STATISTICAL NOTATION IN ROMAN ALPHABET

N_g — common group size when $J \geq 2$ groups include equal numbers of observations

n — number of distinct values in collection of observations

P — probability function or operator, as in $P(A)$, the probability of Event A, or $P(x_i)$, the probability of the value x_i

p — a specific probability value. May denote probability of success in a binomial experiment, significance probability, or population proportion

$\mathbb{P}$ — cumulative probability

q — probability of a failure in a binomial experiment, equal to $(1 - p)$

R — number of rows in a two-way classification of data

r (or r_{XY}) — correlation of X and Y for N pairs of observations

s^2 — variance of a collection of observations

s — standard deviation of a collection of observations; $\sqrt{s^2}$

Sk — skew

Sk_P — Pearson's coefficient of skew

SS — abbreviation for *sum of squares*, as in sum of squares within groups (denoted SS_W)

t — any variable distributed as Student's t

$t(\nu)$ — distribution of Student's t for ν degrees of freedom

V — a random variable or set

v (or v_i) — a value of the random variable V

$V(X)$ — variance of the random variable X

W — a random variable or set; the generalized statistic

w (or w_i) — a value of the random variable (or statistic) W

X — a random variable or set

X/N — the binomial proportion

x (or x_i) — a value of the random variable X

$\tilde{x}$ — the median for a collection of observations; the sample median

$\bar{x}$ — the mean for a collection of observations; the sample mean

Y — a random variable or set

y (or y_i) — a value of the random variable Y

$\hat{y}$ — value of Y predicted by linear regression

Z — any random variable in standard form

z (or z_i) — a particular value of Z

(continued on back endpapers)

STATISTICS

STATISTICS
CONCEPTS AND APPLICATIONS

Harry Frank
Department of Psychology
The University of Michigan – Flint

Steven C. Althoen
Department of Mathematics
The University of Michigan – Flint

CAMBRIDGE
UNIVERSITY PRESS

Published by the Press Syndicate of the University of Cambridge
The Pitt Building, Trumpington Street, Cambridge CB2 1RP
40 West 20th Street, New York, NY 10011-4211, USA
10 Stamford Road, Oakleigh, Melbourne 3166, Australia

First published 1994

Printed in the United States of America

10 9 8 7 6 5 4 3 2 1

Library of Congress Cataloging-in-Publication Data
Frank, Harry, 1942–
Statistics: concepts and applications / Harry Frank, Steven C.
Althoen.
p. cm.
ISBN 0-521-44554-X
1. Statistics. I. Althoen, Steven C. II. Title.
QA276.12.F73 1994
519.5 – dc20   93-27502

A catalog record for this book is available from the British Library.

ISBN 0–521–44554–X hardback

To

Dr. David A. August, M.D.,
　　Division of Surgical Oncology
Dr. Robert L. Cody, M.D., and Dr. Max S. Wicha, M.D.,
　　Division of Hematology and Oncology,
　　Department of Internal Medicine
Dr. Allen S. Lichter, M.D.,
　　Department of Radiation Oncology
　The University of Michigan Medical School

frontline soldiers in the war on cancer, where statistical
decisions are truly matters of life and death

Harry Frank

To

Marcia, Michael, and Tosca
for all their patience during the preparation of this text

Steven C. Althoen

Contents

Appendices

Preface

It is trite and a little pompous to describe writing a book as a journey of discovery. In the present case it happens to be literally true. The authors didn't set out to write a book. We set out to be running partners. In the course of our weekly 10-mile runs, we discovered that despite our wide differences in training and teaching experience, we had come to very similar conclusions about how undergraduate statistics should be taught. Much of the book was written, or at least drafted, during our runs. Running does strange things to the mind. Lack of oxygen forces a runner toward mental simplicity. (Try a few simple arithmetic problems after you've run three or four miles; it is a humbling exercise.) Sometimes this leaves the runner in a hopeless muddle. Sometimes it results in startling clarity by paring away everything that is not essential. We hope that the final result—this book—has captured the clarity and excised the muddle.

TO THE INSTRUCTOR

A textbook is written for students, but students don't have a chance to read it unless the instructor buys it. That's why the traditional note "To the Instructor" usually precedes the note "To the Student." It is the authors' opportunity to make a genteel sales pitch directly to the potential buyer.

Since you are teaching undergraduate courses, you probably don't have enough time to examine all the new books that cross your desk. We'll make things easy. This book is probably *not* for you if you want a book

1. for students who have *not* had at least one good, solid course in college algebra,
2. that is tied to a specific academic discipline,
3. with minimal coverage of probability theory.

If any of the preceding apply to you, give this complimentary copy to a colleague or a talented student or return it to the publisher. Better yet, send it to one of the authors. We have lots of relatives who are expecting gift copies, and (unlike the publisher) *we* will refund your postage.

———

If you are still reading, you are probably teaching the sort of course we had in mind when we wrote this book.

The course in which the book evolved is housed in a department of psychology, but about a third of the students who enroll are majoring in nursing, health care, biology, chemistry, or computer science. The book is most ideally suited, therefore, to the sort of introductory general statistics course frequently offered by departments of statistics or mathematics for students in the behavioral and natural sciences.

Student preparation

The expected level of mathematical preparation is modest, but uncompromising. Your students should be familiar with roots, powers, absolute values, inequalities, graphs, and equations. It is helpful if they have had exposure to functions, variables, and factorial notation. Familiarity with calculus is neither assumed nor required. In Chapter 7 (Box 7.1, p. 241) we *briefly* explain the integral sign $\int$, but only as a sidebar to show that the definitions of expected value, variance, and cumulative probability for continuous random variables are analogous to the definitions given for the discrete case.

Notation

Contemporary statistical notation derives from a host of traditions, disciplines, and historical precedents, and as a consequence there is only a handful of symbols that may truly be thought of as "standard notation." Where such standard notation exists, we use it. In cases where conventions vary widely and we had to make arbitrary decisions, we tried to choose notation that is most consistent with the established conventions and, at the same time, tried to avoid notational distinctions that are difficult to render on a blackboard.

Unfortunately, even widely used notation is occasionally ambiguous, inconsistent, contrary to more general convention, or conceptually misleading. In such cases we have sometimes chosen notation that may be less familiar, but is by no means unique to this text.

A glossary of the most frequently used symbols appears in the end papers, and our few departures from common practice are as follows:

1. The number of *observations*, or experimental trials, is denoted with a capital N. The number of *distinct values* in a collection of observations is denoted with a lowercase n. We find that this notational distinction helps avoid confusion between summations for raw data and summations using frequency notation. By reserv-

ing N for the former, we give precedence to the most traditional notation for the number of observations (trials) in binomial experiments.

2. Random variables, or measurements, are sometimes treated as value sets and unless otherwise indicated are therefore denoted with capital letters (e.g., V, W, X, Y). Corresponding lowercase letters (v, w, x, y) are *values* of random variables.

3. We observe the standard practice of denoting sample statistics with Roman letters and denoting population values and parameters with Greek letters.[1] The quantity

$$\frac{\sum (x - \bar{x})^2}{N}$$

is the variance of a collection of N observations and is therefore denoted s^2. The quantity

$$\frac{\sum (x - \bar{x})^2}{N - 1}$$

is the estimate of the *population* variance and is therefore denoted $\hat{\sigma}^2$.

4. We have found that introductory statistics students often have difficulty with the notion that *probability* in a confidence statement, such as

$$P(x - 1.96\sigma \leq \mu \leq x + 1.96\sigma) = .95$$

ceases to be meaningful as soon as the value of x is specified. To avoid awkward and often counterintuitive semantics, we therefore use $\mathbb{C}$ (for confidence), rather than P. For example,

$$\mathbb{C}(x - 1.96\sigma \leq \mu \leq x + 1.96\sigma) = .95$$

5. Nomenclature for sums (and means) of squares in analysis of variance now differs widely. We retain the traditional sum of squares *within* groups (rather than adopting the sum of squares *error* favored by some authors), but we have taken advantage of the recent liberalization to eliminate a long-standing grammatical flaw: We refer to the sum of squares *among* groups (denoted SS_A), rather than the sum of squares *between* groups.

Probability theory

We devote 123 pages to our three chapters on probability. This is a sizable fraction of the book, but it must be emphasized that much of this is exercises, worked-out examples, and graphic illustrations. Since many of the students for whom the text is intended find probability uncomfortably abstract, we present the material at a more gradual pace than might

[1] There are two notable exceptions. We use *cov* to denote the population covariance. We also follow the nearly universal convention of denoting binomial probability of success as p.

be optimal for a more mathematically oriented readership, developing concepts in a methodical, step-by-step fashion and building in ample redundancy.

You will also find that many of the probability "story problems" involve scenarios from the popular fantasy role-playing game, *Advanced Dungeons & Dragons*®. No familiarity with the game is necessary. We use these problems only because the game makes use of a variety of polyhedral dice (4-sided, 6-sided, 8-sided, etc.) in combinations that generate a refreshing variety of discrete probability distributions. Our students have found these exercises a welcome respite from the usual run of balls-and-urns scenarios or gambling problems. Incidentally, most hobby shops sell packets of these dice for a few dollars. They can be used in class to generate data on the spot.

Boxed displays

Thin-line boxes are used to highlight important definitions and definitional equations. Bold-line, numbered boxes are used for a number of proofs and derivations, supplementary explanations, and other such material that we consider too important to bury in an appendix, but which would be intrusive if incorporated into the main flow of text. We have included these in the same spirit in which hikers and cross-country skiers tuck an extra pair of socks in the outside pocket of a day pack: Better to have them and not need them than to need them and not have them.

Topical coverage and instructional strategies

Our selection of topics is predicated on the belief that an introductory general statistics course should not be a superficial survey or compendium of techniques. It should teach *less* and do it *thoroughly*, with deliberate emphasis on fundamental principles. Consequently, a number of topics that have largely disappeared from introductory texts are covered in some detail. Our treatment of statistical description, for example, is more extensive than you might expect in a contemporary book. We do this for three reasons. First, many instructors simply want more emphasis on description. If you are looking for a "traditional" treatment of topics like skew and kurtosis you will find it. If you feel that these topics are a waste of time, they comprise a self-contained unit (at the end of Chapter 2), and you can omit them with no loss of continuity. The same is true for the treatment of percentiles (Chapter 3).

Second, a strictly descriptive context allows earlier introduction of some topics, such as correlation and regression, that would have to be deferred if description and inference were interwoven.

Third, relative frequency distributions are tangible and real and therefore make it easier to grasp many concepts that can be very elusive if first introduced in the context of probability distributions. For example, even though the availability of computers has made it unnecessary as a practical matter to calculate the mean using frequency notation, early introduction of this technique makes it easier to present the expected value later on. Likewise, our rather detailed coverage of the median with grouped data paves the way for introducing the notion of probability as the area under a density function.

(Incidentally, our treatment of descriptive statistics as a topic in its own right means that you will not find anything about *sampling* under the rubric of description. In Part I we talk about "data collections," not "samples." A sample implies a parent population, which begs the question of inference. Random sampling is therefore introduced with inference in Chapter 8.)

Because of its heuristic usefulness, discrete probability is also covered more deliberately than has become customary. Since discrete variables are more palpable than continuous variables, we find that fewer students "hit the wall" when introduced to methods based on the normal curve if much of the architecture of hypothesis testing is first developed in the context of the binomial distribution. From there, it is an easy progression to normal approximation methods for testing hypotheses about proportions and, then, to hypotheses about means.

The other side of this pedagogical coin is that there are many things you will *not* find in this book. We exclude a number of introductory-level topics that might be desirable in books geared to specific disciplines. For example, there is no discussion of time series, which is a standard topic in business statistics, or measurement models, a staple of educational statistics.

We have also limited our presentation of more advanced material. Our treatment of experimental design (Appendix VI) is confined to fundamentals. There is no discussion of such topics as randomized blocks, Latin squares, nested factors, partial F tests, and so on. Nor do we discuss multiple correlation or multiple regression, and there is no coverage of nonparametric statistics per se. Nevertheless, Chapters 13–15 offer a selection of those advanced topics that are fundamental to the widest range of research traditions and which follow with least elaboration from the core material in Chapters 1–12 (e.g., without introducing new distributions).

Furthermore, Chapters 13–15 are organized to permit either a broad survey or narrower, in-depth coverage of selected techniques. None of the last three chapters assumes background beyond Chapter 12, and more specialized applications are treated at the end of each chapter.

Examples, exercises, and computer software packages

As suggested by the title, one of the principle goals of this text is mastery of applications. Almost every new topic is therefore illustrated with worked-through examples. In addition:

- Exercises are placed at the end of each topical section in order to promote step-by-step consolidation of problem-solving skills.
- Routine calculational drill problems comprise 30 to 40 percent of the exercises. This is no longer fashionable in disciplines outside of mathematics, but it lets students become familiar with the basic

mechanics of problem solving without the additional task of trans-
lating textual situations into numbers. The balance between drill
exercises and content problems changes in favor of content problems
as one proceeds through each set of exercises, as one proceeds
through the exercise sets in any particular chapter, and as one
proceeds from chapter to chapter through the book.

- To demonstrate the diversity of content to which statistical techniques are applied and to appeal to a broad range of student
interests, we have drawn content problems and examples from a
wide variety of academic areas (e.g., paleoanthropology, agriculture,
epidemiology, psychology, animal behavior, and environmental
chemistry) as well as from the everyday statistical environment—
sports, public opinion surveys, census data, civil rights litigation,
etc.

- The personal computer has become the basic tool of the statistical
workplace, and use of statistical software packages is becoming a
common feature of introductory statistics curricula. Accordingly, we
have prepared a companion workbook (*User-Friendly*) that includes
a tutorial for the ASP (A Statistical Package for Business, Economics and the Social Sciences©) system and supplementary exercises. In addition, many of the content problem in *this* book include
data sets that can be used easily with ASP or any other software
package. However, the text is intended to stand independently of the
software-driven workbook.

    Instead of gearing the text to computer protocols, we develop and
provide calculational formulae in terms of basic computational
quantities (e.g., sums of squares and sums of products) that can be
obtained with a single entry of data values on most "student"
models of hand calculators. We have taken this approach because it
is our experience that immersion in the mechanics of calculation has
fundamental instructional importance. One cannot become a world-
class tennis player by watching videotapes of Boris Becker or become
a superb downhill racer by watching replays of Franz Klammer defy
gravity at Innsbruck. You have to hit the ball and you have to take
your falls. Similarly, one cannot acquire a statistical sense for data
without *doing* statistics. *Watching* numbers come up on a computer
screen is no substitute. The computational algorithms eliminate
much of the drudgery of routine arithmetic without depriving the
student of the instructional benefits that derive from working
through the intermediate steps and seeing raw data evolve into
interpretable statistical results.

- Appendix IX provides the *solutions* to the odd-numbered problems,
not just the *answers*. The detail of explanation for the more challenging exercises is about the same level as the explanations given in
the text for worked-out examples. We consider the solutions to be an
integral elaboration of the text. Furthermore, the odd- and even-
numbered problems are usually paired in such a way that the
solutions to odd-numbered exercises offer implicit guidelines for
solving neighboring even-numbered exercises.

Permission to photocopy materials from this book

A number of the appendices and chapter sections are self-contained presentations that might be useful additional reading for students in other courses. We have asked Cambridge University Press to grant all requests to photocopy materials from this book if copies are to be distributed free of charge to students.

TO THE STUDENT

Galileo is supposed to have said that mathematics is the language in which God wrote the universe. Since Galileo was a fluent speaker of mathematics, his claim may seem a little self-serving. Nevertheless, it is probably safe to say that mathematics—or that body of mathematical applications called statistics—is the language in which Man *reads* the universe. Statistics in one form or another has become the common language by which agronomists, chemists, educators, geneticists, medical researchers, political scientists, psychologists, and sociologists—not to mention attorneys, corporate executives, baseball managers, public policy advocates, political strategists, military planners, and the insurance industry—read some portion of the world in which we all live.

Statistics is a language with a numerical vocabulary, a mathematical grammar, and, like any language, its own, distinctive way of shaping the speaker's view of the world. In many respects an introductory statistics course is therefore like an introductory course in German, Hebrew, Japanese, or any other language. Language texts provide glossaries for quick reference to unfamiliar words. A glossary of mathematical and statistical notation likewise appears in the end papers of this book. A quick glance at the glossary reveals that the symbol system of statistics shares an important property with the symbol system of other languages: Meaning often depends on context. For example, the lowercase Greek alpha (α) can mean the probability of a Type I error or it can mean the y-intercept of a regression equation. Native speakers of English are seldom confused about the meaning of the word "draw," but the *Oxford English Dictionary* lists over 75 meanings (and does not include the American English use of "draw" as a noun meaning a shallow gully).

Textbooks for introductory language courses almost invariably advise the student to buy a dictionary. Similarly, you will need a calculator for the exercises in this book. The text provides computational formulae that reduce both the time required to perform calculations and the chances of making a mistake. To use most of these formulae, you need a calculator that computes sums, sums of squares, and sums of products. These functions are found on almost all but the least expensive calculators, and the symbols that most commonly appear on calculator keys to indicate these operations are Σx, Σy, Σx^2, Σy^2, and Σxy. More specialized (and only slightly more expensive) calculators allow you to enter your data and then obtain a variety of statistics directly, each with a single keystroke. Check with your instructor before purchasing a calculator. He or she may want you to use a model that has (or *doesn't* have!) certain capabilities.

Every statistics course, like every language course, also presupposes some particular level of background experience. If you have read the preface section "To the Instructor" you know that this book assumes that you have more or less mastered the fundamentals of college-level algebra. Courses in mathematics beyond the prerequisite level are helpful only insofar they leave a residue of mathematical maturity. There are very few places in the book where past experience with the material found in, say, a calculus or analytic geometry course will be helpful.

Not surprisingly, the first encounter with statistics is often like the first encounter with any foreign language. It *can* be fraught with apprehension, frustration, and bewilderment. Statistics can also be richly rewarding, offering fresh perspectives on daily experience and aesthetic appreciation of the elegant logic of scientific reasoning. Professional golfer Walter Hagen once said, "We pass this way only once, so don't hurry, don't worry, and be sure to smell the flowers on the way."

We can't improve on this advice.

Tips for using this book

By the time you have been in school long enough to be reading this book, you have already had many years to develop your own highly individual strategy for using textbooks and completing assignments. On the other hand, authors are also individuals and necessarily approach the task of writing books with certain assumptions about how their books can be used most effectively. These assumptions begin with authors' own experience, but it is very difficult to recall with any authenticity one's first encounter with something that has become familiar. (Try conjuring up a vivid recollection of the first time you rode a bicycle or drove a car or tasted a hot fudge sundae.) Authors therefore subject their own students to endless manuscript drafts before a book is published, and they find that their assumptions must inevitably be molded and adapted to the reality of student experience. The following advice is based on how we planned for the book to be used and on feedback from our many students who have described the strategies they found successful—and not so successful.

- Don't try to work backward. The backward approach goes something like this: Begin with the exercises. When you are stumped by a problem, scramble madly through the chapter to find an example that looks like the problem you're working on. If you don't find one, look in the back of the book for the solution to a similar problem. If all else fails, read just enough of the chapter to find out how to do the problem.
- Read the entire chapter first. This ordinarily means that you should try to stay ahead of your instructor's lectures, and this requires a measure of discipline. Very few people (and the authors are *not* numbered among this lucky few) can absorb mathematics from a book alone. Most people need, in addition, to have mathematics *told* to them and *demonstrated*, so there is some natural reluctance to

venture alone into unknown mathematical territory. The lectures will be much more enlightening if you have already read the material. Trust us.

- Use a variety of reading strategies. We have said that introductory statistics is like an introductory language course. The fundamental skills in learning a language are *reading* skills. The same is true for statistics. One important reading skill is to use different strategies at different stages of the learning process.

  The first time you read new material, read for *passive understanding*, or *familiarity*. Don't try to put it all together. Read through the exercises, but don't do them. Skip the boxed proofs and derivations. *Follow* the examples, but don't work them out unless you can't see how the authors got their answers. Just make sure you understand each sentence *as you read it*. This reading serves a number of purposes. First, it gives you confidence. If you can understand every sentence, then it can truly be said that there's nothing in the chapter that you *don't* understand. Second, even a light reading begins the unconscious process of organizing the new material—making connections with material you've learned earlier and creating the mental pegs on which you'll hang material you pick up on the next reading and the points explained by your instructor. Third, it shows where the chapter leads. Getting there is easier if you know where you're going.

  The second time you read is for *comprehension*. This time, you actively work toward putting the pieces together, seeing how the various sections and subsections (and even sentences) form a single body of information and how the new topics relate to earlier topics. Start with the outline at the beginning of the chapter *and* with the summary at the end of the chapter. Then, read with a pencil in your hand and work through the examples and the derivations. Taking a crack at some of the *drill* exercises can be useful now. You've *seen* the material, you've probably had some of it *told* to you; now it's time to *do* it.

  The third reading is for *mastery*. At this stage of the game you can get maximum benefit from the more challenging content exercises (a euphemism for story problems). Students have repeatedly told us that if they read the material thoroughly *before* attempting the exercises, they actually spend less *total* time (reading plus exercises) than they spend on *just the problems* when they try doing the exercises cold (i.e., when they do things backward). More important, *this* is the stage of learning when the exercises serve to consolidate your *understanding* of what you've been reading.

- How to approach story problems without fear and loathing. The routine, computational drill exercises at the beginning of almost every exercise set are intended to help you master the mechanics of doing problems, but in the real world, there is no divine agency (or instructor or author) to set up problems for you. You have to figure out what to do before you set about applying the mechanics of doing it. This is why we have story problems.

When you do computational drills, you begin with the specifics of the problem and apply general techniques. With story problems it's better to begin with general principles and move to the specifics. That is, *don't* start with the problem. There is an infinity of potential problems out there, and if you approach each one individually, all you will do is get lost in the details, drink too much coffee, and get an ulcer. On the other hand, there are only a few *kinds* of problems, and for each kind of problem, there are only a few *principles* to be applied. So, make it easy on yourself. Read the story, sit back, and ask yourself *what* you are supposed to do (e.g., describe a collection of data, compute a probability, estimate a parameter, test an hypothesis, etc.). Then, ask yourself what are the general principles involved in this sort of task. What information do you need, what assumptions have to be satisfied, etc.? *Then* get down to the specifics. Ask yourself how the principles can be applied to this particular problem. Where is the information you need given or implied in the story, what assumptions can you make, what operations do you perform on the data to get your answer, etc.?

- We have saved the most important tip for last: *It is almost always easier to figure out how to get from what you know to what you want if you draw a picture*.

P.S. Don't forget the flowers.

TO ALL READERS

The corrected proof pages that we returned to the publisher were, of course, perfect. However, it is a well-known fact that typographical and other errors are spontaneously generated by drafty warehouses. If you find any such errors, no matter how trivial, please let us know by snail mail or e-mail:

salthoen@umich.edu

Your help will be acknowledged in the next edition, and we will keep you on an e-mailing list and send you cumulative errata as they are detected by other readers. Errors are sometimes corrected between printings, so please indicate the printing of your copy, as shown on the copyright page —the final digit in the sequence beginning with 10.

Acknowledgments

Two names appear on the cover of this book, but many people contributed to what is between the covers. First, we want to express appreciation to the University of Michigan–Flint undergraduate students who, with relative good cheer, used various versions of the manuscript as their principal text in Psychology 301. Most of you detected errors in the exercises, many of you identified passages that were perfectly sensible to the writers but incomprehensible to anyone else, and some of you accepted our challenge to rewrite sections you didn't like. All of you can rest easier knowing that the single most nearly universal complaint about the manuscript has been rectified: The published text *does* have an index!

We also want to thank the Psychology 301 laboratory/teaching assistants, the book's intellectual midwives: Mark Stefanski, Tom Bowyer, Mark Siefert, and Amy Collins Siefert. The last two, first as students and later as assistants, saw the text through at least three full revisions and undertook to rewrite a number of especially troublesome passages. The present treatment of the binomial random variable and binomial experiments in Chapter 6 is the only slightly revised work of these two remarkable students.

Family, friends, and University of Michigan–Flint (UM–F) colleagues enriched the content and improved our presentation in many ways. Melvin J. Warrick, former Associate Director of the Human Engineering Division, USAF Aerospace Medical Laboratory, read Part I with scrupulous attention, furnished the references from which we obtained the anthropometry data in Exercises 8.2, and made a number of valuable suggestions that inspired our Postscript to Part I. Martha G. Frank patiently explained how chemists use statistics, set up several of the analytic chemistry exercises, and furnished suitable data. Dave Dvorak, Deputy Principal of Mott Adult High School (and running partner), used the manuscript to supplement the statistics course required in his Ph.D. program and copyedited every line. Krista Hansen (UM–F Department of Mathematics) simplified cumbersome algebra in several places, and

Ed Bowden and Paul Bronstein (UM-F Department of Psychology) furnished raw data from their own research for exercises.

Two former colleagues also contributed to the project. Cay Horstmann taught mathematics at UM-F while attending graduate school and now combines an academic career with private enterprise. He is the father of ChiWriter™, the multifont word processor in which we wrote the manuscript and produced both the companion workbook, *User-Friendly*, and the instructor's manual. George Blackford, who once chaired the UM-F Department of Economics, left academia altogether for the world of software. He is the creator of ASP®, the statistical software package bundled with *User-Friendly*. His generous collaboration in this project prompted a number of changes in the present text.

A word about our reviewers: Reviewers are underpaid, underloved, unacknowledged collaborators. Underpaid because publishers could not possibly afford fair market value for their time and expertise; underloved because, if they do their job well, they make work for the authors and delay production; and unacknowledged because they traditionally work in anonymity. At our request, Cambridge University Press departed from this tradition and invited open reviews. We can therefore thank George Poole (Eastern Tennessee State University, Department of Mathematics) by name. Our treatment of probability runs a briefer and smoother course because of his thoughtful and detailed comments, and his suggestions concerning percentiles of ungrouped data furnished a much-needed link to our treatment of percentiles in *User Friendly*. Editors, of course, are never anonymous. Nor are they ordinarily qualified to act as reviewers. Alan Harvey is an exception. He is a mathematician, and his initial comments persuaded us to draft him for full-scale review duties. It was a good choice. He has a keen eye for devilish flaws that are really quite serious but so subtle that readers—and authors—glide over them with only a vague sense of uneasiness.

PART I

Organization and description of data

CHAPTER 1

The organization of data

CHAPTER OUTLINE

A. THE MEANING OF DATA

The word "data" appears in many contexts and frequently is used in ordinary conversation. Although the word carries something of an aura of scientific mystique, its meaning is quite simple and mundane. It is Latin for "those that are given" (the singular form is "datum"). Data may therefore be thought of as the *results of observation*. Data are collected in many aspects of everyday life. Statements given to a police officer or physician or psychologist during an interview are data. So are the correct and incorrect answers given by a student on a final examination. Almost any athletic event produces data: the time required by a runner to complete a marathon, the number of errors committed by a baseball team in nine innings of play, the number of shots on goal during a period of hockey. And, of course, data are obtained in the course of scientific inquiry: the positions of artifacts and fossils in an archaeological site, the number of interactions between two members of an animal colony during a period of observation, or the spectral composition of light emitted by a star.

1. Quantitative data

Data may be used in a variety of ways to reach conclusions or generate interpretations. For example, the symptoms reported to a physician or clinical psychologist may lead to a diagnosis. The discovery of a particular tool or other artifact in close proximity to fossilized skeletal remains may lead an anthropologist to conclude that the remains are those of a particular species of hominid, e.g., Neanderthal man or Modern man. However, if data are to be treated *statistically*, the observations must be expressed in *numerical* form. That is, they must be *quantitative*. Data appropriate to statistical analysis would therefore include *scores* on the schizophrenia scale of the Minnesota Multiphasic Personality Inventory, the *number* of correct answers on a multiple-choice final examination, the *vertical distance* between a fossilized bone and a flint cutting tool, the *number* of hours, minutes, and seconds required to run a marathon. Consequently, *statistical* data always consist of a *collection of numbers*.

2. Types of quantitative data

The numbers with which a statistician begins are ordinarily of two types, *measurements*, or *scores*, and *frequencies*. When the phenomena that a scientist observes are expressed as or translated into numerical values, these data are called *measurements*. Height expressed in feet and inches is a *measurement* of linear body size. Weight expressed in kilograms is a *measurement* of body mass. An IQ *score* is a *measurement* of academic aptitude. Employee ratings are *measurements* of work performance. In the general case, measurements are represented by letters falling toward the end of the alphabet, usually x, y, or z. When events are expressed as numbers, the term *observations* is understood to mean *numerical* observations, i.e., *measurements*.

Some phenomena cannot be translated into numbers. That is, some observations are *qualitative*, or *categorical*, rather than *quantitative*. Gender (male versus female), political party affiliation (Democrat versus Republican), and diagnostic classification (schizophrenia, depression, psychopathy) are examples of qualitative observations. Whether the events one observes are quantitative (and expressed as numbers) or qualitative, one can record the *number of times* each event occurs. Data of this sort are called *frequencies*. Examples of frequencies would include the *number of students* who obtained a particular score of an examination, the *number of women* who favor a particular candidate or ballot proposal, the *number of students* who miss a particular examination item, the *number of times* a particular member of a baboon troop displays dominance or threat toward other members of the troop, and so forth. In the tables that follow, and subsequently throughout this text, a lowercase *f* will be use to indicate the frequency with which an event occurs.

B. REPRESENTATIONS OF DATA

Despite the fact that an 18-month-old baby is capable of mental activity that challenges the ability of even the most sophisticated computer, the human mind is very limited in some ways. Unlike a computer, for example, it can keep track of only a very small number of unorganized bits of information. Try this experiment on yourself: *Without putting them in any particular geometric pattern*, imagine first a single point in space, then two points, then three points, four points, and so on. Chances are that unless you "cheated" and organized the points systematically (e.g., located them at the corners of imaginary squares or arranged them in evenly spaced rows and columns) you lost track somewhere between five and nine.

Even a small-scale piece of research may involve dozens of observations, and studies that yield several thousand observations are not uncommon. Before a scientist can even contemplate the task of interpretation, therefore, data must be organized so that they are reduced to manageable proportions. The most common organizational schemes are *tabular* and *graphic* representations.

1. Tabular representation: the frequency table

Example 1.1. Let us suppose that a statistics teacher has 25 students each toss four coins and record the number of heads. These experiments might yield the following 25 results:

2 2 0 1 1 2 3 1 2 2 3 3 4 1 3 2 3 3 1 2 2 1 2 2

It is obvious that even with only 25 observations the data call for a more comprehensible display. One very simple expedient would be to array them from the smallest value to the largest value:

0 1 1 1 1 1 1 1 2 2 2 2 2 2 2 2 2 2 3 3 3 3 3 3 4

This sort of presentation *organizes* data, but it does not *reduce* the quantity of data. However, it does reveal something interesting: Although the experiments produced 25 *observations*, these data comprise only 5 different, or distinguishable, numerical *values*. Every value is equal to 0, 1, 2, 3, or 4. Consequently, the display of data is condensed significantly with no loss of information if each observed value is paired with its *frequency*, as given in Table 1.1.

Table 1.1. Frequency distribution for number of heads obtained in toss of four coins with experiment repeated 25 times

Observation (x)	0	1	2	3	4
Frequency (f)	1	7	10	6	1

Because it indicates how frequencies are distributed among the outcomes of an experiment, a representation of data in which every observed value is paired with its frequency is called a *frequency distribution*.

In Example 1.1 the numerical value 2 appears 10 times, which constitutes 40 percent of the total number of observations (25). If, however, the value 2 had appeared 10 times in 100 observations, this would amount to only 10 percent of the observations. Sometimes, then, the *frequency* with which a particular value is observed is less revealing than is its *relative* frequency, that is, the frequency divided by the number of observations. A representation of data in which every observed value is paired with its relative frequency is called a *relative frequency distribution*. Customarily, the number of observations is denoted by the capital letter N, so relative frequency becomes f/N. The relative frequency distribution for the data in our imaginary coin-toss experiment is given in Table 1.2.

Another useful presentation of data is the *cumulative* frequency distribution. The cumulative frequency of a value is its frequency *plus the frequencies of all smaller values*. Accordingly, the cumulative frequencies given in Table 1.3 indicate that 18 students obtained two or *fewer* heads

Table 1.2. Relative frequency distribution for number of heads obtained in toss of four coins with experiment repeated 25 times

Observation (x)	0	1	2	3	4
Relative Frequency (f/N)	.04	.28	.40	.24	.04

Table 1.3. Cumulative frequency distribution and relative frequency distribution for number of heads obtained in toss of four coins with experiment repeated 25 times

Observation (x)	0	1	2	3	4
Cumulative f	1	8	18	24	25
Cumulative f/N	.04	.32	.72	.96	1.00

in their four tosses, 24 students obtained three or *fewer* heads, and all of the students obtained four or fewer heads.

The cumulative *relative* frequency of any value is similarly defined as the relative frequency of the value plus the relative frequencies of all smaller values. From Table 1.2 we know that the relative frequency of 0 heads is .04 and the relative frequency of 1 head is .28. As shown in the bottom row of Table 1.3, the cumulative relative frequency of 1 head is .04 + .28 = .32. It is also apparent from inspection of Table 1.3 that the cumulative relative frequency of any value is equal to its cumulative frequency divided by the total number of observations, N. The cumulative relative frequency of the largest observed value (e.g., 4) is, therefore, always equal to 1.0.

a. Tabular representation of grouped data. In the preceding example only five possible numbers could be observed. In some experiments, the number of values that it is possible to observe may be so large that even the sorts of frequency distributions discussed above become unwieldy.

Example 1.2.1. Suppose 1,000 persons each toss 100 coins and count the number of heads. In this experiment 101 different values might be observed (0 heads, 1 head, ..., 100 heads), which means that a distribution that represented *every* value would be very cumbersome indeed. Furthermore, even with a large number of observations, it is likely that *some* of the values (e.g., 0 heads or 100 heads) would not turn up at all.

Under these conditions it is customary to *group* the values into intervals, or classes, and to tabulate the frequency (or relative frequency, or cumulative frequency, etc.) of observations in each class. Table 1.4 is a

Table 1.4. Frequency distribution for number of heads obtained in toss of 100 coins with experiment repeated 1,000 times

x	0–9	10–19	20–29	30–39	40–49	50–59	60–69	70–79	80–89	90–99
f	13	41	93	147	240	200	160	67	13	26

grouped frequency distribution for 1,000 people tossing 100 coins and counting the number of heads.

A grouped distribution organizes *and* reduces data, but there is a loss of information. In this example we know that 240 persons obtained some number between 40 and 49, but we cannot tell from the table how many persons observed *exactly* 40 heads, how many times *exactly* 41 heads turned up, and so forth. Nevertheless, the benefits of comprehensibility are generally assumed to offset the disadvantage of information loss, especially if some of the *x*-values account for only a small percentage of the total observations.

In setting up a grouped distribution, one must first decide how many intervals to use and how wide the intervals will be. The *fewer* intervals one uses, the wider each interval and the *greater the information loss*; the *more* intervals one uses, the narrower each interval and the *less comprehensible* the display. Comprehensibility is largely in the eye of the beholder, but 6 to 20 intervals is usually workable. In addition, the number of intervals n should be small enough so that the *average* interval frequency (N/n) is greater than 5.

To calculate interval width, call the smallest value that will appear in the table x_a and the largest value x_b. If all n intervals are to be of the same width, then for integer data

$$\frac{x_b - x_a + 1}{n} = \text{interval width} \qquad \text{[1.1]}$$

The table is easier to construct and interpret if the interval width [1.1] is a whole number. This usually can be achieved either by adjusting n or by letting x_a be smaller than the smallest observed value or letting x_b be larger than the largest observed value.

Example 1.2.2. To see how this works, let us suppose that 1 and 99 were the smallest and largest values actually obtained in Example 1.2.1 and that we wanted to organize the data into 10 intervals $(n = 10)$. Then $x_a = 1$ and $x_b = 99$, and by equation [1.1] the interval width is 9.9. This is not a whole number, so we have two choices. We could have used 11 intervals instead of 10. By equation [1.1] this results in an interval width of 9. The other option was either to let $x_b = 100$ or to let $x_a = 0$ (as illustrated in Table 1.4). Either adjustment yields an interval width of 10.

Then one must determine the lower and upper *class limits* of each interval, that is, the smallest and largest value, respectively, to be included in each class. Interval width may be thought of as the difference between the *lower* class limit of any interval and the *lower* class limit of the *next* interval and is therefore the most obvious consideration in calculating class limits. Once it was decided in Example 1.2.2 that the interval width was to be 10 and that $x_a = 0$, the lower class limit of the second interval was determined: $x_a + 10 = 10$.

A less obvious consideration is the precision of one's measurements. In Example 1.2.1, the numerical observations could assume only integer values. That is, they could be expressed only as whole numbers. One cannot, after all, obtain 9.5 heads in 100 tosses of a coin. Since it is impossible to obtain any value *between* 9 and 10, there is no potential ambiguity created by setting the upper class limit of the first interval equal to 9 and the lower class limit of the second interval equal to 10. The situation is different if observations can take fractional values.

Example 1.3.1. Suppose that the *x*-values in Table 1.4 are *lengths* in inches. Depending on the precision of the measuring instrument, it is perfectly feasible to expect a measurement of, say, 9.5 in. or 9.75 in. or 9.0000001 in. all of which fall "between" the first and second classes defined in Table 1.4. If, on the other hand, we defined our classes as 0 to 10, 10 to 20, and so forth, an observation of exactly 10 inches would fall in *two* classes.

To avoid this dilemma statisticians have adopted the convention of extending class limits *half a measurement unit above the largest* observable value in the class and *half a measurement unit below the smallest* observable value in the class. The intervals in Table 1.4 would thus become − .5 to 9.5, 9.5 to 19.5, 19.5 to 29.5, and so forth. Whether the *x*-values represent measurements taken to the nearest inch or the number of heads obtained in 100 tosses of a coin, every possible observation can fall in one and *only one* interval. The principle is simple, but it can be difficult to put the principle into practice if observations are recorded in *multiples* of the units in which measurements were taken.

Example 1.3.2. Suppose once more that the *x*-values in Table 1.4 represent lengths in inches, but let us now imagine that our instrument is accurate to the nearest *tenth* of an inch and that our smallest measurement is 0.1 inches and our largest is 99.9 inches.

One immediate difficulty is that 0.1 and 9.9 are *decimal* values, and we said earlier that equation [1.1] is for *integer* data. The requirements for equation [1.1] are actually more specific and less restrictive: The 1 in the numerator represents 1 *unit of measurement*, so x_a and x_b must be *whole numbers* of units of measurement. By *units of measurement*, we mean the *most precise* units that the measurement instrument records. In this example, the unit of measurement is the *tenth* of an inch. Since the smallest and largest values are given in inches, we multiply them by 10 to express them in *tenths* of inches. That is, $x_a = 1$ *tenth* and $x_b = 999$ *tenths*. Then for $n = 10$ intervals, equation [1.1] gives an interval width of 99.9 *tenths*. If we set $x_a = 0$, the interval width calculated from equation [1.1] is once again a whole number:

$$\frac{x_b = x_a + 1}{n} = \frac{999 - 0 + 1}{10} = 100$$

The first interval therefore includes all measured lengths from 0 to 99 *tenths* of an inch, and the class limits are − .5 to 99.5 tenths. The second interval includes all measured lengths from 100 to 199 *tenths* of an inch, and the class limits are 99.5 to 199.5 tenths, and so on. For a table with lengths expressed in *inches*, simply divide the class limits by 10, which gives us − .05 to 9.95, 9.95 to 19.95, 19.95 to 29.95, and so on.

If the observations are in *hundredths* of a unit, one multiplies x_a and x_b by 100, if in *thousandths*, by 1000, and so on. This approach also works if data are in *integer* (rather than fractional) multiples of the unit of measurement.

BOX 1.1

Calculating class limits for grouped data

Step 1. Find the smallest observation and the largest observation and express these values in *units* of measurement. That is, if observations are recorded in *tenths* of units, multiply by 10; if observations are recorded in *tens* of units, multiply by .1, and so on.

Step 2. Decide how many intervals you want. Call this number n.

Step 3. Calculate the interval width:

$$\frac{x_b - x_a + 1}{n}$$

where x_a is the smallest and x_b is the largest observable value represented in the table. As a first approximation, let x_a equal the smallest observation and let x_b equal the largest observation.

Step 4. If the interval width calculated in Step 3 is a whole number, go to Step 7. If the interval width is not a whole number, round it *up* to the nearest integer.

Step 5. Multiply the *rounded-up* interval width by n.

Step 6. Change n or adjust x_a or x_b (or both) so that $x_b - x_a + 1$ equals the value calculated in Step 5. Make x_a *less than* or equal to the smallest observation and x_b equal to or *greater than* the largest observation.

Step 7. Calculate *lower* class limits. The lower limit of the first interval is $x_a - .5$. The lower limit of the second interval is obtained by adding the interval width to the lower limit of the first interval, etc.

Step 8. Calculate *upper* class limits. The upper limit of an interval is equal to the lower limit of the next interval. The upper limit of the last interval is its lower limit plus the interval width.

Step 9. Express class limits in the same scale as the observations.

Example 1.4. Suppose that the data in Table 1.4 are annual incomes measured to the nearest thousand dollars and that the smallest observed value is $1,000 and the largest is $99,000. The 1 in equation [1.1] therefore represents 1 thousand dollars, and we multiply 1,000 and 99,000 by .001 to obtain x_b and x_b.

The procedure discussed above (and outlined in Box 1.1) assumes that intervals are to be of equal width. Although this is the most common practice, it is not unusual to leave the first or last interval "open ended." For example, if only one or two persons in Example 1.2.1 had obtained 0 to 9 heads (or 90 to 99 heads), the first interval might have been defined as all values below 20 (or the last interval as all values above 79).

2. Graphic representation: polygons and histograms

Tabular distributions provide both economy of representation and organization. The adage that "One picture is worth a thousand words," however, has a certain measure of truth in the representation of quantitative data. That is, data are often better understood if represented in graphic form than in tabular form. If we represent the values of x on the horizontal axis of a graph and the frequencies associated with these values on the vertical axis, then every point represents a numerical value of x and the frequency with which it is observed, f_x. The five points in Figure 1.1, for example, represent the number of heads observed in four tosses of a fair coin, each paired with its frequency of observation, as given in Table 1.1.

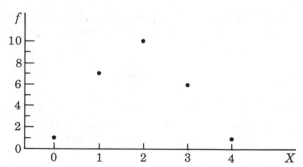

Figure 1.1. Point graph representing frequency distribution in Table 1.1.

When many points are involved, it is sometimes difficult to identify each point with its appropriate x-value. A point graph can be made more readable by connecting the points and creating a *frequency polygon*, as shown in Figure 1.2.

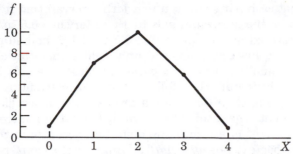

Figure 1.2. Frequency polygon representing distribution in Table 1.1.

An alternative way to improve the readability of a point graph is to drop perpendicular lines from each point to the horizontal axis. Each point therefore becomes a vertical line, the height of which represents the frequency associated with the numerical value from which it originates (Figure 1.3).

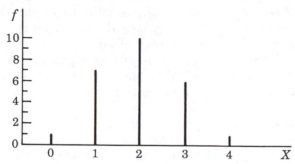

Figure 1.3. Line graph representing frequency distribution in Table 1.1.

While this type of graphic representation is mathematically accurate, it is still not as easily interpretable as we might wish. To create a more readable visual display, we generally substitute solid rectangles for our vertical lines. This familiar "bar graph" is known as a *histogram* and is probably the most popular method of representing frequency distributions. The data in Table 1.1 are again represented in Figure 1.4 below. It

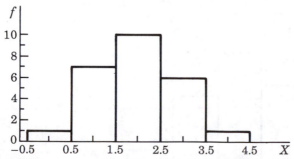

Figure 1.4. Histogram representing frequency distribution in Table 1.1.

will be noted that each bar spans a unit-width interval; that is, the value 0 is represented on the horizontal axis by the interval -0.5 to $+0.5$, the value 1 is represented by the interval $+0.5$ to $+1.5$, and so forth. The reader is reminded, however, that the only values that can actually be observed are those falling at the *midpoints* of these intervals (0, 1, etc.). Construction of a histogram thus follows the same convention discussed above for establishing class limits for a grouped frequency distribution: Any observation that is quantified by counting (as in the number of heads observed in four tosses of a coin) is by definition accurate to the nearest integer, so each bar extends *one half a measurement unit* beyond the actual precision of measurement.

a. The relative frequency histogram. In addition to representing *frequency* distributions, histograms can be used to represent *relative* frequency distributions. In Figure 1.5, the vertical axis has been rescaled to represent the values 0 to 0.40 in order to display graphically the data presented in Table 1.2.

It might also be noted that since the bars in Figure 1.5 are one unit wide, the area of each bar is numerically equal to its height, that is, the relative frequency of the value over which the bar is centered. Accordingly, the *cumulative* relative frequency of any value x is equal to the *sum of the areas* for all bars representing values equal to or less than x. The shaded portion of the relative frequency histogram in Figure 1.6 thus represents the cumulative relative frequency of $x = 1$, as given in Table 1.3.

Even if the bars of a histogram are *not* one unit wide, so long as they are of *equal width*, the area of each bar is still *proportional* to the relative frequency of observations falling in its interval, and *cumulative* relative frequency of any x is likewise equal to the *proportion* of the area to the left of x. These apparently trivial observations will assume considerable importance later in the text.

b. Cumulative frequency and cumulative relative frequency histograms. As an alternative to representing cumulative relative frequency as the area of one or more bars of a relative frequency histogram, we can represent either cumulative frequencies or cumulative relative frequencies on the vertical axis of a histogram, as shown in Figure 1.7. Reading

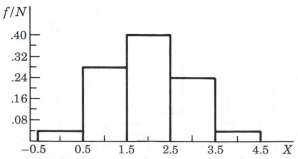

Figure 1.5. Histogram representing relative frequency distribution in Table 1.2.

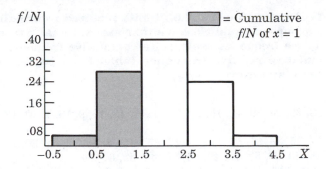

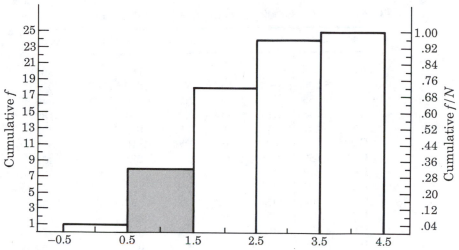

Figure 1.7. Histogram representing cumulative frequency and relative frequency distributions in Table 1.3.

from the right-hand scale (Cumulative f/N), the height of the bar over the interval 0.5 to 1.5 in Figure 1.7 is therefore numerically equal to the shaded area in Figure 1.6.

EXERCISES 1.1

In Exercises 1–6 organize the given data in (a) a frequency table, (b) a relative frequency table, (c) a cumulative frequency table, and (d) a cumulative relative frequency table.

1. 1, 1, 1, 2, 2, 3, 3, 3, 4, 5
2. 1, 3, 4, 4, 5, 5, 5, 6, 6, 7
3. 4, 3, 2, 6, 1, 5, 2, 3, 1, 2
4. 6, 1, 5, 4, 3, 2, 5, 3, 3, 3
5. 0.3, 0.3, 0.2, 0.5, 0.3, 0.4, 0.3, 0.1, 0.5, 0.2, 0.3, 0.4, 0.1, 0.4, 0.3, 0.4, 0.1, 0.2, 0.3, 0.4, 0.2, 0.3, 0.3, 0.2, 0.3
6. 180, 170, 140, 150, 180, 140, 180, 180, 140, 150, 140, 140, 160, 180, 160, 140, 160, 170, 140, 170, 180, 170, 180, 170, 180

In Exercises 7–12 determine the class limits required to organize the data into 6–20 groups of equal class width. Next, write (a) a frequency table, (b) a relative frequency table, (c) a cumulative frequency table, and (d) a cumulative relative frequency table. Keep copies of your answers for use in later exercise sets.

7. 18, 23, 3, 7, 8, 29, 24, 30, 12, 15, 1, 17, 13, 19, 12, 12, 9, 18, 7, 12, 13, 18, 17, 22, 25
8. 46, 21, 89, 47, 35, 36, 67, 53, 42, 75, 47, 85, 40, 73, 48, 32, 41, 20, 75, 48, 48, 32, 52, 61, 49, 50, 69, 59, 30, 40, 31, 25, 43, 52, 62, 50
9. 43.8, 51.8, 53.4, 40.1, 36.8, 34.8, 37.5, 31.6, 44.7, 51.5, 45.3, 45.3, 46.0, 39.9, 42.3, 41.5, 47.8, 44.0, 41.5, 46.2, 48.0, 55.9, 50.5, 43.9, 44.3
10. .81, 1.00, .72, 1.78, .18, 1.81, .15, 1.94, .17, 1.65, .32, 1.80, 1.05, 1.67, 1.71, 1.32, .61, .62, .63, 1.24, .42, 1.72, 1.72, .23, .40, .27, .43, .44, 1.93, .50

11. Per Capita Outstanding Debt by State for 1982 (in dollars)

State	Debt	State	Debt	State	Debt	State	Debt
AL	503	IN	187	NE	188	SC	986
AK	9,206	IA	157	NV	810	SD	1,026
AZ	88	KS	168	NH	1,380	TN	346
AR	243	KY	727	NJ	1,194	TX	182
CA	437	LA	986	NM	632	UT	495
CO	257	ME	765	NY	1,470	VT	1,318
CT	1,487	MD	999	NC	271	VA	489
DE	2,289	MA	1,119	ND	497	WA	603
FL	307	MI	417	OH	463	WV	925
GA	301	MN	652	OK	397	WI	563
HI	2,186	MS	297	OR	2,363	WY	1,134
ID	453	MO	374	PA	526		
IL	654	MT	507	RI	2,098		

Source: U.S. Bureau of the Census (1985). *Statistical Abstract of the United States* (105th ed.), pp. 276–277.

12. Average Secondary Teacher's Salary by State for 1984 (x $1,000)

State	Salary	State	Salary	State	Salary	State	Salary
AL	18.0	IN	22.0	NE	19.6	SC	18.2
AK	36.8	IA	20.8	NV	23.0	SD	16.7
AZ	21.8	KS	19.7	NH	17.4	TN	18.0
AR	17.4	KY	20.7	NJ	23.6	TX	20.8
CA	27.2	LA	19.6	NM	21.8	UT	21.4
CO	23.8	ME	18.2	NY	27.5	VT	18.5
CT	23.0	MD	24.5	NC	18.5	VA	20.8
DE	21.4	MA	22.7	ND	21.2	WA	25.2
FL	19.2	MI	29.1	OH	22.1	WV	18.5
GA	18.9	MN	25.2	OK	19.0	WI	23.6
HI	24.0	MS	16.2	OR	23.7	WY	25.1
ID	19.4	MO	19.8	PA	22.9		
IL	25.3	MT	21.8	RI	23.6		

Source: U.S. Bureau of the Census (1985). *Statistical Abstract of the United States* (105th ed.), p. 141.

In Exercises 13 and 14 draw (a) the point graph, (b) the frequency polygon, (c) the line graph, and (d) the histogram for the given frequency table and relative frequency table.

13.

Observation (x)	0	1	2	3	4	5
Frequency (f)	4	14	30	31	15	2
Relative Frequency (f/N)	.04	.15	.31	.32	.16	.02

14.

Observation (x)	-3	-2	-1	0	1	2
Frequency (f)	7	13	14	8	5	3
Relative Frequency (f/N)	.14	.26	.28	.16	.10	.06

In Exercises 15 and 16 draw histograms that represent the given frequency and relative frequency tables.

15.

Weight (pounds)	130.5– 140.5	140.5– 150.5	150.5– 160.5	160.5– 170.5	170.5– 180.5	180.5– 190.5
Frequency (f)	6	26	68	68	26	6
f/N	.03	.13	.34	.34	.13	.03

16.

Height (inches)	61.5– 63.5	63.5– 65.5	65.5– 67.5	67.5– 69.5	69.5– 71.5	71.5– 73.5
Frequency (f)	10	50	20	10	5	5
f/N	.10	.50	.20	.10	.05	.05

In Exercises 17–28 draw histograms that represent (a) the frequency distribution, (b) the relative frequency distribution, (c) the cumulative frequency distribution, and (d) the cumulative relative frequency distribution for the data in the previous exercise indicated.

17. Exercise 1.1.1
19. Exercise 1.1.3
21. Exercise 1.1.5
23. Exercise 1.1.7
25. Exercise 1.1.9
27. Exercise 1.1.11

18. Exercise 1.1.2
20. Exercise 1.1.4
22. Exercise 1.1.6
24. Exercise 1.1.8
26. Exercise 1.1.10
28. Exercise 1.1.12

29. Flip 5 coins 25 times and record the number of heads obtained. Organize your data in a frequency table, a histogram, and a cumulative frequency histogram.
30. While watching television, time the first 25 *individual* program interruptions (commercials, station breaks, etc.) to the nearest 10 seconds. Organize the data in a frequency table and a histogram.

C. SUMMARY

Data are the results of observations and can be *qualitative* or *quantitative*. Statistical techniques can be applied only to quantitative, or numerical, data. A *measurement* (or *score*) is a numerical representation of an observed event. A *frequency* is the number of times an event occurs.

The most basic convention for organizing data is some form of *frequency distribution*. A frequency distribution is a representation in which every observed value x is paired with the frequency f with which it is observed. A closely related representation is the *relative frequency distribution*, in which the frequency is divided by N, the total number of observations.

Frequency distributions and relative frequency distributions can also be given in *cumulative* form.

- The cumulative frequency of x is the frequency of x plus the frequencies of all values smaller than x.
- The cumulative frequency of the largest observed value must always be N.
- The cumulative *relative* frequency of x is the relative frequency of x plus the relative frequencies of all values smaller than x.
- The cumulative relative frequency of the largest observed value must always be 1.0.

If more than 8 to 10 different values of x are observed, data are sometimes *grouped* into classes or intervals. In a grouped distribution the frequency or relative frequency is reported for each class instead of each individual x-value. One way to make sure that no observations fall between classes is to define a common boundary between adjacent intervals. To make sure that no observation falls on a boundary between intervals, the *limits* of each class are ordinarily reported to one-half a unit beyond the most precise unit that the measurement instrument records.

Distributions of data can be presented in *tabular* form or in *graphic* form. Tabular representations include frequency tables, relative frequency tables, cumulative frequency tables, and cumulative relative frequency tables.

Graphic representations include point graphs, frequency polygons, line graphs, and histograms (bar graphs).

CHAPTER 2

Describing distributions

CHAPTER OUTLINE

A. DESCRIPTIVE STATISTICS

Desirable properties of descriptive statistics
- Single-valued
- Algebraically tractable
- Considers every observed value
- Considers the frequency of every observed value

Statistical shorthand
- Summation notation
- The Algebra of Summations

Exercises 2.1

B. DESCRIBING THE LOCATION OF A DISTRIBUTION

The mode: Mo.

The median: $\tilde{x}$
- The median for ungrouped data
- The median for grouped data

Exercises 2.2

The mean: $\bar{x}$
- The mean as "center of mass" of a distribution

Exercises 2.3

C. DESCRIBING THE DISPERSION OF A DISTRIBUTION

The range

The mean difference among observations

The mean deviation from the mean
The variance and standard deviation
- Computational formulae for the variance and standard deviation
- The coefficient of variation
Exercises 2.4

D. DESCRIBING THE SHAPE OF A DISTRIBUTION
Skewness
Kurtosis
Exercises 2.5

E. SUMMARY

A. DESCRIPTIVE STATISTICS

At the beginning of Chapter 1, we said that data are the results of observations that one uses to reach conclusions or generate interpretations. Conclusions and interpretations, however, are not based on the *raw* numerical observations themselves, but on the *information* they contain: The analytic chemist who measures the concentration of lead in 100 samples of groundwater taken from the site of a proposed housing development is not focally concerned with these 100 numbers themselves, no matter how comprehensibly organized. The chemist is concerned with whether or not future residents of the site are going to suffer lead poisoning. Organization, therefore, is only the first step that a researcher takes when trying to extract information from observations. The second step is to *describe* or *summarize* one or more properties of the numerical data that have been collected. To this end statisticians have developed a body of conventional indices called *descriptive statistics*:

> A descriptive statistic is a numerical index that describes or summarizes some characteristic of a frequency or relative frequency distribution.

In any collection of data, there are a number of properties that may be quantitatively described in this manner. Suppose, for example, that a college instructor administers an examination with 100 possible points. If the instructor wants to be sure that the test is pitched at an appropriate level of difficulty for this particular class, it is important to know whether the majority of students scored toward the low end of the total possible range of points, somewhere in the middle, or toward the high end. This information is summarized in a statistic that describes the *central tendency* or *location* of a set of observations.

Location is not the only descriptive property that might be of interest to a teacher. One purpose of any examination is to distinguish among students with greater and lesser mastery of the subject matter and thus provide a basis for the assignment of grades. If scores ran from the teens to the nineties, the instructor might at least feel confident that assignment of extreme grades reflected wide differences in test performance. If, however, the highest and lowest examination scores were separated by only a few points, there is reason to question the meaningfulness of *any* grades the instructor might assign. In evaluating an examination, therefore, a teacher might very conceivably wish to determine the extent to which test scores are spread across their total possible range of values. This property is captured by statistics that describe *dispersion*, or *variability*.

In addition, the instructor might be concerned with whether or not the test results agreed with other measures of student performance, such as a term paper. That is, did the students who received high scores on the examination also do well on the term paper? In this regard, the teacher might suspect that the two tasks involved different skills or even incompatible skills. The degree of agreement between any two measures is summarized by statistics that describe the property of *covariation*.

1. Desirable properties of descriptive statistics

As the reader might imagine, it is possible to conceive of a number of alternative mathematical expressions to describe any particular property of a collection of data. Some of these alternatives, however, incorporate more information or are easier to handle than others. The descriptive statistics that are most commonly used have, at least in part, evolved by the process of elimination or modification to meet the following criteria:

a. A descriptive statistic should be single-valued. That is, a descriptive property of a collection of data should be represented by a single number.

b. A descriptive statistic should be algebraically tractable. One should be able to calculate, transform, or otherwise manipulate a descriptive statistic using ordinary arithmetic operations, such as addition, subtraction, multiplication, roots, and powers.

c. A descriptive statistic should consider every observed value. By definition, a summary (numerical or otherwise) involves some loss of information. This loss is reduced if every numerical value that appears in the collection of measurements is represented in the calculation of one's descriptive statistic.

d. A descriptive statistic should consider the frequency of every observed value. As we saw in our discussion of frequency distributions, a collection of data is likely to include some numbers that appear more often than others. If, for example, a large number of people each tossed 100 coins and counted the number of heads, we would probably find that 2 heads or 3 heads showed up less frequently than 49 heads or 50 heads. Insofar as a descriptive statistic should somehow characterize the *entire collection* of data, each numerical value should therefore contribute to the statistic in proportion to the frequency with which it is observed.

2. Statistical shorthand

a. Summation notation. In the preceding section it was pointed out that descriptive statistics should be algebraically tractable. The most common arithmetic operation used in the calculation of descriptive statistics is addition. Given seven observations yielding the numbers 0, 9, 10, 10, 3, 9, and 15, we ordinarily express their sum as

$$0 + 9 + 10 + 10 + 3 + 9 + 15$$

This, however, is a sum of seven particular values. To express the idea of a sum of *any* seven values, more general notation is required. One way to do this would be to let x (or y or z, etc.) represent the measurement and to use a subscript indicator to distinguish among specific observations. A sum of *any* seven numerical observations would thus be represented as

$$x_1 + x_2 + x_3 + x_4 + x_5 + x_6 + x_7$$

Even this representation, however, is not completely general. It is limited to cases in which *exactly* seven observations are made. In the last chapter we pointed out that the number of observations in a collection of data is denoted by N. The general case of a sum of N observations can therefore be represented as

$$x_1 + x_2 + \cdots + x_N$$

where the ellipsis ($\cdots$) stands for the observations with subscripts between 2 and N.

Expressing a sum in this fashion provides generality, but it is not very compact; it takes up a good deal of space on a page, and if several sums are involved, expressing them in this way can produce a bewildering array. To simplify matters, mathematicians and statisticians have established the convention of using Σ (the Greek capital letter *sigma*, for *sum*) to symbolize the addition of *all* the numbers in a data set. The most streamlined form of *sigma* notation is

$$\sum x$$

This may look a little foreign, but all it means is *add up all the x-values*.

This is simple, but sometimes it can be ambiguous. In the example given above, seven observations were made, but only five different numerical values were observed, 0, 3, 9, 10, and 15. Consequently, it is unclear whether "all the x-values" refers to all seven *observations* or whether it means *all* five of the *numerical values* that were observed. When such ambiguity is possible, the *upper limit of summation* appears above the *sigma* to indicate the number of quantities that are to be added together.

The notation

$$\sum^{N} x$$

says unambiguously to add all of the N observations. In our example,

$$\sum^{N} x = \sum^{7} x = 0 + 9 + 10 + 10 + 3 + 9 + 15$$

In expressions that refer to *any* set of N numbers, $x_1, x_2, \ldots, x_N$, the general value x in the summation is subscripted with an *index*, usually i, j, or k. The *lower limit of summation* may also be indicated.

$$\sum^{N} x_i = \sum_{i=1}^{N} x_i = x_1 + x_2 + \cdots + x_N$$

Just as the capital letter N is used to indicate the number of *observations*, the lowercase letter n is used to indicate the number of distinct *values* that the observations assume (or, in some instances *can* assume). For the seven numbers 0, 9, 10, 10, 3, 9, and 15,

$$\sum^{n} x$$

tells the reader to add the n (e.g., five) numerical values:

$$\sum^{n} x = 0 + 3 + 9 + 10 + 15$$

To see how this convention is used, suppose that an experimenter had 1,000 persons each toss four coins and count the number of heads. This would yield 1,000 observations ($N = 1,000$), each of which could assume only five values ($n = 5$). There are two ways to compute the total number of heads. The experimenter could simply add the 1,000 numbers:

$$x_1 + x_2 + \cdots + x_{1,000}$$

where x_1 is the first observation, x_2 is the second observation, and so forth. However, the same result is obtained with many fewer calculations if each observed value is multiplied by its frequency and the five products are then summed:

$$x_1 f_1 + x_2 f_2 + x_3 f_3 + x_4 f_4 + x_5 f_5$$

where x_1 is the first *observed value*, f_1 is the number of times x_1 is observed, and so forth. In this expression the x-value and the f-value in each product are assigned the same numerical subscript. This makes it clear that x_1 is to be multiplied only by its own frequency, f_1, that x_2 is to be multiplied by f_2, and so on. When a sum of products is expressed in *sigma* notation, both factors (e.g., x and f) are subscripted with the *same* index:

$$\sum^{n} x_i f_i \qquad \left(\text{or } \sum_{i=1}^{n} x_i f_i \right)$$

b. The Algebra of Summations. The following rules are commonly called the *Algebra of Summations*. The proofs given in Appendix I are mathematically trivial, but it is our experience that no one (including the authors!) really *understands* the rules without the sort of drill that demonstrates how they work with real numbers. The problems at the end of this section are intended to provide that essential drill.

Rule 1: If c is a constant, then

$$\sum^{N} c = Nc$$

Think of $\sum^{N} c$ as $\sum^{N} x$ where every x-value is equal to the same number, c. For example, if one makes five observations (i.e., $N = 5$) and every observation is equal to 6 (i.e., $c = 6$), then

$$\sum^{N} x = 6 + 6 + 6 + 6 + 6 = 5(6)$$

Rule 2: If c is a constant and $x_1, x_2, \ldots, x_N$ is any collection of N numbers, then

$$\sum^{N} cx = c \sum^{N} x$$

Suppose $x_1 = 5$, $x_2 = 11$, and $x_3 = 16$, and we multiply each observation by 3, that is, $c = 3$. Then

$$\sum^{N} cx = 3(5) + 3(11) + 3(15)$$

$$= 3 \underbrace{(5 + 11 + 15)}$$

$$= c \qquad \sum^{N} x$$

In illustrating Rules 1 and 2, we let c be a positive number, but it should be understood that c may be either positive or negative.

Rule 3: If $x_1, x_2, \ldots, x_N$ is any set of N numbers and $y_1, y_2, \ldots, y_N$ is another set of N numbers, then

$$\sum^{N} x_i y_i = x_1 y_1 + x_2 y_2 + \cdots + x_N y_N$$

Let $x_1 = 5$, $x_2 = 11$, and $x_3 = 16$, and let $y_1 = 6$, $y_2 = 12$, and $y_3 = 17$. Then,

$$\sum^{N} x_i y_i = (5)(6) + (11)(12) + (16)(17)$$

Rule 4: If $x_1, x_2, \ldots, x_N$ is any set of N numbers and $y_1, y_2, \ldots, y_N$ is another set of N numbers, then

$$\sum_{i}^{N}(x_i + y_i) = \sum_{i}^{N} x_i + \sum_{i}^{N} y_i$$

Let $x_1 = 5$, $x_2 = 11$, and $x_3 = 16$, and let $y_1 = 6$, $y_2 = 12$, and $y_3 = 17$. Then,

$$
\begin{aligned}
\sum^{N}(x_i + y_i) &= (5 + 6) + (11 + 12) + (16 + 17) \\
&= \underbrace{(5 + 11 + 16)}_{\sum^{N} x_i} + \underbrace{(6 + 12 + 17)}_{\sum^{N} y_i} \\
&= \sum^{N} x_i \quad + \quad \sum^{N} y_i
\end{aligned}
$$

EXERCISES 2.1

1. Suppose $N = 5$ and $x_1 = 3$, $x_2 = 2$, $x_3 = 5$, $x_4 = 7$, $x_5 = 8$. Then

$$\sum x = \underline{\hspace{2cm}} \quad \text{and} \quad \frac{1}{N} \sum x = \underline{\hspace{2cm}}$$

2. $\sum x = \underline{\hspace{2cm}}$ and $\frac{1}{N} \sum x = \underline{\hspace{2cm}}$
for the numbers 4, 5, 8, 2, 1, 3, and 12.

3. $\sum^{6} x = \underline{\hspace{2cm}}$
for the numbers $x_1 = 2$, $x_2 = 2$, $x_3 = 2$, $x_4 = 2$, $x_5 = 2$, and $x_6 = 2$.

4. Let $N = 100$ and $x_i = 4$ for $i = 1, 2, \ldots, 100$. Then

$$\sum^{100} x_i = \underline{\hspace{2cm}}$$

5. Suppose $N = 6$ and $x_i = i$ for $i = 1, 2, 3, 4, 5, 6$. Then

$$\sum^{N} x_i = \underline{\hspace{2cm}}$$

6. Suppose $N = 6$ and $x_i = i^2$ for $i = 1, 2, 3, 4, 5, 6$. Then,

$$\sum^{N} x_i = \underline{\hspace{2cm}}$$

7. Suppose $N = 5$ and $x_1 = 3$, $x_2 = 2$, $x_3 = 2$, $x_4 = 4$, $x_5 = 3$. Then,

$$\sum^{N} x_i = \underline{\hspace{2cm}} \quad \text{and} \quad \sum^{n} x_i = \underline{\hspace{2cm}}$$

8. Suppose $N = 5$ and $x_i = i^2 - 3i + 3$ for $i = 1, 2, 3, 4, 5$. Then,

$$\sum_{i}^{N} x_i = \underline{\qquad} \quad \text{and} \quad \sum_{i}^{n} x_i = \underline{\qquad}$$

9. Suppose $x_1 = 1$, $x_2 = 2$, $x_3 = 4$, $x_4 = 8$, $x_5 = 16$, and $y_1 = 2$, $y_2 = 4$, $y_3 = 2$, $y_4 = 1$, $y_5 = 3$. Then

$$\sum x = \underline{\qquad}, \quad \sum y = \underline{\qquad}, \quad \text{and} \quad \sum xy = \underline{\qquad}$$

Does $(\sum x)(\sum y) = \sum xy$?

10. With the data of Exercise 2.1.9,

$$\sum x + \sum y = \underline{\qquad} \quad \text{and} \quad \sum (x + y) = \underline{\qquad}$$

The following x-values are scores obtained by 15 statistics students on the first midterm examination, and the y-values are scores obtained on the second midterm examination.

Student	x	y	Student	x	y	Student	x	y
1	42	22	6	55	31	11	51	28
2	35	18	7	45	24	12	51	28
3	51	28	8	43	23	13	54	30
4	48	26	9	50	25	14	43	23
5	60	34	10	48	26	15	54	30

11. For the first midterm, $N = \underline{\qquad}$.
12. For the second midterm, $N = \underline{\qquad}$.

13. $\sum^{N} x = \underline{\qquad}$

14. $\sum^{N} y = \underline{\qquad}$

Arrange the scores for the two midterms in ungrouped frequency distributions.

15. For the first midterm, $n = \underline{\qquad}$.
16. For the second midterm, $n = \underline{\qquad}$.
17. For the first midterm, $f_3 = \underline{\qquad}$.
18. For the second midterm, $f_7 = \underline{\qquad}$.

19. $\sum^{n} f_i x_i = \underline{\qquad}$.

20. $\sum^{n} f_i y_i = \underline{\qquad}$

21. What can you say about your answers to Exercises 2.1.13 and 2.1.19?
22. What can you say about your answers to Exercises 2.1.14 and 2.1.20?
23. Suppose the first examination was supposed to have 100 possible points, but in a fit of creative enthusiasm the instructor came up with questions that totaled 125 possible points. He was so proud of his efforts that he couldn't bear to part with any of them and decided instead to decrease everyone's score by one fifth. That is, he decided to multiply everyone's score on the first examination by 4/5, or .80. What is the sum of these adjusted scores?

24. Since all examinations in the course were intended to contribute equally to students' grades, the second examination was also intended to have 100 possible points, but the instructor came up with questions that totaled only 80 possible points. Since $80 + (1/4)80 = 100$, he therefore decided to increase everyone's score by one-fourth. That is, he multiplied every y-score by 1.25. What is the sum of these adjusted scores?

25. Compare your answers to Exercises 2.1.13 and 2.1.23. How are they related? State the rule given in the Algebra of Summations that accounts for this result.

26. Compare your answers to Exercises 2.1.14 and 2.1.24. How are they related? State the rule given in the Algebra of Summations that accounts for this result.

27. Add the score on the first examination to the score on the second examination for each student. What is the sum of the 15 combined scores? How does this result compare with the sum of your answers to 2.1.13 and 2.1.14? State the definition, rule, or principle by which you know that this is true in general?

28. Subtract each student's score on the second examination from his score on the first examination. What is the sum of the 15 differences? How does this result compare with the *difference* between your answers to 2.1.13 and 2.1.14? State this relationship in summation notation.

29. Suppose $x_1 = 1$, $x_2 = 1$, $y_1 = 1$, and $y_2 = 1$. Calculate Σxy and $\Sigma x \Sigma y$. What can you conclude?

30. Suppose $x_1 = 2$ and $x_2 = 3$. Calculate Σx^2 and $(\Sigma x)^2$. What can you conclude?

31. Multiply the score on the first examination by the score on the second examination for each student. What is the sum of these products?

32. Square each student's score on the first examination. What is the sum of these squares?

33. Multiply your answer to Exercise 2.1.13 by your answer to Exercise 2.1.14 and compare the result with your answer to Exercise 2.1.31. What does this tell you about Σxy and $\Sigma x \Sigma y$?

34. Square your answer to Exercise 2.1.13 and compare it with your answer to Exercise 2.1.32. What does this tell you about Σx^2 and $(\Sigma x)^2$?

B. DESCRIBING THE LOCATION OF A DISTRIBUTION

One of the most dramatic aspects of the Apollo 11 moon landing in 1969 was the live television coverage that allowed viewers around the world to witness Neil Armstrong's foot touch the lunar surface even as it happened. Today, satellite communication links and computer networks make available to the electronic and print news media an unprecedented wealth and immediacy of cultural, economic, and political information. Nonetheless, American television networks and newspapers devote more space to coverage of sports than to any other feature of contemporary life. For this reason, most of us have grown up in an environment saturated with the statistics that describe various aspects of competitive athletics. Although we may not always know exactly what they mean or how they are computed, most of us therefore realize that earned-run averages, golf handicaps, free-throw percentages, and the like tell us something about an individual athlete's *typical* performance.

These statistics summarize many observations and therefore incorporate considerable information about an athlete's overall skill. Like any summary, however, they sacrifice a certain amount of information for the sake of brevity. Roger Maris's lifetime batting average of .260 simply tells us that over his entire major league career, 26 percent of his turns at bat resulted in base hits. It does not tell us that he had a hot streak with the Yankees in 1961 and, while breaking Babe Ruth's record for home runs hit in a single season, averaged .269. Nor does it tell us that in 1966 he hit the ball only 81 times in 348 times at bat and was traded to St. Louis. More generally, these statistics tell us nothing about the *extreme* values in one's collection of numerical observations. Rather, they tell us where the *typical* observation is *located*. As we shall see below, the "typical observation" may not actually be observed; the hockey goalie who in the course of a season gave up an average of four goals per game may not have had a single game in which *exactly* four goals were scored against him. Thus, it is sometimes more helpful to think of such measures as indicating where observations *tend to be* con*centrated*. Location statistics may therefore be understood to describe the *cen*tral *tend*ency of a distribution.

1. The mode: *Mo.*

In everyday speech, something is "in the mode" if it is fashionable or popular. In statistics "popularity" refers to frequency of observation, and the most frequently observed value in a collection of observations is therefore called the *mode*. In Chapter 1 we proposed an imaginary coin-toss experiment that yielded the distribution shown in Table 2.1. In this distribution two heads is the most frequently observed value, so

$$Mo. = 2$$

The mode is easily determined and simple to interpret, but it has several undesirable properties. First, it is not necessarily single-valued. Two or more values may be observed with equal frequency. If some of the people in our experiment were tossing coins biased in favor of heads and some were tossing coins biased in favor of tails, we might expect to obtain a distribution with *two* modes (called a *bimodal* distribution), as shown in Table 2.2.

Table 2.1. Frequency distribution for number of heads obtained in toss of four coins with experiment repeated 25 times

Observation (x)	0	1	2	3	4
Frequency (f)	1	7	10	6	1

Table 2.2. Bimodal frequency distribution
(number of heads obtained in toss of four
coins with experiment repeated 25 times)

Observation (x)	0	1	2	3	4
Frequency (f)	3	7	6	7	2

OJ

Moreover, the mode (or modes) is determined entirely by inspection; it cannot be calculated arithmetically. Finally, the mode tells us only about the most popular value(s) in the distribution and considers neither the other values nor their frequencies.

2. The median: $\tilde{x}$

A more commonly used statistic is the *median*. As the name suggests, the median is the value in the *middle* of the distribution, by which we mean that there are exactly as many observations *above* the median as there are *below* the median.

a. The median for ungrouped data. To find the median for any collection of ungrouped data, one must first order the N observations from the smallest to the largest, so that the smallest observed value occupies position 1, the next smallest position 2, and so on. Then, if N is odd, the median is the value in the middle position, that is, position $(N + 1)/2$. For example, if one has 13 observations,

$$1, 1, 2, 3, 4, 4, 5, 6, 8, 10, 14, 15, 17$$

the median falls in position $(13 + 1)/2 = 7$. The value in the seventh position is 5, so $\tilde{x} = 5$.

If N is odd, the median is the value in position

$$\frac{N + 1}{2}$$

If N is even, the situation is not quite so clear. Suppose we add one more observation, say $x = 2$, to the data collection given above. We now have 14 numbers in our ordered list:

$$1, 1, 2, \mathbf{2}, 3, 4, 4, 5, 6, 8, 10, 14, 15, 17$$

Because 14 is an even number, there is no "middle" position. When N is even, the median is therefore defined by convention to fall halfway between the middle *two* observations. The two scores in the middle of the distribution are those in positions $N/2$ and $(N/2) + 1$, so for $N = 14$, the middle *two* are the seventh and eighth numbers. The number 4 occupies position seven in our list of 14 observations, and the number occupying position eight is 5, so $\tilde{x} = (4 + 5)/2 = 4.5$.

If N is even, the median is the *average* of the values falling in positions

$$\frac{N}{2} \quad \text{and} \quad \frac{N}{2} + 1$$

In the last example, the median ($\tilde{x} = 4.5$) did not correspond to any value that was actually included in the collection of observations. When N is even the median can be an observed value *only* if the middle pair of observations happen to be equal. This is why it is so often the case that *no* individual in a large corporation earns the median salary for all employees, *no* student receives the median score on an examination, and so forth.

b. The median for grouped data. In the last section we saw that median is defined differently when N is odd than when N is even and—in the latter case—that the median is itself an observed value only when the middle pair are equal. Clearly, then, it not possible to define the median for grouped data in a way that agrees with the median for ungrouped data in all three of these situations. For reasons that go beyond the scope of this book, the median for grouped data is based on the ungrouped case in which N is *even* and the middle pair are *unequal*.

When data are grouped, the median $\tilde{x}$ is that value at or below which exactly 50 percent of the observations fall.

That is, the sum of the relative frequencies for all x less than or equal to the median is .50:

$$\sum f_i / N = .50$$

for values $x_i \leq \tilde{x}$. Or, in terms of cumulative relative frequency,

$$\text{Cumulative} \frac{f_{\tilde{x}}}{N} = .50$$

To see that this agrees with the median for ungrouped data when N is even and the middle pair are unequal, consider again the 14 observations,

$$1, 1, 2, 2, 3, 4, 4, 5, 6, 8, 10, 14, 15, 17$$

The median for these data was determined to be 4.5, and it is seen that exactly half (i.e., seven) of the observations fall at or below this value.[1]

To see what happens if the middle pair are equal, suppose we substitute another 1 for the 5 in our data set. The median is then equal to 4, and 8 of the 14 observations fall at or below the median. To see what happens if N is odd, suppose we drop 17 from the data set, leaving only 13 observations. The median is 4, and 7 of the 13 observations fall at or below the median.

Even if we treat grouped data as if N is even and the middle pair are unequal, it is not ordinarily possible to identify the middle value(s), but a close approximation can be obtained by *linear interpolation*[2] *if we are willing to assume that all of the observations in any interval are evenly distributed* from the lower class limit to the upper class limit—that is, if we assume that all of the values in the interval have the *same* relative frequency.

To see how this assumption allows us to determine the median for a collection of grouped data, recall from Chapter 1 that a relative frequency distribution can be represented in the form of a histogram where the relative frequency of every *value* is represented by the area of a bar. *If we are willing to assume that all of the values in an interval x_a to x_b are observed with the same frequency* (and therefore have the same histogram height), we can likewise represent the relative frequency of every *interval* in a grouped distribution as the area of a histogram bar.

Example 2.1.1. Interviewers for a particular market survey organization routinely ask the following question at the end of each interview: To the nearest thousand dollars, how much money did your household earn last year? Did you earn $4,000 or less? $5,000 to $9,000? $10,000 to $14,000? $15,000 to $19,000? or $20,000 or more? Suppose that after conducting 20 interviews, one of the interviewers tabulates the data reported in Table 2.3 and represented in the histogram shown in Figure 2.1. Please note that the interviewer has established class limits according to the

Table 2.3. Annual household income ($\times \$1,000$) for 20 interview respondents

Income (x)	4.5 or less	4.5–9.5	9.5–14.5	14.5–19.5	19.5 or more
Frequency (f)	3	1	8	6	2
f/n	.15	.05	.40	.30	.10

[1] Although the procedure for determining the median for ungrouped observations when N is even and the middle values are unequal always yields a value at or below which 50 percent of the observations fall, the converse is not true. In the example given above, 50 percent of the observations lie on either side of *any* value between 4 and 5, e.g., 4.1 or 4.973.

[2] A discussion of linear interpolation is presented in Appendix II.

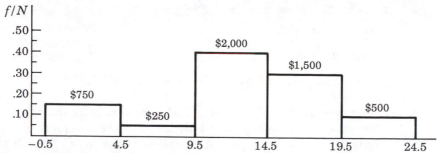

Figure 2.1. Relative frequency histogram for incomes (× $1,000) for 20 interview respondents. Number above each bar is its area.

convention discussed in Chapter 1. Note also that we have "closed" the end intervals in Figure 2.1 so that the bars will each have a width of $5,000.

Recall, too, that the sum of the areas for bars representing *all values equal to or less than x* is proportional to the *cumulative* relative frequency of x. If the area of a relative frequency histogram for grouped data were divided precisely in half by a vertical line, the line would therefore cross the horizontal axis at the value for which cumulative relative frequency is .50. By the definition given above, this point is the *median* of the distribution.

Example 2.1.2. The area for each bar of the histogram in Figure 2.1 is its height (f/N) multiplied by its width ($5,000), and the total area for the histogram is the sum of these areas:

.15($5,000) + 0.5($5,000) + .40($5,000) + .30($5,000) + .10($5,000)

= 1.0($5,000) = $5,000

The median is therefore the point below which the area of the histogram is half of $5,000 or $2,500.

The area of the first two bars is

.15($5,000) + .05($5,000) = $1,000

which is less than $2,500, and the area of the first three bars is

.15($5,000) + .05($5,000) + .40($5,000) = $3,000

which is more than $2,500. The median must therefore lie somewhere in the third interval, which is called the *median interval*. Specifically, the median is that point in the third interval that would add exactly $1,500 to the area of the first two bars. The area of the third bar is

.40($5,000) = $2,000

and the $1,500 we need is 3/4 of this:

$$\frac{\$1,500}{\$2,000} = \frac{3}{4}$$

The median is therefore the point that cuts off the lower 3/4 of the third bar.

Because the top of the bar is flat (which reflects the assumption that observations are evenly distributed across the interval), the point that would cut off the lower 3/4 of its area is simply 3/4 of the distance from the lower class limit of the interval ($x_a = \$9,500$) to the upper class limit ($x_b = \$14,500$):

$$x_b - x_a = \$14,500 - \$9,500 = \$5,000$$

so

$$.75(x_b - x_a) = \$3,750$$

and the median is obtained by adding $3,750 to the lower class limit (x_a) of the median interval:

$$\tilde{x} = \$9,500 + \$3,750 = \$13,250$$

In Figure 2.2 the area below the median ($\tilde{x} = \$13,250$) is indicated by shading.

Example 2.1.3. Now suppose that the survey included one more observation, say another income between $14,500 and $19,500. The frequency and relative frequency distribution would be as given in Table 2.4.

If we were once again to construct a histogram for our data, the area for the first two bars would be

$$.143(\$5,000) + .048(\$5,000) = \$955.00$$

which is $1,545 short of 50 percent of the total area. The third bar would have an area of .381($5,000) = $1,905, and the point that would bisect the area of the graph would therefore be $1,545/$1,905, or approxi-

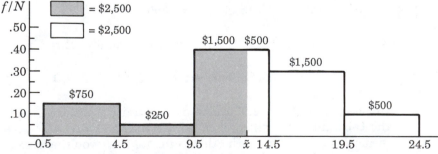

Figure 2.2. Relative frequency histogram for incomes (×$1,000) for 20 interview respondents. Number above each bar is area, and shaded portion represents area below median.

Table 2.4. Annual household income ($\times$ $1,000) for 21 interview respondents

Income (x)	4.5 or less	4.5–9.5	9.5–14.5	14.5–19.5	19.5 or more
Frequency (f)	3	1	8	7	2
f/N	.143	.048	.381	.333	.095

mately 81 percent of the distance from $9,500 to $14,500:

$$.81(\$14,\!500 - \$9,\!500) = .81(\$5,\!000) = \$4,\!050$$

so

$$\tilde{x} = \$9,\!500 + \$4,\!050 = \$13,\!550$$

The strategy for determining the median with grouped data can therefore be applied whether N is even or odd, and is expressed computationally as follows:

$$\tilde{x} = x_a + \frac{.50N - \text{cumulative } f_a}{f_m}(x_b - x_a) \qquad \textbf{[2.1]}$$

where

(**1**) x_a is the *lower* class limit of the median interval,
(**2**) x_b is the *upper* class limit of the median interval,
(**3**) cumulative f_a is the number of observations *below* the median interval, and
(**4**) f_m is the number of observations *in* the median interval, i.e., f_m = cumulative f_b − cumulative f_a.

Of course, the class containing the median must be obtained by inspection. Also, if f_m happens to be zero, this formula fails, since division by zero is meaningless. (See Example 2.3.)

Example 2.2. The basic strategy in Example 2.1 was based on interpreting the median as the point that separates the upper half of the area of a relative frequency histogram from the lower half. This required that all class intervals be of equal width. In the present example we see that the formula given above can be applied even when class widths are unequal, though in this case the area interpretation of the median is not valid. Table 2.5 lists the populations of all California cities with populations of at least 100,000 in 1982.

There are 27 observations. The first two intervals account for fewer than half the observations (12 < 13.5), and the first three account for more than half (17 > 13.5). The median interval is therefore the third

Table 2.5. Populations of California cities to nearest 1,000

Population (x)	110.5 or less	110.5–130.5	130.5–200.5
Frequency (f)	5	7	5

Population (x)	200.5–300.5	300.5–500.5	500.5 or more
Frequency (f)	4	2	4

class, 130.5–200.5. In terms of our formula, we have

$.50N = .50(27) = 13.5$

$x_b - x_a = 200.5 - 130.5 = 70$

Cumulative $f_a = 12$

$f_m = 5$

The median is therefore

$$130.5 + \frac{13.5 - 12}{5}(70) = 130.5 + \frac{1.5}{5}(70) = 151.5$$

That is, $\tilde{x} = 151,500$.

Example 2.3. As indicated above, the formula for the median fails if $f_m = 0$. Here is what is conventionally done in that case. The U.S. coin collection owned by one of the authors has the face values summarized in Table 2.6:

Table 2.6. Face values of coin collection $(\times 1\text{¢})$

Value (x)	4.5 or less	4.5– 9.5	9.5– 14.5	14.5– 19.5	19.5– 24.5	24.5– 29.5	29.5 or more
f	10	17	23	0	0	47	3

The median for grouped data derives from the ungrouped case where N is even and the middle pair are unequal. With 100 observations, the median therefore falls halfway between the 50th value and the 51st value. The cumulative frequency of the third interval is 50, and since grouped observations are assumed to be evenly distributed within every interval, we think of the 50th observation as lying at the *upper* limit of the third interval, 14.5. Similarly, we think of the 51st observation as lying at the *lower* limit of the next class that *includes any observations*. Since 14.5–19.5 and 19.5–24.5 are empty, the 51st observation lies at the lower end of the interval 24.5–29.5, and

$$\tilde{x} = \frac{14.5 + 24.5}{2} = 19.5$$

That is, we combine all consecutive empty intervals and take the median to be the midpoint of this *collapsed* interval.

If the collection included one more coin with face value equal to 25 cents or more, the median would fall in the interval 24.5–29.5 and would be calculated by formula [2.1], ignoring the empty intervals.

Example 2.4. In Chapter 1 we pointed out that in tabulating data it is customary to extend the class limits one half a unit beyond one's precision of measurement even in cases where an experiment yields integer data and each interval includes only one observable value. That is, *ungrouped* data are customarily treated as if they were *grouped* data. One advantage of this convention is that the median can become a more meaningful index of central tendency. Suppose that the numbers given below are ratings of teacher performance (1 = poor; 5 = outstanding) received by two instructors:

Instructor 1:   1 1 2 2 3 4 4 4 4 4 5

Instructor 2:   1 2 3 4 4 4 5 5 5 5 5

If we were to determine median ratings for both instructors according to the procedures given for ungrouped data, both would receive median ratings of 4. One feels, however, that Instructor 2 typically received higher ratings and that this should somehow be reflected in numerical summaries of their performance. If the data are tabulated treating each x-value at a unit-width interval (Table 2.7), the procedure for computing a median from *grouped* data will capture the difference in their ratings. Since $N = 11$ for both instructors, their medians both fall in the class 3.5–4.5. For Instructor 1

$$.50N = .50(11) = 5.5$$

$$x_b - x_a = 4.5 - 3.5 = 1.0$$

Cumulative $f_a = 5$

$$f_m = 5$$

Table 2.7. Student ratings of two instructors

Rating (x)	$-0.5 - 1.5$	1.5–2.5	2.5–3.5	3.5–4.5	4.5–5.5
Instructor 1 (f)	2	2	1	5	1
Instructor 2 (f)	1	1	1	3	5

His median is therefore

$$3.5 + \frac{5.5 - 5}{5}(1.0) = 3.5 + \frac{.5}{5}(1.0) = 3.6$$

For Instructor 2

$.50N = .50(11) = 5.5$

$x_b - x_a = 4.5 - 3.5 = 1.0$

Cumulative $f_a = 3$

$f_m = 3$

His median is therefore

$$3.5 + \frac{5.5 - 3}{3}(1.0) = 3.5 + \frac{2.5}{3}(1.0) \doteq 4.33$$

The symbol $\doteq$ means that the right side is accurate to as many decimal places shown. The median is actually $4\frac{1}{3}$.

The median satisfies two of our criteria for a good statistic: The median is always single-valued, and since the number of observations above the median equals the number of observations below the median, the median takes into account the frequencies of all values in the distribution.

Although it considers the frequencies of all observed values, the median does not consider the values per se. That is, x-values are considered to be either greater than or less than the median, but their relative magnitudes are not distinguished. Thus, an extreme x-value and a moderate x-value make the same contribution to the median if both are observed with equal frequency. This is illustrated in the two distributions given in Table 2.8. Even though the distribution for the second group includes much smaller numerical observations than the distribution for the first group, both have the same median, $\tilde{x} = 100$.

Table 2.8. Two frequency distributions with the same median

x		2	3	4	85	90	92	100	107	111	116
f	(Group 1)	0	0	0	1	2	2	1	2	2	1
f	(Group 2)	1	2	2	0	0	0	1	2	2	1

It should be emphasized that this property is not necessarily a deficiency. When we discuss *skewness* later in the chapter, we will see that location is sometimes better described by an index that is resistant to extreme values.

The median's most serious defect is that it is nonalgebraic. This is sometimes difficult to see, because the median can be expressed in terms of ordinary arithmetic operations. For example, the formula for finding the median from grouped observations involves only addition, subtraction, and division. However, *all* procedures for calculating the median require that observations be ordered, either individually or by interval, and such orderings require comparisons that are *logical* rather than *arithmetic*.

EXERCISES 2.2

Find the mode for the data collections in Exercises 1–4.

1. 1, 3, 4, 4, 2, 3, 5, 1, 3, 3, 5, 4, 2, 2, 2, 3, 3, 4, 4, 5

2. 3, 3, 10, 3, 12, 14, 17, 20, 14, 3, 12, 3, 10, 3, 12, 12, 10, 3, 12, 14

3. Number of Marriages per 1,000 of U.S. Population for the years 1965 – 1981

Year	Rate	Year	Rate	Year	Rate
1965	9.3	1971	10.6	1977	9.9
1966	9.5	1972	10.9	1978	10.3
1967	9.7	1973	10.8	1979	10.4
1968	10.4	1974	10.5	1980	10.6
1969	10.6	1975	10.0	1981	10.6
1970	10.6	1976	9.9		

Source: U.S. Bureau of the Census (1985). *Statistical Abstract of the United States* (105th ed.), p. 57.

4. Number of Deaths per 1,000 of U.S. Population for the Years 1960 and 1965 – 1981

Year	Rate	Year	Rate	Year	Rate
1960	9.5	1970	9.5	1976	8.8
1965	9.4	1971	9.3	1977	8.6
1966	9.5	1972	9.4	1978	8.7
1967	9.4	1973	9.3	1979	8.5
1968	9.7	1974	9.1	1980	8.8
1969	9.5	1975	8.8	1981	8.6

Source: U.S. Bureau of the Census (1985). *Statistical Abstract of the United States* (105th ed.), p. 57.

In Exercises 5 and 6 calculate the mode for each data collection. Can the mode be regarded in any sensible way as the "most popular" value?

5. 1, 1.01, 1.02, 1.03, 1.04, 1.05, 1.06, 1.07, 1.08, 1.09, 1,000, 1,000
6. 1, 2, 3, 4, 5, 6, 7, 8, 9, 10

In Exercises 7–24 find the median.

7. 1, 1, 2, 2, 3, 4, 5, 7, 7, 7, 9
8. 2, 2, 2, 3, 3, 4, 7, 8, 9, 10, 15
9. 7, 8, 1, 1, 9, 19, 11, 2, 3, 4, 8
10. 7, 3, 15, 3, 12, 3, 4, 5, 10, 9, 12
11. 1, 1, 1, 2, 2, 3, 3, 3, 4, 5
12. 1, 3, 4, 4, 5, 5, 5, 6, 6, 7
13. 7, 4, 8, 9, 3, 9, 2, 1, 3, 5
14. 5, 11, 3, 1, 5, 7, 8, 9, 10, 8
15. Frequency Distribution of Weight (pounds) for 100 Persons

Weight	130.5–140.5	140.5–150.5	150.5–160.5	160.5–170.5	170.5–180.5	180.5–190.5
Frequency	10	20	30	20	10	10

16. Frequency Distribution of Height (inches) for 100 Persons

Height	61.5–63.5	63.5–65.5	65.5–67.5	67.5–69.5	69.5–71.5	71.5–73.5
Frequency	10	20	30	20	10	10

17. Frequency Distribution of Maze-Running Times (recorded to nearest second) for 100 "Maze-Bright" Rats

Time	30.5–36.5	36.5–42.5	42.5–48.5	48.5–54.5	54.5–60.5
Frequency	4	10	14	27	45

18. Frequency Distribution of Maze-Running Times (recorded to nearest second) for 100 "Maze-Dull" Rats

Time	60.5–67.5	67.5–74.5	74.5–81.5	81.5–88.5	88.5–95.5	95.5–102.5
Frequency	23	38	17	12	6	4

19. The data in Exercise 1.1.9 on p. 14 (a) as they are presented in the text and (b) as you grouped them in Exercise 1.1.9.
20. The data in Exercise 1.1.10 on p. 14 (a) as they are presented in the text and (b) as *you* grouped them in Exercise 1.1.10.

Exercises 21 and 22 are evaluations of teachers on a 1–5 rating scale. Assume that ratings are actually grouped in intervals one unit in width. For example, the rating "4" represents a value somewhere between 3.5 and 4.5.

21. Instructor 1

Rating (x)	5	4	3	2	1
Frequency (f)	3	6	2	1	1

22. Instructor 2

Rating (x)	5	4	3	2	1
Frequency (f)	8	3	2	0	0

23. Highest recorded temperatures (°F) for 70 selected U.S. cities for various periods, all ending in 1982. (Source: U.S. Bureau of the Census, ibid., p. 209.)

118, 115, 114, 113, 113, 112, 111, 110, 110, 109, 109, 108, 108, 108, 107, 107, 107, 107, 107, 107, 107, 106, 106, 106, 106, 106, 105, 105, 105, 105, 105, 105, 105, 105, 104, 104, 104, 104, 104, 104, 104, 104, 104, 103, 103, 103, 103, 102, 102, 102, 102, 102, 102, 102, 102, 102, 101, 101, 100, 100, 99, 99, 99, 98, 98, 98, 97, 93, 90

24. Lowest recorded temperatures (°F) for 70 selected U.S. cities for various periods, all ending in 1982. (Source: U.S. Bureau of the Census, ibid., p. 210.)

60, 53, 31, 23, 20, 20, 17, 14, 12, 12, 7, 5, 4, 4, 3, 0, −1, −3, −3, −4, −5, −5, −5, −7, −7, −8, −10, −11, −12, −12, −13, −13, −13, −15, −15, −15, −16, −16, −17, −18, −19, −19, −19, −20, −20, −21, −21, −22, −23, −23, −24, −25, −25, −25, −26, −26, −26, −28, −30, −30, −30, −34, −34, −36, −36, −37, −39, −39, −43, −44

3. The mean: $\bar{x}$

Sometime in the course of your education you were taught to compute an average. The average, you were told, is simply the sum of several values divided by the number of values comprising the sum. In the intervening years you may have used this simple formula to compute such things as your average monthly income, the average cost of books per class, or perhaps your average highway speed maintained on a long trip. Certainly very few of you have escaped the omnipresent gradepoint average. This childhood friend is the most widely used of all measures of location and is about to renew your acquaintance as the arithmetic *mean*, $\bar{x}$.

Let us suppose that we have made the following observations, 5, 4, 2. The average, or mean, of these values is computed by taking their sum and dividing by the number of observations:

$$\bar{x} = \frac{5 + 4 + 2}{3} = \frac{11}{3} \doteq 3.67$$

In the general case of N observations, this may be expressed by the definitional formula

$$\bar{x} = \frac{\sum\limits^{N} x}{N} \qquad\qquad [2.2]$$

The arithmetic mean of N observations is the sum of the observations divided by N

$$\bar{x} = \frac{\sum\limits^{N} x}{N}$$

In real life, statistical problems rarely involve as few as three observations, and use of the definitional formula can be very tedious. Suppose, for example, that we have nine observations, 5, 5, 4, 4, 4, 4, 2, 2, 2. Using the definitional formula, the mean is

$$\frac{\sum\limits^{9} x}{9} = \frac{5 + 5 + 4 + 4 + 4 + 4 + 2 + 2 + 2}{9} \doteq 3.56$$

These data, however, consist of repeated observations of only three values, 5, 4, and 2, and as discussed on pages 22–23,

$$\sum\limits^{N} x_1 = \sum\limits^{n} x_i f_i$$

so

$$\bar{x} = \frac{\sum\limits^{3} x_i f_i}{9} = \frac{5(2) + 4(4) + 2(3)}{9} \doteq 3.56$$

When many of one's observations assume the same numerical value, it is therefore easier to use the following computational formula:

$$\bar{x} = \frac{\sum\limits^{n} x_i f_i}{N} \qquad\qquad [2.3]$$

The mean is single-valued, and from the computational formula given above we can see that it also satisfies all of our other criteria for a good statistic. Calculation involves only the operations of addition, multiplication, and division, so the mean is algebraically tractable. The summation runs from x_1 to x_n, and every observed value is therefore included in the calculation. Finally, we see that the frequency of every x-value f_i is also included in the computation.

These last two criteria are compromised only if one must calculate the mean from grouped data. When data are grouped in n intervals it is assumed that all observations in the interval x_a to x_b are concentrated at the midpoint of the interval (sometimes called the *class mark*),

$$\frac{x_b + x_a}{2}$$

Then, in the expression

$$\bar{x} = \frac{\sum\limits_{}^{n} x_i f_i}{N}$$

the value x_i is the midpoint of the ith interval and f_i is the frequency of observations in the ith interval.

Example 2.5. To see how this works, let us calculate the mean income associated with the data given in Example 2.1.1 (p. 30) and shown again in Table 2.9. First we must close the first and last intervals in some sensible way, perhaps as we did to compute the median. Then,

$$x_1 = \frac{-0.5 + 4.5}{2} = 2$$

$$x_2 = \frac{4.5 + 9.5}{2} = 7$$

$$x_3 = \frac{9.5 + 14.5}{2} = 12$$

$$x_4 = \frac{14.5 + 19.5}{2} = 17$$

$$x_5 = \frac{19.5 + 24.5}{2} = 22$$

Table 2.9. Annual household income ($\times$ $1,000) for 20 interview respondents

Income (x)	4.5 or less	4.5–9.5	9.5–14.5	14.5–19.5	19.5 or more
Frequency (f)	3	1	8	6	2
f/N	.15	.05	.40	.30	.10

and the mean is then

$$\bar{x} = \frac{3(2) + 1(7) + 8(12) + 6(17) + 2(22)}{20} = 12.75$$

or $12,750.00.

a. The mean as "center of mass" of a distribution. As discussed earlier, the median is the *middle* value of a frequency distribution in the sense that the number of observations above the median equals the number of observations below the median. The sense in which the arithmetic mean falls at the middle of the distribution is captured by a *physical* model of a frequency distribution. Suppose an experiment yielded the following observations: 2, 2, 2, 4, 4, 4, 4, 5, 5. To construct a model of this distribution we could use an ordinary ruler to represent the horizontal axis. Since all of our x-values are positive, they correspond, respectively, to the 2-, 4-, and 5-in. markings on the ruler. At each of these points on the "axis" we could place a wood block or a stack of coins or some other weight proportional to the frequency with which that value appears. These weights would then be analogous to the bars of a histogram. If we were now to locate a fulcrum at the point B where the weighted ruler is perfectly balanced (see Figure 2.3), we would find that B is equal to the mean, $\bar{x} \doteq 3.56$. The mean, therefore, lies in the "middle" of the distribution in the sense that it is the exact *center of mass* of the distribution.

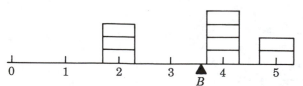

Figure 2.3. Physical model of frequency distribution illustrating mean as center of mass.

To understand why this is so, it might be helpful to review some basic high school physics. In Figure 2.4 we have a lever, a fulcrum, and a weight of 100 lbs suspended from one end of the lever at a distance of 1 ft from the fulcrum. If the other end of the lever is 10 ft from the fulcrum, how much weight must we apply to balance the 100-lb weight? We all know from personal experience that when we apply weight to one end of a lever it causes the other end to move in the opposite direction. Some 2,000 years ago a waterlogged Greek named $A\rho\chi\iota\mu\eta\delta\eta s$ informed the world that this force is equal to the weight applied to the lever multiplied by its distance from the fulcrum. That is,

Force = weight × distance

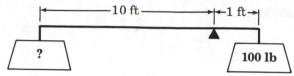

Figure 2.4. 100-lb weight 1 ft from fulcrum counterbalanced by unknown weight 10 ft from fulcrum.

The 100-lb weight therefore exerts a force of 100 ft-lbs and would be counterbalanced by a 10-lb weight suspended from the 10-ft lever arm:

100 lbs × 1 ft = 10 lbs × 10 ft

Now, think of the model in Figure 2.3 as a system comprised of a lever, a fulcrum, and various weights. Because it is balanced, the forces on both sides of the fulcrum must be equal. However, the weights to the right of the fulcrum tend to pull the lever *clockwise*, while the weights to the left of the fulcrum pull the lever *counterclockwise*. When represented algebraically the two forces therefore take opposite signs.

(weight × distance)$_{\text{right}}$ = − (weight × distance)$_{\text{left}}$

or

(weight × distance)$_{\text{right}}$ + (weight × distance)$_{\text{left}}$ = 0   **[2.4]**

That is, the sum of *all* forces is equal to zero.

Recall that each "weight" represents the frequency f_i associated with a particular value x_i and that fulcrum is at point B. The distance of any weight f_i from the fulcrum must therefore be $(x_i - B)$, so the force exerted by any weight f_i is

$$f_i(x_i - B)$$

From Equation **[2.4]**, then,

$$\sum^{n} f_i(x_i - B) = 0$$

If we convert this expression from frequency notation and sum across all N observations,

$$\sum^{N} (x_i - B) = 0$$ **[2.5]**

By the rules of summation,

$$\sum_{i}^{N}(x_i - B) = \sum_{i}^{N}x_i - \sum_{i}^{N}B$$

so from [2.5]

$$\sum_{i}^{N}x_i - \sum_{i}^{N}B = 0$$

and

$$\sum_{i}^{N}x_i = \sum_{i}^{N}B$$

Since B is a constant,

$$\sum_{i}^{N}x_i = NB$$

Therefore,

$$B = \frac{\sum_{i}^{N}x_i}{N} = \bar{x}$$

which means that the balance point B falls at the mean.

If we substitute $\bar{x}$ for B in equation [2.5] we obtain

$$\sum_{i}^{N}(x_i - \bar{x}) = 0 \qquad\qquad [2.6]$$

The difference $x_i - \bar{x}$ is defined as the *deviation* of x_i from the mean. The mean is therefore that point about which the sum of deviations is zero.

Formula [2.6] can be derived directly from the Algebra of Summations without recourse to mechanical analogs and plays an important role in many of the proofs and derivations found throughout this text. Because the mean itself can be represented in a variety of ways, the sum of deviations about the mean does not always appear in its most obvious form, $\Sigma(x - \bar{x})$, and it takes a bit of practice to recognize it. In the exercises below, the reader will be introduced to some of the diverse forms and contexts in which this expression appears.

──────── EXERCISES 2.3 ────────

In Exercises 1–12 calculate the mean of each data collection
   1. 1, 1, 2, 2, 3, 4, 5, 7, 7, 7, 9
   2. 2, 2, 2, 3, 3, 4, 7, 8, 9, 10, 15
   3. 7, 8, 1, 1, 9, 19, 11, 2, 3, 4, 8
   4. 7, 3, 15, 3, 12, 3, 4, 5, 10, 9, 12

5. Frequencies of Course Evaluation Ratings

Rating (x)	5	4	3	2	1
Frequency (f)	3	6	2	1	1

6. Frequencies of Course Evaluation Ratings

Rating (x)	5	4	3	2	1
Frequency (f)	8	3	2	0	0

7. Frequency Distribution of Weight (pounds) for 100 Persons

Weight	130.5–140.5	140.5–150.5	150.5–160.5	160.5–170.5	170.5–180.5	180.5–190.5
Frequency	10	20	30	20	10	10

8. Frequency Distribution of Height (inches) for 100 Persons

Height	61.5–63.5	63.5–65.5	65.5–67.5	67.5–69.5	69.5–71.5	71.5–73.5
Frequency	10	20	30	20	10	10

9. Frequency Distribution of Maze-Running Times (recorded to nearest second) for 100 "Maze-Bright" Rats

Time	30.5–36.5	36.5–42.5	42.5–48.5	48.5–54.5	54.5–60.5
Frequency	4	10	14	27	45

10. Frequency Distribution of Maze-Running Times (recorded to nearest second) for 100 "Maze-Dull" Rats

Time	60.5–67.5	67.5–74.5	74.5–81.5	81.5–88.5	88.5–95.5	95.5–102.5
Frequency	23	38	17	12	6	4

11. (a) Calculate the *ungrouped* mean for the data set in Exercise 1.1.9 on p. 14.
    (b) Calculate the *grouped* mean for these data using the intervals you constructed in your answer to Exercise 1.1.9.
12. (a) Calculate the *ungrouped* mean for the data set in Exercise 1.1.10 on p. 14.
    (b) Calculate the *grouped* mean for these data using the intervals you constructed in your answer to Exercise 1.1.10.

Exercises 13 and 14 refer to the following table. Fill in all the blanks, letting $c = 2$.

x	cx	$(x+c)$	$(x-c)$	y	$(x+y)$
3	___	___	___	9	___
4	___	___	___	13	___
6	___	___	___	15	___
8	___	___	___	19	___
9	___	___	___	24	___
Σ	___	___	___	___	___	___
$\dfrac{\Sigma}{N}$	___	___	___	___	___	___

13. Calculate $c\bar{x}$, $\bar{x} + c$, $\bar{x} - c$, and $\bar{x} + \bar{y}$. How do these compare with the entries in the last row of the table, $\overline{cx}$, $\overline{(x+c)}$, $\overline{(x-c)}$, and $\overline{(x+y)}$?
14. Complete the following statements in terms of c, $\bar{x}$, and $\bar{y}$:
    (a) If a constant c is multiplied by each of N numbers, $x_1, x_2, \ldots, x_N$, the mean of the products $\overline{cx} = $ _____.
    (b) If a constant c is added to each of N numbers, $x_1, x_2, \ldots, x_N$, the mean of the sums $\overline{(x+c)} = $ _____.
    (c) If a constant c is subtracted from each of N numbers, $x_1, x_2, \ldots, x_N$, the mean of the differences $\overline{(x-c)} = $ _____.
    (d) If $x_1, x_2, \ldots, x_N$ is one collection of N numbers and $y_1, y_2, \ldots, y_N$ is another collection of N numbers and $(x+y)_i = (x_i + y_i)$, the mean of the sums $\overline{(x+y)} = $ _____.

Exercises 15–18 assume the following:

c is a constant.
$\bar{x}$ is the mean of N numbers, $x_1, x_2, \ldots, x_N$.
$\bar{y}$ is the mean of N numbers, $y_1, y_2, \ldots, y_N$.

15. Use the Algebra of Summations to prove that if

$$x_1 = x_2 = \cdots = x_N = c, \text{ then } \bar{x} = c \text{ (i.e., prove } \bar{c} = c)$$

16. Use the Algebra of Summations to prove that $\overline{x+c} = \bar{x} + c$.
17. Use the Algebra of Summations to prove that $\overline{cx} = c\bar{x}$.

18. Use the Algebra of Summations to prove that if

$$(x + y)_i = (x_i + y_i), \text{ then } \overline{x + y} = \bar{x} + \bar{y}$$

The Canadian Ski Marathon is a two-day event in which racers complete 40 mi the first day, camp overnight, and complete another 60 mi on the second day. All skiers are required to carry 15-lb packs. The packs are weighed at the end of each day, and a penalty of 30 sec per mile is added to a racer's time for every pound that his or her pack is light. Seven skiers from the Riverbend Gliders Ski Club complete the first day in a combined time of 1,421 min, but every skier's pack is discovered to weigh only 14 lbs. They complete the second day in a combined time of 2,093 min, but on the second day, every skier's pack turns out to weigh only 13 lbs.

19. (a) *Not counting penalties*, what is the average time per skier on the first day? Calculate the answer in both minutes and hours.
  (b) *Not counting penalties*, what is the average time per skier on the second day? Calculate the answer in both minutes and hours.
  (c) What is the average time per skier on the first day *including penalties*? Calculate the answer in both minutes and hours.
  (d) What is the average time per skier on the second day *including penalties*? Calculate the answer in both minutes and hours.
20. (a) What is the average time (in minutes) per skier for the entire two-day race, *not counting the penalties*?
  (b) What is the average time (in hours) per skier for the entire two-day race, *not counting the penalties*?
  (c) What is the average time (in minutes) per skier for the entire two-day race, *including penalties*?
  (d) What is the average time (in hours) per skier for the entire two-day race, *including penalties*?
21. A teacher grades a set of examinations and finds that the average score is 75.5. The next day she discovers a 25th test with a score of 80. What is the average of the 25 examination scores?
22. The average score on a set of 25 tests is 75.8. A student who received a score of 47 drops the course. What is the average of the remaining 24 examinations?

In the year 10 B.C. (Before Calculators) the following trick was used to calculate the mean by hand:

  (1) Guess a value c near $\bar{x}$.
  (2) Calculate the mean $\overline{x - c}$.
  (3) Add c to this result.

Use this technique to compute $\bar{x}$ for the data in Exercises 23 and 24.

23. 80, 78, 75, 70, 66, 65, 64, 60, 58, 55
24. 10, 10, 10, 13, 13, 15, 17, 18, 18, 18

In Exercises 25 and 26 calculate the mode, median, and mean for the data sets given in each problem.

25. Data Set 1: 2, 2, 3, 3, 3, 4, 4, 5, 5, 6
Data Set 2: 2, 2, 3, 3, 3, 4, 4, 5, 5, 6, 100
26. Data Set 1: 50, 52, 53, 53, 53, 55, 55, 56, 58, 60
Data Set 2: 2, 50, 52, 53, 53, 53, 55, 55, 56, 58, 60

Draw histograms to represent the data in Exercises 27 and 28, and mark the mode, median, and mean on the horizontal axis.

27. 1, 3, 3, 3, 3, 3, 3, 4, 5, 6
28. 1, 2, 3, 4, 4, 5, 5, 5, 5, 6
29. Let y_i be defined as $(x_i - \bar{x})$. Calculate $\bar{y}$ for the data in Exercise 2.3.27.
30. Let y_i be defined as $(x_i - \bar{x})$. Calculate $\bar{y}$ for the data in Exercise 2.3.28.
31. The following x-values are examination scores earned by 25 students in a college algebra class:

$$43, \quad 29, \quad 51, \quad 31, \quad 40, \quad 53, \quad 43, \quad 55, \quad 39, \quad 50, \quad 11, \quad 41, \quad 46,$$
$$50, \quad 42, \quad 41, \quad 40, \quad 46, \quad 46, \quad 43, \quad 16, \quad 54, \quad 33, \quad 37, \quad 41$$

The instructor adds a bonus of $b = 3$ points to every examination. Calculate

(a) $\sum^N \dfrac{(x + b)}{N}$ **(b)** $\sum^N \dfrac{(x + b - \bar{x})}{N}$

32. The following y-values are heights of students in a high school physical education class:

$$65, \quad 67, \quad 73, \quad 62, \quad 63, \quad 66, \quad 66, \quad 67, \quad 71, \quad 70, \quad 70, \quad 69, \quad 68, \quad 74,$$
$$54 \quad 56, \quad 68, \quad 71, \quad 58, \quad 69, \quad 70, \quad 63, \quad 62, \quad 65, \quad 59, \quad 72, \quad 73, \quad 48$$

Students were measured with their shoes on. The coach estimates that shoes add about 3/4 in. to each measurement and therefore subtracts a constant h (for "heel") = .75 from each height. Calculate

(a) $\sum^N \dfrac{f_i(y_i - h)}{N}$ **(b)** $\sum^N \dfrac{f_i(y_i - h - \bar{y})}{N}$

C. DESCRIBING THE DISPERSION OF A DISTRIBUTION

Suppose you are a college golf coach and must choose one of two players for a team tournament. The players shot identical scores in qualifying play and they have the same handicap. The only other data you have are the distances for several hundred drives that they shot from the same practice tee over a period of several weeks. The distributions of these distances are given in Figure 2.5. Both players drive an average of 220 yards. Which would you choose?

Now, suppose you are the director of admissions at a university and you are to recommend to the faculty that they adopt one of two entrance examinations. The two distributions for the same large national sample of high school seniors are given in Figure 2.6. Both tests yield the same

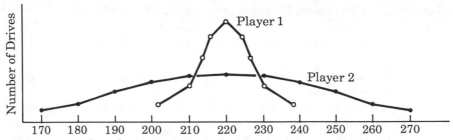

Figure 2.5. Distances (yards) of drives by two college golfers.

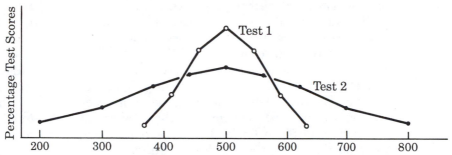

Figure 2.6. Distribution of entrance examination scores for two tests administered to the same national sample of high school seniors.

average score ($\bar{x} = 500$). Which test would you recommend?

In both of the examples given above, a decision is to be made on the basis of distributions that have identical means (as well as identical modes and medians). Decisions must therefore be based on distribution properties *other than* location. In the first example the coach must decide whether to choose a golfer who seems to have a pretty fair chance of driving over 250 yards but who is equally likely to drive less than 190 yards (Player 2) or a golfer who is unlikely to drive *either* more than 240 yards *or* less than 200 yards (Player 1). In the second example the admissions officer must decide whether it is more likely that Test 2 is *exaggerating* differences in college potential or that Test 1 is *obscuring* such differences. Insofar as a sensible decision can be made from the available data in either case, it must be based on considerations of the *dispersion* or *variability* in the data.

Just as statisticians have developed a number of statistics that describe location, so have they developed a variety of indices that describe the dispersion or variability of a distribution. Some of these, however, are more consistent with our criteria for a "good" statistic than are others.

1. The range

Perhaps the most obvious measure of dispersion is simply the difference between the largest observation and the smallest observation. This statistic is known as the *range* and is given by the formula

$$\text{Range} = x_{\max} - x_{\min} \qquad\qquad [2.7]$$

Given the observations 73, 70, 87, 80, 90, and 85, the range is therefore $90 - 70 = 20$.

Like the mode, the range is easy to interpret and simple to compute, but unfortunately it satisfies only one of our criteria for a good statistic. It is single-valued. This value, however, is obtained entirely by inspection. Determination of the range requires that we first identify the minimum and maximum values in our data collection. This means that each observation must be compared with every other observation, and the reader will recall that such comparisons are logical rather than algebraic. Second, the range utilizes only two values. The other values make no contribution to its computation. Finally, the frequencies of even the maximum and minimum x-values are not considered in calculation of the range. If we examine the two frequency distributions in Table 2.10, it is evident that these deficiencies can give a misleading assessment of variability. In the first distribution the observations are uniformly distributed from 50 to 65, and the 15-point range therefore provides an index of dispersion that, at least intuitively, seems meaningful. In the second distribution, however, all but three of the observations lie between 56 and 59. Nonetheless, these three atypical, or outlying, observations give this distribution also a range of 15, even though we have no observations between 51 and 55 or between 60 and 64.

Table 2.10. Two frequency distributions with the same range

x	50	51	52	53	54	55	56	57	58	59	60	61	62	63	64	65
f (Group 1)	2	2	2	2	2	2	2	2	2	2	2	2	2	2	2	2
f (Group 2)	1	0	0	0	0	0	5	8	9	7	0	0	0	0	0	2

2. The mean difference among observations

As indicated above, the range depends on the difference between just two values, $x_{\max}$ and $x_{\min}$, which may or may not be typical of the entire sample. To get better representation of the observed values and surmount the difficulties illustrated in the last example, we might therefore consider taking the average difference among *all* observations. To compute this statistic we subtract each of the N observations x_j from each of the other $N - 1$ observations x_i, giving us a total of $N(N - 1)$ differences to average:

$$\frac{\sum (x_i - x_j)}{N(N - 1)}$$

If we take *every* difference, however, we necessarily include *both* $(x_i - x_j)$ *and* $(x_j - x_i)$. These two values will, of course, be of equal magnitude but of opposite sign, giving us a net sum of zero.

We could avoid this problem by considering only *one* difference for every pair of values, thus giving us a total of $N(N - 1)/2$ terms. But even if we considered only half of the differences this would still involve a prohibitively large number of calculations. With 100 observations, for example, we would have 4,950 differences to compute and sum together. Moreover, the contribution of each pair, x_i and x_j, depends upon their *order*. For example, 5 and 2 will contribute $+3$ to the sum if their difference is calculated as $(5 - 2)$, but they will contribute -3 to the sum of differences if their difference is calculated as $(2 - 5)$. In principle one could eliminate negative values by stipulating that $x_i \geq x_j$ for every difference in $\Sigma(x_i - x_j)$. However, this convention would implicitly require that the N observations be ordered such that $x_1 \geq x_2 \geq \cdots \geq x_N$, and, as indicated earlier, ordering is a nonalgebraic operation.

Despite its obvious attraction, the strategy of developing a measure of dispersion based on differences among observations is freighted with problems. An alternative approach is to base a dispersion statistic on differences between observations and some fixed value.

3. The mean deviation from the mean

The possibility of capturing dispersion by looking at the deviation of every observation from some constant immediately begs the question of selecting an appropriate constant. It makes intuitive sense that the value of our constant should be typical of the values observed in our data. We might therefore select the most frequently observed value, the mode, but if we want our index of variability to reflect the average difference among *all* observations, our constant should itself reflect all of the observations. On this basis an appropriate candidate is the arithmetic mean, $\bar{x}$.

The deviation of score x from the mean $\bar{x}$ tells us how much x differs from the "typical" score, so we might suppose that the average deviation

$$\frac{\sum\limits_{}^{N}(x - \bar{x})}{N}$$

summarizes the variation in the entire data collection. However, it will be recalled that the sum of deviations about the mean is always zero. Therefore, the numerator in this expression is zero, and the mean deviation from the mean must likewise equal zero irrespective of the variability in the distribution. This is a singularly unattractive property in a statistic designed to measure variability.

In our discussion of the mean as the center of mass of a distribution, it was shown that the sum of deviations for x-values to the right of mean is equal to the *negative* of the sum of deviations for x-values to the left of the mean and that the sum of *all* deviations from the mean is therefore zero. However, if all deviations were treated as *positive* values, this property would disappear, and the mean deviation would become a meaningful index of variability. When we consider only the *magnitude* of a number and disregard its *sign*, we speak of the *absolute* value of the number. We signify the absolute value by enclosing the number between vertical lines. That is,

$$|x| = \begin{cases} x \text{ if } x \geq 0 \\ -x \text{ if } x < 0 \end{cases}$$

Thus, in absolute terms,

$$| + 10| = | - 10| = 10$$

We might therefore define our index of dispersion as the average, or mean, *absolute* deviation from the mean:

$$\frac{\sum\limits_{}^{N} (|x - \bar{x}|)}{N}$$

The average absolute deviation from the mean is sometimes called the *mean deviation*. It captures variability in a way that is easily interpreted and plays a role in a number of specialized statistical methods that are not covered in this text. However, we obtain what turns out to be a more useful statistic if we employ a different tactic to convert negative values to positive values.

4. The variance and standard deviation

Recall from elementary algebra that the product of two negative numbers is always positive. We can therefore remove all negative signs by *squaring* our deviations. Substituting squared deviations for absolute deviations yields

$$\frac{\sum\limits_{}^{N} (x - \bar{x})^2}{N} \hspace{3cm} [2.8]$$

This is the mean *squared* deviation from the mean and is defined as the variance, s^2.

The variance of N observations is the mean of squared deviations $(x - \bar{x})$

$$s^2 = \frac{\sum\limits_{}^{N} (x - \bar{x})^2}{N}$$

Example 2.6. Consider the data

$$2, 4, 5, 8, 11$$

Statistical calculators are designed to calculate means, variances, and other statistics quickly and easily. If you do not have such a calculator it is convenient to arrange the calculations as in Table 2.11:

Table 2.11. Calculation of variance for data in example 2.6

x	$x - \bar{x}$	$(x - \bar{x})^2$
2	-4	16
4	-2	4
5	-1	1
8	2	4
11	5	25
$\sum\limits^{5}$ 30		50

So, $\bar{x} = 30/5 = 6$, and $s^2 = 50/5 = 10$.

Recall from earlier discussion that

$$\sum_{}^{N} x_i = \sum_{}^{n} x_i f_i$$

Similarly,

$$\sum_{}^{N} (x_i - \bar{x})^2 = \sum_{}^{n} (x_i - \bar{x})^2 f_i$$

so expression [**2.8**] is equivalent to

$$s^2 = \frac{\sum\limits^{n} (x_i - \bar{x})^2 f_i}{N} \qquad [\textbf{2.9}]$$

From [**2.9**] it can be seen that, unlike the other possible dispersion indices we have considered, the variance satisfies all four criteria for a good statistic. It is single-valued, and, as indicated above, it involves only ordinary arithmetic operations. Moreover, one can see from the limits of summation that the variance incorporates all n observed values of x. Finally, each squared deviation $(x_i - \bar{x})$ enters the sum with the same frequency f_i as the x-value from which it is computed.

Example 2.7. Recall the coin-toss data in Example 1.1 and shown again in Table 2.12:

Table 2.12. Frequency distribution for number of heads obtained in toss of four coins with experiment repeated 25 times

Observation (x)	0	1	2	3	4
Frequency (f)	1	7	10	6	1

Once again a tabular approach is convenient for setting up the calculation of variance (see Table 2.13):

Table 2.13. Calculation of variance for data in Example 2.7

x	f	xf	$x - \bar{x}$	$(x - \bar{x})^2$	$(x - \bar{x})^2 f$
0	1	0	−1.96	3.8416	3.8416
1	7	7	−.96	.9216	6.4512
2	10	20	.04	.0016	.0160
3	6	18	1.04	1.0816	6.4896
4	1	4	2.04	4.1616	4.1616
$\sum_{}^{5}$	25	49			20.96

Since $\Sigma f = 25 = N$,

$$\bar{x} = 49/25 = 1.96$$

and

$$s^2 = 20.96/25 = .8384$$

Equation **[2.9]** can also be used to compute the variance from grouped data. As with the analogous formula for the mean, however, n is the number of *intervals* rather than the number of x-values, x_i is the midpoint of the ith interval, and f_i is the number of observations falling in the ith interval.

Example 2.8. Table 2.14 shows the distribution of annual household incomes from Example 2.1:

Table 2.14. Annual household income ($\times$ $1,000) for 20 interview respondents

Income (x)	4.5 or less	4.5–9.5	9.5–14.5	14.5–19.5	19.5 or more
Frequency (f)	3	1	8	6	2
f/N	.15	.05	.40	.30	.10

Using the ($n = 5$) interval midpoints calculated in Example 2.5 for our x-values gives us the computational layout shown in Table 2.15:

Table 2.15. Calculation of variance for data in Example 2.8

x	f	xf	$x - \bar{x}$	$(x - \bar{x})^2$	$(x - \bar{x})^2 f$
2	3	6	-10.75	115.5625	346.6875
7	1	7	-5.75	33.0625	33.0625
12	8	96	$-.75$	.5625	4.5000
17	6	102	4.25	18.0625	108.3750
22	2	44	9.25	85.5625	171.1250
$\sum^{5}$	20	255			663.7500

Since $\Sigma f = 20 = N$,

$$\bar{x} = 255/20 = 12.75$$

and

$$s^2 = 663.7500/20 = 33.1875$$

The variance satisfies our criteria for a good statistic, but it can be difficult to interpret. The other dispersion statistics that we considered are in the same units as the values from which they are computed. If our data were heights (expressed in inches) of incoming freshmen, for example, a range of 24 would indicate that there was a 24-in. difference between the tallest freshman and the shortest. When we compute the variance, however, each observation (or, more precisely, the deviation of each observation) is squared, and the variance is therefore given in units squared. In the present example, a variance $s^2 = 16$ would represent 16 *square* inches, which has very little meaning in the measurement of height.

BOX 2.1

Computational formulae for variance and standard deviation

$$s^2 \equiv \frac{\sum\limits_{}^{N}(x-\bar{x})^2}{N} = \frac{\sum\limits_{}^{N}(x^2 - 2x\bar{x} + \bar{x}^2)}{N}$$

$$= \frac{1}{N}\sum\limits_{}^{N}(x^2 - 2x\bar{x} + \bar{x}^2)$$

$$= \frac{1}{N}\left(\sum\limits_{}^{N}x^2 - \sum\limits_{}^{N}2x\bar{x} + \sum\limits_{}^{N}\bar{x}^2\right)$$

$$= \frac{1}{N}\sum\limits_{}^{N}x^2 - \frac{1}{N}\sum\limits_{}^{N}2x\bar{x} + \frac{1}{N}\sum\limits_{}^{N}\bar{x}^2$$

Since 2 and $\bar{x}$ are constants, this expression becomes

$$\frac{1}{N}\sum\limits_{}^{N}x^2 - 2\bar{x}\left(\frac{1}{N}\sum\limits_{}^{N}x\right) + \frac{1}{N}(N\bar{x}^2)$$

and since $(1/N)\Sigma x = \bar{x}$,

$$s^2 = \frac{\sum\limits_{}^{N}x^2}{N} - 2\bar{x}^2 + \bar{x}^2$$

$$= \frac{\sum\limits_{}^{N}x^2}{N} - \bar{x}^2$$

or, in frequency notation,

$$s^2 = \frac{\sum\limits_{}^{n}x_i^2 f_i}{N} - \bar{x}^2$$

The standard deviation $s = \sqrt{s^2}$, so it follows that

$$s = \sqrt{\frac{\sum\limits_{}^{N}x^2}{N} - \bar{x}^2} = \sqrt{\frac{\sum\limits_{}^{n}x_i^2 f_i}{N} - \bar{x}^2}$$

For purposes of description, therefore, we frequently take the square root of the variance, which is defined as the *standard deviation*

$$s = \sqrt{\frac{\sum\limits^{N}(x - \bar{x})^2}{N}} = \sqrt{\frac{\sum\limits^{n}(x_i - \bar{x})^2 f_i}{N}} \qquad\qquad [2.10]$$

which gives us an index of dispersion expressed in the same units as the observations from which it is calculated.

a. Computational formulae for the variance and standard deviation. All of the formulae given above require that one calculate the deviation of every x-value from the mean. This involves a minimum of n computations, each of which represents the opportunity for a calculational error. Formulae [2.11] and [2.12] permit calculation of s^2 and s with fewer computations and are particularly useful with hand calculators.

$$s^2 = \frac{\sum\limits^{N}x^2}{N} - \bar{x}^2 = \frac{\sum\limits^{n}x_i^2 f_i}{N} - \bar{x}^2 \qquad\qquad [2.11]$$

Similarly,

$$s = \sqrt{\frac{\sum\limits^{N}x^2}{N} - \bar{x}^2} = \sqrt{\frac{\sum\limits^{n}x_i^2 f_i}{N} - \bar{x}^2} \qquad\qquad [2.12]$$

Derivations for these expressions are given in Box 2.1.

Example 2.9. Tabulating data to use the computational formula for variance (or standard deviation) is a little different from the tabulation format we have used to compute variance from the definitional formula. Consider once again the coin-toss data in Example 2.7, as shown in Table 2.16:

Table 2.16. Calculation of variance for data in Example 2.7 using computational formula

x	f	xf	x^2	$x^2 f$
0	1	0	0	0
1	7	7	1	7
2	10	20	4	40
3	6	18	9	54
4	1	4	16	16
$\sum\limits^{5}$	25	49		117

Since $\Sigma f = 25 = N$,

$$\frac{\Sigma x^2 f}{N} = \frac{117}{25} = 4.68$$

and

$$\bar{x}^2 = \left(\frac{49}{25}\right)^2 = (1.96)^2 = 3.8416$$

so $s^2 = 4.68 - 3.8416 = .8384$, which is exactly what we obtained using the definitional formula for variance in Example 2.7. Furthermore, it can be seen by examining the tabulation of intermediate values that the computational formula involves fewer and simpler calculations.

b. The coefficient of variation. We have said that the standard deviation is more interpretable than the variance, because it is expressed in the same units of measurement as the observations. Even so, the meaning of the standard deviation is elusive. First, variability is intrinsically a more abstract notion than location. Location statistics summarize measurements; in one way or another, a location statistic represents the "typical" observation. But variation statistics capture *differences among* measurements and are therefore removed from the observations by a layer of arithmetic (subtraction). Second, the standard deviation does not easily translate into nonmathematical language, as do the range and even the mean deviation. Consequently, it is often difficult to say *how much* variability is indicated by any particular numerical value of s.

To illustrate this problem, suppose that the standard deviation of weekly incomes for six individuals is 8.165. Given only this information, it is almost impossible to say whether the incomes exhibit little variability, a modest degree of variability, or a great deal of variability. However, let us further suppose that these "incomes" are weekly allowances for a group of 12-year-old children: $5, $10, $10, $15, $20, and $30. *Looking* at the raw data, the variability seems substantial. The child with the largest allowance receives six times as much as the child with the smallest allowance. We might therefore conclude that a standard deviation of 8.165 represents a high degree of variation.

On the other hand, let's imagine that the six individuals are corporate executives with annual incomes of $500,005, $500,010, $500,010, $500,015, $500,020, and $500,030. The largest income is only six-thousandths of 1 percent more than the smallest, and variability therefore seems negligible. For these data, then, the standard deviation 8.165 represents very little dispersion.

The executive incomes in this example were obtained by adding $500,000 to each of the allowances, and the two distributions therefore differ only in location: The average allowance is $15 and the average executive income is $500,015. Thus, we might say that a standard deviation of 8.165 indicates a high degree of variability *in relation to* a mean of 15 but indicates little variability *in relation to* a mean of 500,015. Accordingly, it is sometimes easier to get a feel for the variability in a distribution if one divides the standard deviation by the mean, $s/\bar{x}$.

In our income example we looked at the *percentage* by which the largest value exceeded the smallest value. Likewise, the ratio $s/\bar{x}$ can be multiplied by 100 to express the standard deviation as a percentage of the mean. This statistic is called the *coefficient of variation*, denoted *cv*.

$$cv = 100(s/\bar{x}) \hspace{4cm} [2.13]$$

The coefficient of variation is 100 times the ratio of the standard deviation to the mean

$$cv = 100\left(\frac{s}{\bar{x}}\right)$$

Although the standard deviation is the same for both sets of incomes, the coefficients of variation accord with our sense that the allowances are more variable than the executive incomes. For the allowances,

$$cv \doteq 100\left(\frac{8.165}{15}\right) \doteq 54.4$$

For the executive salaries,

$$cv \doteq 100\left(\frac{8.165}{500{,}015}\right) \doteq .0016$$

The coefficient of variation is particularly useful in two circumstances: (1) comparing dispersion in data sets that have markedly different means and (2) comparing dispersion in sets of data that are expressed in different units of measurement.

EXERCISES 2.4

In Exercises 1–4 calculate the range for each data collection.

1. 1, 1, 1, 2, 2, 3, 3, 3, 4, 5
2. 1, 3, 4, 4, 5, 5, 5, 6, 6, 7
3. 4, 3, 2, 6, 1, 5, 2, 3, 1, 2
4. 6, 1, 5, 4, 3, 2, 5, 3, 3, 3

Each data set in Exercises 5–10 has a mean of 3. Calculate the variance and standard deviation of each set. Draw a histogram using the same scale on the horizontal axis for each set.

5. 1, 2, 3, 4, 5
6. 0, 2, 3, 4, 6
7. −5, −4, 3, 10, 1
8. −8, −1, 4, 9, 11
9. 1, 2, 3, 3, 4, 5
10. 2, 3, 3, 3, 3, 4

For each data set in Exercises 11–22 calculate the variance, the standard deviation, and the coefficient of variation. If you already calculated the means for these data in Exercises 2.3 (pp. 44–45), you may wish to use the computational formulae derived in Box 2.1 (p. 56).

11. 1, 1, 2, 2, 3, 4, 5, 7, 7, 7, 9
12. 2, 2, 2, 3, 3, 4, 7, 8, 9, 10, 15
13. 7, 8, 1, 1, 9, 19, 11, 2, 3, 4, 8
14. 7, 3, 15, 3, 12, 3, 4, 5, 10, 9, 12
15. 43.8, 51.8, 53.4, 40.1, 36.8, 34.8, 37.5, 31.6, 44.7, 51.5, 45.3, 45.3, 46.0, 39.9, 42.3, 41.5, 47.8, 44.0, 41.5, 46.2, 48.0, 55.9, 50.5, 43.9, 44.3
16. .81, 1.00, .72, 1.78, .18, 1.81, .15, 1.94, .17, 1.65, .32, 1.80, 1.05, 1.67, 1.71, 1.32, .61, .62, .63, 1.24, .42, 1.72, 1.72, .23, .40, .27, .43, .44, 1.93, .50

17. Frequencies of Course Evaluation Ratings

Rating (x)	5	4	3	2	1
Frequency (f)	3	6	2	1	1

18. Frequencies of Course Evaluation Ratings

Rating (x)	5	4	3	2	1
Frequency (f)	8	3	2	0	0

19. Frequency Distribution of Weight (pounds) for 100 Persons

Weight	130.5– 140.5	140.5– 150.5	150.5– 160.5	160.5– 170.5	170.5– 180.5	180.5– 190.5
Frequency	10	20	30	20	10	10

20. Frequency Distribution of Height (inches) for 100 Persons

Height	61.5– 63.5	63.5– 65.5	65.5– 67.5	67.5– 69.5	69.5– 71.5	71.5– 73.5
Frequency	10	20	30	20	10	10

21. Frequency Distribution of Maze-Running Times (recorded to nearest second) for 100 "Maze-Bright" Rats

Time	30.5–36.5	36.5–42.5	42.5–48.5	48.5–54.5	54.5–60.5
Frequency	4	10	14	27	45

22. Frequency Distribution of Maze-Running Times (recorded to nearest second) for 100 "Maze-Dull" Rats

Time	60.5–67.5	67.5–74.5	74.5–81.5	81.5–88.5	88.5–95.5	95.5–102.5
Frequency	23	38	17	12	6	4

Exercise 23 refers to the following table. Fill in all the blanks, letting $c = 2$.

x	cx	$(x+c)$	$(x-c)$	y	$(x+y)$
3	——	——	——	9	——
4	——	——	——	13	——
6	——	——	——	15	——
8	——	——	——	19	——
9	——	——	——	24	——
Σ	——	——	——	——	——
$\dfrac{\Sigma}{N}$	——	——	——	——	——

23. Calculate the variance and standard deviation of cx, $(x + c)$, and $(x - c)$. How do these compare with s_X^2, the variance of x, and s_X, the standard deviation of x? Calculate the variance of $(x + y)$. How does it compare with s_X^2 and s_Y^2?

24. Repeat Exercise 2.4.23, letting $c = -3$.

25. Use the Algebra of Summations to verify the following:
  (a) If $x_1 = x_2 = \cdots = x_N = c$, then $s_X^2 = 0$ (i.e., prove $s_C^2 = 0$).
  (b) The variance of $(x + c) = s_X^2$.
  (c) The variance of $cx = c^2 s_X^2$.

26. (a) What does the fact stated in Exercise 2.4.25(b) tell you about the histogram of x and the histogram of $x - c$?

(b) To determine the variance of $x + y$, write

$$s^2 = \sum^N \frac{\left[(x + y) - (\overline{x + y})\right]^2}{N}$$

$$= \sum^N \frac{\left[x + y - \bar{x} - \bar{y})\right]^2}{N}$$

$$= \sum^N \frac{\left[(x - \bar{x}) + (y - \bar{y})\right]^2}{N}$$

For this to equal the variance of x plus the variance of y, what sum must be zero?

Here are the actual times (in minutes) from the Canadian Ski Marathon mentioned in Exercises 2.3.19 and 2.3.20 (p. 47).

Skier	First Day (40 mi)	Second Day (60 mi)	Total Time
Chris Wineslurp	251	377	628
Robin Verycool	267	372	639
Ronald Verycool	267	372	639
Lou Howdy	264	304	568
Ule Muslimsen	229	324	553
J. Tiberius Kirk*	−240	−240	−480
Scott Saltlick	383	585	968

27. (a) What is the standard deviation of skiers' times on the first day? Calculate the answer in both minutes and hours.
    (b) What is the standard deviation of skiers' times on the second day? Calculate the answer in both minutes and hours.
    (c) If 30 seconds per mile is added to each skier's time on the first day, what is the club variance for the first day? Calculate the answer in both minutes and hours.
    (d) If one minute per mile is added to each skier's time on the second day, what is the club variance for the second day? Calculate the answer in both minutes and hours.
28. (a) What is the standard deviation (in minutes) for the entire two-day race?
    (b) What is the standard deviation (in hours) for the entire two-day race?

D. DESCRIBING THE SHAPE OF A DISTRIBUTION

Location and dispersion are *quantitative* concepts. If one looks at a frequency polygon it is not possible to gain any impression of either the

---

Note: When James Kirk arrived at the finish line four hours before the race started, the judges became suspicious. They inspected his equipment and discovered miniature warp-drive engines in his bindings. He was disqualified, and you may exclude his time from the data.

location or the dispersion of the distribution unless the graph includes numbers on the *x*-axis. Even without numbers, however, one can get a sense of the general form or shape of a frequency polygon. The *graphic* features of a distribution are most commonly described in terms of two properties called *skewness* and *kurtosis*. It should be noted, however, that because "shape" is essentially a graphic, rather than quantitative, notion, the statistics used to summarize skewness and kurtosis do not always capture these features as sensibly as, say, the mean and standard deviation summarize location and dispersion. This can be especially true for small samples. Furthermore, it is uncertain what these graphic properties mean when applied to polygons with irregular shapes, e.g., those that are serrate or multimodal. Although some of our examples demonstrate computational procedures with small samples of ungrouped data, these statistics are most meaningful when applied to large collections of grouped data for which the frequency polygons exhibit fairly regular continuity from point to point.

1. Skewness

When the portion of a frequency polygon to the right of the mean is the mirror image of the portion to the left, as in Figure 2.7a, the distribution is said to be *symmetrical*. If more observations lie on one side of the

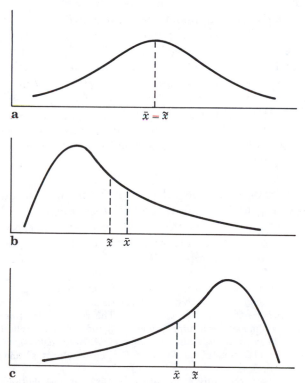

Figure 2.7. (a) Symmetrical distribution (b) distribution skewed to right, (c) distribution skewed to left.

mean than on the other side, the distribution is said to be *skewed*. If more observations lie to the left of the mean, the distribution is typically balanced (at the mean) by a long "tail" to the right, and the distribution is said to *positively skewed*, or skewed to the right, as illustrated in Figure 2.7(b). If more observations lie to the right of the mean, as in Figure 2.7(c), the longer tail extends to the left, and the distribution is said to be *negatively skewed*, or skewed to the left.

To say that *more* of the observations lie to the left than to the right of the mean implies that *over half* of the observations lie to the left. For a positively skewed distribution, the median (as defined for grouped data!) must therefore lie to the left of the mean. Conversely, the median of a negatively skewed distribution falls to the right of the mean. This interpretation of skew is captured in *Pearson's Second Coefficient of Skewness* or, more simply, Pearson's skewness coefficient, which we denote Sk_p.

$$Sk_p = \frac{3(\bar{x} - \tilde{x})}{s}$$
[2.14]

When the mean and median are equal, as in the symmetrical distribution illustrated in Figure 2.7(a), Pearson's skewness coefficient must equal zero. When a distribution is skewed to the right, the mean is larger than the median, so Sk_p is positive for a positively skewed distribution. The mean is smaller than the median when a distribution is skewed to the left, so Sk_p is negative for a negatively skewed distribution.[3]

The magnitude or degree of skew is also reflected in Pearson's coefficient. The greater the departure from symmetry, the larger the absolute value of Sk_p. This is because the mean, *but not the median*, considers every observed value and is therefore more sensitive to the extreme, or outlying, values in the tail of the distribution. Thus, with skewed data the median ordinarily lies closer to the concentration of observations than does the mean and, as indicated earlier in the chapter, is often the preferred index of location.

Example 2.10. In Example 2.1 we calculated the median annual household income for a group of 20 interview respondents to be 13.25 ($\times$$1,000). The frequency distribution for these data is reproduced in Table 2.17. The mean for these data was calculated in Example 2.5 to be 12.75 ($\times$$1,000), and in Example 2.8 the variance was found to be

---

[3] The sign of Sk_p depends only on the numerator, because s is positive. The standard deviation in [2.14] is a scaling constant that makes it possible to compare distributions of observations taken in different units of measurement. The rationale for this practice is developed in our discussion of standardization (Chapter 3), and a similar application is seen in the development of the correlation coefficient (Chapter 4). Note that s is raised to the first power (i.e., $s = s^1$) and is thus of the form s^k. Scaling constants s^3 and s^4 will be appearing shortly.

Pearson initially proposed to capture skew with $(\bar{x} - Mo.)$. For distributions much like those in Figures 2.7(b) and (c), he noticed that $|\bar{x} - Mo.| \cong 3|\bar{x} - \tilde{x}|$, whence the 3 in [2.14].

Table 2.17. Annual household income ($\times$ \$1,000) for 20 interview respondents

Income (x)	8.5 or less	4.5–9.4	9.5–14.5	14.5–19.5	19.5 or more
Frequency (f)	3	1	8	6	2
f/N	.15	.05	.40	.30	.10

33.1875. The value of Pearson's skewness coefficient is therefore

$$Sk_p = \frac{3(12.75 - 13.25)}{5.761} \doteq -.26037$$

indicating the slight, negative departure from symmetry seen in the histogram for these data (Figure 2.1, p. 31).

Pearson's skewness coefficient is intuitively appealing and—if the mean and median are available—is quickly calculated. Moreover, it is single-valued, and because Sk_p is computed from the mean, every observed value x_i enters the statistic multiplied by its frequency, f_i. However, Pearson's skewness coefficient requires nonalgebraic operations because it involves computation of the median.

A more algebraically tractable index of skewness is

$$Sk = \frac{\sum\limits_{}^{N}(x - \bar{x})^3}{Ns^3} \qquad\qquad \textbf{[2.15]}$$

or, in frequency notation,

$$Sk = \frac{\sum\limits_{}^{n}(x_i - \bar{x})^3 f_i}{Ns^3} \qquad\qquad \textbf{[2.16]}$$

To see how **[2.15]** captures skewness, consider the sum of cubed deviations

$$\sum\limits_{}^{N}(x - \bar{x})^3 \qquad\qquad \textbf{[2.17]}$$

in the numerator and recall that:

1. The cube of a negative number is negative.
2. The cube of a number greater than 1 is much larger than the number itself.
3. The cube of a number less than -1 is much smaller than the number itself.

Suppose the distribution is skewed right as in Figure 2.7(b):

- Half the observations lie on either side of the median, and the mean lies to the right of the median.
- Therefore, *more* than half of the observations lie to the *left* of the mean, so the number of *negative* deviations $(x - \bar{x})$ is greater than the number of *positive* deviations.
- However, the *sum* of deviations is zero, so the absolute values of the positive deviations must tend to be larger than the absolute values of the negative deviations.

It follows that the absolute values of the positive deviations *cubed* will be *much* larger than the absolute values of the negative deviations cubed, so the *sum* of the cubed deviation scores $(x - \bar{x})^3$ will be positive.

On the other hand, if the distribution is skewed left, as in Figure 2.7(c), the cubed negative deviations predominate to make expression **[2.17]** negative.

Finally, if the distribution is symmetrical, as in Figure 2.7(a), the positive and negative deviations in **[2.17]** will cancel, and the sum will be zero.

Just as with Pearson's skewness coefficient, therefore, the algebraic *sign* of **[2.15]** ordinarily indicates the direction of skew.[4] The sum of cubed deviations $\Sigma(x - \bar{x})^3$ does not, however, capture the magnitude, or degree, of skew. This is because unless a distribution is symmetrical, the (absolute) value of the sum will be larger for a large number of observations than for a small number of observations. The sum of cubed deviations is therefore divided by N to give the *average* cubed deviation from the mean. As in Pearson's skewness coefficient (footnote 3), we divide by the scaling constant s^k, in this case s^3.

Example 2.11. Consider the following collection of numbers:

$$1, 3, 5, 5, 5, 7, 7, 7$$

Calculation of Sk is easier if the data and the various steps are presented in tabular form, as in Table 2.18. From the first column sum (Σx) in Table 2.18 we calculate $\bar{x} = 5$, which permits us to calculate the values in the other columns. Since $N = 8$, the variance $s^2 = 32/8 = 4$, so $s = 2$, and from **[2.15]**

$$Sk = \frac{-48}{8(2^3)} = -\frac{3}{4}$$

That is, the distribution is *negatively* skewed, or skewed left, as seen in Figure 2.8.

---

[4] The sum of cubed deviations taps the direction of skew less directly than does Pearson's skewness coefficient, and with small sets of data, Sk_p and Sk do not always agree in sign, especially if the departure from symmetry is slight. This will be seen in Exercises 2.5.

Table 2.18. Calculation of skewness for data in Example 2.11

x	$x - \bar{x}$	$(x - \bar{x})^2$	$(x - \bar{x})^3$
1	-4	16	-64
3	-2	4	-8
5	0	0	0
5	0	0	0
5	0	0	0
7	2	4	8
7	2	4	8
7	2	4	8
Σ 40	0	32	-48

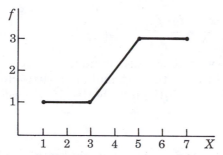

Figure 2.8. Frequency polygon for data in Example 2.11.

Although the data in Example 2.11 are skewed, it will be noted that the mean and the median are both equal to 5. Symmetry implies that the mean and median are equal, but it is clear from this example that equality of mean and median does not necessarily imply symmetry.

For grouped data, we use expression [2.16] to compute Sk and use the midpoints of the intervals for our x-values.

Example 2.12. In Example 2.10 we calculated Pearson's skewness coefficient for a distribution of annual household income reported by 20 interview respondents. We close the open-ended intervals in Table 2.17 to create five classes of equal width and subtract the midpoint (x) of each interval from the mean (12.75) to generate all of the values required to compute Sk. These are given in Table 2.19. From Example 2.8 the variance is equal to 33.1875, so the standard deviation is 5.761, and equation [2.16] yields

$$Sk = \frac{-1,876.875}{20(5.761)^3} \doteq -.4908$$

Table 2.19. Calculation of skewness for data in Example 2.12

x	f	$x - \bar{x}$	$(x - \bar{x})^3$	$(x - \bar{x})^3 f$
2	3	-10.75	$-1{,}242.296875$	$-3{,}726.890625$
7	1	-5.75	-190.109375	-190.109375
12	8	$-.75$	$-.421875$	-3.375
17	6	4.25	76.765625	460.59375
22	2	9.25	791.453125	$1{,}582.90625$
Σ				$-1{,}876.875$

which like Pearson's skewness coefficient in Example 2.10 indicates a slight negative skew.

In summary, the skewness of a distribution

$$Sk = \frac{\sum\limits^{N}(x - \bar{x})^3}{Ns^3} = \frac{\sum\limits^{n}(x_i - \bar{x})^3 f_i}{Ns^3}$$

- Skewness is zero for a symmetrical distribution.
- Skewness is *usually* positive if the distribution is skewed right.
- Skewness is *usually* negative if the distribution is skewed left.

2. Kurtosis

Kurtosis is not a malady characterized by unpredictable episodes of abrupt and rude behavior. It is a term that refers to the "peakedness" or "flatness" of a frequency polygon. A *leptokurtic* distribution is one in which most of the observations are concentrated near the mode and in the tails, as illustrated in Figure 2.9(a). Figure 2.9(b) shows a *mesokurtic* distribution, which has is less peaked and has more observations in the "shoulders" than a leptokurtic distribution. The distribution in Figure 2.9(c) is flat, like a table or *plat*eau, and is described as *platykurtic.*[5]

The algebraic formulae for describing kurtosis are the same as the formulae for skewness, **[2.15]** and **[2.16]**, with both exponents replaced by 4:

$$Kur = \frac{\sum\limits^{N}(x - \bar{x})^4}{Ns^4} \qquad\qquad \textbf{[2.18]}$$

or

$$Kur = \frac{\sum\limits^{n}(x_i - \bar{x})^4 f_i}{Ns^4} \qquad\qquad \textbf{[2.19]}$$

---

[5] All of these words derive from Greek. *Kyrtōsis*, curvature; *leptos*, slender; *mesos*, middle, *platys*, broad.

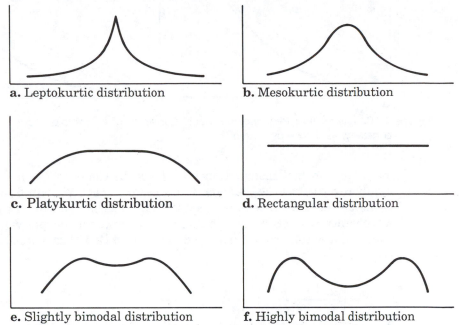

a. Leptokurtic distribution   **b.** Mesokurtic distribution

c. Platykurtic distribution   **d.** Rectangular distribution

e. Slightly bimodal distribution   **f.** Highly bimodal distribution

Figure 2.9. *Distributions exhibiting different degrees of kurtosis and modality.*

The more "peaked" a distribution is, the greater the value of *Kur*; the flatter a distribution is, the smaller the value of *Kur*. Superficially, the notion of kurtosis as "peakedness" seems straightforward, but it is not quite as simple as it appears. First, the notion of kurtosis (at least insofar as it is captured by formula **[2.18]**) makes sense only if a distribution is symmetrical, i.e., if $Sk = 0$. Second, "peakedness" does not simply mean the "height" of a frequency distribution. *Kur* is generally greater to the extent that observations cluster within one standard deviation of the mean, but observations that lie *exactly* at the mean have deviations $(x - \bar{x})$ of zero and contribute little to *Kur*. Kurtosis is more a matter of how abruptly the mode arises, that is, the pitch or—as the Greek root implies—the "curvature" of the rise from tails to mode. As shown in Figures 2.9(a) and 2.9(b), a leptokurtotic distribution has "sunken" or concave "shoulders" and a mesokurtic distribution has "rounded" or convex "shoulders." Accordingly, values in the shoulders of a distribution exert more influence on **[2.18]** than does the peak. The notion of curvature also has implications for the tails of a distribution. Figure 2.10(a) and 2.10(b) are identical, except for the addition of one point in the tail of (b), which gives it deeper concavity. Likewise, distributions with long tails yield higher values of *Kur* than distributions with truncated tails.

When we discussed skewness, the value $Sk = 0$ was a natural reference point, because $Sk > 0$ generally indicates that a distribution is skewed to the right and $Sk < 0$ ordinarily indicates that a distribution is skewed left. There is no such obvious benchmark for kurtosis, because $Kur \geq 1$. However, the bell-shaped distribution in Figure 2.9(b) approxi-

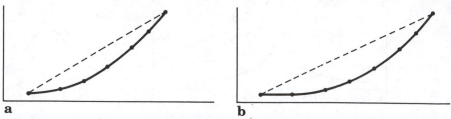

Figure 2.10. Tails of two frequency polygons. Curvature of (b) exhibits more extreme concavity, as shown by dotted lines.

mates the normal distribution introduced in Chapter 7, which is commonly regarded as the "standard" of mesokurtosis. We will find it easier to discuss kurtosis if we use this distribution as the benchmark by which we characterize other distributions as leptokurtic or platykurtic.[6] In addition, it is somewhat easier to describe how [2.18] captures kurtosis if we note that

$$s^4 = \left(\frac{\sum (x - \bar{x})^2}{N} \right)^2 = \frac{\left(\sum (x - \bar{x})^2 \right)^2}{N^2}$$

and see that the "working" part of [2.18] is the ratio

$$\frac{\sum (x - \bar{x})^4}{\left(\sum (x - \bar{x})^2 \right)^2}$$

[2.20]

There are four basic things to note about [2.20]:

1. When observations are *removed* from either the shoulders or the tails of a mesokurtic distribution, the numerator and denominator of [2.20] both *decrease*.
2. When observations are *added* to either the shoulders or the tails of a mesokurtic distribution, the numerator and denominator of [2.20] both *increase*.
3. Changes in the *shoulders* have a greater effect on the *denominator* than on the numerator.
4. Changes in the *tails* have a greater effect on the *numerator* than on the denominator.

Therefore, if one begins with a mesokurtic distribution,

- *Kur* tends to increase if observations are removed from the shoulders or added to the tails, and
- *Kur* tends to decrease if observations are added to the shoulders or removed from the tails.

[6]The normal distribution has *Kur* = 3, and some authorities therefore recommend using *Kur* − 3 to capture kurtosis. Distributions for which (*Kur* − 3) > 0 are characterized as leptokurtic (i.e., more "peaked" than the normal distribution), and distributions for which (*Kur* − 3) < 0 are characterized as platykurtic (i.e., "flatter" than the normal distribution).

We stated earlier that [2.18] is a sensible measure of peakedness only if a distribution is symmetrical. It is further arguable that the very notion of "peakedness" is meaningful only if a distribution has a single mode. An alternative interpretation suggested by Darlington[7] is that kurtosis captures the *modality* of a distribution, with high values of *Kur* indicating that observations tend to cluster sharply about the mode and small values indicating bimodality, the existence of two peaks. If all observations assume the same value, *Kur* is undefined because the denominator of [2.18] is zero. However, it is not hard to show that *Kur* tends to infinity as observations tend to a single value. If observations assume only two values and frequencies are equal, bimodality is maximized, and *Kur* = 1, which is its smallest possible value. Figures 2.9(a) to 2.9(f) illustrate the continuum from a leptokurtic unimodal distribution to a highly bimodal distribution and exhibit a corresponding decrease in *Kur*.

EXERCISES 2.5

Each data set in Exercises 1–12 has a mean of 4 and variance of 9. Calculate Sk_P and Sk for each set. Draw a histogram using the same scale on the horizontal axis for each set.

1. 2, 2, 2, 2, 2, 4, 8, 10   **2.** 0, 2, 2, 2, 4, 6, 6, 10
3. 0, 0, 2, 4, 5, 5, 7, 9   **4.** 0, 0, 3, 3, 4, 6, 7, 9
5. 0, 0, 1, 4, 6, 7, 7, 7   **6.** 0, 0, 1, 5, 5, 6, 7, 8
7. 0, 0, 2, 3, 5, 7, 7, 8   **8.** 0, 0, 2, 4, 4, 6, 8, 8
9. 0, 0, 3, 4, 5, 5, 5, 10   **10.** 0, 2, 2, 2, 3, 7, 7, 9
11. 0, 1, 2, 2, 5, 6, 7, 9   **12.** 0, 1, 2, 3, 3, 7, 8, 8

In Exercises 13–15 verify that Pearson's skewness coefficient (Sk_P) is negative for each data set and then calculate Sk.

13. 1, 1, 6, 6, 6, 10   **14.** 1, 1, 6, 6, 6, 13
15. 1, 1, 6, 6, 6, 7
16. You must fire one of two instructors, and your decision must be justified by their teaching evaluations. The instructors' teaching evaluation scores have the same mean and variance. How can you use skewness to help with your decision?

Calculate *Kur* for each data set in Exercises 17–28. Draw a histogram using the same scale on the horizontal axis for each set.

17. 1, 4, 4, 4, 4, 4, 4, 4, 4, 7
18. 3, 4, 4, 4, 4, 4, 4, 4, 4, 5
19. 0, 2, 3, 4, 5, 5, 5, 6, 6, 6, 6, 6, 6, 6, 7, 7, 7, 8, 9, 10, 12
20. 1, 2, 3, 4, 4, 4, 5, 5, 5, 5, 5, 5, 5, 6, 6, 6, 7, 8, 9
21. 1, 2, 3, 3, 4, 4, 4, 5, 5, 5, 5, 5, 5, 5, 6, 6, 6, 7, 7, 8, 9
22. 1, 2, 3, 3, 3, 4, 4, 4, 4, 4, 5, 5, 5, 6, 7

---

[7]Darlington, R. B. (1970). Is kurtosis really "peakedness"? *American Statistician*, *24(2)*, 19–22. However, see also Balanda, K. P., and MacGillivray, H. L. (1988). Kurtosis: A critical review. *American Statistician*, *42(2)*, 111–119.

23. 1, 2, 3, 4, 4, 5, 5, 5, 6, 6, 6, 6, 6, 7, 7, 7, 8, 8, 9, 10, 11
24. 1, 2, 3, 4, 4, 5, 5, 5, 5, 6, 6, 7, 8, 9
25. 1, 2, 2, 3, 3, 4, 4, 5, 5, 6, 6, 7
26. 1, 1, 2, 2, 3, 3, 4, 4, 5, 5, 6, 6, 7, 7
27. 1, 2, 2, 2, 2, 3, 4, 5, 6, 6, 6, 6, 7
28. 1, 1, 1, 1, 7, 7, 7, 7

In Exercises 29 and 30 calculate the Sk_P, and Sk, and Kur for each set of data.

29. Frequency Distribution of Maze-Running Times (recorded to nearest second) for 100 "Maze-Bright" Rats

Time	30.5–36.5	36.5–42.5	42.5–48.5	48.5–54.5	54.5–60.5
Frequency	4	10	14	27	45

30. Frequency Distribution of Maze-Running Times (recorded to nearest second) for 100 "Maze-Dull" Rats

Time	60.5–67.5	67.5–74.5	74.5–81.5	81.5–88.5	88.5–95.5	95.5–102.5
Frequency	23	38	17	12	6	4

In Exercises 31–33 group the data and then calculate Sk_P, Sk, and Kur.

31. Per Capita Outstanding Debt by State for 1984 (in dollars)

State	Debt	State	Debt	State	Debt	State	Debt
AL	503	IN	187	NE	188	SC	986
AK	9,206	IA	157	NV	810	SD	1,026
AZ	88	KS	168	NH	1,380	TN	346
AR	243	KY	727	NJ	1,194	TX	182
CA	437	LA	986	NM	632	UT	495
CO	257	ME	765	NY	1,470	VT	1,318
CT	1,487	MD	999	NC	271	VA	489
DE	2,289	MA	1,119	ND	497	WA	603
FL	307	MI	417	OH	463	WV	925
GA	301	MN	652	OK	397	WI	563
HI	2,186	MS	297	OR	2,363	WY	1,134
ID	453	MO	374	PA	526		
IL	654	MT	507	RI	2,098		

Source: U.S. Bureau of the Census (1985). *Statistical Abstract of the United States* (105th ed.), pp. 276–277.

32. Average Secondary Teacher's Salary by State for 1984 ($\times$ $1,000)

State	Salary	State	Salary	State	Salary	State	Salary
AL	18.0	IN	22.0	NE	19.6	SC	18.2
AK	36.8	IA	20.8	NV	23.0	SD	16.7
AZ	21.8	KS	19.7	NH	17.4	TN	18.0
AR	17.4	KY	20.7	NJ	23.6	TX	20.8
CA	27.2	LA	19.6	NM	21.8	UT	21.4
CO	23.8	ME	18.2	NY	27.5	VT	18.5
CT	23.0	MD	24.5	NC	18.5	VA	20.8
DE	21.4	MA	22.7	ND	21.2	WA	25.2
FL	19.2	MI	29.1	OH	22.1	WV	18.5
GA	18.9	MN	25.2	OK	19.0	WI	23.6
HI	24.0	MS	16.2	OR	23.7	WY	25.1
ID	19.4	MO	19.8	PA	22.9		
IL	25.3	MT	21.8	RI	23.6		

Source: U.S. Bureau of the Census, ibid., p. 141.

33. Number of Persons ($\times$ 1,000) Below Poverty Level by State for 1979

State	Number	State	Number	State	Number	State	Number
AL	720	IN	516	NE	163	SC	500
AK	42	IA	286	NV	69	SD	113
AZ	351	KS	232	NH	75	TN	736
AR	424	KY	626	NJ	689	TX	2,036
CA	2,627	LA	765	NM	226	UT	148
CO	285	ME	141	NY	2,299	VT	59
CT	243	MD	405	NC	840	VA	611
DE	68	MA	532	ND	79	WA	396
FL	1,287	MI	946	OH	1,089	WV	287
GA	884	MN	375	OK	394	WI	398
HI	92	MS	587	OR	274	WY	36
ID	117	MO	582	PA	1,210		
IL	1,231	MT	94	RI	94		

Source: U.S. Bureau of the Census, ibid., p. 457.

E. SUMMARY

A descriptive statistic is a numerical index that describes or summarizes some characteristic of a frequency or relative frequency distribution. A good descriptive statistic should (1) be single-valued, (2) be algebraically tractable, (3) consider every observed value, and (4) consider the frequency of every observed value.

The most common arithmetic operation used in the calculation of descriptive statistics is addition. The sum of any collection of N numbers, $x_1, \ldots, x_N$, can be denoted

$$\sum^{N} x \quad \text{or} \quad \sum^{N} x_i \quad \text{or} \quad \sum_{i=1}^{N} x_i$$

If these N observations include n different values, then

$$\sum^{N} x_i = \sum^{n} x_i f_i \quad \left(\text{or} \quad \sum_{i=1}^{n} x_i f_i \right)$$

The basic rules for manipulating summations comprise the *Algebra of Summations*. If c is any constant (positive or negative) and if $x_1, \ldots, x_n$ is one collection of N numbers and $y_1, \ldots, y_n$ is another collection of N numbers, then

(1) $\displaystyle\sum^{N} c = Nc$

(2) $\displaystyle\sum^{N} cx = c \sum^{N} x$

(3) $\displaystyle\sum^{N} x_i y_i = x_1 y_1 + \cdots + x_N y_N$

(4) $\displaystyle\sum^{N} (x_i + y_i) = \sum^{N} x_i + \sum^{N} y_i$

Statistics that describe the *location* of a distribution indicate where the bulk of observations fall. A location statistic may be thought of as the "typical" value that is observed.

The most frequently observed value in data collection is called the *mode*, denoted *Mo*. It is simple to interpret, but it meets few criteria for a good statistic, and it is always possible for two or more values to occur with the same frequency. A distribution may therefore be *bi*modal, *tri*modal, or otherwise *multi*modal.

A more useful statistic is the *median* ($\tilde{x}$), which is the value that occupies the middle position in the distribution. To find the median for a collection of *ungrouped* data, they must first be ordered from the smallest value to the largest. If N is odd, the median is the value in position $(N + 1)/2$. If N is even, the median is defined to be halfway between the middle *two* observations. That is, it is the average of the values occupying positions $N/2$ and $(N/2) + 1$.

For *grouped* data, the median is the value at or below which exactly 50 percent of the observations fall and is computed by equation **[2.1]**:

$$\tilde{x} = x_a + \left(\frac{.50N - \text{cumulative } f_a}{f_m} \right)(x_b - x_a)$$

where

- x_a is the lower class limit of the median interval,
- x_b is the upper class limit of the median interval,
- cumulative f_a is the number of observations *below* the median interval, and
- f_m is the number of observations *in* the median interval.

All of the procedures for determining the median require that the observations be arranged in numerical order, smallest to largest. This is a nonalgebraic operation, which is the median's most serious deficiency.

The most widely used location statistic is the *mean*, or average, of the observations. The mean ($\bar{x}$) is defined in equation [2.2] as

$$\frac{\sum\limits^{N} x}{N}$$

where N is the number of observations. When many of the observations assume the same numerical value, it is usually easier to use the computational formula given in [2.3]:

$$\bar{x} = \frac{\sum\limits^{n} x_i f_i}{N}$$

where n is the number of different x-values in the data collection, and f_i is the frequency with which the value x_i occurs.

Statistics that describe the *dispersion* in a distribution indicate how much variation there is among the observations.

The simplest index of dispersion is the difference between the largest and smallest x-values. This statistic is called the *range*. The range considers only two values and can therefore give a very distorted picture of how much variation exists in the entire data collection.

The "typical" degree of departure from the "typical" value in the distribution is captured by the *variance* and the *standard deviation*.

The variation is denoted s^2 and is defined in equation [2.8] as the average of the squared deviations from the mean:

$$\frac{\sum\limits^{N} (x - \bar{x})^2}{N}$$

This expression is equivalent to

$$s^2 = \frac{\sum\limits^{n} (x_i - \bar{x})^2 f_i}{N}$$

as given in equation [2.9].

The variance plays an important role in the inferential techniques presented in Part III of this text, but for descriptive purposes it is often more useful to express variation in the same units as one's observations. This is accomplished by taking the square root of the variance, which defines the standard deviation, s:

$$s = \sqrt{\frac{\sum\limits_{}^{N}(x - \bar{x})^2}{N}} = \sqrt{\frac{\sum\limits_{}^{n}(x_i - \bar{x})^2 f_i}{N}}$$

For purposes of computation, the variance and standard deviation may be calculated by formulae [2.11] and [2.12] derived in Box 2.1:

$$s^2 = \frac{\sum\limits_{}^{N}x^2}{N} - \bar{x}^2 = \frac{\sum\limits_{}^{n}x_i^2 f_i}{N} - \bar{x}^2$$

$$s = \sqrt{\frac{\sum\limits_{}^{N}x^2}{N} - \bar{x}^2} = \sqrt{\frac{\sum\limits_{}^{n}x_i^2 f_i}{N} - \bar{x}^2}$$

Interpreting the degree of dispersion implied by a standard deviation can depend on the location of the distribution, so it is sometimes helpful to compute the *coefficient of variation* as given in formula [2.13]:

$$cv = 100(s/\bar{x})$$

The general form of a distribution, that is, the "shape" of its (relative) frequency polygon, is described by statistics that indicate *skew* and *kurtosis*. The skew of a distribution is its departure from *symmetry* about the mean. A distribution is said to be *positively* skewed if the longer "tail" of the polygon is to the right of the mean and *negatively* skewed if the longer tail is to the left of the mean. Pearson's skewness coefficient [2.14]

$$Sk_{\text{P}} = \frac{3(\bar{x} - \tilde{x})}{s}$$

captures the direction and magnitude of skew because the mean considers every observed value and is therefore more sensitive than the median to extreme values, i.e., values out in the longer tail of the distribution. If data are skewed the median ordinarily lies closer than the mean to the concentration of observations and may therefore be the preferable index of location.

A more commonly used index of skew is given in equations [2.15] and [2.16]:

$$Sk = \frac{\sum\limits_{}^{N}(x - \bar{x})^3}{Ns^3} = \frac{\sum\limits_{}^{n}(x - \bar{x})^3 f_i}{Ns^3}$$

The kurtosis of a distribution indicates the degree to which the frequency polygon is "peaked," and the conventional index for this property is given in equations [2.18] and [2.19]:

$$Kur = \frac{\sum\limits^{N}(x - \bar{x})^4}{Ns^4} = \frac{\sum\limits^{n}(x - \bar{x})^4 f_i}{Ns^4}$$

Large values of [2.18] or [2.19] indicate that most of the observations cluster around the mean and that the distribution is sharply peaked. This is called a *leptokurtic* distribution. Intermediate values (near 3.0) indicate a *mesokurtic* distribution, with the preponderance of observations in the shoulders of the polygon. Small values of *Kur* indicate that *x*-values are more evenly distributed. Such a distribution is described as *platykurtic*. A value of *Kur* near 1 is obtained if a distribution is bimodal.

CHAPTER 3

Describing individuals in distributions: percentiles and standard scores

CHAPTER OUTLINE

A. PERCENTILE RANKS AND PERCENTILES

A special case: The median

Percentiles with ungrouped scores

Exercises 3.1

B. STANDARDIZATION

Measurement in the sciences

Establishing common scales of measurement

- Comparison of physical measurements
- Comparison of behavioral measurements: Standardization

The standard score, *z*

Exercises 3.2

C. SUMMARY

In the last chapter we discussed some of the ways in which statisticians describe distributions of data. In this chapter we consider a somewhat different problem, how statisticians describe the position of an *individual observation* in a distribution. In this context, we will frequently refer to our observations as *scores*.

A. PERCENTILE RANKS AND PERCENTILES

Perhaps the most common convention for describing the position of an individual's score in a distribution of scores is the *percentile rank*:

> The percentile rank of any particular score x is the percentage of observations equal to or less than x.

The percentile rank of x is therefore equal to its cumulative relative frequency multiplied by 100:

$$\text{Percentile rank } x_i = 100\left(\text{cumulative}\frac{f_i}{N}\right)$$

Example 3.1. Suppose that the following scores are obtained on a multiple-choice midterm examination:

$$18, 29, 31, 32, 33, 35, 40, 43, 48, 48, 49, 52, 55, 56, 59$$

Six of the fifteen students, or $6/15 = 40$ percent, earned scores equal to or less than 35, so the percentile rank of the student who scored 35 points is 40. That is, for $x = 35$,

Cumulative $f/N = .40$

Percentile rank $= 100\,(\text{cumulative } f/N)$

$= 40$

Similarly, about 66.67 percent of the students (10 of 15) received scores equal to or less than 48, so the percentile rank of $x = 48$ is 66.67.

Although high school teachers and college professors do not always report percentile ranks for scores on classroom examinations, they are almost a universal feature of tests that are administered throughout a nation, state, school district, or some other well-defined population. In such cases, however, the feedback that one receives is often in the form of a table that shows percentile ranks for a selection of test scores that may *not* include the recipient's score.

Example 3.2. Nessa, a student in the Aldamarian School District, earns a score of 960 on a test of clerical skills and is given the conversion chart appearing in Table 3.1 from which to determine her percentile rank.

Table 3.1. Selected scores and percentile ranks for Aldamarian clerical skills examination

Raw Score (x)	650	675	700	725	750	800	850	900	950	1,000	1,050
Percentile Rank	40	45	50	55	60	69	77	84	89	93	96

The tabled scores do not include $x = 960$, but if Nessa is willing to assume that the scores between any two tabled values are evenly distributed, she can determine her percentile rank by linear interpolation:[1] First, she must find the nearest tabled scores *below* 960 and *above* 960. These are 950, which we denote x_a, and 1,000, which we denote x_b. The distance from x_a to x_b, is therefore

$$x_b - x_a = 50$$

Her own score is 10 points higher than x_a, so it occupies a position 1/5 of the distance from x_a to x_b:

$$\frac{960 - 950}{1,000 - 950} = \frac{10}{50} = \frac{1}{5}$$

If scores are evenly distributed from x_a to x_b, then the *percentile rank* of 960 is likewise 1/5 of the distance from the *percentile rank* of x_a to the *percentile rank* of x_b, that is, 1/5 of the distance from 89 to 93. Since

$$\left(\tfrac{1}{5}\right)(93 - 89) = 0.8$$

the percentile rank of 960 is $89 + .8 = 89.8$. That is,

$$\text{Percentile rank} = 89 + \frac{960 - 950}{1,000 - 950}(93 - 89) = 89.8$$

And, in general, if $x_a < x < x_b$, then

Percentile rank $x =$ [3.1]

$$\text{percentile rank } x_a + \frac{x - x_a}{x_b - x_a}(\text{percentile rank } x_b - \text{percentile rank } x_a)$$

Percentile ranks can also be calculated when the raw data are presented in the form of a frequency distribution.

Example 3.3. Let us suppose that 1,100 students in the Aldamarian School District took the clerical skills examination and the data were reported as in Table 3.2. What is the percentile rank of $x = 960$? If we

---

[1] For a more thorough discussion of linear interpolation, see Appendix II.

Table 3.2. Frequency distribution of scores on the Aldamarian clerical skills examination

x	650 and below	651–675	676–700	701–725	726–750	751–800	801–850	851–900	901–950	951–1,000	1,001–1,050	1,051 and up
f	440	55	55	55	55	99	88	77	55	44	33	44

adopt the convention introduced in Chapter 1 of extending class limits one half a measurement unit beyond the actual precision of measurement, $x_a = 950.5$, and $x_b = 1,000.5$.[2] We can see from the table that 979 of the 1,100 students who took the examination scored below 950.5, so as before the percentile rank of x_a is

$$100\left(\frac{979}{1,100}\right) = 89$$

Likewise, the cumulative frequency of scores falling below 1,000.5 is 1,023, and the percentile rank of x_b is

$$100\left(\frac{1,023}{1,100}\right) = 93$$

From equation [3.1]

$$\text{Percentile rank } 960 = 89 + \frac{960 - 950.5}{1,000.50 - 950.5}(93 - 89) = 89.76$$

In Examples 3.1 and 3.2, the problem was to determine the percentage of observations that fall at or below a particular score, x. In many situations the task is reversed. One wishes to find the x-value at or below which fall a specified percentage of observations, K.

> The Kth percentile is the x-value at or below which fall K percent of observations.

Example 3.4. Suppose that Nessa knows that she must score above the 95th percentile to qualify for entrance to the Aldamarian Academy of Necromantic Studies. She therefore wants to calculate the score she must obtain when she (re)takes the examination. She knows from Table 3.1 that the 93rd percentile is 1,000 and that the 96th percentile is 1,050,

[2] This practice introduces some distortion. Strictly speaking, x_a is 950 and x_b is 1,000. However, extending these limits by half a unit of measurement makes determination of percentile ranks more consistent with other, closely related calculational procedures for grouped frequency distributions. Answers to problems in Exercises 3.1 are based on this practice.

and 95 is 2/3 of the distance from 93 to 96. That is,

$$\frac{95 - 93}{96 - 93} = \frac{2}{3}$$

If scores are evenly distributed between the 93rd percentile and the 96th percentile, then the 95th percentile is 2/3 of the distance from 1,000 to 1,050, that is, (2/3) (50) points above 1,000. The cutoff score for entrance into the academy is therefore

$$1,000 + 33.33 = 1,033.33 \cong 1,034$$

and in general,

Kth percentile   [**3.2**]

$$= x_a + \frac{K - \text{percentile rank } x_a}{\text{percentile rank } x_b - \text{percentile rank } x_a}(x_b - x_a)$$

1. A special case: the median

Recall from Chapter 2 that the median for any collection of grouped data is that value at or below which 50 percent of the observations fall. The median is therefore the 50th percentile. By formula [**3.2**], the 50th percentile equals

$$x_a + \frac{50 - \text{percentile rank } x_a}{\text{percentile rank } x_b - \text{percentile rank } x_a}(x_b - x_a)$$

If we divide both numerator and denominator by 100, all percentile ranks become cumulative relative frequencies, so

$$x_a + \frac{.50 - \text{cumulative}\dfrac{f_a}{N}}{\text{cumulative}\dfrac{f_b}{N} - \text{cumulative}\dfrac{f_a}{N}}(x_b - x_a)$$

Multiplying both numerator and denominator by N,

$$x_a + \frac{.50N - \text{cumulative } f_a}{\text{cumulative } f_b - \text{cumulative } f_a}(x_b - x_a)$$

Cumulative f_a is the total number of observations *below* the median interval. Cumulative f_b minus cumulative f_a is the number of observations *in* the median interval, and we therefore see that when $K = 50$, formula **[3.2]** for the Kth percentile is the same as formula **[2.1]** for calculating the median from grouped data.

2. Percentiles with ungrouped scores: a word of caution

We based our discussion of the median as the 50th percentile on the definition of the median for *grouped* data. Since computation of the median is quite different for ungrouped data, it should not be surprising that the 50th percentile in a collection of ungrouped data may not correspond to the median. For example, given the scores 1, 2, 3, and 4, the value at or below which 50 percent of the observations fall is 2, but the median is the average of the middle two scores (2 and 3) and therefore equals 2.5. Nor will the median of an ungrouped collection of data necessarily have a percentile rank of 50. Consider the scores 1, 1, 2, 2, 2. The median is 2, but the percentile rank of 2 is 100.

These anomalies do not arise because percentiles are inadequately defined. The fundamental concept of a single value below which falls some percentage of values is strictly appropriate only when "gaps" between adjacent values are very, very small. The interpolation procedures we use with grouped data implicitly treat the observations in any interval as though they were evenly distributed from the upper class limit to the lower class limit—rather like smearing individual peanuts into an even paste of peanut butter that fills all the gaps. This assumption is not always warranted, so calculating the Kth percentile by interpolation can sometimes yield results that are perplexing, equivocal, or just plain silly. When data are ungrouped, more sensible results can often be obtained if the Kth percentile is defined and calculated as described in Box 3.1. This definition and calculational method are natural extensions of those we presented for calculating the median of ungrouped data. The disadvantage to this approach is that the notion of percentile cannot easily be "turned around" to assign a unique percentile rank to a particular score x.

BOX 3.1

The Kth percentile for ungrouped data

When data are not grouped, the Kth percentile is sometimes defined as that value below which fall *no more* than K percent of the scores and above which fall *no more* than $(100 - K)$ percent of the scores. A unique value that satisfies this definition is obtained by generalizing the procedure for calculating the median of ungrouped data:

- Arrange the N scores in order from smallest to largest.
- Number the *positions* occupied by the scores $1, \ldots, N$.
- Calculate K percent of N. Call this value N_K.
- If N_K is an integer, then the Kth percentile is the average of scores in positions N_K and $N_K + 1$.
- If N_K is not an integer, then the Kth percentile is the *smallest* score with position number greater than N_K.

──────── EXERCISES 3.1 ────────────────────────────

In Exercises 1 and 2 find the percentile rank of the *x*-values indicated in each ungrouped collection of scores.

1. (a) $x = 15$ (b) $x = 19$ (c) $x = 22$
18, 23, 3, 7, 8, 29, 24, 30, 12, 15, 1, 17, 13, 19, 12, 12, 9, 18, 7, 12, 13, 17, 17, 22, 25

2. (a) $x = .32$ (b) $x = .72$ (c) $x = 1.72$
.81, 1.00, .72, 1.78, .18, 1.81, .15, 1.94, .17, 1.65, .32, 1.80, 1.05, 1.67, 1.71, 1.32, .61, .62, .63, 1.24, .42, 1.72, 1.72, .23, .40, .27, .43, .44, 1.93, .50

In Exercises 3 and 4 use formula [3.2] to find the percentiles indicated in each ungrouped collection of scores.

3. (a) 25th percentile   (b) 50th percentile   (c) 90th percentile
46, 21, 89, 47, 35, 36, 67, 53, 42, 75, 47, 85, 40, 73, 48, 32, 41, 20, 75, 48, 48, 32, 52, 61, 49, 50, 69, 59, 30, 40, 31, 25, 43, 52, 62, 50

4. (a) 20th percentile   (b) 50th percentile   (c) 85th percentile
43.8, 51.8, 53.4, 40.1, 36.8, 34.8, 37.5, 31.6, 44.7, 51.5, 45.3, 45.3, 46.0, 39.9, 42.3, 41.5, 47.8, 44.0, 41.5, 46.2, 48.0, 55.9, 50.5, 43.9, 44.3

In Exercises 5 and 6 use the procedures in Box 3.1. Report the percentage of scores that fall below and that fall above the *K*th percentile calculated by this method.

5. Find the percentiles indicated in Exercise 3.1.3.
6. Find the percentiles indicated in Exercise 3.1.4.

Exercises 7–10 refer to the data in Exercise 1.1.12 (p. 14).

7. What is the percentile rank of average salaries paid in 1984 to secondary teachers in

(a) Nebraska (NE)   (b) Michigan (MI)   (c) California (CA)

8. What is the percentile rank of average salaries paid in 1984 to secondary teachers in

(a) Massachusetts (MA)   (b) Indiana (IN)   (c) Oklahoma (OK)

9. In 1984, which states fell at the

(a) 10th   (b) 50th   (c) 80th

percentiles in average teachers' salaries?

10. In 1984, which states fell at the

(a) 25th   (b) 75th   (c) 90th

percentiles in average teachers' salaries?

Exercises 11 and 12 refer to the following percentile ranks for selected scores on the Wechsler Adult Intelligence Scale.

IQ	100	104	108	110	113	116	119	125	131	135
Percentile rank	50	60	70	75	80	85	90	95	98	99

11. What are the percentile ranks for persons with IQ scores of

(a) 102 (b) 127 (c) 133

12. What IQ scores correspond to the following percentiles?

(a) 58th (b) 83rd (c) 97th

Exercises 13 and 14 refer to the following distribution of completion times for the first 200 women finishers in the 1988 Telemark Korteloppet (27-km) cross country ski race.

Time (in minutes)	100–114	115–129	130–144	145–159	160–174	175–189	190–204	205–219	220–234
f	4	5	5	18	18	27	44	40	39

13. What time (to the nearest minute) corresponds to the

(a) 25th percentile (b) median (c) 88th percentile

(Hint: The 100th percentile is the *shortest* completion time.)

14. A total of 583 women finished the race.

(a) What was the percentile rank of the longest time listed above?
(b) What was the percentile rank of Darlene Kozarek, who finished the race in 3 hours 20 minutes?
(c) What was the percentile rank of Valentina Ustilentsva, who completed the race in 2 hours 15 minutes?

B. STANDARDIZATION

It should be apparent from our preceding discussion of percentile ranks and percentiles that the description of an individual score in a distribution implies *comparison* of a particular score x with other scores in the distribution. For example, to say that the percentile rank of x is 40 tells us that 40 percent of the *other scores* in the distribution were at or below x. Another approach to comparing individual scores is *standardization*, but to appreciate the underlying logic of standardization, it is helpful to understand something about the nature of measurement in scientific work.

1. Measurement in the sciences

In Chapter 1 we said that observations expressed as numerical values are called measurements and that measurements provide the basis for scientific generalization. Indeed, it is a truism in the scientific community that basic knowledge develops only as fast as our ability to measure precisely the phenomena that we observe. Thus, classical Newtonian physics was firmly grounded in the empirical experiments of Galileo, who determined the acceleration of gravity at the earth's surface:

$$g = 32 \text{ ft/sec}^2$$

Similarly, modern physics could not have developed without the earlier experiments of Michelson, who precisely measured the speed of light. And, more recently, the molecular structure of DNA was established only after X-ray diffraction techniques provided John Watson and Frances Crick with precise measurements of intramolecular distances in the DNA crystal. The importance of measurement in the sciences is aptly expressed in the following statement by the nineteenth-century British physicist, Lord Kelvin:

> I often say that when you can measure what you are speaking about, and express it in numbers, you know something about it; but when you cannot express it in numbers, your knowledge is of a meager and unsatisfactory kind.

Certainly not even Lord Kelvin would claim that scientific knowledge proceeds *directly* from measurement. What is crucial to the establishment and confirmation of scientific hypotheses is *comparisons* among measurements. Thus, Galileo completely revised the seventeenth-century conception of the physical universe when he compared the times required by light objects and heavy objects to fall the same distance and found them to be identical.

In order to compare two measurements, however, it is necessary that both be expressed in the *same scale of measurement*. This requirement has two conditions. First, both measurements must be taken from the *same reference point*. When measuring physical properties, such as length or volume, it is generally assumed that the reference point is *true* zero. Thus, a length of 15 inches is usually assumed to mean 15 inches *longer than zero inches*. This is not the case for all measurements, however. Consider the measurement of temperature. Which is warmer, 25°C or 293°K? Although 293 is a larger number than 25, the Kelvin and Celsius scales do not have the same zero point; 0°C is equal to 273°K, and 25°C is therefore 5 degrees warmer than 293°K.

The second *requirement* is that our measurements be in *units of equal magnitude*. If object A is 15 units long and object B is only 11 units long, we cannot conclude that A is longer than B unless we know that both are expressed in units of the same length. If A is 15 *inches* long, for example, and B is 11 *yards* long, then B is decidedly longer than A. Before Canada converted to the metric system, a tourist from Michigan discovered that gasoline, which cost $1.00 per gallon in Detroit, averaged $1.20 (U.S. dollars) per gallon in Toronto. The tourist therefore concluded that Canadian gasoline was 20 cents per gallon more expensive than the domestic product. This was based on the implicit assumption that Canadian gallons and United States gallons were of equal volume. At the time, however, Canadian petroleum products were measured in *Imperial* gallons, which are equal to approximately 1.2 United States (or *wine*) gallons, so a tank of gas was just about the same price in Ontario as it would have been in Detroit.

2. Establishing common scales of measurement

a. Comparison of physical measurements. To illustrate the basic problem of comparing incompatible measurements, let us consider a familiar scenario.

Example 3.5. Jim and Joe have planned a hunting trip and are debating which is the shorter of two possible routes. They therefore elect to take two cars, one by each route, and compare the distances recorded on their odometers. They leave from Jim's house the next morning and when they arrive at the trailhead in the afternoon, Jim's odometer reads 16,225 and Joe's reads 5,050. Before they can compare their measured distances, however, they must first subtract the mileage that was *already* on the two cars when they departed that morning. They are, after all, concerned only with the *net* mileage logged on this particular trip and not the total mileage accumulated since the cars were built.

	Jim	**Joe**
Odometer readings at trailhead	16,225	5,050
Odometer readings at Jim's house	16,000	5,000
Net distance traveled	225	50

Note that by subtracting the mileage recorded at the beginning of the trip they have satisfied the first requirement for a meaningful comparison of measurements. They have transformed the two measurements so that both originate from the same reference point, Jim's house. At this juncture it certainly appears that Jim took the long way around. But this assumes that both odometers were recording distance in the same units of measurement. In fact, however, Joe's car is a rare model manufactured in the Black Forest (by Elves), and his odometer is calibrated in *leagues*. Jim's car, on the other hand, is a somewhat better-known model manufactured in Milan (by hysterics), and his odometer is calibrated in *kilometers*. Before they can conclude that Joe's route is shorter, therefore, they must first express their readings in some common unit of measurement and thus satisfy the second requirement for meaningful comparison.

In a spirit on international compromise they agree to convert their measurements to English miles. For Jim, this conversion requires that he divide his net distance, given in kilometers, by the number of kilometers per mile:

$$\frac{\text{Distance in kilometers}}{\text{Number of kilometers per mile}} = \text{distance in miles}$$

Similarly, Joe must divide his recorded distance in leagues by the number of leagues per mile:

$$\frac{\text{Distance in leagues}}{\text{Number of leagues per mile}} = \text{distance in miles}$$

Being well-equipped outdoorsmen they carry the most recent edition of the *Handbook of Chemistry and Physics* (paperback edition), and a quick glance at the appropriate conversion table tells them that a mile equals 1.61 kilometers and that a mile equals .333 leagues. Expressed in miles, therefore, the length of Jim's route is

$$\frac{225 \text{ km}}{1.61 \text{ km per mi}} \doteq 140 \text{ mi}$$

and to the nearest mile, the length of Joe's route is

$$\frac{50 \text{ leagues}}{.333 \text{ leagues per mi}} \doteq 150 \text{ mi}$$

With the two measurements thus converted to a common scale of measurement we can now see that Joe's route was 10 mi longer than Jim's. This transformation was intuitively obvious, but it actually involved several steps, which are summarized in Table 3.3.

b. Comparison of behavioral measurements: standardization. Although we chose an example involving measurement of a physical property, distance, the importance of measurement is certainly not restricted to the physical sciences. As we pointed out in Chapter 1, measurement also plays an important role in the social and behavioral sciences. Economists conduct market surveys to measure consumer demand. Teachers administer examinations to measure acquisition of course material. Psychologists construct tests to measure personality traits. And, as we all know, colleges use entrance examinations to measure scholastic aptitude. As in the physical sciences, these measurements are used to make comparisons upon which decisions are based. College entrance examinations are used to select the most promising applicants, consumer surveys are used to target potential markets for new products, etc.

In making these comparisons the behavioral scientist faces the same problems as our two drivers, Jim and Joe: The quantities that are to be compared must be expressed in common units of measurement relative to a fixed reference point of origin. However, physical measurements are

Table 3.3. Steps required to compare distances measured in different units

	Jim	Joe
(1) Reference point (zero point)	Jim's House	Jim's House
(2) Numerical value of (1)	16,000	5,000
(3) Distance from (1)	225	50
(4) Unit of measurement	Kilometer	League
(5) Transformed unit	Mile	Mile
(6) Number of measurement units per transformed unit	1.61	.333
(7) Distance from (1) expressed in transformed units	140	150

more easily compared than behavioral measurements, because physical properties are measured in units that are defined *independently* of the particular measurements under consideration. When we say that Jim traveled 225 km, this distance may be interpreted in terms of an *external* referent, the standard meter. The situation is quite different when we interpret a behavioral measurement, such as a score of 600 on the Scholastic Aptitude Test. An SAT score assumes meaning *only* in comparison to the scores of other individuals who complete the *same* examination. There are no external standards for behavioral measurements (e.g., no "standard SAT point"), so behavioral scientists use a statistical procedure called *standardization* to convert measurements into a scale that permits meaningful comparisons.

Example 3.6. Suppose a college instructor is especially concerned about the progress of one particular student. The teacher has the student's test results from the first two examinations and wishes to determine whether the student's performance, relative to the rest of the class, improved, declined, or remained constant. The two scores are 63 and 75, and in absolute terms it appears that the student performed better on the second test. However, the instructor has no reason to assume that the two tests were scored in the same scale of measurement. The first examination, for example, might have been more difficult than the second, in which case the student's relative performance might not have improved at all. These raw test scores are therefore analogous to the final odometer readings in the preceding example. Like the hunters, our instructor must express the scores in reference to some common point and convert them to some common unit of measurement.

The hunters in Example 3.5 established Jim's house as their point of reference. That is, they were concerned only with the distance to the trailhead *in relation* to Jim's house, so each driver subtracted from his final odometer reading a value that *described the location* of Jim's house (viz., his initial reading). In the present example, meaningful interpretation requires that the test scores be expressed *in relation* to the scores of other students who took the same examination. Our instructor therefore subtracts from each score a value that *describes the location* of its distribution. From Chapter 2 we know that the location of a distribution is described by its mean $\bar{x}$, so the instructor defines the class mean as the reference point. The class mean for the first examination was 50, and the mean for the second examination was 70. Subtracting these two values from the corresponding test scores, we obtain the following *deviation* scores:

$$x_1 - \bar{x}_1 = 63 - 50 = 13$$

$$x_2 - \bar{x}_2 = 75 - 70 = 5$$

If these deviation scores are to be interpreted as "distances" from a common reference point, the instructor must necessarily assume that the two means, $\bar{x}_1$ and $\bar{x}_2$, represent the *same* point in much the same way as the two initial odometer readings, 16,000 and 5,000, represented a common point of departure in the earlier problem. In the present context this is not an unreasonable assumption. Although the average *score* obtained by a particular group of students will vary from test to test (depending on test length, difficulty, etc.), the average *level of mastery* that underlies test scores is a property of the *group*, and therefore remains relatively constant. This is the basis for the common practice of "grading on the curve."

Even though the student scored 13 points above the mean on the first examination and only 5 points above the mean on the second, the instructor has no guarantee that the two tests were scored in the same units of measurement. If, for example, the first midterm had 150 possible points ar.d the second had only 100 possible points, then the points on the second examination might be "larger" than the points on the first, much as leagues are larger than miles. Consequently, a deviation score of 5 points on the second test might actually represent better performance than a deviation score of 13 on the first, just as 50 leagues is longer than 225 kilometers. The two deviation scores must therefore be converted to some common unit of measurement.

The instructor has no external standard to which the measurements can be referred, but this is not an insurmountable difficulty. A standard is, after all, nothing more than a fixed interval that is expressed in the same units as the various measurements one wishes to compare. In Example 3.5 the drivers were able to transform their measurements into miles, because the mile is an interval of fixed magnitude that their conversion table expressed in both leagues and kilometers. In the present example, therefore, our instructor requires some fixed interval that is expressed in the "points" for which both examinations were scored.

Since it is assumed that student performance is a property of the group, the amount of *variability* in this performance, like the *average* performance, should be the same from test to test. This implies that the standard deviations for his two tests, which appear below, represent intervals of *equal* length.

$$s_1 = 15$$

$$s_2 = 5$$

Because s_1 and s_2 are assumed to represent the *same* degree of variability, it may at first seem puzzling that they are not numerically equal. This is because (as indicated in Chapter 2) the standard deviation is always expressed in the same units as the measurements from which it is calculated. That is,

$$s = 15 \text{ test points on Examination 1}$$

$$s = 5 \text{ test points on Examination 2}$$

This is analogous to the situation in Example 3.5. Two stretches of road, each a mile long, are the same length even though *numerically* equal to .333 if measured in leagues or 1.61 if measured in kilometers. Like the mile, s therefore satisfies our definition of a standard: It is a fixed interval that is expressed in the units (test points, in this example) of the two measurements to be compared, so the instructor may adopt s as the common unit of measurement and transform the two deviation scores to s-units, just as Jim and Joe converted their leagues and kilometers to *miles*. For the first examination,

$$\frac{13 \text{ test points above the mean}}{15 \text{ test points per } s\text{-unit}} \doteq .87 \, s\text{-units above the mean}$$

and for the second examination

$$\frac{5 \text{ test points above the mean}}{5 \text{ test points per } s\text{-unit}} = 1.0 \, s\text{-units above the mean}$$

The instructor can therefore be reasonably confident that the student's performance, relative to other students in the class, improved from the first examination to the second examination.

The steps in this comparison are summarized in Table 3.4 and are seen to be analogous to the comparison of distances summarized in Table 3.3.

Table 3.4. Steps required to compare scores on different tests

	Test 1	Test 2
(1) Reference point (zero point)	Mean ($\bar{x}$)	Mean ($\bar{x}$)
(2) Numerical value of (1)	63	70
(3) Distance from (1)	13	5
(4) Unit of measurement	Test$_1$ point	Test$_2$ point
(5) Transformed unit	s	s
(6) Number of measurement units per transformed unit	15	5
(7) Distance from (1) expressed in transformed units	.87	1.0

3. The standard score, z

Any value x from a distribution with mean $\bar{x}$ and standard deviation s can be transformed to a *standard score*, or z-score, by subtracting $\bar{x}$ and dividing by s. That is,

$$z = \frac{x - \bar{x}}{s}$$

[3.3]

Example 3.7. Calculate the standardized value of $x = 4$ in the data set 3, 3, 4, 5, 7, 7, 9, 12, 13, 21. First we must calculate the mean and standard deviation. The reader should confirm that these are $\bar{x} = 8.4$ and $s \doteq 5.35$. Then,

$$z = \frac{x - \bar{x}}{s} \doteq \frac{4 - 8.4}{5.35} \doteq -0.822$$

Given a particular z-score, one can also work backward and calculate the corresponding value of x.

Example 3.8. For the data set in Example 3.7, find the value of x that has a standardized value $z = 2.36$. First we solve equation **[3.3]** for x:

$$z = \frac{x - \bar{x}}{s}$$

so

$$sz = x - \bar{x}$$

and

$$x = sz + \bar{x}$$

Therefore, for $z = 2.36$,

$$x \doteq 5.35(2.36) + 8.4 \doteq 21$$

If *all* of the values in a collection are standardized, $\bar{x}$ is subtracted from every score, and it might be recalled from the exercises in Chapter 1 that adding a constant to (or subtracting a constant from) every x-value changes the location of the distribution of x. It is easy to see that the mean of a distribution of standard scores is zero, $\bar{z} = 0$: Let the mean for a collection of N observations be $\bar{x}$ and the standard deviation be s. Then, for any value x

$$z = \frac{x - \bar{x}}{s}$$

and

$$\bar{z} = \frac{\sum\limits^{N} z}{N} = \frac{1}{N} \sum\limits^{N} \left(\frac{x - \bar{x}}{s} \right) = \frac{1}{N} \sum\limits^{N} \left(\frac{1}{s}(x - \bar{x}) \right)$$

Since s is a constant for this particular collection of values, we know from our Algebra of Summations that

$$\sum\limits^{N} \left(\frac{1}{s}(x - \bar{x}) \right) = \frac{1}{s} \sum\limits^{N} (x - \bar{x})$$

so

$$\bar{z} = \frac{1}{N} \cdot \frac{1}{s} \sum\limits^{N} (x - \bar{x})$$

However, $\sum\limits^{N} (x - \bar{x})$ is the sum of deviations about the mean and is therefore equal to zero, which gives us

$$\bar{z} = \frac{1}{N} \cdot \frac{1}{s} \cdot 0 = 0 \qquad\qquad \textbf{[Q.E.D.]}$$

As the reader may have anticipated, standardization also affects the variance of a distribution. The problems developed in Examples 3.5 and

3.6 both required that incompatible measurements be converted to some common unit of measurement. In the travel problem the two drivers converted their measurements from leagues and kilometers to miles. That is, the mile became the transformed unit. In this new scale of measurement, therefore,

Mile = 1 unit

That is, if the unit is the mile, then a mile must equal 1 unit. We had an analogous situation in the second problem, where the instructor transformed a student's test scores to standard units. When scores are standardized, the units in which they are expressed are equal to the standard deviation, s. If the transformed unit is equal to s, it is necessarily true that

$s = 1$ unit

For any distribution of standardized scores, the standard deviation is 1, which means that the variance, s^2, must likewise equal 1. This, too, is easily proved. By the definition of variance,

$$s_z^2 = \frac{\Sigma(z - \bar{z})^2}{N}$$

but since we know that $\bar{z} = 0$, this reduces to

$$s_z^2 = \frac{\Sigma(z)^2}{N}$$

$$= \frac{1}{N}\sum_{}^{N}\left(\frac{x - \bar{x}}{s}\right)^2 = \frac{1}{N}\sum_{}^{N}\left(\frac{1}{s^2}(x - \bar{x})^2\right)$$

Once again, for any particular collection of data, s^2 is a constant, so

$$s_z^2 = \frac{1}{s^2} \cdot \frac{1}{N}\sum_{}^{N}(x - \bar{x})^2$$

$$= \frac{1}{s^2} \cdot s^2 = 1 \qquad\qquad\qquad \textbf{[Q.E.D.]}$$

Standardization of any collection of data involves two operations: First, the mean $\bar{x}$ is subtracted from every x-value. This locates the distribution at zero. Then the N deviation scores $(x - \bar{x})$ are divided by the standard deviation s. This sets the variance (and, therefore, the standard deviation) equal to 1. It is therefore true that *any* distribution of standardized values has a mean of 0 and a standard deviation of 1.

―――――――― EXERCISES 3.2 ――――――――――――――――――

In Exercises 1–6 calculate *z*-scores for the **boldface** *x*-values in each data set. You may use the means and standard deviations you calculated for these data in Exercises 2.3 and 2.4.

1. 1, 1, 2, **2**, 3, 4, 5, **7**, 7, 7, 9
2. 2, **2**, 2, 3, 3, 4, 7, 8, **9**, 10, 15
3. 7, 8, 1, 1, 9, 19, **11**, **2**, 3, 4, 8
4. 7, **3**, 15, 3, 12, 3, 4, 5, 10, **9**, 12
5. **43.8**, 51.8, 53.4, 40.1, 36.8, 34.8, 37.5, 31.6, 44.7, 51.5, 45.3, 45.3, 46.0, 39.9, 42.3, 41.5, 47.8, 44.0, 41.5, 46.2, 48.0, 55.9, **50.5**, 43.9, 44.3
6. .81, 1.00, **.72**, 1.78, .18, 1.81, .15, 1.94, .17, 1.65, .32, 1.80, **1.05**, 1.67, 1.71, 1.32, .61, .62, .63, 1.24, .42, 1.72, 1.72, .23, .40, .27, .43, .44, 1.93, .50
7. Using the data in Exercises 2.3.7 (p. 45) and 2.4.19 (p. 60), calculate the *z*-scores for the following weights.

   (a) $x = 136$ lbs   (b) $x = 155$ lbs   (c) $x = 183$ lbs

8. Using the data in Exercises 2.3.8 (p. 45) and 2.4.20 (p. 60), calculate the *z*-scores for the following heights.

   (a) $x = 62$ in.   (b) $x = 67$ in.   (c) $x = 72$ in.

9. Using the data in Exercise 2.4.19 (p. 60), what weights correspond to the following *z*-scores?

   (a) $z = -1.5$ (b) $z = 0$ (c) $z = 1.6$

10. Using the data in Exercise 2.4.20 (p. 60), what heights correspond to the following *z*-scores?

    (a) $z = -3.7$ (b) $z = -.08$ (c) $z = 2.33$

11. Two applicants for the same scholarship have qualifications that are similar in every way except for their scores on the SAT, which they took in different years. One applicant received a raw score of 650 in a year when the mean was 495 and the standard deviation was 91. The other received a score of 610 in a year when the mean was 419 and the standard deviation was 83. What were the applicants' *z*-scores?

12. The cutoff score for an entry-level civil service position is $z = 1.5$ on a test that has a mean of 63 and a variance of 56.25. If five applicants obtain scores of 75, 68, 78, 59, and 80, which of them are successful?

13. Students in an introductory biology class are given the following grade distribution:

Score	30 and below	31–35	36–45	46–51	52 and up
Grade	F	D	C	B	A

Students are given their *z*-scores and are told that the test mean was 43 and the standard deviation was 5.1.

(a) What grade was received by a student with a *z*-score of 1.5?
(b) How many standard deviations from the mean was the score of the person who earned the lowest **C** in the class?

14. A mathematics examination has 86 possible points. The mean on the examination is 51 and the variance is 306. The teacher grades on a "straight" scale and requires a score of 90 percent of possible points for an **A**. One student receives a z-score of 1.6. Did the student receive an **A**?

15. SAT scores are "rescaled" by multiplying every individual's z-score by 100 and adding 500. These are called y-scores. Raw scores of 475, 540, and 734 were obtained by three high school seniors on the mathematics subtest of the SAT in a year when the mean was 493 and the standard deviation was 114.
    (a) Calculate their y-scores.
    (b) Why are these called *rescaled* scores?

16. Scores on a widely used personality test are calculated by multiplying an individual's z-score by 10 and adding 50. These are called T-scores (not to be confused with Student's t–scores, introduced in Chapter 10).
    (a) What is the mean of all T-scores?
    (b) What is the standard deviation of all T-scores?

C. SUMMARY

The most common techniques for describing the position of an *individual observation* in a distribution involve either *percentile* calculations or *standardization*.

The *percentile rank* of any particular score x is the percentage of observations equal to or less than x. If scores are ungrouped, calculating the percentile rank of x is simply a matter of counting all of the observations equal to or less than x, dividing by N, the total number of observations, and multiplying by 100. Typically, however, percentile ranks are tabled in a form that reports percentile ranks for a selection of x-values, and the value in question may not be among those reported. In this situation, one can use formula **[3.1]** to approximate the percentile rank of x by linear interpolation:

Percentile rank $x =$

$$\text{percentile rank } x_a + \frac{x - x_a}{x_b - x_a}(\text{percentile rank } x_b - \text{percentile rank } x_a)$$

where x_a is the nearest tabled x-value that is smaller than x, and x_b is the nearest tabled x-value that is greater than x.

Sometimes the situation is reversed. One begins with a percentile rank (K) and wants to find the corresponding x-value. That is, one wants to find the Kth *percentile*: The Kth percentile is the x-value at or below which fall K percent of observations. Formula **[3.2]** gives a linear interpolation approximation of the Kth percentile:

Kth percentile

$$= x_a + \frac{K - \text{percentile rank } x_a}{\text{percentile rank } x_b - \text{percentile rank } x_a}(x_b - x_a)$$

If $K = 50$, formula [3.2] is identical to formula [2.1] for calculating the median from grouped data. The median is not necessarily equal to the 50th percentile if data are ungrouped.

Describing the position of an individual score in a distribution implies comparison with other scores. This requires that scores be in the same *scale of measurement*. Measurements must be taken from the same reference point, or zero point, and they must be in the same units of measurement. In making comparisons of *physical* measurements, such as length or weight, this is usually implicit in the measurement operations. Physical measurements are taken from an *externally defined* zero point, and *external standards* have been established for conventional units of physical measurement.

Behavioral measurements differ from physical measurements in two important ways. First, a score of zero on a behavioral measure does not imply a complete absence of the underlying attribute. Second, there are no externally defined standard units of measurement.

In comparing measurements of the same attribute, statisticians therefore make the assumption that mean and variance of the *attribute* remain constant, even if the mean and variance for different *measures* of the attribute are not. This makes it possible to use the mean as a common reference point and the standard deviation as the unit of measurement. An individual's standard score is therefore defined as the number of standard deviations above or below the mean of the distribution of scores. This is expressed in equation [3.3]. The *standardized* value of any score x is denoted z:

$$z = \frac{x - \bar{x}}{s}$$

If all of the scores in a collection of data are standardized, the mean $\bar{z}$ of the standardized scores is equal to 0, and the variance s_z^2 of standardized scores is 1. The standard deviation of standardized scores s_z must likewise equal 1.

CHAPTER 4

Describing joint distributions of data: covariance, correlation, and regression

A. THE SCATTER DIAGRAM

In Chapter 2 we introduced the notion of describing distributions of data and developed statistics that summarize a number of descriptive properties (location, dispersion, etc.), and in Chapter 3 we examined some of the ways statisticians describe the position of an individual observation. In both chapters, however, our discussion was confined to situations involving *one* set of observations, or measurements—family income, maze-running time, score on a particular aptitude test, and so on. In the present chapter we extend our treatment of statistical description to properties that characterize the *relationship* between *two* sets of measurements taken on the *same* group of objects or individuals. That is, we consider the statistics that describe a *joint distribution* of data.

To get a feel for the meaning of a joint distribution, let us suppose we have recorded the height (to the nearest inch) and weight (to the nearest 5 lbs) of every student enrolled in a large freshman college course. These data could then be tabulated as shown in Table 4.1.

Table 4.1. Heights and weights for N subjects

Subject	Weight (X)	Height (Y)
001	160 lbs	67 in.
002	220 lbs	76 in.
003	180 lbs	72 in.
⋮	⋮	⋮
N	135 lbs	61 in.

We know from Chapter 1 that such a listing can be more economically displayed in the form of a frequency distribution. Since our observations involve *two* measurements, however, such a display must include *three* things: the frequency of every weight, x; the frequency of every height, y; and the frequency of every *combination* of weight and height, which we call the *joint frequency* of x and y. These frequencies are given in Table 4.2.

Each entry in the body of Table 4.2 represents the joint frequency of persons of a particular weight (x) and a particular height (y). For example, we see that our group includes 2 people who weigh 105 lbs and are 55 in. tall. The values labeled f_y in the right-hand margin of the table are the row sums. Each sum is the number of persons of height y. The number 7 at the end of the third row therefore indicates that 7 people in the group are 56 inches tall. The values labeled f_x in the bottom margin are the column sums. Each sum is the number of persons weighing x pounds. Because they are displayed in the margins of the table, the frequencies of x-values and y-values are called the *marginal* distributions of X and Y. If the marginal frequencies are divided by N, the number of observations (that is, the number of *pairs*), we obtain the marginal relative frequencies.

Table 4.2. Joint and marginal frequencies of weight
and height for N individuals

Height (Y)	Weight in Pounds (X)							f_y
	100	105	110	115	120	$\cdots$	300	
54 in.	1	1	0	0	0	$\cdots$	0	2
55 in.	1	2	2	0	0	$\cdots$	0	5
56 in.	0	0	0	1	2	$\cdots$	0	7
$\cdots$			$\cdots$					$\cdots$
84 in.	0	0	0	0	0	$\cdots$	1	1
f_x	2	5	5	8	12	$\cdots$	1	

We also know from Chapter 1 that a collection of observations can be represented as points on a line. An alternative way to represent the data in Table 4.2 is to let our two measurements, weight and height, be recorded on the X and Y axes of a plane, so that each point in the plane represents the combination, or joint occurrence, of some particular weight and some particular height. Using weight and height values as coordinates, we can then plot a point for each person, as shown in Figure 4.1.[1] This sort of representation is called a *scatter plot* (or scatter diagram or scattergram). The data plotted in Figure 4.1 include no one who weighs less than 100 lbs or who is shorter than 54 in., so representing the complete range of x-values from 0 lbs to 300 lbs and the complete range of y-values from 0 in. to 84 in. (7 feet) leaves a lot of blank space in the scatter plot. For this reason, it is common practice to omit portions of the X and Y axes between the origin $(0, 0)$ and the smallest observed values.

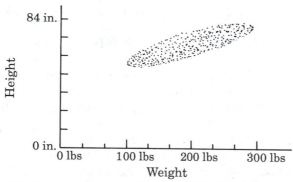

Figure 4.1. Scatter plot of weight and height of N individuals.

---

[1] To represent the repeated values in Table 4.2, for example, the two individuals with coordinates $(105, 55)$, we put *two* points close together near that point in the plane.

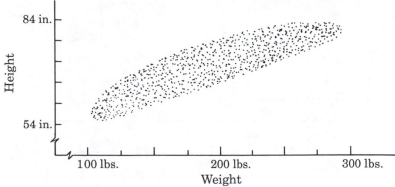

Figure 4.2. *Scatter plot of weight and height of N individuals.*

These discontinuities are represented by breaks (e.g., - ⅄ -) in Figure 4.2, but are not so denoted in subsequent figures.

1. The concept of a statistical relation

If we examine the scattergram of heights and weights we notice various characteristics. First, the plot is roughly elliptical, and the ellipse is markedly longer than it is wide.

This property has a good deal of importance in terms of our ability to estimate the value of one measurement from knowledge about the value of the other measurement. Let us imagine that we draw a subject at random from the list of students in our freshman course and that we want to define a pair of values that are certain to bracket the person's height. In addition, we would like our estimate to be as precise as possible, so we want our two values to define the narrowest possible interval.[2] With no information other than the fact that the person is a student in the class, the best we can say is that the subject's height falls somewhere in the *marginal range* of heights, 54 in. to 84. in., an interval of 30 in. If, on the other hand, we knew that the student's weight was, say, 150 lbs, the scatter plot suggests a more sensible estimate. Although our entire group ranges in height from 54 to 84 in., we see that individuals weighing 150 lbs range only from 60 to 72 in. (Figure 4.3). Thus, knowing our subject's weight has allowed us to reduce the range of our estimate from 30 to just 12 in. This means that our estimate of the person's height has become more precise on the basis of *knowing* the person's *weight*. This is the essence of a statistical relation: Knowing the value of one measurement reduces uncertainty about (or increases the precision of estimate of) the value of a second measurement. More

[2] This was the problem faced by the coffin maker in the samurai movie *Yojimbo*. He expected the hero to kill one of the swordsmen in the *daimyo*'s retinue, but he didn't know which of the swordsmen would be selected for the duel. He wanted to be sure the coffin he built would be neither too short for the shortest man nor longer than needed (and therefore unnecessarily expensive) for the tallest opponent.

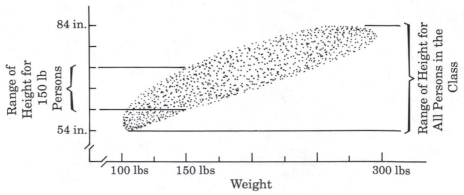

Figure 4.3. Range of heights for entire group compared with range of heights for persons weighing 150 lbs.

formally,

> If (X, Y) are pairs of measurements and if knowing the value of one member x of any pair increases the precision of estimate of the second member y, the two measurements are statistically related.

Exactly what we mean by "increasing the precision of estimate" or "reducing uncertainty" will be explained more fully when we discuss correlation later in the chapter. Even without the formal mathematics, however, it should be apparent that the degree to which knowing x allows us to reduce the range of our estimate for y is related to the *width* of the scatterplot. That is, the narrower the plot, the smaller the range of Y-values associated with any given value of X. Therefore, as the magnitude, or strength, of the statistical relation increases, the width of the scattergram decreases in comparison to its length.

Suppose, for example, that we collect four measurements on a large group of preadolescent, male elementary and middle-school students: grade-point average (X_1), weight (X_2), age (X_3), and height to the nearest inch (Y). If we plot GPA against height, weight against height, and age against height, our three scatter plots might look something like those appearing in Figure 4.4. The range of heights for persons with GPA equal to x_1 is almost the same as the range of heights for the entire group, so the relation between height and GPA is weak, indeed, virtually nonexistent. The range of heights for persons weighing x_2 pounds is somewhat smaller than the range of heights for the entire group, and these two measurements therefore exhibit a moderate degree of statistical relationship. The range of heights for persons of age x_3 is very small compared to the range of heights for the entire group, which we would interpret as indicating a strong relation between height and age for our group of schoolboys.

To see what happens when prediction is *perfect*, suppose we take a second measurement of height (say, in centimeters) and call this X_4. The

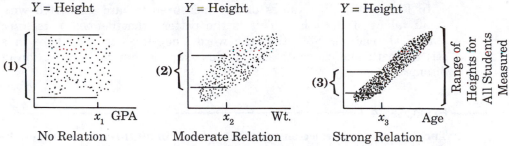

Figure 4.4. Scatter plots for grade-point average, weight, and age against height (Y). (1) represents the range of heights for all persons with GPA of x_1. (2) represents the range of heights for all persons of weight x_2. (3) represents the range of heights for all persons of age x_3.

scatter plot for X_4 and Y would have one and only one value y for every particular x-value (viz., its conversion from inches to centimeters), and our plot would be a straight line. When a plot degenerates to a single line (or other curve), so that each value of X is associated with one and only one value of Y, there is *no* variability in Y among persons with the same value on measurement X.

In addition to *strength*, or *magnitude*, or a statistical relation, the scatter plot indicates the *direction* of the relation. In Figure 4.4, it is seen that height *increases* as weight and age increase. Relations such as these, in which small values of X are associated with small values of Y and large values of X are associated with large values of Y, are said to be *positive*. We can, however, imagine a number of pairs of measurements that might exhibit *negative* relationships (of varying strength). For example, resting heart rate is considered to be a fairly good index of aerobic fitness. Long-distance runners and cross-country skiers typically find their resting pulse rate to be in the range of 45 to 55 beats per minute, while resting heart rates of 75 to 90 are not uncommon in overweight, inactive smokers. Were we to plot number of weekly hours of aerobic exercise against resting heart rate, therefore, we might obtain a scatter plot such as given in Figure 4.5.

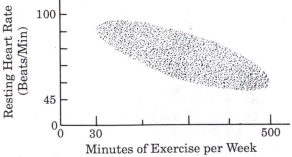

Figure 4.5. Scatter plot of exercise and heart rate.

In Figure 4.5 large values of one measurement are associated with small values of the other. This is the pattern that defines a *negative* statistical relation. Note that the word "negative" is used here in a strictly mathematical sense and, in this example, has *no* connotation of disapproval.

─────── EXERCISES 4.1 ───────

The data sets in these exercises will be used in all three exercise sets in this chapter. Keep copies of the scatter plots for use later. For each data set, unless otherwise noted, (a) construct joint and marginal frequency distribution tables, (b) plot a scatter diagram, (c) state whether or not the data appear to be statistically related and if so, whether the relationship is positive or negative.

1.

Observation	X	Y	Observation	X	Y
1	1	1	5	3	3
2	2	1	6	4	3
3	2	2	7	4	4
4	3	2	8	5	4

2.

Observation	X	Y	Observation	X	Y
1	2	4	5	4	3
2	3	4	6	4	2
3	3	3	7	4	1
4	3	2	8	5	1

3.

Observation	X	Y	Observation	X	Y
1	1	4	6	4	4
2	1	5	7	4	2
3	2	2	8	4	1
4	2	4	9	5	2
5	2	5	10	5	1

4.

Observation	X	Y	Observation	X	Y
1	1	1	6	3	2
2	1	2	7	3	3
3	2	1	8	3	4
4	2	2	9	4	3
5	2	3	10	4	4

5.

Observation	X	Y	Observation	X	Y
1	1	1	5	3	4
2	1	5	6	4	2
3	2	2	7	5	1
4	3	3	8	5	5

6.

Observation	X	Y	Observation	X	Y
1	1	2	5	3	1
2	1	3	6	3	4
3	2	1	7	4	2
4	2	4	8	4	3

7.

Observation	X	Y	Observation	X	Y
1	2	3	7	4	6
2	3	4	8	5	7
3	4	5	9	2	5
4	5	6	10	3	6
5	2	4	11	4	7
6	3	5	12	5	8

8.

Observation	X	Y	Observation	X	Y
1	2	8	7	4	5
2	3	7	8	5	4
3	4	6	9	2	6
4	5	5	10	3	5
5	2	7	11	4	4
6	3	6	12	5	3

9.

Observation	X	Y	Observation	X	Y
1	25	48	5	24	48
2	22	45	6	24	47
3	23	47	7	24	46
4	22	46	8	25	47

10.

Observation	X	Y	Observation	X	Y
1	26	49	5	22	45
2	23	46	6	23	45
3	24	47	7	22	43
4	27	47	8	23	45

For Exercises 11 and 12, do only parts (b) and (c).

11.

Observation	X	Y	Observation	X	Y
1	90	−1.2	6	48	10.0
2	10	38.2	7	54	11.1
3	22	27.4	8	29	30.1
4	37	21.2	9	0	−6.0
5	7	29.6	10	14	29.3

12.

Observation	X	Y	Observation	X	Y
1	91	−12.4	6	−44	0.1
2	39	−2.9	7	−10	−4.9
3	63	−16.6	8	−42	16.6
4	97	−30.8	9	5	6.7
5	1	6.3	10	−18	7.8

13. Times (in minutes) to complete Canadian ski marathon for six skiers (as given in Exercises 2.4.27 and 2.4.28, p. 62)

Skier	First Day (X)	Second Day (Y)
Chris Wineslurp	251	377
Robin Verycool	267	372
Ronald Verycool	267	372
Lou Howdy	264	304
Ule Muslimsen	229	324
Scott Saltlick	383	585

14. Scores obtained by 15 statistics students on first midterm examination (X) and second midterm examination (Y)

Student	x	y	Student	x	y	Student	x	y
1	42	22	6	55	31	11	51	28
2	35	18	7	45	24	12	51	28
3	51	28	8	43	23	13	54	30
4	48	26	9	50	25	14	43	23
5	60	34	10	48	26	15	54	30

15. (x, y): (25, 48), (26, 49), (25, 45), (23, 46), (23, 47), (24, 47), (25, 46), (22, 45), (24, 48), (22, 45), (24, 47), (23, 45), (24, 46), (22, 43), (24, 45), (23, 46), (24, 46), (24, 47), (25, 47), (25, 47)

16. (x, y): (30, 48), (20, 54), (15, 50), (10, 48), (25, 52), (25, 54), (30, 48), (15, 49), (15, 51), (30, 50), (20, 54), (30, 50), (25, 52), (10, 47), (20, 52), (15, 49), (35, 47), (20, 53), (30, 51), (20, 52)

B. QUANTITATIVE DESCRIPTION OF A STATISTICAL RELATION

The scatter plot has allowed us to develop, in an intuitive fashion, the basic notion of a statistical relation. Now we turn to development and discussion of numerical indices that describe the magnitude and directionality of statistical relations. Before we begin, however, we should point out that in all of the statistical relations we have illustrated so far, the points of the scatter plot tend to form or to cluster around a straight line. That is, we have implicitly confined our discussion to *linear* statistical relations. Likewise, it must be understood that the quantitative indices we develop below reflect the direction and magnitude of only the linear relationship between X and Y.

1. The covariance C_{XY}

Let us again consider the four scatter plots discussed above: age and height, weight and height, grade-point average and height, and exercise and heart rate. Now, however, let us suppose that for every pair of values (x, y), we subtract $\bar{x}$ from x, and subtract $\bar{y}$ from y. We know from earlier discussion that the mean of any set of deviation scores is zero. This transformation therefore has the effect of shifting the origin of each plot to $(\bar{x}, \bar{y})$, as illustrated in Figure 4.6. Consequently, each plot is now divided into four quadrants, which we number (counterclockwise from the upper right quadrant) I to IV. If we examine the four plots, it can be seen that the majority of data points in the plots representing positive relationships are in quadrants I and III, the majority of points in the plot representing a negative relationship are in quadrants II and IV, and the points in the plot representing the joint distribution of statistically unrelated measurements are about evenly distributed among all four quadrants. Furthermore, the points for the *strong* positive relationship are *more* concentrated in quadrants I and III than are the points for the *moderate* positive relation.

It should also be noted that since the four quadrants are defined by axes corresponding to $(x - \bar{x}) = 0$ and $(y - \bar{y}) = 0$, the x-deviation values represented in quadrants I and IV are positive, while those in quadrants II and III are negative. Similarly, the y-deviation values in quadrants I and II are positive, while those in quadrants III and IV are negative. Because the product of two values with the same algebraic sign is positive, the product of deviation scores $(x_i - \bar{x})(y_i - \bar{y})$ is therefore *positive* for any individual whose data point falls in *either* quadrant I or quadrant III. On the other hand, positive values multiplied by negative values yield negative products, and $(x_i - \bar{x})(y_i - \bar{y})$ will be *negative* for any data point in either quadrant II or quadrant IV.

When X and Y are positively related, i.e., when data points are concentrated in quadrants I and III, the *sum* of such products for all N individuals, $\Sigma(x_i - \bar{x})(y_i - \bar{y})$, will therefore yield a positive value. The stronger the relationship (i.e., the fewer the points in quadrants II and IV), the fewer the negative products, and the larger this positive value will be. Conversely, if data points are concentrated in quadrants II and IV, as when X and Y are negatively related, this expression will yield a negative value, and the stronger the relation, the smaller the value.[3] If the measurements are not statistically related the points are evenly distributed among the four quadrants, so the positive values will tend to cancel out the negative values, and the sum will be near zero.

The sum of the products of every individual's x-deviation score multiplied by his or her y-deviation score therefore provides a numerical index

---

[3] That is, the *absolute* value of $\Sigma(x - \bar{x})(y - \bar{y})$ becomes *larger* as the strength of relationship increases. Remember, -10 is *smaller* than, say, -9.

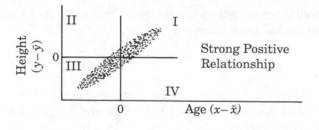

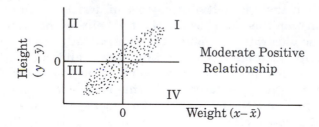

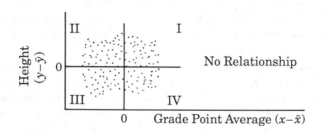

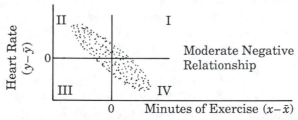

Figure 4.6. Scatter plots of deviation scores $(x - \bar{x})$, $(y - \bar{y})$ for four joint distributions.

that reflects both the direction and magnitude of the statistical relation of X and Y. However, unless two sets of measurements are completely unrelated, the absolute value of the sum tends to increase with sample size. Thus, for example, even if height and weight exhibit exactly the same degree of *positive* relationship in two different groups of persons, $\Sigma(x_i - \bar{x})(y_i - \bar{y})$ should be *larger* for the larger group. For a *negative* relationship, $\Sigma(x_i - \bar{x})(y_i - \bar{y})$ becomes *smaller* as N increases. This troublesome complication can be avoided by dividing the sum by the number of pairs, N. This yields the *covariance* of X and Y, which is denoted C_{XY}:

$$C_{XY} = \frac{\sum\limits_{i}^{N}(x_i - \bar{x})(y_i - \bar{y})}{N} \qquad [4.1]$$

The mean of the products of deviation scores $(x_i - \bar{x})(y_i - \bar{y})$ is the *covariance* of X and Y:

$$C_{XY} = \frac{\sum\limits_{i}^{N}(x_i - \bar{x})(y_i - \bar{y})}{N}$$

Example 4.1. Calculate the covariance C_{XY} for the following pairs of values (x_i, y_i): (25, 48), (26, 49), (24, 46), (23, 46), (27, 47), (21, 43), (24, 47), (27, 48), (22, 43), (24, 46). First, calculate the mean of the x-values and the mean of the y-values. These turn out to be $\bar{x} = 24.3$ and $\bar{y} = 46.3$. Now subtract $\bar{x}$ from every x-value x_i, subtract $\bar{y}$ from every y-value y_i, and average the products $(x_i - \bar{x})(y_i - \bar{y})$. Many calculators allow the user to accumulate such a sum of products, but it is easier to get a feel for the calculations if the steps are laid out in tabular form (Table 4.3), as we did with skew and kurtosis:

Table 4.3. Calculation of covariance for data in Example 4.1

Observation (i)	x_i	$(x_i - \bar{x})$	y_i	$(y_i - \bar{y})$	$(x_i - \bar{x})(y_i - \bar{y})$
1	25	0.7	48	1.7	1.19
2	26	1.7	49	2.7	4.59
3	24	−0.3	46	−0.3	0.09
4	23	−1.3	46	−0.3	0.39
5	27	2.7	47	0.7	1.89
6	21	−3.3	43	−3.3	10.89
7	24	−0.3	47	0.7	−0.21
8	27	2.7	48	1.7	4.59
9	22	−2.3	43	−3.3	7.59
10	24	−0.3	46	−0.3	0.09
Σ					31.10

Dividing $\Sigma(x_i - \bar{x})(y_i - \bar{y}) = 31.10$ by 10, we obtain a covariance $C_{XY} \doteq 3.11$, which indicates a positive statistical relationship.

——————— EXERCISES 4.2 ————————————————————————

Exercises 1–16. For the data sets in Exercises 4.1.1–4.1.16, do the following: (a) calculate $\bar{x}$ and $\bar{y}$, (b) add the vertical line $x = \bar{x}$ and the horizontal line $y = \bar{y}$ to the scatter plots you saved from Exercises 4.1, and (c) calculate the covariance C_{XY}.

17. Calculate the covariance of per capita state debt and teachers' salaries given in Exercises 1.1.11 and 1.1.12 (p. 14). Using just the *extreme* values of X and Y (i.e., the range of X and the range of Y) and what you can tell from the covariance, draw a rough outline of the scatter plot.
18. Turn to Exercise 2.4.26(b) on page 62 and complete the following sentence: The variance of $(X + Y)$ will equal the variance of X plus the variance of Y if _____ is equal to zero.

2. The correlation coefficient r_{XY}

The covariance has a number of properties that make it an important statistic:

1. If X and Y are not statistically related, $C_{XY} = 0$.
2. $|C_{XY}|$ increases as the strength of the relation between X and Y.
3. The sign of C_{XY} corresponds to the direction of the relation between X and Y.
4. C_{XY} satisfies all four criteria for a good descriptive statistic.

Unfortunately, it has one regrettable deficiency: The magnitude of C_{XY} varies with the units of measurement. This may not be immediately obvious, but consider the following example. Let us define X as weight to the nearest pound, Y as height measured in feet, and W as height measured in inches. If we take our measurements on the same N persons, the relation between height and weight is the same irrespective of the units in which our measurements are taken, and a meaningful index of statistical relationship should therefore be expected to yield the same value for weight (X) and height in feet (Y) or for weight and height in inches (W). The covariance of X and Y is

$$C_{XY} = \frac{1}{N} \sum_{i}^{N} (x_i - \bar{x})(y_i - \bar{y})$$

and the covariance of X and W is

$$C_{XW} = \frac{1}{N} \sum_{i}^{N} (x_i - \bar{x})(w_i - \bar{w})$$

Since any individual's height in inches is equal to 12 times his or her height measured in feet, $w_i = 12y_i$. And from our discussion of the mean, we know it must be likewise true that $\bar{w} = 12\bar{y}$. The covariance of X and W can therefore be rewritten as

$$\frac{1}{N} \sum_{i}^{N} (x_i - \bar{x})(12y_i - 12\bar{y}) = \frac{1}{N} \sum_{i}^{N} (x_i - \bar{x})12(y_i - \bar{y})$$

$$= 12\left(\frac{1}{N} \sum_{i}^{N} (x_i - \bar{x})(y_i - \bar{y})\right) = 12C_{XY}$$

That is, the covariance of weight and height measured in inches is 12 times the covariance of weight and height measured in feet. This, alas, substantially reduces the usefulness of the covariance. It might be meaningful to compare covariances of the same measurements (for example, height in inches and weight in pounds) for two different groups (say, men and women), but comparisons among covariances for *different* measurements would yield little information. If, for example, the covariance of height and weight in our fictitious group of schoolchildren were 2,890 and the covariance calculated for their heights and ages were 1,512, we could conclude only that both statistical relations were positive. We could not infer that height and weight were more strongly related than height and age, because we could inflate the latter covariance 12-fold by converting age in years to age in months.

This flaw, as they say in Vienna, is critical but not serious. It is eliminated if we express *all* observations in the same units of measurement. This is precisely what is accomplished when measurements are *standardized*. Before calculating the covariance we therefore transform both sets of measurements to standard units. By the definition given in expression **[4.1]**, the covariance for two standardized variables Z_X and Z_Y is

$$C_{Z_X Z_Y} = \frac{1}{N} \sum_{i}^{N} (z_X - \bar{z}_X)(z_Y - \bar{z}_Y)$$

Since the mean of any collection of standardized measurements is zero, this becomes

$$\frac{1}{N} \sum_{i}^{N} (z_X)(z_Y)$$

Converting from standard units to natural units,

$$C_{Z_X Z_Y} = \frac{1}{N} \sum_{i}^{N} \left(\frac{x_i - \bar{x}}{s_X}\right)\left(\frac{y_i - \bar{y}}{s_Y}\right)$$

$$= \frac{1}{N} \sum_{i}^{N} \frac{1}{s_X} \cdot \frac{1}{s_Y}(x_i - \bar{x})(y_i - \bar{y})$$

$$= \sum_{i}^{N} \frac{(x_i - \bar{x})(y_i - \bar{y})}{N s_X s_Y} \qquad \qquad [4.2]$$

which is defined as the *correlation coefficient*, denoted r_{XY}.

The definitional formula for the Pearson product–moment correlation coefficient is

$$r_{XY} = \sum_{}^{N} \frac{(x_i - \bar{x})(y_i - \bar{y})}{N s_X s_Y}$$

Example 4.2. Geneticists hypothesize that American couples exhibit a tendency toward *assortative mating* with respect to height. Tall men marry tall women, and short men marry short women. Table 4.4 reports

Table 4.4. Heights of husbands and wives

Couple (i)	Husband (x_i)	Wife (y_i)	Couple (i)	Husband (x_i)	Wife (y_i)	Couple (i)	Husband (x_i)	Wife (y_i)
1	76	71	6	71	65	11	68	65
2	75	70	7	71	65	12	68	66
3	75	70	8	70	67	13	67	63
4	72	67	9	68	64	14	67	65
5	72	71	10	68	65	15	62	61

Source: Hays, W. L. (1973). *Statistics for the Social Sciences* (2nd ed.). New York: Holt, Rinehart and Winston, p. 665.

Table 4.5. Calculation of correlation for data in Example 4.2

Couple (i)	x_i	$(x_i - \bar{x})$	y_i	$(y_i - \bar{y})$	$(x_i - \bar{x})(y_i - \bar{y})$
1	76	6	71	4.67	28.02
2	75	5	70	3.67	18.35
3	75	5	70	3.67	18.35
4	72	2	67	.67	1.34
5	72	2	71	4.67	9.34
6	71	1	65	−1.33	−1.33
7	71	1	65	−1.33	−1.33
8	70	0	67	.67	0.00
9	68	−2	64	−2.33	4.66
10	68	−2	65	−1.33	2.66
11	68	−2	65	−1.33	2.66
12	68	−2	66	−0.33	.66
13	67	−3	63	−3.33	9.99
14	67	−3	65	−1.33	3.99
15	62	−8	61	−5.33	42.64
Σ	1,050		995		140
Means	70		66.33		
s	3.6		2.9		

heights (in inches) for 15 American married couples. Calculate the correlation of husbands' and wives' heights. Once again, the easiest way to get started is to lay out the data and the intermediate calculations in tabular form, as shown in Table 4.5. From equation [4.2], the correlation coefficient is $r = 140/[(15)(3.6)(2.9)] \doteq .89$.

a. Computational formula for correlation. The equation given in [4.2] is the *definitional* or *conceptual* formula for correlation, but it involves many calculations. Formula [4.3], which is derived in Box 4.1, requires fewer calculations, or keystrokes on a hand calculator, and therefore introduces fewer opportunities for computational error.

$$r_{XY} = \frac{N\Sigma xy - \Sigma x \Sigma y}{\sqrt{N\Sigma x^2 - (\Sigma x)^2}\sqrt{N\Sigma y^2 - (\Sigma y)^2}} \qquad [4.3]$$

Example 4.3. Use the computational formula [4.3] to calculate the correlation of husbands' and wives' heights from the data in Example 4.2. The computational efficiency of this method is apparent if one compares the layouts of data and intermediate calculations in Tables 4.5 and 4.6. In

Table 4.6. Calculation of correlation for data in Example 4.2 using computational formula

Couple (i)	x_i	x_i^2	y_i	y_i^2	$x_i y_i$
1	76	5,776	71	5,041	5,396
2	75	5,625	70	4,900	5,250
3	75	5,625	70	4,900	5,250
4	72	5,184	67	4,489	4,824
5	72	5,184	71	5,041	5,112
6	71	5,041	65	4,225	4,615
7	71	5,041	65	4,225	4,615
8	70	4,900	67	4,489	4,690
9	68	4,624	64	4,096	4,352
10	68	4,624	65	4,225	4,420
11	68	4,624	65	4,225	4,420
12	68	4,624	66	4,356	4,488
13	67	4,489	63	3,969	4,221
14	67	4,489	65	4,225	4,355
15	62	3,844	61	3,721	3,782
Σ	1,050	73,694	995	66,127	69,790

BOX 4.1

Computational formula for correlation coefficient

$$r_{xy} = \sum_{}^{N} \frac{(x_i - \bar{x})(y_i - \bar{y})}{Ns_X s_Y} = \frac{\sum^{N}(x_i y_i - \bar{x} y_i - \bar{y} x_i + \bar{x}\bar{y})}{Ns_X s_Y}$$

$$= \frac{\dfrac{1}{N}\left(\Sigma xy - \bar{x}\Sigma y - \bar{y}\Sigma x + N\bar{x}\bar{y}\right)}{s_X s_Y} = \frac{\dfrac{\Sigma xy}{N} - \bar{x}\dfrac{\Sigma y}{N} - \bar{y}\dfrac{\Sigma x}{N} + \bar{x}\bar{y}}{s_X s_Y}$$

$$= \frac{\dfrac{\Sigma xy}{N} - \bar{x}\bar{y} - \bar{y}\bar{x} + \bar{x}\bar{y}}{s_X s_Y} = \frac{\dfrac{\Sigma xy}{N} - \bar{x}\bar{y}}{s_X s_Y} = \frac{\dfrac{\Sigma xy}{N} - \dfrac{\Sigma x \Sigma y}{N^2}}{s_X s_Y}$$

$$= \left(\frac{N^2}{N^2}\right)\frac{\left(\dfrac{\Sigma xy}{N} - \dfrac{\Sigma x \Sigma y}{N^2}\right)}{s_X s_Y} = \frac{N\Sigma xy - \Sigma x \Sigma y}{Ns_X Ns_Y}$$

$$= \frac{N\Sigma xy - \Sigma x \Sigma y}{N\sqrt{\dfrac{\Sigma x^2}{N} - \bar{x}^2}\; N\sqrt{\dfrac{\Sigma y^2}{N} - \bar{y}^2}}$$

$$= \frac{N\Sigma xy - \Sigma x \Sigma y}{\sqrt{N^2\left[\dfrac{\Sigma x^2}{N} - \left(\dfrac{\Sigma x}{N}\right)^2\right]}\sqrt{N^2\left[\dfrac{\Sigma y^2}{N} - \left(\dfrac{\Sigma x}{N}\right)^2\right]}}$$

$$= \frac{N\Sigma xy - \Sigma x \Sigma y}{\sqrt{N\Sigma x^2 - (\Sigma x)^2}\sqrt{N\Sigma y^2 - (\Sigma y)^2}}$$

---

addition, many hand calculators provide the intermediate quantities $(\Sigma x, \Sigma x^2, \text{etc.})$ in a single step. From Formula **[4.3]**, the correlation is equal to

$$\frac{15(69{,}790) - (1{,}050)(995)}{\sqrt{15(73{,}694) - (1{,}050)^2}\;\sqrt{15(66{,}127) - (995)^2}} \doteq .90$$

In this example, and in general, the computational formula occasions less rounding error and therefore yields a somewhat more accurate answer than the definitional formula.

b. Interpreting the correlation coefficient. If we compare formulae [4.1] and [4.2], we see that

$$
r_{XY} = \frac{C_{XY}}{s_X s_Y}
$$

When two measurements X and Y are unrelated, therefore, the correlation coefficient must, like the covariance, equal zero. At the other extreme, we can conceive of a situation in which X and Y are *perfectly* related, that is, where any value x_i is associated with one and only one value of y. If the relationship is such that the scatter plot of X and Y is a straight line, it is shown in Box 4.2 that the correlation of X and Y is 1 when the relationship is positive and -1 when the relationship is negative.

Whatever the units in which X and Y are measured, then, a positive r-value indicates that X and Y are positively related, and the closer the value is to $+1$, the stronger the relation. A negative r-value indicates a negative statistical relation, and the closer the value to -1, the stronger the relation.[4]

The correlation coefficient is a useful and versatile statistic, but one must be careful not to confuse correlation with *causation*. For instance, when the Surgeon General reports a high correlation between smoking and lung cancer, one is tempted to suspect that smoking somehow *causes* physiological changes in lung tissue. This is probably true, but in general, a high correlation between X and Y (i.e., a value r_{XY} near 1) does *not* necessarily imply that X causes Y or that Y causes X. It may simply mean that some other factor W, or some combination of factors, influences both.[5] For example, it is a time-tested truism in educational research that children's shoe size (X) is highly correlated with spelling ability (Y). This is because shoe size and spelling performance are both correlated with age (W); older children have bigger feet than younger children and older children spell better. Similarly, per capita milk consumption is highly and positively correlated with a country's incidence of cancer. This is not because milk causes cancer or because cancer produces an appetite for milk. Milk consumption and longevity are both correlated with national wealth, and the older a person becomes, the greater the risk of cancer.

---

[4] A general proof that the values $+1$ and -1 are the maximum and minimum values of the correlation coefficient is beyond the scope of this book.

[5] Hence, tobacco industry claims that some as-yet unidentified genetic factor accounts for *both* susceptibility to lung cancer *and* a disposition to heavy tobacco use.

BOX 4.2

$r_{XY} = 1$ for perfect correlation

Proof: If X and Y are perfectly related and the relationship is *linear*, each value y_i can be expressed as $bx_i + a$. Since the formula for r_{XY} has standard deviations in its denominator we can assume that $s_X > 0$ and $s_Y > 0$.

$$r_{XY} = \sum^N \frac{(x_i - \bar{x})(y_i - \bar{y})}{Ns_X s_Y} = \sum^N \frac{(x_i - \bar{x})\left[(bx_i + a) - (\overline{bx + a})\right]}{Ns_X s_{bX+a}}$$

From exercises in Chapter 2 we know that adding a constant to all the x-values in a collection of observations adds the constant to the mean and that multiplying all of the x-values by a constant multiples the mean by the constant. Therefore, $(\overline{bx + a}) = b\bar{x} + a$. Likewise, we know that adding a constant to all of the values in a collection of observations has no effect on the variance, but that if $y = bx$, then $s_Y^2 = b^2 s_X^2$. Therefore,

$$s_{bX+a}^2 = b^2 s_X^2$$

If we assume $b > 0$, then

$$s_{bX+a} = bs_X$$

Consequently,

$$\sum^N \frac{(x_i - \bar{x})\left[(bx_i + a) - (\overline{bx + a})\right]}{Ns_X s_{bX+a}}$$

$$= \sum^N \frac{(x_i - \bar{x})\left[(bx_i + a) - (b\bar{x} + a)\right]}{Ns_X bs_X} \qquad [1]$$

In the numerator in expression [1],

$$\left[(bx_i + a) - (b\bar{x} + a)\right] = bx_i - b\bar{x} + a - a = b(x_i - \bar{x}),$$

so r_{XY} becomes

$$\sum^N \frac{(x_i - \bar{x})b(x_i - \bar{x})}{Ns_X bs_X} \qquad [2]$$

The b in the numerator cancels the b in the denominator, and

$$r_{XY} = \sum^N \frac{(x_i - \bar{x})(x_i - \bar{x})}{Ns_X s_X} = \sum^N \frac{(x_i - \bar{x})^2}{Ns_X^2}$$

$$= \frac{1}{s_X^2} \sum_N \frac{(x_i - \bar{x})^2}{N} \qquad [3]$$

However, the summation in expression [3] is exactly equal to s_X^2, so when X and Y are perfectly related,

$$r_{XY} = \frac{1}{s_X^2} \cdot s_X^2 = 1 \qquad \text{Q.E.D.}$$

This result assumes that b is positive. If b is negative, then

$$s_{bX+a} = -bs_X$$

and expressions [1] and [2] have $-b$ in the denominator. Thus, the cancellation operation in [2] leaves a -1, and $r_{XY} = -1$.

One must also be cautious in interpreting weak correlations. To be sure, a value of r near zero *may* mean that X and Y are not statistically related. However, even a *perfect* statistical relation will yield $r = \pm 1$ only if the relationship is *linear*. A statistical relationship may also be *curvilinear*. For example, cats that fall from buildings apparently reach terminal velocity after about four or five stories of fall and then, if they have time, extend their limbs spread-eagle, which increases air resistance and slows them down. Consequently, the proportion of cats killed by falls from New York City skyscrapers is about the same for falls of 9 stories as for falls of just 1 story and is much higher for falls of 5 stories.[6] The relationship of proportion of fatalities and distance fallen (see Figure 4.7) is very strong, but the relationship is curvilinear and would yield a correlation near zero.

c. Linearity and "reduction of uncertainty." We began this chapter by explaining the conventions for tabling the joint and marginal distributions of pairs of measurements taken on the same group of N individuals. Subsequent discussion focused on the graphic representation of the joint distribution (the scatter plot), but it is also possible to

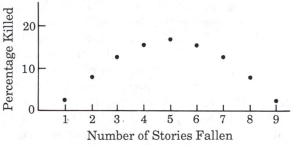

Figure 4.7. Plot of percentage of cats killed in falls by number of stories fallen.

[6] Diamond, J. (1989). How cats survive falls from New York skyscrapers. *Natural History*, August, 20–26.

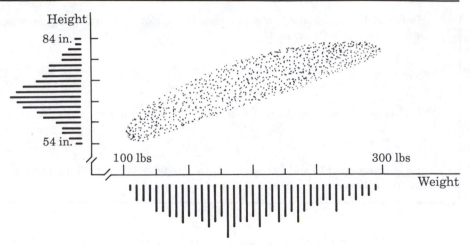

Figure 4.8. Marginal and joint distributions of height and weight.

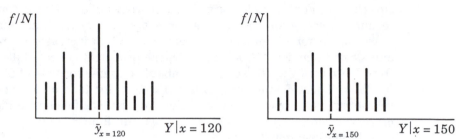

Figure 4.9. Conditional relative frequency distributions of height for persons weighing 120 and 150 pounds.

represent the marginal distributions in graphic form. In Figure 4.8, line graphs representing the marginal relative frequencies of weights (X) and heights (Y) in Table 4.2 have been added to the scatter plot of height and weight that appeared in Figure 4.2.

In addition to plotting the distribution of Y (or X) for the entire group, we could plot frequency or relative frequency distributions of Y for every observed value of X. That is, we could plot the distribution of height for all persons weighing 100 lbs, the distribution of height for all persons weighing 105 lbs, etc. The distribution of y-scores for all persons who obtained a particular score x is called the *conditional* distribution of *Y given x*. In general, "Y given x" is denoted $Y|x$. When x is assigned a specific value, say 100 pounds, we use notation of the form $Y|(x = 100)$. The conditional distributions of height (Y) for persons weighing 120 lbs and for persons weighing 150 lbs are given in Figure 4.9.[7]

---

[7] Likewise, we could plot the distributions of weights for all persons who were 4 ft 6 in. tall, all persons who were 4 ft 7 in. tall, etc. The labels X and Y are arbitrary, so everything we say about the conditional distribution of $Y|x$ applies to the conditional distribution of $X|y$.

In the conditional distribution of heights (Y) for persons weighing 120 lbs [that is, in the distribution of $Y|(x = 120)$ in Figure 4.9] the mean is denoted $\bar{y}_{x=120}$. Likewise, the *conditional* mean of $Y|(x = 150)$ is denoted $\bar{y}_{x=150}$. Conditional means are important to the interpretation of the correlation coefficient, because they are central to what is meant by a *linear* relation. Earlier, we said that a statistical relation is linear if the points of the scatter plot tend to cluster about a straight line. More specifically,

> A statistical relationship is described as linear if the conditional means of $Y|x$ lie on a straight line.

The variances of the conditional distributions are also important. Earlier in the chapter we explained that the essence of a statistical relation is that knowing the value of one member x of a pair of measurements reduces our uncertainty about the value of the other member of the pair, y. Uncertainty was not formally defined, but *degree* of uncertainty was discussed intuitively in terms of the range of y-values observed in the *entire collection of data* compared to the range of y-values observed *among persons who received the same score x on measurement X*. That is, we compared the range of the *marginal* distribution of Y to the range of the *conditional* distribution of $Y|x$. The *range* provided a graphically convenient way to introduce the notion that reduction of uncertainty hinges on the difference between marginal and conditional variation, but a formal treatment of this notion requires an algebraically tractable index of dispersion, viz., the *variance*.

The variance of the *marginal* distribution of Y (that is, s_Y^2) reflects the degree to which we would be uncertain about an estimate of Y in the absence of any information about X, whereas the *conditional* variance of $Y|x$ reflects our degree of uncertainty associated with an estimate of Y when the value of X is known to equal x. If it is assumed that all of the conditional variances are equal, the common conditional variance is denoted $s_{Y|x}^2$. The *percentage* by which uncertainty (as reflected in variance) about an individual's y-score is *reduced* by specifying the individual's x-score is therefore expressed by the ratio

$$\frac{s_Y^2 - s_{Y|x}^2}{s_Y^2} \qquad\qquad [4.4]$$

which is sometimes called the *correlation ratio*.[8]

Example 4.4. Calculate the conditional variance $s_{Y|x}^2$ and the correlation ratio for the following data set.

---

[8] The most common notation for the correlation ratio is η^2, where η is the lowercase Greek *eta*, but notation varies from author to author.

	X		
7	9	11	23
24.0	24.0	27.0	37.0
27.0	30.0	27.8	38.9
Y 23.0	34.1	35.0	34.2
21.0	29.0	33.0	33.2
29.0	29.0	27.0	42.1
20.0	26.1	28.0	34.0

We first calculate the conditional variances for Y for each value of x. The conditional variance of $Y|(x = 7)$ is simply the variance of the six values in the column headed $x = 7$, and the reader can confirm that $s^2_{Y|x=7} \doteq 10.0$. Similarly, $s^2_{Y|x=9} \doteq 10.0$, $s^2_{Y|x=11} \doteq 10.0$, and $s^2_{Y|x=23} \doteq 10.0$. Since all four variances are equal, the common conditional variance

$$s^2_{Y|x} \doteq 10.0$$

To obtain the correlation ratio, we then calculate the variance of all 24 y-values in the table, $s^2_Y \doteq 30.1$. Therefore, the correlation ratio is about

$$\frac{30.1 - 10.0}{30.1} \doteq 0.67$$

Thus, our uncertainty about any individual's y-score is reduced about 67 percent if we know his or her x-score.

When the common conditional variance $s^2_{Y|x}$ is equal to the marginal variance s^2_Y, there is *no* reduction in uncertainty, and it can be seen from equation [**4.4**] that the correlation ratio is equal to 0; when the relation of X and Y is perfect, each x-value is associated with only one y-value, and the conditional variance $s^2_{Y|x}$ is zero, so the correlation ratio equals 1. Because the correlation ratio can never be negative, it is apparent that the correlation *ratio* is not equal to the correlation *coefficient*. However, it can be shown that if the relationship of X and Y is *linear*, the correlation ratio is equal to the *square* of the correlation coefficient, r^2_{XY}, which is called the *coefficient of determination*. If the relationship of X and Y is not linear, it is easily demonstrated that r^2_{XY} is *less than* the correlation ratio.

Example 4.5. Scores for 10 individuals on two measures, X and Y, are presented in both tabular and graphic form in Figure 4.10. In this collection of data the points of the scatter plot tend to cluster about an inverted "V" or "U" defined by the conditional means of $Y|x$ (indicated by o on the graph in Figure 4.10). If we calculate the conditional variance of Y for any particular value x, we find that $s^2_{Y|x} = 4$, and if we calculate

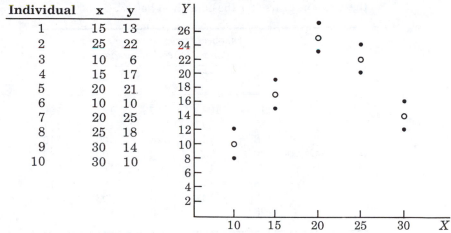

Individual	x	y
1	15	13
2	25	22
3	10	6
4	15	17
5	20	21
6	10	10
7	20	25
8	25	18
9	30	14
10	30	10

Figure 4.10. Scores (x, y) for 10 individuals ($\bullet$) and conditional means of $Y|x$ (o).

the (marginal) variance of Y across all 10 observations, we find that $s_Y^2 = 33.04$. The correlation ratio is therefore

$$\frac{s_Y^2 - s_{Y|x}^2}{s_Y^2} = \frac{33.04 - 4}{33.04} \doteq .88$$

That is to say, almost 90 percent of the variance of Y is eliminated if the value of X is specified—a very strong statistical relation indeed. However, if we calculate the Pearson correlation we find that $r \doteq .32$, which means that $r^2 \doteq .10$.

EXERCISES 4.3

In Exercises 1–10 calculate the correlation for earlier exercises indicated.

1. Exercise 4.1.1 (p. 104)   **2.** Exercise 4.1.2 (p. 104)
3. Exercise 4.1.3 (p. 104)   **4.** Exercise 4.1.4 (p. 104)
5. Exercise 4.1.5 (p. 104)   **6.** Exercise 4.1.6 (p. 105)
7. Exercise 4.1.9 (p. 105)   **8.** Exercise 4.1.10 (p. 105)
9. Exercise 4.1.11 (p. 105)   **10.** Exercise 4.1.12 (p. 106)

11. A traditional issue in the psychology of learning concerns the frequency with which a behavior should be reinforced in order to maximize the strength of the habit. One measure of how frequently a reinforcer is administered is the number of times the subject must perform the behavior before receiving a reward. A measure of habit strength is the number of *trials to extinction*, that is, the number of trials in which the behavior persists after reinforcement is no longer administered. Calculate the correlation between these two measures for the data given below:

	Number of times response must be performed to receive reinforcer (X)				
	5	**10**	**15**	**20**	**25**
	118	107	87	141	106
	87	83	100	130	151
	113	75	126	109	110
Trials to extinction	80	78	122	103	143
(Y)	109	111	96	136	155
	68	82	88	99	112
	110	112	123	103	148
	75	111	130	137	115

12. Another measure of frequency of reinforcement is the time between reinforcements. Calculate the correlation between number of minutes between reinforcements and trials to extinction from the following data:

	Minutes between reinforcements (X)				
	.50	**1**	**5**	**10**	**20**
	109	106	87	141	107
	75	151	100	130	83
	113	148	126	109	75
Trials to extinction	80	143	122	103	78
(Y)	118	155	96	136	111
	68	112	88	99	82
	110	110	123	103	112
	87	115	130	137	111

13. Table 1 of Appendix VIII is a listing of five-digit, computer-generated random numbers. Using any technique whatsoever, use these numbers to make up a fictitious data set of 15 two-digit numbers. Let these numbers be your x-values in the table below. Let a be the month in which you were born (January = 1, February = 2, etc.). Let b be the day of the month on which you were born; if b is odd, multiply it by -1. Use these values to calculate the y-values in the table, and calculate r_{XY}.

x	$bx + a = y$	x	$bx + a = y$	x	$bx + a = y$
——	——	——	——	——	——
——	——	——	——	——	——
——	——	——	——	——	——
——	——	——	——	——	——
——	——	——	——	——	——

14. Calculate the correlation between divorce rate (X) and unemployment rate (Y) for the years given below.

Year	Unemployment	Divorce	Year	Unemployment	Divorce
1960	5.5	2.2	1983	10.9	4.9
1965	4.5	2.5	1984	8.6	5.0
1970	4.9	3.5	1985	8.1	5.0
1975	9.1	4.8	1986	7.9	4.9
1980	7.9	5.2	1988	6.3	4.8
1982	11.0	5.0			

Source: U.S. Bureau of the Census (1990). *Statistical Abstract of the United States* (110th ed.), pp. 86, 89, 396; U.S. Bureau of the Census. *Statistical Abstract of the United States* (bicentennial edition), p. 135.

Many scientists who study animal behavior are interested in the relationship between social dominance and reproductive success. The following data are wins in aggressive encounters, number of cubs born (1978–1982), and number of cubs surviving 1 year for 12 female spotted hyenas observed in the Masai Mara National Reserve in Kenya.

Female	Wins	Cubs Born	Cub Survival	Female	Wins	Cubs Born	Cub Survival
04	63	5	5	30	4	5	3
03	45	6	6	22	3	3	2
63	11	2	No data	11	3	1	No data
N2	10	5	1	44	5	2	No data
KB	3	4	2	16	2	3	3
40	9	5	2	31	3	3	1

Source: Frank, L. G. (1986). Social organization of the spotted hyaena *Crocuta crocuta*. II. Dominance and reproduction. *Animal Behaviour, 34*, 1510–1527.

15. What is the correlation between wins and cubs born?

16. What is the correlation between wins and one-year cub survival?

Exercises 17 and 18 refer to the following heights in inches (X) and weights in pounds (Y) for a fictitious group of 30 college freshmen.

Subject	x	y	Subject	x	y	Subject	x	y
1	67	160	11	78	200	21	70	125
2	76	219	12	72	200	22	66	175
3	72	180	13	60	175	23	73	150
4	60	110	14	70	145	24	71	140
5	84	275	15	75	275	25	69	175
6	71	175	16	66	155	26	63	180
7	66	200	17	66	130	27	61	125
8	81	250	18	69	125	28	69	230
9	57	150	19	63	145	29	57	105
10	72	150	20	70	190	30	61	135

17. (a) Calculate the covariance of heights in inches and weights in pounds for the data given above.
    (b) Calculate the covariance of heights in feet and weights in pounds for the data given above.
18. (a) Calculate the correlation of heights in inches and weights in pounds for the data given above.
    (b) Calculate the correlation in heights in feet and weights in pounds for the data given above.
19. For the data in Exercise 4.3.11 calculate the conditional variance $s^2_{Y|x=10}$.
20. For the data in Exercise 4.3.11 calculate the conditional variance $s^2_{Y|x=20}$.
21. For the data in Exercises 4.3.17 and 4.3.18 calculate the conditional variance $s^2_{Y|x=66}$.
22. For the data in Exercises 4.3.17 and 4.3.18 calculate the conditional variance $s^2_{X|y=175}$.
23. For the data in Exercise 4.1.7 (p. 105) calculate the conditional variance $s^2_{Y|x}$ and the correlation ratio.
24. For the data in Exercise 4.1.8 (p. 105) calculate the conditional variance $s^2_{Y|x}$ and the correlation ratio.
25. The covariance of X and Y in Example 4.1 was calculated to be 3.11. If in addition we know that s_X and s_Y are both equal to 1.9, what is r_{XY}?
26. Consider the data in Example 4.2 (p. 112).
    (a) Calculate the variance of husbands' heights.
    (b) Calculate the variance of husbands of women who are 5 ft 5 in. tall.
    (c) By what percentage is variance of husbands' heights reduced if we consider only husbands of women who are 5 ft 5 in. tall?
    (d) What connection do you see between your answer to 4.3.26(c) and the answer calculated in Example 4.2?

For Data Set 1 and Data Set 2 in Exercises 27 and 28, calculate the following: (a) the marginal variance of Y, (b) the conditional variance and the conditional mean of $Y|x$ for every x-value, (c) the correlation ratio, (d) the coefficient of determination.

27.

Data Set 1				Data Set 2			
X	Y	X	Y	X	Y	X	Y
50	15	170	195	50	75	90	135
130	115	210	235	130	235	90	175
170	215	50	35	210	255	270	115
270	335	240	255	240	195	130	255
90	95	270	315	90	115	50	15
210	295	130	135	210	295	270	75
50	55	270	355	240	215	130	295
130	155	90	135	50	35	170	355
270	295	240	295	240	235	210	275
240	315	130	175	170	295	270	95
170	175	170	235	270	135	170	335
210	255	50	75	210	315	170	315
90	115	210	275	50	55	90	155
240	275	90	75	240	175	130	275

(a) What do you notice about the correlation ratios in Data Sets 1 and 2? What would you say about the strength of the relationship between X and Y in Data Sets 1 and 2?

(b) What do you notice about the coefficients of determination for Data Sets 1 and 2?

(c) From your answers to (a) and (b) what might you conclude about the relationship between X and Y in Data Sets 1 and 2?

(d) Support your answer to (c) by drawing graphs with the conditional means of $Y|x$ represented on the vertical axis and x-values on the horizontal axis for Data Sets 1 and 2.

28.

Data Set 1				Data Set 2			
X	Y	X	Y	X	Y	X	Y
245	10	125	190	245	70	205	130
165	110	85	230	165	230	205	170
125	210	245	30	85	250	25	110
25	330	55	250	55	190	165	250
205	90	25	310	205	110	245	10
85	290	165	130	85	290	25	70
245	50	25	350	55	210	165	290
165	150	205	130	245	30	125	350
25	290	55	290	55	230	85	270
55	310	165	170	125	290	25	90
125	170	125	230	25	130	125	330
85	250	245	70	85	310	125	310
205	110	85	270	245	50	205	150
55	270	205	70	55	170	165	270

(a) What do you notice about the correlation ratios in Data Sets 1 and 2? What would you say about the strength of the relationship between X and Y in Data Sets 1 and 2?

(b) What do you notice about the coefficients of determination for Data Sets 1 and 2?

(c) From your answers to (a) and (b) what might you conclude about the relationship between X and Y in Data Sets 1 and 2?

(d) Support your answer to (c) by drawing graphs with the conditional means of $Y|x$ represented on the vertical axis and x-values on the horizontal axis for Data Sets 1 and 2.

C. LINEAR REGRESSION

Up to now, we have used the terms "reduction of uncertainty" and "precision of estimate" rather interchangeably. In practice, these may address somewhat different questions. For example, a college instructor may be interested in whether students who do well on the term paper also do well on the final examination and whether students who do poorly on one also do poorly on the other. In this situation the only issue is the degree to which the two methods of assessing student performance are in agreement. In other situations the instructor might actually want to predict one score from the other. Suppose, for instance, that a student is hospitalized just before the final examination. If the instructor must assign a course grade based on the student's past performance, one strategy is to predict or estimate the score on the final examination from the score on the term paper.

In the last section we formalized the notion of reduction of uncertainty in terms of conditional and marginal variation and saw that when the relationship of X and Y is linear, reduction of uncertainty is related intimately to the correlation coefficient, r_{XY}. In this section we shall likewise formalize the notion of precision of estimate and demonstrate that precision of estimate, like reduction of uncertainty, is tied to correlation.

1. The regression equation: prediction and error

A friend of ours has a motor home and spends much of the winter visiting cross-country ski areas in the Midwest. He has a Fahrenheit thermometer mounted on the outside of the vehicle near a window, so that he can read the air temperature (and decide which wax to use on his skis) without getting up from the breakfast table. When he goes to Ontario and hears a weather report announce a current temperature of $-10°$ Celsius, our friend can "predict" with absolute certainty the Fahrenheit temperature that he will see on his thermometer. That is because there is an equation

$$F = \tfrac{9}{5}C + 32$$

that relates the two measures. Thus, $-10°C$ corresponds to

$$\tfrac{9}{5}(-10) + 32 = 14°F \qquad [4.5]$$

(or, as our friend interprets such things, *special green wax*).

Equation [4.5] presents what we have called a "perfect" relationship between two measures. Each value of C yields exactly one value of F. Furthermore, equation [4.5] is of the general form

$$y = bx + a \qquad [4.6]$$

You should recall from your algebra course that equation [4.6] is called a *linear* equation (cf. Box 4.2), because its graph in an XY-coordinate system is a straight line.[9] A straight line has the simplest of all equations and linear predictions are, accordingly, simple to calculate and easy to comprehend. However, *none* of the graphs of data in this chapter looks anything like a line (or, for that matter, a curve). Generally, a scatter plot is roughly an ellipse with many different y-values for each x-value. Nevertheless, it is possible to find the linear equation that *best fits* a scattering of data points. In a sense we can think of such an equation as a

---

[9]You may be more familiar with the notation $y = mx + b$. In the context of correlation and regression, b replaces m and a replaces b.

way of organizing and summarizing joint distributions of data that is analogous to the way we used grouping to organize and summarize distributions of one measure.

In Chapter 2 we learned that there are certain criteria by which some statistics do a better job of summarizing such distribution properties as location and dispersion than do other statistics. Similarly, some lines do a better job of summarizing a scatter plot than do other lines. Equation **[4.6]** is determined once we know the values of b and a, so the task is to find the values of a and b that give the *best* summary. The equation that best summarizes the scatter plot of (X, Y) can then be used to predict or estimate the y-value associated with a specified value of X.

To understand what we mean by "best" summary, let us suppose that the data point (x_i, y_i) represents the scores obtained by individual i and that $\hat{y}_i$ is the y-value predicted for individual i by the equation

$$\hat{y} = bx + a$$

In general you expect the prediction to be wrong. That is, we expect that $y_i \neq \hat{y}_i$, because we know that in general any line will miss most data points. The difference $(y_i - \hat{y}_i)$ is the *error of prediction*. The error of prediction for data point i is illustrated in Figure 4.11.

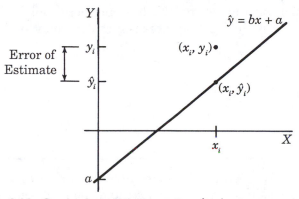

Figure 4.11. Graph of prediction equation $\hat{y} = bx + a$.

Customarily, the line that best fits a joint distribution of observations is taken to be the equation that yields the smallest overall error of estimate. We know that a good descriptive statistic should consider all observed values, so we might try to capture the error associated with our prediction equation by adding all of the prediction errors. This tactic will be confounded by the fact that some positive errors will cancel some negative errors in an uncontrollable way. To avoid this problem we *square* every error $(y_i - \hat{y}_i)$, as we did with variance, and take the overall error of prediction to be the *mean* of the squared errors for all N individuals in the data collection:

$$\frac{\sum\limits_{i}^{N}(y_i - \hat{y}_i)^2}{N}$$

We therefore want the line that yields the smallest possible average of the squared errors, which is sometimes called the *variance of estimate*.[10]

In Appendix III we present a proof that the average squared error of prediction is least when

$$b = r\frac{s_Y}{s_X} \tag{4.7}$$

and

$$a = \bar{y} - b\bar{x} \tag{4.8}$$

From [4.8] we may substitute $\bar{y} - b\bar{x}$ for a in the general expression for the predicted value $\hat{y}$ and we obtain

$$\hat{y} = bx + (\bar{y} - b\bar{x}) \tag{4.9}$$

$$= b(x - \bar{x}) + \bar{y}$$

Accordingly, the best linear prediction of y from a known value of x is obtained from equation [4.9].

The linear equation that yields the best prediction of y from a known value of x is the least squares regression equation

$$\hat{y} = b(x - \bar{x}) + \bar{y}$$

where

$$b = r\frac{s_Y}{s_X}$$

Calculation of b from formula [4.7] requires that one first compute r, s_Y, and s_X. If these values are not available (and you don't especially feel like computing them), it may be quicker to calculate b directly from the raw data using formula [4.10]:

$$b = \frac{N\Sigma xy - \Sigma x\Sigma y}{N\Sigma x^2 - (\Sigma x)^2} \tag{4.10}$$

where N is the total number of observations, and all summations run from 1 to N.

Example 4.6. In 1964 Kitty Genovese was repeatedly stabbed by an assailant in a New York City street. The assault proceeded for over half an hour, and although 38 of the victim's neighbors witnessed the attack

---

[10] Since N is fixed, we can achieve the same goal by making the numerator $\Sigma(y - \hat{y})^2$ as small as possible, thus, the name *least squares* line. This approach was first put forth by C. F. Gauss (1777–1855) and independently by A. M. Legendre (1752–1833).

from their apartment windows, not one person intervened or even telephoned the police. This event raised a question that subsequently became a major research question in social psychology: Under what conditions will people intervene in an emergency? One finding that has weathered many replications is that there tends to be what social psychologists John M. Darley and Bibb Latané call a "diffusion of responsibility." The more people who are present, the more reluctant people are to "become involved."

The following experiment is typical of studies conducted by Darley and Latané and their associates[11] to test the hypothesis that willingness to intervene is related to group size. Each subject is an undergraduate student waiting in the anteroom of a university office to keep an appointment with an academic advisor. When the subject arrives, there are already 1, 2, 4, 5, or 11 other persons also apparently waiting to keep appointments. A minute or two after the subject's arrival, the crash of a heavy bookcase falling over is heard from an adjacent room, followed by calls for help that become increasingly desperate as time passes.

The crash and the victim's calls for help are tape recorded, and the other persons in the anteroom are confederates of the experimenter. The length of time a subject waits before investigating is Y, and X is the number of persons (subject plus confederates) on the scene. Data for this experiment conducted for 25 individuals might look like those in Table 4.7. What score (time) would be predicted for an individual in a group of 4 persons? Of 38 persons? To use formula [4.10] to calculate b, one needs the quantities Σx, Σx^2, Σy, and Σxy:

Table 4.7. Length of time (in seconds) before intervening by group size

	Group size (x)				
	2	3	5	6	12
Time (y)	13	9	20	22	32
	9	11	20	19	26
	10	10	16	20	19
	12	10	17	18	28
	8	8	16	16	31
Σy	52	48	89	95	136

$$\Sigma x = 5(2) + 5(3) + 5(5) + 5(6) + 5(12) = 140$$
$$\Sigma x^2 = 5(4) + 5(9) + 5(25) + 5(36) + 5(144) = 1{,}090$$
$$\Sigma y = 52 + 48 + 89 + 95 + 136 = 420$$

[11] See Latané, B., and Nidas, S. (1981). 10 years of research on group size and helping. *Psychological Bulletin*, 89, 308–324.

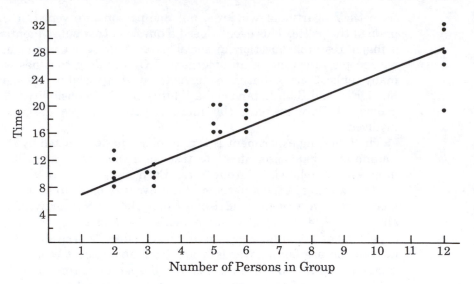

Figure 4.12. Scatter plot and regression line for Example 4.6.

$$\Sigma xy = (2)(13) + (2)(9) + (2)(10) + (2)(12) + (2)(8)$$

$$+ (3)(9) + (3)(11) + (3)(10) + (3)(10) + (3)(8)$$

$$+ (5)(20) + (5)(20) + (5)(16) + (5)(17) + (5)(16)$$

$$+ (6)(22) + (6)(19) + (6)(20) + (6)(18) + (6)(16)$$

$$+ (12)(32) + (12)(26) + (12)(19) + (12)(28) + (12)(31)$$

$$= 2{,}895$$

Therefore,

$$b = \frac{25(2{,}895) - [(140)(420)]}{25(1{,}090) - (140)^2} \doteq 1.77$$

From Σx and Σy, $\bar{x} = 5.6$, and $\bar{y} = 16.8$, so by formula **[4.10]** the predicted response time for a person in a group of 4 persons is

$$\hat{y} = 1.77(4 - 5.6) + 16.8 \doteq 14 \text{ sec}$$

For 38 persons[12]

$$\hat{y} = 1.77(38 - 5.6) + 16.8 \doteq 74 \text{ sec}$$

[12] Of course, extrapolating so far beyond the data is somewhat risky.

In Figure 4.12 we have drawn the scatter plot and shown the regression line

$$\hat{y} = 1.77(x - 5.6) + (16.8)$$

2. Interpreting regression

a. Regression line as locus of conditional means. Earlier, we said that a statistical relation is described as linear if the conditional means of $Y|x$ lie on a straight line. We can now be more explicit: When X and Y are linearly related, the conditional means of $Y|x$ lie on the *regression* line, $\hat{y} = bx + a$. The y-value predicted for any particular value x therefore equals the mean of the distribution of $Y|x$.

b. Correlation and the slope of the regression line. Recall from your algebra course that the coefficient b is called the *slope* of the line $y = bx + a$. That is, b is the number of units that Y changes for every unit change in X. For the least squares regression line we know that

$$b = r\frac{s_Y}{s_X}$$

The slope of the regression line is therefore always proportional to the correlation between X and Y.

The relationship between correlation and the slope of the regression line is most easily seen if we standardize X and Y. Let $\hat{z}_Y$ be the predicted *standardized* value of Y associated with z_X, the *standardized* value of X. The prediction equation then becomes

$$\hat{z}_Y = bz_X + a$$

where

$$b = r\frac{s_{Z_y}}{s_{Z_x}}$$

from equation [4.7] and

$$a = \bar{z}_Y - b\bar{z}_X$$

from equation [4.8]. However, we know that the standard deviation s_Z for any collection of standardized scores is 1, so

$$b = r\frac{1}{1} = r$$

For standardized scores, then, the slope b is equal to the correlation r of X and Y. We also know that the mean $\bar{z}$ of any collection of standardized scores is 0, so

$$a = 0 - b(0) = 0$$

The least squares regression equation for standardized scores therefore reduces to

$$\hat{z}_Y = rz_X$$

Thus, when $r = 1$, we see that

$$\hat{z}_Y = (1)z_X = z_X$$

so the regression line $\hat{z}_Y = rz_X$ rises at an angle of 45 degrees to the horizontal axis, as illustrated in Figure 4.13(a). At the other extreme, when $r = 0$,

$$\hat{z}_Y = (0)z_X = 0$$

for all values of z_X. That is, the regression line $\hat{z}_Y = rz_X$ coincides with the horizontal axis, as illustrated in Figure 4.13(c). As correlation decreases from 1 to 0, the predicted values $\hat{z}_Y$ are therefore said to *regress*[13] to the (marginal) mean, $\bar{z}_Y = 0$.

Figure 4.13 considers only nonnegative correlations. Correlations of -1 and $-.49$ (approximately) would give us mirror images of Figures 4.13(a) and 4.13(b). And in general, if Z_X and Z_Y are negatively correlated, the angle that the regression line makes with the positive horizontal axis is between 135 degrees ($r = -1$) and 180 degrees ($r = 0$).

c. Regression, correlation, and conditional variance. We noted in our discussion of correlation that if X and Y are linearly related, then the coefficient of determination is equal to the correlation ratio,

$$r_{XY}^2 = \frac{s_Y^2 - s_{Y|x}^2}{s_Y^2}$$

[13] This was first observed by Sir Francis Galton (1822–1911). In developing quantitative techniques to study heredity, Galton measured a number of biological attributes (x) in fathers and, from these scores, predicted scores ($\hat{y}$) for their sons. Since relations between fathers' scores and sons' scores were imperfect ($0 < r < 1$), Galton often found that the sons' *predicted* scores were closer to the mean for sons ($\bar{y}$) than were their fathers' *predictor* scores (x) to the mean of fathers ($\bar{x}$). He concluded from this that biological inheritance produced a *regression* toward the mean from one generation to the next. Galton's terminology has fared better than his inferences about inheritance.

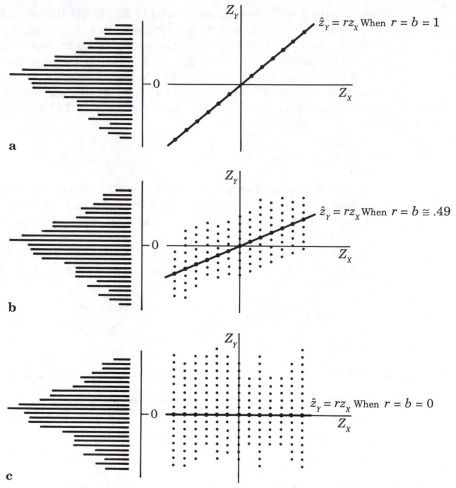

Figure 4.13. Regression line and conditional distributions for (a) $r = 1$, (b) $r \cong .49$, (c) $r = 0$. Points (•) indicate predicted values $\hat{z}_Y$.

so

$$r_{XY}^2 = 1 - \frac{s_{\hat{Y}|x}^2}{s_Y^2}$$

$$1 - r_{XY}^2 = \frac{s_{\hat{Y}|x}^2}{s_Y^2}$$

and

$$s_{\hat{Y}|x}^2 = s_Y^2\left(1 - r_{XY}^2\right) \qquad\qquad [4.11]$$

We know that b is always proportional to r and that for standardized observations $b = r$. From equation **[4.11]** we should therefore expect the variance of $Y|x$ to decrease as b increases. In the case of standardized scores Z_X and Z_Y, this is seen in the marginal distributions of Z_Y and the conditional distributions of $Z_{Y|x}$ illustrated in Figure 4.13.

When $b = 0$, and the regression line is therefore horizontal, the conditional variance equals the marginal variance, as in Figure 4.13(c). In the general case, it is likewise seen from equation **[4.11]** that if $r_{XY} = 0$, then

$$s_Y^2(1 - 0) = s_{Y|x}^2 = s_Y^2$$

As b increases, and the regression line "rotates" from the horizontal the conditional variance diminishes, as in Figure 4.13(b). At 45 degrees (i.e., when $b = 1$), we see in Figure 4.13(a) that the conditional variance vanishes altogether. All of the observed values z_Y associated with any particular value z_X are identical; they all lie on the line $\hat{z}_Y = bz_X$. And from equation **[4.11]** we see that in the general case, it is similarly true that $r = 1$ gives us

$$s_{Y|x}^2 = s_Y^2(1 - 1) = 0$$

EXERCISES 4.4 ———————————————————————————————

1. For each of the data sets below, calculate the values of a and b of the regression line $\hat{y} = bx + a$ and calculate $\hat{y}$ for the x-values indicated.

Data Set 1		Data Set 2		Data Set 3	
X	Y	X	Y	X	Y
14	10	148	151	131	207
10	11	152	159	126	200
13	12	121	167	130	202
13	11	136	150	132	209
11	9	141	164	144	216
9	10	149	152	129	203
11	11	137	160	140	203
7	9	135	170	136	211
12	10	129	160	128	216
13	12	128	163	138	201
		155	154	149	221
				129	210

(a) For Data Set 1, $x = 5$, $x = 8$, $x = \mathbf{9}$
(b) For Data Set 2, $x = 130$, $x = \mathbf{141}$, $x = 150$
(c) For Data Set 3, $x = \mathbf{126}$, $x = 135$, $x = 145$
(d) The boldface x-values in (a), (b), and (c) were *observed* values. Which of the three corresponding $\hat{y}$-values exhibits the least error of estimate, i.e., for which x-value is $\hat{y}$ nearest y? (Since the three data sets are in different

scales of measurement, the errors should be converted to standard units to make this comparison.)

2. For each of the data sets below, calculate the values of a and b of the regression line $\hat{y} = bx + a$ and calculate $\hat{y}$ for the x-values indicated.

Data Set 1		Data Set 2		Data Set 3	
X	Y	X	Y	X	Y
11	24	90	87	58	139
9	17	100	39	54	127
10	17	120	75	57	130
12	18	150	75	59	146
8	16	170	72	60	137
10	20	180	54	53	132
7	12	200	50	51	112
9	16	200	45	58	147
		220	45	62	147
		220	45	56	139
		220	84	55	136
		230	39		

(a) For Data Set 1, $x = 6$, $x = $ **8**, $x = 14$
(b) For Data Set 2, $x = 110$, $x = 160$, $x = $ **230**
(c) For Data Set 3, $x = 48$, $x = $ **53**, $x = 61$
(d) The boldface x-values in (a), (b), and (c) were *observed* values. Which of the three corresponding $\hat{y}$-values exhibits the least error of estimate, i.e., for which x-value is $\hat{y}$ nearest y? (Since the three data sets are in different scales of measurement, the errors should be converted to standard units to make this comparison.)

3. Use the data below from Exercise 4.3.11 (p. 122) to
(a) calculate linear regression coefficients a and b
(b) estimate the habit strength (number of unreinforced trials required for extinction) of a response acquired on a schedule of 1 reinforcement every 35 times the response was performed.

| | | **Frequency of reinforcement** | | | | |
		Number of times response must be performed to receive reinforcer				
		5	**10**	**15**	**20**	**25**
		118	107	87	141	106
		87	83	100	130	151
		113	75	126	109	110
Habit strength	80	78	122	103	143
Trials to extinction	109	111	96	136	155
		68	82	88	99	112
		110	112	123	103	148
		75	111	130	137	115

4. Use the data below from Exercise 4.3.12 (p. 122) to
   (a) calculate linear regression coefficients a and b, and
   (b) estimate the habit strength (number of unreinforced trials required for extinction) of a response acquired on a schedule of 1 reinforcement every 15 minutes.

	Frequency of Reinforcement Minutes between reinforcements				
	.50	1	5	10	20
	109	106	87	141	107
	75	151	100	130	83
	113	148	126	109	75
Habit Strength	80	143	122	103	78
Trials to extinction	118	155	96	136	111
	68	112	88	99	82
	110	110	123	103	112
	87	115	130	137	111

5. Use the data below from Example 4.2 (p. 112) to
   (a) calculate linear regression coefficients a and b, and
   (b) estimate the height of the wife of a man who is 6 ft 1 in. tall.

Couple (i)	Husband (x_i)	Wife (y_i)	Couple (i)	Husband (x_i)	Wife (y_i)	Couple (i)	Husband (x_i)	Wi $(y$
1	76	71	6	71	65	11	68	6!
2	75	70	7	71	65	12	68	6(
3	75	70	8	70	67	13	67	6:
4	72	67	9	68	64	14	67	6!
5	72	71	10	68	65	15	62	6:

6. The data below give the annual inflation rate and the prime interest rate for years 1972–1978.
   (a) calculate linear regression coefficients a and b, and
   (b) predict the prime interest rate for a year in which inflation is 5 percent.

Year	Inflation	Prime Rate	Year	Inflation	Prime Rate
1972	3.3	5.5	1976	5.8	6.8
1973	6.2	8.0	1977	6.5	6.9
1974	11.0	10.8	1978	7.6	9.0
1975	9.1	7.9			

Source: U.S. Bureau of the Census (1980). *Statistical Abstract of the United States* (100th ed.), pp. 419 and 541.

7. Use the data below from Exercise 4.3.14 to
   (a) calculate linear regression coefficients a and b and
   (b) predict the divorce rate when unemployment reaches the depression-level benchmark of 17 percent.

Year	Unemployment	Divorce	Year	Unemployment	Divorce
1960	5.5	2.2	1983	10.9	4.9
1965	4.5	2.5	1984	8.6	5.0
1970	4.9	3.5	1985	8.1	5.0
1975	9.1	4.8	1986	7.9	4.9
1980	7.9	5.2	1988	6.3	4.8
1982	11.0	5.0			

Source: U.S. Bureau of the Census (1990). *Statistical Abstract of the United States* (110th ed.), pp. 86, 89, 396; U.S. Bureau of the Census. *Statistical Abstract of the United States* (bicentennial edition), p. 135.

8. Atomic absorption spectrophotometery is a technique that is used to determine the concentration of certain elements in a sample of groundwater, industrial waste, etc. Light of a particular wavelength is passed through the sample and the spectrophotometer calculates the amount of light that the sample absorbs. This value is given in *atomic absorption units* and is ordinarily proportional to the concentration of the element, but the atomic absorption of many elements changes with age. A chemist wants to be able to predict such changes for a particular element and therefore records the atomic absorption of 15 samples of the same known concentration of the element. Five samples are 3 weeks old, 5 are 5 weeks old, and 5 are 7 weeks old. Atomic absorption readings for the 15 samples are given below.

	Age of sample		
	3 weeks	**5 weeks**	**7 weeks**
	58	40	25
Atomic	60	37	24
absorption	55	42	26
	56	40	23
	61	39	25

Use these data to
(a) calculate linear regression coefficients a and b, and
(b) estimate the absorption of a 6-week-old sample of the same concentration.

D. SUMMARY

When two measurements X and Y are taken on the *same* group of N individuals, the *joint distribution* of X and Y may be represented by letting X and Y be the axes of a plane and letting the *pair* of scores for individual i be represented by the point in the plane at coordinates (x_i, y_i). A representation of this sort is called a *scatter plot* or *scatter diagram*.

If the scatter diagram shows that the y-values associated with any *particular* value of x exhibit less dispersion than the *entire collection* of y values, one can make a more precise estimate of y if the value of x is known than if the value of x is not known. The measurements X and Y are then said to be statistically related. The degree to which precision of estimate of y is increased by knowing the value of x is called the *strength* or *magnitude* of the statistical relation. A statistical relation is also characterized by its *direction*. If large x-values tend to be paired with large y-values and small x-values with small y-values, the direction of the relation is *positive*. If the relationship is *negative*, large values of X tend to be paired with small values of Y and small x-values with large y-values.

One numerical index that describes the magnitude and direction of a statistical relationship is the *covariance*, which is denoted C_{XY} and is defined in expression **[4.1]**.

$$C_{XY} = \frac{\sum\limits_{}^{N}(x_i - \bar{x})(y_i - \bar{y})}{N}$$

The covariance satisfies all of our criteria for a good descriptive statistic and has three other important properties: (1) If X and Y are not statistically related, $C_{XY} = 0$, (2) the absolute value of C_{XY} increases as the strength of the relation between X and Y, (3) the sign C_{XY} corresponds to the direction of the relation between X and Y.

The covariance can be difficult to interpret because it is sensitive to units of measurement. The smaller the units, the larger the covariance. The problem is overcome by transforming X and Y to standard units. The covariance for two *standardized* measurements Z_X and Z_Y is called the *correlation coefficient*, denoted r_{YX}, and is defined by expression **[4.2]**.

$$r_{XY} = \sum\limits_{}^{N} \frac{(x_i - \bar{x})(y_i - \bar{y})}{N s_X s_Y}$$

For computational purposes, the correlation coefficient is more easily determined by formula **[4.3]**:

$$r_{XY} = \frac{N\Sigma xy - \Sigma x \Sigma y}{\sqrt{N\Sigma x^2 - (\Sigma x)^2}\sqrt{N\Sigma y^2 - (\Sigma y)^2}}$$

When the relationship between X and Y is *linear*, and the relationship is perfect, $r_{XY} = 1$ if the relationship is positive, and $r_{XY} = -1$ if the relationship is negative. If X and Y are completely unrelated, $r_{XY} = 0$.

The amount by which uncertainty about an individual's y-score is reduced if the person's x-score is known can be expressed by the *correlation ratio*

$$\frac{s_Y^2 - s_{Y|x}^2}{s_Y^2}$$

where s_Y^2 is the variance of all the y-values and is called the *marginal* variance of Y, and $s_{Y|x}^2$ is the variance of all the y-values associated with a particular x-value and is called the *conditional* variance of $Y|x$. If X and Y are *linearly* related, this ratio is equal to the *square* of the correlation coefficient, r_{XY}^2, which is called the *coefficient of determination*.

The basic task in regression problems is to obtain the values a and b in the linear equation

$$y = bx + a$$

that give the most accurate prediction of Y when the value of X is known. If $\hat{y}_i$ is the *predicted* value for individual i and y_i is the *observed* value for individual i, then accuracy is a matter of reducing the average of squared error

$$\frac{\sum\limits_{i}^{N} (y_i - \hat{y}_i)^2}{N}$$

It can be shown that the average squared error of estimate is smallest when

$$b = r \frac{s_Y}{s_X}$$

and

$$a = \bar{y} - b\bar{x}$$

Substituting these values into the general equation for a straight line, we obtain

$$\hat{y} = bx + (\bar{y} - b\bar{x})$$

which is called the *linear regression* equation or the *least squares* equation.

If the relationship of X and Y is linear, then for any specified value x, the value $\hat{y}$ predicted by the least squares equation is the mean of the conditional distribution $Y|x$.

Correlation and regression techniques are closely related. The slope b of the regression line is always proportional to the correlation coefficient r, and for standardized observations, $b = r$. This is reflected in two ways in the scatter plot of X and Y: First, as the (absolute) value of r increases, the regression line becomes steeper, that is, more nearly vertical. Second, as the slope of the regression line increases, the conditional variances of $Y|x$ becomes smaller. This is most easily seen in the scatter plot for standardized scores. When $b = 0$, conditional variance equals the marginal variance; when $b = 1$, the conditional variance equals zero.

Postscript to part I: numbers, intervals, graphs, and statistics

A. GRAPHS AND STATISTICS: TWO SIDES OF A COIN

In Chapter 1 we said that observations are grouped in order to make a large data collection comprehensible, but what we meant by "comprehensible" was left deliberately vague. We may now say that a distribution of data is comprehensible to the extent that it captures the descriptive properties discussed in Chapter 2, i.e., central tendency, skew, and so on. We also suggested in Chapter 1 that the number of intervals be chosen so that the *average* interval frequency exceeds 5. This is because most descriptive properties of a distribution reflect *differences* among frequencies, and unless the average frequency is greater than 5, the frequencies don't have much room to vary.

This same consideration is one of the basic reasons for grouping data: Unless the number of observations N is more than 5 times the number of *possible* scores, it is very difficult to get an intuitive feel for the descriptive properties of an ungrouped distribution. In Table 1.4 (p. 6), we tabulated fictional data for an experiment in which 1,000 persons each tossed 100 coins and counted the number of heads. We developed this example to illustrate the advantages of grouping, but the number of observations ($N = 1,000$) is almost 10 times the number of possible scores (101), so many of the observations would necessarily yield identical values. Consequently, even if the data were not grouped, we would probably detect a distinct mode near $x = 50$, symmetry about the mode, and a moderate degree of kurtosis. However let us suppose instead that we have only $N = 300$ persons each toss 100 coins and count the number of heads. It is very likely that every x-value would have a frequency of 0, 1, or 2. The central tendency, symmetry, and kurtosis of the distribution would therefore be undetectable unless the data were grouped.

One obvious remedy is to forgo frequency tables or histograms and to proceed directly to calculation of descriptive statistics. Indeed, many current texts take this approach, arguing that the availability of statistical computer packages has largely eliminated the need to group or otherwise reduce large data collections. As the following example illustrates, however, one should be wary of technological shortcuts and keep in mind the observation attributed to Benjamin Disraeli that there are three kinds of lies: lies, damned lies, and statistics.

Example 1. Consider the following data sets:

Data Set 1

0.40,	0.40,	0.40,	0.41,	0.42,	0.42,	0.43,	0.44,
0.44,	0.44,	0.47,	0.49,	1.98,	1.99,	1.99,	2.00,
2.10,	2.11,	2.14,	2.17,	2.20,	2.23,	3.00,	3.10,
3.20,	3.23,	3.30,	3.40,	4.90,	5.00,	5.00,	5.10,
6.60,	6.70,	6.77,	6.80,	6.90,	7.00,	7.77,	7.80,
7.83,	7.86,	7.89,	7.90,	8.00,	8.01,	8.01,	8.02,
9.51,	9.53,	9.56,	9.56,	9.56,	9.57,	9.58,	9.58,
9.59,	9.60,	9.60,	9.60				

Data Set 2

− 13.30,	2.80,	3.09,	3.56,	3.78,	3.98,	4.00,	4.02,
4.05,	4.10,	4.12,	4.30,	4.32,	4.40,	4.45,	4.60,
4.60,	4.67,	4.67,	4.70,	4.80,	4.88,	4.88,	4.89,
4.90,	4.90,	4.95,	4.98,	5.00,	5.00,	5.00,	5.00,
5.02,	5.05,	5.10,	5.10,	5.11,	5.12,	5.12,	5.20,
5.30,	5.33,	5.33,	5.40,	5.40,	5.55,	5.60,	5.68
5.70,	5.88,	5.90,	5.95,	5.98,	6.00,	6.02,	6.22,
6.44,	6.91,	7.20,	23.30				

The following descriptive statistics might lead one to suspect that the two distributions are very similar:

	Data Set 1	Data Set 2
$\tilde{x}$	5	5
$\bar{x}$	5	5
s^2 (approx.)	11.8258	11.8258
Sk	0	0
Sk_P	0	0

However, if we group the data into unit-width intervals, we see from the histograms in Figures 1(a) and 1(b) that the distributions are in fact

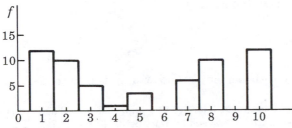

Figure 1(a). Data Set 1.

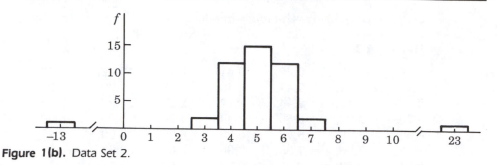

Figure 1(b). Data Set 2.

quite different. The distribution for Data Set 1 is basically U-shaped, and the distribution of Data Set 2 is basically bell-shaped.[1]

1. Outliers

In part, these very different distributions yielded almost identical descriptive statistics because Data Set 2 includes the values -13.30 and 23.30, which—as Figure 1(b) makes evident—lie far from the bulk of observations. Occasionally, a collection of observations includes scores that differ so markedly from the main body of data that descriptive statistics give a distorted picture of the properties they are intended to capture (as in per capita debt for the state of Alaska in Exercise 2.5.31, p. 72). In some cases this results from errors in measurement, e.g., calibration of instruments. In other instances it may indicate factors the experimenter did not consider and which violate the fundamental assumptions of the experiment, e.g., the artificially inflated "boom-town" economy of Alaska during the construction of the Alaskan pipeline. Such observations are called *outliers*, and several criteria have been suggested for identifying outliers so that they can be examined and possibly eliminated from the data.

One approach is to define outliers as values that lie more than two and a half standard deviations from the mean, i.e., values above or below

$$\bar{x} \pm 2.5s$$

This criterion is probably most appropriate when the distribution is basically unimodal and symmetrical. If data are skewed, we know that the median may be a better index of central tendency than the mean. For skewed data, a quick and widely used criterion defines outliers as any values that lie outside the interval

$$\tilde{x} \pm 2(x_{.75} - x_{.25})$$

where $x_{.75}$ is the 75th percentile and $x_{.25}$ is the 25th percentile.[2] This

---

[1] Since Data Set 1 is bimodal and Data Set 2 is unimodal, this difference is reflected in kurtosis. For Data Set 1, $Kur = 1.43$ and for Data Set 2, $Kur = 26.74$.
[2] The difference $(x_{.75} - x_{.25})$ is called the interquartile range.

criterion, however, is sensitive to kurtosis. A more robust criterion defines outliers as values that are *greater* than $x_{.75} + 1.5(x_{.75} - x_{.25})$ or *less* than $x_{.25} - 1.5(x_{.75} - x_{.25})$.

B. INTERVALS AND SCALES

Example 1 tempts us to rewrite Disraeli's epigram and to suggest that there are liars, damned liars, and outliers, but elimination of outliers will not necessarily make a distribution sensible. The data sets in the following example include no outliers.

Example 2. Consider the following data sets:

Data Set 3

1.00,   2.00,   3.00,   4.00,   5.00,
6.00,   7.00,   8.00,   9.00,   10.00

Data Set 4

1.25,   1.72,   2.50,   5.48,   5.50,
5.50,   5.51,   8.50,   9.25,   9.76

Once again, the descriptive statistics for these data sets are essentially identical:

	Data Set 3	Data Set 4
$\tilde{x}$	5.5	5.5
$\bar{x}$	5.5	5.5
s^2 (approx.)	8.25	8.25
Sk	0	0
Sk_P	0	0
Kur (approx.)	1.8	1.8

Moreover, if we examine the data sets, we see that the *ungrouped* distributions are nearly identical. Every value has a frequency of 1, except for the value 5.5, which appears twice in Data Set 4. However, if we group the data with an interval width of 2, differences between the distributions become evident:

	0.995 – 2.995	2.995 – 4.995	4.995 – 6.995	6.995 – 8.995	8.995 – 10.995
Data Set 3 f	2	2	2	2	2
Data Set 4 f	3	0	4	0	3

Although grouping data can give a good idea of the shape of a distribution, one has to be careful. Data can also be grouped to mislead, as the following examples show.

Example 3. Here are the earnings for union members at a small company.

$24,000	$28,000	$48,000	$68,000
$24,000	$40,000	$52,000	$72,000
$28,000	$48,000	$52,000	$72,000
$28,000	$48,000	$60,000	$74,000

The union argues that the bulk of these wages is very low and to support the argument groups the data into four equal intervals of width $20,000.

$10,500 – $30,500	$30,500 – $50,500	$50,500 – $70,500	$70,500 – $90,500
f 5	4	4	3

Management, on the other hand, argues that union wages are top-heavy and that low wage earners are rare in the rank and file. The management position is supported by the following distribution in which salaries are likewise grouped in four equal intervals, but with an interval width of $16,000.

$10,500 – $26,500	$26,500 – $42,500	$42,500 – $58,500	$58,500 – $74,500
f 2	4	5	5

Histograms for the distributions presented by the union and by management are illustrated in Figures 2(a) and 2(b) below.

Choosing intervals to skew the distribution is only one of the tactics by which the data in Figure 2 are represented to support one or another claim. In addition, it should be noted that the initial value on the vertical axis of Figure 2(a) is 2.5 and the initial value of the vertical axis of Figure 2(b) is 1.5. By showing only the tops of each histogram bar, one exaggerates the differences in their heights. This tactic is often used on graphs

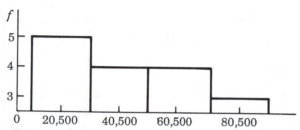

Figure 2(a). Union distribution of annual worker earnings.

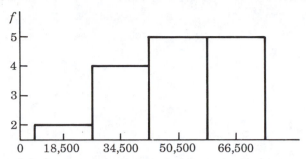

Figure 2(b). Management distribution of annual worker earnings.

that plot such data as sales figures, frequency of homicides, incidence of AIDS or other diseases, or worker injuries over time in order to exaggerate trends. Such plots are sometimes called "gee-whiz" graphs.[3]

In Chapter 4 we noted that two measurements are in the same scale if they have the same zero point and the same units of measure. The "gee-whiz" effect is created by manipulating the zero point. Changing units of measurement can also affect the appearance and overall impression of a distribution. Joint distributions of data are especially vulnerable to this sort of juggling.

Example 4. Consider the following data points:

Subject	1	2	3	4	5	6	7	8	9	10
X	1	2	2	3	3	4	5	5	3	4
Y	6	7	8	7	8	8	9	10	9	9

The scatter plot in Figure 3(a) is scaled so that each y-unit is represented by about 8 mm on the vertical axis and each x-unit is represented by about 18 mm on the horizontal axis.

In Figure 3(b), the Y-axis has been rescaled so that 8 mm corresponds to 5 units. This has the effect of compressing the scatter plot vertically and gives the appearance that the regression line is nearly parallel to the horizontal axis, which suggests that the correlation is near zero.

In Figure 3(c), the X-axis has been rescaled, so that 18 mm represents approximately 5 x-units instead of 1 x-unit as in Figures 3(a) and 3(b). This compresses the scatter plot horizontally and makes the regression line steeper, thereby increasing the appearance of statistical relatedness. (Incidentally, the correlation of X and Y is about .83.)

The moral of these examples is clear: Descriptive statistics may not give the whole picture, and the same data can be pictured differently.

[3] For more examples of such practices, see Huff, D. (1954). *How to Lie with Statistics.* New York: Norton.

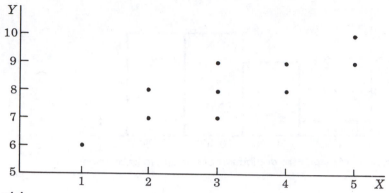

Figure 3(a).

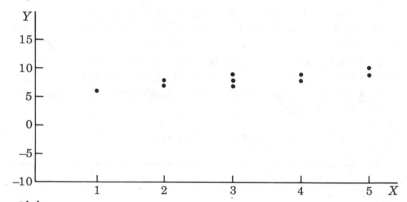

Figure 3(b).

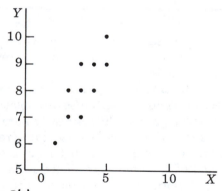

Figure 3(c).

There are no hard and fast rules for presenting (or detecting if you have been presented) the most "honest" picture. Making descriptive sense from a collection of numbers is as much an art as a science, but one can usually avoid the most serious pitfalls if statistical *description* is based on raw data and tabular or graphic *organization* is in a *grouped* format. When one's graphic or tabular display of data seems inconsistent with the properties reflected in the statistics, it's a good idea to examine one's choice of intervals, one's choice of scale, and so on.

C. STATISTICAL DESCRIPTION AS DATA EXPLORATION

Throughout Part I we have examined various techniques for organizing and describing collections of numbers that represent observations. *Data*, it will be recalled, are those which are given. They are the numbers that an experimenter has in hand, so to speak. We have, therefore, implicitly assumed that the experimenter is concerned only about the objects that have actually been observed—where the measurements on these objects tend to cluster on the real line, how much variation they exhibit, how their distribution is shaped, and so on. In Part III we introduce techniques by which an experimenter can use such observations to make inferences about the much larger and *unobserved* universe from which the objects in hand were selected. In the context of statistical *inference*, description has a somewhat different role. Statistical inference is essentially a *confirmatory* enterprise; statistical description is—or can be—an *exploratory* enterprise. Careful and methodical description of data can become a form of "data prospecting" that furnishes insights into the sorts of questions the researcher *should* be asking about the world beyond what has already been observed. This approach to statistical description is sometimes called *exploratory data analysis*[4] and emphasizes the use of visually striking data displays to reveal potential "deposits" of information. The widespread availability of personal computers has made exploratory data analysis a very practical adjunct to traditional statistical description, and several of these techniques are presented in our companion volume of computer exercises, *User Friendly*.

[4] Tukey, J. W. (1977). *Exploratory Data Analysis*. Reading, MA: Addison-Wesley.

PART II

Probability

Introduction to probability

CHAPTER OUTLINE

Introductory statistics students are frequently puzzled when they discover that a preliminary exposure to probability is considered essential to the study of statistics. At the outset, probability seems to be an exercise in pure mathematics that is far removed from concerns of scientific research. Nothing could be further from the truth. *Probability* is a mathematical abstraction. Probabilities are no more real than mathematically perfect rectangles or triangles, but they can serve as *models* for the relative frequencies of real, concrete events.

The use of mathematical concepts as models of real-world events and objects is really a familiar part of everyday life. Given the height and width of a wall, a painter can use the geometric properties of a mathematically ideal rectangle to calculate the amount of paint needed to cover the wall. A manufacturer of roof trusses can use the trigonometric properties of a mathematically ideal triangle to calculate the lengths of joists and rafters a builder will need to construct a roof. As we shall see in Part III, the statistician may likewise use the mathematical properties of probability to make inferences about the location, dispersion, and other distribution properties of yet-unobserved collections of potential observations.

Because mathematical models are idealizations, they are often imperfect reflections of reality. No wall is a perfect rectangle. Drywall and plywood warp and buckle, and even the most skillfully constructed wall seldom has perfect right angles or opposite sides of exactly equal length. Most methods for measuring the pitch of a roof give only approximations of the angle between the joists and the rafters. To the extent that a real-world phenomenon violates the assumptions of its mathematical idealization, predictions will have a certain amount of error: The painter needs another pint, the builder discovers that rafters must be trimmed, and so on. Generalizations and predictions based on probability theory are also imperfect, but we shall see in Part III that probability theory permits the statistician not only to make inferences, but also to express their degree of accuracy in quantitative terms.

Mark Twain is supposed to have said that he didn't know what prose was, and then discovered he'd been speaking it all his life.[1] Descriptive statistics is a lot like that. Most of us have been using numbers to describe tangible, everyday phenomena ever since we learned how to count, and computation of averages was first encountered in grade school. Consequently, many of the concepts developed in Part I were new in a formal sense, but not really unfamiliar. In Part II we will introduce a number of formal mathematical concepts, such as *random variables*, *probability distributions*, and the *expected value*. This should be easier now that the reader already has some familiarity with their more tangible and accessible cousins, measurements, relative frequency distributions, and the mean.

---

[1] If Twain *did* say it, he cribbed it from Molière's *Le Bourgeois Gentilhomme*. (Act II, scene iv).

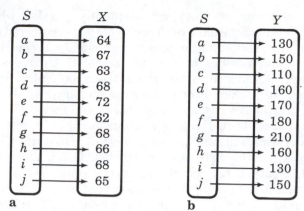

Figure 5.1. Sample space S of 10 individuals and value set of their heights (X) and weights (Y).

A. THE CLASSICAL APPROACH TO PROBABILITY

Suppose you have a group of 10 people, a, b, c, d, e, f, g, h, i, and j, and that you intend to select a person at random from this group. Your random selection can have 10 possible results, which are listed in the collection labeled S in Figures 5.1(a) and 5.1(b). Any action that can have *more than one* possible result is called a *random experiment*, and a representation of every *possible* result of a random experiment is called the *sample space* of the experiment. The collection labeled S is therefore the sample space of the experiment "select a person at random from this group."

Let us further suppose that you have measured the height of each person to the nearest inch and weighed each person to the nearest 10 pounds. The height of each person in S is listed in Figure 5.1(a). This collection of numbers is labeled X. The arrow that connects a value in X to every individual in S represents the operation, "measure the height of each person to the nearest inch." A rule that assigns a numerical value to every result in the sample space of a random experiment is called a *random variable*, and the numbers that are assigned by this rule are called the *value set* of the random variable. In Figure 5.1(b), the collection of numbers labeled Y is the value set of the random variable "weigh each person to the nearest 10 lbs."

Imagine now that the experiment is conducted as follows: You write the name (a, b, etc.) of each person in S on a slip of paper, put the slips in an urn, close your eyes, and pull one slip of paper out of the urn.[2]

Example 5.1. What are the chances that the height of the individual whose slip you draw is 68 in.? There are three individuals (d, g, and i)

---

[2] For arcane reasons that are revealed only to advanced statistics students, a receptacle from which things are drawn in probability problems is called an *urn*. If you don't have an urn, a brown paper bag will work nicely. So will a white paper bag.

who are 68 in. tall and 10 individuals altogether, so the chances are 3 in 10, or $3/10 = .30$. In the somewhat more formal language of mathematics, we say that the *probability* of selecting an individual with height 68 in. is .30.

In mathematics, any well-defined collection of objects (such as S or X) is called a *set*, and the individual objects that belong to a set are called *elements*. A collection made up of *some* of the objects in a set is called a *subset*. The subset of results that satisfy a particular condition or possess a particular property comprise what is called an *event*, denoted E. The outcome "an individual who is 68 in. tall" may therefore be thought of as event E defined by the subset of all persons in S with height 68 in., so the *probability* of event E is $3/10$, which we write

$$P(E) = \frac{3}{10} \quad \text{or} \quad P(E) = .30$$

to indicate that the probability that event E happens is .30. Since E is defined by the subset of persons for whom $x = 68$, this may also be written

$$P(x = 68) = .30$$

Example 5.2. What is $P(y = 130)$? The values in Y denote weight, so this question asks the chances of selecting a person who weighs 130 lbs. In Figure 5.1(b) we find two such individuals, a and i, so this probability is $2/10$. That is,

$$P(y = 130) = \frac{2}{10} = .20$$

The reasoning by which we solved these two simple probability problems suggests an intuitively appealing definition of probability: For an event E in the sample space S,

$$P(E) = \frac{\text{number of results in } E}{\text{number of results in } S}$$

Although it is cast in the modern language of sets and events, this is indeed the essence of the *classical definition of probability* first developed by Blaise Pascal (1623–1662) and Pierre de Fermat (?1601–1665) in their 1653 correspondence.

> If an experiment can produce m different and mutually exclusive results, all of which are equally likely, and if f of these are favorable, the probability of a favorable result is f/m.

Mutual exclusivity will be formally defined later, but what it means in practical terms is that the occurrence of one result precludes the occurrence of any other result. For example, the experiment "select an individual" from the sample space S in Figure 5.1 cannot yield both person a and person b on the same draw. Selection of individual a and selection of individual b are mutually exclusive results.

Probability theory derives from techniques developed in the seventeenth century to calculate the changes of winning in various games of chance, which is why the classical definition speaks of *favorable*, or *successful*, results. Contemporary applications of probability are not limited to gambling, and *favorable* simply means that the result belongs to the event E for which one is calculating probability.

Example 5.3. Let's use the classical definition of probability to calculate $P(x \leq 67)$. Since X denotes height, a favorable result is selecting an individual who is no taller than 67 in. In Figure 5.1(a) we see that this condition is satisfied by individuals a, b, c, f, h, and j. That is, if E is defined as "obtain an individual whose height is less than or equal to 67 in.," then E is the subset a, b, c, f, h, j. Mathematicians list the members of a set in braces, so we may write

$$E = \{a, b, c, f, h, j\}$$

and since there are six individuals in E,

$$P(E) = \frac{6}{10} = .60$$

Example 5.4. What is $P(y \leq 300)$? From Figure 5.1(b) we see that everyone in S weighs less than 300 lbs, so it is absolutely certain that any individual selected will satisfy this condition. The event $y \leq 300$ is called the *certain* event and is assigned a probability of 1. This is easily seen from the classical definition of probability:

$$P(y \leq 300) = \frac{10}{10} = 1$$

The event in this example included every result in S. Since S includes everything that can possibly happen, S—or any event that includes every result in S—is the certain event, and $P(S) = 1$.

Example 5.5. What is $P(x > 73)$? None of the individuals in S is taller than 73 in., so the event $x > 73$ is impossible. The *impossible* event may be thought of as an event with no favorable results, and from the classical definition of probability, it follows that

$$P(x > 73) = \frac{0}{10} = 0$$

The impossible event is the set that contains no elements. This is called the *empty* set and is always assigned a probability of zero.

BOX 5.1

Functions and random variables

Intuitively, a function is simply a rule by which every member of one set is assigned to or paired with one member of another set. More formally,

> Let X and Y be sets. If f is a rule that assigns to every element x in the set X a *unique* element y in the set Y, then f is said to be a function that *maps* X into Y.

Suppose that X is the set comprised of six students in a seminar and that Y is the set of term paper topics handed out by the instructor. Let us further suppose that we define the rule "choose a topic for a term paper" and that student x_1 chose topic y_1, student x_2 chose topic y_2, and students x_3, x_4, x_5, and x_6 all chose the same topic, which we call y_3. The rule "choose a topic for a term paper" satisfies our definition of a function and may be represented as follows:

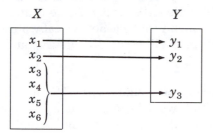

The rule in this example permits more than one student to choose the same topic (four students chose topic y_3), but it does *not* permit a student to choose more than one topic. This is what is meant when it is said that a function assigns a *unique* element in Y to every element in X.

When X and Y are sets of numbers, the function can often be expressed in the form of a mathematical equation. For example, the rule "for every real number x assign a number y equal to 3 more than x" would ordinarily be written

$$y = x + 3$$

Likewise, if X is the set of all integers from 1 to 5, the function defined by the rule "y equals the square of x" would ordinarily be given as

$$y = x^2$$

where $X = \{1, 2, 3, 4, 5\}$. If we let $x = 3$, then $y = 9$. That is, the function f assigns the y-value 9 to the x-value 3. This is denoted $f(3) = 9$, and in general if y is the value that the function f assigns to the value x, then we write $f(x) = y$. This is read, "f of x equals y," *not* "f *times* x equals y."

We know that a random experiment is any action that can yield two or more results and that a *random variable* is a rule that assigns a number to every possible result of a random experiment. A random variable is therefore a *function* and can be formally defined as follows:

> A random variable is a function that assigns a real number to every possible result of a random experiment.

We have seen that the classical definition of probability has many attractive properties. It has both an intuitive simplicity that appeals to our commonsense notion of probability and sufficient theoretical power to generate important implications about the mathematical properties of probability. With the introduction of two more notions from set theory, the classical definition can, in addition, be extended from situations involving *one* event E to situations involving *several* events,

$$E_1, E_2, E_3, \ldots$$

1. The probability of E_1 and E_2: the multiplication theorem

In some problems involving two or more events, the experimenter is concerned with the probability that *all* of the events will occur. That is, what is the probability that the experiment will yield a result that is common to E_1 and E_2 and E_3 and ...? For example, if we define Event 1 as "draw an 8" and Event 2 as "draw a heart," we might easily imagine that a poker player holding a 5, 6, 7, and 9 of hearts has a more than casual interest in the probability that the next card he draws will belong to *both* Event 1 *and* Event 2. Such problems are typically phrased in the following form: What is the probability of E_1 and E_2 and E_3 and ...?

Example 5.6. Select an individual from the sample space S (in Figure 5.1). If E_1 is $x \geq 66$ and E_2 is $y \leq 150$, what is $P(E_1$ and $E_2)$? We may think of (E_1 and E_2) as a *new* event, "choose an individual who possesses *both* of the following properties: (1) is at least 66 in. tall, (2) weighs 150 lbs or less." This new event is comprised of all elements common to both E_1 and E_2 and is called the *intersection*[3] of E_1 and E_2. The intersection of sets is denoted by the symbol ∩, which looks a bit like the n in the word $a \cap d$. The event E_1 is the set $\{b, d, e, g, h, i\}$, and E_2 is the set $\{a, b, c, i, j\}$. The only results common to both sets are b and i, so the intersection $E_1 \cap E_2$ is $\{b, i\}$. This means that only two results are favorable to the condition "is at least 66 in. tall *and* weighs 150 lbs or less," so by the classical definition of probability,

$$P(E_1 \cap E_2) = \frac{2}{10} = .20$$

Intersections do not always include results. If $E_3 = \{c, g, h\}$ and $E_4 = \{a, e, g\}$ and $E_5 = \{b, e, h\}$, then

$$E_3 \cap E_4 \cap E_5$$

is empty, because there is no element common to all three sets, E_3 and E_4 and E_5. This intersection is therefore the impossible event.

---

[3] In general, the intersection of a collection of sets is the set of all elements in common to *all* of the sets in the collection.

a. Conditional probability. In our treatment of statistical relationships in Chapter 4, we introduced the notion that information about *one* property of a result can affect our degree of uncertainty about *other* properties of the result. We pointed out, for instance, that we can make a more precise estimate about someone's height if we know the individual's weight. This was discussed in the context of conditional *variance*. We may also think about this reduction of uncertainty in terms of conditional *probability*.

Example 5.7.1. Select an individual from the sample space S (in Figure 5.1). What is the probability that the person chosen is no more than 67 in. tall? There are six results that satisfy the condition ($x \leq 67$), so

$$P(x \leq 67) = 6/10 = .60$$

Example 5.7.2. Suppose, however, an experimenter draws the slip and announces that the selected individual weighs at least 150 lbs. *Now*, what is the probability that the person selected is no more than 67 in. tall? The announcement tells us that the event

$$E_1 = (y \geq 150) = \{b, d, e, f, g, h, j\}$$

has occurred. The question may therefore be rephrased: "Under the *condition* that event E_1 has occurred, what is the probability of event $E_2 = (x \leq 67) = \{a, b, c, f, h, j\}$?" This probability is called the *conditional probability* of E_2 *given* E_1. Conditional probability is written using the same notational convention we used to indicate conditional variance:

$$P(E_2|E_1)$$

which is read, "P of E_2 given E_1." In Example 5.7.2 it is easy to use the classical definition of probability to compute $P(E_2|E_1)$. Since we know that E_1 has occurred, only seven results are possible: b, d, e, f, g, h, and j. Of these seven, only four, b, f, h, and j, are favorable to the occurrence of E_2. Therefore,

$$P(E_2|E_1) = \frac{4}{7} \doteq .57$$

In general, when the prior occurrence of Event 1 is given as a condition of the experiment, the *possible* results are those in E_1, and the *favorable* results are those results in E_2 which are *also* in E_1. That is, the favorable results are those in $E_1 \cap E_2$. In the language of classical probability, therefore,

$$P(E_2|E_1) = \frac{\text{number of results in } E_1 \cap E_2}{\text{number of results in } E_1}$$

Dividing both the numerator and denominator by the number of results in S, this becomes

$$\frac{(\text{number of results in } E_1 \cap E_2)/(\text{number of results in } S)}{(\text{number of results in } E_1)/(\text{number of results in } S)} \qquad \textbf{[5.1]}$$

Observe that by the classical definition of probability, the numerator in expression **[5.1]** is the probability of the intersection $E_1 \cap E_2$ and the denominator is the probability of E_1. Therefore,

$$P(E_2|E_1) = \frac{P(E_1 \cap E_2)}{P(E_1)} \qquad \textbf{[5.2]}$$

If we multiply both sides of equation **[5.2]** by $P(E_1)$, we get

$$P(E_2 \cap E_1) = P(E_1)P(E_2|E_1) \qquad \textbf{[5.3]}$$

This result is sometimes called the *multiplication theorem*:

> The probability of E_1 and E_2 is $P(E_1)P(E_2|E_1)$.

Example 5.7.3. If an individual is selected from the sample space S in Figure 5.1, what is the probability that the person weighs at least 150 lbs *and* is no more than 67 in. tall? We found in Example 5.7.2 that $P(y \geq 150) = 7/10$ and that $P(x \leq 67|y \geq 150) = 4/7$, so by the multiplication theorem,

$$P(E_1 \text{ and } E_2) = \left(\frac{7}{10}\right)\left(\frac{4}{7}\right) = \frac{4}{10} = .40$$

This result is easily confirmed by the classical definition of probability. The intersection of E_1 and E_2 includes the four results $\{b, f, h, j\}$, so

$$P(E_1 \cap E_2) = \frac{4}{10} = .40$$

b. Independent events. Two events E_1 and E_2 are said to be independent if $P(E_2|E_1) = P(E_2)$. That is, two events are independent if the probability of E_2 is the same whether or not E_1 has occurred. Since independence means that $P(E_2|E_1) = P(E_2)$, it follows that when two events are independent, the multiplication theorem **[5.3]** becomes

$$P(E_1 \cap E_2) = P(E_1)P(E_2) \qquad \textbf{[5.4]}$$

This is called the multiplication theorem for *independent events*:

> If event E_1 and event E_2 are independent, the probability of E_1 and E_2 is $P(E_1)P(E_2)$.

Example 5.8.1. What is the probability that an individual selected from the sample space S in Figure 5.1 weights at least 160 lbs and is less than 64 in. tall? Let E_1 be the event $y \geq 160$ and let E_2 be the event $x < 64$. Event 2 is $\{c, f\}$, and by the classical definition of probability,

$$P(E_2) = \frac{2}{10} = .20$$

Event 1 is $\{d, e, f, g, h\}$, and of these five results only f is favorable to Event 2. Therefore,

$$P(E_2|E_1) = \frac{1}{5} = .20$$

$$= P(E_2)$$

which means that Event 2 is independent of Event 1. By the classical definition of probability, $P(E_1) = 5/10$, so

$$P(E_1 \text{ and } E_2) = P(E_1)P(E_2|E_1) = \left(\frac{5}{10}\right)\left(\frac{1}{5}\right)$$

$$= P(E_1)P(E_2) = \left(\frac{5}{10}\right)\left(\frac{2}{10}\right)$$

$$= .10$$

It is important to understand that independence is symmetrical: If E_2 is independent of E_1, then E_1 is likewise independent of E_2. That is, if $P(E_2|E_1) = P(E_2)$, then $P(E_1|E_2) = P(E_1)$. This is easily proved and is demonstrated in the following example.

Example 5.8.2. In Example 5.8.1 we let E_1 be $y \geq 160$ and E_2 be $x < 64$, and we showed that $P(E_2|E_1) = P(E_2)$. It is easily seen that

$$P(E_1|E_2) = P(E_1).$$

The probability of E_2 is $2/10$, and since E_1 and E_2 have only result f in common, the probability of $(E_1 \cap E_2)$ must be $1/10$. Therefore, from formula **[5.2]**,

$$P(E_1|E_2) = \frac{P(E_1 \cap E_2)}{P(E_2)} = \frac{1/10}{2/10} = \frac{1}{2} = P(E_1)$$

A good statistician develops a "feel" for independence. For example, if we roll two dice we would expect the events "get a 4 on the first die" and "get a 3 on the second die" to be independent, since the first event has no effect on the second event. Of course, one can sometimes get into deep trouble by trusting statistical intuition. In 1936, pollsters for *Literary Digest* falsely assumed that the events E_1 "has a telephone" and E_2 "votes republican" were independent. Thus, when they used results of weekly telephone surveys to predict the presidential election, they assumed $P(E_2|E_1)$ was equal to $P(E_2)$. However, a disproportionate number of democratic voters lived in economically depressed rural areas and had no telephone service, so $P(E_2)$ was really much smaller than $P(E_2|E_1)$. Alf Landon did *not* win the election.

2. The probability of E_1 or E_2: the addition theorem

Another type of question that frequently arises when several events are involved is the probability that *at least one* of the events will occur. The notion of *at least one* is usually signified by the word "or." The poker player holding a 5, 6, 7, and an 8, not all of which are of the same suit, doesn't care whether the next card is a 4 or a 9. The issue that determines the wager is the probability of drawing either a 4 *or* a 9.

Example 5.9. Select an individual from S (in Figure 5.1). If we define E_1 as $x \leq 65$ and E_2 as $y \geq 160$, what is $P(E_1 \text{ or } E_2)$? Once more, we have defined a new event: "Choose an individual who possesses *at least* one of the following properties: (1) is less than or equal to 65 in. tall, (2) weighs at least 160 lbs." In the language of set theory, this new event is called the *union* of E_1 and E_2. That is, the union of E_1 and E_2 is the set of elements that belong to at least one of the events.[4] The union of events is denoted by the symbol $\cup$, a stylized capital U, as in $\cup$nion. In the present example we see from Figure 5.1(a) that $E_1 = \{a, c, f, j\}$ and from Figure 5.1(b) $E_2 = \{d, e, f, g, h\}$. The union $E_1 \cup E_2$ includes all elements in either E_1 or E_2 (or in both[5]), so

$$E_1 \cup E_2 = \{a, c, d, e, f, g, h, j\}$$

Similarly, if $E_3 = \{a, b, c\}$ and $E_4 = \{d, e, f\}$, then

$$E_3 \cup E_4 = \{a, b, c, d, e, f\}$$

The notion of a union is by no means limited to two sets. If $E_5 = \{a, b, c\}$ and $E_6 = \{c, f, g\}$ and $E_7 = \{f, h\}$, then

$$E_5 \cup E_6 \cup E_7 = \{a, b, c, f, g, h\}$$

---

[4] In general, the union of a collection of sets is the set of all elements that belong to at least *one* set in the collection.

[5] In mathematics and in this text the word "or" is always inclusive. That is, when we say, for example, "Harriet *or* Al will go," we mean to allow that *both* Harriet and Al might go.

To calculate the probability of a union of sets, we simply treat the union as a single event, and the probability follows directly from the classical definition. In Example 5.9 the union $E_1 \cup E_2$ includes 8 elements, so

$$P(E_1 \cup E_2) = \frac{8}{10} = .80$$

It is important to note in this example that $P(E_1 \cup E_2)$ does *not* equal $P(E_1) + P(E_2)$: The probability of E_1 is $4/10 = .40$, and the probability of E_2 is $5/10 = .50$, so $P(E_1) + P(E_2) = .90$. This happens because individual f is included in both events; the person is only 62 in. tall *and* weights 180 lbs. A simple tally of the individuals included in E_1 plus the number of individuals in E_2 therefore counts f twice, so to calculate $P(E_1 \cup E_2)$, we must *subtract* $P\{f\}$ from the sum $P(E_1) + P(E_2)$:

$$P(E_1 \cup E_2) = P(E_1) + P(E_2) - P\{f\}$$

And, in general, the probability associated with *all* results common to both E_1 and E_2 must be subtracted from $P(E_1) + P(E_2)$ to obtain $P(E_1 \cup E_2)$. That is,

$$P(E_1 \cup E_2) = P(E_1) + P(E_2) - P(E_1 \cap E_2) \qquad [5.5]$$

Equation [5.5] is commonly presented in the form of the *addition theorem*:

> The probability of E_1 or E_2 is $P(E_1) + P(E_2) - P(E_1$ and $E_2)$.

a. Mutually exclusive events. We know from the discussion following Example 5.6 (p. 156) that events sometimes have no results in common. Such events are said to be *mutually exclusive*.

Example 5.10. From Figure 5.1(a) let E_1 be $x \leq 65$, and from Figure 5.1(b) let E_2 be $y = 160$. What is $P(E_1$ or $E_2)$? Event $E_1 = \{a, c, f, j\}$, and event $E_2 = \{d, h\}$. We have $P(E_1) = 4/10$ and $P(E_2) = 2/10$, and since the two events have no results in common, $P(E_1 \cap E_2) = 0$. From equation [5.5], then,

$$P(E_1 \cup E_2) = \frac{4}{10} + \frac{2}{10} - 0 = \frac{6}{10} = .60$$

And in general, to say that two events E_1 and E_2 are mutually exclusive means that their intersection $E_1 \cap E_2$ is the empty set and that $P(E_1 \cap E_2)$ is therefore equal to zero. When $P(E_1 \cap E_2) = 0$, equation [5.5] becomes

$$P(E_1 \cup E_2) = P(E_1) + P(E_2) \qquad [5.6]$$

Equation **[5.6]** is often called the addition theorem *for mutually exclusive* events:

> If event E_1 and event E_2 are mutually exclusive, the probability of E_1 or E_2 is $P(E_1) + P(E_2)$.

In Example 5.10 sets E_1 and E_2 were mutually exclusive by happenstance; there is no logical necessity that everyone weighing 160 lbs should be taller than 65 in. In other situations, mutual exclusivity may be implicit in the procedures that define the experiment and the properties that define the events. That is, the occurrence of one necessarily prevents the occurrence of the other. Obtaining a head *and* obtaining a tail are mutually exclusive on *one* toss of a coin. Likewise, one cannot draw *one* card that is *both* an ace *and* a king.

Interrelationships among probability, union and intersection are sometimes better understood if events are illustrated in a spatial representation called a Venn diagram.[6] From Example 5.9, let E_1 be the event $x \leq 65$ and E_2 be the event $y \geq 160$. Then E_1 is $\{a, c, f, j\}$ and E_2 is $\{d, e, f, g, h\}$. These two events, their union, and their intersection are identified in Figure 5.2.

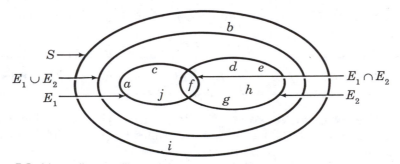

Figure 5.2. Venn diagram illustrating union and intersection of two events.

As represented in this diagram, we see that

E_1 has 4 elements, which gave us $P(E_1) = .40$,
E_2 has 5 elements, which gave us $P(E_2) = .50$,
$E_1 \cap E_2$ has 1 element, which gave us $P(E_1 \text{ and } E_2) = .10$,

and

$E_1 \cup E_2$ has 8 elements, which gave us

$$P(E_1 \text{ or } E_2) = .80 = P(E_1) + P(E_2) - P(E_1 \text{ and } E_2)$$

[6]Venn diagrams are named for the English mathematician John Venn (1834–1923), although Leonhard Euler (1707–1773) originated the use of such diagrams.

─────── EXERCISES 5.1 ───────

1. An urn contains four white balls, two red balls, and two green balls. Use the classical definition of probability to find the probability of
   (a) drawing a red ball on one draw from the urn,
   (b) drawing a black ball on one draw from the urn,
   (c) drawing at least one white ball if five balls are drawn simultaneously.

2. A European roulette wheel has 37 numbers, 0 to 36. Except for 0, which is green, half the numbers are red and half are black. Use the classical definition of probability to find
   (a) the probability of hitting number 18,
   (b) the probability of hitting a red number,
   (c) the probability of hitting an even number.

3. If two cards are drawn (without replacement) from an ordinary playing deck, what is the probability that the first will be the 7 of hearts and the second will be the 5 of diamonds?

In the game of blackjack, two cards are dealt to each player, one face down and one face up. Kings, queens, and jacks count as 10, aces count as 1 or 11 at the discretion of the possessor, and all other cards count as face value. Players can take additional cards. The object is to get a higher total than the dealer, but not over 21. The dealer has the ace of clubs showing. Player 1 has the 5 of hearts showing. Player 2 has the 9 of spades showing. Player 3 has the 10 of diamonds showing. You have the king of clubs and the 3 of hearts. Draw one more card and let X be the total for all three of your cards. *Note: In calculating probabilities for card games, all cards that are not showing (or in your own hand) are treated as though they are still in the deck.*

4. (a) What cards comprise the event "draw a card that forces your total over 21"?
   (b) If you draw one of the cards you listed in (a), what is the value set of X?
   (c) If you draw one of the cards you listed in (a), what is $P(x \leq 22)$?
   (d) If you draw one of the cards you listed in (a), what is $P(x = 21 | \text{third card is a diamond})$?
   (e) If you draw one of the cards you listed in (a), what is $P(x > 21 | \text{third card is a diamond})$?

5. What is the probability that your third card will give you exactly 21?

6. Consider the following experiment and random variable: Roll an ordinary 6-sided die and count the spots on the uppermost face.
   (a) What is the value set for this random variable?
   (b) Let E_1 be "obtain a prime number." (Note: A prime number is an integer greater than 1 that is divisible only by itself and 1.) What is $P(E_1)$?
   (c) Let E_2 be "obtain a 5." What is $P(E_2)$?
   (d) What is $P(E_2 | E_1)$?
   (e) Are these two events independent? Why or why not?
   (f) Use the multiplication theorem to find the probability of obtaining both a prime number and a 5.

7. Consider the following experiment: Toss four coins, one at a time, and list the results in order. Define the random variable "count the number of heads."
   (a) What is the value set for this random variable?
   (b) For any *particular* coin, use the classical definition of probability to determine the probability of the event "obtain a head."

    (c) Use the multiplication theorem to find the probability of obtaining the sequence, head, tail, head, tail.

    (d) Generalizing from your answer to (c), use the addition theorem to find the probability of zero heads in four tosses, the probability of one head in four tosses, the probability of two heads in four tosses, the probability of three heads in four tosses, and the probability of four heads in four tosses.

8. Suppose there is a 60 percent chance of snow and a 15 percent chance that your employer will give you the day off. To determine the probability of one or the other, you should _____ the two probabilities, but then you must _____ the probability that there will be snow *and* and that you will get the day off. If the two events are independent, the probability that both events will occur is _____. Therefore, the probability of either (or both) events occurring is _____ + _____ − _____ = _____. If you did not subtract the probability of both events, the answer would be incorrect, because _____.

9. A traveler one day was making his way
      through a woods both wild and deep.
The road split in two and he muttered, "*Mon Dieu!*
      Where in the world shall I sleep?"
To the south lay an inn most noted for sin,
      but a constable shrewd and upright.
To the north (said the sign), a village benign,
      a safer abode for the night.
The village, it seemed, was a traveler's dream,
      for thieves chose the inn, as a rule,
But the sheriff in town had earned wide renown
      as a lax and incompetent fool.
Suppose that the probability of being robbed is .60 at the inn and .20 in the village. On the other hand, the probability that the constable at the inn will recover the traveler's money is .70, but the probability that the sheriff in the village will recover the traveler's money is only .10. Which is the better choice?

10. What is the probability of drawing *either* a heart *or* a face card (king, queen, or jack) on one drawing from a 52-card deck?

11. *Advanced Dungeons & Dragons* ® (*AD & D* ®) is a role-playing game in which simulated combat is resolved by rolling dice. A 20-sided die (with faces numbered 1 to 20) is rolled to determine whether or not a player's attack is successful (i.e., whether or not the player's character "hits" an opponent). If the attack is successful, other dice are rolled to determine how much damage is inflicted by the attack. Suppose you need a 17 or better "to hit" an opponent, and suppose that your character is using a longsword, for which damage is determined by rolling an 8-sided die (with faces numbered 1 to 8).

    (a) Let E_1 be "hits opponent." What is $P(E_1)$?

    (b) On a successful attack, let E_2 be "inflicts 5 or more points of damage." What is $P(E_2)$?

    (c) If you need 5 or more points to kill your opponent, what is the probability of killing your opponent on one attack?

    (d) If you roll a 14 on your 20-sided die, what is the probability of killing your opponent on that attack? The event "kills opponent on that attack" is therefore the _____ event.

    (e) If you need only 1 point of damage to kill your opponent and roll an 18 on your 20-sided die, what is the probability of killing your opponent on that attack? The event "kills opponent on that attack" is therefore the _____ event.

12. Some $AD \& D^{®}$ characters can attack with two weapons. A 20-sided die is rolled for *each* attack, and damage is rolled separately for each *successful* attack. The total damage inflicted is the sum of the "damage" rolls. Suppose that Zlog is a warrior who fights with a sword in one hand and a dagger in the other. Against a particular opponent he needs to roll 11 or better on a 20-sided die to "hit" with the sword and needs 13 or better to "hit" with the dagger. His sword damage is determined by rolling an 8-sided die, and dagger damage is determined by rolling a 4-sided die.
   (a) If Zlog attacks with both weapons, what is the probability that Zlog will hit with *just* the sword?
   (b) If Zlog hits with just the sword, what is the probability that he will do at least 5 points of damage?
   (c) What is the probability that Zlog will hit with *both* weapons?
   (d) If Zlog hits with both weapons, what is the probability that he will do at least 5 points of damage?
   (e) What is the overall probability that Zlog will do at least 5 points of damage?

13. A family consists of a mother, a father, their daughter, Drusilla, and their sons, Claudius, Gaius, and Tiberius. Select two members of this family.
   (a) Let E_1 be "includes two males." What is $P(E_1)$?
   (b) Let E_2 be "includes Drusilla." What is $P(E_2)$?
   (c) What is $P(E_2|E_1)$?
   (d) From your answer to (c) you know that E_2 and E_1 are _____ .
   (e) What is $P(E_1)P(E_2)$?
   (f) What is $P(E_1)P(E_2|E_1)$?
   (g) From your answers to (e) and (f) you know that E_1 and E_2 are _____ .
   (h) What do your answers to (d) and (g) suggest about mutual exclusivity and independence of events?

3. Counting results of more complicated experiments

By the classical definition of probability we know that if event E is a subset of equally likely and mutually exclusive results in the sample space S, then

$$P(E) = \frac{\text{number of results in } E}{\text{number of results in } S}$$

Much of our discussion of classical probability was based on the experiment "select an individual" from the sample space of 10 persons in Figure 5.1. Since there are only 10 elements in the sample space it was a simple matter to count the results in any event E that we defined. For more complicated experiments we need more sophisticated methods to count results. The basis for all our counting techniques is the *Principle of Multiplication in Counting*.[7]

---

[7] In honor of William Allen Whitworth we present a direct quote from his classic text, *Choice and Chance* (4th ed.), Cambridge, 1886 p. 59. Despite the age of this work it is still perhaps the best detailed introduction to counting.

If one operation can be performed in m ways, and (when it has been performed in any way) a second operation can then be performed in n ways, there will be mn ways of performing the two operations.

Example 5.11. If one rolls an ordinary die and flips a coin, how many results are possible? There are six possible results $(1, 2, 3, 4, 5, 6)$ from rolling the die and two (H, T) from tossing the coin, so there are $6 \cdot 2 = 12$ results altogether. These are

$1H$	$2H$	$3H$	$4H$	$5H$	$6H$
$1T$	$2T$	$3T$	$4T$	$5T$	$6T$

Example 5.12. Until recently, Michigan automobile license plates displayed three letters followed by three digits, e.g., TKE 583. Assuming that it was permissible to repeat letters and numbers, how many different license plates were possible? The number of letter sequences is $26 \cdot 26 \cdot 26 = 17{,}576$, so there are 17,576 ways to perform the first "operation." Similarly, there are $10 \cdot 10 \cdot 10 = 1{,}000$ ways to perform the second "operation." By the Principle of Multiplication, the number of possible license plates for Michigan automobiles was therefore

$$17{,}576 \cdot 1{,}000 = 17{,}576{,}000.$$

(Actually, the number was somewhat smaller, because Michigan law prohibits letter combinations that produce obscenities or ethnic slurs.)

The Principle of Multiplication can be extended to any number of operations. Whitworth's *Extended Principle of Multiplication*:

If one operation can be performed in m ways, and then a second can be performed in n ways, and then a third in r ways, and then a fourth in s ways (and so on), the number of ways of performing the operations will be $m \times n \times r \times s \times \& c$.

Example 5.13. How many ways can the letters A, B, C, D be arranged in order (without repetition)? Imagine four blanks:

——————   ——————   ——————   ——————

There are four ways to fill the first blank. No matter how that blank is filled, there are three letters left to fill the second blank. Now, no matter which letters are chosen to fill the first two blanks, there are two letters left for the third blank. After the third blank is filled there is only one letter remaining for the last blank. By the Extended Principle of Multiplication there are $4 \cdot 3 \cdot 2 \cdot 1 = 24$ ways to arrange our four letters:

ABCD	*BACD*	*CABD*	*DABC*
ABDC	*BADC*	*CADB*	*DACB*
ACDB	*BCDA*	*CBDA*	*DBCA*
ACBD	*BCAD*	*CBAD*	*DBAC*
ADBC	*BDAC*	*CDAB*	*DCAB*
ADCB	*BDCA*	*CDBA*	*DCBA*

The tree diagram in Box 5.2 is another way to visualize this process.

Making lists or tree diagrams may be time consuming, but it is the best way to understand counting. Even if a complete list is too long to make or to diagram, a partial list can give a good idea of the answer.

Example 5.14. How many 3-letter words can be made using our 26-letter alphabet without repeating letters in any of the words? In this example a "word" is just a series of letters in order. Think of three blanks,

_____   _____   _____

BOX 5.2

Tree diagram for Example 5.13

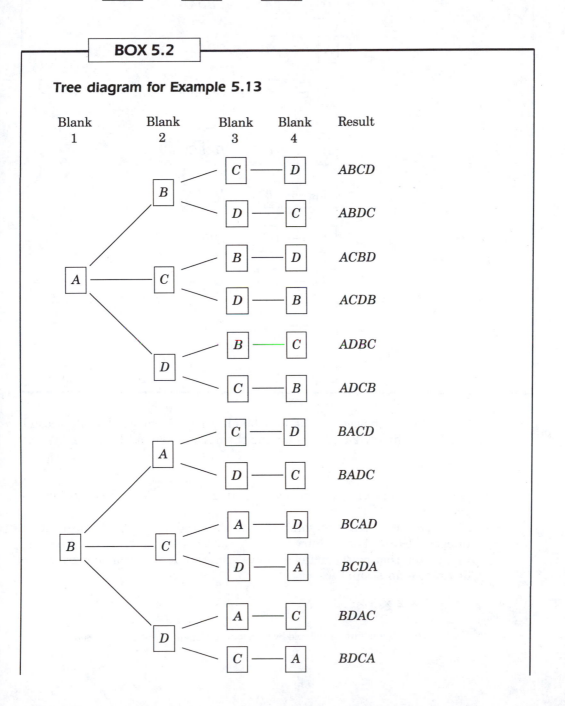

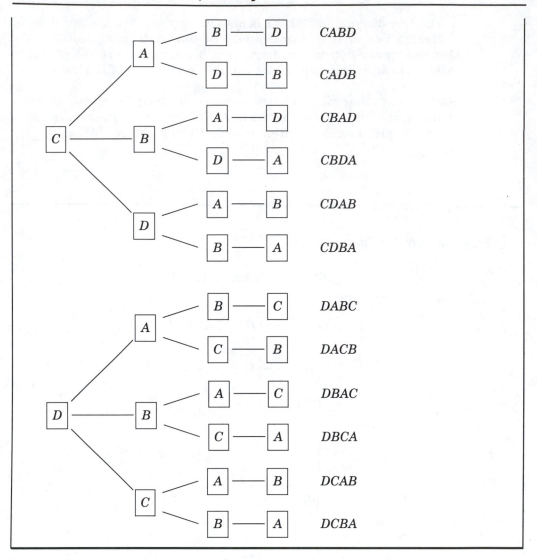

There are 26 choices for the first blank, then 25 choices for the second blank, and finally 24 choices for the third blank. Thus, there are

$$26 \cdot 25 \cdot 24 = 15{,}600$$

words. We can't draw a complete tree diagram, but Box 5.3 shows how to get started.

In Example 5.13, the solution was the product of the 4 consecutive integers, 1 to 4. The product of the N consecutive integers 1 to N comes up so often that mathematicians have named it *N-factorial* and given it a distinctive notation:[8]

$$N! = 1 \times 2 \times 3 \times \cdots \times N$$

[8]The notation $N!$ was first introduced by Christian Kramp of Strasbourg in 1808.

BOX 5.3

Partial tree diagram for Example 5.14

BOX 5.4

The first 10 factorials

$1! = 1$

$2! = 1 \cdot 2 = 2$

$3! = 1 \cdot 2 \cdot 3 = 2! \cdot 3 = 6$

$4! = 1 \cdot 2 \cdot 3 \cdot 4 = 3! \cdot 4 = 24$

$5! = 1 \cdot 2 \cdot 3 \cdot 4 \cdot 5 = 4! \cdot 5 = 120$

$6! = 1 \cdot 2 \cdot 3 \cdot 4 \cdot 5 \cdot 6 = 5! \cdot 6 = 720$

$7! = 1 \cdot 2 \cdot 3 \cdot 4 \cdot 5 \cdot 6 \cdot 7 = 6! \cdot 7 = 5,040$

$8! = 1 \cdot 2 \cdot 3 \cdot 4 \cdot 5 \cdot 6 \cdot 7 \cdot 8 = 7! \cdot 8 = 40,320$

$9! = 1 \cdot 2 \cdot 3 \cdot 4 \cdot 5 \cdot 6 \cdot 7 \cdot 8 \cdot 9 = 8! \cdot 9 = 362,880$

$10! = 1 \cdot 2 \cdot 3 \cdot 4 \cdot 5 \cdot 6 \cdot 7 \cdot 8 \cdot 9 \cdot 10 = 9! \cdot 10 = 3,628,800$

Thus,

$$5! = 1 \times 2 \times 3 \times 4 \times 5 = 120$$

or

$$5! = 1 \cdot 2 \cdot 3 \cdot 4 \cdot 5 = 120$$

Please note that

$$5! = 4! \cdot 5 = 3! \cdot 4 \cdot 5$$

The first 10 factorials are given in Box 5.4.

In order to make certain formulae more compact, it is very convenient to take $0! = 1$, even though the idea of $0!$ makes no sense by any arithmetic definition.

a. Permutations. A *permutation* is an arrangement or configuration of objects. It is therefore characterized by the *identity* of the elements it includes *and* by the *order* in which they appear. In Example 5.13 we listed all 24 permutations of the letters A, B, C, and D. Suppose we have N distinct objects from which we choose r distinct objects. Each ordered listing of the r objects chosen from the collection of N objects is called a *permutation of N objects taken r at a time*. The total number of possible permutations of N objects, r at a time, is denoted $_N P_r$.[9]

---

[9]This notation for the number of permutations was first introduced by Harvey Goodwin of Cambridge before 1869.

Example 5.15. In how many ways can four $(N = 4)$ people be seated in two $(r = 2)$ different seats (with two people left standing)? By the Principle of Multiplication there are four persons who can occupy the first seat and, when the first seat is filled, three persons who can take the second seat. Therefore, there are $4 \times 3 = 12$ permutations of four things taken two at a time:

$$_4 P_2 = 12$$

In Example 5.13 we arranged the four letters A, B, C, D in order and discovered that there were $24 = {_4 P_4}$ ways to do so. In Box 5.2 we began a list of the 15,600 permutations of 26 letters, 3 at a time, and discovered:

$$_{26} P_3 = 15,600$$

The general formula for $_N P_r$ is obtained directly from the Extended Principle of Multiplication. Once again, let us represent our r possible choices as r blanks to be filled.

$$\frac{N}{1\text{st}} \quad \frac{N-1}{2\text{nd}} \quad \frac{N-2}{3\text{rd}} \quad \cdots \quad \frac{?}{r\text{th}}$$

There are N ways to fill the first blank, then $N - 1$ ways to fill the second, etc. By the Extended Principle of Multiplication, the number of ways to fill all r blanks is therefore

$$_N P_r = N(N - 1)(N - 2) \cdots (?)$$

The number of ways to fill the last blank is denoted with a question mark (?) because it is not immediately obvious, except in the case where we want to arrange all N objects:

$$\frac{N}{1\text{st}} \quad \frac{N-1}{2\text{nd}} \quad \frac{N-2}{3\text{rd}} \quad \cdots \quad \frac{1}{N\text{th}}$$

By the Extended Principle of Multiplication, the number of permutations of all N objects must be

$$N(N - 1)(N - 2) \cdots 1 = N!$$

Now, consider the situation that presents itself when we have filled the first r of the N blanks: We have $N - r$ objects remaining, so the next blank $(r + 1)$ can be filled $N - r$ ways. When that blank is filled, we have $(N - r - 1)$ objects remaining, and so on down to the last blank, which can be filled by only 1 object. If we examine our array of N blanks we see that we have filled them from right to left with the integers 1 to N, so the question mark (?) in the rth blank must be $N - r + 1$.

$$
\underbrace{\frac{N}{\text{1st}} \ \frac{N-1}{\text{2nd}} \cdots \frac{N-r+1}{r\text{th}}}_{\text{first } r \text{ blanks}} \ \underbrace{\frac{N-r}{(r+1)\text{th}} \ \frac{N-r-1}{} \cdots \frac{3}{} \ \frac{2}{} \ \frac{1}{N\text{th}}}_{\text{last } N-r \text{ blanks}}
$$

By the Extended Principle of Multiplication, the number of ways to fill the first r blanks is therefore

$$N(N-1)(N-2)\cdots(N-r+1)$$

However, it will be observed that the number of ways to fill the *first r* blanks is exactly the same as the number of ways to fill *only r* blanks. Therefore,

$$_N P_r = N(N-1)(N-2)\cdots(N-r+1) \qquad [5.7]$$

For example, let us suppose that $N = 7$ and $r = 3$, then $(N - r + 1) = 5$, and

$$_N P_r = {_7 P_3} = 7 \cdot 6 \cdot 5 = 210$$

It is likewise seen from the Extended Principle of Multiplication that the number of ways to fill the last $N - r$ blanks is

$$(N-r)(N-r-1)\cdots 1 = (N-r)!$$

We already know that there are $N!$ ways to fill all N blanks, and by the Principle of Multiplication we know that this must equal the number of ways to fill the first r blanks multiplied by the number of ways to fill the last $(N - r)$ blanks. Therefore,

$$_N P_r (N-r)! = N!$$

Dividing both sides by $(N - r)!$, we obtain

$$_N P_r = \frac{N!}{(N-r)!} \qquad [5.8]$$

The number of permutations or arrangements of N objects taken r at a time is denoted $_N P_r$ and is equal to

$$N(N-1)(N-2)\cdots(N-r+1) = \frac{N!}{(N-r)!}$$

For example, let us once again suppose that $N = 7$ and $r = 3$. Then,

$$_7P_3(7 - 3)! = 7!$$

and

$$_7P_3 = \frac{7!}{(7 - 3)!} = \frac{7!}{4!} = \frac{7 \cdot 6 \cdot 5 \cdot 4 \cdot 3 \cdot 2 \cdot 1}{4 \cdot 3 \cdot 2 \cdot 1} = 7 \cdot 6 \cdot 5 = 210$$

Example 5.16. Players in a state lottery can bet on any sequence of four one-digit numbers, 0 to 9, and no digit can be repeated. What is the probability that the winning number will be 1234? Every possible result is one permutation of 4 digits chosen from 10 digits. The number of results in the sample space is therefore

$$_{10}P_4 = \frac{10!}{(10 - 4)!} = 10 \cdot 9 \cdot 8 \cdot 7 = 5,040$$

There is only one favorable result (the permutation 1234), so the probability is

$$\frac{1}{5,040} \doteq .000198$$

b. Combinations. A *combination* is a selection of distinct objects *without regard* to order. For example, if we select three letters from among A, B, C, D, then ABC and ABD are different combinations because they are not comprised of the same three letters. In contrast,

$$ABD, ADB, BAD, BDA, DAB, DBA$$

are six permutations of the *same* combination, since all six include the same three letters, A, B, and D.

We use the notation $_NC_r$ to denote the number of combinations of N distinct objects taken r at a time.[10] To understand how we calculate $_NC_r$, consider all permutations of A, B, C, and D taken three at a time: From formula **[5.7]** there are $4 \cdot 3 \cdot 2 = 24$ such permutations, as follows:

ABC	ABD	ACD	BCD
ACB	ADB	ADC	BDC
BAC	BAD	CAD	CBD
BCA	BDA	CDA	CDB
CAB	DAB	DAC	DBC
CBA	DBA	DCA	DCB

[10] This notation was probably first introduced by Robert Potts in 1880, although George Peacock used the notation C_r in 1830. Mathematicians denote $_NC_r$ by $\binom{N}{r}$, a notation originally devised by Leonhard Euler in 1778. The idea of reading $_NC_r$ as "N choose r" is attributed to H. S. M. Coxeter. This won't contribute much to your understanding of statistics, but it might do wonders for your *Trivial Pursuit* game.

Note that we have arranged for each column to include all of the possible permutations of *one* combination. The first column includes all permutations of the combination A, B, and C. The second column includes all permutations of the combination A, B, and D, and so forth. Since each combination has three letters, there are $_3P_3 = 3! = 6$ permutations in each column. The total number of permutations is therefore equal to the number of combinations multiplied by the number of ways to order each combination:

$$_4P_3 = (_4C_3)(_3P_3) = (4)(6) = 24$$

This result reflects the mechanics by which we actually obtain permutations of objects: First, we choose r objects from a collection of N objects. There are $_NC_r$ ways to perform this operation. Then, we arrange all of the r objects we have chosen in all possible orders. From earlier discussion, we know that the number of ways to perform this operation is

$$r(r - 1)(r - 2) \cdots 1 = r!$$

By the Principle of Multiplication, the total number of ways to order N objects taken r at a time must therefore be the product $_NC_r$ by $r!$:

$$_NP_r = (_NC_r)r!$$

so

$$_NC_r = \frac{_NP_r}{r!} \qquad [5.9]$$

From equation [5.8],

$$_NP_r = \frac{N!}{(N - r)!}$$

so equation [5.9] can be rewritten

$$_NC_r = \frac{\dfrac{N!}{(N - r)!}}{r!} = \frac{N!}{r!(N - r)!} \qquad [5.10]$$

The number of combinations of N objects taken r at a time is denoted $_NC_r$ and is equal to

$$\frac{N!}{r!(N - r)!}$$

Example 5.17. Calculate the number of combinations of 8 objects chosen 6 at a time, $_8C_6$. If we use equation [5.9] directly,

$$_8C_6 = \frac{_8P_6}{6!} = \frac{8 \cdot 7 \cdot \cancel{6} \cdot \cancel{5} \cdot 4 \cdot \cancel{3}}{\cancel{6} \cdot \cancel{5} \cdot 4 \cdot \cancel{3} \cdot 2 \cdot 1} = \frac{56}{2} = 28$$

Or we can be devious and lazy (which in statistics usually means intelligent!) and realize that every time we choose 6 of the 8 objects, we reject 2 of the objects. Thus, the number of ways to *include* 6 objects is exactly equal to the number of ways to *exclude* 2 objects. Thus,

$$_8C_6 = {_8C_2} = \frac{_8P_2}{2!} = \frac{8 \cdot 7}{2 \cdot 1} = 28$$

and in general,

$$_NC_r = {_NC_{N-r}} \qquad\qquad [5.11]$$

Note that by equation [5.9], $_NC_N = 1$. There is only one way to select N objects from N objects: Take them all. The number of ways to select zero objects from N objects cannot be obtained from [5.9], but it falls out of [5.11] very nicely:

$$_NC_0 = {_NC_{N-0}} = {_NC_N} = 1$$

There is only one way to select 0 objects: Do nothing.

Example 5.18. A graduate seminar includes five students. The instructor assigns two **A**s and three **B**s, but on the day that he must submit his grade list to the college, he leaves his grade book at home. How many ways can he report two **A**s and three **B**s? This can be solved two ways, both of which give the same result. The instructor can choose two names to assign **A**s (and the rest **B**s):

$$_5C_2 = \frac{5!}{2!(5-2)!} = \frac{5!}{2!3!} = \frac{5!}{(2)(6)} = 10$$

or he can choose three names to assign **B**s (and the rest **A**s):

$$_5C_3 = \frac{5!}{3!(5-3)!} = \frac{5!}{3!2!} = \frac{5!}{(6)(2)} = 10$$

Example 5.19.1. The long, slow pitch: A steering committee of 7 is to be chosen at random from a club with 40 members. How many committees can be formed? From equation [5.9],

$$_{40}C_7 = \frac{40 \cdot 39 \cdot 38 \cdot 37 \cdot 36 \cdot 35 \cdot 34}{7 \cdot 6 \cdot 5 \cdot 4 \cdot 3 \cdot 2 \cdot 1} = 18{,}643{,}560$$

Example 5.19.2. The fast-breaking curve: Exactly 10 members of the club want the club disbanded. If it takes a unanimous vote of the steering committee to disband the club, what is the probability that the club will be disbanded? That is, what is $P(E)$ where E is the event "7 of the 10 members who favor disbanding the club are selected for the steering committee"? For E to occur, the steering committee must be comprised of 7 of the 10 disbanders, which can happen

$$_{10}C_7 = \left(\frac{10 \cdot 9 \cdot 8}{3 \cdot 2 \cdot 1} \right) = 120$$

ways. From Example 5.19.1 we know that there are 18,643,560 possible results of the experiment "select 7 members for the steering committee," so by the classical definition of probability,

$$P(E) = \frac{120}{18,643,560} \doteq .0000064$$

Example 5.19.3. Let's suppose that it requires only a majority vote of the steering committee to disband the club. What is the probability that the club will be disbanded? That is, what is the probability that 4 or 5 or 6 or 7 of the 10 club members who favor disbanding will be chosen for the steering committee? We know by the addition theorem that

$$P(4 \text{ or } 5 \text{ or } 6 \text{ or } 7) = P(4) + P(5) + P(6) + P(7)$$

First, let's puzzle out $P(4)$, the probability of selecting exactly 4 disbanders. The key to this kind of problem is to keep in mind that the committee must be comprised of 4 members who favor disbanding *and* 3 members who oppose disbanding. The number of ways to choose 4 of the 10 disbanders is

$$_{10}C_4 = \frac{10 \cdot 9 \cdot 8 \cdot 7}{4 \cdot 3 \cdot 2 \cdot 1} = 210$$

The number of ways to choose 3 of the 30 who oppose disbanding is

$$_{30}C_3 = \frac{30 \cdot 29 \cdot 28}{3 \cdot 2 \cdot 1} = 4,060$$

By the Principle of Multiplication, the number of ways to form the committee is therefore

$$210 \cdot 4,060 = 852,600$$

and by the classical definition of probability, the probability of obtaining a steering committee comprised of 4 who favor and 3 who oppose disbanding is

$$\frac{852,600}{18,643,560} \doteq .0457316$$

The probability of selecting 5 disbanders and the probability of selecting 6 disbanders are calculated in the same way, and the probability of selecting 7 disbanders is calculated in Example 5.19.1. The reader can confirm that the probability of choosing a steering committee that includes a majority who favor disbanding is approximately

$$.0457316 + .0058798 + .0003379 + .0000064 \doteq .052$$

——————— EXERCISES 5.2 ———————————————

All probabilities in the following exercises are to be calculated by the classical definition of probability.

1. There are three routes from a person's home to her place of work. There are four parking lots where she works, three entrances into her building, two elevators to her floor, and one route from each elevator to her office door.
   (a) How many ways can she go from her home to her office?
   (b) If she makes her various choices (route, entrance, etc.) at random, what is the probability that she will take Morningside Drive, park in lot A, use the south entrance, and take elevator 1?
   (c) As she starts her car one morning, she recalls that parking lots A and B are closed for repairs. What is the probability that she will take Industrial Avenue, park in lot D, use the north entrance, and take elevator 2?
2. An experiment consists of the following series of activities: roll an ordinary six-sided die, draw a card from a standard playing deck, flip a dime, and flip a quarter.
   (a) How many outcomes are possible?
   (b) What is the probability of the following outcome: roll a 1 on the die, draw the 3 of diamonds, get "heads" on both coins?
   (c) What is the probability of the following outcome: roll a 6 on the die, draw the ace of hearts, get "heads" on one coin and "tails" on the other?

3 Calculate the following:

   (a) 6! (d) (6 × 5 × 4)3! (g) (6!)/(4!)
   (b) 12! (e) (7 × 6 × 5)4! (h) (100!)/(96!)
   (c) 0! (f) 7 × 6! (i) (87!)/[(77)!(10!)]

4. Caluclate the following:

   (a) 7! (d) 13! (g) (10!)/(6!)
   (b) 5! × 6 × 7 (e) 11! × 12 × 13 (h) 10 × 9 × 8 × 7
   (c) 6! × 7 (f) 12! × 13 (i) 10 × 9 × ⋯ × (10 − 4 + 1)

5. Calculate the following:

(a) $_3P_2$

(b) $_4P_4$

(c) $_4P_1$

(d) $_7P_5$

(e) $(7!)/(7-5)!$

(f) $7 \times 6 \times \cdots \times (7-5+1)$

6. Calculate the following:

(a) $(_3P_2)/(_2P_2)$

(b) $(_7P_4)/(_4P_4)$

(c) $(7!)/(4!)(3!)$

(d) $_{11}P_6$

(e) $(11!)/(11-6)!$

(f) $11 \times 10 \times \cdots \times (11-6+1)$

7. In stud poker, the order in which cards are dealt is important, because players bet after each deal. In seven-card stud poker,

(a) how many different hands can be dealt from an ordinary 52-card deck?

(b) what is the probability of being dealt the ace followed by the king followed by the queen followed by the jack followed by the 10 of the same suit?

8. A pipe smoker has a rack with a row of spaces for nine pipes.

(a) If he owns only seven pipes, in how many ways can they be arranged in the rack?

(b) If he uses one of the empty spaces for pipe cleaners and the other for his pipe tamper, in how many ways can pipes, cleaners, and tamper be arranged?

(c) If he places his pipes and other paraphernalia entirely at random, what is the probability that on any given day the tamper and the cleaners will occupy adjacent spaces?

9. An eclectic and somewhat fussy colleague of ours has the books in her office grouped by topic. One of her bookcases includes 15 books on animal behavior, 11 works of fiction (most of it pretty trashy), 7 philosophy books, 6 volumes of poetry, and 22 statistics books (most of which she's never read). If all of the books are kept together by topic, how many ways can they be arranged in the bookcase

(a) if the five topics are arranged in alphabetical order?

(b) if the five topics can be in any order?

(c) If the topics can be in any order, what is the probability that they *happen* to be in alphabetical order?

(d) If the topics can be arranged in any order, what is the probability that they happen to be in alphabetical order and that the books in each group are ordered alphabetically by the last name of the author?

10. Four men and three women are to be seated at a lunch counter that has only five stools.

(a) How many ways can the customers be arranged at the counter?

(b) How many ways can they be arranged at the counter if all the women are to be seated?

(c) How many ways can they be arranged at the counter if all the women are to be seated and if men occupy the first stool and the last stool?

(d) If customers take seats at random, what is the probability that all of the women are seated and that men occupy the first stool and the last stool?

(e) If customers take seats at random, what is the probability that all of the men are seated and that a woman occupies the middle stool?

11. Calculate the following:

(a) $_8C_6$

(b) $(_8P_6)/(_6P_6)$

(c) $_7C_5$

(d) $(7!)/[(5!)(2!)]$

12. Calculate the following:

    (a) $_5C_5$ (c) $_5C_3$ (e) $_5C_1$
    (b) $_5C_4$ (d) $_5C_2$ (f) $_5C_0$

13. A janitor finds three out-of-order signs on the floor in front of a row of seven telephone booths. If he hangs them up on three telephone booths at random, what is the probability that he will place them correctly?

14. Three customers at a crowded restaurant find only one available table. The table has four empty chairs, but as they are about to sit down, a waiter cautions them that three of the chair seats are freshly painted. Without the waiter's warning, what is the probability that all three would have sat down on fresh paint?

15. A public health official tells the owner of a restaurant that 25 percent of the restaurant's tables must be reserved for nonsmokers.
    (a) If there are 32 tables, how many ways can the owner satisfy the letter of the law?

    Suppose that 8 of the tables seat two people, 16 of the tables seat four people, and 8 of the tables seat six people. If the owner assigns tables to the nonsmoking section at random, what is the probability that

    (b) all two-person tables will be reserved for nonsmokers?
    (c) 4 two-person tables and 4 six-person tables will be reserved for nonsmokers?

16. Once a week, homework exercises in an advanced statistics class are explained by a group of four students selected at random the preceding week.
    (a) If there are 24 students in the class, how many different groups can be formed?

    Four of the students received **A**s on the first midterm, 5 received **B**s, 11 received **C**s, and 4 received **D**s.

    (b) What is the probability that next week's group will be made up entirely of students who received **A**s on the midterm?
    (c) What is the probability that next week's group will include exactly one person who received a **D**?

At the end of the Professional Ski Instructors of America fall (Nordic) workshop last year, the 22 instructors were divided into two groups (A and B) of 11 skiers each for various competitive events. The last event was a 4-skier relay race.

17. (a) In how many different orders could Group A send off 4 skiers from the entire group of 11 to cover the first, second, third, and fourth legs of a relay?
    (b) How many different relay teams could Group A select?
    (c) In how many ways could Group A order its 4-skier relay team to cover the first, second, third, and fourth legs of the relay?
    (d) What is the relationship among your answers to (a), (b), and (c)?

18. (a) Assume that each group selects its team at random and that each team's racing order is determined at random. If individual a is in Group A and individual b is in Group B, what is the probability that they will race against one another?
    (b) If individuals a, b, c, and d are in Group A and if individuals e, f, g, and h are in Group B, what is the probability that a will race against e, b against f, c against g, and d against h?

B. OTHER APPROACHES TO PROBABILITY

1. The relative frequency approach to probability

The classical definition of probability made it possible for us to develop methods for calculating probabilities of intersections of events, conditional events, and unions of events, but direct application of the classical definition is really quite limited. Situations in which the number of possible results is known in advance, the number of favorable results is known in advance, and all results are equally likely are typical only of games of chance.

Furthermore, even in games of chance, probability provides only an imperfect model of the *relative frequency* with which particular events occur. No die is so perfectly balanced that all six faces are *exactly* equally likely to turn up; no roulette wheel is so mathematically true that every number is *exactly* as likely as every other. But, insofar as games and gaming apparatus are designed to meet the assumptions of probability theory, the mathematics of probability permits very accurate prediction of the relative frequency with which such gambling events as "obtain a six on the roll of a single die" or "obtain an odd number on the spin of a roulette wheel" will occur. Conversely, insofar as experiments are designed to meet the assumptions of probability theory, probabilities can often be *estimated* by making empirical observations and calculating the *relative frequencies* with which the events occur.

The following experiment possesses the usual trappings of probability problems—urns, balls, etc.—but it is far more typical of the real-life problems faced by statisticians.

Example 5.20. *The compulsive colleagues.* An experimenter has an urn containing *m* balls, where *m* is a large, but *unknown* number. Some of the balls *may* be red. The experimenter defines the event *E* as "obtain a red ball" and wishes to assign a probability *p* to this event. Since the number of results is unknown and the number of favorable results is unknown, the experimenter cannot apply the classical definition of probability. He therefore decides to draw a *sample*[11] of balls from the urn—making sure that every ball in the urn is as likely to be selected as every other—and use the relative frequency of red balls in his sample to *estimate p*. The experimenter draws a total of 10 balls from the urn, and 4 of them are red, so $f/N = .40$. He returns the balls to the urn, and as he sits down to write[12]

$$\frac{f}{N} = \frac{4}{10} = .40 \cong p$$

---

[11] The notion of a *sample* will be elaborated in Chapter 8. For the moment, think of a sample as a subset of *N* results chosen blindly from a sample space containing *m* possible results.

[12] The symbol $\cong$ means "approximately equal to." We use it instead of $\doteq$ when the approximation does not result from rounding.

a co-worker wanders in to borrow a book and looks over the experimenter's shoulder. "Gee," says the visitor, "10 is a pretty small number of observations. If I were you, I think I'd be more confident if I looked at 100 balls."

The experimenter sighs, shakes the urn vigorously, stirs the balls around, and draws out 100 balls. Thirty-eight of them are red, so he erases what he has written and jots down

$$\frac{f}{N} = \frac{38}{100} = .380 \cong p$$

At that very moment, his research assistant stops by to pour herself a cup of coffee. "Hmmm," she says. "That's a pretty big urn. You're the principal investigator on this project, but if it were up to me, I think I'd look at 1,000 balls."

Not wanting to inhibit his assistant from making suggestions, the experimenter returns his 100 balls to the urn, mixes them thoroughly, and draws a sample of 1,000 balls, of which 388 are red. Erasing his much frayed paper with care he writes

$$\frac{f}{N} = \frac{388}{1,000} = .388 \cong p$$

As fate would have it, he finishes his writing just as his department head drops by to ask advice on setting up the office World Series pool. "Only a *thousand?*" he queries. "In *my* day we wouldn't settle for a sample smaller than *ten* thousand!"

Despite an overwhelming urge to suggest that his chairman couldn't count to 10,000 on his best day, the experimenter crumples up his paper, tosses his bucket of 1,000 balls back into the urn, rolls the urn around the floor (inadvertently crushing the chairman's Gucci® shoe and the foot inside it), and decants a sample of 10,000 balls, of which 3,840 are red. He writes,

$$\frac{f}{N} = \frac{3,840}{10,000} = .384 \cong p$$

and a grim thought suddenly occurs to him: *With my luck the next person through that door will be the Dean of the college. I better draw 100,000 balls just to be safe.* After many tedious hours (long after his co-worker, research assistant, department head, and Dean—who spent her entire day on the golf course—had gone home for the day) the experimenter completes his tally. Of the 100,000 balls in his sample, 38,500 are red.

He picks up the urn to return the balls and discovers that it is empty. He immediately realizes that he has now satisfied the conditions necessary to use the classical definition and writes down

$$\frac{f}{N} = \frac{f}{m} = \frac{38,500}{100,000} = .385 = p$$

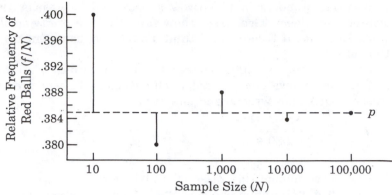

Figure 5.3. Relative frequency of red balls based on samples of 10, 100, 1000, 10,000, and 100,000 observations. Vertical lines represent $|(f/N) - p|$ where $p = .385$.

The results of these experiments are plotted in Figure 5.3, which illustrates two important conclusions to be drawn from the example. First, when the "sample" includes all m possible results, $f/N = p$. Second, as N increases, the absolute difference between relative frequency (f/N) and probability (p) decreases. That is, f/N gets closer to p. Of course, it doesn't always happen like that. A sample of 1,000 balls drawn from our fictitious urn could include 200 red balls (or, for that matter, zero red balls or 1,000 red balls)—in which case the relative frequency .40 based on $N = 10$ would have been much closer to $p = .385$ than the relative frequency based on $N = 1,000$. Nonetheless, as N increases, probability gets closer to 1 that $|(f/N) - p|$ is smaller than *any* positive value one cares to choose. This is sometimes called the *Law of Large Numbers*. Although it is not a "law" in the sense that $|(f/N) - p|$ must decrease as N increases, it is a property (which we shall discuss in Chapter 8) that makes f/N a good estimator of p, particularly for large samples.

2. The mathematical definition of probability

The preceding example demonstrated an approach to *estimating* probabilities when the requirements of the classical definition cannot be satisfied. This does not eliminate the *definitional* limitations inherent in the classical approach to probability; it merely gets us around them. If probability is to serve as a model for experiments in general, and not just those that emulate the logic and symmetry of gaming, we therefore need a more general conception of probability.

The conditions that limit application of the classical definition of probability concern the number and nature of the individual *results* in the sample space. If we focus instead on the probabilities of *events*, these restrictions cease to be troublesome. In this connection, it will be noted that the various techniques we have developed for calculating probabilities have something very fundamental in common: They are rules by which one assigns numbers to events. Probability may therefore be thought of as a *function* (see Box 5.1, p.155) that assigns a number to the outcome of an experiment. Furthermore, from what we have already learned about probability, we can deduce several things about these numbers.

First, we know that the probability of the *impossible* event is 0 and that the probability of the *certain* event is 1. The probability of any event E is therefore a value between 0 and 1.

The second property we can ferret out comes from the addition theorem. The classical definition of probability requires that results be mutually exclusive. Suppose we have a sample space with m mutually exclusive results. Now, let us define m events in such a way that each event includes one and only one result. Individual results of an experiment are generally denoted e (for *e*lement), so $E_1 = \{e_1\}$, $E_2 = \{e_2\}$, and so on. If f of our results

$$\{e_1, e_2, \ldots, e_f\}$$

are "successful," then events

$$\{E_1, E_2, \ldots, E_f\}$$

are successful. By the addition theorem for mutually exclusive events, the probability of a success is therefore

$$P(E_1) + P(E_2) + \cdots + P(E_f)$$

This may seem almost trivially obvious, but it liberates our concept of probability from the gambling model in a very important respect: Defining the probability of success no longer depends on the *equal likelihood* of successful results. (Since "equally likely" means "equally probable," this approach also gets us out of the sticky position of defining probability in terms of probability.) Of course, *if* it happens that all events are in fact equally likely, then $P(E_i) = 1/m$ and

$$P(E_1) + P(E_2) + \cdots + P(E_f) = \frac{f}{m}$$

but

$$P(E_1) + P(E_2) + \cdots + P(E_f) = P(\text{success})$$

in any case.

Finally, we know that since S includes every possible result of the experiment and *one* of the possible results is certain to occur,

$$P(S) = 1$$

These properties give us the *mathematical* definition of probability:

Let S be the sample space of an experiment. Then *probability* is a function P that assigns real numbers to events in such a way that

(1) $0 \le P(E) \le 1$ for any event E,

(2) $P(E_1 \cup E_2 \cup \cdots) = P(E_1) + P(E_2) + \cdots$ for any collection of mutually exclusive events $\{E_1, E_2, \ldots\}$,

and

(3) $P(S) = 1$.

C. SUMMARY

Probability is a pure mathematical concept. Probability is no more real than a mathematically ideal circle or rectangle. However, probability can be used as a *model* for the relative frequency of real events. This makes it possible to predict the relative frequencies of events that have not yet been observed.

The *classical definition* of probability was formulated to predict outcomes of gambling games: If an experiment can produce m different and mutually exclusive results, all of which are equally likely, and if f of these are favorable, the probability of a favorable result is f/m.

The rules for calculating the probability of two or more events are formulated in terms of the *intersection* and the *union* of the events. The intersection (denoted $\cap$) of two events is the subset of all results that belong to *both* of the events. The union (denoted $\cup$) of two events is the subset of results that belong to *at least one* of the events.

The *conditional* probability of Event 2 *given* Event 1 is the probability that Event 2 will occur *if* Event 1 has already occurred. This is denoted $P(E_2|E_1)$, and

$$P(E_2|E_1) = \frac{P(E_1 \cap E_2)}{P(E_1)}$$

The probability that *both* events E_1 and E_2 will occur is the probability of the intersection $E_1 \cap E_2$. It follows from the formula for conditional probability that

$$P(E_1 \text{ and } E_2) = P(E_1)P(E_2|E_1)$$

which is called *the multiplication theorem*.

Two events are said to be *independent* if the occurrence of E_1 has no effect on the probability of E_2. This means that $P(E_2|E_1) = P(E_2)$.

Therefore, if E_1 and E_2 are independent,

$$P(E_1 \text{ and } E_2) = P(E_1)P(E_2)$$

The probability that *either* Event 1 *or* Event 2 will occur is the probability of their union and is given by the *addition theorem*:

$$P(E_1 \text{ or } E_2) = P(E_1) + P(E_2) - P(E_1 \text{ and } E_2)$$

Two events are said to be *mutually exclusive* if they have no results in common. This means that for mutually exclusive events,

$$P(E_1 \text{ and } E_2) = 0,$$

and

$$P(E_1 \text{ or } E_2) = P(E_1) + P(E_2)$$

Results of many complicated experiments are either *permutations* or *combinations* of objects.

A *permutation* is an ordered arrangement of r objects taken from a collection of N objects. The number of possible permutations of N objects taken r at a time is denoted $_NP_r$:

$$_NP_r = N(N-1)(N-2) \cdots (N-r+1) = \frac{N!}{(N-r)!}$$

A *combination* is an unordered collection of r objects taken from a collection of N objects. The number of possible combinations of N objects taken r at a time is denoted $_NC_r$:

$$_NC_r = \frac{N!}{r!(N-r)!}$$

When the requirements of the classical approach to probability cannot be satisfied, one can sometimes conduct an experiment and *estimate* the probability p of an event by the relative frequency f/N of the event. This is called the *relative frequency* approach to probability and is possible because $|(f/N) - p|$ tends to decrease as sample size N increases.

The *mathematical definition* of probability provides a more general model for the relative frequencies of events than does the classical definition: Let S be the sample space of an experiment. Then probability is a function P that assigns real numbers to events in such a way that $0 \le P(E) \le 1$ for any event E,

$$P(E_1 \cup E_2 \cup \cdots) = P(E_1) + P(E_2) + \cdots$$

for any collection of mutually exclusive events $\{E_1, E_2, \ldots\}$, and $P(S) = 1$.

CHAPTER **6**

Discrete probability distributions

CHAPTER OUTLINE

A. INTRODUCTION: THE NOTION OF A PROBABILITY DISTRIBUTION

In the last chapter we developed techniques for calculating probability and developed the notion that the probability of an event is a mathematical idealization, or model, of the relative frequency with which the event occurs. If we can determine the probability of *every possible* outcome of an experiment, we can model the relative frequency distribution for the experiment. A model of this sort is called a *probability distribution*.

Example 6.1. Recall the very first example of an experiment discussed in this text: Have 25 persons each toss four coins and record the number of heads. The (fictitious) relative frequency distribution presented for this experiment was given in Table 1.2 and is reproduced in Table 6.1(a). The experiment "toss four coins" was encountered again in Exercise 5.1.7, and the student was called on to calculate the probability of every value of the random variable "count the number of heads." These probabilities appear in Table 6.1(b). The relative frequencies in Table 6.1(a) differ slightly from the probabilities in 6.1(b), but the relative frequencies were based on only 25 observations. From Chapter 5, we would expect the discrepancies to be much smaller if the experiment were repeated 250 or 2,500 or 25,000 times.

1. Discrete and continuous random variables

In Chapter 5 we defined a random variable as a rule that assigns a real number to every result of an experiment. In Example 6.1 this rule was "count the number of heads." The value set of this random variable must therefore have a *countable* number of values—as, of course, must any random variable that explicitly assigns values to results by counting. By countable, we mean that the values can be put into one-to-one

Table 6.1 (a). Relative frequency distribution for number of heads obtained in toss of four coins with experiment repeated 25 times

Observation (x)	0	1	2	3	4
Relative Frequency (f/N)	.04	.28	.40	.24	.04

Table 6.1 (b). Probability distribution modeling relative frequency distribution for number of heads obtained in toss of four coins

x_i	0	1	2	3	4
$P(x_i)$	.0625	.25	.375	.25	.0625

correspondence with a subset of the natural numbers $\{1, 2, 3, 4, \ldots\}$. The nature of most such experiments places a finite limit on such a value set, but we can imagine some that do not. For example, toss a fair coin until you get a run of 1,000,000,000 heads in a row and count the number of tosses. The *smallest* value this random variable can take is 1,000,000,000, but there is *no* largest value. Although the value set for this random variable is therefore infinite, it is still countable. Random variables that can assume only a countable number of values are called *discrete* random variables. Probability models for discrete random variables assign a probability to every value that the random variable can assume.

The experiment illustrated in Figure 5.1 (p. 152) is, in principle, quite different. The sample space S included 10 individuals, and the random variable X was each person's height. *Measurements* of height (or any other physical property) are necessarily discrete because of limitations on the precision of measuring instruments. For example, the experiment said that height was measured *to the nearest inch*. But height itself is *not* a discrete dimension. If (as in Figure 5.1) the shortest person in a group is 62 in. tall and the tallest person is 72 in. tall, the heights of the remaining persons—if not the *measurements* of their heights—can assume *any* values between 62 and 72. Random variables of this sort are said to be *continuous*. It is not possible to assign a nonzero probability to every *value* of a continuous random variable, so continuous and discrete random variables are treated quite differently.

The variety of phenomena—both continuous and discrete—that are subjected to statistical investigation is immense, but the student of elementary statistics can take comfort in noting that most of these variables can be modeled by only seven probability distributions.[1] They are listed in Box 6.1. Only the first two of these are discussed in the present chapter, the third is developed in the next chapter, and the rest will be introduced as they are required in the context of statistical inference, which occupies Parts III and IV of this book.

BOX 6.1

List of all the probability distributions needed for most common statistical applications

> Uniform distribution (Chapter 6 §B)
> Binomial distribution (Chapter 6 §C)
> Normal distribution (Chapter 7 §B)
> Student's t distribution (Chapter 10 §B)
> *Chi*-square distribution (Chapter 11)
> F distribution (Chapter 11)
> Poisson distribution (Appendix VIII)

---

[1] Actually, if you continue your study of statistics beyond this text, you might encounter a few other distributions, such as the *gamma*, hypergeometric, or Cauchy distribution, but applications of these in most scientific work are infrequent.

B. THE UNIFORM DISTRIBUTION

A random variable X that takes on n values, each of which has the same probability, $1/n$, is called a *uniformly distributed* random variable, which we denote by writing $X : U(n)$. For example, if X is a uniformly distributed random variable that can assume the 10 integer values 0 to 9, we write $X : U(10)$ and have

$$P(x = 0) = P(x = 1) = \cdots = P(x = 9) = .10$$

If X can assume integer values from 0 to 99 and $X : U(100)$,

$$P(x = 0) = P(x = 1) = P(x = 2) = \cdots = P(x = 99) = .01$$

and so forth.

The table of random numbers (Table 1) in Appendix VIII approximates a uniform probability distribution. One of the properties of randomly generated numbers is that every one-digit number appears with the same relative frequency $(1/10)$ as every other one-digit number, every two-digit number has the same relative frequency $(1/100)$ as every other two-digit number, and so on. Accordingly, the uniform distribution is also called the *equally likely* distribution.

Example 6.2. Consider the experiment "roll one ordinary 6-sided die" and the random variable "count the number of spots on the uppermost face." If the die is assumed to be a perfect cube and the material of which it is made is of consistent density throughout, each of the six outcomes, 1, 2, 3, 4, 5, and 6, is equally likely, and the uniform probability distribution $X : U(6)$, where $P(x) = 1/6$, would seem an appropriate model for this experiment. However, if you actually roll a 6-sided die, say, 36 times, the results might be as recorded in Table 6.2. As in Example 6.1, most of the relative frequencies differ a bit from the probabilities, but one expects that the more often one repeats the experiment, the closer the relative frequencies will get to 1/6.

1. Describing probability distributions

In Chapter 2 we developed quantitative indices that describe the location and dispersion of a relative frequency distribution. There are similar indices that describe probability distributions, and the simplicity of the uniform probability distribution makes it a convenient vehicle for introducing them.

a. Location: the expected value of a random variable. Let's calculate the mean number of spots that came up in the experiment recorded in Table

Table 6.2. Results of rolling a 6-sided die 36 times

Outcome	x	f	f / N	P(x)
[⚀]	1	5	$\frac{5}{36}$	$\frac{1}{6}$
[⚁]	2	6	$\frac{6}{36}$	$\frac{1}{6}$
[⚂]	3	8	$\frac{8}{36}$	$\frac{1}{6}$
[⚃]	4	7	$\frac{7}{36}$	$\frac{1}{6}$
[⚄]	5	4	$\frac{4}{36}$	$\frac{1}{6}$
[⚅]	6	6	$\frac{6}{36}$	$\frac{1}{6}$

6.2. From formula **[2.3]**

$$\bar{x} = \frac{\sum_{i}^{n} x_i f_i}{N} = \frac{1(5) + 2(6) + 3(8) + 4(7) + 5(4) + 6(6)}{36}$$

$$= \tfrac{1}{36}\left[1(5) + 2(6) + 3(8) + 4(7) + 5(4) + 6(6)\right]$$

$$= 1\left(\tfrac{5}{36}\right) + 2\left(\tfrac{6}{36}\right) + 3\left(\tfrac{8}{36}\right) + 4\left(\tfrac{7}{36}\right) + 5\left(\tfrac{4}{36}\right) + 6\left(\tfrac{6}{36}\right)$$

$$\doteq 3.472$$

The fractions in parentheses are the relative frequencies (f/N) given in the fourth column of Table 6.2, and in general,

$$\bar{x} = \frac{\sum_{i}^{n} x_i f_i}{N}$$

$$= \sum_{i}^{n} x_i \frac{f_i}{N} \tag{6.1}$$

Now, suppose someone says "*Imagine* you throw this die 100,000,000 times." What will be the mean number of spots that show up? Since you probably don't have time to run the experiment,[2] you would have to *model* the experiment. So, instead of using relative frequencies, you are forced to use the probability values listed in the column $P(x)$ in Table

---

[2] If you worked 8 hours a day, 7 days a week and required one second to roll a die and record the result, this experiment would take a little over $9\frac{1}{2}$ years. So you better get started.

6.2. These values remain the same no matter how often the die is rolled. If we substitute $P(x_i)$ for f_i/N in equation [6.1] we obtain

$$\sum_{}^{n} x_i P(x_i) \qquad\qquad [6.2]$$

If the values $P(x_i)$ are obtained by calculations on the sample space of an experiment, we know from our discussion of Example 5.20 that probability and relative frequency are conceptually and arithmetically identical. Under such circumstances, expressions [6.1] and [6.2] are equivalent, so [6.2] is the mean. In the present example, however, the sample space is not available. The experiment was never conducted, and the probabilities are derived on the basis of theoretical or conceptual considerations. Consequently, [6.2] is called the *expected value* or the *expectation*:

The expected value of a discrete random variable X is

$$E(X) = \sum_{i=1}^{n} x_i P(x_i) \qquad\qquad [6.3]$$

The expected value $E(X)$ may be thought of as the mean that would be obtained in a mathematically ideal experiment, that is, an experiment that conformed in every way to the assumptions of probability theory. Consequently, the *numerical value* of $\sum x_i P(x_i)$ is often denoted μ_X or μ, the lowercase Greek *mu*, which corresponds to m (for *mean*) in the Roman alphabet;[3]

$$\sum x_i P(x_i) = \mu$$

Example 6.3. The uniformly distributed variable in Example 6.2 can take six values, each of which has been assigned the probability $1/6$. If we use equation [6.3] to calculate the expectation of X for this example, we find that

$$E(X) = 1\left(\tfrac{1}{6}\right) + 2\left(\tfrac{1}{6}\right) + 3\left(\tfrac{1}{6}\right) + 4\left(\tfrac{1}{6}\right) + 5\left(\tfrac{1}{6}\right) + 6\left(\tfrac{1}{6}\right) = 3.5$$

You are well aware that the mean of a collection of data is rarely equal to a datum in the collection. In the same way, the expected value of a random variable is rarely an actual value in the set X. Thus, when we say, "the expected value of the roll of one 6-sided die is 3.5" we do *not*

---

[3] Since $E(X)$ is *numerically* equal to μ, the two notations are largely interchangeable, and both are used throughout the book. Frequently, the choice is purely a matter of simplifying notation, e.g., getting rid of parentheses or subscripts. Otherwise, we will ordinarily use $E(X)$ when we are emphasizing the *function* by which the value μ is calculated, and we will use μ when we are emphasizing the *value* obtained by our computations.

mean that we expect to get a 3.5. You cannot get a 3.5 on the roll of a die. However, if you roll the die a large number of times and average the results, you can expect that average to be 3.5. For the data in Table 6.2, we calculated the average to be about 3.472.

Like so many concepts in probability theory, the notion of expected value has its roots in games of chance.

Example 6.4. Suppose that you are asked by a community service organization to buy a raffle ticket for $1.00. Unless you simply consider the dollar to be a charitable donation, you might try to determine the investment value of your purchase. That is, what is the ticket actually worth in terms of its potential returns compared with the certain forfeiture of $1.00. This determination depends on two factors. First you must know the dollar value of the prize. If the prize were worth less than $1.00 there is no way that the ticket could yield a net gain, but if the prize were, say, an $80.00 wristwatch, you have a chance to make a $79.00 profit on your investment.

Even with a potential gain of $79.00, the value of the investment depends on how many tickets are sold. If only one ticket is sold, you are certain to win, and the purchase is worth $79.00, i.e., $80.00 in merchandise less the $1.00 price of the ticket. If 100 tickets are sold, there is a .99 probability of losing $1.00 and only a .01 probability of making a $79.00 profit. Consequently, the worth of the ticket can be evaluated by multiplying the probability of winning by the net value of the win and subtracting from this the probability of losing multiplied by the cost of the loss.

$$.01(\$79.00) - .99(1.00) = -\$.20$$

This can be expressed as the expected value of the ticket:

$$\sum x_i P(x_i)$$

where X represents the net value of the two possible outcomes, "win" and "don't win." The value associated with a win is $x_1 = 79$, and $P(x_1) = .01$. The value associated with not winning is $x_2 = -1$, and $P(x_2) = .99$. Accordingly,

$$E(X) = (79)(.01) + (-1)(.99) = -.20$$

As in Example 6.3, you *cannot* lose 20¢ on this raffle; you can either lose $1.00 or gain $79.00, but if you participated in many such raffles, you could expect to lose an average of 20¢ per ticket.

The Algebra of Expectations. The Algebra of Summations made it possible for us to demonstrate many important properties of relative frequency distributions, and the rules for calculating expectations make it likewise possible to demonstrate important properties of probability distributions. These rules are proved formally in Appendix V, but expected values are very similar to means, and many of the rules for calculating expectations are completely analogous to principles we have already developed for calculating means. For example, in Exercises 2.3.15–2.3.17 it was proved that if c is a constant (positive or negative) and $\{x_1, \ldots, x_N\}$ is any set of numbers with mean $\bar{x}$, then

(1) $\bar{c} = c$
(2) the mean of sums $(x + c) = \bar{x} + c$
(3) the mean of products $(cx) = c\bar{x}$.

In the Algebra of Expectations we find, similarly:

Rule 1: If c is a constant, then

$$E(c) = c$$

Rule 2: If c is a constant and X is a random variable with expectation $E(X) = \mu$, then

$$E(X + c) = \mu + c$$

Rule 3: If c is a constant and X is a random variable with expectation $E(X) = \mu$, then

$$E(cX) = c\mu$$

In Exercise 2.3.18 it was proved that if $(x + y)_i = (x_i + y_i)$, then $\overline{(x + y)} = \bar{x} + \bar{y}$. In the Algebra of Expectations we find a similar rule for calculating the expectation of the sum of variables $(X + Y)$:

Rule 4: If X is a random variable with expectation $E(X) = \mu_X$ and Y is a random variable with expectation $E(Y) = \mu_Y$, then

$$E(X + Y) = \mu_X + \mu_Y$$

The reader is cautioned, however, that the proof of Rule 4 does *not* follow by analogy from the proof in Exercise 2.3.18. The mean $\overline{(x + y)}$ is calculated over the N sums $(x_i + y_i)$. The expected value of $(X + Y)$ is strictly analogous to the mean calculated over all NM sums $(x_i + y_j)$ for the sets $\{x_1, \ldots, x_N\}$ and $\{y_1, \ldots, y_M\}$. The proof of Rule 4 requires an introduction to joint probability distributions, which are covered in Chapter 14. For the moment, it must suffice to *define* $E(X + Y)$ as the sum $\mu_X + \mu_Y$.

b. Dispersion: the variance of a random variable. We know that the dispersion in a relative frequency distribution of N observations can be described by the variance of the observations, which is defined in equation [2.9] as

$$s^2 = \frac{\sum\limits_{i}^{n} (x_i - \bar{x})^2 f_i}{N}$$

By the same algebraic rearrangement that led to [6.1], this is identical to

$$s^2 = \sum\limits_{i}^{n} (x_i - \bar{x})^2 \frac{f_i}{N}$$

In an *idealized* experiment—that is to say, an experiment that has not been (or for practical reasons cannot be) run—we can *model* the variance of X by substituting $P(x_i)$ for f_i/N and by substituting μ for $\bar{x}$:

$$\sum\limits_{i}^{n} (x_i - \mu)^2 P(x_i)$$

This expression is the *variance* or *expected variance* of the random variable X.

> The variance of a discrete random variable X is
>
> $$V(X) = \sum\limits_{i=1}^{n} (x_i - \mu)^2 P(x_i)$$
>
> [6.4]

The numerical value of $V(X)$ is often denoted σ^2 (or σ_X^2), the lowercase Greek *sigma* squared:

$$\sum (x_i - \mu)^2 P(x_i) = \sigma^2$$

We use the Greek letter to distinguish the variance of a *random variable* from the variance of a *collection of observations*, s^2.

Example 6.5. We know that $E(X)$ for the uniformly distributed random variable in Table 6.2 is 3.5, so from formula [6.4] the variance of X is

$$(1 - 3.5)^2\left(\tfrac{1}{6}\right) + (2 - 3.5)^2\left(\tfrac{1}{6}\right) + (3 - 3.5)^2\left(\tfrac{1}{6}\right)$$

$$+ (4 - 3.5)^2\left(\tfrac{1}{6}\right) + (5 - 3.5)^2\left(\tfrac{1}{6}\right) + (6 - 3.5)^2\left(\tfrac{1}{6}\right) \doteq 2.92$$

Computational formulae for V(X). We know from Chapter 2 that the variance s^2 of any collection of observations can be calculated using the following computational formula [2.11]:

$$s^2 = \frac{\sum\limits_{}^{n} x_i^2 f_i}{N} - \bar{x}^2$$

By the same argument we have used above, the denominator N can be moved into the summation in [2.11], which gives us

$$s^2 = \sum_{}^{n} x_i^2 \frac{f_i}{N} - \bar{x}^2$$

and if we substitute $P(x_i)$ for f_i/N and μ for $\bar{x}$, we obtain an analogous expression for the expected variance:

$$V(X) = \sum_{}^{n} x_i^2 P(x_i) - \mu^2 \qquad\qquad [6.5]$$

For the values in Table 6.2, equation [6.5] gives us

$$(1^2)(\tfrac{1}{6}) + (2^2)(\tfrac{1}{6}) + (3^2)(\tfrac{1}{6}) + (4^2)(\tfrac{1}{6}) + (5^2)(\tfrac{1}{6}) + (6^2)(\tfrac{1}{6}) - (3.5)^2$$

$$\doteq 2.92$$

which is same value we obtained by calculating the variance from expression [6.4].

In Chapter 2 we pointed out that the variance of a collection of observations,

$$s^2 = \frac{\sum\limits_{}^{N} (x - \bar{x})^2}{N}$$

is the *mean* of the squared deviations from the mean, $\bar{x}$. Likewise, the variance of a random variable,

$$V(X) = \sum (x_i - \mu)^2 P(x_i)$$

is the *expected value* of the squared deviations from the expected value, μ. That is,

$$V(X) = E\big[(X - \mu)^2\big] \qquad\qquad [6.6]$$

If we expand the square on the right-hand side of [6.6],

$$V(X) = E(X^2 - 2X\mu + \mu^2)$$

$$= E(X^2) - E(2X\mu) + E(\mu^2)$$

The values 2, μ, and μ^2 are constants, so by Rules 2 and 3 of the Algebra of Expectations,

$$V(X) = E(X^2) - 2\mu E(X) + \mu^2$$

and since $E(X) = \mu$,

$$V(X) = E(X^2) - 2\mu^2 + \mu^2$$

$$= E(X^2) - \mu^2 \qquad\qquad [6.7]$$

The Algebra of Variances. Since the variance may be thought of as the expected value of $(X - \mu)^2$, the rules that govern the behavior of variances are numbered sequentially with our Algebra of Expectations, both below and in Appendix V. Once again, however, many of the principles for calculating the variances of random variables are analogous to principles we have already established for calculating the variances of data collections. In Exercise 2.4.25, for example, it was proved that if c is a constant and $\{x_1, \ldots, x_N\}$ is a set of numbers with variance s_X^2, then

(1) $s_c^2 = 0$
(2) the variance of the sums $(x + c)$ is s_X^2
(3) the variance of the products (cx) is $c^2 s_X^2$

Similarly:

Rule 6: If c is a constant, then

$$V(c) = 0$$

Rule 7: If c is a constant and X is a random variable with variance $V(X) = \sigma^2$, then

$$V(X + c) = \sigma^2$$

Rule 8: If c is a constant and X is a random variable with variance $V(X) = \sigma^2$, then

$$V(cX) = c^2\sigma^2$$

In Exercises 2.4.26 and 4.2.18 we explained that if $X = \{x_1, \ldots, x_N\}$ and $Y = \{y_1, \ldots, y_N\}$ are such that the covariance of X and Y is zero, then the variance of the sums $(x + y)$ equals the sum of the variances $s_X^2 + s_Y^2$. A similar relationship holds true for the sum of random variables, $X + Y$.

Rule 9: If X and Y are independent random variables with variances $V(X) = \sigma_X^2$ and $V(Y) = \sigma_Y^2$, then

$$V(X + Y) = \sigma_X^2 + \sigma_Y^2$$

BOX 6.2

Independence and covariance of X and Y.

- The notion of independence of events is extended in Chapter 14 to define independence of random variables. We know from Chapter 5 that events E_1 and E_2 are independent if

$$P(E_2|E_1) = P(E_2)$$

Likewise, discrete random variables X and Y are independent if

$$P(y = y_k|x = x_j) = P(y_k)$$

for every x-value x_j and every y-value y_k.
- $cov(X, Y)$ describes the strength and direction of linear relationship for random variables X and Y much as C_{XY} captures these properties for two sets of measurements.

The derivation in Box 4.1 is easily adapted to obtain the following computational formula for covariance:

$$C_{XY} = \frac{N\Sigma xy - \Sigma x \Sigma y}{N^2}$$

Therefore,

$$C_{XY} = \frac{\Sigma xy}{N} - \frac{\Sigma x \Sigma y}{N^2} = \frac{\Sigma xy}{N} - \left(\frac{\Sigma x}{N}\right)\left(\frac{\Sigma y}{N}\right) = \overline{xy} - (\bar{x})(\bar{y})$$

The covariance of random variables X and Y takes the same form as the last expression, with expected values in place of means

$$cov(X, Y) = E(XY) - E(X)E(Y)$$

Although $cov(X, Y)$ and C_{XY} are analogous in form and meaning, $E(XY)$ is defined in terms of joint probability and is not, therefore, strictly analogous to $\overline{xy}$. See Rule 5 in Appendix V.

It will be noted that Rule 9 speaks to *independent* random variables and makes no mention of the *covariance* of random variables X and Y, which we denote $cov(X, Y)$. The notions of independence and covariance of random variables are previewed briefly in Box 6.2, and it is shown in Appendix V (corollary to Rule 5 of the Algebra of Expectations) that $cov(X, Y) = 0$ when X and Y are independent. Hence the analogy of Rule 9 to our earlier observation that $s^2_{X+Y} = s^2_X + s^2_Y$ when $C_{XY} = 0$. As with Rule 4, the reader is cautioned that the variance s^2 of the N sums $(x_i + y_i)$ is *not* an analog for $V(X + Y)$. The variance of $(X + Y)$ is strictly analogous to s^2 calculated over all NM sums $(x_i + y_j)$ for the sets $\{x_1, \ldots, x_N\}$ and $\{y_1, \ldots, y_M\}$ and involves joint probabilities, which are discussed later in the text.

2. Standardized random variables

From Chapter 3 we know that any score x in a distribution with mean $\bar{x}$ and variance s^2 can be standardized by subtracting $\bar{x}$ and dividing by s:

$$z = \frac{x - \bar{x}}{s}$$

The standard deviation of a *random variable*, like the standard deviation of a set of measurements, is the square root of its variance. Any value of a random variable can therefore be standardized by subtracting μ and dividing by σ:

$$z = \frac{x - \mu}{\sigma}$$

where $E(X) = \mu$ and $V(X) = \sigma^2$.

Example 6.6. For the uniformly distributed random variable in Table 6.2, what is the standardized value of $x = 3$? We know from Example 6.3 that $E(X) = 3.5$, and we know from Example 6.5 that $V(X) \doteq 2.92$, so

$$z = \frac{x - \mu}{\sigma} \doteq \frac{3 - 3.5}{\sqrt{2.92}} \doteq -.29$$

Standardization can be thought of as a function that assigns a unique value z to every value of X. Accordingly

$$\frac{X - \mu}{\sigma}$$

is itself a random variable, which is customarily denoted Z. We know from Chapter 3 that for any collection of standardized measurements $\{z_1, z_2, \ldots, z_N\}$, the mean $\bar{z} = 0$, and the variance $s_Z^2 = 1$. Likewise, it is proved in Box 6.3 that $E(Z) = 0$ and $V(Z) = 1$. That is, if X has an expected value of μ and a variance of σ^2, then the random variable

$$Z = \frac{X - \mu}{\sigma}$$

has an expected value of 0 and a variance of 1 *no matter what the values of μ and σ^2*. Hence, Z is called the *standardized* random variable, and

$$\frac{X - \mu}{\sigma}$$

is sometimes called the *standard form* of the random variable X.

BOX 6.3

Expected value and variance of a standardized random variable

Let X be a random variable with $E(X) = \mu$ and $V(X) = \sigma^2$. Then, the standardized random variable

$$Z = \frac{X - \mu}{\sigma}$$

1. $E(Z) = E\left(\frac{X - \mu}{\sigma}\right)$

The expectation μ and variance σ^2 are constants, so by Rule 3 of the Algebra of Expectations

$$E\left(\frac{X - \mu}{\sigma}\right) = \frac{1}{\sigma}E(X - \mu)$$

and by Rule 2 of the Algebra of Expectations

$$\frac{1}{\sigma}E(X - \mu) = \frac{1}{\sigma}[E(X) - \mu] = \frac{1}{\sigma}(\mu - \mu) = 0$$

2. $V(Z) = V\left(\frac{X - \mu}{\sigma}\right)$

By Rule 8 of the Algebra of Expectations

$$V\left(\frac{X - \mu}{\sigma}\right) = \frac{1}{\sigma^2}V(X - \mu)$$

and by Rule 7 of the Algebra of Expectations,

$$\frac{1}{\sigma^2}V(X - \mu) = \frac{1}{\sigma^2}V(X) = \frac{1}{\sigma^2}(\sigma^2) = 1$$

It is important to understand that standardization does not change the basic shape or form of a probability distribution. If X is uniformly distributed, then Z is uniformly distributed; if the distribution of X has positive skew, the distribution of Z has positive skew, and so on. Nor does standardization affect probability relationships. That is, if

$$z_a = \frac{x_a - \mu}{\sigma} \quad \text{and} \quad z_b = \frac{x_b - \mu}{\sigma}$$

then $P(x_a \leq x \leq x_b) = P(z_a \leq z \leq z_b)$. Standardization simply relocates the distribution so that the expected value falls at 0 and rescales the values of the random variable so that they are expressed in units equal to σ.

─────── EXERCISES 6.1 ───────

In Exercises 1 and 2, the set X is all of the values associated with the possible outcomes in each of eight different experiments. Use the classical definition of probability to calculate the probability of the x-values in each set and then calculate the expected value and variance for each set.

1. (a) $X = \{1, 1, 1, 2, 2, 2, 3, 3, 3, 6, 6, 6, 10, 10, 10\}$
   (b) $X = \{1, 1, 3, 3, 3, 3, 5, 5, 6, 7, 8, 8, 9, 9, 9, 9\}$
   (c) $X = \{-1, -1, -1, -1, -1, -1, -1, -1, -1, 5\}$
   (d) $X = \{5, 5, 5, 5, 100\}$
2. (a) $X = \{1, 2, 2, 3, 3, 3, 4, 4, 4, 5, 5, 6\}$
   (b) $X = \{8, 9, 10, 11, 12, 12, 15, 15, 15, 15, 20, 20, 20, 20, 20, 25, 25, 25, 25,$
        $25, 25, 25, 25, 25, 25\}$
   (c) $X = \{1.5, 1.5, .25, .25, .25, .25, 1.0, 1.0, .5, .5\}$
   (d) $X = \{.6, .6, .8, .8, .8, 1.2, 1.2, 1.2, 2.6, 2.6, 3.2, 3.2, 3.8, 3.8, 4.2\}$
3. $X : U(3)$ and can take the values $1, 3, 7$.
   (a) Calculate the expected value and the variance of X.
   (b) Calculate the expected value and the variance of X^2.
   (c) What can you say in general about $[E(X)]^2$ and $E(X^2)$?
   (d) What can you say in general about $[V(X)]^2$ and $V(X^2)$?
4. Let X be the score rolled on a 6-sided die and let Y be the score rolled on a 4-sided die. Calculate all possible values of $(X - Y)$ and the probability of every value $(x - y)$.

   (a) $E(X) =$ _____   (b) $E(Y) =$ _____   (c) $E(X - Y) =$ _____
   (d) What can you say about $E(X - Y)$ and $E(X)$ and $E(Y)$_____
   (e) $V(X) =$ _____   (f) $V(Y) =$ _____   (g) $V(X - Y) =$ _____
   (h) Keeping in mind that scores obtained on two dice are independent, what can you say about $V(X - Y)$ and $V(X)$ and $V(Y)$?

5. A random variable is distributed as follows:

x	1	2	3	4	5	6
$P(x)$	$\dfrac{1}{12}$	$\dfrac{1}{12}$	$\dfrac{1}{6}$	$\dfrac{1}{6}$	$\dfrac{1}{6}$	$\dfrac{1}{3}$

   (a) $E(X) =$ _____   (b) $V(X) =$ _____
   (c) From (a) and (b), calculate $E(X^2)$. Hint: See formula **[6.7]**.

6. A random variable is distributed as follows:

x	0	1	2	3	4	5	6	7	8
$P(x)$	.001	.008	.041	.124	.232	.278	.209	.090	.017

   (a) $E(X) =$ _____   (b) $V(X) =$ _____
   (c) From (a) and (b), calculate $E(X^2)$.

7. A random variable can take the values $1, 2, 3, 4$. It is distributed in such a way that $P(x) = kx$, where $k = P(x = 1)$.
   (a) Find the probability distribution of X.
   (b) $E(X) = $ _____ (c) $V(X) = $ _____

8. A random variable can assume only odd integer values between 0 and 12. It is distributed in such a way that $P(x) = x/\overset{n}{\Sigma}x$.
   (a) Find the probability distribution of X.
   (b) $E(X) = $ _____ (c) $V(X) = $ _____

9. Let X be the score obtained on one roll of a 20-sided die.
   (a) How is X distributed? (b) What is the expected value of X?
   (c) What is the variance of X?

10. Let X be all randomly generated five-digit numbers, a sample of which appear in Table 1 of Appendix VIII.
    (a) How is X distributed?
    (b) What is the expected value of X?

11. At a carnival ring-toss game the following events can occur: (A) win nothing, (B) win an ice cream cone worth 75¢, (C) win a stuffed animal worth $10.00, ($D$) win a roller coaster ride worth $1.00, or ($E$) win a balloon worth 20¢. The probabilities of these events are as follows.

Event	A	B	C	D	E
Probability	.50	.14	.01	.05	.30

If one gets three throws for a dollar, what is the expected value of each throw?

12. Michigan Instant Lottery tickets cost $1.00 each. Tickets are printed in batches of 1,000,800 and each batch includes the following distribution of winners:

Number of Tickets	Prize
10	$10,000
25	$ 1,000
100	$ 100
100	$ 50
22,680	$ 5
84,670	$ 2

Source: Michigan Lottery Commission.

(a) What is the expected value of a lottery ticket?
(b) Mr. Jones buys five Michigan Instant Lottery tickets every day. Mr. Brown buys none, but he spends two weeks every year in Las Vegas. Mr. Brown spends $700.00 on travel, lodging and entertainment and every day in Las Vegas he gambles $1,000 at roulette, where he plays the number 7 on every bet. Roulette played in this way has the *worst* payoff of any casino game. The expected value of such a bet is -2¢ for every dollar wagered. Over the long haul, which of the two men spends more money on gambling?

As explained in Exercises 5.1, the amount of damage inflicted in *Advanced Dungeons & Dragons* ® is determined by rolling dice. Different weapons do different amounts of damage. In Exercises 13–16 (a) represent the probability distribution (as either a graph or a table) for damage inflicted by the weapon described, (b) calculate the expected value of one successful attack for the weapon described, and (c) calculate the variance for successful attacks with the weapon described.

13. A hunting arrow inflicts 1 to 6 points of damage, which is determined by rolling a 6-sided die with faces numbered from 1 to 6.

14. A dagger inflicts 1 to 4 points of damage, which is determined by rolling a 4-sided die with faces numbered from 1 to 4.

15. If wielded two-handed, a bastard sword* inflicts 2 to 8 points of damage, which is calculated by rolling two 4-sided dice, each numbered 1 to 4, and summing the results of the two rolls.

16. A longsword inflicts 1 to 8 points of damage, which is determined by rolling a single 8-sided die with faces numbered 1 to 8.

17. A rather scruffy barbarian who makes occasional appearances in our *AD&D*® games has exceptional strength that allows him to add one point to the damage he rolls on the dice.
    (a) What is the expected damage for a successful attack by this character if he uses a bastard sword? (Assume he is wielding it two-handed.)
    (b) State the rule in the Algebra of Expectations that allows you to determine the answer to 6.1.17(a) from the answer to 6.1.15(b).
    (c) What is the expected variance of successful bastard sword attacks?
    (d) State the rule in the Algebra of Variances that allows you to determine the answer to 6.1.17(c) from the answer to 6.1.15(c).
    (e) Let $y = (x - 6)/\sqrt{2.5}$ where X is bastard-sword damage inflicted by the barbarian. Then $E(Y) = $ _____ and $V(Y) = $ _____. This illustrates that _____.

18. A character with sufficient dexterity is permitted to attack with two weapons (i.e., with one weapon in each hand). Suppose that such a character is using a longsword in one hand and a dagger in the other.
    (a) What is the expected damage if the character hits with *both* weapons?
    (b) State the rule in the Algebra of Expectations that allows you to determine the answer to 6.1.18(a) from your answers to 6.1.14(b) and 6.1.16(b).
    (c) What is the expected variance of successful attacks with both weapons?
    (d) State the rule in the Algebra of Variances that allows you to determine the answer to 6.1.18(c) from the answers to 6.1.14(c) and 6.1.16(c).

19. An archer can shoot two arrows in a single round of combat.
    (a) If an archer uses hunting arrows, what is the expected value of damage inflicted in one round of combat if we assume that both shots always hit?
    (b) State the rule in the Algebra of Expectations that allows you to determine the answer to 6.1.19(a) from your answer to 6.1.13(b).
    (c) What variance can an archer expect for damage inflicted in one round of combat if we assume that all attacks are successful?
    (d) State the rule in the Algebra of Variances that allows you to determine the answer to 6.1.19(c) from the answer to 6.1.13(c).

20. Hansufel Axe-Beard, the celebrated warrior and tavern brawler, is equally handy wielding a bastard sword with two hands or using it one-handed while using a dagger in the other hand. Unfortunately, Hansufel is none too bright and can't figure out whether he should expect to do more damage using his sword two-handed or fighting with two weapons. He journeys to the great walled city of Umf and seeks advice from the bearded sage Al'tin. If a bastard

---

*A generic label for several medieval sword types distinguished by grips that are "too long for one hand, too short for two hands." Also called hand-and-a-half swords.

sword wielded in one hand inflicts the same damage as a longsword, what does the sage tell him?

21. An *AD&D*® player requires a 17 or higher on a 20-sided die "to hit" a particular opponent with a longsword. If his attack is successful, he will then roll to determine the damage he has inflicted, as described in Exercise 6.1.16. What is the expected value of a longsword attack against this opponent?

22. Suppose that Hansufel Axe-Beard requires a 10 or better on a 20-sided die "to hit" an opponent if he wields his bastard sword with both hands, but that he requires a 12 or better "to hit" if he uses it one-handed and a 16 or better "to hit" with a dagger (since he must wield it in his left hand). Calculate the expected values that the sage would compute in order to give appropriate advice.

23. Two *AD&D*® characters are armed with bastard swords. Bastard-sword damage against *large* creatures is calculated by throwing two 8-sided dice (and adding the results of the two rolls). In addition, the first character's sword is magical, allowing him to add 2 points to the total damage he rolls on the dice, and because he has unusual strength he is allowed to add 1 more point of damage.

(a) Calculate the variance that each character should expect in damage inflicted on large creatures.

(b) State the rule in the Algebra of Variances that accounts for the difference between the variance calculated for the two characters. Suppose, now, that the second character's sword is "Giant Slayer," which does *double* the damage rolled on the dice against giants (definitely large creatures!).

(c) Calculate the variance he should expect in the damage he inflicts on giants.

(d) State the rule in the Algebra of Variances that accounts for the difference between the variance you calculated for this character in 6.1.23(a) and in 6.1.23(c).

C. THE BINOMIAL DISTRIBUTION

For the experiment "flip three coins" let the random variable X denote the number of heads obtained. Then, X takes on four values: 0, 1, 2, or 3 (heads). Table 6.3 presents three of infinitely many possible probability distributions we could use as a model for the relative frequency distribution of X.

Before reading ahead, find three coins. We'll wait. .... OK, now flip the three coins and record the number of heads. Repeat this experiment 16

Table 6.3. Probability distributions modeling the experiment toss three coins

	Distribution 1	Distribution 2	Distribution 3
x	$P(x)$	$P(x)$	$P(x)$
0	.25	.125	0
1	.25	.375	0
2	.25	.375	0
3	.25	.125	1

times and then calculate the relative frequency of each of the four possible outcomes, $x = 0$, $x = 1$, $x = 2$, and $x = 3$. Unless you are tossing three two-headed coins, Distribution 2 should most closely model the relative frequency distribution you obtained. Distribution 2 is an example of a *binomial* distribution. We will begin our discussion of the binomial distribution by constructing a mathematical model of the experiment, much as we did in Example 6.2. We will then "run" this ideal experiment to obtain the distribution.

1. The binomial experiment

An experiment with exactly two possible results is called a *trial*. Although "guilty" and "not guilty" come to mind as good names for these two results, they are always designated *success* and *failure* in probability theory.[4] The most obvious example of a trial is the toss of a coin, designating one side a success and the other a failure, but any experiment can be turned into a trial by defining one or more possible results as successes and *all other* possible results as failures. For example, in a roll of a fair die, one could call "obtain a 6" a success and "obtain any other value" a failure. Similarly, in Example 5.3 we let X denote height and calculated the probability that the height of an individual selected from a sample space S (Figure 5.1) is less than or equal to 67 in. This entire experiment can be thought of as a single trial in which "success" is the subset of persons in S for whom $x \leq 67$ and "failure" is the subset of persons for whom $x > 67$. By custom, the probability of a success is denoted p, and the probability of failure is denoted $q = 1 - p$. In some situations, such as games of chance and coin tossing, the logic and symmetry of the experiment make it possible to calculate p by the classical definition of probability. In other situations, p is estimated by relative frequency.

A *binomial experiment*[5] is any experiment that can be regarded as a sequence of N trials in which

(1) N is defined before the experiment begins.
(2) The result of every trial can be classified into one of two mutually exclusive categories, *success* and *failure*.
(3) The result of any trial is independent of the results of all other trials.
(4) The probability of success p does not change from trial to trial.

Condition (4) implies that the probability of failure q must likewise remain constant from trial to trial, since $q = 1 - p$.

---

[4] These terms are entirely arbitrary and do not characterize the value or desirability of either result. For example, a sports medicine physician comparing the probability of injury in two different sports would nominally record injuries as "successes."
[5] Also called a *Bernoulli trials* experiment, after Jacques Bernoulli (1654–1705), who was the first to describe the binomial distribution.

Our coin-toss experiment may therefore be thought of as 16 repetitions of a binomial experiment: We specified in advance that ($N = 3$) coins were to be tossed, and each toss had two possible results, "coin lands heads up" and "coin lands tails up." By instructing you to record the number of heads, we implicitly defined "coin lands heads up" as a success, and the result of any one toss is in no way influenced by the results for the other coins, so $P(Head) = p$ is always $1/2$.

A result of a binomial experiment is a sequence of N trials, each of which is either a success or a failure. Following the strategy we developed in Chapter 5, we can represent each trial as a blank and use the Extended Principle of Multiplication to determine the number of possible results of a binomial experiment. Since there are exactly two possible results for each trial, there are exactly two ways to fill in each of the N blanks:

$$\underbrace{\underline{2} \quad \underline{2} \quad \underline{2} \quad \ldots \quad \underline{2}}_{N \text{ blanks}}$$

Thus, there are 2^N possible results in the sample space of a binomial experiment. In the experiment "toss three coins" the number of results is $2^3 = 8$, and these are listed below.

$$TTT \quad \text{--- 0 successes}$$

$$\left.\begin{array}{l} HTT \\ THT \\ TTH \end{array}\right\} \text{--- 1 success}$$

$$\left.\begin{array}{l} THH \\ HTH \\ HHT \end{array}\right\} \text{--- 2 successes}$$

$$HHH \quad \text{--- 3 successes}$$

(Perhaps you begin to see why Distribution 2 in Table 6.3 is the right model: $.125 = 1/8$ and $.375 = 3/8$.)

A *binomial random variable* is the number of successes in N trials and can therefore assume the $N + 1$ integer values $0, 1, 2, \ldots, N$. We shall see below that the probability distribution of a binomial random variable X is completely determined by the number of trials N and the probability of success p, and these values must be provided if one is to calculate (or look up) the probability of any x-value. However, instead of writing, "X is a binomial random variable for a series of N trials with probability p of success on each trial," statisticians use the more succinct notation, $X : B(N, p)$. For example, in our coin-toss experiment, $X : B(3, 0.5)$.

We have said that every result of a binomial experiment is a sequence of failures and successes. Since the trials in a binomial experiment are independent, we can use the multiplication theorem for independent events given in Chapter 5 to determine the probability of any such

---

BOX 6.4

Probabilities of results of binomial experiment and values of $X : B(4, p)$

Result	Probability	Number of Successes
SSSS	$pppp = p^4$	- - - 4
SSSF	$pppq = p^3 q^1$	
SSFS	$ppqp = p^3 q^1$	
SFSS	$pqpp = p^3 q^1$	- - - 3
FSSS	$qppp = p^3 q^1$	
SSFF	$ppqq = p^2 q^2$	
SFSF	$pqpq = p^2 q^2$	
SFFS	$pqqp = p^2 q^2$	
FSSF	$qppq = p^2 q^2$	- - - 2
FSFS	$qpqp = p^2 q^2$	
FFSS	$qqpp = p^2 q^2$	
FFFS	$qqqp = p^1 q^3$	
FFSF	$qqpq = p^1 q^3$	
FSFF	$qpqq = p^1 q^3$	- - - 1
SFFF	$pqqq = p^1 q^3$	
FFFF	$qqqq = q^4$	- - - 0

---

sequence. For example, in a binomial experiment with six trials, the result

Success, Success, Failure, Success, Failure, Success

has probability

$$ppqpqp = p^4 q^2$$

The probabilities of all of the results of a binomial experiment with four trials and all values of the random variable $X : B(4, p)$ are given in Box 6.4.

The results listed in Box 6.4 are necessarily mutually exclusive. One cannot, for example, obtain *SSFS* and *SFSS* in the *same* sequence of trials, because *SSFS* calls for a success on the second trial and *SFSS* calls for a failure on the second trial, and "success" and "failure" are mutually exclusive categories. By the addition theorem for mutually exclusive results, therefore, the probability of obtaining exactly x successes in N trials is the sum of the probabilities of all results with x successes (and, therefore, $N - x$ failures). However, it will be noted from Box 6.4 that all results with exactly x successes have the same probability. For example, every result for which $x = 3$ successes has probability

p^3q^1. And, in general, it is true that if $X : B(N, p)$ the probability of *any* result with exactly x successes (and $N - x$ failures) is

$$p^x q^{N-x}$$

Therefore, the probability of obtaining x successes must equal

$$(\text{Number of ways to obtain } x \text{ successes}) p^x q^{N-x}$$

To deduce the number of ways to obtain x successes in four trials, look at the entries in Box 6.4 corresponding to three successes. Each result in which one obtains three successes in four trials corresponds to one possible *combination* of four trials taken three at a time:

Result	Occurs if successes are obtained on
SSSF	Trial 1 Trial 2 Trial 3
SSFS	Trial 1 Trial 2 Trial 4
SFSS	Trial 1 Trial 3 Trial 4
FSSS	Trial 2 Trial 3 Trial 4

The *number* of ways to obtain exactly three successes in four trials must therefore be the *number* of combinations of four trials taken three at a time, $_4C_3$. By equation **[5.10]** this is equal to

$$\frac{4!}{3!(4-3)!} = \frac{4 \cdot 3 \cdot 2 \cdot 1}{(3 \cdot 2 \cdot 1)(1)} = 4$$

This reasoning applies to any number of trials N and any number of successes x: The number of ways to obtain x successes in N trials is $_NC_x$.

These observations provide a model for the distribution of binomial random variables. We know that the probability of every event "obtain (exactly) x successes in N trials" is $p^x q^{N-x}$ and that there are $_NC_x$ such events. Therefore,

$$P(x) = {_NC_x} p^x q^{N-x} \qquad \text{[6.8]}$$

If X is a binomially distributed random variable with probability of success p, then the probability of obtaining x successes in N trials is

$$_NC_x p^x q^{N-x}$$

The numbers $_NC_x$ are sometimes called *binomial coefficients* because they are the coefficients in the expansion of $(x + y)^N$. The presence of the binomial coefficient $_NC_x$ in **[6.8]** gives this distribution its name. If you remember the binomial theorem from some other mathematics course, it

---

BOX 6.5

The binomial theorem and binomial probability

The binomial theorem states,

$$(p + q)^N = \sum_{x=0}^{N} {}_NC_x p^x q^{N-x}$$

Since $(p + q) = 1$,

$$\sum_{x=0}^{N} {}_NC_x p^x q^{N-x} = 1^N = 1$$

---

is really easy to see that the sum of all the assigned probabilities in the binomial distribution is 1. This is shown in Box 6.5

Example 6.7. In a sequence of 4 trials with probability of success $p = 1/3$, what is the probability that there will be exactly 2 successes?

Since $p = 1/3$, we have $q = (1 - p) = 2/3$. There are ${}_4C_2 = 6$ ways to choose two successes in four trials. Each way has probability $ppqq = p^2 q^2 = (1/3)^2(2/3)^2 = 4/81$. Thus, the probability of obtaining exactly two successes is $6(4/81) = 8/27$.

Example 6.8. In a binomial experiment with $N = 7$ and $p = 3/4$, what is the probability of getting no more than 2 successes?

Since $p = 3/4$, we have $q = 1 - (3/4) = 1/4$. We want the probability of 0, 1, or 2 successes in 7 trials. There is only 1 way to get 0 successes (namely, get all failures). Accordingly, ${}_7C_0 = 1$. There are ${}_7C_1 = 7$ ways to get exactly 1 success; and there are ${}_7C_2 = 21$ ways to get exactly 2 successes. The events "obtain 0 successes," "obtain exactly 1 success," and "obtain exactly 2 successes" are mutually exclusive, so by the mathematical definition of probability, $P(0, 1, \text{ or } 2 \text{ successes})$ is obtained by addition:

$$ {}_7C_0 p^0 q^7 + {}_7C_1 p^1 q^6 + {}_7C_2 p^2 q^5 $$

$$ = (1)\left(\tfrac{3}{4}\right)^0\left(\tfrac{1}{4}\right)^7 + (7)\left(\tfrac{3}{4}\right)^1\left(\tfrac{1}{4}\right)^6 + (21)\left(\tfrac{3}{4}\right)^2\left(\tfrac{1}{4}\right)^5 $$

$$ = \frac{1}{16,384} + \frac{(7)(3)}{16,384} + \frac{(21)(9)}{16,384} = \frac{211}{16,384} $$

Example 6.9. In a binomial experiment with $N = 10$ and $p = 3/5$, what is the probability of getting no more than 9 successes?

Apparently we are to calculate the sum of the probabilities of getting 0, 1, 2, 3, 4, 5, 6, 7, 8, or 9 successes. However, in this experiment we must certainly get $0, 1, 2, \ldots, 9$, or 10 successes, all of which are mutually exclusive. Thus,

$$P(0 \text{ or } 1 \text{ or} \ldots \text{or } 9 \text{ successes}) + P(10 \text{ successes}) = 1$$

so

$$P(0 \text{ or } 1 \text{ or} \ldots \text{or } 9 \text{ successes}) = 1 - P(10 \text{ successes})$$

There is only one way to get 10 successes:

$$P(10 \text{ successes}) = p^{10} = (3/5)^{10} \doteq .00605$$

Therefore,

$$P(0 \text{ or } 1 \text{ or} \ldots \text{or } 9 \text{ successes}) \doteq 1 - .00605 = .99395$$

2. The expected value and variance of a binomial random variable

Counting the number of successes in a series of N trials can be thought of as the following function: Assign the value 1 to every success and 0 to every failure and total the results across all N trials. The result of each trial is therefore represented as a random variable that can assume only the values 0 and 1. Such a variable is called an *indicator* random variable and is denoted I.

$$I = \{0, 1\}$$

Since $P(\text{success}) = p$ and $P(\text{failure}) = (1 - p) = q$, it must be true that

$$P(i = 1) = p$$

and

$$P(i = 0) = q$$

From the definition of expected value in equation **[6.3]**,[6] then

$$E(I) = \sum_{}^{2} i_j P(i_j) = 0(q) + 1(p) = p \qquad \textbf{[6.9]}$$

If every trial is quantified as an indicator random variable, then a binomial random variable can be thought of as the sum of N indicator

[6]We use the subscript j instead of i in expression **[6.9]** because otherwise the summation would go from $i = 0$ to $i = 1$. That is, the index (subscript) would take the values of the random variable I.

random variables. That is, if $X : B(N, p)$, then

$$X = I_1 + I_2 + \cdots + I_N$$

Therefore,

$$E(X) = E(I_1 + I_2 + \cdots + I_N)$$

From the Algebra of Expectations we know that the expectation of a sum of random variables $(X + Y)$ is equal to the sum of their individual expectations $\mu_X + \mu_Y$. This result extends naturally to the expected value of the sum of any number of variables. Thus, if $E(I_i) = \mu_i$, then,

$$E(X) = \mu_1 + \mu_2 + \cdots + \mu_N$$

In a binomial experiment, p remains constant from trial to trial. That is,

$$p_1 = p_2 = \cdots = p_N = p$$

so by equation [6.9],

$$\mu_1 = \mu_2 = \cdots = \mu_N = p$$

Consequently,

$$E(X) = \underbrace{p + p + \cdots + p}_{N \text{ times}}$$

That is,

$$E(X) = Np \qquad\qquad\qquad [6.10]$$

Most people find $E(X)$ to be obvious and intuitively sensible. For example, if the probability of success is $1/3$ how many successes would you expect in, say, 12 trials? You expect success one third of the time, so $E(X) = 12(1/3) = 4$.

If $X : B(N, p)$, then $E(X) = Np$.

The easiest way to derive the variance of a binomial random variable is from the computational formula given in [6.7].

$$V(I) = E(I^2) - \mu_I^2$$

The random variable I^2 can assume only the values $0^2 = 0$ and $1^2 = 1$. Moreover, since $i^2 = 0$ if and only if $i = 0$,

$$P(i^2 = 0) = P(i = 0) = q$$

Likewise, $i^2 = 1$ if and only if $i = 1$, so

$$P(i^2 = 1) = P(i = 1) = p$$

From formula [6.3] for expected value, it follows that

$$E(I^2) = 0(q) + 1(p) = p$$

In addition, we know from [6.9] that the expected value of I is p, that is, $\mu_I = p$, so

$$(\mu_I)^2 = p^2$$

Therefore,

$$V(I) = E(I^2) - (\mu_I)^2$$

$$= p - p^2 = p(1 - p) = pq \qquad\qquad \text{[6.11]}$$

As before, the binomial random variable $X: B(N, p)$ is expressed as the sum of N indicator random variables,

$$X = I_1 + I_2 + \cdots + I_N$$

so

$$V(X) = V(I_1 + I_2 + \cdots + I_N)$$

The trials of a binomial experiment are independent, and we know from the Algebra of Variances that if X and Y are independent variables, the variance of $(X + Y)$ equals the variance of X plus the variance of Y. As before, this extends to the case of N variables, so

$$V(X) = V(I_1 + I_2 + \cdots + I_N)$$

$$= \underbrace{pq + pq + \cdots + pq}_{N \text{ times}}$$

That is,

$$V(X) = Npq \qquad\qquad \text{[6.12]}$$

$$\boxed{\text{If } X: B(N, p), \text{ then } V(X) = Npq.}$$

For reference purposes, the expected value and variance of the binomial random variable are given in Box 6.6.

Example 6.10. For the experiment "toss four coins and count the number of heads," what is the variance of X? The number of heads in 4 tosses of a coin is distributed $B(4, 0.5)$. Therefore,

$$V(X) = (4)(.5)(1 - .5) = (4)(.5)(.5) = 1$$

Example 6.11. Roll a fair 6-sided die 20 times and count the number of times that 6 turns up. What is the standard deviation of your random variable? In this experiment 6 is a success and any other value is a failure. Therefore, $p = 1/6$ and $q = 1 - (1/6) = 5/6$. For 20 rolls, the variance of X is $(20)(1/6)(5/6) = 100/36$. The standard deviation is the square root of the variance and is therefore equal to $10/6 \doteq 1.67$.

a. The expected value and variance of the binomial proportion X/N.
Questions about binomial experiments are sometimes asked in terms of the *proportion* of successes (X/N) rather than the *number* of successes, X. The problem in Example 6.7, for example, was to find the probability of exactly two successes when X is binomially distributed with $N = 4$ and $p = 1/3$. It is easily seen that this is identical to asking for the probability that two-fourths of the trials are successful. So, in Example 6.7,

$$P(x = 2) = P\left(\frac{x}{N} = \frac{2}{4}\right)$$

and, in general, if $X : B(N, p)$, then the probability of any particular value x_i is equal to the probability of the binomial proportion x_i/N. That is,

$$P(x = x_i) = P\left(\frac{x}{N} = \frac{x_i}{N}\right)$$

The expected value and variance of the proportion X/N are easily derived from the expected value and variance of the binomial random variable. First, we note that the proportion X/N is the random variable X multiplied by the constant $(1/N)$. That is,

$$\frac{X}{N} = \frac{1}{N} \cdot X$$

so

$$E\left(\frac{X}{N}\right) = E\left(\frac{1}{N} \cdot X\right)$$

From the Algebra of Expectations, then,

$$E\left(\frac{X}{N}\right) = \frac{1}{N}E(X)$$

From expression [6.10] we know that $E(X) = Np$. Therefore,

$$E\left(\frac{X}{N}\right) = \frac{1}{N}Np = p \qquad\qquad [6.13]$$

Similarly,

$$V\left(\frac{X}{N}\right) = V\left(\frac{1}{N}\cdot X\right)$$

and from the Algebra of Variances,

$$V\left(\frac{X}{N}\right) = \left(\frac{1}{N}\right)^2 V(X)$$

Since $V(X) = Npq$,

$$V\left(\frac{X}{N}\right) = \left(\frac{1}{N}\right)^2 Npq = \frac{Npq}{N^2}$$

so

$$V\left(\frac{X}{N}\right) = \frac{pq}{N} \qquad\qquad [6.14]$$

The expected value [6.13] and variance [6.14] of a proportion are included in Box 6.6.

3. The standardized binomial random variable

We know that the standard form for the value x of any random variable is

$$z = \frac{x - \mu}{\sigma}$$

where $\mu = E(X)$ and $\sigma = \sqrt{V(X)}$. The value x of a binomial random variable is therefore converted to standard units by the formula

$$z = \frac{x - Np}{\sqrt{Npq}} \qquad\qquad [6.15]$$

Example 6.12. In Example 6.8 we calculated the probability of obtaining two successes for $X: B(7, 3/4)$. What is the standardized value of $x = 2$? The expected value of X is $7(3/4) = 5.25$. The variance of X is

$7(3/4)(1/4) = 1.3125$. Therefore,

$$z = \frac{2 - 5.25}{\sqrt{1.3125}} \doteq -2.84$$

Example 6.13. For the random variable in Example 6.9 (p. 208), what is the probability that $z \leq 1.6667$? For $X : B(10, 3/5)$,

$$E(X) = Np = 10(3/5) = 6$$

and

$$V(X) = Npq = 10(3/5)(2/5) = 2.4$$

Therefore,

$$1.6667 = \frac{x - Np}{\sqrt{Npq}} = \frac{x - 6}{\sqrt{2.4}}$$

and $x = (1.6667\sqrt{2.4}) + 6 \doteq 9$. From Example 6.9 (pp. 208–9), we know that $P(x \leq 9) \doteq .99395$.

BOX 6.6

Expected value, variance, and standard deviation of a binomial random variable and a binomial proportion

If $X : B(N, p)$, then

$$E(X) = \mu_X = Np \qquad\qquad [6.10]$$

$$V(X) = \sigma_X^2 = Npq \qquad\qquad [6.12]$$

$$\sigma_X = \sqrt{Npq}$$

$$E\left(\frac{X}{N}\right) = \mu_{X/N} = p \qquad\qquad [6.13]$$

$$V\left(\frac{X}{N}\right) = \sigma_{X/N}^2 = \frac{pq}{N} \qquad\qquad [6.14]$$

$$\sigma_{X/N} = \sqrt{\frac{pq}{N}}$$

──────── EXERCISES 6.2 ────────────────────────────────

1. Which of the following experiments is not a binomial experiment, and why not?
   (a) Toss a fair coin until 5 heads come up in a row.
   (b) Toss a fair coin 20 times and record the number of heads.

(c) Draw a card from an ordinary playing deck and record whether or not it is a heart. Repeat the experiment 10 times without replacing the cards after each draw.

(d) Roll three dice 10 times and record the number of times that the total number of spots is 3.

2. Which requirement(s) for a binomial experiment is (are) violated in each of the following?

(a) Roll an ordinary 6-sided die 10 times. On each roll call a 1 or a 2 "Joe"; call a 3 or a 4 "Pete"; call a 5 or a 6 "Throckmorton."

(b) Roll three ordinary 6-sided dice 12 times and record the number of spots that come up on each roll (so X can take any value from 3 to 18). Repeat the experiment until you obtain at least 6 scores of 18 in your 12 rolls.

(c) Go to the table of random numbers (Table 1 of Appendix VIII). Pick a number at random. Let the first two digits be R and the next two be C. Let x be the five-digit number located at the intersection of row R and column C. If the first digit of x is a 7 call it a success; if it is not a 7 call it a failure. If your trial is successful, cross out x and do not choose it again. When you have made 10 permissible choices, record the number of successes.

3. The expression

$$_{15}C_5(.82)^5(.18)^{10} + {}_{15}C_6(.82)^6(.18)^9 + {}_{15}C_7(.82)^7(.18)^8$$

is the probability that _____ for a random variable that is distributed _____ . Calculate the expected value and variance of this random variable.

4. Let $X : B(100, \frac{1}{6})$.

(a) Use binomial coefficients to write a general expression for the probability that $(85 \le x \le 88)$.

(b) Calculate the expected value and variance of X.

5. In a binomial experiment with 5 trials, what is the probability of obtaining exactly 3 successes if $p = .12$?

6. In a binomial experiment with 5 trials, what is the probability of obtaining 1, 2, 3, 4, or 5 successes if $p = .42$?

7. Consider the experiment "roll a 6-sided die." If the experiment is repeated three times and X is the number of 6s, what is

(a) $P(x = 0)$ (b) $P(x \ge 2)$

(c) the expected number of 6s

(d) the expected variance in number of 6s obtained in three rolls

(e) the expected proportion of 6s

(f) the expected variance in the proportion of 6s obtained in three rolls

8. Let $X : B(10, .23)$. What is

(a) $P(x = 5)$ (b) $P(x \ne 5)$ (c) $P(x \le 3)$

(d) $E(X)$ (e) $V(X)$ (f) $E\left(\dfrac{X}{N}\right)$ (g) $V\left(\dfrac{X}{N}\right)$

9. Consider the experiment in Exercise 6.2.7 and let a success be "obtain a value *other than* a 6." Calculate

(a) the expected number of successes in three rolls of the die

(b) the expected variance in the number of successes in three rolls of the die

(c) the standardized value of $x = 2$ successes

(d) the expected proportion of successes in three tosses of the die

(e) the expected variance in the proportion of successes in three tosses of the die

(f) the standardized score corresponding to an observed proportion of successes $= 2/3$

10. Consider the random variable in Exercise 6.2.8. Calculate

    (a) the z-score corresponding to 5 successes

    (b) the z-score corresponding to 50 percent successes

    (c) the z-score corresponding to 20 percent successes

11. A genetics class includes 5 men and 17 women, and each student is as likely to attend class on any day as any other student. If on one particular day 18 students attend class, what is the probability that this group will have between 13 and 16 women, inclusive?

12. An introductory sociology class is given a multiple-choice quiz with 15 items. Each item has 5 choices for every answer. If a passing score is 70 percent, what is the probability that a student will earn a passing grade if he guesses at random?

13. The rejection rate for a certain journal is 45 percent. If the journal accepts articles at random, how many articles would someone have to submit to have a probability of .75 of getting at least one article accepted?

14. Go to the table of random numbers in the back of the book and select 20 five-digit numbers. Record the last digit for each of the 20 numbers. Let "obtain a 9" be a success. Model this experiment with the appropriate binomial random variable. What is the probability of obtaining as many 9s as you actually obtained, or *fewer*?

15. A pharmaceuticals company claims to have developed a vaccine that is 95 percent effective against parvovirus in dogs. To test this claim the veterinary school at ESU (Enormous State University) injects 25 dogs with the vaccine and then exposes them to parvovirus. Five of the dogs develop the disease. What is the probability of 20 or fewer successful immunizations if the company's claim is accurate? What does this suggest about the company's claim?

16. A missile manufacturer claims that its missiles are 90 percent effective. The Lower Slobovian Air Force tests this claim by firing 10 missiles at Upper Slobovia. It obtains 6 hits. What is the probability of obtaining *6 or fewer* successes if indeed $p = .90$?

17. A plastics manufacturer claims that its clay-lined waste dump prevents benzene from getting into the local ground water, but residents are suspicious. A random survey of 100 persons living within a quarter of a mile of the plant turns up 4 people with a form of cancer known to be correlated with benzene pollution. The base rate for this form of cancer in the U.S. population is .001. What is the probability of observing 4 or more cases if the plant is *not* polluting the local water supply?

18. Consider the experiment "roll three 6-sided dice" and the random variable "count the total number of spots on the uppermost faces of the dice." Let "obtain an 18" be a success. If the experiment is repeated 12 times, what is

    (a) $P(x = 0)$ (b) $P(x = 1)$ (c) $P(x = 2)$ (d) $P(x \geq 2)$

Hint: $p = P(\text{success}) = P(18) = P(\text{obtaining three 6s in three rolls of a die.})$

19. Consider the experiment "roll four 6-sided dice" and the random variable "ignore the lowest die and count the number of spots on the uppermost faces of the other three." Let "obtain an 18" be a success. If the experiment is

repeated 6 times, what is

(a) $P(x = 0)$ (b) $P(x = 1)$ (c) $P(x = 2)$ (d) $P(x \geq 2)$

Hint: $p \equiv P(\text{success}) = P(18) = P(\text{obtaining three 6s in four rolls of a die}).$

D. REPRESENTATION OF PROBABILITY DISTRIBUTIONS OF DISCRETE RANDOM VARIABLES

There are three basic ways to display or represent a discrete probability distribution, and two of these—tabular and graphic representation—are already familiar. They have been used since Chapter 1 for representation of data.

1. Tabular representation

In everyday statistical work the most common representation is the probability table, a simple list of all the values that the variable can assume and the probability associated with each value. When the random variable can assume only a small number of values, the typical probability table is laid out very much like a relative frequency table for ungrouped data. Table 6.2 (p. 190) gives the probability values for $X : U(6)$, and Distribution 2 in Table 6.3 (p. 203) gives the probability values for $X : B(3, 0.5)$.

A more extensive compilation of binomial probabilities is given in Table 2 of Appendix VIII. Almost all calculators compute factorials (many calculate $_N C_r$ directly) and powers, which makes it a very simple task to calculate the binomial probability of a single value. But to determine the probability that x is larger than or smaller than a particular value or that x lies between two values can require that many individual probabilities be calculated. Table 2 can be used to obtain the probability of any value of x if $N \leq 25$ and if $p = .01, .05, .10, \ldots$, or $.99$. The number of trials N is given in the far left column, and for every N-value there are $N + 1$ rows, corresponding to $x = 0, 1, \ldots, N$. The columns represent the probability of success, p. For example, the probabilities for $X : B(3, 0.50)$ appear in the block of four entries for $N = 3$ under the column $p = .50$:

N	x	$\cdots$	.50
3	0	$\cdots$	125
	1	$\cdots$	375
	2	$\cdots$	375
	3	$\cdots$	125

These values (except for decimal points, which are excluded from Table 2 to save space) are the same as those in Distribution 2 of Table 6.3. Likewise, the entries for $N = 4$ under the column $p = .50$ give the values for $X : B(4, 0.50)$, which are the same as those in Table 6.1(b) on p. 187.

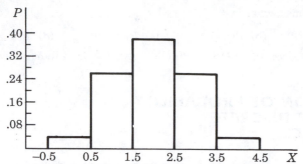

Figure 6.1. Histogram representing probability distribution in Table 6.4, $X : B(4, 0.5)$.

2. Graphic representation

In Chapter 1 we showed how relative frequencies can be represented graphically, either as histograms or as (relative) frequency polygons. This convention can likewise be used for probability distributions. For example, the relative frequency distribution in Table 6.1(a) was presented as a histogram in Figure 1.5. The corresponding probability distribution [Table 6.1(b)], can be similarly represented, as in Figure 6.1. We have replaced the f/N in Figure 1.5 with a P in Figure 6.1 to indicate that the vertical axis represents probability values instead of relative frequencies. The histogram in Figure 6.2 represents the uniform probability distribution in Table 6.2.

It will be noted in both Figures 6.1 and 6.2 that we have observed the convention of representing each possible value of X by an interval that is one unit wide. In Figure 6.2, for example, the value $x = 0$ spans the interval -0.5 to 0.5. This convention was introduced in Chapter 1 as a graphic convenience, but we also pointed out that since the area enclosed by any rectangle is equal to its height multiplied by its width, the *area* of each unit-width bar must be numerically equal to its height. We later exploited this property in our discussion of the median to represent relative frequency as the *area* enclosed by each bar.

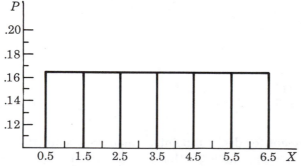

Figure 6.2. Histogram representing probability distribution in Table 6.2, $X : U(6)$.

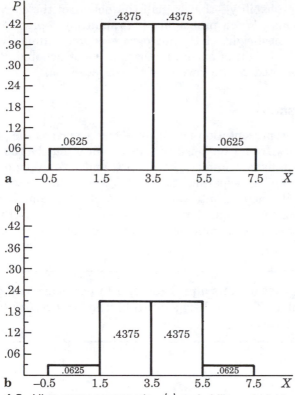

Figure 6.3. Histograms representing (a) probability as height and (b) probability as area for distribution of $X : B(7, 0.5)$.

It is similarly true that the area enclosed by each of the bars in Figures 6.1 and 6.2 is equal to the probability x, where x is represented by the interval $x \pm 0.5$. This means, of course, that the total area enclosed by the histogram is 1.0. This is most readily apparent in Figure 6.2, where the height of each bar is $.16666\ldots = 1/6$, and the base of the entire histogram spans 6 units. The area is therefore $(1/6)(6) = 1$.

For reasons that will become apparent when we begin our discussion of continuous probability, histographic representations of probability distributions conventionally represent probability as the *area* of each bar, rather than the *height* of each bar. We have pointed out that such representations are essentially identical when each bar is one unit wide. But, when the values of X are grouped, so that each bar represents two or more values of X, the representations will be quite different. In Figure 6.3 we present histograms for the distribution of $X : B(7, 0.5)$ where each bar represents the probability for two values 0 and 1, 2 and 3, etc. In Figure 6.3(a) probability is represented as height, and in Figure 6.3(b) probability is represented as area. Since the area of any rectangle is equal to its height multiplied by its width, the height of any rectangle is equal to its area divided by its width. The bars in Figure 6.3 are each *two* units wide, so the height of each bar in Figure 6.3(b) must be numerically equal to *one-half* its probability—that is, half the height of the corresponding bar in Figure 6.3(a). When probability is graphically represented as the area, therefore, the height of the histogram is *not* ordinarily equal to probability, so the height of such a histogram is not denoted P. Instead, it is generally denoted ϕ, the lowercase Greek letter *phi*.

3. Functional representation

The only type of representation not encountered earlier is the probability function. We know from Chapter 5 that probability may be thought of as a function that assigns a number between 0 and 1 to every outcome in the sample space of an experiment. We also know that outcomes of an experiment can be mapped into the value set of a random variable and that probabilities are assigned to these values. Functions that assign probabilities to values of a random variable are called *probability functions*.

> If P is a rule that assigns a probability $P(x)$ to every value of the random variable X, then P is said to be the *probability function* of the random variable.

Thus, the function

$$P(x) = {}_N C_x p^x q^{N-x}$$

is the probability function for the binomial random variable $X: B(N, p)$. Please note that the probability function in this general form does not permit probabilities to be calculated. The probability of any particular value x can be determined only if N and p are assigned numerical values, e.g., 10 and 0.5. Constants that must be assigned specific values in order to calculate $P(x)$ are called the *parameters* of the probability function P. When the appropriate function is identified *and* the parameter values are given, a distribution is said to be completely *specified*. For this reason, a statement of the form $X: B(10, 0.5)$ is sometimes called the *specification statement* of the distribution.

——————— EXERCISES 6.3 ———————————————————

1. The expression $X: U(1{,}000)$ is the _____ statement for the _____ random variable with _____ $n =$ _____ and with $P(x) =$ _____.

2. The expression $X: B(13, 0.267)$ is the _____ statement for the _____ random variable with _____ $N =$ _____ and _____ $= .267$. Calculate the expected value and variance of X.

3. For the random variable in Exercise 6.1.7 (p. 201)
   (a) construct a probability table.
   (b) construct a probability histogram, where each bar is one unit wide and the vertical axis represents probability.
   (c) write the probability function.

4. For the random variable in Exercise 6.1.8 (p. 201)
   (a) construct a probability table.
   (b) construct a probability histogram, where each bar is two units wide and the vertical axis represents probability.
   (c) write the probability function.
5. For the random variable in Exercise 6.1.7 (p. 201)
   (a) construct a probability histogram, where each bar is one unit wide and the areas of the bars represent probability.
   (b) construct a probability histogram, where each bar is two units wide and the vertical axis represents probability.
   (c) using the same horizontal scale as in Exercise 6.3.5(b) construct a probability histogram, where each bar is two units wide and the areas of the bars represent probability.
   (d) compare the heights of the bars in Exercises 6.3.5(a) and 6.3.3(b).
   (e) compare the heights of the bars in Exercises 6.3.5(b) and 6.3.5(c).
6. For the random variable in Exercise 6.1.8 (p. 201), construct a histogram where each bar is two units wide and the areas of the bars represent probability. Use the same horizontal scale as in Exercise 6.3.4(b). Compare the heights of the bars in Exercises 6.3.6 and 6.3.4(b).

E. SUMMARY

A *probability distribution* is a mathematical idealization, or model, of the relative frequency distribution of outcomes of a random experiment. If a random variable can assign only a *countable* number of values to the result of a random experiment, it is said to be a *discrete* random variable. Probability models for discrete random variables assign a nonzero probability to every value that the random variable can assume.

A random variable that takes on n values, each of which is assigned the same probability $(1/n)$, is said to be uniformly distributed and is denoted $X : U(n)$.

Properties like location and dispersion are summarized for probability distributions by quantitative indices that are very similar to the descriptive statistics that describe these properties for relative frequency distributions.

The location of a probability distribution can be described by the *expected value* of the random variable, denoted $E(X)$. The expected value of a discrete random variable is defined as

$$E(X) = \sum x_i P(x_i)$$

where $P(x_i) = P(x = x_i)$. The actual value of $E(X)$ is frequently denoted μ.

The dispersion of a probability distribution can be described by the *variance* or *expected variance* of the random variable, denoted $V(X)$. The variance of a discrete random variable is defined as

$$V(X) = \sum (x_i - \mu)^2 P(x_i)$$

where $\mu = E(X)$ and $P(x_i) = P(x = x_i)$. The variance of X can sometimes be calculated more easily by the formula

$$V(X) = E(X^2) - \mu^2$$

The actual value of $V(X)$ is often denoted σ^2.

The standard deviation of a random variable X is the square root of its variance,

$$\sigma = \sqrt{V(X)}$$

A random variable can be standardized by subtracting the expected value μ from every x-value and dividing by the standard deviation σ:

$$z = \frac{x - \mu}{\sigma}$$

If Z is a standardized random variable, $E(Z) = 0$ and $V(Z) = 1$.

An experiment with only two possible outcomes is called a *trial*, or a *Bernoulli trial*, and the two possible outcomes are customarily designated *success* and *failure*. The probability of success is denoted p, and the probability of failure is denoted q. For any trial, therefore, $p + q = 1$, so $q = 1 - p$.

A *binomial experiment* (also called a *Bernoulli trials experiment*) is a sequence of N trials in which (1) N is defined before the experiment begins, (2) the result of any trial is independent of the results of all other trials, and (3) the probability of success p remains constant from trial to trial.

The result of a binomial experiment is a sequence of N successes and failures, and a *binomial random variable* is the *number* of successes obtained. The notation $X : B(N, p)$ is a compact way to say that X is a binomial random variable in a Bernoulli trials experiment with N trials and a probability of success p on every trial.

If $X : B(N, p)$, the probability of *any* result with x successes (and, therefore, $N - x$ failures) is

$$p^x q^{N-x}$$

In a binomial experiment the *number of ways* one can obtain exactly x successes in N trials is equal to the number of combinations of N things taken x at a time,

$$_N C_x = \frac{N!}{x!(N - x)!}$$

For $X : B(N, p)$ the probability of x successes is equal to the sum of the probabilities of all results with x successes. All such results have probability $p^x q^{N-x}$, so the probability of x successes is

$$P(x) = {}_N C_x p^x q^{N-x}$$

For a binomial random variable $X : B(N, p)$,

$$E(X) = Np \quad \text{and} \quad V(X) = Npq$$

If $X : B(N, p)$, then for the proportion of successes, X/N,

$$E\left(\frac{X}{N}\right) = p \quad \text{and} \quad V\left(\frac{X}{N}\right) = \frac{pq}{N}$$

Like any random variable, the binomial random variable can be standardized. If $X : B(N, p)$, then for any particular value x

$$z = \frac{x - \mu}{\sigma} = \frac{x - Np}{\sqrt{Npq}}$$

If Z is the standardized binomial random variable, then (as always) $E(Z) = 0$ and $V(Z) = 1$.

Discrete probability distributions can be represented in *tabular* form, in *graphic* form, or in *functional* form.

A probability table is a list of the values that the random variable can assume together with the probability associated with each value. If X can take many values, a probability table may include only a selection of these.

The most common form of probability graph is the probability histogram. Every value x of a discrete random variable corresponds to a single point on the real line, but in a probability histogram each value is typically represented as an *interval* extending from $x - .5$ to $x + .5$. The probability associated with the value x may then be represented by a bar for which this interval is the base. If probability is represented by the *height* of the bar, the vertical axis is ordinarily labeled P. If the probability $P(x)$ is represented by the *area* of the bar, the vertical axis is labeled ϕ.

The third convention for representing a discrete probability distribution is the *probability function*. If P is a rule that assigns a probability to every value of the random variable X, then P is said to be the probability function of the random variable. Accordingly,

$$P(x) = {}_N C_x p^x q^{N-x}$$

is the probability function of the random variable $X : B(N, p)$. Any probability function has certain *parameter* values that must be specified in order to calculate $P(x)$. For the binomial probability function these are N and p. A statement of the form $X : B(36, 0.5)$ or $X : U(10)$ is therefore called the *specification statement* of the distribution.

CHAPTER 7

Continuous probability distributions

C. REPRESENTATION OF PROBABILITY DISTRIBUTIONS OF CONTINUOUS RANDOM VARIABLES
Tabular representation
Graphic representation
Functional representation

D. SUMMARY

In Chapter 6 we developed the uniform and binomial probability distributions and showed that they were appropriate models, or idealizations, for the relative frequency distributions we might obtain in certain kinds of experiments. All of the experiments we modeled, however, had one thing in common: The observations involved *discrete* phenomena. That is, all of the experiments involved *counting* or some other operation that could in principle assign one of the natural numbers

$$\{1, 2, 3, 4, \dots\}$$

to each observation. We pointed out, though, that not all phenomena are in one-to-one correspondence with the natural numbers. Some can assume *any* real number value, e.g., $1\frac{1}{2}$, 5.333, $\sqrt{17}, \dots$, etc. These are *continuous* phenomena. In this chapter we develop the logic and the conventions that statisticians use to model the relative frequency distributions associated with *continuous* phenomena, such as weight, length, height, and most other dimensions by which we describe physical and behavioral properties.

Apart from the fact that distributions of discrete and continuous variables are mathematically different in some important ways, the machinery for handling continuous variables is important for another reason: Even when an experiment involves discrete phenomena, the appropriate discrete model can be very difficult to use if the number of observations is large. In such cases, it is often convenient to use a continuous model to *approximate* the more exact discrete model.

The key to understanding continuous probability distributions and how they differ from discrete probability distributions lies in thinking about the representation of probability as the *area* of a histogram, and it is this topic with which we begin.

A. THE UNIFORM DISTRIBUTION REVISITED: PROBABILITY AS AREA AND THE NOTION OF PROBABILITY DENSITY ϕ

We introduced discrete probability distributions in Chapter 6 with a uniformly distributed random variable with only six possible values, the number of spots obtained on one toss of a fair die. To begin our discussion of continuous probability distributions, let us define a discrete random variable that can take only two possible values.

Example 7.1. Suppose that we have two individuals, A and B, and we have recorded their weights to the nearest pound:

Individual	Weight (to nearest pound)
A	177
B	178

Suppose further that we write A on one slip of paper and B on another slip of paper and draw one slip at random. If this experiment were repeated many times (replacing the slip after each draw) we would expect to choose someone weighing 177 pounds on about half of our draws and someone weighing 178 on about half of our draws. The appropriate model for this relative frequency distribution is the discrete uniform probability distribution: We have two possible values, each of which is assigned the same probability, $p = 1/2$.

From Chapter 6 we know that we can represent this probability distribution in the form of a histogram, where each observable value is represented by a bar and the probability associated with each weight is represented by the *area* of the bar. The probability histogram for this experiment is illustrated in Figure 7.1. Since our unit of measurement is the pound, our customary practice of extending the bases of our histogram bars half a measurement unit above and half a unit below the observable values gives us one bar associated with the interval 176.5 to 177.5 and another bar associated with the interval 177.5 to 178.5. The base of each bar is therefore one unit wide. The height of any rectangle is equal to its area divided by its base, and since the probability represented by the area of each bar is $1/2$, the height of each bar must be $(1/2)/1 = 1/2$. This means, of course, that the total area of the histogram is 1, the probability associated with the set of all possible results of the experiment.

Example 7.1.1. Now suppose we have a scale accurate to the nearest *half* pound and that we have three individuals:

Individual	Weight (to nearest 0.5 lb)
A	177.0
B	177.5
C	178.0

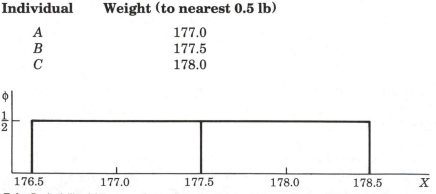

Figure 7.1. Probability histogram for uniformly distributed random variable with 2 values.

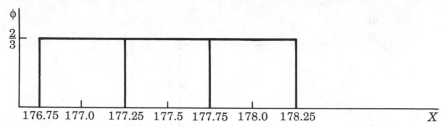

Figure 7.2. Probability histogram for uniformly distributed random variable with 3 values.

If we perform the same experiment as in Example 7.1, the appropriate model for our relative frequencies is the histogram appearing in Figure 7.2, which is drawn so that the areas of the bars represent the probability (1/3) associated with each weight. Since our scale is accurate to the nearest *half* pound, each bar covers an interval spanning a *quarter* pound above and below each observable value. That is, 177 is represented by the interval 176.75 to 177.25; the value 177.5 is represented by the interval 177.25 to 177.75; and 178 is represented by the interval 177.75 to 178.25. The base of each bar is therefore an interval one half pound in width. The height of each rectangle is once again determined by dividing its area by its base, so

$$\text{Height} = \frac{1/3}{1/2} = \frac{2}{3}$$

Because the three possible results are equally likely, the entire histogram is itself a rectangle, and the height of the histogram is everywhere equal to the height of any one bar, 2/3. This is confirmed if we divide the total area ($= 1$) by the entire base. The base is $178.25 - 176.75 = 1.5 = 3/2$, so

$$\text{Height} = \frac{1}{3/2} = \frac{2}{3}$$

Example 7.1.2. Finally, let us suppose that we have recorded the weights of 101 individuals using a scale accurate to the nearest *hundredth* of a pound.

Individual	**Weight (to nearest .01 lb)**
A	177.00
B	177.01
C	177.02
⋮	⋮
Elrond	177.98
Imrahir	177.99
Glorfindel	178.00

The histogram for this collection of values has too many bars for us to

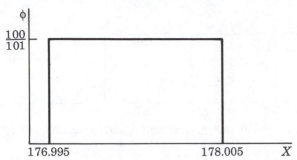

Figure 7.3. Probability histogram for uniformly distributed random variable with 101 values.

represent, but we can tell where it begins and where it ends. The unit of measurement is .01. One half of that is .005. Consequently, the histogram begins at 177.00 − .005 = 176.995 and ends at 178.005. The width of the entire histogram is therefore 178.005 − 176.995 = 1.01, nearly 1. So, the height must be 1/1.01 = 100/101, which of course is also nearly 1.

Looking at Figures 7.1, 7.2, and 7.3 we see that increasing the precision of measurement has several interrelated effects on the histogram for our discrete uniform random variable: (1) the width of the base tends toward 1, (2) the height of the histogram tends toward 1, (3) the number of bars increases, and (4) the width of each bar decreases. We could further increase our precision of measurement and weigh individuals to the nearest .001 pound or .0001 pound and so on. The limit of this process is the *continuous* uniform distribution, which models *absolute* precision of measurement. If X is the continuous random variable representing weight, then a statement like $x = 177.03$ means that the individual weighs *exactly* 177.03 lbs—not 177.0299993 or 177.030001—but exactly 177.03 lbs. If we drew a histogram to represent *exact* weights, there would be infinitely many infinitely thin bars between 177 and 178. In Figure 7.4, we show only the tops of those bars, which form an unbroken line. Because the value of ϕ for every point x between 177 and 178 is 1, the equation for this line is $\phi(x) = 1$. The function ϕ is called

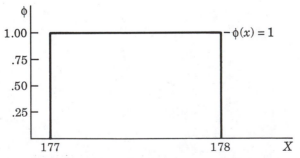

Figure 7.4. Density function for a continuous random variable uniformly distributed over the interval [177, 178].

the *probability density function*, or, more simply, the *density function* for the distribution.

$\phi(X)$ is said to be the probability density function for the random variable X if the area under ϕ between x_a and x_b is equal to $P(x_a \leq x \leq x_b)$.

The notion of "area under" the graph of a function is formalized a little later in the chapter (Box 7.1) and incorporated into our definition of a density function.

Example 7.2. Consider the experiment "pick a real number at random from the interval of real numbers between 1 and 5 inclusive." Since any number is as likely to be drawn as any other number, the random variable is uniformly distributed, and its density function can be represented as in Figure 7.5.

Figure 7.5 can be viewed as infinitely many infinitely thin histogram bars distributed between 1 and 5. Since the base is $5 - 1 = 4$ and the total area must be 1, the height must be $1/4$. Let x denote the number you select. Note that x can be *any* real number value between 1 and 5. Of course, x could be one of the integers 1, 2, 3, 4, or 5. But x could also be one of infinitely many fractions like $2\frac{1}{3}$ or $27/8$ or $113/30$, etc., or one of infinitely many irrational numbers like π, $\sqrt{2}$, $\sqrt[3]{61}$, etc.

Example 7.2.1. What is the probability that the number selected is between 1 and 4? That is, what is $P(1 < x < 4)$? Since three-fourths of the numbers in the interval $[1, 5]$ are between 1 and 4, we have $P(1 < x < 4) = 3/4$. You can *and should* think of this probability as the area shaded in

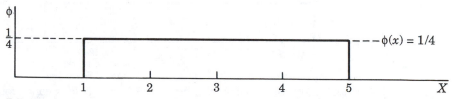

Figure 7.5. Density function for a continuous random variable uniformly distributed over interval $[1, 5]$

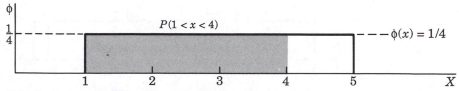

Figure 7.6. Proability of selecting a number between 1 and 4 for a continuous random variable uniformly distributed over the interval $[1, 5]$.

Figure 7.6. Thus, probability equals the height of the shaded rectangle multiplied by the base:

$$\frac{1}{4} \times 3 = \frac{3}{4}$$

Example 7.2.2. What is the probability that the number selected is between 3.7 and 4.2? Since $4.2 - 3.7 = .5$, the interval 3.7 to 4.2 is half a unit wide and must include one eighth of the values between 1 and 4. Therefore, $P(3.7 < x < 4.2) = 1/8$. Once again, this may be thought of as the shaded area of Figure 7.7. Multiplying the height times the base of the shaded rectangle,

$$\frac{1}{4} \times 0.5 = \frac{1}{8}$$

Example 7.2.3. What is the probability that the value selected is *exactly* 2.3? This problem leads us to one of the most important differences between discrete and continuous probability. If we think of 2.3 as a "rectangle" over the interval 2.3 to 2.3, we see that our "rectangle" has a width of zero. That is, the rectangle has degenerated to a line segment, as seen in Figure 7.8. Accordingly, it has an area of zero, and $P(x = 2.3) = 0$.

This may come as a surprise. Here we have an event, "select 2.3," which is entirely possible, since 2.3 *is* in the interval 1 to 5. Yet, the probability of this eminently possible event is zero. This is a characteristic of *all* continuous random variables: The probability of achieving any particular value is zero.

Since probability zero is assigned to each individual value, we can be a bit sloppy with inequalities. For example, in Example 7.5 it really doesn't

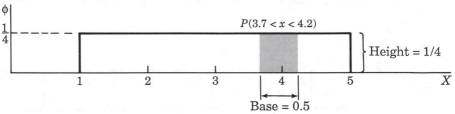

Figure 7.7. Probability of selecting a number between 3.7 and 4.2 for a continuous random variable uniformly distributed over the interval [1, 5].

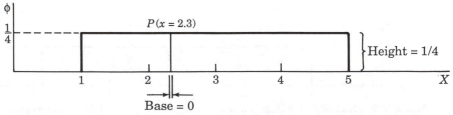

Figure 7.8. Probability that $x = 2.3$ where X is a continuous random variable uniformly distributed over the interval [1, 5].

matter if we include or exclude the endpoints of the intervals. Thus,

$$P(1 < x < 4) = P(1 \le x < 4) = P(1 < x \le 4) = P(1 \le x \le 4)$$

1. Cumulative probability $\mathbb{P}$

In Chapter 1 we explained that the cumulative relative frequency of any value is the relative frequency of the value plus the relative frequencies of all smaller values. That is, the cumulative relative frequency of any value x_i is the sum of the relative frequencies for all x-values less than or equal to x_i. The notion of cumulative probability is completely analogous:

> The cumulative probability of any value x_i is defined as $P(x \le x_i)$

The cumulative probability of x_i is in this text denoted $\mathbb{P}(x_i)$, that is,

$$\mathbb{P}(x_i) = P(x \le x_i)$$

This definition applies to both discrete and continuous random variables. In the discrete case the cumulative probability of x_i can be obtained by calculating the sum

$$\sum P(x)$$

for all $x \le x_i$. For a continuous random variable X with density function ϕ, the cumulative probability of x_i is represented as the area under ϕ to the left of x_i. In Example 7.2.1, for instance, the smallest value the x can take is 1, so

$$\mathbb{P}(4) = P(x \le 4) = P(1 \le x \le 4) = \frac{3}{4}$$

and is represented by the shaded area in Figure 7.6. Likewise, the cumulative probability of $x = 2.5$ in Example 7.2 is the area to the left of 2.5 under the graph of ϕ in Figure 7.5. The interval 1 to 2.5 is one and a half units wide, and the height of the histogram is $1/4$, so

$$\mathbb{P}(2.5) = P(x \le 2.5) = P(1 \le x \le 2.5) = \left(\frac{1}{4} \times 1.5 \right) = \frac{3}{8}$$

──────── EXERCISES 7.1 ──

1. Graph the density function for the random variable X: "pick a real number between 2 and 12 at random."
  (a) On the graph of this density function shade the area that represents $P(4 \le x \le 7)$. Calculate $P(4 \le x \le 7)$.
  (b) On the graph of this density function shade the area that represents $P(x \le 9)$. Calculate $P(x \le 9)$.
  (c) Shade the area corresponding to $\mathbb{P}(5)$. Evaluate $\mathbb{P}(5)$.
  (d) Shade the area for $P(4 < x)$. Calculate $P(4 < x)$.

2. Graph the density function for the random variable X: "pick a real number at random between -4 and 8."
   (a) On the graph of this density function shade the area that represents $P(-1 \leq x \leq 7)$. Calculate $P(-1 \leq x \leq 7)$.
   (b) On the graph of this density function shade the area that represents $P(x \leq 3)$. Calculate $P(x \leq 3)$.
   (c) Shade the area corresponding to $\mathbb{P}(0)$. Calculate $\mathbb{P}(0)$.
   (d) Shade the area for $P(-2 < x)$. Calculate $P(-2 < x)$.

B. THE NORMAL DISTRIBUTION

If a fair coin is flipped 1,000,000 times, what are the chances that it will come up heads between 499,200 and 500,500 times (inclusive)? Although you may not be particularly interested in this question, having recently seen the binomial distribution, you recognize at once that this is a question about a random variable $X: B(1,000,000, 0.5)$. We are asking for

$$P(499,200 \leq x \leq 500,500)$$

In principle, this problem is no different from

$$P(3 \leq y \leq 5)$$

for a random variable $Y: B(7, 0.5)$. In practice, the first problem is considerably more difficult because $N = 1,000,000$ is a very large number. You could look up the answer to the second problem in Table 2 of Appendix VIII:

$$P(3 \leq y \leq 5) \doteq .2734 + .2734 + .1641 = .7109$$

or with a hand calculator, you could obtain the exact result in a minute or two by computing

$$_7C_3\left(\tfrac{1}{2}\right)^3\left(\tfrac{1}{2}\right)^4 + {}_7C_4\left(\tfrac{1}{2}\right)^4\left(\tfrac{1}{2}\right)^3 + {}_7C_5\left(\tfrac{1}{2}\right)^5\left(\tfrac{1}{2}\right)^2 = \frac{91}{128}$$

Table 2 does not go up to $N = 1,000,000$, and there are at least two reasons you would not want to use the binomial probability function to calculate

$$P(499,200 \leq x \leq 500,500)$$

First, you would need to sum 1,301 terms. Second, each of those terms involves the product of a very large binomial coefficient multiplied by a very small probability factor, for example,

$$_{1,000,000}C_{499,281} \cong 10^{300,000}$$

(that's 1 followed by 300,000 zeros), multiplied by

$$\left(\tfrac{1}{2}\right)^{499,281}\left(\tfrac{1}{2}\right)^{500,719} = \left(\tfrac{1}{2}\right)^{1,000,000}$$

which is practically zero. The product of a very large and very small number causes problems for a calculator, or even for a computer. Consider:

$$(1,000,000) \times (.000001) = 1$$

but

$$(1,000,000) \times (.0001) = 100$$

Note that a difference of only .000099 (approximately 1 in 10,000) in the second factor yields a 100-fold change in the product. Since computers and calculators must necessarily round off to do arithmetic, it is not difficult to see that rounding errors in the calculation of very small probability values can produce immense errors when the probabilities are then multiplied by large binomial coefficients. Even if your computer makes an error of only .0008 in each of the 1,301 terms, its answer could be off by as much as ± 1.0. That is particularly devastating, since we know that the answer itself is between 0 and 1.

Our difficulty here seems as intractable as it is trivial; after all, who really cares about flipping coins? The resolution of the difficulty, however, allows us to develop the world's most important probability distribution, which has significant applications in all of the sciences and in mathematics. What we are calling the "world's most important probability distribution" is the normal distribution,[1] and we approach it by way of the binomial distribution by much the same logic as we generated the continuous uniform distribution from the discrete uniform distribution. The route to the normal distribution has a number of detours, however, so it is essential that we keep in mind three important notions we developed in the last section: First, for any discrete random variable X

$$P(x_a \le x \le x_b)$$

can be represented as the area of histogram bars corresponding to the x-values from x_a to x_b. Second, the total area under the histogram represents the probability of all possible results and must therefore always equal 1. Third, we have seen in Examples 7.1 and 7.2 that as more and more values are crowded into an interval of (more or less) fixed

[1] Sir Francis Galton was probably the first (1877) to use the term "normal" to characterize this distribution, apparently as a synonym for "usual," "common," etc. See Kruskal, W. (1978). Formulas, numbers, words: Statistics in prose. *The American Scholar*, 47, 223–229. Many natural phenomena are "normally" distributed for reasons discussed in Box 8.1 on p. 285.

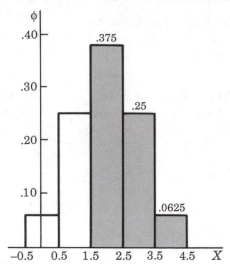

Figure 7.9. Probability histogram for $X : B(4, 0.5)$.

width, x_a to x_b, the histogram bars become infinitely narrow, and the tops of the "bars" become an unbroken line, the density function of the distribution.

1. The effects of sample size on the histogram for the binomial distribution

Instead of contemplating the daunting problem of modeling the number of heads in 1 million tosses of a coin, we shall begin with a more modest task. Let X be the number of heads in 4 tosses of a coin, and let us suppose that we are interested in the probability of obtaining 2 to 4 heads, inclusive. From our discussion of basic probability, we know that

$$P(2 \le x \le 4) = P(x = 2) + P(x = 3) + P(x = 4)$$

If we represent each integer value x of the random variable as the interval $x - .5$ to $x + .5$ and we represent the probability associated with x by the area of a bar constructed over its interval, then $P(2 \le x \le 4)$ is represented by the shaded area in Figure 7.9, that is, the area between 1.5 and 4.5.

Now let us suppose we are still interested in the probability of obtaining 2 to 4 heads but that we increase the number of tosses from $N = 4$ to $N = 16$. The probability histogram for $X : B(16, 0.5)$ is shown in Figure 7.10, and $P(2 \le x \le 4)$ is the area from 1.5 to 4.5. The histogram illustrated in Figure 7.10 is much flatter than that of $X : B(4, 0.5)$. This is because the *total* area enclosed by the bars must be equal to 1, and this area is spread over a wider range of x-values.

This increase in spread is reflected in the variance. From [**6.12**] we know that for any binomial random variable, $V(X) = Npq$. All other things (viz., the value of p) being equal, the variance of a binomial

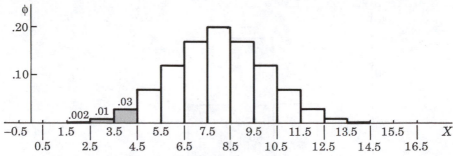

Figure 7.10. Probability histogram for $X : B(16, 0.5)$.

random variable therefore increases as N increases. For $X : B(4, 0.5)$, variance equals $(4)(.5)(.5) = 1$. For $X : B(16, 0.5)$, variance equals $(16)(.5)(.5) = 4$.

Insofar as we might like to represent $P(2 \leq x \leq 4)$ as the area between $x = 1.5$ and $x = 4.5$, it is seen that the effects of increasing sample size have an unfortunate consequence: The bar associated with $x = 2$ is shorter than the thickness of the horizontal axis. If we were to increase the sample size to 36, as illustrated in Figure 7.11, we would find that the bars associated with all three of our values, 2, 3, and 4, are essentially invisible.

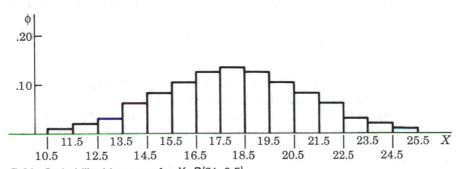

Figure 7.11. Probability histogram for $X : B(36, 0.5)$.

And for $X : B(1,000,000, 0.5)$, *none* of the bars of the histogram would be visible.

If the key to solving our coin-toss problem hinges on the conception of probability as "area" under the histogram, we obviously need some approach that prevents the histogram from disappearing into the X-axis. The tactic turns out to be *standardization*.

2. THE PROBABILITY HISTOGRAM FOR THE STANDARDIZED BINOMIAL RANDOM VARIABLE

We know from expression **[6.15]** that the standardized binomial random variable is

$$Z = \frac{X - Np}{\sqrt{Npq}}$$

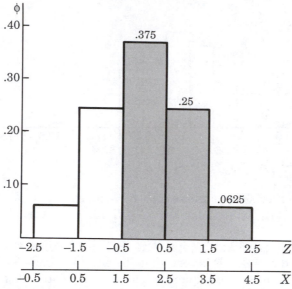

Figure 7.12. Probability histogram for $Z = (X - Np) / \sqrt{Npq}$ where $X : B$ (4, 0.5).

If $X : B(4, 0.5)$, then $Np = (4)(.5) = 2$, and $Npq = (4)(.5)(.5) = 1$, so

$$Z = \frac{X - 2}{1}$$

The probability histogram for Z is given in Figure 7.12. Figure 7.12 looks exactly the same as the histogram for X illustrated in Figure 7.9, but it is centered at zero. As before, $P(2 \leq x \leq 4)$ is represented as the area between $x = 1.5$ and $x = 4.5$. In terms of Z, this translates to the area between

$$z = \frac{1.5 - 2}{1} = -.5$$

and

$$z = \frac{4.5 - 2}{1} = 2.5$$

When we increase the number of tosses to 16, the expected value becomes $(16)(.5) = 8$, and the variance becomes $Npq = (16)(.5)(.5) = 4$, so

$$Z = \frac{X - 8}{2}$$

The histogram for Z when $X : B(16, 0.5)$ is given in Figure 7.13.

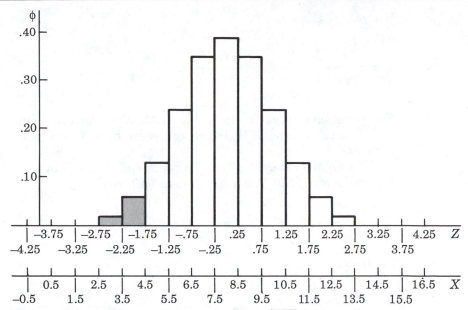

Figure 7.13. Probability histogram for $Z = (X - Np) / \sqrt{Npq}$ where $X : B(16, 0.5)$.

If we compare the histograms for Z when $N = 4$ and when $N = 16$, we see that increasing the number of observations has a very different effect than it had on the corresponding histograms for X. First, the histogram of Z does not seem to spread out as much. The range of z-values over which we constructed bars for $N = 4$ was -2.5 to $+2.5$. For $N = 16$, the range of values for which the bars are perceptibly taller than the thickness of the horizontal axis is -2.75 to $+2.75$. This is because the variance of *any* standardized random variable is equal to 1. As N gets larger, the variance of X increases, but variance of Z does not.

Second, we note that the histogram of Z does not flatten out. This, too, is because the variance of Z is unaffected by sample size. As we've already seen, the width of the histogram of Z increases slowly as N increases. Since the total area of the histogram is always equal to 1, the height of the histogram must likewise change very little. We see, for example, that the height of the bar erected over the modal interval in Figure 7.12 ($N = 4$) and the height of the modal bar in Figure 7.13 ($N = 16$) are both just under .40.

Third, we notice that the histogram bars in Figure 7.13 are narrower than the bars in Figure 7.12. For $N = 16$ we have 17 bars (corresponding to x-values $0, 1, 2, \ldots, 16$) that occupy much the same span of z-values as the 5 bars (corresponding to x-values 0, 1, 2, 3, and 4) in the histogram of Z for $N = 4$. As N increases, the width of each bar must therefore decrease.

To understand how this happens, recall that if X is a binomial random variable, then any x-value in standard units is

$$z = \frac{x - Np}{\sqrt{Npq}}$$

For $p = .5$, then,

$$z = \frac{x - N(.5)}{\sqrt{N(.5)(.5)}} = \frac{x - N(.5)}{.5\sqrt{N}}$$

For any given interval x_a to x_b, the width of the corresponding standardized interval z_a to z_b therefore decreases by a factor of $.5\sqrt{N}$ as N increases.

Consider the interval 2.5 to 3.5 representing $x = 3$ in the probability histogram for $X: B(N, 0.5)$. In Figure 7.12, $N = 4$, so

$$\mu = Np = (4)(.5) = 2 \quad \text{and} \quad \sigma = \sqrt{Npq} = \sqrt{(4)(.5)(.5)} = 1.0.$$

Therefore,

$$z_a = \frac{2.5 - 2}{1} = .5 \quad \text{and} \quad z_b = \frac{3.5 - 2}{1} = 1.5$$

and the width of the bar is $1.5 - .5 = 1.0$.

If $N = 16$, as in Figure 7.13, then

$$\mu = Np = (16)(.5) = 8 \quad \text{and} \quad \sigma = \sqrt{Npq} = \sqrt{(16)(.5)(.5)} = 2.0.$$

Therefore,

$$z_a = \frac{2.5 - 8}{2} = -2.75 \quad \text{and} \quad z_b = \frac{3.5 - 8}{2} = -2.25$$

so the width of the bar is $-2.25 - (-2.75) = .50$, half as wide as when $N = 4$.

For $N = 36$ we have

$$\mu = Np = (36)(.5) = 18 \quad \text{and} \quad \sigma = \sqrt{(36)(.5)(.5)} = 3,$$

so the bar representing $x = 3$ extends from

$$z_a = \frac{2.5 - 18}{3} \doteq -5.17 \quad \text{to} \quad z_b = \frac{3.5 - 18}{3} \doteq -4.83$$

as is illustrated in Figure 7.14. The width of the bar is therefore $-5.17 - (-4.83) \doteq .333$, so it is now only one-third as wide as when $N = 4$.

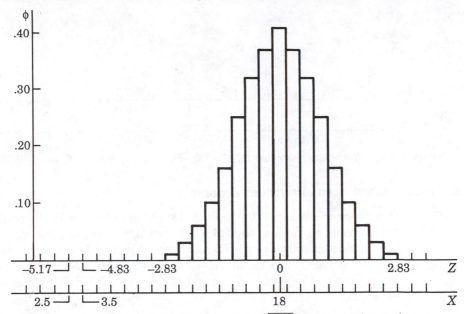

Figure 7.14. Probability histogram for $Z = (X - Np) / \sqrt{Npq}$ where $X : B(36, 0.5)$.

For $X : B(1,000,000, 0.5)$, each bar has a width of only .002, so the bars are narrower than the thickness of the lines used to represent them. There are so many bars jammed together that their tops appear to form a smooth curve instead of the familiar stairsteps. Figure 7.15 shows the histogram for Z when $X : B(1,000,000, 0.5)$. Recall that our inquiry began with the question of calculating $P(499,200 \leq x \leq 500,500)$. In Figure 7.15 this probability is the shaded area between $x = 499,199.5$ and $x = 500,500.5$. For $X : B(1,000,000, 0.5)$, we have

$$\mu = Np = (1,000,000)(.5) = 500,000$$

and

$$\sigma = \sqrt{Npq} = \sqrt{(1,000,000)(.5)(.5)} = 500$$

so the corresponding z-values are

$$\frac{499,199.5 - 500,000}{500} \doteq -1.60$$

and

$$\frac{500,500.5 - 500,000}{500} \doteq 1.00$$

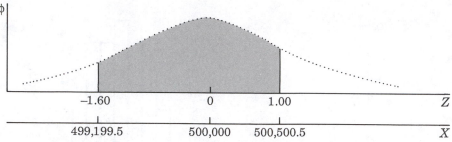

Figure 7.15. Probability histogram for $Z = (X - Np)/\sqrt{Npq}$ where $X : B(1,000,000, 0.5)$.

3. The standard normal random variable

Beginning in about 1730 Abraham De Moivre (1667–1754) wrestled with the problem of finding the function that approximates the standardized binomial histogram for $p = .5$ and large N.[2] The function he discovered in 1733 is called the *standard normal probability density function*—or, more simply, the standard normal curve—and is given by the following expression:

$$\phi(z) = \frac{1}{\sqrt{2\pi}} e^{-z^2/2} \qquad\qquad [7.1]$$

where $\pi \doteq 3.141592654\ldots$ and should be familiar to most readers from plane geometry as the ratio of the circumference of a circle to its diameter, and $e \doteq 2.718281828459045\ldots$ is another important constant in mathematics, the base of the natural logarithm system.

If ϕ is the density function of a random variable X, we know that $P(x_a \le x \le x_b)$ is the area under the graph of ϕ between x_a and x_b. If the graph of ϕ is a histogram, area can be calculated by addition, but if the graph of ϕ is a curve, area is calculated by a technique called *integration*, which is studied in calculus. Areas under the graph of the standard normal curve appear in Table 3 of Appendix VIII, so one does not need to use calculus to use the standard normal density function as a probability model. However, a number of important concepts that are defined in terms of summation for discrete random variables are defined in terms of integration for continuous random variables. If one is at least familiar with calculus *notation*, it is easily seen that the definitions are analogous. Integral notation is introduced in Box 7.1.

We stated earlier that ϕ is a density function if the area under ϕ from x_a to x_b is equal to $P(x_a \le x \le x_b)$. Our introduction to integral notation in Box 7.1 allows us to formalize this definition: $\phi(x)$ is said to be the probability density function for the random variable X if

$$\int_{x_a}^{x_b} \phi(x)\, dx = P(x_a \le x \le x_b)$$

[2] In 1738 De Moivre's own translation of his 1733 work appeared. It begins: "Altho' the solution of problems of chance often require that several terms of the binomial $(a + b)^N$ be added together, nevertheless in very high powers the thing appears so laborious, and of so great a difficulty, that few people have undertaken the task; for besides James [sic] and Nicolas Bernoulli, two great mathematicians, I know of no body that has attempted it ... ."

| BOX 7.1 |

Areas under density functions as integrals

Figure 7.1.1 shows the graph of a density function ϕ and a shaded area under the graph between vertical lines at x_a and x_b for $x_a < x_b$.

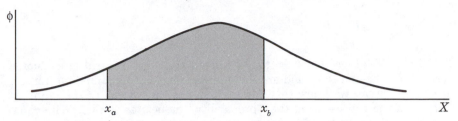

Figure 7.1.1. Area under the graph of a density function ϕ between x_a and x_b.

We can think of this area as the *sum* of the areas of infinitely many infinitely narrow histogram bars. An "infinitely narrow" bar has zero area, but integral calculus makes sense of the notion of summing infinitely many zeros. The result is not really a sum, so we don't denote area with a Σ; but to remind us that the area problem originates as a sum, we use $\int$, an elongated S. This symbol is called an *integral sign*.

With Σ-notation we often denote the limits of summation. We know from Chapter 2, for example, that

$$\sum_{i=a}^{b}$$

indicates that the sum begins with x_a and ends with x_b. With an integral sign, we also indicate the lower and upper limits, for example

$$\int_{x_a}^{x_b}$$

This is read, "the integral from x_a to x_b" and indicates that the area is between vertical lines at x_a and x_b.

For any x-value, the height of the graph in Figure 7.1.1 is $\phi(x)$. The area of a rectangle is its base times its height, so we need a base to go along with the height $\phi(x)$. Mathematicians use dx to denote this infinitely thin base. (Please note that dx is to be regarded as a single symbol, *not* as d times x.) Therefore, the shaded area in Figure 7.1.1, representing $P(x_a \leq x \leq x_b)$, is denoted

$$\int_{x_b}^{x_b}\phi(x)\,dx$$

which is read, "the integral from x_a to x_b of ϕ of $x\,dx$."

The area of interest is not always *between* two values. In Figure 7.1.1 the area under ϕ *below* x_b is $\mathbb{P}(x_b)$, the cumulative probability of x_b. The area for all real number values less than x_b is indicated with the notation

$$\int_{-\infty}^{x_b}\phi(x)\,dx$$

which is read, "the integral from minus infinity to x_b of ϕ of $x\,dx$."

BOX 7.2

Expectation and variance of a continuous random variable

The expected value of a continuous random variable with density function ϕ is

$$E(X) = \int_{-\infty}^{+\infty} x\phi(x)\,dx$$

The limits of integration ($-\infty$ to $+\infty$) mean that the integral is evaluated over all possible x-values and are therefore analogous to summation limits 1 to n. In addition, we know that $\phi(x)\,dx$ denotes the "area" of the infinitely thin rectangle representing the "probability" associated with x. The equation given above is therefore analogous to equation [**6.3**], which defines the expected value of a discrete random variable:

$$E(X) = \sum_{i=1}^{n} x_i P(x_i)$$

Similarly, the variance of a continuous random variable with density function ϕ is

$$V(X) = \int_{-\infty}^{+\infty} (x - \mu)^2 \phi(X)\,dx$$

which is analogous to the expression given in [**6.4**] for the variance of a discrete random variable:

$$V(X) = \sum_{i=1}^{n} (x_i - \mu)^2 P(x_i)$$

a. Expectation and variance of the standard normal random variable. In Chapter 6 we fully developed the notions of the expected value and variance for discrete random variables. As you might expect, expected value and variance are also defined for continuous random variables. The definitions (which are given in Box 7.2) are entirely analogous to their discrete counterparts, essentially by replacing summation notation (Σ) with integral notation ($\int$).

Although one needs calculus to compute expected value and variance of a continuous random variable, they still capture the same properties as in the discrete case. The expected value falls in the "middle" of the distribution in the sense that it is the "center of mass." Likewise, the variance of a continuous distribution expresses the degree of dispersion or spread.

Since the standard normal distribution is obtained from the standardized binomial distribution by letting N increase, it should not be surprising that the expected value of the standard normal distribution is 0 and the variance of the standard normal distribution is 1.

b. Properties of the standard normal density function. As we indicated earlier, Table 3 in Appendix VIII makes it possible to look up the probability associated with any interval of standard normal values, for example, the interval -1.60 to 1.00 in Figure 7.15. But use of the table is easier to understand if the reader is familiar with certain properties of the standard normal curve. These can be deduced fairly easily from formula **[7.1]**.

First, the area under the whole curve represents the probability that z takes on *some* value. This is certain, so the area under the entire curve is 1.

Second, the curve is symmetrical about zero: For any particular z-value, $\phi(z) = \phi(-z)$. For example, $\phi(1.2) = \phi(-1.2)$, $\phi(-.0683) = \phi(.0683)$, etc. This is because z is first *squared* before any of the other operations in expression **[7.1]** are performed. We know that $z^2 = (-z)^2$, so any operations on z^2 must give the same value as the operations performed on $(-z)^2$.

We said in Chapter 2 that a curve is symmetrical if the portion to the right of the mean is the mirror image of the portion to the left. Since the mean of any standardized variable is zero, half of its area must therefore lie above (i.e., to the right of) $z = 0$ and half must lie below (i.e., to the left of) $z = 0$. That is,

$$P(z \le 0) = P(z \ge 0) = .5$$

Third, the curve has a maximum value at $z = 0$. That is, the *mode* of the standard normal random variable occurs at $z = 0$. This is easily seen if **[7.1]** is rewritten in the form

$$\phi(z) = \frac{1}{\sqrt{2\pi}} \cdot \frac{1}{e^{z^2/2}} \qquad\qquad \textbf{[7.2]}$$

Because z^2 appears in the denominator, $\phi(z)$ must *increase* as z^2 *decreases*. However, z^2 can never be negative, so the denominator decreases to its smallest possible value—and $\phi(z)$ reaches its largest value—when $z = 0$.

The value of $\phi(z)$ for $z = 0$ is easily calculated. When z is set equal to zero,

$$e^{z^2/2} = e^0 = 1$$

Therefore,

$$\phi(0) = \frac{1}{\sqrt{2\pi}} \left(\frac{1}{1}\right) = \frac{1}{\sqrt{2\pi}} \doteq .39894228$$

$$\cong .4$$

as we discovered for the binomial example, which the normal approximates.

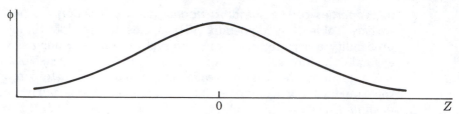

Figure 7.16. The standard normal curve.

Fourth, $\phi(z)$ is said to be *asymptotic* to zero. We already know that $\phi(z)$ increases as z^2 decreases. Conversely, we can see from equation **[7.2]** that $\phi(z)$ decreases as z^2 increases. However, we can also see from equation **[7.2]** that $\phi(z)$ is always positive. As z departs from zero, the graph of $\phi(z)$ therefore gets closer to the horizontal axis ($\phi = 0$) but never quite reaches it. This is what mathematicians mean by the expression "asymptotic to zero."

These properties give the standard normal curve its distinctive symmetrical, bell-shaped form (see Figure 7.16): The standard normal curve lies entirely above the horizontal axis; it has a single mode, which occurs at $z = 0$; it is symmetrical about its mode (which is also equal to its expected value); as z moves away from zero in either the positive or negative direction, the tails of the curve, "flatten out," nearly touching the horizontal axis.

4. The table of cumulative probabilities for the standard normal curve

In our discussion of the continuous uniform distribution we indicated that for any continuous random variable X with density function ϕ, the cumulative probability for the value x_i is the area that lies to the left of x_i under the graph of ϕ. This convention is central to understanding how to use Table 3 in Appendix VIII to calculate probabilities for the standard normal random variable.

The entries in Tables 3 and 3A of Appendix VIII are *cumulative* probabilities for values of the standard normal random variable expressed to two decimal places. Table 3 includes all such z-values from -3.09 to $+3.09$, and Table 3A includes selected values of $z > 3.09$. To use Table 3, find the row that corresponds to your z-value expressed to the *first* decimal place. For example, to find the cumulative probability of $z = 1.25$, we first find 1.2 in the first column (the column headed z). This number identifies the row that includes the cumulative probabilities for the z-values $1.20, 1.21, \ldots, 1.29$.

The second step is to find the column that gives the number in the *second* decimal place of our z-value. For $z = 1.25$, the column is .05 (because $1.25 = 1.2 + .05$). The entry at the intersection of this row and column is the cumulative probability of $z = 1.25$ and is equal to .8944. That is, according to Table 3,

$$\mathbb{P}(1.25) = .8944$$

or

$$P(z \leq 1.25) = .8944$$

or, since the probability of any *one* value of a continuous random variable is zero,

$$P(z < 1.25) = .8944$$

This value can be interpreted as the *area* under the standard normal curve to the left of 1.25.

Example 7.3.1. If Z is the standard normal random variable, what is the cumulative probability of $z = 1.00$? The cumulative probability of 1.00 is obtained by entering Table 3 at the row labeled 1.0 and the column headed .00. This is found to be $\mathbb{P}(1.00) = .8413$.

Example 7.3.2. What is $P(z \leq 1.07)$ if Z is the standard normal random variable? We know that $P(z \leq 1.07) = \mathbb{P}(1.07)$ so we once again enter Table 3 at row 1.00, but this time we move across the page to the column headed .07, and we find that $P(z \leq 1.07) = .8577$.

Example 7.3.3. Use Table 3 to find $\mathbb{P}(2.45)$. The cumulative probability for $z = 2.45$ is opposite 2.4 and under .05. Here we find the entry $.9^2 2857$. What does the 9^2 mean? It is space-saving shorthand for 99. The cumulative probability associated with 2.45 is therefore .992857.

Example 7.3.4. If Z is the standard normal random variable, what is the probability that z is less than -1.87? For a continuous random variable, $P(z < -1.87)$ is equal to $P(z \leq -1.87)$, so this is simply another way of asking about the cumulative probability of -1.87. The value .03074 is found at the intersection of the row labeled -1.8 and the column headed .07, so $P(z < -1.87) = .03074$.

Example 7.3.5. What is the cumulative probability of $z = 0.00$? The entry lying at row .0 and column .00 is .5000. Since we have already established that half the area under the normal curve lies above $z = 0$ and half the area lies below $z = 0$, this should come as no surprise.

If you dutifully worked all of these examples (instead of just reading and taking things on faith), use of Table 3 probably became rather routine. Nevertheless, the examples serve to emphasize a number of points.

- The entries in Table 3 (and, indeed, all of the subsequent tables in Appendix VIII) are *cumulative* probabilities.
- Graphically, the entry for any value z_i in Table 3 is the area under the standard normal curve to the left of z_i.

- All of the tabled values except .5000 are approximations.[3] However, they are expressed to at least four decimal places, and that, as the saying goes, is close enough for government work.
- In Example 7.3.4 we saw that the cumulative probability of $z = -1.87$ was just a little greater than .03, and in Example 7.3.3 we saw that the cumulative probability of 2.45 was .99-plus. Even though the standard normal random variable can assume any real value, the curve drops very quickly as z departs from zero. The total area under the curve (from $-\infty$ to $+\infty$) is, of course, 1, but more than 99.7 percent of this lies between -3.00 and $+3.00$. This is why we can get away with tabling cumulative probability values for such a small interval in the domain of all possible z-values.
- The standard normal random variable is continuous, so for any particular value z_i it is always the case that $P(z \le z_i) = P(z < z_i)$.

Thus far, we have used Table 3 only to find the probability that z is *less than* some particular value z_i. As indicated earlier, cumulative probabilities can also be used to find the probability that z lies *between* two values, z_a and z_b. If $z_a < z_b$, then

$$P(z_a \le z \le z_b) = \mathbb{P}(z_b) - \mathbb{P}(z_a)$$

Example 7.4.1. Use Table 3 to find $P(-1.87 \le z \le 1.07)$. The basic method for using cumulative probabilities to determine the probability associated with an interval, z_a to z_b, is best understood when illustrated in terms of area. In Figure 7.17(a), the shaded area under the curve to the left of 1.07 represents $\mathbb{P}(1.07)$; similarly, the shaded area under the curve to the left of -1.87 in Figure 7.17(b) represents $\mathbb{P}(-1.87)$. If the area in Figure 7.18(b) is subtracted from the area in Figure 7.17(a), the difference is the shaded band of area in Figure 7.17(c) *between* -1.87 and 1.07. We know from earlier discussion that this area represents

$$P(-1.87 \le z \le 1.07)$$

so it follows that

$$P(-1.87 \le z \le 1.07) = \mathbb{P}(1.07) - \mathbb{P}(-1.87)$$

From Example 7.3.2 we know that $\mathbb{P}(1.07) \doteq .8577$, and in Example 7.3.4 we found that $\mathbb{P}(-1.87) \doteq .03074$. Therefore,

$$P(-1.87 \le z \le 1.07) \doteq .8577 - .03074 = .82696$$

Example 7.4.2. If Z is the standard normal random variable, what is the probability that z lies between the values 1.5 and 1.64? We use Table 3 just as we did in Examples 7.3 to locate the entries corresponding to

---

[3] Most cumulative probabilities in the remainder of the text are therefore written with the symbol $\doteq$ instead of $=$.

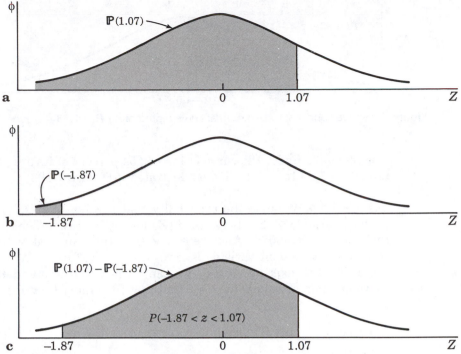

Figure 7.17. (a) Cumulative probability of $z = 1.07$, (b) cumulative probability of $z = -1.87$, (c) $P(-1.87 \leq z \leq 1.07)$

$z = 1.50$ and $z = 1.64$, and we find that $\mathbb{P}(1.50) \doteq .93319$ and $\mathbb{P}(1.64) \doteq .94950$. Therefore,

$$P(1.50 \leq z \leq 1.64) \doteq .94950 - .93319 = .01631$$

This probability corresponds to the narrow shaded strip in Figure 7.18.

Example 7.4.3. What percentage of the area under the standard normal density function lies between $z = -1.93$ and $z = -.12$? From Table 3 we find that $\mathbb{P}(-1.93) \doteq .02680$ and $\mathbb{P}(-.12) \doteq .4522$. Therefore, the area under $\phi(z)$ between -1.93 and $-.12$ is about

$$.4522 - .02680 = .4254$$

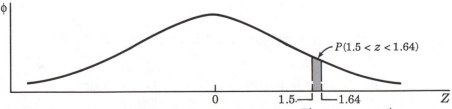

Figure 7.18. Area under standard normal curve representing $P(1.5 \leq z \leq 1.64)$

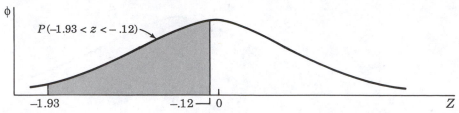

Figure 7.19. Area under standard normal curve representing $P(-1.93 \leq z \leq -.12)$.

The total area under the curve is 1, so 42.54 percent of its area lies in the interval -1.93 to $-.12$. This is illustrated in Figure 7.19.

Example 7.4.4. What is the probability that a value of standard normal random variable Z lies between $E(Z)$ and 1.76? From Table 3 we find that $\mathbb{P}(1.76) \doteq .96080$. And we know that the expected value of any standardized random variable is zero, that is, $E(Z) = 0$. We can either use Table 3 to look up $\mathbb{P}(0)$ or we can recall that the standard normal curve is symmetrical about $z = 0$ and that $\mathbb{P}(0)$ must therefore equal .5, so

$$P(0 \leq z \leq 1.76) \doteq .96080 - .5 = .4608$$

The table of cumulative probabilities for the standard normal curve can also be used to obtain the probability that z is larger than any particular value z_i. That is, we can find the probability that z lies in the "upper tail" of the standard normal curve.

Example 7.5. If Z is the standard normal random variable, what is the probability that z is greater than or equal to 1.07? This problem can be solved in two ways. Since we have tabled both positive and negative z-values and since the standard normal curve is symmetrical, it is seen that $P(z \geq 1.07) = P(z \leq -1.07)$. So we might simply look up the cumulative probability of -1.07 in Table 3 and find that $\mathbb{P}(-1.07) \doteq .1423$.

If you find yourself using some other statistics book, however, you may discover that the author has tabled only positive z-values. This presents no real difficulty, but the explanation is easier if you turn back to Figure 7.17 (p. 247). The shaded area in Figure 7.17(a) represents the cumulative probability of $z = 1.07$, that is, $P(z \leq 1.07)$. Since the standard normal random variable is continuous, the unshaded area is $P(z \geq 1.07)$. The total area under the curve is 1, so

$$P(z \geq 1.07) = 1 - \mathbb{P}(1.07)$$

In Table 3 (or from Example 7.3.2) we find that $\mathbb{P}(1.07) \doteq .8577$. Therefore,

$$P(z \geq 1.07) = 1 - \mathbb{P}(1.07) \doteq 1 - .8577 = .1423$$

And in general, $P(z \geq z_i) = 1 - \mathbb{P}(z_i)$.

Many of the problems involving use of probability density functions require us to work "backward," that is, to begin with a cumulative probability and to find the corresponding value of the random variable. By now we have presented enough examples so that you probably realize that the cumulative probabilities in Table 3 get larger (i.e., closer to 1) as you work down the table and as you move across the table—from left to right for positive z-values and from right to left for negative z-values.

The strategy for finding a z-value that corresponds to a given cumulative probability $\mathbb{P}$ is therefore very straightforward: Begin at the top of the column headed .00 and move down the page until you find the two consecutive values that span $\mathbb{P}$. Enter the table in the row that is *closer* to $z = 0.00$ and move across the page until you find either the exact value of $\mathbb{P}$ (which is very unlikely) or until you once again find two consecutive values that span $\mathbb{P}$. In the latter case, choose the tabled value that is closer to $\mathbb{P}$ and read the corresponding z-value from the row and column headings.[4]

Example 7.6.1. What z-value in Table 3 corresponds to a cumulative probability of .97381? Beginning at the top of the column headed .00 and moving down, we eventually find that $\mathbb{P} = .97128$ is the first entry in row 1.9 and that $\mathbb{P} = .97725$ is the first entry in row 2.0. Therefore, we know that the $\mathbb{P}$-value we want is in row 1.9. Moving across row 1.9 we find that $\mathbb{P} = .97381$ lies in the column headed .04. Therefore, the z-value for which $\mathbb{P} = .97381$ is 1.94.

Example 7.6.2. Find z such that $\mathbb{P}(z) = .95$. Using the same strategy we first find that the row we want is 1.6. Moving across the row, however, we do not find $\mathbb{P} = .95$. We find, instead, that the entry in column .04 is $\mathbb{P} = .94950$ and that the entry in column .05 is .95053. Since .94950 is closer to $\mathbb{P} = .95$ than is .95053, the desired z-value is 1.64.

Example 7.6.3. Find z_i such that $P(z \geq z_i) = .005$. This is another form of the "upper tail" problem. Therefore, we can either find $-z_i$ such that $\mathbb{P}(-z_i) = .005$ or we can find z_i such that $\mathbb{P}(z_i) = 1 - .005$.

If we employ the first strategy, we find that the $\mathbb{P}$-value in the .00 column of row -2.6 is $.0^24661$, or .004661, and we find that the $\mathbb{P}$-value entered in the first column of row -2.5 is $.0^26210$, or .006210. We therefore know that $-z_i$ is -2.5-something. Moving across the row, we find that $\mathbb{P}(-2.57) \doteq .005085$ and that $\mathbb{P}(-2.58) \doteq .004940$. The cumulative probability of $z = -2.58$ is closer to .005 than is the cumulative probability of $z = -2.57$, so $-z_i = -2.58$ and $z_i = 2.58$.

The second strategy calls for us to find the z-value for which $\mathbb{P} = 1 - .005 = .995$. Moving down column .00 we find that $\mathbb{P}(2.5)$ is .993790 (which is smaller than .995) and that $\mathbb{P}(2.6)$ is .995339 (which is

---

[4] Some books tell you to take the larger of the two z-values. Some books suggest you interpolate (see Appendix I). This seems a little finicky to us, but if your $\mathbb{P}$-value falls "exactly" halfway between two tabled values, and you fancy yourself a person who is devoted to precision, feel free to average the two z-values. We don't mind.

larger than .995). Moving across row 2.5, we find that $\mathbb{P}(2.57) \doteq .994915$ and that $\mathbb{P}(2.58) \doteq .995060$. This approach therefore gives us the same answer as before, $z_i = 2.58$.

──────── EXERCISES 7.2 ────────────────────────────────

In Exercises 1 and 2 use Table 3 (or 3A) of Appendix VIII to find the indicated cumulative probability for the standard normal random variable.

1. (a) $\mathbb{P}(1)$ (b) $\mathbb{P}(-2)$ (c) $\mathbb{P}(1.5)$
      (d) $\mathbb{P}(-.23)$ (e) $\mathbb{P}(3.50)$ (f) $\mathbb{P}(-4)$

2. (a) $\mathbb{P}(2)$ (b) $\mathbb{P}(-1)$ (c) $\mathbb{P}(-1.5)$
      (d) $\mathbb{P}(.45)$ (e) $\mathbb{P}(3.2)$ (f) $\mathbb{P}(-3.5)$

In Exercises 3–12 the random variable Z is the standard normal random variable.

3. Find $P(z_a \leq z < z_b)$ where

   (a) $z_a = -1$ and $z_b = 1$ (b) $z_a = -1.64$ and $z_b = 1.64$
   (c) $z_a = 1.35$ and $z_b = 2.55$ (d) $z_a = -3.2$ and $z_b = -1.5$
   (e) $z_a = -.86$ and $z_b = 1.41$ (f) $z_a = 0$ and $z_b = 1.76$

4. Find $P(z_a \leq z < z_b)$ where

   (a) $z_a = -2$ and $z_b = 2$ (b) $z_a = -1.96$ and $z_b = 1.96$
   (c) $z_a = 1.05$ and $z_b = 1.82$ (d) $z_a = -1.45$ and $z_b = -.16$
   (e) $z_a = -1.23$ and $z_b = .67$ (f) $z_a = -3.9$ and $z_b = 3.9$

5. Use Table 3 (or 3A) of Appendix VIII to find the following:

   (a) $P(z \leq -.83)$ (b) $P(z \geq 1.64)$
   (c) $P(z \geq 1.96)$ (d) $1 - \mathbb{P}(-1.96)$
   (e) $P(z < -1.0)$ (f) $P(z > 2.33)$
   (g) $1 - \mathbb{P}(2.33)$ (h) $P(z \geq 2 \text{ or } z \leq -2)$

6. Use Table 3 (or 3A) of Appendix VIII to find the following:

   (a) $P(z > 2.0)$ (b) $P(z > -.54)$
   (c) $P(z > 0.00)$ (d) $P(z \leq 1.64)$
   (e) $1 - \mathbb{P}(-3.02)$ (f) $P(z > -3.02)$
   (g) $1 - \mathbb{P}(1.0)$ (h) $P(z < -1.64)$
   (i) $P(z \geq 1 \text{ or } z \leq -2)$

7. Find the value of z for which $\mathbb{P}(z)$ is approximately

   (a) .07353   (b) .995975   (c) .004661   (d) .4761
   (e) .5040   (f) .9997   (g) .00005   (h) .001114

8. Find the value of z for which $\mathbb{P}(z)$ is approximately

   (a) .91149   (b) .7019   (c) .3409   (d) .999
   (e) .93319   (f) .97725   (g) .009903   (h) .5000

9. Find the value of z in Table 3 or 3A for which $\mathbb{P}(z)$ is *most nearly* equal to

(a) .001   (b) .005   (c) .025
(d) .9994   (e) .90   (f) .975

10. Find the value of z in Table 3 or 3A for which $\mathbb{P}(z)$ is *most nearly* equal to

(a) .999   (b) .05   (c) .98
(d) .0005   (e) .99   (f) .9975

11. Find z_i such that $P(z \geq z_i)$ is approximately

(a) .50   (b) .05   (c) .02   (d) .005   (e) .0005

12. Find z_i such that $P(z \geq z_i)$ is approximately

(a) .025   (b) .01   (c) .0025   (d) .001   (e) .00125

13. What percentage of the area under the standard normal density function lies between $z = 0$ and $z = .5$?

14. What percentage of the area under the standard normal density function lies between $z = .5$ and $z = 1.0$?

15. What percentage of the area under the standard normal density function lies within one standard deviation of $E(Z)$? Hint: What is the numerical value of the standard deviation of the standard normal? Now rewrite this exercise in the form $(-z_i \leq z \leq z_i)$ for the appropriate z_i.

16. What percentage of the area under the standard normal density function lies within two standard deviations of $E(Z)$? Hint: See Exercise 7.2.15.

17. Find an interval centered at 0 over which the area under the standard normal density function is approximately 0.5. Express this answer in terms of probability.

18. Find an interval centered at 0 over which the area under the standard normal density function is approximately 0.9. Express this answer in terms of probability.

19. Find two different intervals over which the area under the standard normal density function is approximately 0.8.

20. Find two different intervals over which the area under the standard normal density function is approximately 0.2.

5. The normal approximation to the binomial

We introduced our discussion of the normal distribution with a deceptively innocent question: If a fair coin is flipped 1,000,000 times, what are the chances that it will come up heads between 499,200 and 500,500 times (inclusive)? That is, what is

$$P(499{,}200 \leq x \leq 500{,}500)$$

for $X : B(1{,}000{,}000, 0.5)$? We learned that the methods developed in Chapter 6 for calculating binomial probabilities are not easily applied when N is very large. However, we have also discovered that when N is large, the *standardized* binomial distribution approaches the standard normal distribution. Accordingly, the standard normal distribution can be used to *approximate* binomial probabilities.

If the probability distribution for $X: B(1{,}000{,}000, 0.5)$ were represented as a histogram, $P(499{,}200 \leq x \leq 500{,}500)$ would be the area of the bars from $x_a - .5 = 499{,}199.5$ to $x_b + .5 = 500{,}500.5$. Earlier in the chapter we calculated the z-values corresponding to x_a and x_b,

$$z_a = \frac{499{,}199.5 - 500{,}000}{500} \doteq -1.60$$

and

$$z_b = \frac{500{,}500.5 - 500{,}000}{500} \doteq 1.00$$

The probability of obtaining between 499,200 and 500,500 heads in 1,000,000 tosses of a fair coin is therefore approximately equal to $P(-1.60 \leq z \leq 1.00)$ under the standard normal curve. From Table 3 in Appendix VIII, $\mathbb{P}(1.00) \doteq .8413$ and $\mathbb{P}(-1.60) \doteq .05480$. Therefore,

$$P(499{,}200 \leq x \leq 500{,}500) \cong .8413 - .05480 = .7865$$

Thus, there is about a 79 percent chance of getting between 499,200 and 500,500 heads in 1,000,000 flips of a fair coin. This is the shaded area illustrated in Figure 7.15.

In real-life problems (as opposed to problems contrived by textbook authors for instructional purposes) sample sizes of a million are pretty unusual. One of the attractive features of the normal approximation is that N really does not need to be very large for one to obtain a good approximation to $X: B(N, 0.5)$.

Example 7.7. Use the standard normal to approximate $P(4 \leq x \leq 7)$ for $X: B(12, 0.5)$. The histogram for this distribution is shown in Figure 7.20. We have shaded the area that represents $P(4 \leq x \leq 7)$. That area includes histogram bars that extend from $4 - .5 = 3.5$ to $7 + .5 = 7.5$. For $X: B(12, 0.5)$ we have

$$E(X) = Np = (12)(.5) = 6$$

and

$$V(X) = Npq = (12)(.5)(.5) = 3$$

So the relevant endpoints for the *standardized* binomial are

$$\frac{3.5 - 6}{\sqrt{3}} \doteq -1.44 \quad \text{and} \quad \frac{7.5 - 6}{\sqrt{3}} \doteq 0.87$$

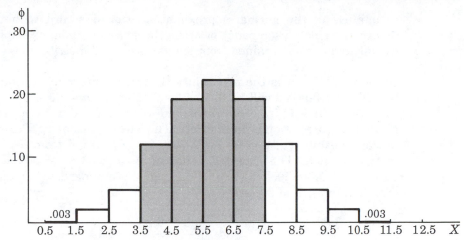

Figure 7.20. Probability histogram for $X : B(12, 0.5)$.

From Table 3 of Appendix VIII we see that $\mathbb{P}(0.87) \doteq .8078$ and that $\mathbb{P}(-1.44) \doteq .07493$. Therefore the area under the standard normal curve between -1.44 and $.87$ is approximately

$$.8078 - .07493 = .73287$$

Since $N = 12$ is small, we can calculate the exact answer by hand and compare it with our approximation:

$$_{12}C_4\left(\tfrac{1}{2}\right)^4\left(\tfrac{1}{2}\right)^8 + {}_{12}C_5\left(\tfrac{1}{2}\right)^5\left(\tfrac{1}{2}\right)^7 + {}_{12}C_6\left(\tfrac{1}{2}\right)^6\left(\tfrac{1}{2}\right)^6 + {}_{12}C_7\left(\tfrac{1}{2}\right)^7\left(\tfrac{1}{2}\right)^5$$

$$= 495\left(\tfrac{1}{2}\right)^{12} + 792\left(\tfrac{1}{2}\right)^{12} + 924\left(\tfrac{1}{2}\right)^{12} + 792\left(\tfrac{1}{2}\right)^{12}$$

$$= \frac{3003}{4096} \doteq .73315$$

The error in using the normal approximation when $N = 12$ is, therefore, merely .00028.

The standard normal distribution can also be used to approximate the binomial when p is not .5. However, as p becomes either larger or smaller than .5, the standardized binomial distribution becomes increasingly skewed and therefore less and less like the standard normal. This skewness can be reduced by increasing the size of the sample. The rule of thumb says that you may use the normal approximation when $Np \geq 5$ and $Nq \geq 5.$[5] In Example 7.7 we had

$$Np = 12(.5) = Nq = 12(.5) = 6 > 5$$

[5] If $Np < 5$ or $Nq < 5$, the Poisson distribution can be used to approximate the binomial. See the note following Table 7 of Appendix VIII.

so our use of the normal approximation was in accord with the rule (hence, the small discrepancy between the exact binomial probability and the approximation obtained from standard normal distribution).

Example 7.8. What is the probability that the number of 3s, obtained in 50 rolls of a fair die will be 6, 7, 8, or 9? Here, $Np = 50(1/6) = 8\frac{1}{3}$ and $Nq = 50(5/6) = 41\frac{2}{3}$, so both Np and Nq are greater than 5 and the normal approximation is appropriate. The bars representing $P(6 \le x \le 9)$ in the probability histogram for $X: B(50, 1/6)$ extend from $6 - .5 = 5.5$ to $9 + .5 = 9.5$. The expected value of X is $50(1/6) \doteq 8.33$, and the variance of X is $50(1/6)(5/6) \doteq 6.94$, so the standardized scores corresponding to 5.5 and 9.5 are

$$\frac{5.5 - 8.33}{\sqrt{6.94}} \doteq -1.07 \quad \text{and} \quad \frac{9.5 - 8.33}{\sqrt{6.94}} \doteq 0.44$$

In Table 3 of Appendix VIII we find $\mathbb{P}(.44) \doteq .6700$ and $\mathbb{P}(-1.07) \doteq .1423$. Therefore,

$$P(6 \le x \le 9) \cong .6700 - .1423 = .5254$$

The exact value calculated to five decimal places using the binomial probability function is .54422. Even with a value of p as small as $1/6$, the normal approximation produces an error of only about .017.

Example 7.9. If one of us flips a coin 64 times, what is the probability that he will get *exactly* 32 heads? In the histogram for $X: B(64, 0.5)$, this probability is represented by the area of the bar centered at $x = 32$. This bar extends from $32 - .5 = 31.5$ to $32 + .5 = 32.5$. We have $E(X) = 64(.5) = 32$ and $V(X) = 64(.5)(.5) = 16$, so the standard scores are therefore

$$\frac{31.5 - 32}{4} = -.125 \quad \text{and} \quad \frac{32.5 - 32}{4} = .125$$

These values fall halfway between the nearest z-values shown in Table 3, so to gain a little bit more accuracy we will average $\mathbb{P}(-.12)$ and $\mathbb{P}(-.13)$ for the cumulative probability of $-.125$ and we will average $\mathbb{P}(.12)$ and $\mathbb{P}(.13)$ for the cumulative probability of $.125$. From Table 3 we find $\mathbb{P}(-.12) \doteq .4522$ and $\mathbb{P}(-.13) \doteq .4483$, so $\mathbb{P}(-.125) \cong .45025$. Likewise, $\mathbb{P}(.12) \doteq .5478$ and $\mathbb{P}(.13) \doteq .5517$, so $\mathbb{P}(.125) \cong .54975$. Therefore,

$$P(x = 32) \cong .54975 - .45025 = .0995$$

The binomial probability to five decimal places is .09934.

Example 7.10. On an automobile assembly line with 100 workers, each worker performs a task independently of the other workers, and each has a 90 percent chance of making no errors. That is, $p = .9$ that a worker will perform the task at hand correctly. What is the probability that an automobile will come off the line without a single defect? Let the random variable X denote the number of tasks performed correctly. We need $P(x = 100)$ for $X : B(100, 0.90)$.

This random variable can assume any value from 0 to 100, and since $E(X) = 100(.90) = 90$, we see that it must be very skewed. Nevertheless, $Np = 90$ and $Nq = 100(.10) = 10$, and since these are both greater than 5, the normal approximation is acceptable. In the probability histogram for $X : B(100, 0.90)$ the area representing $P(x = 100)$ is the last bar, which begins at 99.5 and extends to 100.5. We have already noted that the expected value of X is 90, and $V(X) = 100(.90)(.10) = 9$. Therefore, the standard scores we submit to the table of cumulative probabilities for the standard normal random variable are

$$\frac{99.5 - 90}{3} \doteq 3.17 \quad \text{and} \quad \frac{100.5 - 90}{3} = 3.5$$

Values larger than $z = 3.09$ are not shown in Table 3, but in Table 3A we find that $\mathbb{P}(3.5) \doteq .9998$. Because the standard curve is so "flat" in the tails, cumulative probabilities change very slowly, so z-values are given only to one decimal place. Therefore, we let $\mathbb{P}(3.17) \cong \mathbb{P}(3.2) \doteq .9993$, and for $X : B(100, 0.90)$,

$$P(x = 100) \cong .9998 - .9993 = .0005$$

Even if each worker stands a 90 percent chance of performing his or her task without error, it is almost inconceivable that an automobile will come off the line without a single assembly error.

Incidentally, it is actually easier to calculate the exact probability in the present example than to use the normal approximation:

$$P(x = 100) = {}_{100}C_{100}(.90)^{100}(.10)^0 = (1)(.90)^{100}(1)$$

$$= (.90)^{100} \doteq .000027$$

which tells us that our approximation, even with p as large as .90, was in error only by about .00047.

Example 7.11. A student is supposed to take a 100-question multiple-choice test. Each question will have five choices, and a passing grade is 60 percent or better. She decides to sleep late, so she asks her roommate, who knows nothing about the material, to take the test for her. She fails to mention that all of the questions will be in Japanese, so the roommate won't even be able to read the questions. What is the probability that the student will receive a passing grade?

The best the roommate will be able to do is take random guesses, so the probability of getting any particular answer correct is $1/5$. The probability of getting a passing grade is therefore $P(x \geq 60)$ where

$X : B(100, 0.2)$. With these parameters, $E(X) = (100)(.2) = 20$ and $V(X) = (100)(.2)(.8) = 16$. The normal approximation to $P(x \geq 60)$ is therefore the area under the standard normal curve above

$$\frac{59.5 - 20}{4} \doteq 9.88$$

There is no point in looking up that value, since even Table 3A goes only to $z = 5$. The probability of obtaining a passing score by random guessing is very nearly zero. (On the other hand, the probability that the student will be expelled from school is nearly 1.)

a. The distribution of a binomial proportion X/N. Let us suppose that X is a binomially distributed random variable, $X : B(N, p)$. We know from equation **[6.13]** that the expected value of a binomial proportional binomial proportion

$$E\left(\frac{X}{N}\right) = p$$

and from **[6.14]** that the variance of a binomial proportion

$$V\left(\frac{X}{N}\right) = \frac{pq}{N}$$

Therefore, the *standardized* binomial proportion $Z_{X/N}$ must be

$$Z_{X/N} = \frac{\dfrac{X}{N} - p}{\sqrt{\dfrac{pq}{N}}} \qquad\qquad \textbf{[7.3]}$$

It will be noted that

$$\frac{\dfrac{X}{N} - p}{\sqrt{\dfrac{pq}{N}}} = \frac{(X - Np)\dfrac{1}{N}}{(\sqrt{Npq})\dfrac{1}{N}}$$

$$= \frac{X - Np}{\sqrt{Npq}}$$

is the standardized binomial random variable. This means that for any value x of a binomial random variable, standardizing the proportion x/N will give the same value as standardizing the value x. For example, we saw earlier (Figure 7.12) that if $X: B(4, 0.5)$, then for $x = 1.5$,

$$z = \frac{1.5 - 2}{1} = -.5$$

For the proportion $(1.5)/4$

$$z_{X/N} = \frac{\dfrac{1.5}{4} - .5}{\sqrt{\dfrac{(.5)(.5)}{4}}} = -.5$$

Since the probability distribution for the standardized binomial random variable approximates the standardized normal distribution when $Np \geq 5$ and $Nq \geq 5$, the standardized proportion

$$Z_{X/N} = \frac{\dfrac{X}{N} - p}{\sqrt{\dfrac{pq}{N}}}$$

must likewise have an approximately standard normal distribution when $Np \geq 5$ and $Nq \geq 5$.[6] We can therefore use the standard normal distribution to approximate probabilities associated with binomial proportions. The procedure follows directly from the normal approximation to the binomial, so for any interval

$$\frac{x_a}{N} \quad \text{to} \quad \frac{x_b}{N}$$

we standardize the values

$$\frac{x_a - .5}{N} \quad \text{and} \quad \frac{x_b + .5}{N}$$

Example 7.12. We believe that 40 percent of the students at our school are female. There is only one walkway to our cafeteria, and everyone must pass through it at lunchtime. We observe the first 120 students to come through the doors and record the gender (M or F) of every fifth student, giving us a tally of $N = 24$. Assuming that our belief is correct, what is the probability that the proportion of females on our list will be

[6] If $Np < 5$ or $Nq < 5$, the Poisson distribution can be used to approximate the distribution of proportions. See the note following Table 7 of Appendix VIII.

equal to or greater than 50 percent? That is, if $X : B(24, 0.40)$, what is

$$P\left(\frac{x}{24} \geq .50\right)$$

If $(x/24) = .50$, then $x = (.50)(24) = 12$, so

$$z_{X/N} = \frac{\dfrac{12 - .5}{N} - p}{\sqrt{\dfrac{pq}{N}}} = \frac{\dfrac{12 - .5}{24} - .40}{\sqrt{\dfrac{(.40)(.60)}{(24)}}}$$

$$= \frac{\dfrac{11.5}{24} - .40}{\sqrt{.01}} \doteq .79$$

In Table 3 of Appendix VIII we find that

$$\mathbb{P}(.79) = P(z \leq .79) \doteq .7852$$

Therefore,

$$P\left(\frac{x}{N} \geq .5\right) \cong P(z > .79) \doteq 1 - .7852 = .2148$$

This is easily confirmed from Table 2 of Appendix VIII. We know that if $(x/24) \geq .50$, then $x \geq 12$, and in the distribution of $X : B(24, 0.40)$ we find that

$$P(x = 12) + \cdots + P(x = 24) \doteq .213$$

Example 7.13. Suppose that in Example 7.12 we tallied the genders of $N = 96$ students instead of $N = 24$. Now what is the probability that at least 50 percent of the people we observe are female? In this problem we want to find $P[(x/96) \geq .50]$ where $X : B(96, 0.40)$.

If $(x/96) = .50$, then $x = (.50)(96) = 48$, so

$$z_{X/N} = \frac{\dfrac{48 - .5}{96} - p}{\sqrt{\dfrac{pq}{N}}}$$

As before, $p = .40$, but

$$\frac{pq}{N} = \frac{(.40)(.60)}{96} = .0025$$

so

$$z_{X/N} = \frac{\dfrac{47.5}{96} - .40}{\sqrt{.0025}} \doteq 1.9$$

From Table 3 we find $\mathbb{P}(1.9) \doteq .97128$, so

$$P\left(\frac{x}{96} \geq .50\right) \cong P(z \geq 1.9) \doteq 1 - .97128 \doteq .029$$

This means that the probability of the event "obtain 48 or more females in 96 observations on a population that is 40 percent female" is only about 2.9 percent. If we actually conducted the experiment described in Example 7.13 and found 48 or more females on our list, we would probably reject our belief that 40 percent of our student population is female. Of course, we might also look for other reasons why we observed so many females. Perhaps a women's physical education class was dismissed just before noon and poured through the doors all at once.

One can also use the normal approximation to find the proportion x/N that corresponds to some specific cumulative probability or cuts off some specific percentage in the upper tail of the distribution of X/N. We will see in Chapter 9 that this procedure is very important in some types of statistical inference, but it reveals a problem we have not encountered in other examples: Since the standard normal random variable Z is continuous, there is a z-value that corresponds to any observable value of x/N. *But*, X/N is discrete, so it is not possible to find an observable value of x/N that corresponds to every value of z.

Example 7.14. If $X : B(100, 0.2)$ what proportion cuts off the top 10 percent of the distribution of X/N? What we want is the proportion x/N corresponding to z_i, where $P(z \geq z_i) = .10$, which means that $\mathbb{P}(z_i) = 1 - .10 = .90$. The nearest $\mathbb{P}$-value to .90 in Table 3 is $\mathbb{P} = .8997$, for which $z = 1.28$. To find the corresponding value of x/N, we therefore set

$$\frac{\dfrac{x}{N} - p}{\sqrt{\dfrac{pq}{N}}} = \frac{\dfrac{x}{100} - .2}{\sqrt{\dfrac{(.2)(.8)}{100}}} = 1.28$$

so

$$\frac{x}{100} = 1.28\sqrt{\frac{(.2)(.8)}{100}} + .20 = .2512$$

However, it will be noted that if $(x/100) = .2512$, then

$$x = (.2512)(100) = 25.12$$

Since X is binomially distributed it can take only integer values (0, 1, ..., 100), so the nearest observable x-value is 25. When a binomial random variable is represented in the form of a histogram, the bar representing $x = 25$ extends from 24.5 to 25.5. If we take $25/100$ as the cutoff value, then by the normal approximation our cumulative probability is $\mathbb{P}(z)$, where

$$z = \frac{\dfrac{24.5}{100} - .2}{\sqrt{\dfrac{(.2)(.8)}{100}}} \doteq 1.13$$

From Table 3 we find that $\mathbb{P}(1.13) \doteq .8708$, so the probability that x/N is equal to or greater than $25/100$ is approximately

$$1 - .8708 = .1292 > .10$$

On the other hand, if we take $26/100$ as our cutoff value, then $z \doteq 1.38$, and $\mathbb{P}(1.38) \doteq .91621$. The probability that x/N is equal to or greater than $26/100$ is therefore approximately

$$1 - .91621 = .08379 < .10$$

For $X : B(100, 0.2)$ there is *no* observable x-value that cuts off exactly the top 10 percent of the distribution of x/N.

6. The nonstandard normal random variable

In Chapter 6 we explained that the distribution of a discrete random variable is specified by its probability function and the values of its parameters. We have subsequently used expressions of the form

$$X : B(N, p)$$

as a convenient shorthand to indicate that a random variable X is binomially distributed, that the sample size is N, and that the probability of success on any trial is p.

The convention for specifying the distribution of a continuous random variable is similar, except that we identify the *density* function rather than a *probability* function. We have already discussed the density function for the standard normal random variable (expression [**7.1**]), but we have said nothing about its parameters. The parameters for the standard normal random variable Z are the expected value of Z, which is 0, and the variance of Z, which is 1. The specification statement for the standard normal random variable is therefore

$$Z : N(0, 1)$$

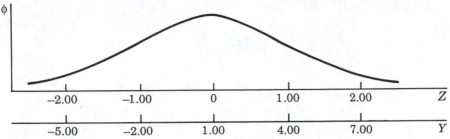

Figure 7.21. Distributions for $Z:N(0, 1)$ and $Y:N(1, 9)$.

Let us suppose that Z is the standard normal random variable, $Z: N(0, 1)$, and let us define a second random variable Y as $3Z + 1$. That is, the y-value corresponding to any particular z-value is calculated by multiplying z by 3 and adding 1:

$$y = 3z + 1$$

We know from the Algebra of Expectations that

$$E(Y) = 3\mu_Z + 1 = 3(0) + 1 = 1$$

and from the Algebra of Variances that

$$V(Y) = 3^2\sigma_Z^2 = 9(1) = 9$$

As indicated in Figure 7.21, this transformation has no effect on the shape of the distribution. Our manipulations only shift the distribution one unit to the left (so that it is centered at 1) and change the size of the units. Consequently, our new variable Y is still said to be normally distributed, but its parameter values are 1 and 9. Since $E(Y) \neq 0$ and $V(Y) \neq 1$, the variable Y is not the *standard* normal random variable. It is, rather, one of *infinitely many* normal random variables, each of which is determined by a different pair of values μ and σ^2. In the general case, the specification statement for a normally distributed random variable is therefore

$$X: N(\mu, \sigma^2)$$

where $E(X) = \mu$ and $V(X) = \sigma^2$. In our present example, then,

$$Y: N(1, 9)$$

This is why we have always referred to *the* standard normal random variable, rather than *a* standard normal random variable. The *standard* normal random variable is *the* normally distributed random variable with expectation equal to zero and variance equal to one.

In the general case, the formula for the density function of the normally distributed random variable X with expectation μ and variance

σ^2 is

$$\phi(x) = \frac{1}{\sqrt{2\pi\sigma^2}} e^{-\frac{1}{2}\left(\frac{x-\mu}{\sigma}\right)^2} \qquad\qquad [7.4]$$

Since $E(X) = \mu$ and $V(X) = \sigma^2$, we see that

$$\frac{x-\mu}{\sigma} = z$$

so expression [7.4] becomes

$$\phi(x) = \frac{1}{\sqrt{2\pi\sigma^2}} e^{-z^2/2}$$

And, if $\sigma^2 = 1$,

$$\phi(x) = \frac{1}{\sqrt{2\pi}} e^{-z^2/2}$$

which is identical to expression [7.1], the density function for the standard normal random variable.

It should therefore be no surprise that the graph of the (nonstandard) normal density function has all the properties of its standardized cousin:

- The total area under the whole graph is 1.
- The curve is symmetrical about its expected value, μ.
- The curve has a maximum value of

$$\frac{1}{\sqrt{2\pi\sigma^2}}$$

  when $x = E(X) = \mu$.
- The curve is asymptotic to zero.

Consequently, the graphs of all the nonstandard normal density functions are bell-shaped curves, similar to those illustrated throughout this chapter.

If you check Appendix VIII for a table of cumulative probabilities for any nonstandard normally distributed random variable, you will discover that none has been included. This is not because nonstandard normal random variables are unimportant; nor is it because the publisher is cheap. It is because the table for cumulative probabilities of the standard normal random variable (Table 3) can be used to calculate probabilities for *any* normally distributed random variable. The key is that if

$$X : N(\mu, \sigma^2)$$

then

$$Z = \frac{X - \mu}{\sigma}$$

is distributed as the standard normal random variable. That is,

$$Z : N(0, 1)$$

This means that $P(x_a \leq x \leq x_b)$ can always be obtained by calculating

$$z_a = \frac{x_a - \mu}{\sigma} \quad \text{and} \quad z_b = \frac{x_b - \mu}{\sigma}$$

and then using Table 3 to find $P(z_a \leq z \leq z_b)$.

Example 7.15. Suppose $X : N(485, 625)$. What is the probability that $x \leq 460$; that is, what is $\mathbb{P}(460)$? Since $E(X) = 485$ and $V(X) = 625$,

$$Z = \frac{X - 485}{\sqrt{625}}$$

and $Z : N(0, 1)$. The value $x = 460$ corresponds to

$$z = \frac{460 - 485}{\sqrt{625}} = \frac{460 - 485}{25} = -1.0$$

so $P(x \leq 460) = P(z \leq -1.0) = \mathbb{P}(-1.0)$. From Table 3 we find that $\mathbb{P}(-1.0) \doteq .1587$.

Example 7.16. Suppose that X is a normally distributed random variable with $E(X) = 500$ and $V(X) = 20$. Find $P(490 \leq x \leq 510)$. Standardizing our x-values we obtain

$$z_a = \frac{490 - 500}{\sqrt{20}} \doteq \frac{490 - 500}{4.47} \doteq -2.24$$

and

$$z_b = \frac{510 - 500}{\sqrt{20}} \doteq \frac{510 - 500}{4.47} \doteq 2.24$$

In Table 3 of Appendix VIII we find that $\mathbb{P}(2.24) \doteq .98745$ and that $\mathbb{P}(-.24) \doteq .01255$. Therefore,

$$P(490 \leq x \leq 510) = P(-2.24 \leq z \leq 2.24)$$

$$= \mathbb{P}(2.24) - \mathbb{P}(-2.24)$$

$$\doteq .98745 - .01255 = .9749$$

Example 7.17.1. Let X be IQ scores and suppose that $X : N(100, 225)$. What is the probability that the IQ score of a person selected at random falls between 80 and 90? Using Table 3 in Appendix VIII we first calculate

$$z_a = \frac{80 - 100}{15} \doteq -1.33 \quad \text{and} \quad z_b = \frac{90 - 100}{15} \doteq -0.67$$

We then find that $\mathbb{P}(-1.33) \doteq .09176$ and $\mathbb{P}(-0.67) \doteq .2514$, so

$$P(80 \le x \le 90) = P(-1.33 \le z \le -0.67)$$

$$\doteq .2514 - .09176$$

$$= .15964$$

In Examples 7.6 we learned how to work backward, that is, to begin with a particular cumulative probability and find the corresponding value of the standard normal random variable. We can do the same thing with nonstandard normally distributed random variables.

Example 7.17.2. What IQ score falls at the 95th percentile? That is, what is the value x_i for which $P(x \le x_i) = .95$? We know that $P(x \le x_i) = \mathbb{P}(x_i)$, but to find the x-value for which $\mathbb{P} = .95$, we first find the z-value for which $\mathbb{P} = .95$. In Table 3 we find that $\mathbb{P}(1.64) \doteq .94950$ and $\mathbb{P}(1.65) \doteq .95053$, so we let $z \cong 1.64$. Since it is true by definition that

$$z = \frac{x - \mu}{\sigma}$$

the x-value that corresponds to $z = 1.64$ is easily calculated:

$$z_i = \frac{x_i - \mu}{\sigma}$$

so

$$x_i = \mu + z_i \sigma$$

$$= 100 + 1.64(15)$$

$$= 124.6 \cong 125$$

———— EXERCISES 7.3 ————

All of the binomial probabilities in the following exercises are to be obtained using the normal approximation unless the exercise requires the exact probability.

1. Find
    (a) $P(18 \le x \le 24)$ for $X : B(54, 0.4)$
    (b) $P(1 \le x \le 4)$ for $X : B(25, 0.2)$

(c) $P(650 \le x)$ for $X : B(1,800, 1/3)$

(d) $P(x \le 2,000)$ for $X : B(2,500, 4/5)$

(e) $P(x = 7,280)$ for $X : B(8,100, 0.9)$

2. Find

(a) $P(29 \le x \le 35)$ for $X : B(54, 0.6)$

(b) $P(21 \le x \le 25)$ for $X : B(25, 0.8)$

(c) $P(1,190 \le x)$ for $X : B(1,800, 2/3)$

(d) $P(x \le 500)$ for $X : B(2,500, 1/5)$

(e) $P(x = 800)$ for $X : B(8,100, 0.1)$

3. To calculate $P(x = 50)$ using the normal approximation to $X : B(100, 0.5)$, we approximate the area of the binomial histogram bar centered at _____. This bar actually extends from _____ to _____. Now, calculate $P(x = 50)$ for

(a) $X : B(100, 0.5)$ (b) $X : N(50, 25)$.

4. To calculate $P(x = 200)$ using the normal approximation to $X : B(400, 0.5)$, we approximate the area of the binomial histogram bar centered at _____. This bar actually extends from _____ to _____. Now, calculate $P(x = 200)$ for

(a) $X : B(400, 0.5)$ (b) $X : N(200, 100)$

5. Calculate $P(948 \le x \le 972)$ for

(a) $X : B(2,400, 0.4)$ (b) $X : N(960, 576)$

6. Calculate $P(88 \le x \le 92)$ for

(a) $X : B(100, 0.9)$ (b) $X : N(90, 9)$

7. If a fair coin is flipped 50 times, what is the probability it comes up heads 25 times?

8. If a fair coin is flipped 50 times, what is the probability it comes up heads 24 times?

9. If an ordinary 6-sided die is rolled 60 times and Event A is defined as "obtain either a 2 or a 3," what is the probability that Event A will occur on 20 of the rolls?

10. What is the probability that 2 will be obtained exactly 10 times in 60 rolls of an ordinary 6-sided die?

11. If an ordinary 6-sided die is rolled 60 times and Event A is defined as "obtain either a 5 or a 6," what is the probability that Event A will occur on 16, 17, 18, 19, or 20 rolls?

12. What is the probability that the number of 2s obtained in 60 rolls of an ordinary 6-sided die will be 11, 12, or 13?

13. If an experimenter flips a fair coin 1,000 times he can be 90 percent certain of obtaining 500 $\pm$ how many heads?

14. If an experimenter flips a fair coin 1,000 times she can be 95 percent certain of obtaining 500 $\pm$ how many heads?

15. In the game "Telephone" 1,000 people are seated in a row. A message is whispered to the first person at the left, who then whispers it to the person to his or her right, etc. If the probability is only 1 percent that any person introduces an error into the message and if we assume that no one accidentally *corrects* an error, what is the probability that the 1,000th person reports a completely correct message?

16. Repeat Exercise 7.3.15 if the probability of introducing an error is 5 percent.

17. Suppose you have an urn containing one red ball and one white ball. Conduct the following experiment 10 times: Draw a ball at random from the urn, record its color, and return it to the urn. Let X be the number of times a white ball is drawn.
    (a) Use the binomial distribution to determine the exact probability of drawing 4 or 5 or 6 or 7 white balls.
    (b) Calculate the probability of drawing 4 or 5 or 6 or 7 white balls using the normal approximation to the binomial.
    Suppose you now add another red ball to the urn.
    (c) Use the binomial probability distribution to determine the exact probability of drawing 4 or 5 or 6 or 7 white balls.
    (d) Calculate the probability of drawing 4 or 5 or 6 or 7 white balls using the normal approximation to the binomial.
    (e) Why does (b) approximate (a) better than (d) approximates (c)?

18. You have a coin that you suspect is biased against tails, and you decide to conduct an experiment to test this suspicion by tossing the coin 100 times and recording the number of heads. You make an agreement with yourself that you will *assume* that the coin is fair, but if (given this assumption)

    P(the number of heads you obtain and all larger numbers)

    is less than .20, you will conclude that the coin is, in fact, biased. You toss the coin and obtain 55 heads.
    (a) *If* the coin is fair, what is the probability of obtaining 55 heads?
    (b) *If* the coin is fair, what is the probability of obtaining 55 *or more* heads?
    (c) Should you conclude that the coin is fair or that the coin is biased?
    (d) What is the probability that your conclusion is wrong?

19. An ordinary 6-sided die is rolled 60 times. If Event A is defined as "obtain either a 2 or a 3," what is the probability that Event A will occur between 26.67 and 33.33 percent of the time?

20. If an ordinary 6-sided die is rolled 60 times, what is the probability of getting a 2 between 18.33 and 21.67 percent of the time?

21. We believe that 60 percent of our workers favor unionization. We sample 100 workers at random. Assuming that our belief is correct, what is the probability that fewer than half will favor unionization?

22. We believe that 50 percent of American workers favor a smoke-free workplace. We sample 100 workers at random. Assuming that our belief is correct, what is the probability that more than 60 will favor a smoke-free workplace?

23. Q'mog is a novice thief and after serving an undistinguished apprenticeship in the local Thieves' Guild has a 30 percent chance of picking a pocket without detection. As his final examination, the Guild Master assigns him the task of picking the pockets of 50 people at a crowded street festival. What is the probability that Q'mog will be successful on 20 to 35 percent of his attempts?

24. Q'mog's brother Darrel is a journeyman thief and has a 45 percent chance of successfully picking a lock. The Guild Master of the East Lankmar Thieves' Guild (Local 299) asks Darrel to demonstrate his skills for an apprentice class. He presents Darrel with a collection of 75 locks that he has accumulated over the years and tells the journeyman that if he picks 40 percent or fewer of them, the Master will fine him 5 gold pieces. However, if he picks at least 55 percent of them, the Master will give him a bonus of 10 gold pieces.
    (a) What is the probability that Darrel will lose 5 gold pieces?
    (b) What is the probability that Darrel will earn 10 gold pieces?

(c) How many gold pieces does Darrel expect to gain (or lose) as a result of his demonstration?

In Exercises 25–28 use Table 3 (or 3A) of Appendix VIII to find the indicated probability.

25. (a) $P(3 < x < 5)$ for $X : N(4, 9)$
    (b) $P(10.3 < x < 12.5)$ for $X : N(9, 9)$
    (c) $P(25 < x)$ for $X : N(20, 25)$
    (d) $P(x < 40)$ for $X : N(50, 49)$

26. (a) $P(25 < x < 30)$ for $X : N(27, 9)$
    (b) $P(-1.2 < x < 2.5)$ for $X : N(1, 2)$
    (c) $P(5 < x)$ for $X : N(4, 9)$
    (d) $P(x < 4)$ for $X : N(3, 1)$

27. Let $X : N(9, 9)$. Find the value of x_i such that $P(9 - x_i < x < 9 + x_i) = .9$.

28. Let $X : N(4, 4)$. Find the value of x_i such that $P(4 - x_i < x < 4 + x_i) = .95$.

29. The Wechsler Intelligence Scale for Children (WISC) has a mean of 100 and a variance of 225 for the general population. If a child falls in the lowest 10 percent of the population, a certain clinical psychologist routinely urges the child's parents to have the child tested for possible brain damage. Assuming that WISC scores are normally distributed, what is the cut-off score that the psychologist uses?

30. The Verbal subtest of the Scholastic Aptitude Test and the Mathematics subtest each have a mean of 500 and a standard deviation of 100 based on the scores obtained by high school seniors. The ATY (Academically Talented Youth) program uses SAT scores of 11- and 12-year-olds as criteria for eligibility. Any youngster who obtains *either* a Mathematics score of at least 720 *or* a composite (i.e., Mathematics plus Verbal) score of at least 1,280 is accepted.

    (a) If the Mathematics score is normally distributed, what percentage of high school seniors would be expected to score below an applicant who was barely eligible based on his Mathematics score?

    (b) If the subtest scores are independent and the composite score is normally distributed, what percentage of high school seniors would be expected to score below an applicant who was barely eligible based on her composite score?

31. Midterm scores for the first author's introductory psychology course are almost always approximately normally distributed. One year, two sections of the course produced the following means and standard deviations:

	μ	σ
Section 1	29	4
Section 2	20	6

One student did not include the section number on the test and the name was illegible. The student's score was 25.

    (a) What is the probability of any student in Section 1 receiving a score as low or lower than 25?

    (b) What is the probability of any student in Section 2 receiving a score as high or higher than 25?

    (c) To which section should the instructor guess that the student belongs?

C. REPRESENTATION OF PROBABILITY DISTRIBUTIONS OF CONTINUOUS RANDOM VARIABLES

We concluded Chapter 6 with a discussion of tabular, graphic, and functional representations of probability distributions for discrete random variables. These conventions are also used to represent probability distributions of continuous random variables, but they are somewhat different from their discrete counterparts, and the differences highlight some of the important distinctions between discrete and continuous probability.

1. Tabular representation

Most of the examples in this chapter required the reader to use Table 3 of Appendix VIII to obtain probability values associated with normally distributed random variables. So, the reader is surely aware that statisticians use tables to represent probability distributions for continuous random variables. Such tables are different, however, from discrete probability tables in at least two ways.

First, since the probability associated with any particular value x of a continuous random variable is 0, tabled values are *cumulative* probabilities. For example, in Table 3 the value .97725 associated with $z = 2.00$ is $\mathbb{P}(2.00) = P(z \leq 2.00)$. Contrast this with the value .375 associated with $x = 2$ in Table 2 for $X : B(4, 0.5)$. Here, $.375 = P(x = 2)$.

Second, the probability values tabled for discrete random variables are often *exact* values. For example, if $X : B(4, 0.5)$,

$$P(x = 2) = .375$$

In contrast, most of the cumulative probabilities tabled for continuous random variables are *approximations*. Thus, if $Z : N(0, 1)$,

$$P(z \leq 2.00) \doteq .97725$$

This is likewise true for the values of other continuous random variables for which we have tabled cumulative probabilities in Appendix VIII.

2. Graphic representation

Like discrete probability distributions, continuous distributions can be represented histographically. Again, however, there are differences. At the end of Chapter 6 we pointed out that the *height* of any bar of a discrete probability histogram most generally represents the probability of the value over which the bar is centered. In a representation of this sort, the height of the bar centered over the value x is denoted $P(x)$. We also noted that discrete probability can be represented as the *area* of the bar. In a representation of this sort, the height of a continuous probability histogram is *always* denoted ϕ.

In Figures 7.1 to 7.3 we saw that for discrete random variables, the observable values included between x_a and x_b are represented on the X-axis of a probability histogram by the interval extending from $x_a - .5$ units of measurement to $x_b + .5$ units of measurement.

As we have mentioned before, this is in part a graphic convenience. It is also a reflection of one of the realities of measurement: We necessarily use *discrete* measures to obtain values of *continuous* phenomena. Any observable value x therefore represents the myriad real *but unmeasurable* numbers between $x + .5$ and $x - .5$. In Figure 7.2, for example, the *observable* value 177.5 lbs actually represents *all* weights between 177.25 and 177.75 lbs, and the set of observable measurements {177.0, 177.5, 178.0} represents *all* weights from 176.75 to 178.25 lbs.

The X-axis of a continuous distribution may be thought of as a representation of the phenomenon itself, rather than our imperfect measurements of the phenomenon. This means that any particular value x represents only the *exact* quantity x. This was seen in Figure 7.4, where the values included between *exactly* 177 lbs and *exactly* 178 lbs were represented on the X-axis by the interval $x_a = 177$ to $x_b = 178$.

3. Functional representation

The probability distribution for a discrete random variable, like the binomial, can often be represented by a rule that assigns a probability $P(x)$ to every x-value. To obtain $P(x_a \leq x \leq x_b)$ one can therefore calculate the *sum*

$$\sum_{i=a}^{b} P(x_i)$$

The distribution for a continuous random variable can be represented by a rule that assigns a probability *density* $\phi(x)$ to every x-value. To obtain $P(x_a \leq x \leq x_b)$ one must therefore calculate the *integral*

$$\int_{x_a}^{x_b} \phi(x)\, dx$$

or use tables of cumulative probabilities, which have been calculated by integration. Some calculators and most statistical software packages calculate cumulative probabilities for a variety of density functions.

D. SUMMARY

A random variable that can assume *any* real value is called a *continuous* random variable. Probability distributions for continuous random variables are important for two reasons. First, many physical and behavioral properties studied by scientists are continuous. Second, continuous probability models can often be used to obtain approximations of discrete probabilities.

The logic and conventions that statisticians use to model relative frequency distributions of continuous random variables are easier to understand if we think of probability as the *area* of a probability histogram. If X is a *discrete* random variable, $P(x_a \leq x \leq x_b)$ corresponds to the area of the bars centered at $x_a, \ldots, x_b$. This interval begins at x_a *minus* half a unit of measurement and extends to x_b *plus* half a unit of measurement. As the unit of measurement becomes smaller, the bars become narrower. For a *continuous* random variable, the bars are infinitesimally narrow, and $P(x_a \leq x \leq x_b)$ is represented by the area of the histogram over the interval that begins *exactly* at x_a and ends *exactly* at x_b. If $P(x_a \leq x \leq x_b)$ is represented as the area under the probability graph between x_a and x_b, the graph is called the *probability density function* of x and is denoted ϕ.

The *cumulative probability* of any value x_i is defined as $P(x \leq x_i)$ and is denoted $\mathbb{P}(x_i)$. If probability is represented by the graph of the density function, the cumulative probability of x_i is the area under the graph to the *left* of x_i.

For large values of N, the density function for the *standardized* binomial random variable loses its "staircase" appearance and becomes a smooth, bell-shaped curve. This curve is called the *standard normal probability density function* or the *standard normal curve* and is given by the following expression [**7.1**]:

$$\phi(z) = \frac{1}{\sqrt{2\pi}} e^{-z^2/2}$$

where $\pi \doteq 3.1416$ and $e \doteq 2.718$ and $z = (x - \mu)/\sigma$.

Expression [**7.1**] allows us to deduce a number of important properties of the standard normal curve:

- $\phi(z)$ is symmetrical about zero.
- $\phi(z)$ has a maximum value at $z = 0$.
- $\phi(z)$ is asymptotic to zero.

If Z is the standard normal random variable and if $z_a < z_b$, the probability that z falls between z_a and z_b is the cumulative probability of z_b minus the cumulative probability of z_a:

$$P(z_a \leq z \leq z_b) = \mathbb{P}(z_b) - \mathbb{P}(z_a)$$

Cumulative probabilities for all two-place standard normal values from $z = -3.09$ to $z = +3.09$ are given in Table 3 of Appendix VIII. To obtain $P(z_a \leq z \leq z_b)$ look up $\mathbb{P}(z_b)$ and $\mathbb{P}(z_a)$ and calculate the difference, $\mathbb{P}(z_b) - \mathbb{P}(z_a)$. Cumulative probabilities for selected values of $z > 3.09$ appear in Table 3A.

If $X: B(N, p)$ and if $Np \geq 5$ and if $Nq \geq 5$, then the standard normal distribution *approximates* the probability distribution for the *standard-*

ized binomial random variable

$$Z = \frac{X - Np}{\sqrt{Npq}}$$

Accordingly, one can obtain a close approximation to $P(x_a \leq x \leq x_b)$ by calculating

$$z_a = \frac{(x_a - .5) - Np}{\sqrt{Npq}} \quad \text{and} \quad z_b = \frac{(x_b + .5) - Np}{\sqrt{Npq}}$$

finding $\mathbb{P}(z_a)$ and $\mathbb{P}(z_b)$ in the table of cumulative probabilities for the standard normal curve (Table 3 of Appendix VIII), and calculating $\mathbb{P}(z_b) - \mathbb{P}(z_a)$.

One can also use the standard normal curve to obtain approximate probabilities associated with the *proportion* of successes obtained in a binomial experiment, X/N. If $X : B(N, p)$, it is easily shown that the standardized value of the binomial proportion x/N

$$\frac{\frac{x}{N} - p}{\sqrt{\frac{pq}{N}}}$$

is exactly equal to the standardized value of x. When $Np \geq 5$ and $Nq \geq 5$, one can therefore obtain a close approximation of

$$p\left(\frac{x_a}{N} \leq \frac{x}{N} \leq \frac{x_b}{N}\right)$$

by calculating

$$z_a = \frac{\frac{x_a - .5}{N} - p}{\sqrt{\frac{pq}{N}}} \quad \text{and} \quad z_b = \frac{\frac{x_b + .5}{N} - p}{\sqrt{\frac{pq}{N}}}$$

and calculating $\mathbb{P}(z_b) - \mathbb{P}(z_a)$ from entries found in the table for cumulative probabilities for the standard normal random variable.

A random variable X is said to be normally distributed if its density function [7.4] is

$$\phi(x) = \frac{1}{\sqrt{2\pi\sigma^2}} e^{-\frac{1}{2}\left(\frac{x-\mu}{\sigma}\right)^2}$$

If X is a *standardized* random variable, then $\mu = 0$ and $\sigma^2 = 1$, and [7.4] reduces to [7.1]. Hence, the *standard* normal density function is the density function for a *standardized* random variable that is normally distributed. This means that if X is a normally distributed random variable with $E(X) = \mu$ and $V(X) = \sigma^2$, then

$$P(x_a \leq x \leq x_b) = P(z_a \leq z \leq z_b)$$

where

$$z_a = \frac{x_a - \mu}{\sigma} \quad \text{and} \quad z_b = \frac{x_b - \mu}{\sigma}$$

The conventions for tabular, graphic, and functional representation of continuous probability distributions highlight some of the important differences between continuous and discrete probability.

The probability of any particular value x of a continuous random variable is zero. Therefore, probability distributions for continuous random variables are always tabulated cumulatively. That is, for any value x_i of a continuous random variable, the entry in a probability table is $\mathbb{P}(x_i) = \mathbb{P}(x \leq x_i)$.

In a graphic representation of a continuous probability distribution, the probability $P(x_a \leq x \leq x_b)$ is the area under the graph between x_a and x_b. Although probability distributions for discrete random variables can also be represented in terms of area, it is important to note that in the discrete case the interval begins at $x_a - .5$ and ends at $x_b + .5$.

The probability distribution for a continuous random variable can also be represented by specifying its probability density function ϕ and the values of the appropriate parameters. The parameters for the normal distribution are the expected value μ and the variance σ^2. The specification statement for any normally distributed random variable is $X : N(\mu, \sigma^2)$. The expected value of any standardized random variable Z is 0 and the variance is 1. Therefore, the specification statement for the standard normal random variable is $Z : N(0, 1)$.

PART III

Introduction to statistical inference

CHAPTER 8

Sampling distributions and estimation

CHAPTER OUTLINE

A. POPULATIONS AND SAMPLES

In Part I we developed the computational machinery and many of the concepts involved in the organization and quantitative description of collections of data. The importance of description notwithstanding, the scientific philosophy of our times places much greater emphasis on explanation, prediction, and generalization than on pure description.

Throughout most of his life, the Dutch astonomer Tycho Brahe (1546–1601) carefully recorded the positions of the known planets. These observations, however, were only empirical curiosities until Johannes Kepler (1571–1630) incorporated them into an explanatory model of the solar system that permitted prediction of planetary motion and generalization to planets not yet discovered. In a similar vein, we pointed out at the beginning of Chapter 1 that statistical data are collected in order to reach conclusions or generate interpretations. The interview data described in Example 2.1 might reveal interesting relationships between, say, annual family income and recognition of the product or service marketed by the client who commissioned the survey. But these relationships are ultimately important to the client only to the extent that they apply to *all* persons who meet the criteria by which this particular group of 20 respondents was selected.

In this spirit, contemporary statistics is more concerned with statistical *inference* than statistical *description*.

> *Inferential statistics* is a body of quantitative techniques that enable the scientist to make appropriate generalizations from limited observations.

Many of the features that distinguish problems of inference from problems of description revolve around the difference between a *sample* and a *population*, and these are illustrated in the following example.

Example 8.1. Suppose that the manufacturer of extended-life light bulbs claims that his bulbs burn an average of 1,500 hours. To maintain this quality, he selects one bulb at random from every batch of 20 and records burn time. Suppose further that a production run of 2,000 light bulbs generates the observations summarized in Table 8.1. Superficially, this

Table 8.1. Frequency distribution for burn time of 100 light bulbs

Burn Time (in hours)	1375.55 or less	1375.5– 1425.5	1425.5– 1475.5	1475.5– 1525.5	1525.5– 1575.5	1575.5– 1625.5	1625.5 or more
f	2	5	8	60	19	3	3

$\bar{x} = 1,505.5$ hr

scenario is very similar to the sorts of description problems we encountered in Part I: An experimenter (scientist, manufacturer, etc.) has a collection of quantitative observations, has organized them into a frequency distribution, and has described the location of the distribution by calculating the mean. The problem differs, however, in one crucial respect: The experimenter is interested in the distribution of burn time in the 100 bulbs he has observed *only insofar* as it may be representative of the distribution of burn time for the *larger pool* of 2,000 bulbs from which the observations were drawn (or perhaps, the considerably larger pool consisting of *all* bulbs he has ever manufactured or will in the future manufacture). In general, the universe of potential observations about which the experimenter wishes to make some general statement, or *inference*, is called the *population* (or parent population), and the inferences concern the same properties of relative frequency distributions we have already encountered in the context of statistical description, viz., location, variability, etc.

Depending upon the size and accessibility of the parent population, it may be impossible to take measurements on all of its members. For instance, if the manufacturer in Example 8.1 defined his population as all bulbs ever manufactured, many of these would no longer be accessible.[1] The manufacturer must therefore make *inferences* about the distribution of burn time in his parent population based on a collection of data that constitutes a relatively small fraction of the population. This collection of data is called a *sample*. A sample therefore differs from any other collection of observations only in the intent of the observer. If the manufacturer could be certain that his sample reflected the distribution characteristics of the parent population with absolute fidelity, he might conclude that the average burn life of the bulbs he manufactures is 1,505.5 hours. He knows from experience, however, that average burn life differs from sample to sample. The computational procedures appropriate to description of data are, therefore, not sufficient to the task of making inferences about populations. The leap from sample to population crosses the chasm from what is *observed* to what is *unobserved*. The chasm is bridged by *unobservable* probability distributions and application of probability theory developed in Part II, as diagramed below.

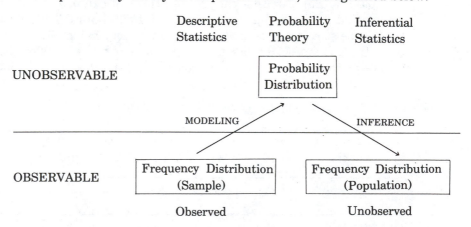

As we know from Chapter 1 a collection of measurements, such as those that comprise a sample, satisfy the definition of *data*, those which are given. In contrast, the numerical values assigned to members of a population are *potential*, rather than actual. Although the numbers themselves do not yet exist in fact (and so cannot be thought of as *given*) the experimenter always knows the procedures by which they are to be generated. For example, select a bulb at random and record how long it burns. Selection of a bulb is an activity that can have more than one result and therefore satisfies the definition of a random experiment given in Chapter 5. Accordingly, the rule "record how long it burns" must be a random variable. In the context of inferential statistics, (potential) measurements are referred to as values of random variables, rather than data.

In Chapter 1 we developed the notion of pairing relative frequency f/N with every value (or group of values) in a collection of observations. Relative frequencies can be calculated only *after* measurements have been obtained. Inasmuch as a population is understood to be a pool of *potential* observations, measurements *have not yet been obtained*, so relative frequencies cannot be calculated. But we know from Part II that values of *random variables* can be associated with probabilities. Whereas data collections are typically organized into relative frequency distributions that are characterized by descriptive statistics, populations are modeled by *probability distributions* that are characterized by *parameters*.

As we know, descriptive statistics are generally denoted by Roman letters, e.g., s^2 for the variance, r for correlation, and so on. The corresponding *population values* are modeled by the parameters of probability distributions and are generally denoted by the Greek letters used to denote these parameters (σ^2 for the value of the population variance, ρ the value of the correlation coefficient, etc.). The parallels between samples (or collections of measurements in general) and populations can be summarized as follows:

	Samples	**Populations**
Numerical values	Data	Values of random variables
Distribution	Relative frequency	Probability
Characterized by	Descriptive statistics	Population values (or parameters)
Notation	Roman (s, r, etc.)	Greek (σ, ρ, etc.)

Beginning later in this chapter, we sometimes find it useful to let W denote an unspecified or "generic" statistic. Likewise, we use θ (the lowercase Greek *theta*) to denote an unspecified parameter or unknown population value.

In some situations the population of interest is relatively small and it is possible to take measurements on all members. The experiment is said to *exhaust* the population. In this circumstance the probability of each value is, by the classical definition, exactly equal to its relative frequency, and the descriptive statistics that characterize the "sample" *are* the population values.

1. Sampling as a random experiment

We pointed out in Example 8.1 that selection of an individual member of a population is a random experiment. It is likewise true that selection of the N individuals that comprise one's sample is a random experiment.

Example 8.2. Let us suppose that a developmental physiologist is concerned with the distribution of heights in the population of freshmen women at some university. The freshman class includes 4,000 women, so it is not feasible to measure every one. She therefore decides to estimate various properties of the population distribution from a sample of 100 incoming women. The number of possible samples (that is, the number of possible *groups* of 100 freshman women) is the number of combinations of 4,000 women taken 100 at a time. That is, there are $_{4,000}C_{100}$ possible *different* groups that she might select for observation. Drawing a sample of N observations from a population of K individuals is therefore a *random experiment*, and if each possible sample is thought of as one possible result, such an experiment generates a sample space with $m = {_K}C_N$ elements. For $K = 4,000$ and $N = 100$ this sample space is larger than you might think. In fact, m is about equal to 5×10^{201}, a 5 followed by 201 zeros. This number is considerably larger than the number of atoms in the known universe.

There are many ways that our fictitious physiologist (or any scientist) might select her sample. She might simply take every 40th name from an alphabetical roster of women provided by the office of admissions; she might take the first 100 or the last 100; if the list includes students' social security numbers she might take the lowest (or highest) 100 numbers; she might get a listing of names on a computer disk, have the computer shuffle them, and then take the first 100. Whatever the procedure by which her observations are chosen, the important thing is that the relative frequency distribution they generate faithfully reflects the distribution of values in the parent population. A sample is of little use, for example, if it includes a high frequency of values that are relatively rare in the parent population.

To illustrate how this might happen, let us suppose that our physiologist knows that during the week preceding registration all freshmen undergo physical examinations at the student health center. To obtain her sample, she therefore goes to the health center and records the heights of the first 100 women to complete their examinations. Suppose, however, that the women's basketball coach has arranged with the health center to schedule all freshmen basketball players for the first day of physical examinations. Given this circumstance, the experimenter's sample would probably include a disproportionate representation of basketball players, and it is therefore likely that the relative frequency of heights exceeding, say, 70 in. would be considerably greater in her sample than in the population of freshman women. To avoid this sort of sampling error, statisticians generally require that samples be generated by *random sampling*:

> Random sampling is defined as any method of selection that guarantees that all *possible* samples are equally likely to be drawn.

For a sample of 100 selected from a population of 4,000 potential observations, random sampling would give each possible sample a probability of approximately $1/(5 \times 10^{201})$ of being drawn. Drawing the names from a hat would ensure randomness of the sample. Taking the 100 individuals with the lowest social security numbers would not ensure random sampling of *individuals*, but since social security numbers are assigned wholly independently of height, a sample drawn in this fashion would almost certainly be random with respect to the variable under consideration—though not, perhaps, with respect to some other variable, such as age.

2. Statistics as random variables

In Example 8.1 we presented a scenario in which a manufacturer of light bulbs tested a sample of bulbs from each production run. We now know that drawing such a sample constitutes a random experiment and that for any given production run, each of the m possible samples he might draw can be represented as an element e in a sample space S, as illustrated in Figure 8.1.

Example 8.3. To simplify matters, let us suppose that the manufacturer decides to sample 10 percent of every production run and to produce bulbs in lots of 50 instead of 2,000. That is, he will draw a sample of 5 bulbs from every production run of 50 bulbs. In this example, then, the population consists of 50 bulbs, the sample size is 5, and the number of possible samples m is therefore $_{50}C_5 = 2{,}118{,}760$.

After the five bulbs are drawn the manufacturer lights them with standard 110-volt line current until they burn out and records their burn

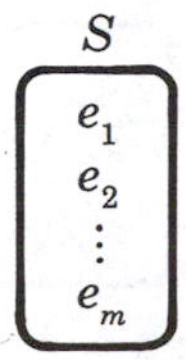

S

$$\begin{array}{|l|} \hline e_1 \\ e_2 \\ \vdots \\ e_m \\ \hline \end{array}$$

Figure 8.1. *Sample space for the experiment "draw a sample of light bulbs."*

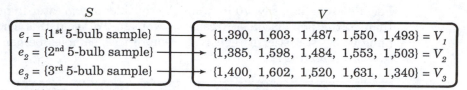

Figure 8.2. Graphic mapping of measurement rule that assigns value sets to each of three samples.

life. For example, a particular sample of five bulbs might turn out to have burn times of 1,390, 1,603, 1,487, 1,550, and 1,493 hr, respectively. As we know from earlier discussion, this test is a random variable—a function that assigns a numerical value (burn time) to each bulb.

The function mapping three samples, e_1, e_2, and e_3, into three sets of numbers is illustrated in Figure 8.2. The first five-bulb sample, e_1, yields the five burn times listed in the set labeled V_1. The second and third samples, e_2 and e_3, have the burn times listed in sets V_2 and V_3.

Now let us define a statistic that assigns a number w to each value set V. In Example 8.1 this statistic was the mean, $\bar{x}$, but could as easily be the median, the variance, Pearson's skewness coefficient, etc. The rule assigning a value of $\bar{x}$ to each of the three samples in Figure 8.2 is diagramed in Figure 8.3.

Figure 8.3 illustrates two important points about the whole business of drawing samples, assigning numbers to the observations, and calculating statistics. First, every sample is comprised of a different combination of observations, and it is therefore unlikely that any two value sets V_i and V_j contain exactly the same numbers. Nonetheless, it is entirely possible for two value sets to be assigned the same value of the statistic. In Figure 8.3, for example, it is seen that the first two five-bulb samples have the same mean, 1,504.6 hr. In general it must therefore be true that n, the number of w-values, is equal to *or less than* the number of samples, m.

Second, we note that even though the same value w may be assigned to two or more samples, e.g., the mean calculated for samples e_1 and e_2 in Figure 8.3, each sample is assigned only one value of w. Since each sample is the result of a random experiment, any single-valued statistic is therefore a *random variable*.

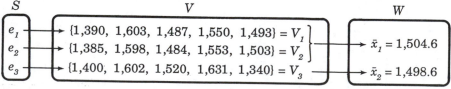

Figure 8.3. Graphic mapping of three sets of observations into two values of $\bar{X}$.

—————— EXERCISES 8.1 ——————————————————————

In Exercises 1–4 identify the population, sample, and random variable in each experiment.

1. Many colleges and universities admit new students before the end of their senior year in high school. A university admissions officer wants to obtain an academic profile on the incoming freshman class. He takes 30 names from the roster of incoming freshmen and writes to their high schools to obtain their final high school grade-point averages.
2. An automobile manufacturer wants to estimate the average cost of maintenance in the first year of service for a new model. A survey is sent to 1,000 persons who purchased the new car asking them to report the amount they spent on maintenance.
3. A municipal civil service is developing standards for a new entrance-level examination for police department applicants. The current group of applicants is administered the examination. Their scores are sealed and will be correlated with the performance evaluations they receive after six months on the job.
4. A player of *Advanced Dungeons & Dragons* ® believes that her favorite 10-sided die is biased against high numbers. To test the die she rolls it 200 times and records the number of times that "10" shows up. She divides this number by 200.
5. Devise a method to ensure that the sample in Exercise 8.1.1 is random.
6. Devise a method to ensure that the sample in Exercise 8.1.2 is random.
7. List the names of every member of your immediate family, including yourself.
   (a) How many samples of two individuals can you draw from this list of names?
   (b) List all of the samples along with the age of each person in every sample.
   (c) Calculate the median age and the range of ages for every sample.
   (d) Are all the medians in your answer to (c) the same?
   (e) Are all the ranges in your answer to (c) the same?
   (f) Your answers to (d) and (e) illustrate the point made in the text that sample statistics are _____ _____ .
8. Using any method you wish, draw five two-digit numbers from Table 1 of Appendix VIII. Repeat this experiment five times. Calculate the mean $\bar{x}$ and variance s^2 for each of your five samples.
   (a) Are all the means the same?
   (b) Are all the variances the same?
   (c) Your answers to (a) and (b) indicate that means and variances of samples are _____ _____ .

B. SAMPLING DISTRIBUTIONS

In Part I of this text we became accustomed to thinking of a good descriptive statistic, such as the mean $\bar{x}$ or the variance s^2, as having a *single* value, which we compute from N observations. The key to thinking of statistics as *variables* is to keep in mind that one can draw *many* samples of N observations from the same population and that the value of any particular statistic W (the mean, the variance, etc.) can differ from sample to sample.

Given the set of values $W = \{w_1, w_2, \ldots, w_n\}$ one can, in addition, calculate the relative frequency of every value w_i. If one does this for *all possible* samples of size N, these relative frequencies f_i/m are, by the classical definition, probabilities. In principle, therefore, one can always assign a probability to every value w in the set W:

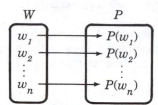

Figure 8.4. Graphic mapping of n values of the statistic W into a set of probability values.

The probability distribution defined by the function illustrated in Figure 8.4 is called the *sampling distribution* of the statistic W, and in general

> The probability distribution associated with any statistic is called the *sampling distribution* of that statistic.

In practice, of course, we do not draw all possible samples of size N and calculate the w-value for each. The sampling distribution of W is therefore a probability *model* for the relative frequency distribution of the statistic W. Throughout most of the rest of this book we shall be dealing with statistics as random variables, and many—if not most—of the statistical techniques we develop will focus on the sampling distributions of these statistics.

1. The sampling distribution of the mean $\bar{X}$

From Chapter 2 we know that one of the more *descriptively* useful properties of a relative frequency distribution is its location, and that for many types of data the mean $\bar{x}$ is the most useful statistic for describing this property. We shall see in the course of Part III that much of inferential statistics is also concerned with the mean. In part this is because the location of a distribution is such a fundamental piece of information. In part it is because the sampling distribution of the mean $\bar{X}$ has properties that allow us to base our inferences on some very convenient mathematical assumptions. These assumptions concern the probability density function ϕ to be used as a model for the relative frequency distribution of $\overline{X}$ and the parameters appropriate to that function.

a. The density function of $\overline{X}$: the Central Limit Theorem. Let us suppose that a researcher proposes to draw a sample of N mathematics test scores from the population of K persons who completed the Scholastic Aptitude Test in a particular year. Let us further suppose that there are n values, $x_1, x_2, \ldots, x_n$, distributed among the K individual test scores and that the observations are to be independent of one another.

As indicated at the beginning of the chapter, each observation may be thought of as a random variable, which in this case can potentially yield any one of the n test scores, $x_1, x_2, \ldots, x_n$. When we draw a random sample of size N from this population we make N observations and thus obtain N random variables:

X_1:   First observation

X_2:   Second observation

$\vdots$

X_N: Nth observation

Consider the first observation, X_1. The set of scores from which the experimenter will make his selection is the entire population. The distribution of x-values for the first observation is, therefore, precisely the same as the distribution of x-values for the parent population. Since the scores are to be drawn at random, this is likewise true for all subsequent observations. The random variables $X_1, X_2, \ldots, X_N$ are therefore both *independent* and *identically distributed*.

We know from Chapter 2 that the mean of a collection of N values $x_1, x_2, \ldots, x_N$ is equal to

$$\bar{x} = \frac{\sum\limits^{N} x_i}{N}$$

or

$$\bar{x} = \frac{1}{N}(x_1 + x_2 + \cdots + x_N)$$

Likewise, for a set of N random variables $X_1, X_2, \ldots, X_N$

$$\overline{X} = \frac{1}{N}(X_1 + X_2 + \cdots + X_N)$$

$$= \frac{1}{N}(X_1) + \frac{1}{N}(X_2) + \cdots + \frac{1}{N}(X_N) \qquad [8.1]$$

Letting $W_i = \frac{1}{N}(X_i)$, the variable $\bar{X}$ may be rewritten as

$$\bar{X} = W_1 + W_2 + \cdots + W_N \qquad [8.2]$$

If $X_1, X_2, \ldots, X_N$ are independent and identically distributed, as when observations are independent, $W_1, W_2, \ldots, W_N$ are likewise independent and identically distributed. The mean $\bar{X}$ can therefore be expressed as the *sum* of N independent, identically distributed random variables, W_i.

In this connection, one of the most important theorems in statistics is the Central Limit Theorem, which tells us that if a random variable Y is *itself* the sum of N independent and identically distributed random variables, Y becomes normally distributed as N increases. The formal statement of this theorem is given below:

The Central Limit Theorem. If $W_1, W_2, \ldots, W_N$ are independent, identically distributed random variables and $Y = W_1 + W_2 + \cdots + W_N$, the probability density function of Y approaches

$$\frac{1}{\sqrt{2\pi\sigma_Y^2}} e^{-\frac{1}{2}\left(\frac{y-\mu_Y}{\sigma_Y}\right)^2}$$

as N approaches infinity.

From equation [8.2], we may therefore conclude that the distribution of sample means becomes increasingly normal as the number of observations N increases. The proof of the Central Limit Theorem is beyond the scope of this text, but its importance would be difficult to overstate. The Central Limit Theorem underlies much of the repertoire of statistical inference and the applicability of these methods to an extraordinary range of scientific problems (see Box 8.1).

We have emphasized several times that, discounting manual solutions, the computation of any probability $P(x_a \leq x \leq x_b)$ requires the probability function associated with X. In many scientific problems, of course, it is unrealistic to expect that the population distribution of our random variable will conform precisely to some convenient mathematical model —even the ubiquitous normal curve. Irrespective of the population distribution of X, however, we can often invoke the Central Limit Theorem and assume that *means* of large samples drawn from the population are normally distributed. We may thus rely on the normal distribution to determine probabilities associated with intervals $[\bar{x}_a, \bar{x}_b]$.

As with the binomial approximation to the normal, moreover, surprisingly small samples are often large enough to permit confident use of the normal distribution to compute such probabilities. To illustrate this, Bailey (1971) used a computer to generate sampling distributions of $\bar{X}$ based on samples of various sizes for random variables with uniform,

BOX 8.1

The pervasive Central Limit Theorem

We have already encountered one of the most important implications of the Central Limit Theorem, the normal approximation to the binomial. In Chapter 6 we demonstrated that a binomial random variable may be represented as the sum of N independent indicator random variables. That is, if $X : B(N, p)$, then $X = I_1 + I_2 + \cdots + I_N$, where $I = \{0, 1\}$. As N increases, the binomial random variable satisfies the conditions of the Central Limit Theorem and therefore approaches the normal. Similarly, we shall see in Chapter 11 that the χ^2 (chi-square) distribution is also defined by a sum of N independent random variables and, like the binomial distribution, becomes normal as N increases.

The Central Limit Theorem also explains why the normal curve so closely approximates the distributions of many economic, biological, sociological, and geophysical phenomena that can themselves be thought of as the sum of many other independent variables. For example, biologists assume that weight, body measurements, and other characteristics that exhibit *quantitative* differences—rather than *qualitative* differences, like blood type—result from the additive effects of many genes. These *polygenes* most commonly exist in two forms, one of which contributes a measurable increment to the attribute and one of which does not. The potential contribution of any particular gene may therefore be represented by either a 1 or a 0, so each gene can be quantified as an indicator variable. Sexual reproduction assures that any individual receives an independent assortment of parental chromosomes, each of which is likely to include both contributing ($= 1$) and noncontributing ($= 0$) genes. The sum of these genetic "values" determines the amount of the attribute possessed by the offspring, and like the sum of any large number of independent indicator variables, most polygenic characteristics are approximately normally distributed.

J-shaped, and U-shaped probability distributions. These computer simulations are illustrated in Figure 8.5, and it is evident that even when the distribution of X is far from normal, one requires only modest sample sizes for the distribution of $\overline{X}$ to assume the symmetrical, bell-shaped form that characterizes the normal distribution.

For samples of 50 or more observations, sampling experiments show that the distribution of $\overline{X}$ is approximately normal even if the distribution of X is highly irregular (discontinuous, multimodal, etc.). Fewer observations are required with random variables that are more nearly normally distributed. If, for example, the distribution of X is approximately symmetrical about the mean and has only one mode, the table of cumulative normal probabilities yields very accurate estimates of $P(\overline{x}_a \leq \overline{x} \leq \overline{x}_b)$ for samples of 10 or more observations. Indeed, if the distribution of X is *exactly* normal, then $\overline{X}$ is normal for *any* size sample. This is readily apparent if we consider that for the smallest possible sample ($N = 1$), the observed value x is the mean of the sample, and the distributions of X and $\overline{X}$ must therefore be identical. For most

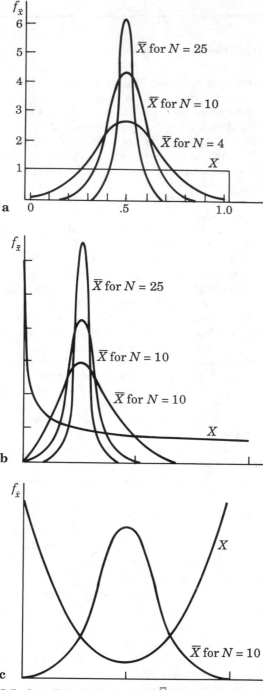

Figure 8.5. Sampling distributions of $\bar{X}$ for various sample sizes, where distribution of X is uniform (a), J-shaped (b), and U-shaped (c). (Adapted from Bailey, D.E. (1971), Probability and Statistics: Models for Research. New York: Wiley)

variables studied in the behavioral and biological sciences, sample sizes of 30 or more are generally considered adequate to justify the assumption that the sampling distribution of $\bar{X}$ is approximately normal.

b. The expected value and variance of $\bar{X}$. From Chapter 7 we know that the parameters of the normal distribution are μ, the expected value, and σ^2, the variance.

The expected value of $\bar{X}$ is easily derived from equation [8.1]:

$$\bar{X} = \frac{1}{N}(X_1 + X_2 + \cdots + X_N)$$

Therefore,

$$E(\bar{X}) = E\left(\frac{1}{N}(X_1 + X_2 + \cdots + X_N)\right)$$

However, since $1/N$ is a constant, it may be factored out of the expectation, so

$$E(\bar{X}) = \frac{1}{N}E(X_1 + X_2 + \cdots + X_N)$$

We also know from the Algebra of Expectations that the expected value of a sum is equal to the sum of the expected values. That is, if $E(X_1) = \mu_1$ and $E(X_2) = \mu_2$ and $\cdots E(X_N) = \mu_N$, then

$$E(\bar{X}) = \frac{1}{N}(\mu_1 + \mu_2 + \cdots + \mu_N) \tag{8.3}$$

Let us now suppose that the population mean is μ. Since the distributions of $X_1, X_2, \ldots, X_N$ are all identical to the population distribution, it follows that all N random variables have the same expected value, μ, the population mean. That is,

$$\mu_1 = \mu_2 = \mu_3 = \cdots = \mu_N = \mu \tag{8.4}$$

Equation [8.3] may therefore be rewritten,

$$E(\bar{X}) = \frac{1}{N}(\mu + \mu + \cdots + \mu)$$

$$= \frac{1}{N}(N\mu)$$

so

$$E(\bar{X}) = \mu \tag{8.5}$$

For any sample of N observations, the expected value of the sample mean is equal to the population mean, $\mu_{\bar{X}} = \mu$.

The variance of the sampling distribution of $\bar{X}$ can be derived in a very similar fashion. Let us suppose that the variance of scores in the population is σ^2. From equation [8.1],

$$\bar{X} = \frac{1}{N}(X_1 + X_2 + \cdots + X_N)$$

Therefore,

$$V(\bar{X}) = V\left(\frac{1}{N}(X_1 + X_2 + \cdots + X_N)\right)$$

Once again, $1/N$ is a constant, but when a constant is factored out of a variance, recall that it is *squared*, so

$$V(\bar{X}) = \frac{1}{N^2}V(X_1 + X_2 + \cdots + X_N)$$

Since $X_1 + X_2 + \cdots + X_N$ is the sum of N *independent* random variables, it follows from Rule 9 of the Algebra of Variances that the variance of this sum is equal to the sum of the individual variances: If $V(X_1) = \sigma_1^2$ and $V(X_2) = \sigma_2^2$ and $\cdots$ and $V(X_N) = \sigma_N^2$, then

$$V(X_1 + X_2 + \cdots + X_N) = \sigma_1^2 + \sigma_2^2 + \cdots + \sigma_N^2$$

Therefore,

$$V(\bar{X}) = \frac{1}{N^2}(\sigma_1^2 + \sigma_2^2 + \cdots + \sigma_N^2) \qquad \text{[8.6]}$$

And, since the variables representing our N observations all have the same distribution as the parent population, they must all have the same variance, σ^2, the population variance. Equation [8.6] therefore becomes

$$V(\bar{X}) = \frac{1}{N^2}(N\sigma^2)$$

so

$$V(\bar{X}) = \frac{\sigma^2}{N} \qquad \text{[8.7]}$$

For samples of N independent observations, the variance of the sample means is equal to the population variance σ^2 *divided by* the sample size N, that is

$$\sigma_{\bar{X}}^2 = \frac{\sigma^2}{N}$$

Likewise, the standard deviation of the sample means $\sigma_{\bar{X}}$ is

$$\sqrt{\frac{\sigma^2}{N}} = \frac{\sigma}{\sqrt{N}}$$

and is sometimes called the *standard error of the mean*.

We see from [**8.7**] that as sample sizes *increase*, i.e., as N becomes larger, the variability among sample means *decreases*. This is an extraordinarily important phenomenon. Indeed, many of the procedures involved in drawing inferences about means is based on this. It is also easy to forget. Don't.

The preceding discussion has established that as sample size N increases, the sampling distribution of means becomes increasingly normally distributed. Furthermore, the expected value of $\bar{X}$ is μ and the variance of $\bar{X}$ is σ^2/N, where μ and σ^2 are the population mean and variance of the random variable X. In the following examples, we begin to see how this might be of use in drawing inferences about populations based on sample data.

Example 8.4. To illustrate the application of the Central Limit Theorem to the sampling distribution of $\bar{X}$, let us suppose that in one particular year the mean mathematics score for all persons taking the SAT was 485 and the variance was 18,750.

Now, let us suppose that the experimenter discussed earlier proposed to draw a sample of 30 observations and wanted to calculate the probability that the mean of his sample will be equal to or less than 460. That is, he wishes to determine $P(\bar{x} \le 460) = \mathbb{P}(460)$. He doesn't know how the SAT scores themselves are distributed, but his sample size is large enough to assume on the basis of the Central Limit Theorem that the *means* of such samples will be (approximately) normally distributed. Recall from Chapter 7 that if a random variable X is normally distributed, the cumulative probability of any value x_i may be determined by standardizing the value and submitting it to a table of cumulative probabilities of the standard normal curve. That is,

$$\mathbb{P}(x_i) = \mathbb{P}(z_i)$$

where

$$z_i = \frac{x_i - \mu_X}{\sigma_X}$$

To determine $P(\bar{x} \leq 460)$ the first step is to calculate

$$\frac{460 - \mu_{\bar{X}}}{\sigma_{\bar{X}}}$$

Since the mean μ of the population of SAT scores is equal to 485, the expected value $\mu_{\bar{X}}$ of all possible means must also equal 485. Similarly, the variance $\sigma_{\bar{X}}^2$ for all means of sample size N is equal to σ^2/N, which in this example is $18{,}750/30 = 625$. Therefore,

$$\frac{460 - \mu_{\bar{X}}}{\sigma_{\bar{X}}} = \frac{460 - 485}{\sqrt{625}} = \frac{-25}{25} = -1.0$$

From Table 3 of Appendix VIII, we see that $\mathbb{P}(-1.0) \doteq .1587$, and the probability that the mean of any sample of 25 observations will be less than or equal to 460 is therefore (approximately) 0.16.

Example 8.5. The mean length of unsolicited, promotional copies of introductory statistics textbooks received by the authors is 500 pages. Textbook editors in many publishing firms come from the sales ranks, so they often know nothing about statistics, and their major concern is that any book they contract be as much like all the other books as possible. Consequently, page length does not vary much ($\sigma^2 = 100$) and is normally distributed. Given this (not completely apocryphal) information, what is the probability that the mean number of pages for a sample of five textbooks would be between 490 and 510 pages?

Since the random variable, page length, is normally distributed, means from samples of *any* size will likewise be normally distributed. Therefore,

$$P(490 \leq \bar{x} \leq 510) = P\left(\frac{490 - \mu_{\bar{X}}}{\sigma_{\bar{X}}} \leq z \leq \frac{510 - \mu_{\bar{X}}}{\sigma_{\bar{X}}}\right)$$

under the standard normal curve. Since $\mu = 500$, it must also be true that $\mu_{\bar{X}} = 500$ and if $\sigma^2 = 100$, then $\sigma_{\bar{X}}^2 = 100/5 = 20$, which means that $\sigma_{\bar{X}} = \sqrt{20} \doteq 4.47$. So,

$$P(490 \leq \bar{x} \leq 510) \doteq P\left(\frac{490 - 500}{4.47} \leq z \leq \frac{510 - 500}{4.47}\right)$$

$$\doteq P(-2.24 \leq z \leq 2.24)$$

$$= \mathbb{P}(2.24) - \mathbb{P}(-2.24)$$

From Table 3 of Appendix VIII, $\mathbb{P}(2.24) \doteq .98745$ and $\mathbb{P}(-2.24) \doteq .01255$. Therefore,

$$P(490 \leq \bar{x} \leq 510) \doteq .98745 - .01255 = .9749$$

so the probability is about .975 that the average page length for a sample of five books would fall between 490 and 510 pages.

Example 8.6. Now, let us suppose that page length is not really normally distributed, but is distributed unimodally and symmetrically, so that means from samples of, say, size 20 could safely be assumed to be normally distributed. Assuming that the authors, sensitive souls that they are, had the fortitude to examine 20 of the dreadful things, what is the probability that the mean page length would fall between 495 and 505 pages?

As in the preceding example, the limits of the interval, 495 and 505, are first standardized:

$$\frac{495 - \mu_{\bar{X}}}{\sigma_{\bar{X}}} = \frac{495 - 500}{\sqrt{100/20}} = \frac{-5}{\sqrt{5}} \doteq \frac{-5}{2.24} \doteq -2.24$$

Similarly,

$$\frac{505 - \mu_{\bar{X}}}{\sigma_{\bar{X}}} \doteq \frac{5}{2.24} \doteq 2.24$$

These are the same standardized values we obtained in Example 8.5, and the probability associated with the interval [495, 505] is, likewise, .975. In this example, however, the interval is *half* as *wide* as the interval in Example 8.5, [490, 510]. Because sample size in Example 8.6 was four times larger than the sample size in Example 8.5, the standard error of the mean $\sigma_{\bar{X}} = \sqrt{\sigma^2/N}$ was half as large. This allowed us to "trap" the same amount of probability in an interval only half as wide (see Figure 8.6). The significance of this will become apparent when we discuss estimation later in the chapter.

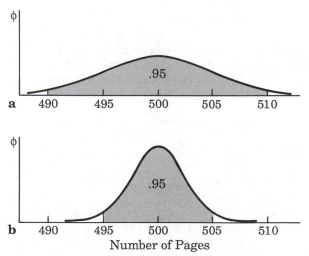

Figure 8.6. Probability density functions for $\bar{X}$ when (a) $N = 5$ and (b) $N = 20$. See Examples 8.5 and 8.6.

—————— EXERCISES 8.2 ——————————————————————————

In Exercises 1–4 we report the mean μ and variance σ^2 of a random variable X in a specific population. Calculate the expected value $E(\bar{X})$ of the mean for a random sample of N observations and the variance $V(\bar{X})$ of the means of all samples of size N, for the values of N given in each exercise.

1. In a population of 12,200 East German civilians (aged 30 to 39) the mean height in inches was $\mu = 67.3$ and the variance was $\sigma^2 = 8$. [Source: Hertzberg, H. T. E. (1972). Engineering anthropology. In H. Van Cott and R. Kinkage (Eds.). *Human Engineering Guide to Equipment Design*. Washington, DC: U.S. Government, p. 482.]

   (a) $N = 4$ (b) $N = 8$
   (c) $N = 50$ (d) $N = 100$

2. In a population of 5,765 20-year-old Norwegian military recruits, the mean weight in pounds was $\mu = 154.2$ and the variance σ^2 was 268.96. (Source: Hertzberg, H. T. E. ibid., p. 489.)

   (a) $N = 4$ (b) $N = 8$
   (c) $N = 16$ (d) $N = 100$

3. In a population of 1,905 Air Force women, the mean knee circumference in inches was $\mu = 14.29$ and the variance $\sigma^2 = 0.7921$. [Source: Clauser, C. et al. (1972). *Anthropometry of Air Force Women*. Wright-Patterson Air Force Base, OH: Aerospace Medical Research Laboratory, p. 149.]

   (a) $N = 2$ (b) $N = 10$
   (c) $N = 50$ (d) $N = 100$

4. In a population of 1,905 Air Force women, the mean hand breadth in inches was $\mu = 2.97$ and the variance was $\sigma^2 = 0.0225$. (Source: Clauser, C. et al., ibid., p. 241.)

   (a) $N = 2$ (b) $N = 10$
   (c) $N = 50$ (d) $N = 100$

Exercises 5–8 refer to the populations in Exercises 1–4. Calculate the probability that the mean $\bar{x}$ of a random sample of the given size N will satisfy the inequalities given. In Exercises 5 and 6 calculate the width of the interval defined by the inequalities.

5. The random sample is drawn from the population in Exercise 1.
   (a) $P(66.41 \leq \bar{x} \leq 68.19)$ for $N = 10$
   (b) $P(66.41 \leq \bar{x} \leq 68.19)$ for $N = 40$
   (c) $P(66.41 \leq \bar{x} \leq 68.19)$ for $N = 90$
   (d) $P(66.85 \leq \bar{x} \leq 67.75)$ for $N = 40$
   (e) $P(67.00 \leq \bar{x} \leq 67.60)$ for $N = 90$
6. The random sample is drawn from the population in Exercise 2.
   (a) $P(152.08 \leq \bar{x} \leq 157.37)$ for $N = 15$
   (b) $P(152.08 \leq \bar{x} \leq 157.37)$ for $N = 60$
   (c) $P(152.08 \leq \bar{x} \leq 157.37)$ for $N = 135$
   (d) $P(153.14 \leq \bar{x} \leq 155.79)$ for $N = 60$
   (e) $P(153.49 \leq \bar{x} \leq 155.26)$ for $N = 135$

7. The random sample is drawn from the population in Exercise 3.
  (a) $P(\bar{x} \le 14.49)$ for $N = 20$
  (b) $P(\bar{x} \le 14.49)$ for $N = 40$
  (c) $P(\bar{x} \le 14.49)$ for $N = 80$

8. The random sample is drawn from the population in Exercise 4.
  (a) $P(\bar{x} \ge 2.99)$ for $N = 15$
  (b) $P(\bar{x} \ge 2.99)$ for $N = 30$
  (c) $P(\bar{x} \ge 2.99)$ for $N = 60$

In order to understand sampling distributions, it helps to see what happens for populations about which you already know everything. Imagine that since the dawn of time, the Astral Accounting Office has recorded the results of every roll of every ordinary six-sided die ever rolled by anyone on this planet. The list of results contains billions of 1s, 2s, 3s, 4s, 5s, and 6s. The relative frequency distribution for this population is modeled by the probability distribution of $X:U(6)$ for $X = \{1, 2, 3, 4, 5, 6\}$, for which we calculated $E(X) = 3.5$ and $V(X) \doteq 2.92$ in Examples 6.3 and 6.5. The samples discussed in Exercises 9 and 10 may be thought of as drawn from this population.

9. Here are 10 samples of size 10. Rolls of the die were simulated by drawing 10 1-digit numbers (discarding draws of 7, 8, 9, and 0) for each sample from Table 1 of Appendix VIII.

				Sample					
1	**2**	**3**	**4**	**5**	**6**	**7**	**8**	**9**	**10**
1	1	1	3	3	2	1	6	2	6
2	5	1	6	6	5	4	6	5	3
2	2	3	4	2	2	5	4	3	3
4	5	4	1	6	6	6	6	4	4
3	1	1	4	2	4	2	2	3	5
2	5	2	1	3	1	6	1	4	6
6	6	2	3	3	6	5	5	3	2
2	1	6	3	6	6	3	2	2	4
5	4	5	2	5	3	5	4	1	2
5	2	1	2	6	4	6	1	3	1

$\bar{x}$

  (a) Calculate the means of the 10 samples, and organize them in a grouped frequency distribution. How does the "shape" of the distribution of means compare with the "shape" of the distribution of the random variable X?
  (b) Calculate the mean of the 10 sample means. What is the difference between this value and $E(X)$?
  (c) Calculate the variance of your 10 sample means. How is this variance related to $V(X)$?

10. Here are five samples of size 20 from our population. Draw five more samples of size 20. Collect data either by rolling a die 100 times or by simulation, using Table 1 of Appendix VIII as described in Exercise 9.

	Sample									
	1	**2**	**3**	**4**	**5**	**6**	**7**	**8**	**9**	**10**
	1	1	3	3	2	1	6	2		6
	2	5	1	6	6	5	4	6	5	3
	2	2	3	4	2	2	5	4	3	3
	4	5	4	1	6	6	6	6	4	4
	3	1	1	4	2	4	2	2	3	5
	2	5	2	1	3	1	6	1	4	6
	6	6	2	3	3	6	5	5	3	2
	2	1	6	3	6	6	3	2	2	4
	5	4	5	2	5	3	5	4	1	2
	5	2	1	2	6	4	6	1	3	1
$\bar{x}$										

(a) Calculate the means of the 10 samples, and organize them in a grouped frequency distribution. How does the "shape" of the distribution of means compare with the "shape" of the distribution of the random variable X? How does the "shape" of the distribution of means compare with the "shape" of the distribution of means in Exercise 9?

(b) Calculate the mean of the 10 sample means. What is the difference between this value and $E(X)$? Is the mean closer to or more distant from $E(X)$ than the mean of the 10 sample means calculated in Exercise 9?

(c) What does your answer to (b) suggest about the relationship of sample size to the variability of the sample means?

(d) Support your answer to (c) by comparing the variance of the 10 sample means in these two exercises with $V(X)$.

Last year the second author took advantage of a TV offer and bought a 3-million volume list of all past and future coin flips in the United States. This handsome, exclusive leather-bound compendium has four columns: date and place of flip; type of coin; name of person flipping; result. Exercises 11 and 12 involve samples from the population consisting of all results on his list.

11. Here are 20 samples of size five of the many trillions of samples of size five we could have taken from the population. The number (X) of heads and the proportion $\left(\frac{X}{N}\right)$ of heads are recorded for each sample.

	Sample																			
	1	**2**	**3**	**4**	**5**	**6**	**7**	**8**	**9**	**10**	**11**	**12**	**13**	**14**	**15**	**16**	**17**	**18**	**19**	**20**
	T	H	H	H	H	T	H	H	H	H	H	H	T	T	T	T	T	T	T	H
	T	H	T	H	T	H	H	H	H	T	T	T	T	T	T	T	H	H	H	H
	H	H	T	H	H	H	T	H	H	H	T	H	H	T	H	H	T	T	T	H
	H	T	T	T	T	H	H	H	T	T	T	H	T	T	H	T	T	H	T	T
	H	H	T	H	T	T	T	H	T	H	T	T	H	T	T	H	H	T	T	H
X	3	4	1	4	2	3	3	5	2	3	1	2	2	0	3	3	2	2	2	3
$\dfrac{X}{N}$	.6	.8	.2	.8	.4	.6	.6	1.	.4	.6	.2	.4	.4	0	.6	.6	.4	.4	.4	.6

(a) Calculate the mean and standard deviation of the 20 proportions. Use these values to determine the values $\left(\frac{x}{N}\right)_a$ and $\left(\frac{x}{N}\right)_b$ that correspond to standardized values $z_a = -0.9$ and $z_b = +0.9$.

(b) What is the width of the interval $\left[\left(\frac{x}{N}\right)_a, \left(\frac{x}{N}\right)_b\right]$?

(c) Use Table 3 of Appendix VIII to calculate the percentage of observations you should expect between $z = -0.9$ and $z = +0.9$ if Z is the standard normal random variable.

(d) What percentage of the 20 sample proportions actually fall between $\left(\frac{x}{N}\right)_a$ and $\left(\frac{x}{N}\right)_b$?

12. Toss 10 coins and record both the number X and the proportion $\frac{X}{N}$ of heads. Repeat the experiment 20 times.

(a) Calculate the mean and standard deviation of your 20 proportions. Use these values to determine values $\left(\frac{x}{N}\right)_a$ and $\left(\frac{x}{N}\right)_b$ that correspond to standardized values $z_a = -0.9$ and $z_b = +0.9$.

(b) What is the width of the interval $\left[\left(\frac{x}{N}\right)_a, \left(\frac{x}{N}\right)_b\right]$?

(c) Use Table 3 of Appendix VIII to calculate the percentage of observations you should expect between $z = -0.9$ and $z = +0.9$ if Z is the standard normal random variable.

(d) What percentage of your 20 sample proportions actually fall between $\left(\frac{x}{N}\right)_a$ and $\left(\frac{x}{N}\right)_b$?

(e) Why did you manage to capture (approximately) the same percentage of your sample proportions as in Exercise 11 in an interval that is only about 70 percent as wide?

13. Ability scores for *Advanced Dungeons & Dragons* ® characters are determined by rolling three 6-sided dice and adding up their values.

(a) How are the values on a single six-sided die distributed?

(b) The *Dungeon Masters Guide* (1st edition) claims that the distribution of ability scores is "bell-shaped." Given your answer to (a) can this be correct? If so, state the law, theorem, rule, etc., that justifies the claim.

14. In *Advanced Dungeons & Dragons* ® damage inflicted by the magic spell *Fireball* is calculated by throwing one 6-sided die for each proficiency level of the spell caster and adding the spots (or numbers) that appear. For example, a sixth-level magic-user rolls six dice, a seventh level magic-user rolls seven dice, etc. Is a sixth-level magic user as justified in assuming the normality of his *Fireball* damage as, say, a fifteenth-level magic-user? Why or why not?

15. The *Advanced Dungeons & Dragons* ® spell *Lightning* also does damage calculated by throwing one 6-sided die per proficiency level of the caster.

(a) How many points of damage *per die* should a sixth-level magic user expect to inflict?

(b) How many points of damage *per die* should a fifteenth-level magic user expect to inflict?

16. An *AD & D* player notices that the average *Lightning* damage inflicted per die by his sixth-level magic user departs more from the expected value (per die) than does the damage inflicted by a fifteenth-level magic user. Explain this in terms of the expected variance per die for a sixth- and for a fifteenth-level magic user.

C. ESTIMATION

One of the ways in which statisticians make inferences about population distributions is to *estimate* the population values that describe various distribution properties. For example, the manufacturer of extended-life light bulbs in Example 8.1 might want to *estimate* the mean hours of burn time for all bulbs in a single production run; the client who commissioned the survey from which we obtained income data might be interested in *estimating* the population correlation between income and years of education; the developmental physiologist in Example 8.2 might want to *estimate* the variance of heights in university freshman women.

Statisticians have developed two general methods that permit such estimation. The first is *point estimation*, which involves only a single value, or point. The developmental physiologist, for example, might estimate the population variance of heights in freshman women to be 34.23.

Point estimates are intuitively appealing, but if a variable is continuous, many of the parameters (including the mean and variance) that describe its population distribution are also continuous. The probability is zero that the physiologist's estimate is *exactly* equal to a parameter that can, in principle, be calculated to an infinite number of decimal places. Alternatively, therefore, she might estimate that the value of her parameter is included in an *interval*, such as [26.16, 45.66]. This would permit her to assign a nonzero probability to her estimate and is an example of *interval* estimation.

1. Point estimation

> A point estimator $\hat{\theta}$ is a single-valued statistic that approximates the value of a population parameter θ.

We know from Part I that calculation of sample statistics requires only the observations at hand, but we saw in Part II that determination of parameter values often requires considerably more information. For example, calculation of the expected value

$$\mu = \sum x_i P(x_i)$$

of a discrete random variable requires that we know every value that the random variable can take and the probability associated with each of these values. Calculation of the variance

$$\sigma^2 = \sum (x_i - \mu)^2 P(x_i)$$

requires, in addition, the expected value, μ. Such information is generally implicit in a *probability* problem, but in *statistical* problems, where probability distributions are *models* for population relative frequency distributions, such information is seldom available: If the statistician knew the population values, there would be no need to make inferences about them! Most often, therefore, we must rely on point estimates based on sample data to *approximate* the population values.

At the beginning of the chapter we said that Roman letters, such as s and r, are used to indicate statistics and that Greek letters, such as σ and ρ, indicate population values. Because an estimate is calculated from sample data, it is a type of statistic. However, point estimates are often substituted for unknown parameter values, so we shall denote estimators with Greek letters. To distinguish estimators from parameters, they will be identified with a circumflex (^), or "hat." Thus, an estimator of the population mean would be indicated by $\hat{\mu}$, which is read "estimate of mu" or, more simply, "mu hat."

a. Desirable properties of point estimators. When we introduced the notion of a descriptive statistic in Chapter 2, we indicated the possibility of formulating many different expressions to describe any particular property of a distribution. Similarly, we could probably conceive of many ways to estimate any population value. As with descriptive statistics, however, there are various criteria that make some estimators more acceptable than others.

The most desirable feature of any estimator is that it yield accurate estimates. This is true whether we are talking about the methods an auto mechanic uses to estimate repair costs, the methods by which the Office of Management and Budget determines its projections of federal revenues, or the calculations a wildlife biologist performs to estimate the population of lions in the Serengeti Plain. In all cases we want the value of the estimator $\hat{\theta}$ to be as close as possible to the population value θ that it estimates.

An *estimate* is a particular value calculated from a particular sample of observations, but an *estimator*, like any statistic, is a random variable. If one were to draw *many* samples and calculate an estimate of the parameter θ from each sample, one would probably obtain many different values for θ. An estimator therefore has a sampling distribution characterized by various properties, such as location and dispersion. These properties generate a number of different ways to interpret the notion of "accuracy" and therefore play an important role in the criteria that make one estimator preferable to another.

i. An estimator should be unbiased. If we think in terms of drawing many samples, one way to define accuracy is to say that *on the average*, the value of $\hat{\theta}$ should equal the population value θ that it estimates. An estimator that possesses this property is said to be *unbiased*.

$\hat{\theta}$ is an *unbiased* estimator of the parameter θ if the expected value of $\hat{\theta}$ is equal to θ. That is, $\hat{\theta}$ is unbiased if $E(\hat{\theta}) = \theta$.

---

ii. *An estimator should be consistent.* Another way to think about the accuracy of an estimator is in terms of the *absolute* difference between the estimate and the population value, $|\hat{\theta} - \theta|$. If we use only *one* member of a large population to determine an estimate $\hat{\theta}$, we wouldn't ordinarily expect it to be very accurate. On the other hand, if we observe *every* member of the population, we can calculate the exact value of θ. Intuitively, we therefore expect an estimator to become more accurate as sample size increases. That is, as N increases we expect that $|\hat{\theta} - \theta|$ should become smaller.

Realistically, of course, one can never be certain that a large sample will *always* give us a smaller absolute difference $|\hat{\theta} - \theta|$ than a smaller sample. Even with all the safeguards of random sampling an experimenter might draw a very large sample that yields an estimate $\hat{\theta}$ that is wildly different from θ. What we *can* require is that such an event be *unlikely*. That is, we can require that the *probability* that $\hat{\theta}$ is close to θ tend to increase as sample size increases.

The notion that $\hat{\theta}$ is "close to" θ is mathematically expressed by saying that the absolute difference $|\hat{\theta} - \theta|$ is smaller than an arbitrarily chosen small positive value, ordinarily denoted ε (the lowercase Greek *epsilon* for "*e*rror tolerance"). What we want, then, is the following: For any value of ε we choose, the probability that $|\hat{\theta} - \theta|$ is less than ε tends to increase with sample size. If $P(|\hat{\theta} - \theta|) < \varepsilon$ approaches 1 as N increases, $\hat{\theta}$ is a *consistent* estimator of θ.[2]

$\hat{\theta}$ is said to be a *consistent* estimator of θ if for any positive value ε,

$$P(|\hat{\theta} - \theta| < \varepsilon)$$

tends to certainty ($p = 1$) as sample size increases.

Since ε is always a *positive* number, the inequality that defines consistency means

$$P(-\varepsilon < \hat{\theta} - \theta < \varepsilon)$$

which is equivalent to

$$P\left[-\varepsilon < (\hat{\theta} - \theta) \quad \text{and} \quad (\hat{\theta} - \theta) < \varepsilon\right]$$

[2] It must be understood that consistency does not imply that $P(|\hat{\theta} - \theta| < \varepsilon)$ is *always* greater for the larger of any two particular sample sizes. To say that $\hat{\theta}$ is consistent means only that for any positive value ε and any probability value $p < 1$, there is *some* sample size N beyond which $P(|\hat{\theta} - \theta| < \varepsilon)$ is always greater than p.

or, adding θ to both sides of each inequality,

$$P(\theta - \varepsilon < \hat{\theta} \quad \text{and} \quad \hat{\theta} < \theta + \varepsilon)$$

It follows that $\hat{\theta}$ is a consistent estimator of θ if

$$P(\theta - \varepsilon < \hat{\theta} < \theta + \varepsilon)$$

tends to 1 as sample size increases.

An estimator that is accurate by one criterion is not necessarily accurate by another. For example, let us suppose that a lazy researcher wants to estimate the mean μ of heights of students in his department. He has his secretary invite a random sample of N students to his office. The worker measures the first student who arrives and then quits for the day, telling his secretary to use the one height as the estimate and to send everyone else home. If all of the students are instructed to show up at the same time, every student is equally likely to be the first in the door. Thus, the expected height of the first student

$$E(\hat{\theta}) = E(X) = \mu$$

so $\hat{\theta}$ is unbiased, but the estimate $\hat{\theta}$ is no more likely to be near μ whether the secretary invites a sample of 10 students or 100 students or 1,000 students. Although the estimator is unbiased it is not, therefore, consistent.

It is also the case that a consistent estimator can be biased. In the next section, for example, we shall see that the variance s^2 of a sample of N observations is a biased estimator of the population variance σ^2 but is nevertheless consistent.

iii. An estimator should be efficient. Since an estimator is a random variable, it is necessarily characterized by a certain amount of variability, and some estimators may be more variable than others. Figure 8.7, for example, shows the sampling distributions of two unbiased estimators, $\hat{\theta}_1$ and $\hat{\theta}_2$. The sampling distribution of $\hat{\theta}_1$ is considerably more variable than the sampling distribution of $\hat{\theta}_2$, so for any particular sample size N,

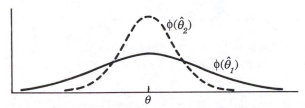

Figure 8.7. Sampling distributions for two estimators $\hat{\theta}_1$ and $\hat{\theta}_2$ of the parameter θ.

the estimator $\hat{\theta}_2$ has the greater likelihood of being near θ. Accuracy in this sense is called *efficiency*.

> The most *efficient* of all unbiased estimators of the parameter θ is the one with the smallest variance for any given sample size.

The *relative* efficiency of two unbiased[3] estimators $\hat{\theta}_1$ and $\hat{\theta}_2$ can therefore be calculated by taking the ratio

$$\frac{V(\hat{\theta}_1)}{V(\hat{\theta}_2)}$$

b. Estimating μ: properties of the sample mean. Earlier in the chapter we proved that $E(\overline{X}) = \mu$. The sample mean is therefore an unbiased estimator of the population mean.

It was also demonstrated that $\sigma_{\overline{X}}^2 = \sigma^2/N$. Therefore, as N increases, $\sigma_{\overline{X}}^2$ tends to zero, and we saw in Examples 8.5 and 8.6 that the interval required to "trap" any amount of probability under the distribution of $\overline{X}$ consequently becomes narrower as sample size gets larger (see Figure 8.6, p. 291). Conversely, if we choose any two *fixed* points, $\mu - \varepsilon$ and $\mu + \varepsilon$, the area between them must *increase* as sample size gets larger (see Figure 8.8). That is, as N increases,

$$P(\mu - \varepsilon \leq \overline{x} \leq \mu + \varepsilon)$$

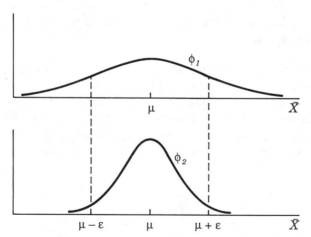

Figure 8.8. Area between $\mu - \varepsilon$ and $\mu + \varepsilon$ for distribution of $\overline{X}$ when N is small (ϕ_1) and when N is large (ϕ_2).

---

[3] We have confined our definition of efficiency to unbiased estimators. If the notion is to be generalized to all estimators, then the most efficient estimator is the one for which $E[(\hat{\theta} - \theta)^2]$ is smallest. It is only when $\hat{\theta}$ is unbiased that $E[(\hat{\theta} - \theta)^2] = V(\hat{\theta})$.

increases, which means that $\overline{X}$ satisfies our definition of a consistent estimator.

The matter of relative efficiency is more complicated, since it requires us to compare all possible unbiased estimators of μ. It is possible to show that among all "reasonable" candidates for estimators of μ, the sample mean is the most efficient.[4]

A special case: relative frequency as an estimator of p. In Chapter 5 we developed an example (5.20) to illustrate that $|(f/N) - p|$ tends to decrease as the sample size N increases. We can now see that this so-called Law of Large Numbers is simply an informal way of saying that relative frequency is a consistent estimator of probability.

To understand why this is so, consider again the problem faced by the experimenter in Example 5.20. He had an urn with m balls of various colors and wanted to estimate p, the probability of the event "obtain a red ball." He therefore drew a sample of N balls, let f be the number of red balls in his sample, and let $f/N = \hat{p}$. One way to calculate f is to assign a value of 1 to each red ball, assign a value of 0 to each non-red ball, and let f equal the sum of the values assigned by this rule. That is, each observation can be thought of as an indicator random variable and the frequency f of the event "obtain a red ball" as the sum of N indicator random variables:

$$f = \sum^{N} I$$

Therefore,

$$\frac{f}{N} = \frac{\sum^{N} I}{N} = \overline{I}$$

From Chapter 6, we know that the expected value of each of these indicator variables is equal to p, the probability of drawing a red ball. The relative frequency f/N can therefore be thought of as the *sample mean* of N observations on a random variable with expected value $\mu = p$. We know that the sample mean is an unbiased and consistent estimator of μ, so relative frequency must be an unbiased and consistent estimator of p.

c. Estimating σ^2. The value set of the random variable $\overline{X}$ is the set of all means $\bar{x}$ for samples of a given size N. Similarly, the value set of the random variable S^2 is the set of all variances s^2 for samples of a given size N. For a finite population of size K, we know that the variance s^2 of an exhaustive sample is the same as the population variance σ^2. Intuitively, it might therefore seem that S^2 should be an unbiased estimator

---

[4] "Unreasonable" estimators include all functions in which the x-values are squared, cubed, or otherwise raised to a power other than 1. Hence, the sample mean is more efficient than any other *linear* estimator of μ.

| BOX 8.2 |

S^2 a biased estimator of population variance

Proof: If $V(X) = \sigma^2$ and $E(X) = \mu$, then from equation [6.7]

$$\sigma^2 = E(X^2) - [E(X)]^2$$

so

$$E(X^2) = \sigma^2 + [E(X)]^2$$

$$= \sigma^2 + \mu^2 \qquad [1]$$

Now let $X_1, X_2, \ldots, X_N$ represent N independent observations on the random variable X, where $E(X) = \mu$ and $V(X) = \sigma^2$. From the computational formula for variance given in equation [2.11]:

$$S^2 = \frac{\sum\limits_{}^{N} X^2}{N} - \bar{X}^2$$

Therefore,

$$E(S^2) = E\left(\frac{\sum\limits_{}^{N} X^2}{N} - \bar{X}^2 \right)$$

$$= E\left(\frac{\sum\limits_{}^{N} X^2}{N} \right) - E(\bar{X}^2) \qquad [2]$$

The constant $1/N$ can be factored out of the first term, so

$$E\left(\frac{\sum\limits_{}^{N} X^2}{N} \right) = \frac{1}{N} E\left(\sum\limits_{}^{N} X^2 \right)$$

$$= \frac{1}{N} E(X_1^2 + X_2^2 + \cdots + X_N^2)$$

$$= \frac{1}{N} [E(X_1^2) + E(X_2^2) + \cdots + E(X_N^2)] \qquad [3]$$

Since $X_1, X_2, \ldots, X_N$ are a random sample, each has a distribution identical to that of X. Thus,

$$E(X_1^2) = E(X_2^2) = \cdots = E(X_N^2) = E(X^2)$$

$$= \sigma^2 + \mu^2 \text{ by equation } [1]$$

Therefore, equation [3] becomes

$$E\left(\frac{\sum\limits_{}^{N} X^2}{N}\right) = \frac{1}{N}\left[(\sigma^2 + \mu^2) + (\sigma^2 + \mu^2) + \cdots + (\sigma^2 + \mu^2)\right]$$

$$= \frac{1}{N}\left[N(\sigma^2 + \mu^2)\right]$$

$$= \sigma^2 + \mu^2 \qquad\qquad [4]$$

Now, let us turn our attention to the second term in equation [2], $E(\bar{X}^2)$.

If we apply [1] to the random variable $\bar{X}$, we see that $E(\bar{X}^2) = \sigma_{\bar{X}}^2 + \mu_{\bar{X}}^2$ and by equation [8.7] we know that $\sigma_{\bar{X}}^2 = \sigma^2/N$. Therefore,

$$E(\bar{X}^2) = \sigma_{\bar{X}}^2 + \mu_{\bar{X}}^2$$

$$= \frac{\sigma^2}{N} + \mu^2 \qquad\qquad [5]$$

Combining equations [4] and [5], equation [2] becomes

$$E(S^2) = (\sigma^2 + \mu^2) - \left(\frac{\sigma^2}{N} + \mu^2\right)$$

$$= \sigma^2 - \frac{\sigma^2}{N} \qquad\qquad [6]$$

of σ^2, but it is demonstrated in Box 8.2 that

$$E(S^2) = \sigma^2 - \frac{\sigma^2}{N} \qquad\qquad [8.8]$$

Since $E(S^2) \neq \sigma^2$, the statistic S^2 is not an unbiased estimator of the population variance σ^2. However, equation [8.8] allows us to derive an estimator of σ^2 that *is* unbiased:

$$E(S^2) = \sigma^2 - \frac{\sigma^2}{N}$$

$$= \sigma^2\left(1 - \frac{1}{N}\right) = \sigma^2\left(\frac{N-1}{N}\right)$$

Therefore,

$$\left(\frac{N}{N-1}\right)E(S^2) = \sigma^2$$

Since $N/(N-1)$ is a constant,

$$\left(\frac{N}{N-1}\right)E(S^2) = E\left(S^2\frac{N}{N-1}\right)$$

and

$$E\left(S^2\frac{N}{N-1}\right) = \sigma^2$$

Therefore,

$$S^2\left(\frac{N}{N-1}\right)$$

is an unbiased estimator of σ^2. For any particular sample of N observations,

$$s^2\left(\frac{N}{N-1}\right) = \frac{\sum\limits^{N}(x-\bar{x})^2}{N} \cdot \frac{N}{N-1}$$

$$= \frac{\sum\limits^{N}(x-\bar{x})^2}{N-1} \qquad\qquad [\textbf{8.9}]$$

Expression [**8.9**] is the definitional formula for the *unbiased estimate of the population variance* and is denoted $\hat{\sigma}^2$ throughout the remainder of the text.[5]

> The statistic
>
> $$\hat{\sigma}^2 = \frac{\sum\limits^{N}(x-\bar{x})^2}{N-1}$$
>
> is an unbiased estimate of the population variance σ^2.

[5] Expression [**8.9**] is sometimes called the "sample estimate of population variance" or, more simply, the "sample variance." We have avoided the latter term because "sample variance" is easily misunderstood to mean the variance of the sample. The variance of a sample (or any collection of N observations) is expression [**2.8**],

$$s^2 = \frac{\sum(x-\bar{x})^2}{N}$$

Similarly, we use $\hat{\sigma}$ to denote

$$\sqrt{\frac{\sum\limits_{N}(x - \bar{x})^2}{N - 1}}$$

though it should be noted that $\hat{\sigma}$ is *not* an unbiased estimate of the population standard deviation σ.

It can also be shown that $\hat{\sigma}^2$ is consistent as well as unbiased. And, since

$$\hat{\sigma}^2 = s^2\left(\frac{N}{N - 1}\right)$$

it is apparent that as N increases, the difference between s^2 and $\hat{\sigma}^2$ must decrease. Consequently, S^2 is a consistent estimator of σ^2, even though it is biased.

2. Interval estimation

One of the more familiar applications of inferential statistics in everyday life is survey research. Several national organizations take weekly polls in which respondents are asked to agree or disagree with statements about current issues. Results are regularly reported in newspapers and on National Public Radio. Pollsters typically report the number of persons participating in the survey (sample size) and the percentage (f/N) of participants who agreed with each statement. In addition, most surveys include a statement like "*The margin of error is plus or minus two percentage points.*" This tells the reader or listener that (1) each percentage f/N is an *estimate $\hat{p}$* of the proportion of persons in the population who would be expected to agree with the statement and that (2) the population value p lies between $\hat{p} + .02$ and $\hat{p} - .02$. The "two percentage points" in this example is what statisticians call the *maximum error of estimate*, and the range of values (e.g., $\hat{p} \pm .02$) that is expected to include the population value p is called a *confidence interval*.

> A confidence interval is a range of values, w_a to w_b, that is expected to include the population parameter θ.

Clearly, a report that 50 percent of one's sample agreed with a particular statement and that the maximum error of estimate was .50 would not be very useful. One does not need a survey to know that the percentage of the population that approves of this public policy or that political candidate lies somewhere between 0 percent and 100 percent. The narrower the interval, the more precise the estimate.

However, a *precise* estimate is not necessarily *accurate*. A prediction that candidate X is expected to receive between 50.51 and 50.52 percent of the popular vote is very precise, but if the pollster who published the prediction had in the past been correct only 1 time in 20, one would not place much confidence in the accuracy of the prediction.

In the matter of predicting election results, methods of sampling have become so sophisticated in recent years that relatively small samples can now yield estimates that are both precise (narrow) and accurate (characterized by high confidence). Exit polls taken during the 1984 presidential election made it possible for the three major television networks to project Ronald Reagan's victory less than 15 minutes after polling places closed on the east coast, and Congress considered legislation prohibiting announcement of network projections until 11:00 pm EST for fear that projections might influence election results in more westerly time zones.

The methods that allow a survey researcher or an election-day pollster to calculate a suitably narrow interval and to quantify the likelihood, or confidence, that the population parameter p is included in that interval are the business of interval estimation.

a. Interval estimates of μ for large samples. Confidence intervals may be determined for a variety of population values besides p, including the population variance σ^2 (see Box 11.2, p. 467) and correlation coefficient ρ. In this section we shall consider interval estimation of the population mean μ.

For any standardized normal random variable Z, we know from Table 3 in Appendix VIII that $\mathbb{P}(-1.96) \doteq .025$ and that $\mathbb{P}(1.96) \doteq .975$, so

$$P(-1.96 \leq z \leq 1.96) \doteq .95$$

or,

$$P\left(-1.96 \leq \frac{x - \mu}{\sigma} \leq 1.96\right) \doteq .95 \qquad \textbf{[8.10]}$$

Consider the left-hand inequality:

$$-1.96 \leq \frac{x - \mu}{\sigma}$$

so,

$$-1.96\sigma \leq x - \mu$$

and

$$-x - 1.96\sigma \leq -\mu \qquad \textbf{[8.11]}$$

Multiplying both sides of equation **[8.11]** by -1,

$$\mu \leq x + 1.96\sigma \qquad \textbf{[8.12]}$$

Now, consider the right-hand inequality in equation [8.10]:

$$\frac{x - \mu}{\sigma} \leq 1.96$$

so,

$$x - \mu \leq 1.96\sigma$$

and

$$-\mu \leq -x + 1.96\sigma \qquad\qquad [8.13]$$

Multiplying both sides of [8.13] by -1,

$$\mu \geq x - 1.96\sigma$$

which is the same as

$$x - 1.96\sigma \leq \mu \qquad\qquad [8.14]$$

From equations [8.12] and [8.14], equation [8.10] can be rewritten:

$$P(x - 1.96\sigma \leq \mu \leq x + 1.96\sigma) \doteq .95 \qquad\qquad [8.15]$$

Equation [8.15] says that for any normally distributed random variable X, the probability is .95 that μ lies between $x - 1.96\sigma$ and $x + 1.96\sigma$. That is, if one were to draw a large number of x-values from the population and calculate the interval $[(x - 1.96\sigma), (x + 1.96\sigma)]$ for *every* x-value, one would expect 95 percent of these intervals to include μ.

Strictly speaking, then, equation [8.15] is only meaningful *until* a value of x is specified. As soon as x assumes a specific numerical value, the limits $x \pm 1.96\sigma$ define a *fixed* interval, which either includes the constant μ (in which case the probability in [8.15] is 1) or does not include μ (in which case the probability in [8.15] is 0). This restriction does not, of course, extend to *subjective* certainty. Let us suppose that X is a normally distributed random variable with $E(X) = \mu$ and variance $V(X) = 100$, and let us suppose that an experimenter draws a single observation $x = 485$ and calculates interval limits

$$x - 1.96\sigma = 485 - 1.96(10) = 465.4$$

and

$$x + 1.96\sigma = 485 + 1.96(10) = 504.6$$

Because 95 percent of *all* such intervals include the population mean, the experimenter could be 95 percent certain that μ lies somewhere in the interval [465.4, 504.6]. This subjective, probabilistic certainty is called *confidence*. We use the symbol $\mathbb{C}$ to denote confidence, and confidence statements are written in much the same way as probability statements.

BOX 8.3

Confidence and cumulative probability

If $X : N(\mu, \sigma^2)$, then

$$\mathbb{C}(x - z_b\sigma \leq \mu \leq x + z_b\sigma) = P(-z_b \leq z \leq z_b)$$

under the standard normal curve. That is, $\mathbb{C}$ is equal to the probability that z is in the interval $[-z_b, z_b]$. The probability that z falls *outside* the interval $[-z_b, z_b]$ must therefore equal $1 - \mathbb{C}$. That is, $1 - \mathbb{C}$ equals

$$P(z < -z_b \quad \text{or} \quad z > z_b) = P(z < -z_b) + P(z > z_b)$$

Because the standard normal curve is symmetrical about zero, $P(z < -z_b) = P(z > z_b)$.

Therefore,

$$1 - \mathbb{C} = 2P(z > z_b)$$

From our discussion of cumulative probability we know that $P(z > z_b) = 1 - \mathbb{P}(z_b)$, so

$$1 - \mathbb{C} = 2[1 - \mathbb{P}(z_b)]$$

and

$$\frac{1 - \mathbb{C}}{2} = 1 - \mathbb{P}(z_b)$$

Therefore,

$$\mathbb{P}(z_b) = \frac{1 + \mathbb{C}}{2}$$

In this example, for instance, we would write $\mathbb{C}(465.4 \leq \mu \leq 504.6) = .95$ or $\mathbb{C} = .95$ that $(465 \leq \mu \leq 504.6)$.

The strategy that led to equation **[8.15]** can be used to calculate any confidence interval for μ when the random variable X is normally distributed: Once the degree of confidence $\mathbb{C}$ has been established, one can use a table of standard normal values to find $-z_b$ and z_b such that $P(-z_b \leq z \leq z_b) = \mathbb{C}$. For any observed value x, the appropriate confidence limits will therefore be $x - z_b\sigma$ and $x + z_b\sigma$. The values $-z_b$ and z_b are called the confidence *coefficients*, and in Box 8.3 we show that for any confidence level $\mathbb{C}$, the coefficient z_b is such that

$$\mathbb{P}(z_b) = \frac{1 + \mathbb{C}}{2} \tag{8.16}$$

The confidence coefficient z_b is therefore the z-value for which $\mathbb{P} = (1 + \mathbb{C})/2$ in Table 3 of Appendix VIII.

Example 8.7. Suppose an experimenter wants a 90 percent confidence interval for the mean of a normally distributed random variable X with variance $\sigma^2 = 225$. The coefficient z_b is the z-value for which cumulative probability $\mathbb{P}$ equals

$$\frac{1 + .90}{2} = .95$$

From Table 3, $\mathbb{P}(1.64) \doteq .95$, so $P(-1.64 \le z \le 1.64) \doteq .90$. If a single observation on the random variable yields, for example, the value $x = 55$, the experimenter can therefore be 90 percent certain that the population mean μ lies between

$$55 - 1.64\sigma = 55 - 1.64(15) = 30.4$$

and

$$55 + 1.64\sigma = 55 + 1.64(15) = 79.6$$

A 90 percent confidence interval for μ is therefore [30.4, 79.6].

Example 8.8. In Example 8.5 we said that the variance for page length of introductory statistics books is 100 and that page length is normally distributed. If, in addition, we know that one particular book is 479 pages long, how wide an interval would we need to be 99 percent certain that the interval includes the mean page length of the entire population of introductory statistics books? The confidence coefficients are the values $\pm z_b$, where $P(-z_b \le z \le z_b) = .99$. From equation [8.16],

$$\mathbb{P}(z_b) = \frac{1 + .99}{2} = .995$$

and in Table 3 we find that $\mathbb{P}(2.58) \doteq .995$. Therefore, we are 99 percent confident that mean page length μ for the population of introductory statistics textbooks is no less than

$$479 - 2.58(10) = 453.2$$

and no greater than

$$479 + 2.58(10) = 504.8$$

The width of this interval is about 52 pages.

Although one can use a single observation x to compute an interval estimate for μ to any desired degree of confidence, this approach has two inconvenient limitations. First, it requires that X be normally distributed. In the case of page length, this assumption is difficult to justify.

Second, the actual width of the interval depends entirely on the standard deviation σ. When σ is large, any interval we compute might be too wide to be useful. For example, even if we were 99 percent certain that the mean length for all introductory statistics books is between 453 and 505, an interval this imprecise doesn't tell us very much.

In many situations, however, both of these problems can be overcome if we base our interval estimate for μ on the sampling distribution of $\bar{X}$. First, we know that $\mu_{\bar{X}} = \mu$. Second, we know from the Central Limit Theorem that however X may be distributed, the sample mean, at least for large samples, is normally distributed. Therefore, if $E(X) = \mu$ and $V(X) = \sigma^2$, then confidence equals $\mathbb{C}$ that

$$(\bar{x} - z_b\sigma_{\bar{X}}) \le \mu \le (\bar{x} + z_b\sigma_{\bar{X}}) \qquad\qquad [8.17]$$

where

$$\mathbb{P}(z_b) = \frac{1 + \mathbb{C}}{2} \quad \text{and} \quad \sigma_{\bar{X}}^2 = \frac{\sigma^2}{N}$$

Example 8.9. Let us suppose that instead of recording the page length of only *one* introductory statistics book, the authors examined 20 books, as described in Example 8.6, and found the mean page length for this sample to be 479. Once again, let us suppose that we want a 99 percent confidence interval for the mean μ of all introductory statistics books. We know from Example 8.8 that for any normally distributed random variable X, with expected value $E(X) = \mu$ and variance $V(X) = \sigma^2$, then $\mathbb{C} = .99$ that

$$(x - 2.58\sigma) \le \mu \le (x + 2.58\sigma)$$

It must therefore be true that confidence is .99 that

$$(\bar{x} - 2.58\sigma_{\bar{X}}) \le \mu_{\bar{X}} \le (\bar{x} + 2.58\sigma_{\bar{X}})$$

or, because $\mu_{\bar{X}} = \mu$ and $\sigma_{\bar{X}} = \sigma/\sqrt{N}$, that

$$\left(\bar{x} - 2.58\frac{\sigma}{\sqrt{N}}\right) \le \mu \le \left(\bar{x} + 2.58\frac{\sigma}{\sqrt{N}}\right)$$

Since $\sigma^2 = 100$, the standard error of the mean $\sigma/\sqrt{N}$ equals

$$10/\sqrt{20} \doteq 10/(4.47) \doteq 2.24$$

and the limits of our 99 percent confidence interval are

$$479 - (2.58)2.24 \doteq 473.2$$

and

$$479 + (2.58)2.24 \doteq 484.8$$

In Example 8.8 the width of the 99 percent confidence interval based on a single observation was about 52 pages. In Example 8.9, where $N = 20$, the 99 percent confidence interval decreased to about 12 pages. This is because $\sigma_{\bar{X}}^2 = \sigma^2/N$, so the standard error of the mean $\sigma_{\bar{X}}$ becomes smaller as sample size increases. For any level of confidence $\mathbb{C}$ we can therefore calculate the sample size required to make the interval $\bar{x} \pm z_b \sigma_{\bar{X}}$ as narrow as we want by setting the maximum error of estimate equal to $z_b \sigma_{\bar{X}}$ and solving for N.

Example 8.10. How many books would we have to read to reduce our 99 percent confidence interval to $\bar{x} \pm 2$ pages? The maximum error of estimate is 2, and since $\mathbb{C} = .99$ we know that $z_b = 2.58$. Therefore,

$$2.58\sigma_{\bar{X}} = 2.58\frac{\sigma}{\sqrt{N}} = 2$$

We know that $\sigma^2 = 100$, so $\sigma = 10$. Therefore,

$$2.58\frac{10}{\sqrt{N}} = 2$$

and

$$2.58(10) = 2\sqrt{N}$$

which give us

$$N = \left(\frac{25.8}{2}\right)^2 \doteq 166$$

Despite the obvious advantages of calculating interval estimates of μ based on the sampling distribution of $\bar{X}$, the calculations still require that the experimenter provide a value for σ, the standard deviation of X. It is unrealistic to expect that an experimenter who must *estimate* the population mean μ will *know* the population standard deviation σ. If, however, $N \geq 30$, one may substitute $\hat{\sigma}$ for σ with relatively little error.[6]

[6] The confidence coefficient

$$\frac{\bar{x} - \mu}{\hat{\sigma}/\sqrt{N}}$$

with cumulative probability equal to $(1 + \mathbb{C})/2$ is obtained from *Student's t* distribution, which is introduced in Chapter 10 (see Box 10.3). For $N \geq 30$ observations, the normal distribution provides a very satisfactory approximation of *Student's t*.

Example 8.11. Thirty-six seniors from a particular high school take the mathematics placement test given by the local community college. The average score for these students is 28 and the variance is 446. Calculate a 95 percent confidence interval for the mean of all seniors from this high school.

From equation [8.16] the confidence coefficient z_b is the value of z for which $\mathbb{P} = (1 + .95)/2 = .975$. From Table 3 we find that $z_b = 1.96$. The desired interval is therefore $28 - 1.96\sigma_{\bar{X}}$ to $28 + 1.96\sigma_{\bar{X}}$. However, we can't calculate $\sigma_{\bar{X}}$ because we don't know the value of σ, so we calculate the unbiased estimator. From [8.9],

$$\hat{\sigma}^2 = s^2\left(\frac{N}{N-1}\right) = 446\left(\frac{36}{35}\right) \doteq 458.7$$

and

$$\hat{\sigma}_{\bar{X}}^2 \doteq \frac{458.7}{36} \doteq 12.74$$

which means that

$$\hat{\sigma}_{\bar{X}} \doteq \sqrt{12.74} \doteq 3.57$$

and that our confidence limits are

$$28 - 1.96(3.57) \doteq 21.00$$

and

$$28 + 1.96(3.57) \doteq 35.00$$

We can therefore be approximately 95 percent confident that the interval $[21, 35]$ includes the population mean μ. And, in general,

If $E(X) = \mu$ and $V(X) = \sigma^2$, then for large N, confidence is approximately $\mathbb{C}$ that

$$\left(\bar{x} - z_b\hat{\sigma}_{\bar{X}}\right) \leq \mu \leq \left(\bar{x} + z_b\hat{\sigma}_{\bar{X}}\right) \qquad [8.18]$$

where $\mathbb{P}(z_b) = \dfrac{1 + \mathbb{C}}{2}$ and $\hat{\sigma}_{\bar{X}}^2 = \dfrac{\hat{\sigma}^2}{N}$

If one begins with a particular interval $[x_a, x_b]$ that is symmetrical about the sample mean $\bar{x}$, one can work backward, so to speak, and determine the level of confidence that the interval includes the population mean μ. From equation [8.16] we know that $\mathbb{P}(z_b) = (\mathbb{C} + 1)/2$.

Consequently,

$$C = 2\mathbb{P}(z_b) - 1 \qquad\qquad [8.19]$$

One can therefore standardize x_b and use equation [8.19] to calculate $\mathbb{C}$.

Example 8.12. A psychologist is developing a test for manual dexterity. She administers the test to 30 job applicants and finds that $\bar{x} = 40$ and $s^2 = 250$. How confident can she be that the population mean μ falls between 35 and 45? For these data,

$$z_b = \frac{45 - \mu}{\sigma_{\bar{X}}}$$

but since neither μ nor σ is known, each must be estimated. We therefore let $\hat{\mu} = \bar{x} = 40$ and

$$\hat{\sigma}_{\bar{X}} = \frac{\hat{\sigma}}{\sqrt{30}} \doteq \frac{\sqrt{250\left(\dfrac{30}{29}\right)}}{\sqrt{30}} \doteq 2.94$$

Therefore,

$$z_b \doteq \frac{45 - 40}{2.94} \doteq 1.70$$

From Table 3 in Appendix VIII we find that $\mathbb{P}(1.70) \doteq .95543$, and from Equation [8.19] the psychologist can have confidence

$$2(.95543) - 1 \cong .91$$

that the interval [35, 45] spans the population mean μ.

We began our discussion of calculating interval estimates for μ with the statement that for a standard normal random variable Z,

$$P(-1.96 \le z \le 1.96) \doteq .95$$

This is, of course, absolutely correct. It is also true that there are an infinite number of *other* intervals z_a to z_b for which $P(z_a \le z \le z_b) = .95$. For example, $P(-1.68 \le z \le 2.97) \doteq .95$ and $P(-1.64 \le z \le 3.09) \doteq .95$. We could therefore define a 95 percent confidence interval for μ as $(\bar{x} - 1.68\sigma_{\bar{X}})$ to $(\bar{x} + 2.97\sigma_{\bar{X}})$ or as $(\bar{x} - 1.64\sigma_{\bar{X}})$ to $(\bar{x} + 3.09\sigma_{\bar{X}})$. For the normal distribution, however, it can be shown that the *smallest* of all such intervals (i.e., the one that yields the most precise estimate of μ) is the one that is symmetrical about $\bar{x}$, that is, $\bar{x} \pm 1.96\sigma_{\bar{X}}$.

b. Interval estimates of p. From Chapter 7 we know that the distribution of the standardized binomial proportion

$$Z_{X/N} = \frac{\dfrac{X}{N} - p}{\sqrt{\dfrac{pq}{N}}}$$

is approximated by $Z : N(0, 1)$ when $Np \geq 5$ and $Nq \geq 5$. Since we know $E(X/N) = p$, the machinery we have developed for using the standard normal distribution to construct interval estimates of μ can be applied to estimates of p.

In our discussion of relative frequency as an estimator of p, we pointed out that the relative frequency of successes f/N may be thought of as the sample mean of N indicator random variables. For a binomial random variable $X : B(N, p)$, the *proportion* of successes in N trials is therefore the *mean* number of successes. That is, if x is the number of successes,

$$\frac{x}{N} = \bar{x}$$

If Np and Nq satisfy our requirement for the normal approximation, then confidence is approximately $\mathbb{C}$ that

$$\left(\frac{x}{N} - z_b\sigma_{X/N}\right) \leq p \leq \left(\frac{x}{N} + z_b\sigma_{X/N}\right)$$

where $\mathbb{P}(z_b) = (\mathbb{C} + 1)/2$. This is completely analogous to expression [8.17], but the situation is different in a subtle but important way: If $X : N(\mu, \sigma^2)$, it is at least conceivable that σ^2 is known even if μ is unknown. However, we know from [6.14] that

$$\sigma^2_{X/N} = \frac{pq}{N}$$

and if we knew the value of p and q that we need to calculate $\sigma_{X/N}$, we wouldn't be calculating an interval *estimate* of p. Therefore, a confidence interval for p always requires that $\sigma_{X/N}$ be estimated.

Example 8.13. The Environmental Protection Agency (EPA) requires that all laboratories conducting EPA analyses provide annual estimates of the proportion of errors made in reporting results obtained by chemists. The quality control officer of a chemistry laboratory therefore pulls a random sample of 2,000 analyses from the preceding year, compares reported results with the chemists' bench notes, and discovers 9 errors. Calculate a 99.5 percent confidence interval for the proportion of errors.

First, we note that since p and q are unknown, we cannot apply our customary criterion (Np and Nq both ≥ 5) to decide whether or not the distribution of $Z_{X/N}$ is approximated by the standard normal distribution. But, we *do* know that *relative frequency* is an unbiased and consistent estimator of p and that the relative frequency of errors was found to be

$$\frac{f}{N} = \frac{9}{2,000} = .0045$$

We therefore let $.0045 = \hat{p}$. And, since $q = 1 - p$, we can let $\hat{q} = (1 - .0045) = .9955$. Accordingly, we find that $N\hat{p} = 9$ and $N\hat{q} = 1,991$.[7]

From equation **[8.16]** the confidence coefficient z_b is that value for which $\mathbb{P} = (1 + .995)/2 = .9975$. In Table 3 we find that the entry nearest to .9975 is .997445, for which $z = 2.81$. The desired interval is therefore

$$\frac{9}{2,000} - 2.81\sigma_{X/N} \quad \text{to} \quad \frac{9}{2,000} + 2.81\sigma_{X/N}$$

For the proportion $X/2,000$,

$$\sigma_{X/N} = \sqrt{\frac{pq}{2,000}}$$

The values of p and q are not known, so we once again let $\hat{p} = .0045$ and let $\hat{q} = (1 - .0045) = .9955$, which gives us an estimate of $\sigma_{X/N}$:

$$\hat{\sigma}_{X/N} = \sqrt{\frac{(.0045)(.9955)}{2,000}} \doteq .0015$$

The limits of the .995 confidence interval are therefore

$$\frac{9}{2,000} - 2.81(.0015) \doteq .00029$$

and

$$\frac{9}{2,000} + 2.81(.0015) \doteq .0087$$

The quality control officer can report to the EPA with 99.5 percent confidence that the proportion of transcription errors in this laboratory

[7] It is easily seen that $N\hat{p} = x$ and $N\hat{q} = N - x$. The standard normal distribution is therefore a suitable approximation if the number of observed successes and the number of observed failures are both greater than or equal to five.

falls in the interval [.00029, .0087]. And, in general,

If N is large, then confidence is approximately $\mathbb{C}$ that

$$\left(\frac{x}{N} - z_b\hat{\sigma}_{X/N}\right) \le p \le \left(\frac{x}{N} + z_b\hat{\sigma}_{X/N}\right)$$

where $\mathbb{P}(z_b) = \dfrac{1 + \mathbb{C}}{2}$ and $\hat{\sigma}_{X/N}^2 = \dfrac{\hat{p}\hat{q}}{N} = \dfrac{\left(\dfrac{x}{N}\right)\left(1 - \dfrac{x}{N}\right)}{N}$ **[8.20]**

Example 8.14. Suppose that 124 respondents in a survey of 200 adults with checking or savings accounts are dissatisfied with the services provided by their financial institutions (bank, savings and loan, etc.). The survey research firm can be 99.9 percent certain that the proportion of all adults who feel this way falls between what two values?

We let $\hat{p} = f/N = 124/200 = .62$, and $\hat{q} = 1 - \hat{p} = .38$. Since $N\hat{p}$ and $N\hat{p}$ are both greater than 5, the proportion (X/N) can be treated as a normally distributed random variable. We therefore look for z_b in Table 3 of Appendix VIII, such that

$$\mathbb{P}(z_b) = \frac{1 + .999}{2} = .9995$$

Table 3 does not include z-values for $\mathbb{P} > .9990$, but in Table 3A we find that $\mathbb{P}(3.30) = .9995$, so

$$\mathbb{C}\left(\frac{x}{N} - 3.30\sigma_{X/N} \le p \le \frac{x}{N} + 3.30\sigma_{X/N}\right) = .999$$

We have already calculated the observed proportion of dissatisfied respondents,

$$\frac{x}{N} = \frac{124}{200} = .62$$

and estimating $\sigma_{X/N}$, we obtain

$$\hat{\sigma}_{X/N} = \sqrt{\frac{\hat{p}\hat{q}}{N}} = \sqrt{\frac{(.62)(.38)}{200}} \doteq \sqrt{.001178} \doteq .034$$

The survey firm can therefore be 99.9 percent confident that the population proportion p of dissatisfied banking customers is between $.62 - (3.30)(.034) \doteq .508$ and $.62 + (3.30)(.034) \doteq .732$.

Expression **[8.20]** can also be used to calculate the sample size required to yield a specified maximum error of estimate.

Example 8.15. In Example 8.14 the population proportion of dissatisfied customers is estimated to be .62, plus or minus about .11. How large a sample would the survey firm need in order to reduce the maximum error of estimate from .11 to .02 and still be 99.9 percent confident that the interval traps p?

As in Example 8.10, we simply set $z_b \hat{\sigma}_{X/N}$ equal to the desired maximum error of estimate and solve for N:

$$3.3\sqrt{\frac{(.62)(.38)}{N}} = .02$$

Squaring both sides and multiplying both sides by N, we get

$$(3.3)^2(.62)(.38) = (.02)^2 N$$

and N is therefore equal to approximately 6,414.

The alert reader may wonder why expression **[8.20]** does not incorporate the convention we established in Chapter 7 for the normal approximation and set the interval limits equal to

$$\frac{x - .5}{N} - z_b \hat{\sigma}_{X/N} \quad \text{and} \quad \frac{x + .5}{N} + z_b \hat{\sigma}_{X/N}$$

The reason is that $E(X/N)$ and $V(X/N)$ are both dependent on p, so expression **[8.20]** is a suitable approximation only when N is so large that the correction $(\pm .5)$ is trivial. For small values of N, even when $N\hat{p}$ and $N\hat{q}$ are both greater than or equal to five, a more conservative strategy is to assume that $\sigma_{X/N}$ is the *largest* value possible. This occurs when $p = .5$ and $q = .5$ and $\sigma^2_{X/N}$ is therefore

$$\frac{(.5)(.5)}{N} = \frac{.25}{N}$$

Under this circumstance one should probably, in addition, correct for continuity, in which case expression **[8.20]** becomes

$$\left(\frac{x - .5}{N} - z_b \frac{.5}{\sqrt{N}}\right) \le p \le \left(\frac{x + .5}{N} + z_b \frac{.5}{\sqrt{N}}\right) \qquad \text{[8.21]}$$

Example 8.16. Suppose that a poll is taken among 25 faculty members at a university and that 60 percent of those polled favor a longer break between semesters (and a shorter summer vacation) and the other 40 percent do not. If p is the proportion of the entire faculty that favors the longer break, one can be 95 percent confident that p falls between what two numbers?

Since $(.60)(25) = 15 > 5$ and $(.40)(25) = 10 > 5$, we may use the normal approximation to the distribution of $Z_{X/N}$. However, since N is only 25 we shall base the upper limit of our interval on $15 + .5 = 15.5$ and the lower limit of our interval on $15 - .5 = 14.5$. And, instead of estimating $\sigma_{X/N}$ from $\hat{p}$ and $\hat{q}$, we shall calculate our interval on the conservative assumption that $\sigma_{X/N}$ is as large as possible. That is,

$$\sigma_{X/N} = \frac{.5}{\sqrt{N}} = \frac{.5}{5} = .1$$

From equation **[8.16]** the confidence coefficient z_b is that value for which $\mathbb{P} = (1 + .95)/2 = .975$, and in Table 3 we find that $\mathbb{P}(1.96) \doteq .975$. The limits of our .95 confidence interval are therefore

$$\frac{14.5}{25} - 1.96(.1) \doteq .38$$

and

$$\frac{15.5}{25} + 1.96(.1) \doteq .82$$

If $\sigma_{X/N}^2 < .25$, the 95 percent confidence interval will be narrower, so we can be *at least* 95 percent confident that between 38 percent and 82 percent of the faculty favor the longer mid-semester break and the shorter summer recess.

─────── EXERCISES 8.3 ───────

1. The following is a sample of 30 scores obtained on the Graduate Record Examination Advanced Test in Psychology (GREATP) by graduates with bachelors degrees in psychology from a state university.

520	470	410	480	440	500	460	510	390	500
470	480	450	480	370	460	480	450	550	490
470	520	450	500	470	520	440	490	460	500

   (a) What are the mean $\bar{x}$ and variance s^2 for this sample?
   (b) Calculate point estimates for μ the expected value and σ^2 the variance of GREATP scores for the entire population of psychology graduates from this institution who have taken this test.
   (c) What desirable properties do your estimates possess?
   (d) Suppose we took the third observation, 410, as an estimate of the population mean μ. What desirable properties would it possess?

2. A manufacturer of television sets collected the following data on picture-tube lifetime (hours × 100) in continuous operation.

450.5– 500.5	500.5– 550.5	550.5– 600.5	600.5– 650.5	650.5– 700.5	700.5– 750.5	750.5– 800.5	800.5– 850.5	850.5– 900.5
4	3	7	21	19	28	39	51	28

(a) Calculate unbiased and consistent point estimates for the expected value and variance of picture-tube lifetime for the entire product line.

(b) Suppose the data had been grouped as follows:

500.5 or less	500.5 – 550.5	550.5 – 600.5	600.5 – 650.5	650.5 – 700.5	700.5 – 750.5	750.5 – 800.5	800.5 or more
4	3	7	21	19	28	39	79

Would the manufacturer be interested in estimating the same population parameters?

3. A coin is tossed 20 times with the following result:

$$H\,T\,T\,T\,H\,T\,T\,H\,T\,T\,T\,H\,H\,T\,H\,T\,T\,H\,T\,H$$

Estimate p, the probability of obtaining a head.

4. A certain pheromone is known to attract male fruit flies of two species, A and B. Ten traps baited with small samples of the pheromone are placed in an orchard. Use the data below to calculate an unbiased and consistent point estimate of the proportion of male fruit flies of species A.

Trap	1	2	3	4	5	6	7	8	9	10
f_A	14	4	8	42	4	7	26	19	2	19
f_B	15	10	8	12	8	8	9	9	5	7

5. Let X be a normally distributed random variable. A sample of N observations yields a mean of 30 and the population variance is known to be 100. Calculate confidence intervals for $\mathbb{C} = .95$ assuming the following sample sizes:

(a) 1   (b) 4   (c) 16

6. Let X be a normally distributed random variable. A sample of N observations yields a mean of 100 and a variance of 222.75. Calculate interval estimates of μ for $\mathbb{C} = .99$ assuming the following sample sizes:

(a) 30   (b) 50   (c) 100

7. Let X be a normally distributed random variable. A sample of 50 observations yields a mean of 30 and a variance of 98. Calculate interval estimates of μ for the following levels of confidence:

(a) $\mathbb{C} = .90$ (b) $\mathbb{C} = .975$ (c) $\mathbb{C} = .995$

8. Let X be a normally distributed random variable. A sample of 9 observations yields a mean of 25, and the population variance is known to be 24. Calculate interval estimates for μ for the following levels of confidence:

(a) $\mathbb{C} = .75$ (b) $\mathbb{C} = .866$ (c) $\mathbb{C} = .98$

9. The following sample of 10 scores was obtained on a measure that is known to be normally distributed:

   $$12, 14, 15, 15, 22, 17, 19, 18, 18, 23$$

   If the population variance is known to be 12, use these data to calculate interval estimates for μ for the following levels of confidence:

   (a) $\mathbb{C} = .95$ (b) $\mathbb{C} = .975$ (c) $\mathbb{C} = .99$

10. The following sample of 9 scores was obtained on a measure that is known to be normally distributed:

    $$-.31, -.11, -.16, .20, .31, .56, .87, .93, 1.26$$

    Assume that the population variance is known to be .30, and use these data to calculate interval estimates for μ for the following levels of confidence:

    (a) $\mathbb{C} = .50$ (b) $\mathbb{C} = .75$ (c) $\mathbb{C} = .90$

11. (a) Consider again the Graduate Record Examination scores in Exercise 8.3.1. The chair of a graduate admissions committee could be 97.5 percent certain that the mean μ of scores for all graduates with bachelors degrees in psychology from this institution falls between _____ and _____.

    (b) How large a sample would he need to be 97.5 percent certain that the population mean μ was no more than 5 points higher or 5 points lower than the sample mean?

12. The following concentrations (parts per billion) of mercury were detected in 30 water samples taken from a proposed housing development site:

0.25	0.22	15.00	12.00	25.00	0.30	0.90	0.85	11.00	9.00
0.30	25.00	13.00	8.00	7.00	8.00	10.00	16.00	9.00	6.00
30.00	10.00	4.00	3.00	6.00	9.00	.50	2.00	8.00	8.00

    (a) An analytical chemist can be 99 percent certain that the concentration of mercury for the entire site is between _____ and _____.

    (b) If the Environmental Protection Agency requires that the developer be 99 percent certain that the concentration does not exceed 10 parts per billion, what size would the sample have to be for the laboratory to achieve this certainty?

13. Let X be a normally distributed random variable. A sample of 40 observations yields a mean of 15, and the population variance is known to be 14. Calculate the confidence $\mathbb{C}$ that μ lies in the following intervals:

    (a) [14.03, 15.97]   (b) [13.47, 16.53]   (c) [14.2, 15.8]

14. Let X be a normally distributed random variable. A sample of 200 observations yields a mean of 83 and a variance of 5.6. Calculate the confidence $\mathbb{C}$ that the population mean μ lies in the following intervals:

    (a) [82.67, 83.33]   (b) [82.61, 83.39]   (c) [82.5, 83.5]

15. Many *AD & D*® players become compulsive dice testers. In odd moments you can catch these poor wretches rolling dice dozens of times to ensure that they are fair. One player rolled his favorite 6-sided die 30 times and obtained the following results:

Score	1	2	3	4	5	6
Frequency	8	5	8	2	3	4

He can be 95 percent certain that the expected value of a single roll falls between _____ and _____.

16. To determine which group of characters in *AD & D*® "attacks" first, the players roll a 10-sided die, and the game master rolls a 10-sided die. The side with the lower number gets the initiative. A certain game master seems to get the initiative more than his players expect, so the players commandeer his 10-sided die and roll it 200 times. The results are listed below.

Score	1	2	3	4	5	6	7	8	9	10
Frequency	16	15	25	19	20	22	12	23	26	22

Calculate a 90 percent confidence interval for the expected value of any roll the game master makes.

17. Use the normal approximation to calculate a 95 percent confidence interval for the proportion of Species A in the population of male fruit flies in the orchard described in Exercise 8.3.4.

18. In a game of *AD & D*®, a player rolls 65 attacks with his 20-sided die, and 43 of these are successful. Construct a 98 percent confidence interval for p, where p is the probability of a successful attack using this particular die.

19. In a 1986 survey, Louis Harris reported that in a poll of 1,500 Americans, 57 percent approved of raising taxes to reduce the national deficit. He stated further that the *actual* percentage for the *entire nation* might vary from that figure by 2.5 percent (i.e., the population proportion might be anywhere from .545 to .595). What is the confidence level for this estimate?

20. A political pollster reports to the candidate who hired him that in a sample of 100 registered voters, the candidate was favored by 58 percent of the respondents. He claims to be "very confident" that between half and 66 percent of the electorate favors the candidate. How confident is "very" confident?

D. SUMMARY

Inferential statistics is a body of quantitative techniques that enable the scientist to make generalizations from limited observations. A statistical generalization, or inference, is a statement about the relative frequency distribution of measurements in a well-defined *population*. The limited observations are called a *sample*. Bridging the gap between the relative frequency of *observed* measurements in the sample and the *unobserved* population distribution is the central role of probability theory in statistics:

1. Use the relative frequency distribution for a sample of measurements to specify a probability distribution for the random variable defined by the measurements.
2. Use the mathematics of probability to draw conclusions about the probability distribution.
3. Apply these conclusions to the relative frequency distribution in the population.

The unobserved measurements in the population are thought of as *values of a random variable*, rather than data, because they are not "given." Likewise, indices of various properties of the population distribution are called *parameters*, rather than statistics, and are denoted by *Greek* letters, rather than Roman letters.

Drawing a sample is a form of random experiment, and most probability models are appropriate only to sample data that are obtained in such a way that all possible samples are equally likely to be drawn. This is called *random sampling*. The number of possible random samples of size N that can be drawn from a population of K individuals is $_KC_N$. Any particular statistic W assigns a number to each of these many samples, and these numbers can differ from sample to sample. A statistic is therefore a random variable. If W is a statistic, the probabilities associated with W define the *sampling distribution of W*.

Like any probability distribution, the sampling distribution of a statistic is specified by its probability or probability density function and its parameters. The density function for the sampling distribution of the mean $\overline{X}$ is closely approximated by the normal distribution when $N \geq 30$. This is a consequence of the *Central Limit Theorem*: If $W_1, \ldots, W_N$ are independent, identically distributed random variables and if $Y = W_1 + \cdots + W_N$, then $\phi(Y)$ approaches the normal density function as N approaches infinity.

If the random variable X has $E(X) = \mu$ and $V(X) = \sigma^2$, then

$$E(\overline{X}) = \mu_{\overline{X}} = \mu$$

and for samples of size N

$$V(\overline{X}) = \sigma_{\overline{X}}^2 = \frac{\sigma^2}{N}$$

Therefore, the distribution of $\overline{X}$ is approximately $N(\mu, \sigma^2/N)$.

One of the ways in which statisticians make inferences about population distributions is to *estimate* their parameter values.

A point estimator is a function that yields a single value, or point. Three desirable properties of a point estimator are *unbiasedness*, *consistency*, and *efficiency*. Let $\hat{\theta}$ be a point estimator of the parameter θ:

1. $\hat{\theta}$ is unbiased if $E(\hat{\theta}) = \theta$.
2. $\hat{\theta}$ is consistent if the probability that $\hat{\theta}$ is arbitrarily close to θ tends to 1 as sample size increases.
3. The most efficient of all unbiased estimators of θ is the one with the smallest variance for any given sample size.

For samples of N observations, the sample mean $\bar{X}$ is an unbiased and consistent estimator of the population mean μ, and it is the most efficient of all linear estimators of μ. The relative frequency f/N of event E in N observations can be thought of as a special case of the sample mean, where the population mean μ is equal to p. The relative frequency is therefore an unbiased and consistent estimator of p.

An unbiased and consistent estimate of the population variance σ^2 is

$$ s^2 \left(\frac{N}{N-1} \right) = \frac{\sum\limits_{}^{N}(x - \bar{x})^2}{N-1} $$

which is denoted $\hat{\sigma}^2$.

Interval estimation yields values w_a and w_b, where the probability that the interval $[w_a, w_b]$ includes θ is specified by the experimenter. The likelihood that θ falls in the interval is called the *confidence* of the estimate, denoted $\mathbb{C}$, and $[w_a, w_b]$ is called the *confidence interval*.

If X is known to be normally distributed, and $V(X) = \sigma^2$, then confidence equals $\mathbb{C}$ that μ falls in the interval

$$ \left[(\bar{x} - z_b \sigma_{\bar{X}}), (\bar{x} + z_b \sigma_{\bar{X}}) \right] $$

where $\mathbb{P}(z_b) = (\mathbb{C} + 1)/2$ under the standard normal curve, and

$$ \sigma_{\bar{X}} = \sqrt{\sigma^2/N} $$

For any given level of confidence, one can therefore make the confidence interval for μ as narrow as desired by letting N increase.

If X is not known to be normally distributed, the Central Limit Theorem nevertheless permits the assumption the $\bar{X}$ is normally distributed if $N \geq 30$.

If $V(X)$ is unknown, one can use $\hat{\sigma}^2$ in place of σ^2 if $N \geq 30$.

The normal approximation to the binomial allows us to extend these procedures to interval estimates for the expected value of the binomial proportion, X/N. If $X: B(N, p)$ and $N\hat{p} \geq 5$ and $N\hat{q} \geq 5$, where $\hat{p} = f/N$ and $\hat{q} = 1 - \hat{p}$, then when N is large, confidence is approximately equal to $\mathbb{C}$ that $E(X/N) = p$ falls in the interval

$$\left[\left(\frac{X}{N} - z_b\hat{\sigma}_{X/N}\right), \left(\frac{X}{N} + z_b\hat{\sigma}_{X/N}\right)\right]$$

where

$$\hat{\sigma}_{X/N} = \sqrt{\frac{\hat{p}\hat{q}}{N}}$$

For small values of N a more conservative estimate (i.e., a wider interval) is obtained by assuming that $\sigma_{X/N}$ takes the largest possible value. For confidence level $\mathbb{C}$, this interval is

$$\left[\left(\frac{x - .5}{N} - z_b\frac{.5}{\sqrt{N}}\right), \left(\frac{x + .5}{N} + z_b\frac{.5}{\sqrt{N}}\right)\right]$$

Summary of frequently used confidence coefficients

Confidence Level ($\mathbb{C}$)	$\mathbb{P}(z) = \dfrac{1 + \mathbb{C}}{2}$	$\pm z$
.90	.95	± 1.64
.95	.975	± 1.96
.975	.9875	± 2.24
.98	.99	± 2.33
.99	.995	± 2.58
.995	.9975	± 2.81
.999	.9995	± 3.30

Hypothesis testing

CHAPTER OUTLINE

C. EVALUATION OF THE STATISTICAL TEST

The concept of power

The power of a test against a composite alternative
hypothesis

Scientific versus statistical significance

D. POSTSCRIPT: DECISIONS CONCERNING HYPOTHESES ABOUT PROPORTIONS

Test statistics: Z_X and $Z_{X/N}$

Upper-tailed tests, lower-tailed tests, and two-tailed tests

Significance level and adjustments for continuity

Exercises 9.3

E. SUMMARY

Like estimation, hypothesis testing permits the scientist to make generalizations about populations from sample data. As the reader might expect, these generalizations are inferences about values of population parameters, and the sample data are used to calculate statistics on which the inferences are based. These *test statistics* are often functions of estimators, but the inferences we base on test statistics are somewhat different from the inferences we base on estimators. An estimator is used to answer the question, "what is the value of parameter θ?" A test statistic is used to answer the question, "is θ_0 the value of parameter θ?"

Another feature that hypothesis tests share with estimation is that they permit us to make probability statements that express the certainty that characterizes our inferences. In the context of estimation, certainty is expressed in terms of confidence; in hypothesis testing, the certainty—or uncertainty— of one's inference is called *significance*.

A. OUTLINE AND EXAMPLE OF HYPOTHESIS TESTING

In ordinary language, an hypothesis is a proposition about the world around us. Whether it takes the form of a belief, a suspicion, a deduction, or an expectation, one quality that sets hypotheses apart from other propositions is that hypotheses are *provisional*. We acknowledge—either implicitly or explicitly—that our decision to accept the proposition as true or reject it as false must await the collection of further evidence.

Sometimes we look for evidence that *confirms* our hypotheses. If Gwen's car is hard to start on cold mornings and she suspects that this is due to a defective battery, she will probably make a mental note of every electrical oddity in the car's performance. In other situations, the logic of testing hypotheses against evidence calls for *disconfirmation*. We formulate *two* mutually exclusive propositions, *A* and *B*, and our decision takes the form of *rejecting* one proposition in favor of the other. The logic of disconfirmation is the cornerstone of statistical hypothesis testing. At first, the logic may seem a little foreign to our everyday modes of reasoning, perhaps even a little tortuous, but it is, after all, the logic that underlies one of our most familiar institutions, criminal trials.

A prosecutor suspects that a defendant is guilty of a crime, but under the American system of criminal justice the accused person is presumed innocent. In order to obtain a conviction, the prosecutor must therefore *disconfirm* the "hypothesis" of innocence in favor of the alternative proposition that the defendant is guilty. To accomplish this, the prosecution presents *factual* evidence that is *both* implausible under the hypothesis of innocence *and* consistent with the hypothesis of guilt. If the prosecution fails, the defendant is found "not guilty." That is, the jury *rejects* the hypothesis of *guilt*. This does not in any way imply that the jury *accepts* the hypothesis of *innocence*. It means only that the evidence is insufficient to sustain a conviction.

To get some feeling for how this logic operates in the context of statistical hypothesis testing, we shall briefly outline the steps involved in a statistical test and illustrate them with a simple example.

1. Formulation

The first step in conducting an hypothesis test is to formulate a proposition about the state of nature. This proposition may concern any observable phenomenon, but it must imply a statement about the population distribution of some random variable. Actually, the investigator formulates two such statements. The first is the *test hypothesis* (H_0), which always states that some parameter θ equals a *specific* value, θ_0. The second is the *alternative hypothesis* (H_1), which specifies the value (or values) that θ is expected to take if the test hypothesis is incorrect.[1]

Example 9.1. Suppose one of the authors believes that a coin is biased in favor of "heads" and decides to test his suspicion by tossing the coin 10 times and counting the number of heads. This experiment satisfies the conditions of a binomial experiment, and if X is defined as the number of heads and p as the probability of obtaining a head on any particular toss, two possible "states of nature" give rise to two different statements about the distribution of X:

State 1, the coin is fair: $X : B(10, 0.5)$
State 2, the coin is biased: $X : B(10, p_1)$

where all we can say about p_1 is that it is greater than .5.

---

[1] One can also formulate hypotheses about *nonparametric* features of the population distribution, e.g., that the distribution is normal or not normal, uniform or not uniform, etc. Hypotheses of this sort are introduced in Chapter 15.

These two statements differ in only one respect, the value of the parameter p. Furthermore, a *specific* value of p is implied only by the first proposition, "the coin is fair." Since the test hypothesis *must* specify a single value for the parameter in question, our two hypotheses are

$H_0: p = .5$

$H_1: p > .5$

The test hypothesis (H_0) in this example is like the presumption of innocence; it describes the distribution of the random variable *if* the investigator's suspicion is groundless. The alternative hypothesis (H_1) describes the distribution of the random variable if the experimenter's suspicion is correct and is therefore analogous to the case argued by the prosecution.

2. Decisions

After the hypotheses are formulated, an experimenter must decide what sort of evidence would *disconfirm* one hypothesis and provide *support* for the other hypothesis. This actually involves two decisions: The experimenter must select a *test statistic* and must define a *rejection rule*.

> A test statistic W is a numerical index that is expected to take the value w_0 if H_0 is correct and is expected to take some other value if H_1 is correct.

In Example 9.1 the random variable itself—that is, the number of heads —can be used as a test statistic. Under the test hypothesis $E(X) = 5$, and the author should expect to get about 5 heads in 10 tosses. Under the alternative he should expect more than 5 heads.

The problem here is that "about" and "more than" are vague terms. To make things more explicit, the experimenter decides *before* conducting the experiment what results (of all possible results) will require that the test hypothesis be rejected and what results will require that the alternative hypothesis be rejected. This is called the rejection rule. For example, the author with his suspicious coin might decide that he will reject H_0 if he obtains 8 or more heads and that he will reject H_1 if he obtains fewer than 8 heads.

The "results" that lead to rejection of H_0 are values of the test statistic that are *unlikely* if H_0 is correct but relatively *likely* if H_1 is correct. These values are called the *rejection region* or *critical region*. If the test statistic falls in the critical region, the test hypothesis is rejected *in favor of* the alternative hypothesis; otherwise, the alternative hypothesis is rejected. This is why the rejection region is made up of values that are more likely under the alternative hypothesis than under the test hypothesis and are, therefore, said to "favor" the alternative hypothesis.

The value in the critical region that is *closest* to the value expected under the test hypothesis (i.e., the value that is most favorable to H_0) is called the *critical value*. In this example, then, the critical region is $\{8, 9, 10\}$ and the critical value is 8.

3. Data collection

In this example the appropriate data are collected by tossing the coin 10 times and counting the number of heads. Let us suppose that the coin comes up heads 7 times.

4. Conclusions

The test itself is conducted by simple inspection. If the observed value of the test statistic falls in the critical region, the test hypothesis is abandoned in favor of the alternative hypothesis. If the statistic does not fall in the critical region, the alternative hypothesis is rejected. In Example 9.1 the observed value of the test statistic X is 7, which does not fall in the critical region, so under his rejection rule the experimenter must reject the alternative hypothesis. Note that we do *not* say that the experimenter *accepts* the *test* hypothesis. The coin is not judged *innocent* of bias. We conclude only that the coin cannot be judged guilty on the basis of the evidence presented.

B. FOUNDATIONS OF HYPOTHESIS TESTING

In the preceding section we outlined a statistical test, but we did not discuss any of the philosophical or statistical considerations on which these procedures are based. In this section we shall examine the four principal aspects of the statistical test from the perspectives of both scientific epistemology and probability theory.

1. Formulation

a. Theory. We've said that a statistical test begins with the formulation of a testable proposition about the state of nature. Typically, such a formulation starts with some sort of conceptualization that expresses relationships among observable phenomena or underlying, hypothetical processes. This conceptualization may be a grand theory, such as Darwin's theory of natural selection, or it may be a "mini-theory" of more limited scope. But, it is likely that the overwhelming majority of hypotheses derive either from prior empirical observation or from that combination of playful speculation, scientific hunches, serendipity, unbridled conjecture, and whimsical inspiration, affectionately known in some scientific circles as SWIG.[2]

[2] Scientific Wild Intuitive Guesswork!

Example 9.2. Several years ago one of the authors conducted a series of experiments comparing learning and problem solving in domestic dogs and timber wolves. Experimenters had to remain silent while testing dogs, because we discovered that their performance was influenced by verbal expression of approval or disapproval. This observation led the author to speculate that sensitivity to human communicative signals, particularly verbal cues, may have been an incidental consequence of domestication. It is not difficult to imagine that responsive companion animals spark more affection in humans than do indifferent animals, and that animals we are attached to receive better care, live longer, and are consequently more reproductively successful.

Exploratory data analysis. An experienced researcher can sometimes (as in the preceding example) see patterns in naturalistic observations and formulate testable hypotheses that bring these patterns into high relief. At other times, patterns and relationships begin to emerge as the investigator proceeds with the preliminary tasks of organizing and describing data. As we indicated in the Postscript to Part I, statistical description can do more than simply reduce a welter of information to manageable proportions. Methodical and creative description is the essence of exploratory data analysis, which can yield clues to unexpected relationships among observed phenomena and the processes that influence these relationships. Such clues are the stuff of which hypotheses are made. Exploratory data analysis can therefore be especially fruitful when one approaches a data collection with no preconceptions whatsoever about how phenomena are related or even *whether* they are related, as when using a large bank of demographic or public opinion data.

b. Model. If a theory is to generate statistically testable hypotheses, it must imply at least a partial description of the population distribution of some measurement, X. In rare instances the theory may fully specify the probability distribution that models the population distribution of X. It is more often the case that the theory implies only a partial specification statement or a particular population value, such as μ or σ^2.

Example 9.3. If the evolutionary "Just-So" story in Example 9.2 has some validity, the dog's nearest wild cousin, the wolf, should *not* be especially sensitive to the spoken word. This implication could be tested by attempting to train wolves to perform a simple discrimination task, such as turning right or left, in response to a verbal command. One traditional apparatus used for such experiments is a *T-maze*, a runway (the stem of the "*T*") intersected at one end by a cross-corridor (which forms the arms of the "*T*").

Suppose that untrained wolves released into the runway of a *T*-maze are observed to turn right as often as they turn left. Now imagine that 20 wolves are each given a series of training trials in which the experimenter calls out either "Right" or "Left" on every trial. If the wolf turns into the arm signaled by the experimenter, it finds the door at the end open and receives a tidbit of food. An incorrect turn leads to a closed door.

After the training period, each wolf is given *one* test trial, which results in either *success* or *failure*, and the total number of successes is recorded for the group.

If, as the author proposed, wolves are insensitive to verbal cues, the training should not affect performance, so for the test trial of any one wolf

$$P(\text{left turn}) = P(\text{right turn}) = .5$$

This experiment (like our earlier coin-toss experiment) satisfies the conditions of a binomial experiment, and if we define X as the number of correct turns, then $X: B(20, 0.5)$.

We have therefore moved from speculation about natural phenomena (evolutionary processes) to a formal experiment that yields numerical results (number of successes) to a probability model for the distribution of that variable (the binomial distribution with parameters $N = 20$ and $p = .5$).

c. Hypotheses. Our model says that if wolves cannot learn verbal cues, then the number of successes in our experiment is distributed $B(20, 0.5)$. If, on the other hand, wolves *do* learn verbal cues then the probability of success on the test trial should be higher than .5. This logic gives us two hypotheses: $p = .5$ and $p > .5$.

The hypothesis $p = .5$ is called a *simple* hypothesis, because it specifies a single value, .5, for the parameter p. We know that any test statistic W must have a specified expected value w_0 under the test hypothesis. Furthermore, the expected value of a test statistic ordinarily depends on the value of the parameter. Therefore, H_0 must always be a simple hypothesis. In the present example, then,

$$H_0: p = .5$$

and in general notation the test hypothesis about the parameter θ is written

$$H_0: \theta = \theta_0$$

Since the test hypothesis states that there is *no* difference between the *true* value θ of the parameter and the *hypothesized* value θ_0, the test hypothesis is sometimes called the *null* hypothesis.

The hypothesis $p > .5$ states that the parameter p may take any of many values and is therefore called a *composite* hypothesis. In the present example, this is the alternative hypothesis:

$$H_1: p > .5$$

In this example, the hypothesis favored by the experimenter happens to be the simple hypothesis and is therefore the test hypothesis. However, when only *one* of the competing propositions about the state of nature implies a simple hypothesis, it is usually the proposition that *contradicts* the experimenter's expectations. Thus, H_0 is ordinarily the hypothesis that the experimenter hopes to reject, as in Example 9.1.

Sometimes the experimenter has enough information to formulate *two* simple hypotheses. For example, had we first trained a large sample of dogs and found that 75 percent of them learned to respond to the verbal cues, our hypotheses would have been $p = .50$ and $p = .75$. If one has *two* simple hypotheses, either may in principle be cast as the test hypothesis. Nonetheless, we shall see later in the chapter that there are statistical, logical, and philosophical reasons to *test* the hypothesis that the experimenter hopes to reject and cast the hypothesis that is favored by the experimenter as the *alternative*.[3]

Even if a simple alternative cannot be formulated, there may at least be enough information to say whether the parameter under H_1 is *larger* or *smaller* than the value specified in H_0. This was the case in our wolf scenario, which permitted us to formulate a *directional* alternative hypothesis, H_1: $p > .5$. Suppose, however, that when we pretested the wolves we found that some prefer to turn right and others prefer to turn left. In order to determine whether or not "left-turning" wolves and "right-turning" wolves are equally common we arrange things so that the test trial for all wolves calls for a left turn. If there are more left-turning wolves, the proportion of correct responses should be greater than .5; if there are more right-turning wolves, the proportion of correct responses should be less than .5. Accordingly, our alternative hypothesis would be *nondirectional*, H_1: $p \neq .5$.

Thus, the alternative may be simple or composite, and if composite may be directional or nondirectional. An alternative hypothesis about the parameter θ is therefore written in one of the following four forms:

$$H_1: \theta = \theta_1 \qquad \text{Simple}$$

$$\left.\begin{array}{l} H_1: \theta > \theta_0 \\ H_1: \theta < \theta_0 \end{array}\right\} \text{Directional}$$
$$\left.H_1: \theta \neq \theta_0 \quad \text{Nondirectional}\right\} \text{Composite}$$

A researcher generally knows whether the alternative hypothesis is simple or composite and whether a composite alternative is directional or nondirectional. It is not always so obvious to a statistical consultant, who may be unfamiliar with the underlying research issue. Formulating the appropriate alternative hypothesis can also be puzzling for the statistics student trying to interpret a story problem. The key generally lies in the "theory." The reasoning that leads a researcher to hypothesize that θ is equal to a particular value θ_0 will generally suggest what to expect if θ is *not* equal to θ_0.

- If the experimenter's thinking furnishes two *specific* values (θ_0 and θ_1), the alternative hypothesis is *simple*.
- If only *one* specific value (θ_0) is available, the alternative is *composite*.

---

[3] Some books therefore refer to the alternative hypothesis as the *experimental* hypothesis.

- If the alternative is composite and the experimenter suspects that the subject population possesses *more* of some attribute or *less* of some attribute than the modeled population, H_1 is probably *directional*.
- If the alternative is composite but the experimenter suspects *only* that the distribution of X in the subject population is *different* from the model, H_1 is *nondirectional*.

2. Decisions

a. Test statistic. Test statistics are the third branch of the statistics family and, like their cousins, descriptive statistics and estimators, must possess certain properties. The basic requirement of a test statistic is that its sampling distribution depend on which hypothesis is correct. If the statistic W is to be used as a test statistic, the distribution of W when H_0 is correct must be *different* from the distribution of W when H_1 is correct. The more one can deduce about the sampling distribution of W under the two hypotheses, the more useful the test statistic, but in any event, we know from our earlier definition that

- the test hypothesis and the alternative hypothesis must imply different *expected values* for W, that is,

  $E(W|H_0 \text{ correct}) \neq E(W|H_1 \text{ correct})$

- the experimenter must be able to state the expected value of W under the *test* hypothesis.

In addition, it must be possible to determine the probabilities associated with values of W when the test hypothesis is correct. In most circumstances this means that the experimenter must be able to say that if H_0 is correct, then some *particular* probability function or density function is an appropriate model for the sampling distribution of W. It also means that the experimenter must specify the parameters appropriate to this function, either by implication from the test hypothesis or by estimation. Finally, as a practical matter, the distribution must be one for which cumulative probabilities are readily available. It is not very useful to know how W is distributed if one cannot look up (or easily calculate) probabilities associated with values of W. The distribution of W implied by the test hypothesis is denoted ϕ_0.

Example 9.4.1. The hypotheses in Example 9.1 concerned one particular coin, and the test statistic was based on a sample of N observations on that coin. In the present example our hypotheses concern the population of *all* wolves, and our test statistic should therefore be based on a representative sample of wolves. Accordingly, we might define our test statistic as the *number* of successes X achieved by N wolves, each of which performed the task once. Under the test hypothesis, the number of successes X is distributed $B(20, 0.5)$, so $E(X) = 20(.5) = 10$. Under the alternative hypothesis, $E(X) = 20(p) > 10$.

Example 9.4.2. Alternatively, we might use the binomial proportion X/N as our test statistic. From Chapter 6 (see Box 6.6.) we know that under H_0

$$E\left(\frac{X}{N}\right) = p = .5$$

and under H_1

$$E\left(\frac{X}{N}\right) = p_1 > .5$$

Appendix VIII does not have tables of probabilities or cumulative probabilities for the binomial proportion, so $x/20$ must be transformed in one of two ways. One option is to multiply $X/20$ by 20, in which case the statistic is reduced to the binomial random variable $X: B(20, p)$ as discussed above.

The second approach is to use the normal approximation to the binomial proportion. We know from Chapter 7 (see pp. 256–257) that when $Np \geq 5$ and $Nq \geq 5$, the distribution of the standardized binomial proportion

$$Z_{X/N} = \frac{\dfrac{X}{N} - p}{\sqrt{\dfrac{pq}{N}}}$$

is approximated by the standard normal distribution, $Z: N(0, 1)$. The number of successes in our wolf experiment is distributed $B(20, p)$, and under the test hypothesis, $p = .5$, so $q = (1 - p) = .5$. Therefore, $Z_{X/N}$ is

$$\frac{\dfrac{X}{20} - .5}{\sqrt{\dfrac{(.5)(.5)}{20}}}$$

[9.1]

and, since $Np = Nq = 10 > 5$, the distribution of $Z_{X/N}$ is (approximately) normal with an expected value of 0 and a variance of 1. If the alternative hypothesis, $p > .5$, is correct, the expected value of X/N is greater than .5, and the expected value of $Z_{X/N}$ is therefore greater than 0.[4]

---

[4] Under the alternative hypothesis we cannot assume that the normal curve provides a good approximation for the distribution of $Z_{X/N}$. Since $p = (1 - q)$ and $p > .5$, it follows that $q \leq .5$, and Nq might be less than 5. For example, if $p = .8$, then $q = .2$, and for 20 observations $Nq = 4$.

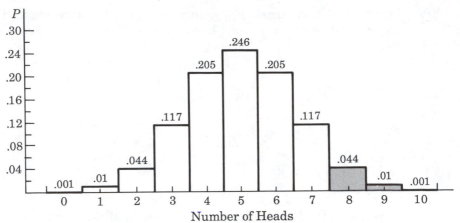

Figure 9.1. Probability distribution for number of heads in 10 tosses if p(Head) $= .5$. Shaded area is probability of 8 heads or 9 heads or 10 heads.

b. Rejection rule. The reasoning that underlies the selection of a critical region is the linchpin to understanding hypothesis testing. Once it is understood—not just memorized, but *understood* at a gut level—all hypothesis testing is just a variation on the same theme.

We have defined the critical region as a set of values that is *less likely* to include the observed value of the test statistic if the test hypothesis is correct than if the alternative hypothesis is correct. This definition hinges on the fundamental uncertainty that characterizes all statistical inference: One observes only a *sample* of the population, and sampling is a random experiment. A test statistic is therefore a random variable, and even if the test hypothesis is correct, the *observed* value can differ considerably from the *expected* value under H_0. However, the sampling distribution implied by the test hypothesis makes it possible to calculate the probability of any such difference. Recall Example 9.1. We stated that if H_0 is correct (the coin is fair), the experimenter should expect to observe *about* 5 heads in 10 tosses, but even with a perfectly fair coin, there is at least a small probability ($\cong .001$) that 10 tosses will all come up heads. The probability of obtaining 9 or more heads is larger ($\cong .011$), and the probability of obtaining 8 or more heads is larger still ($\cong .055$). (See Figure 9.1.)

Since random sampling can yield unlikely values of the test statistic, the statistician must decide just *how* unlikely a result must be in order to conclude that the test hypothesis is incorrect. This is the decision that drives the rejection rule. If the experimenter in Example 9.1 decides to reject H_0 if he obtains 8 or more heads, then H_0 will be rejected if the probability of the test statistic and *all values that are even more favorable to the alternative hypothesis* is less than .055. This probability is called the *significance level* of the test and is denoted by the lowercase Greek letter α (alpha).

Example 9.5.1. If we use the binomial random variable X as the test statistic in our wolf experiment, we know that $E(X) = 10$ if the test hypothesis is correct and that $E(X) > 10$ if the alternative hypothesis is correct. Values of X that are *greater* than 10 therefore favor the alternative hypothesis, so the appropriate critical region is the set of values *greater* than some critical value x_α under the distribution $X : B(20, 0.5)$. We can use Table 2 of Appendix VIII to find the critical value for any level of significance that we might choose.

Suppose we don't want to reject the test hypothesis if the probability exceeds .05 that our test statistic x will fall in the critical region when H_0 is correct. That is, we want to choose x_α in such a way that the following conditional probability holds true:

$$P(x \geq x_\alpha | H_0 \text{ correct}) \leq .05$$

In Table 2 of Appendix VIII we find that for $N = 20$ and $p = .5$,

$$P(x > 17) \doteq .000$$

$$P(x = 17) \doteq .001$$

$$P(x = 16) \doteq .005$$

$$P(x = 15) \doteq .015$$

$$P(x = 14) \doteq .037$$

Therefore,

$$P(x \geq 17) \doteq .001$$

$$P(x \geq 16) \doteq .001 + .005 = .006$$

$$P(x \geq 15) \doteq .001 + .005 + .015 = .021$$

and

$$P(x \geq 14) \doteq .001 + .005 + .015 + .037 = .058$$

If $P(x \geq x_\alpha | p = .5)$ is not to exceed .05, then x_α can be no smaller than 15. The critical value is therefore $x = 15$, and the rejection rule is: If $x \geq 15$, reject the hypothesis that $p = .5$ in favor of the hypothesis that $p > .5$. If $x < 15$, reject the hypothesis that $p > .50$.

Example 9.5.2. Now let's see what happens if we decide to use the standardized binomial proportion $Z_{X/N}$ as the test statistic. We established in Example 9.4.2 that $E(Z_{X/N}) = 0$ if the test hypothesis is correct and that $E(Z_{X/N}) > 0$ if the alternative hypothesis is correct. That is, positive values of Z favor the alternative hypothesis. We also know that if H_0 is correct, the distribution of the test statistic is approximated by $Z : N(0, 1)$. The rejection region is therefore the set of values *greater than* some critical value z_α under the standard normal distribution. Once

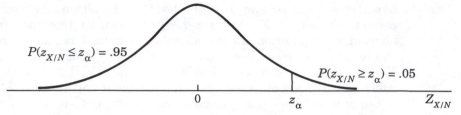

Figure 9.2. Normal approximation to distribution of $Z_{X/N}$ and critical value z_α in Example 9.5.2.

again, the critical value depends on significance level. For $\alpha = .05$ we want z_α such that

$$P\left(z_{X/N} \geq z_\alpha | H_0 \text{ correct}\right) = .05$$

Total area under the standard normal curve is 1, so if $P(z_{X/N} \geq z_\alpha) = .05$, then $P(z_{X/N} \leq z_\alpha)$ must be $1 - .05 = .95$ as shown in Figure 9.2. Therefore, the critical value is the z-value for which $\mathbb{P}(z) = .95$. In Table 3 of Appendix VIII we find that $\mathbb{P}(1.64) \doteq .95$, so if our test statistic is

$$z_{X/N} = \frac{\dfrac{x}{20} - .5}{\sqrt{\dfrac{(.5)(.5)}{20}}} \qquad \qquad [9.2]$$

our rejection rule is: If $z_{X/N} \geq 1.64$, reject the hypothesis that $p = .5$ in favor of the hypothesis that $p > .5$. If $z_{X/N} < 1.64$, reject the hypothesis that $p > .5$.

Because the normal approximation *is* an approximation, it doesn't always yield exactly the same critical x-value as the binomial itself. In Example 9.5.2 we found the critical value z_α to be 1.64, and the corresponding x-value is easily calculated from equation [9.2]:

$$\frac{\dfrac{x}{20} - .5}{\sqrt{\dfrac{(.5)(.5)}{20}}} = 1.64$$

Therefore,

$$\frac{x}{20} = .5 + 1.64\sqrt{\frac{(.5)(.5)}{20}}$$

$$\doteq .68$$

and $x = (.68)(20) = 13.6$. Since X is discrete, this means that H_0 should be rejected if $x \geq 14$. The critical x-value obtained by the normal approximation in Example 9.5.2 is therefore $x = 14$, whereas the binomial solution in Example 9.5.1 gave us a critical value of $x = 15$.

Significance as "unexpectedness." In ordinary conversation an event is said to be *significant* if it *signifies* something or has meaning. And in ordinary life we generally attribute more meaning to events that are unexpected than to events that are commonplace. When a host asks dinner guests if the wine is satisfactory and everyone nods agreeably, we attach little meaning to their behavior. It is conventional; it is expected. Any other reply would be ungracious. However, if a guest replies that the wine too astringent and requests another vintage, the very unexpectedness of the response disposes us to consider it meaningful. That is, we are inclined to regard the statement as signifying something about the wine (or about the guest!). In hypothesis testing as in dining-table conversation, unexpectedness lies at the heart of what we call significance: The *smaller* the significance level α, the *more unexpected* our result must be in order to reject H_0. That is, the greater the difference must be between the *observed* value of the test statistic and the *expected* value under the test hypothesis.

Example 9.5.3. In Example 9.5.2, we found that a significance level $\alpha = .05$ led to a rejection rule under which the test hypothesis is rejected if $x \geq 14$. However, we are using the normal approximation, and $x = 14$ is represented by the continuous interval 13.5 to 14.5. The test hypothesis will therefore be rejected if

$$\frac{x}{N} \geq \frac{13.5}{20} = .675$$

The rejection region and the area under the standard normal curve corresponding to α are illustrated in Figure 9.3(a).

Now let us suppose that instead of setting $\alpha = .05$, we decide that we want a more conservative test and set $\alpha = .01$. To find z_α we refer once again to the table of cumulative probabilities for the standard normal distribution and find the z-value with $\mathbb{P}(z) = 1 - \alpha$. From Table 3, we find that $\mathbb{P}(2.33) \doteq .990097$, so $P(z \geq 2.33) \doteq .01$. If we set $z_\alpha = 2.33$, then from equation **[9.2]**

$$\frac{\dfrac{x}{20} - .5}{\sqrt{\dfrac{(.5)(.5)}{20}}} = 2.33$$

and

$$\frac{x}{20} = .5 + 2.33 \sqrt{\frac{(.5)(.5)}{20}}$$

$$\doteq .76$$

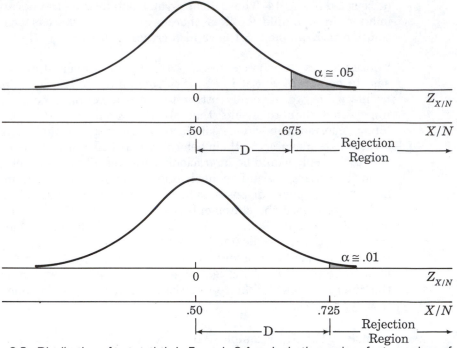

Figure 9.3. Distribution of test statistic in Example 9.4 and rejection regions for two values of α. D indicates distance of critical value from expected value of test statistic under H_0.

The critical number of successes is therefore $(.76)(20) = 15.2$. This value falls in the interval 14.5 to 15.5, which represents 15 successes, so we will reject H_0 if

$$\frac{x}{N} \geq \frac{14.5}{20} = .725$$

For a test with significance level $\alpha = .05$, the critical proportion of successes was .675, but $\alpha = .01$ is smaller than $\alpha = .05$, so the values that allow rejection of H_0 are *less probable* (i.e., more unexpected) than the values defining the rejection region when $\alpha = .05$. The critical value of the test statistic therefore lies *farther* from the value expected under the test hypothesis. This is shown graphically in Figure 9.3.

Confidence and significance. At the very beginning of the chapter, we said that estimation addresses the question, "what is the value of parameter θ?" and reminded the reader that the degree of certainty attached to the answer is expressed in terms of confidence. We also said that hypothesis tests address the question, "is θ_0 the value of the parameter θ?" and that the *un* certainty of the answer is expressed in terms of significance. These inferences and the certainty we attach to them are intimately related: If the value θ_0 falls *outside* a confidence interval for

$$\mathbb{C} = 1 - \alpha$$

then the test hypothesis

$H_0: \theta = \theta_0$

will be rejected at significance level α.

──────── EXERCISES 9.1[5] ──

1. A random variable X can have only two results, A and B. The probability of Event A is denoted p, and a particular theory suggests that $p = .65$.
   (a) For any sample of N observations, the theory implies that $X: B(N, 0.65)$. This specification statement is called the _____.
   (b) $p = .65$ is called the _____.
   (c) $p \neq .65$ is called the _____.
   (d) If one makes 30 observations and X is the number of times that Event A occurs, then X: _____ if H_0 is correct.
   (e) If the test hypothesis is correct, then for 30 observations, $E(X) =$ _____, but if the alternative hypothesis is correct, then $E(X)$_____ 19.5.
   (f) Your answers to (d) and (e) indicate that X would be an appropriate _____.

2. An experiment can have only two results, Event A and Event B. One theory implies that $P(A) = .30$. A researcher favors a competing theory, which implies that $P(B) < .70$. Let Y be the number of times A occurs in N observations.
   (a) Formulate a mathematical model for the frequency distribution of Y.
   (b) What is the test hypothesis?
   (c) What is the alternative hypothesis?
   (d) The experimenter decides to make 20 observations and record the number of times A occurs. How is this test statistic distributed if the test hypothesis is correct?
   (e) If the researcher decides that the test hypothesis will be rejected only if the probability of doing so incorrectly is less than 5 percent, what is the rejection rule?

3. An experiment can yield only two results, A and B. According to one theory the probability of Event A is .50. According to another theory, the probability of Event A is .70. Let X be the number of times A is observed, let p denote the probability of A, and let

   $H_0: p = .50$

   $H_1: p = .70$

   Use the binomial distribution (Table 2 of Appendix VIII) to determine the probability that x falls in each of the following rejection regions when H_0 is correct.
   (a) $N = 10$; rejection region is $\{9, 10\}$
   (b) $N = 20$; rejection region is $x \geq 17$
   (c) $N = 15$; rejection region is $(11 \leq x \leq 15)$
   (d) $N = 25$; rejection region is $x > 18$

──────────────

[5]A number of these exercises require use of Table 2 of Appendix VIII. Instructions for the use of this table are given on p. 217.

4. An experiment can yield only two results, A and B. One theory implies that the probability of Event A is .30. According to another theory, the probability of Event A is greater than .30. Let X be the number of times A is observed, let p denote the probability of A, and let

$H_0: p = .30$

$H_1: p > .30$

Use the binomial distribution (Table 2 of Appendix VIII) to determine the probability that x falls in each of the following rejection regions when H_0 is correct.
(a) $N = 12$; rejection region is $\{9, 10, 11, 12\}$
(b) $N = 18$; rejection region is $x \geq 11$
(c) $N = 22$; rejection region is $(13 \leq x \leq 20)$
(d) $N = 25$; rejection region is $x \geq 14$

5. An experiment can yield only two results, A occurs and A does not occur. According to one theory, the probability of Event A is .80. If the theory is flawed, Event A should occur less than half the time. Let X be the number of times A is observed, let p denote the probability of A, and let

$H_0: p = .80$

$H_1: p < .80$

Use the binomial distribution (Table 2 of Appendix VIII) to determine the probability that each of the following rejection rules will lead the experimenter to reject H_0 when H_0 is in fact correct.
(a) $N = 25$; reject H_0 if $x \leq 14$
(b) $N = 15$; reject H_0 if $x = 0, 1, 2, 3, 4, 5, 6, 7, 8,$ or 9
(c) $N = 10$; reject H_0 if $(0 \leq x \leq 6)$
(d) $N = 20$; reject H_0 if $x \leq 11$

6. An experiment can yield only two results, A and B. According to one theory the probability of Event A is .60. According to another theory, the probability of Event A is less than .60. Let X be the number of times A is observed, let p denote the probability of A, and let

$H_0: p = .60$

$H_1: p < .60$

Use the binomial distribution (Table 2 of Appendix VIII) to determine the probability that each of the following rejection rules will lead the experimenter to reject H_0 when H_0 is in fact correct.
(a) $N = 10$; reject H_0 if $x = 0, 1,$ or 2
(b) $N = 20$; reject H_0 if $x \leq 7$
(c) $N = 15$; reject H_0 if $(0 \leq x \leq 4)$
(d) $N = 25$; reject H_0 if $x = 0, 1, 2, 3, 4, 5,$ or 6

7. Suppose the competing theory in Exercise 9.1.3 implied only that $p \neq .50$. That is,

$H_0: p = .50$

$H_1: p \neq .50$

The experimenter therefore decides that she must reject H_0 if the number times A occurs is *either* much greater *or* much smaller than she expects under the test hypothesis. Use the binomial distribution (Table 2 of Appendix VIII) to determine the probability that she will incorrectly reject H_0 for each of the following sample sizes and rejection rules:

(a) $N = 10$; reject H_0 if $x = 0$ or 10
(b) $N = 18$; reject H_0 if $x = 0, 1, 2, 3, 4, 14, 15, 16, 17,$ or 18

8. Suppose the competing theory in Exercise 9.1.4 implied only that $p \neq .30$. That is,

$H_0: p = .30$

$H_1: p \neq .30$

The experimenter therefore decides that he will reject H_0 if the number times A occurs is *either* much greater *or* much smaller than he expects under the test hypothesis. Use the binomial distribution (Table 2 of Appendix VIII) to determine the probability that he will incorrectly reject H_0 for each of the following sample sizes and rejection rules:

(a) $N = 22$; reject H_0 if $x = 0, 1, 13, 14, 15, 16, 17, 18, 19, 20, 21,$ or 22
(b) $N = 16$; reject H_0 if $x = 0, 1, 9, 10, 11, 12, 13, 14, 15,$ or 16

9. For the hypotheses given in Exercise 9.1.3, use the binomial distribution (Table 2 of Appendix VIII) to determine the rejection region for each of the following:

(a) $\alpha \doteq .06$, $N = 10$ (b) $\alpha \doteq .006$, $N = 20$
(c) $\alpha \doteq .02$, $N = 15$ (d) $\alpha \doteq .05$, $N = 25$

10. For the hypotheses given in Exercise 9.1.6, use the binomial distribution (Table 2 of Appendix VIII) to determine the rejection rule for each of the following:

(a) $\alpha \doteq .02$, $N = 20$ (b) $\alpha \doteq .01$, $N = 8$
(c) $\alpha \doteq .03$, $N = 15$ (d) $\alpha \doteq .05$, $N = 24$

11. Suppose the experimenter in Exercise 9.1.5 has a table of binomial probabilities that includes entries only for $p \leq .50$. Since $P(A \text{ occurs})$ is hypothesized to equal .80, he cannot obtain probabilities for X, the number of times A occurs, unless he wants to use the binomial probability function (Chapter 6, p. 207). However, if he lets Y be the number of times A *fails* to occur and lets q denote $P(A \text{ fails to occur})$, then $p = .80$ implies $q = .20$, and $p < .80$ implies $q > .20$. He therefore formulates the hypotheses

$H_0: q = .20$

$H_1: q > .20$

Use the binomial distribution (Table 2 of Appendix VIII) to determine the rejection region for each of the following:

(a) $\alpha \doteq .001$, $N = 25$ (b) $\alpha \doteq .02$, $N = 15$
(c) $\alpha \doteq .05$, $N = 18$ (d) $\alpha \doteq .03$, $N = 20$

12. Imagine that the experimenter in Exercise 9.1.6 has a table of binomial probabilities that includes entries only for $p \leq .50$. Since the probability of Event A is hypothesized to equal .60, she cannot obtain probabilities for X,

the number of times A occurs, unless she wants to use the binomial probability function (Chapter 6, p. 207). However, if she lets Y be the number of times Event B occurs and lets q denote P(Event B), then $p = .60$ implies $q = .40$, and $p < .60$ implies $q > .40$. She therefore formulates the hypotheses

$H_0: q = .40$

$H_1: q > .40$

Use the binomial distribution (Table 2 of Appendix VIII) to determine the rejection region for each of the following:

(a) $\alpha \doteq .025$, $N = 24$ (b) $\alpha \doteq .06$, $N = 12$

(c) $\alpha \doteq .01$, $N = 8$ (d) $\alpha \doteq .005$, $N = 20$

13. One of the authors is a regular listener of National Public Radio's *Car Talk*, featuring "Click and Clack, The Tappet Brothers." After many careful hours of observation, he hypothesizes that "Click" answers more than 60 percent of the questions that listeners phone in. The author's wife (the mechanic in the family) is also a fan of Click and Clack, and she insists that Click answers only 60 percent of the questions and no more.

Let N be the number of questions asked during any particular broadcast and let X be the number answered by Click.

(a) Write a specification statement for X that will permit formulation of a test hypothesis.

(b) Formulate the hypotheses suggested by this marital dispute. The hypotheses should be appropriate to your model.

Determine the rejection rule for each of the following tests, and state the rule in terms of the proportion X/N. If $Np < 5$ or $Nq < 5$, use the binomial distribution (Table 2 of Appendix VIII); if $Np \geq 5$ and $Nq \geq 5$, use the normal approximation to the binomial.

(c) $\alpha \doteq .02$, $N = 12$ (d) $\alpha \doteq .01$, $N = 18$

(e) $\alpha \doteq .11$, $N = 8$ (f) $\alpha \doteq .04$, $N = 36$

14. A friend of ours was a splendid amateur chef until he discovered microwave cookery. Now, the only spices he uses are rosemary and thyme. In fact, every jar in his spice cabinet contains either rosemary or thyme. Unfortunately, the jars are opaque and are not labeled. The friend's dog got frisky one day and toppled a number of jars onto the kitchen counter, where they spilled. This "experiment" suggested that 40 percent of the jars are filled with rosemary. A week later, the dog repeated his experiment, and this time it appeared that 65 percent of the jars were filled with rosemary. We propose to open N of the jars and let X be the number we find filled with rosemary.

(a) Write a specification statement for the distribution of X based on the first experiment.

(b) Write the test hypothesis implied by the model (a) and the alternative hypothesis implied by the second experiment.

Determine the rejection rule for each of the following tests, and state the rule in terms of the proportion X/N. If $Np < 5$ or $Nq < 5$, use the binomial distribution (Table 2 of Appendix VIII); if $Np \geq 5$ and $Nq \geq 5$, use the normal approximation to the binomial.

(c) $\alpha \doteq .055$, $N = 12$ (d) $\alpha \doteq .001$, $N = 20$

(e) $\alpha \doteq .01$, $N = 5$ (f) $\alpha \doteq .02$, $N = 35$

15. Jim and John are candidates for governor. The local barber guesses that 65 percent of his patrons favor John. The barber's partner thinks that at least 70 percent of shop patrons favor John. They poll the next N patrons and find that x of these patrons favor John. Let p be the proportion of *all* their patrons who favor John, and let

H_0: $p = .65$

H_1: $p = .70$

Use the normal approximation to the binomial to calculate approximate critical regions for the values of N and the significance levels α given below. Express your answers in terms of both X and X/N.

(a) $N = 50$, $\alpha = .05$ (b) $N = 50$, $\alpha = .01$
(c) $N = 100$, $\alpha = .05$ (d) $N = 100$, $\alpha = .01$

16. The authors are faithful readers of Bill Watterson's newspaper comic strip, "Calvin and Hobbes." One of the authors notes that during a one-month period, about 75 percent of the fights between Calvin and Hobbes are instigated by Hobbes. The other author thinks that Hobbes usually starts a much higher proportion of their scraps. To test their competing theories, they go down to the local newspaper and collect N comic strips in which Calvin and Hobbes get into a fight. Let X be the number of these strips in which Hobbes starts the fight.

(a) Write a specification statement for X that will permit formulation of a test hypothesis.
(b) Write the hypotheses appropriate to your model.

Use the normal approximation to the binomial to calculate approximate critical regions for the values of N and the significance levels α given below. Express your answers in terms of both X and X/N.

(c) $N = 50$, $\alpha = .05$ (d) $N = 50$, $\alpha = .01$
(e) $N = 100$, $\alpha = .05$ (f) $N = 100$, $\alpha = .01$

17. A player of *Advanced Dungeons & Dragons* ® has a 6-sided die she believes is biased in favor of 1s. That is, she thinks the number 1 comes up more often than it should, and if she decides this is the case she intends to throw the die away. Let X be the number of times that 1 comes up in N rolls.

(a) What is the model implied by the player's hunch?
(b) She decides to test her hunch by rolling the die 60 times. What are her hypotheses?
(c) If she wants to be 95 percent certain that she won't throw away a perfectly good die, what should the critical region of her test be?
(d) If she wants to be 99 percent certain that she won't throw away a perfectly good die, what should the critical region of her test be?

18. A statistician is flipping coins to collect "real" data for a lecture. He believes that one of his coins is biased in favor of heads, that heads comes up more often than it should. If this is the case, he will drop it back in his coin jug and it will eventually wind up in a parking meter. Let X be the number of heads in 100 tosses of the coin.

(a) What model is suggested by the statistician's suspicion?
(b) He decides to test his suspicion by flipping the coin. What are his hypotheses?

(c) If he wants to be 95 percent certain that he won't discard a nice, fair coin, what should the critical region of his test be?

(d) If he wants to be 99 percent certain that he won't feed a parking meter with a fair coin, what should the critical region of his test be?

19. A signal-detection theory predicts that observers can utilize cues without conscious awareness. If the prediction is correct, the percentage of errors made by subjects on a particular task should not exceed .25.

(a) Define a random variable X appropriate to the theory.

(b) Formulate a model for the distribution of X that implies a *simple* test hypothesis.

(c) What is the appropriate test hypothesis?

(d) What is the appropriate alternative hypothesis?

(e) The experimenter wants to look at the proportion X/N in 40 observations to test his hypothesis. If the experimenter has only the tables in Appendix VIII, what specification statement can he write for the distribution of $X/40$?

(f) Why can't the experimenter use $X/40$ as a test statistic?

(g) If $x = 14$, what is the value of the *appropriate* test statistic?

20. An experiment can yield 5 different results. A number of observations lead an experimenter to suspect that result 1 occurs at least half the time.

(a) Define a random variable X appropriate to the theory.

(b) Write a specification statement for X.

(c) Write an appropriate test hypothesis.

(d) Write an appropriate alternative hypothesis.

(e) The experimenter decides to make 25 observations and to let X be the test statistic. What values will lead to rejection of the test hypothesis if the probability is to be less than .05 that x will fall in the critical region when H_0 is correct?

(f) If the experimenter decides to make 50 observations and uses the normal approximation to the distribution of X/N, what values of x will lead to rejection of the test hypothesis if $\alpha = .05$? (Hint: Remember, X is *discrete*.)

Types of errors and decisions about decisions. In Examples 9.5.1 and 9.5.2 the values $\alpha = .05$ and $\alpha = .01$ were chosen arbitrarily for the sake of illustration. And it has become common practice among many scientists to select, in a similarly arbitrary—or, at least, ritualistic—fashion, some conventional significance level, such as .01, .025, or .05. In the everyday business of designing experiments and reporting research results, it is easier to rely on customary practice than to consider what the statistical significance of one's test really implies. The significance level of a test is, as we have indicated, a conditional probability: *Assuming that the test hypothesis is correct*, significance level is the probability that the observed value of the test statistic will fall in the rejection region. If the test statistic falls in the rejection region, the test hypothesis is rejected, so it follows that the significance level is the probability of rejecting the test hypothesis *when the test hypothesis is correct*. That is,

$$\alpha = P(\text{rejecting } H_0 | H_0 \text{ correct}) \tag{9.3}$$

Rejecting the test hypothesis when it is in fact correct is called a *Type I error*, so

$$\alpha = P(\text{Type I error}) \tag{9.4}$$

A Type I error is rejecting the test hypothesis when it is correct. The significance level of a test is the *probability* of making a Type I error.

$$\alpha = P(\text{Type I error}) = P(\text{rejecting } H_0 | H_0 \text{ correct})$$

Clearly, the choice of a significance level α depends, at least in part, on how important it is to avoid a Type I error. Indeed, if incorrect rejection of H_0 were the only possible error one could make, the choice of a significance level would be simple: Set α equal to some arbitrarily small number, for example, .01 or .001 or even .0001. As the reader has probably anticipated, however, it is also possible to reject the *alternative* hypothesis incorrectly. Rejecting the alternative hypothesis when it is in fact correct is called a *Type II error*, and P (Type II error) can therefore be expressed as the conditional probability

$$\beta = P(\text{rejecting } H_1 | H_1 \text{ correct}) \tag{9.5}$$

where β is the lowercase Greek letter *beta*. The problem with making α arbitrarily small is that for any particular sample size, α and β are inversely related. The smaller the probability of a Type I error, the larger the probability of a Type II error.

To see how this works, it is necessary to discuss the mechanics of calculating β. In this connection, it is very important to keep in mind that the conditions in expressions [9.3] and [9.5] represent two *alternative and mutually exclusive* states of affairs. To say that H_0 is correct implies that the test statistic is distributed as ϕ_0; to say that H_1 is correct implies that the test statistic has some other distribution, ϕ_1. This means that the probabilities α and β are calculated under *different* probability distributions: α is the probability that the test statistic falls *in* the rejection region *given that* the test statistic is distributed as ϕ_0, and β is the probability that the test statistic falls *outside of* the rejection region *given that* the test statistic is distributed as ϕ_1. To calculate β the experimenter must therefore specify a *particular* distribution for the test statistic, which means that H_1 must be a *simple* alternative hypothesis.

A Type II error is rejecting the alternative hypothesis when it is correct. The probability of making a Type II error is denoted β,

$$\beta = P(\text{Type II error}) = P(\text{rejecting } H_1 | H_1 \text{ correct})$$

and is calculated under the distribution specified by the *alternative* hypothesis.

Example 9.6.1. Meanwhile, back in the wolf woods .... If our speculations about domestication (Example 9.2) are wrong and domestication really hasn't made dogs any more sensitive than wolves to human vocal cues, then wolves should be able to learn verbal signals as well as dogs. This was the basis for our alternative hypothesis, $p > 5$. Suppose, however, that we had at some earlier time conducted the experiment described in Example 9.3 with a large sample of domestic dogs and found that 75 percent of them made the correct choice on the test trial. This would allow us to formulate a simple alternative hypothesis: If wolves are no less sensitive to verbal cues than dogs, then 75 percent of the wolves should be successful,

$$H_1: p = .75$$

This means that if H_1 is correct, the number of successes should be distributed $B(20, 0.75)$.

For $\alpha = .01$, we established in Example 9.5.2 that the test hypothesis is to be rejected if the proportion of successes x/N is greater than or equal to .725. Since $(.725)(20) = 14.5$, this means that the test hypothesis is rejected if we observe 15 or more successes and that the alternative is rejected if we observe fewer than 15 successes. The probability of a Type II error is therefore the conditional probability

$$\beta = P\left(\frac{x}{N} < \frac{15}{20}\middle| X : B(20, 0.75)\right)$$

We note that $Np = (.75)(20) = 15 > 5$ and $Nq = (1 - .75)(20) = 5$, so under the alternative hypothesis that $p = .75$, the distribution of

$$Z_{X/N} = \frac{\dfrac{X}{N} - p}{\sqrt{\dfrac{pq}{N}}}$$

is approximately $N(0, 1)$, and the probability β can be determined by standardizing the critical proportion, $(x/N) = .725$,

$$z_\beta = \frac{.725 - p}{\sqrt{\dfrac{pq}{20}}} \qquad\qquad [9.6]$$

and submitting z_β to the standard normal distribution. It is important to emphasize once again that β is conditioned on the assumption that the *alternative* hypothesis is correct. This means that calculation of **[9.6]** is based on $p = .75$, as specified in H_1:

$$z_\beta = \frac{.725 - .75}{\sqrt{\dfrac{(.75)(.25)}{20}}} \doteq -.26 \qquad\qquad [9.7]$$

and the probability of *incorrectly* rejecting H_1 is therefore approximated by $P(z < -.26)$ under the standard normal distribution. From Table 3 of Appendix VIII we find that $\mathbb{P}(-.26) \doteq .3974$. Thus, if we set $\alpha = .01$, which implies a critical value of $x/N = .725$, then $\beta \doteq .40$. The distribution of the test statistic under H_0 and under H_1, along with the areas corresponding to α and β, are illustrated in Figure 9.4(a).

To calculate β,

- Calculate the critical proportion x_α/N.
- Standardize the critical proportion x_α/N under the distribution specified by the *alternative* hypothesis:

$$z_\beta = \frac{\dfrac{x_\alpha}{N} - p_1}{\sqrt{\dfrac{p_1 q_1}{N}}}$$

- Find the probability that $z_{X/N}$ does *not* fall in the rejection region, *which is now expressed in terms of* z_β.

Example 9.6.2. Assuming that a 40 percent chance of committing a Type II error is unacceptably high, let us see what happens to β if we establish a more modest significance level. From Example 9.5.1 we know that for $\alpha = .05$ the test hypothesis will be rejected if the proportion of successes equals or exceeds .675, and the alternative hypothesis will be rejected if the proportion of successes is less than .675. Therefore, $\beta \cong P(z < z_\beta)$ under the standard normal distribution, where

$$z_\beta = \frac{.675 - .75}{\sqrt{\dfrac{(.75)(.25)}{20}}} \doteq -.77$$

From Table 3 of Appendix VIII, we find that $\mathbb{P}(-.77) = .2206$, so $\beta \doteq .22$. The distribution of the test statistic under H_0 and under H_1 and the areas corresponding to α and β are illustrated in Figure 9.4(b).

From Examples 9.6.1 and 9.6.2 (and the areas representing α and β in Figure 9.4) we see that any rejection rule implies a trade-off between the probability of committing a Type I error and the probability of committing a Type II error. Each type of error carries some sort of consequence or cost, and in some types of applied research it is possible to express these costs in dollars, kilowatt-hours of power consumption, worker-days of labor, or some other economic unit. When assignment of such values is possible, the investigator may bring to bear a branch of applied mathematics called *decision theory* in choosing a rejection rule that quantitatively balances the probabilities of Type I and Type II errors against their potential costs.

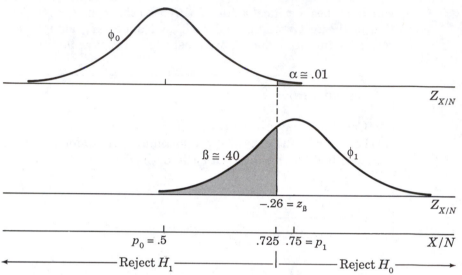

Figure 9.4(a). Distributions of $Z_{X/N}$ under test hypothesis and alternative hypothesis for $\alpha \cong .01$.

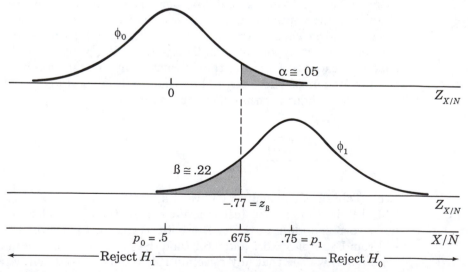

Figure 9.4(b). Distributions of $Z_{X/N}$ under test hypothesis and alternative hypothesis for $\alpha \cong .05$.

There are a number of decision theory algorithms, but in general they are based on the expected value of every conclusion that may arise under any particular rejection rule. The simplest criterion is to select the rule that yields the *least expensive* worst-case scenario. This is called the *minimax loss* criterion.

We know by definition that the probability of a Type I error is α. If, in addition, we let C_I represent the *cost* of a Type I error, then over many repetitions of the experiment, every rejection (correct as well as incorrect) of H_0 would incur an *average* loss equal to the product $\alpha(C_I)$. This quantity must therefore represent the *expected loss* associated with the rejection rule when H_0 is correct. That is,

$$\alpha(C_I) = E(\text{loss}|H_0 \text{ correct})$$

Assuming that the experimenter has a simple alternative hypothesis, the value of α will determine the value of β. If the cost of a Type II error is C_{II}, the experimenter can then compute the product

$$\beta(C_{II}) = E(\text{loss}|H_1 \text{ correct}).$$

The larger of these two expected values is the *maximum* expected loss for this particular rejection rule. In this fashion the experimenter can calculate the maximum expected loss for all of the rejection rules under consideration and choose the rule with the *smallest* maximum loss.

Example 9.7. To put some meat on this abstract skeleton, imagine that our wolf experiment has some dollarable consequences. Let us suppose that if we claim *incorrectly* that wolves and dogs are different, the research team will be professionally discredited and will lose a $10,000 research grant. The cost C_I of a Type I error would therefore be $10,000. On the other hand, suppose that an advertising agency is using wolves in a TV commercial. If wolves must be taught to respond to nonverbal signals, the producer is willing to pay the team $1,600 for one day of consultation with the agency's animal trainer. If we *incorrectly* conclude that wolves and dogs are equally sensitive to verbal cues, the cost of this error, C_{II}, is therefore $1,600. The expected losses for the rejection rules presented in Examples 9.5.1 and 9.5.2 are given in Table 9.1. The maximum loss under each rule is indicated with an asterisk (*), and we

Table 9.1. Expected losses for two rejection rules

Rule	Expected loss given that	
	H_0 Correct	H_1 Correct
I	.05($10,000) = $500*	.22($1,600) = $352
II	.01($10,000) = $100	.40($1,600) = $640*

see that the smaller of the two is $500, generated by Rule I. By the minimax loss criterion, therefore, we should select the first rule, in which the level of significance is set equal to .05 and the probability of a Type II error is .22. It should be understood that some other rule, which was not considered in this example, might have yielded an even smaller maximum loss.

Formal decision theory enjoys more widespread application in economic decision making than in science, because it is seldom possible for a scientist to assign numerical values to the costs of errors (or the payoffs for correct decisions). However, even if a scientist cannot assign an *exact* cost to the consequences of an incorrect conclusion, it is often possible to state which of the two possible errors would be the *more* serious. Analytic chemists testing water samples collected near toxic waste dumps are frequently faced with hypotheses of the type

H_0: Contaminant levels are safe.

H_1: Contaminant levels are dangerous.

Incorrect rejection of H_0 may lead to needless expenditure of hundreds of thousands of dollars in cleanup costs, implementation of new disposal technology, etc. Incorrect rejection of H_1 may lead to incalculable costs in human suffering and possible loss of life compounded by millions of dollars in civil damages and criminal fines. Clearly, it is more important in this instance to avoid a Type II error than to avoid a Type I error.

In *pure* research, as opposed to *applied* research, costs may be largely intangible, though no less real—loss of professional stature, wasted time and effort invested in a particular line of inquiry, or the abandonment of a theory that offered the promise of new understanding. When such intangibles lie in the balance of our conclusions, the best rejection rule is usually one in which the two errors are equally probable.

3. Data collection

a. Random sampling. We have indicated in several places that inferential statistics finds application in a wide variety of scientific disciplines. It is generally immaterial whether a particular random variable represents biological, chemical, anthropological, psychological, or geographical measurements. However, we shall see in later chapters that the choice of a test statistic and the probability model for the sampling distribution of that statistic often depend on the basic research *strategy* by which the data were obtained. In this connection, we have used the term "experiment" throughout this book to indicate any activity that can yield more than one result, and we have used the term "observation" to mean assignment of numerical values to these results. This is what experiment and observation mean in the context of probability. In the context of research methodology these terms are used instead to distinguish between fundamentally different ways of collecting data.

Whether data are experimental or observational, it must always be assumed that they reflect the actual state of nature. If we are to place such absolute reliance on our data, we must exert every effort to ensure that they are as representative as possible of the population(s) about which we hope to draw a conclusion. A scientist's subjects, be they white rats, samples of groundwater, or cancer patients, should therefore be chosen by *random sampling*. Basic distinctions among experimental research and two types of observational research (*quasi-experimental* and *correlational* designs) and the random sampling considerations appropriate to these designs are discussed in Appendix VI.

The fundamental assumption of random sampling has an important implication for testing hypotheses that are based on exploratory data analysis: When data *suggest* hypotheses, one cannot use the *same* data to *test* those hypotheses. In Example 9.1, we began with the supposition that one of the authors believed a coin was biased in favor of heads and formulated the hypotheses

$$H_0: p = .50$$

$$H_1: p > .50$$

where p is P(head). Let's further suppose that the author first became suspicious of the coin when he tossed it five times and obtained five heads. There are $2^5 = 32$ possible samples (cf. p. 205) that might be drawn from the sample space of the experiment "toss a coin five times and record the sequence of heads and tails." To satisfy the definition of random sampling given in Chapter 8, the probability of selecting each sample must therefore be $1/32$. However, if the author uses his original five tosses to test his hypotheses, the sample $HHHHH$ has already been selected and therefore has a probability of 1, so the probability of selecting any other sample is zero. In order to base any inference on random sampling, the experimenter must therefore collect additional data. This caveat must be observed whenever hypotheses are generated by exploratory data analysis.

b. Sample size. The principles of research design emphasize consideration of *how* data are to be collected. In addition, the scientist must also consider *how much* data should be collected, because the *number* of observations in one's sample(s) can bear heavily on both the results and the meaningfulness of a statistical test.

At the end of our discussion of rejection rules, the reader was probably left with the impression that β is determined entirely by the critical value of the test statistic and ultimately, therefore, by the significance level α. As we shall soon discover, however, the probability of a Type II error also depends on the variability of the test statistic. Furthermore, the variability of *every* test statistic developed in Part III is a function of sample size. As sample size *increases*, variability *decreases*. For tests of proportions, the relationships among sample size, variability, and β are exceptionally clear, and our wolf experiment is therefore a particularly good vehicle for introducing these considerations.

We know that the variance of a proportion $V(X/N)$ is equal to $(pq)/N$, and throughout our discussion of Example 9.2 we have assumed that $N = 20$. When we determined the critical value of X/N to be .725 in Example 9.5.2, we began by setting z_α equal to 2.33 and solving

$$\frac{\dfrac{x}{20} - .5}{\sqrt{\dfrac{(.5)(.5)}{20}}} = 2.33$$

for x. It should be obvious that for any sample size *other* than 20, the standard deviation

$$\sqrt{\frac{(.5)(.5)}{N}}$$

would be different, and the critical value of X/N would likewise be different.

Example 9.8. Suppose, as in Example 9.5.2, that $\alpha = .01$, but let us now imagine that we had decided to test 50 wolves instead of 20. If $N = 50$, then z_α equals

$$\frac{\dfrac{x}{50} - .5}{\sqrt{\dfrac{(.5)(.5)}{50}}} = 2.33$$

so

$$\frac{x}{N} = .5 + 2.33\sqrt{\frac{(.5)(.5)}{50}} \doteq .66$$

and $x = (.66)(50) = 33$. Since $x = 33$ corresponds to the interval 32.5 to 33.5, we will reject H_1 if

$$\frac{x}{N} < \frac{33.5}{50} = .67$$

We see in Figure 9.5(a) that the critical value of X/N is *closer to* the expected value of X/N under H_0 with a sample size of 50 than with a sample size of only 20. Conversely, it is *farther from* the value of X/N expected under H_1. This is illustrated in Figure 9.5(b). Increasing sample size therefore has the same effect on β as increasing α.

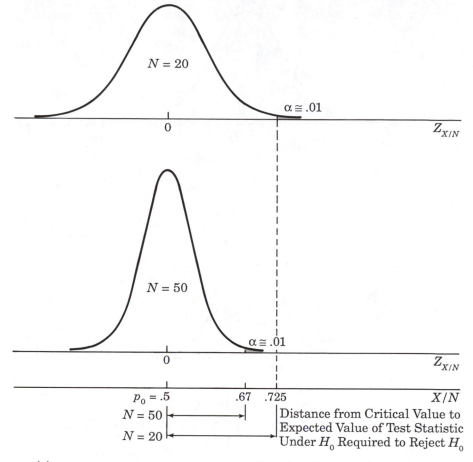

Figure 9.5(a). Distribution under test hypothesis of $Z_{X/N}$ for $N = 20$ and $N = 50$.

To calculate β, we once again standardize our critical value using the values of p and q specified in the *alternative* hypothesis, just as we did in Example 9.6.1:

$$z_\beta = \frac{.67 - .75}{\sqrt{\dfrac{(.75)(.25)}{50}}} \doteq -1.31$$

Consequently, $\beta \doteq P(z < -1.31)$ under the standard normal distribution. In Table 3 of Appendix VIII we find that $\mathbb{P}(-1.31) \doteq .09510$. For $\alpha = .01$ and $N = 50$, $\beta \doteq .10$. Contrast this with $\beta \doteq .40$, which we calculated in Example 9.6.1 when $\alpha = .01$ and $N = 20$.

The reason that β in Example 9.8 is so much smaller than β in Example 9.6.1 is that *increasing* the sample size N *decreased* the variance of X/N. The effect on β of reducing variance is seen in Figure 9.5(b). In general, anything that reduces the variability of the test statistic will reduce the probability of a Type II error. If the variability of

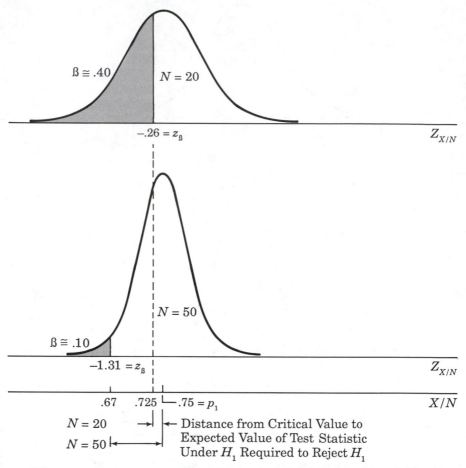

Figure 9.5(b). Distribution under alternative hypothesis of $Z_{X/N}$ for $N = 20$ and $N = 50$.

a test statistic depends on sample size, β can therefore be reduced *independently* of α by increasing the number of observations. For some statistical tests there are procedures for calculating the sample size that yields desired values of α and β. Two such techniques are presented in Chapter 10.

4. Conclusions: rejecting hypotheses versus accepting hypotheses

At the beginning of the chapter we explained that statistical hypotheses are tested by *disconfirmation*, and we have seen that statistical conclusions are consequently stated in terms of *rejecting* hypotheses rather than *accepting* hypotheses. The basis for this practice becomes apparent if we return once again to our wolf experiment.

Example 9.9. In a spirit of eager enthusiasm—fueled by the prospect of a $1,600 consulting fee—we have conducted our wolf experiment with 20 subjects, and 13 of them were able to learn the verbal cue. That is, $13/20 = 65$ percent of the wolves were successful. This falls short of the critical proportion $x/N = .675$ required to reject the test hypothesis at the .05 level. However, since $E(X/N) = p$, the proportion of successes x/N is an unbiased estimate of p. On the basis of our data, therefore, the most likely value of p is .65, not .5 as specified in H_0, and although we must reject H_1, we cannot reasonably accept the test hypothesis.

The curious reality is that it is generally impossible to *prove* an empirical proposition of the form A is B, where A is a class of objects or phenomena and B is some quality or attribute. Consider the proposition, "All swans are white."[6] No matter how many white swans one observes, the truth of the proposition remains in doubt. Unless one observes *all* swans in existence it is possible that only 99 percent of all swans are white—or 99.9 percent or 99.99 percent. On the other hand, it takes only one black swan to *disprove* the proposition. Statistical hypotheses are propositions of the form A is B (e.g., the value of p is greater than .5), and are therefore subject only to disproof.

a. Significance probability. Significance *level* is a property of one's *test*. It indicates "how improbable" a result must be in order to reject H_0. Significance *probability* is a property of your *data*. It indicates "how improbable" your obtained result really is. More formally, we know that α is the probability (assuming H_0 is correct) of obtaining a value of the test statistic that is *at least* as far from the value expected under H_0 *in the direction favored by H_1* as the critical value. That is, significance level is the probability of obtaining a value of the test statistic that is at least as favorable to H_1 as the critical value if H_0 is correct. The significance probability, or p-value, of the result is defined in much the same way:

> Significance probability is the probability of obtaining a value of the test statistic that is *at least* as favorable to H_1 as the value actually obtained *if H_0 is correct*.

In Exercises 9.1.3 to 9.1.6 you were given the value of the population parameter p and the critical region and asked to calculate the probability of rejecting H_0. Implicitly these exercises asked you to calculate the significance *level* of the test. Significance *probability* is ordinarily calculated in the same way, except that you use the *obtained* value of the test statistic instead of the *critical* value.

Example 9.10.1. Suppose that instead of recording 13 successes in our wolf experiment, as in the preceding example, 16 wolves made the correct turn on their test trials. That is, $x/N = 16/20$. The alternative hypothesis is $p > .5$, so the alternative hypothesis favors values of X/N in the *upper* tail of the distribution specified under the test hypothesis. The significance probability of this result is therefore the probability of 16 *or*

---

[6] This proposition was believed by most naturalists from the time of Aristotle until black swans were discovered in New Zealand.

BOX 9.1

Reporting significance probability

Because there are gaps in most cumulative probability tables, the obtained value of the test statistic seldom corresponds to a tabled value. Therefore, significance probability is ordinarily reported in the form of an inequality.

Small significance probabilities favor the alternative hypothesis, so the statement that p is *less than* some value implies rejection of H_0. Small p-values (e.g., $p < .05$) are usually rounded to either one or two significant figures. One common practice is to round small p-values up to the nearest value with a final digit of 0, 1, or 5. A significance probability between .009642 and .009903 might therefore be reported as $p < .0099$ or as $p < .01$.

Large significance probabilities favor the test hypothesis, so the statement that p is *greater than* some value implies rejection of H_1. Large p-values (e.g., $p > .05$) are usually reported to two places and should be *truncated* rather than rounded up, but it is usually acceptable to round *down* to the nearest value with a final digit of 0, 1, or 5. A significance probability between .3669 and .3707 might therefore be reported as $p > .36$ (truncated) or as $p > .35$ (rounded down).

If α is established in advance, it may be necessary to report the p-value to the same number of places as α. Thus, if $\alpha = .0125$, a significance probability slightly less than .01222 might be reported as $p < .0123$; a p-value slightly greater than .00126 might be reported as $p > .0125$.

---

more wolves making the correct turn if H_0 is correct,

$$P\left(\frac{x}{N} \geq \frac{16}{20}\middle| X : B(20, 0.5)\right)$$

and the normal approximation to this probability is $P(z > z_{X/N})$, where

$$z_{X/N} = \frac{\dfrac{15.5}{20} - .5}{\sqrt{\dfrac{(.5)(.5)}{20}}} \doteq 2.46$$

From Table 3 of Appendix VIII we see that $\mathbb{P}(2.46) \doteq .993053$. *If* the test hypothesis is correct, the probability of observing 16 or more successes in 20 trials is therefore $1 - .993503$, which is a little less than .007.

In Example 9.10.1 the observed value of the test statistic corresponds to a tabled value ($z = 2.46$), and the significance probability may therefore be reported either as an approximation (e.g., $p \doteq .007$) or in the form of an inequality ($p < .007$). If the observed statistic does not correspond to a tabled value, the p-value is expressed as an inequality, as discussed in Box 9.1.

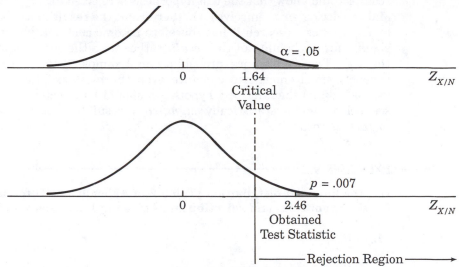

Figure 9.6. Critical region, α, and p-value for Example 9.10.1.

If the test hypothesis is rejected, we know that the obtained value of the test statistic is either equal to the critical value or lies even farther out into the rejection region than the critical value. If α is set in advance, this means that the p-value of any test statistic that falls in the rejection region must be less than or equal to α. These relationships are illustrated for Example 9.10.1 in Figure 9.6 and suggest a somewhat different way to state the rejection rule: If p-value $\leq \alpha$, reject H_0; if p-value $> \alpha$, reject H_1.

The p-value ordinarily corresponds to values of the test statistic that lie *farther* from the value expected under H_0 than the value actually observed. This is an appealing way to think about significance probability, but it holds true only if the test statistic favors the *alternative* hypothesis, as in Example 9.10.1. This doesn't always happen.

Example 9.10.2. Suppose that only 4 of the 20 wolves in Example 9.10.1 had been successful. We calculate the significance probability of this result the same way as in Example 9.10.1:

$$P\left(\frac{x}{N} \geq \frac{4}{20} \,\middle|\, X : B(20, 0.5)\right) \cong P(z \geq -2.46) \doteq .993053$$

In this example, however, the observed value departs from the value expected under H_0 in the direction *unfavorable* to H_1; that is, it falls in the *lower* tail of $X : B(20, 0.5)$. Consequently, most of the values captured by the significance probability ($x = 5, 6, \ldots, 15$) are *closer* than $x = 4$ to the value expected under H_0 (10 successes).

We have said that if α is set in advance, then H_0 is rejected if p-value $\leq \alpha$. Even if α is not set in advance, readers evaluating the test results would know that the test hypothesis is rejected for *any* $\alpha \geq p$-value and can decide for themselves whether or not the result justifies rejection of H_0. In most research it has therefore become customary to report the significance probability of the *result*, rather than the significance level of the *test*. The significance probability in Example 9.10.1 ($p < .007$) is probably small enough to convince even the most skeptical reader (or journal editor) that the test hypothesis should be rejected, but we shall see below that a statistically *significant* result is not necessarily an *important* result.

───────── EXERCISES 9.2 ────────────────────────────────

Use the binomial distribution (Table 2 of Appendix VIII) to calculate β for the hypotheses and rejection rules given in Exercises 1 and 2.

1. H_0: $p = .50$

    H_1: $p = .70$

    (a) $N = 10$; rejection region is $\{9, 10\}$
    (b) $N = 20$; rejection region is $x \geq 17$
    (c) $N = 15$; rejection region is $(11 \leq x \leq 15)$
    (d) $N = 25$; rejection region is $x > 18$
2. An experiment can yield only two results, A and B. According to one theory the probability of Event A is .80. According to another theory, the probability of Event A is .90. Let

    H_0: $p = .80$

    H_1: $p = .90$

    (a) Reject H_0 if $x/N \geq .90$, where $N = 10$.
    (b) Reject H_0 if $x/N \geq .95$, where $N = 20$.
    (c) Reject H_0 if $x/N = 1.00$, where $N = 15$.
    (d) Reject H_0 if $x/N \geq .96$, where $N = 25$.

Use the binomial distribution (Table 2 of Appendix VIII) to calculate β for the hypotheses, sample sizes, and significance levels in Exercises 3 and 4.

3. An experiment can yield only two results, A and B. One theory implies that the probability of Event A is .30. Another theory implies that probability of Event A is .50. Let

    H_0: $p = .30$

    H_1: $p = .50$

    (a) $N = 20$, $\alpha \doteq .113$ (d) $N = 25$, $\alpha \doteq .10$
    (b) $N = 20$, $\alpha \doteq .048$ (e) $N = 25$, $\alpha \doteq .044$
    (c) $N = 20$, $\alpha \doteq .005$ (f) $N = 25$, $\alpha \doteq .005$

4. An experiment can yield only two results, A and B. According to one theory the probability of Event A is .30. According to another theory, the probability of Event A is .40. Let

   H_0: $p = .30$

   H_1: $p = .40$

   (a) $N = 20$, $\alpha \doteq .113$
   (b) $N = 20$, $\alpha \doteq .048$
   (c) $N = 20$, $\alpha \doteq .005$

   (d) $N = 25$, $\alpha \doteq .10$
   (e) $N = 25$, $\alpha \doteq .044$
   (f) $N = 25$, $\alpha \doteq .005$

Calculate β for the hypotheses, sample sizes, and significance levels in Exercises 5 and 6. If $Np \geq 5$ and $Nq \geq 5$ under both the test and the alternative hypotheses, use the normal approximation to the binomial (as illustrated in Example 9.6.1). Otherwise use the binomial distribution.

5. An experiment can yield only two results, A and B. According to one theory the probability of Event A is .50. According to another theory, the probability of Event A is .60. Let X be the number of times A is observed, let p denote the probability of A, and let

   H_0: $p = .50$

   H_1: $p = .60$

   (a) $\alpha \doteq .003$; $N = 12$
       $\alpha \doteq .003$; $N = 30$

   (b) $\alpha \doteq .01$; $N = 10$
       $\alpha \doteq .01$; $N = 25$

6. An experiment can yield only two results, A and B. According to one theory the probability of Event A is .35. According to another theory, the probability of Event A is .75. Let X be the number of times A is observed, let p denote the probability of A, and let

   H_0: $p = .35$

   H_1: $p = .75$

   (a) $\alpha \doteq .04$; $N = 15$
       $\alpha \doteq .04$; $N = 20$

   (b) $\alpha \doteq .01$; $N = 15$
       $\alpha \doteq .01$; $N = 25$

7. An experiment can yield only two results, A and B. According to one theory the probability of Event A is .24. According to another theory, the probability of Event A is .32. Let X be the number of times A is observed, let p denote the probability of A, and let

   H_0: $p = .24$

   H_1: $p = .32$

Find β for each of the following:

(a)	(b)	(c)
$\alpha = .10;\ N = 50$	$\alpha = .05;\ N = 50$	$\alpha = .01;\ N = 50$
$\alpha = .10;\ N = 100$	$\alpha = .05;\ N = 100$	$\alpha = .01;\ N = 100$
$\alpha = .10;\ N = 200$	$\alpha = .05;\ N = 200$	$\alpha = .01;\ N = 200$

8. An experiment can yield only two results, A and B. According to one theory the probability of Event A is .65. According to another theory, the probability of Event A is .75. Let X be the number of times A is observed, let p denote the probability of A, and let

$$H_0:\ p = .65$$

$$H_1:\ p = .75$$

Find β for each of the following:

(a)	(b)	(c)
$\alpha = .10;\ N = 50$	$\alpha = .05;\ N = 50$	$\alpha = .01;\ N = 50$
$\alpha = .10;\ N = 100$	$\alpha = .05;\ N = 100$	$\alpha = .01;\ N = 100$
$\alpha = .10;\ N = 200$	$\alpha = .05;\ N = 200$	$\alpha = .01;\ N = 200$

9. Recall the following hypotheses from Exercise 9.1.3.

$$H_0:\ p = .50$$

$$H_1:\ p = .70$$

   (a) What is the significance probability (p-value) of the result $x = 9$ if the experimenter makes 10 observations?
   (b) What kind of error is possible in (a) if $\alpha \doteq .011$?
   (c) What is the significance probability (p-value) of the result $x = 9$ if the experimenter makes 15 observations?
   (d) What kind of error is possible in (c) if $\alpha \doteq .001$?

10. The experimenter in Exercise 9.1.12 formulated the following hypotheses about the binomial random variable Y:

$$H_0:\ q = .40$$

$$H_1:\ q > .40$$

   (a) What is the significance probability (p-value) of the result $y = 9$ if the experimenter makes 10 observations?
   (b) What kind of error is possible in (a) if $\alpha \doteq .013$?
   (c) What is the significance probability (p-value) of the result $y = 9$ if the experimenter makes 20 observations?
   (d) What kind of error is possible in (c) if $\alpha \doteq .001$?

11. What conclusion should the experimenter in Exercise 9.2.10 draw under each of the following circumstances, and what is the p-value of the result?

   (a) $\alpha \doteq .057,\ N = 18,\ y = 12$ (b) $\alpha \doteq .021,\ N = 20,\ y = 10$
   (c) $\alpha = .05,\ N = 50,\ y = 30$ (d) $\alpha \doteq .10,\ N = 100,\ y = 45$

12. A theory implies that $X: B(N, 0.70)$ for the results of a particular binomial experiment. A competing theory implies that $X: B(N, 0.80)$. An experimenter tests the hypotheses

H_0: $p = .70$

H_1: $p > .70$

What conclusion should he draw under each of the following circumstances, and what is the p-value of his result?

(a) $\alpha = .05$, $N = 15$, $x = 12$ (b) $\alpha = .01$, $N = 20$, $x = 19$
(c) $\alpha = .01$, $N = 50$, $x = 40$ (d) $\alpha = .05$, $N = 100$, $x = 80$

13. Steve has a 6-sided die that is biased against 3s and 4s. Either a 3 or a 4 comes up only about 25 percent of the time. Steve also knows that after about 7 miles of their weekly 10-mile run, Harry has the functional intelligence of a pet rock and can be tempted to bet on anything. Steve proposes to pay Harry $2 for every 3 or 4, provided that Harry pays Steve $1 for every other result. At about 8 miles they agree on a game of 1,000 rolls. Steve arrives smugly at his office a few minutes before game time, and a colleague says, "I saw Harry messing around with your dice while you were in class." Now, Steve knows that Harry has a die that looks exactly like Steve's but is biased *in favor of* 3s and 4s. For Harry's die, $P(3 \text{ or } 4) = .4$. Steve has just enough time to roll his die 24 times to test the hypothesis that Harry didn't switch dice. If his test leads Steve to conclude that a switch was made, he will not go through with the bet. A quick glance at Table 2 of Appendix VIII convinces Steve to pick a significance level of either $\alpha = .120$ or $\alpha = .053$. According to the minimax loss criterion, which is the better choice? (Hint: The first step is to calculate the expected value of the wager if H_0 is correct and the expected value of the wager if H_1 is correct.)

14. Each voter in the city favors either Caligula or Livia for Mayor. With just two weeks to go until the election, Caligula thinks that he has 60 percent of the vote, but Livia also claims to have 60 percent of the vote. This late in the race, Caligula will certainly lose the election if his opponent is correct, and he will drop out of the race. He decides to test the hypothesis that he has only 40 percent of the vote against the alternative that he has 60 percent of the vote. He will sample 100 voters and use a significance level of $\alpha = .10$ or $\alpha = .05$ or $\alpha = .01$. It will cost him about $1,000,000 to continue the campaign, but if he's elected he will receive $10,000,000 in kickbacks from the chariot makers' guild. Use the minimax loss criterion to decide which significance level is best for Caligula to use.

C. EVALUATION OF THE STATISTICAL TEST

1. The concept of power

Let us assume that our statistical test has a simple alternative hypothesis, H_1: $\theta = \theta_1$. We know from our discussion of statistical errors that

$$P(\text{rejecting } H_1 | H_1 \text{ correct}) = \beta$$

That is, β is the probability that a statistical test will *fail* to detect that the true value of the population parameter is θ_1. A rejection rule permits only two possible results. The experimenter must either reject H_0 or reject H_1. Therefore,

$$P(\text{rejecting } H_0) = 1 - P(\text{rejecting } H_1)$$

and if the alternative hypothesis is correct,

$$P(\text{rejecting } H_0 | H_1 \text{ correct}) = 1 - \beta \qquad [9.8]$$

That is, $1 - \beta$ is the probability that the test *will* detect that the population parameter is equal to θ_1. This value is therefore a measure of the sensitivity of the test and, like the sensitivity of an optical microscope, is defined as *power*.

> The power of a statistical test is the probability of correctly rejecting the test hypothesis. That is,
>
> $$\text{Power} = P(\text{rejecting } H_0 | H_1 \text{ correct}) = 1 - \beta$$

It follows from equation [9.8] that any reduction in β increases the power of the test. In our discussion of decision theory, for instance, we established that β decreases as α increases. So, a statistical test may be made more powerful by increasing the level of significance. In our discussion of data collection we further established that β may be reduced by decreasing the variability of the test statistic. This is typically accomplished by increasing sample size. Accordingly, statistical tests based on large samples are more powerful than tests based on small samples.

2. The power of a test against a composite alternative hypothesis

In order to calculate the probability of a Type II error in Examples 9.6.1 and 9.6.2 it was necessary to formulate a simple alternative hypothesis, $p = .75$. This is because the value of β depends upon the *specific* alternative value assigned to the parameter θ. If one has only sufficient information to formulate a composite alternative hypothesis, then β will assume a different value for every possible value that θ is allowed under the alternative. For example, the alternative hypothesis in our wolf experiment was

$$H_1: p = p_1 > .5$$

so β will have a different value for every value of p_1, where $.5 < p_1 \leq 1$. That is, β is a *function* of the alternative parameter value θ_1. Since power is equal to $1 - \beta$, it is likewise true that power varies as a function of θ_1.

Example 9.11. In Example 9.8 we calculated $\beta \doteq .10$ for the alternative hypothesis, H_1: $p = .75$. The power of our test against this alternative was therefore approximately $1 - .10 = .90$. Suppose instead that our dog performance data had suggested that the appropriate alternative hypothesis was $p = .80$. The critical value $x/N = .67$ is determined by the *test* hypotheses and therefore remains the same. So under our new alternative hypothesis,

$$\beta = P\left(\frac{x}{N} < .67 \middle| X : B(50, 0.80)\right)$$

and using the normal approximation

$$z_\beta = \frac{.67 - .80}{\sqrt{\dfrac{(.80)(.20)}{50}}} \doteq -2.30$$

In Table 3 of Appendix VIII we find that $\mathbb{P}(-2.30) \doteq .01072$. So $\beta \doteq .01$, and power is about $1 - .01 = .99$.

For $p_1 = .75$, power is .90; for $p_1 = .80$, power is .99. Were we to calculate power for $p_1 = .85$, $p_1 = .90$, etc., we would find that power increases as the difference between p_0 and p_1 increases. Remember that β and power are *conditional* probabilities; they are probabilities that have meaning *only* under the assumption that the alternative hypothesis is correct, that the *true* value of p is p_1.

What Example 9.11 shows, therefore, is that as the difference between the *hypothesized* parameter value p_0 and the *true* value of p increases, the more likely it is that the test will detect this difference. And in general, if θ is the true value of a population parameter, then the greater the absolute difference $|\theta - \theta_0|$, the more likely it is that the difference will be revealed by the statistical test. In this respect a statistical test is similar to an optical instrument: The farther apart two lines inscribed on a microscope slide, the more likely it is that someone looking through a microscope will discern the separation; the farther apart two stars in a binary system, the more likely it is that an astronomer will be able to see they are not a single star.

We have already pointed out that there are at least two ways to increase the power of a statistical test: (1) Increase α or (2) increase sample size in order to reduce the variability of the test statistic. To these we may now add a third strategy: (3) Increase the difference between θ_1 and θ_0. That is, formulate a different alternative hypothesis—one that specifies a parameter value θ_1 that is farther from the value θ_0 specified in the test hypothesis. (This seems a bit vague, but we'll say more about it in the next section.) The effects on power illustrated in the examples cited in this section are summarized in Table 9.2.

Table 9.2. Effects of α, N, and H_1 on power

Example	α	N	p_1	Power $= 1 - \beta$
9.6.1	.05	20	.75	.60 $= 1 - .40$
9.6.2	.01	20	.75	.78 $= 1 - .22$
9.8	.01	50	.75	.90 $= 1 - .10$
9.11	.01	50	.80	.99 $= 1 - .01$

The sampling distribution under the alternative hypothesis is difficult to specify for some test statistics, so the effects on β of, say, increasing sample size may be difficult to calculate. When control of β is uncertain, the prudent course for the experimenter with two simple hypotheses is to test the hypothesis that has the *higher* cost of error. In the world of science, the cost of incorrectly abandoning established, orthodox theory is generally considered to be higher than the cost of incorrectly dismissing revolutionary or maverick theories. This is not mindless conservatism. It is, for example, our best safeguard against abandonment of established medical treatments in favor of unconventional remedies that may at best be useless and may at worst entice a patient to forgo lifesaving, conventional treatment. Since most researchers are in the business of advancing scientific thought, experiments are generally devised to test propositions that challenge orthodox thinking. Accordingly, the customary practice is to cast the hypothesis favored by the experimenter as the *alternative* hypothesis.

Furthermore, when costs are difficult to quantify, and an experimenter cannot bring to bear the machinery of decision theory, it is all too easy to let hopes and aspirations color the interpretation of data. The difference between what is expected under H_0 and what is observed has a way of looking more significant when the difference favors one's pet (alternative) hypothesis. This is why it's a good idea to establish a rejection rule *before* data are collected, even if you intend to report only the significance probability (p-value) of the result instead of the significance level (α) of the test.

3. Scientific versus statistical significance

Earlier we said that selection of a critical region is the linchpin of hypothesis testing. This decision is the most concrete answer to the fundamental question, "How much difference must there be between what we *actually* observe and what the test hypothesis tells us to *expect* in order to justify abandonment of the test hypothesis?" When we select a significance level for a test we implicitly answer this question in probability terms: The difference between the observed and expected values of the test statistic must be so large that the probability of such a difference is less than or equal to α.

In addition to looking at the *probability* of this difference, one can look at the size of the difference itself. In Example 9.5.2, with $\alpha = .01$ and $N = 20$, the critical proportion of successes turned out to be .725. This means that the difference between the *observed* proportion of successes and the proportion *expected* under H_0 had to be at least $(.725 - .5) = .225$ to reject H_0. In Example 9.8 we again set $\alpha = .01$, but we increased sample size from 20 to 50, and we discovered that the critical value became .67. The critical value .67 is *closer* to the value expected under the test hypothesis (.50) than is .725, so increasing the power of the test made it possible to reject the test hypothesis on the basis of a *smaller* difference $(.67 - .50 = .17)$ between the observed value and the value expected under H_0. This is the central analogy between the resolving power of an optical microscope and the power of a statistical test: The more powerful the microscope, the smaller the spaces it can detect between grid lines on a slide; the more powerful the statistical test, the smaller the differences it can detect (i.e., find to be statistically significant) between the observed and expected values of a test statistic. With a large enough sample we could generate a test so powerful that an observed proportion of 56 or .51 or even .50000001 would allow us to reject the test hypothesis with a significance of .01.

Insofar as the observed proportion x/N is an unbiased estimate of the *true* population proportion p, this begs an obvious question: If the true proportion of successes in the wolf population were, say, .51, would that be different in any scientifically important way from the value expected under the test hypothesis, .50? If not, would we want to reject the hypothesis that $p = .50$ when, in fact, $p = .51$? In more general terms, is it useful to conduct a test that is likely to yield *statistical significance* when the difference between θ and θ_0 is *scientifically trivial*?

This problem reveals a basic flaw in the prevailing scientific practice of making one's samples as large as possible. The fallacy is even more apparent when one has a simple alternative hypothesis. In Example 9.6.1 we tested H_0: $p = .5$ against the alternative, H_1: $p = .75$. Suppose once again that $\alpha = .01$ and suppose that N is large enough that $x/N = .55$ falls in the critical region. Since

$$E\left(\frac{X}{N}\middle| H_0 \text{ correct}\right) = .5$$

and

$$E\left(\frac{X}{N}\middle| H_1 \text{ correct}\right) = .75$$

such a test permits us to reject H_0 on the basis of an observed test statistic that is *closer* to the value expected under H_0 than it is to the value expected under H_1.

A statistical test that is too powerful is, simply, one for which β is too small. Accordingly, the power of a test can be reduced by any manipulation that increases β, for example, selecting a smaller sample or reducing α. If one has a composite alternative hypothesis, power can sometimes be controlled by formulating a *synthetic* alternative hypothesis based on considerations of *scientific* significance. The experimenter must first decide how *small* a difference between θ_0 and the *true* population value θ is scientifically important. Call this difference δ (lowercase Greek *delta*, for difference). That is,

$$|\theta - \theta_0| = \delta$$

Then formulate a *synthetic* alternative hypothesis,

$$H_1: \theta = \theta_0 + \delta$$

or

$$H_1: \theta = \theta_0 - \delta$$

or

$$H_1: \theta = \theta_0 \pm \delta$$

according to the directionality or nondirectionality of the alternative. One can then use the value(s) specified by the synthetic alternative hypothesis to evaluate β and power. If β is disproportionate to the cost of a Type II error, one has the option of specifying some other synthetic value for θ_1, which achieves a more desirable balance between power and scientific significance.

D. POSTSCRIPT: DECISIONS CONCERNING HYPOTHESES ABOUT PROPORTIONS

1. Test statistics: Z_X and $Z_{X/N}$

Many of the principles of hypothesis testing were illustrated throughout this chapter with examples surrounding the hypotheses about the proportion of successes

$$H_0: p = p_0$$

$$H_1: p > p_0$$

in N trials of a binomial experiment. The test statistic in all of these examples was

$$Z_{X/N} = \frac{\dfrac{X}{N} - p_0}{\sqrt{\dfrac{p_0(1 - p_0)}{N}}}$$

which under the hypothesis that $p = p_0$ is distributed approximately $N(0, 1)$. For $X : B(N, p)$, we know that $Z = Z_{X/N}$, so we could just as well have used the statistic

$$Z = \frac{X - Np_0}{\sqrt{Np_0(1 - p_0)}}$$

and formulated our questions in terms of the *number* of successes instead of the *proportion* of successes. Our calculations of critical values, power, and so on would have been identical.

2. Upper-tailed tests, lower-tailed tests, and two-tailed tests

We have emphasized throughout the chapter that a rejection region includes values of the test statistic that are less likely if H_0 is correct than if H_1 is correct. The alternative hypothesis in our wolf example stated that $p > p_0$, so $E(Z_{X/N}|H_1$ correct$) > 0$, and all of the rejection regions discussed in the chapter were in the *upper* tail of the standard normal curve. For example, when we set $\alpha = .05$, we defined the critical region to be all $z \geq z_\alpha$, where $\mathbb{P}(z_\alpha) = 1 - \alpha$. In other experiments, of course, the appropriate alternative hypothesis is that $p < p_0$ (or that $p = p_1$, where $p_1 < p_0$). An alternative hypotheses of this sort calls for a *lower*-tailed test. If $p < p_0$, then $E(Z_{X/N}) < 0$, so values of $Z_{X/N}$ that favor the alternative hypothesis are *negative*, and the rejection region lies in the *lower* tail of the standard normal curve. For significance level α, the critical region for a lower-tailed test therefore includes all $z \leq z_\alpha$, where $\mathbb{P}(z_\alpha) = \alpha$.

Example 9.12. In Example 9.5.1 we tested the hypothesis $p = .5$ against the alternative hypothesis $p > .5$, where the random variable was the number of *successes* in 20 trials. If, instead, we define X as the number of *errors*, then the alternative hypothesis becomes $p < .5$. The critical region therefore falls in the lower tail of the standard normal curve, and if we want α to equal, say, .02, the critical value of our test statistic is the z-value for which $\mathbb{P} = .02$. In Table 3 of Appendix VIII, we find that $\mathbb{P}(-2.05) \doteq .02$, so the critical value z_α is -2.05.

Example 9.12 raises an important notational point. For one-tailed tests, we use z_α to denote the value that cuts off an area equal to α in *either* the upper tail *or* the lower tail of the standard normal curve, depending on the direction of the alternative hypothesis. That is, if the rejection region lies in the upper tail, $\mathbb{P}(z_\alpha) = 1 - \alpha$, but if the rejection region lies in the lower tail, then $\mathbb{P}(z_\alpha) = \alpha$.

When the alternative hypothesis is nondirectional, $H_1: p \neq p_0$, then $E(Z_{X/N}|H_1$ correct$)$ can be *either* larger *or* smaller than zero, and one must therefore partition the rejection region into *two* intervals, one in the lower tail and one in the upper tail of the standard normal curve. This is called a *two-tailed* test. It is customary to make these two regions equally probable, so for a specific level of significance α, we select two critical values, z_a and z_b, such that critical values, z_a and z_b, such that

$$P(z \leq z_a | H_0 \text{ correct}) = \frac{\alpha}{2}$$

and

$$P(z \geq z_b | H_0 \text{ correct}) = \frac{\alpha}{2}$$

From earlier discussion we know that the critical value in the upper tail must be the z-value for which $\mathbb{P}(z) = 1 - (\alpha/2)$, which we denote $z_{\alpha/2}$. Since the standard normal curve is symmetrical about zero, the critical value in the lower tail must therefore be $-z_{\alpha/2}$. With a two-tailed test of proportions, H_0 is rejected if $z_{X/N}$ is either *greater* than (or equal to) $z_{\alpha/2}$ or *less* than (or equal to) $-z_{\alpha/2}$.

Example 9.13. In our discussion of Example 9.3 we said that we could test the hypothesis that "left-turning" and "right-turning" wolves are not equally common by setting our test hypothesis $p = .5$ against the alternative hypothesis $p \neq .5$. If we choose $\alpha = .05$, then $z_{\alpha/2}$ is such that $P(z \geq z_{\alpha/2}) = .025$ under the standard normal curve. In Table 3 we find that $\mathbb{P}(1.96) \doteq .975$, so $P(z \geq 1.96) \doteq .025$, and $P(z \leq -1.96) \doteq .025$. Our two-tailed critical region is therefore all z-values greater than or equal to 1.96 and all z-values less than or equal to -1.96.

3. Significance level and adjustments for continuity

When N is small (even if large enough to make calculation of binomial probabilities a tedious chore) deciding on an appropriate critical region is further complicated by the problem first discussed at the end of Chapter 7: Since X/N is a discrete random variable, it is not always possible to observe a value x/N that corresponds exactly to the critical z-value that one wants.

One approach to this problem is to *correct for continuity* by adding .5 to or subtracting .5 from one's *observed* value of x before standardizing. One then tests the standardized corrected value against the critical z-value. An alternative approach is to adjust the *critical* value so that it corresponds to some *observable* x-value plus or minus .5:[7] Let x_α be the

---

[7] This might be called a *correction for discontinuity*, since it involves adjusting the continuous Z to conform to the discrete X, rather than adjusting the discrete X to conform to the continuous Z.

value that corresponds *exactly* to z_α. That is,

$$z_\alpha = \frac{\dfrac{x_\alpha}{N} - p_0}{\sqrt{\dfrac{p_0(1 - p_0)}{N}}}$$

Solve for x_α and round off to the nearest integer. Call this rounded-off value x_c (x-critical). For an upper-tailed test, the critical value is z corresponding to $(x_c - .5)$ and for a lower-tailed test the critical region is z corresponding to $(x_c + .5)$.

This tactic was first seen in Example 9.5.1. We set $\alpha = .05$, which gave us a critical z-value of 1.64. Solving for the corresponding x-value, we found that $x_\alpha = 13.6$ and rounded it off to the nearest (observable) integer, $x_c = 14$. Since we were using the normal approximation, $x_c = 14$ was represented by the continuous interval 13.5 to 14.5, and the critical region therefore became all z-values greater than or equal to

$$\frac{\dfrac{13.5}{20} - .5}{\sqrt{\dfrac{(.5)(.5)}{20}}} \doteq 1.57 \qquad\qquad [9.9]$$

For the lower-tailed test in Example 9.12 we set $\alpha = .02$, which gave us $z_\alpha = -2.05$. Under the test hypothesis, $p_0 = .5$, so $(1 - p_0)$ also equals .5. If $N = 20$ (as in Example 9.5.1), then

$$-2.05 = \frac{\dfrac{x_\alpha}{20} - .5}{\sqrt{\dfrac{(.5)(.5)}{20}}}$$

and $x_\alpha = 5.42$. Our critical region is therefore all z-values corresponding to proportions equal to or less than 5/20. Under the normal approximation 5 is represented by the interval 4.5 to 5.5, so the critical value of z is

$$\frac{\dfrac{5.5}{20} - .5}{\sqrt{\dfrac{(.5)(.5)}{20}}} \doteq -2.01$$

For a two-tailed test, the calculations are performed as if one were conducting two tests, one for the alternative $p > p_0$ and one for the alternative $p < p_0$, each for significance level $\alpha/2$. The reader can confirm that in Example 9.13 the values $\pm z_{\alpha/2} = \pm 1.96$ yield approximate x-values of 14.38 and 5.62, which—as before—we round to the nearest integers, 14 and 6. Thus, the test hypothesis will be rejected if *either*

$$\frac{x}{N} \geq \frac{13.5}{20} = .675$$

or

$$\frac{x}{N} \leq \frac{6.5}{20} = .325$$

These considerations reveal a fundamental risk involved in using the normal approximation for small values of N. It *tends to increase the probability of a Type I error*. Equation **[9.9]** reminded us that the critical region for $\alpha = .05$ in Example 9.5.1 was $z \geq 1.57$. However, under the standard normal curve, $P(z \leq 1.57) \doteq .94179$, so the probability of a Type I error is approximately $1 - .94179 \doteq .058$. If .058 is unacceptably high, we could elect the more conservative rejection rule, "reject H_0 if $x \geq 15$," in which case the test hypothesis is rejected if

$$z \geq \frac{\dfrac{14.5}{20} - .5}{\sqrt{\dfrac{(.5)(.5)}{20}}} \doteq 2.01$$

The probability of a Type I error is now approximately $1 - .97778 \doteq .022$, which is well below $\alpha = .05$. We know, however, that by decreasing α we increase β, so the choice really depends on the relative importance of avoiding Type I and Type II errors.

If N is very large, the distortion in α (and β) created by the normal approximation is usually trivial, and the critical value of z determined by α and by the direction of the alternative hypothesis may be used without correction.

─────── EXERCISES 9.3 ───────────────────────────────

1. Calculate the power of each of the statistical tests in Exercise 9.2.1 (p. 360).
2. Calculate the power for each of the statistical tests in Exercise 9.2.2 (p. 360).
3. An experiment can yield only two results, A and B. According to one theory the probability of Event A is .50, and we therefore wish to test the hypothesis H_0: $p = .50$. The competing theory suggests only that $p < .50$ and is therefore satisfied by all of the alternative hypotheses listed below. For each alternative hypothesis use Table 2 of Appendix VIII to calculate the power of the statistical test at the $\alpha = .02$ level for $N = 15$ observations.

(a) H_1: $p = .10$ (b) H_1: $p = .20$
(c) H_1: $p = .30$ (d) H_1: $p = .40$

4. An experiment can yield only two results, A and B. According to one theory the probability of Event A is .30, and we therefore wish to test the hypothesis H_0: $p = .30$. The competing theory suggests only that $p > .30$ and is therefore satisfied by all of the alternative hypotheses listed below. For each alternative hypothesis use Table 2 of Appendix VIII to calculate the power of the statistical test at the $\alpha = .10$ level for $N = 25$ observations.

(a) H_1: $p = .40$ (b) H_1: $p = .50$
(c) H_1: $p = .70$ (d) H_1: $p = .90$

5. An experiment can yield only two results, A and B. According to one theory the probability of Event A is .60, and we therefore wish to test the hypothesis H_0: $p = .60$. The competing theory suggests only that $p > .60$ and is therefore satisfied by the alternative hypotheses listed below. For each alternative calculate the power of the statistical test at the $\alpha = .05$ level for $N = 100$ observations.

(a) H_1: $p = .70$ (b) H_1: $p = .65$
(c) H_1: $p = .62$ (d) H_1: $p = .61$

6. An experiment can yield only two results, A and B. According to one theory the probability of Event A is .40, and we therefore wish to test the hypothesis H_0: $p = .40$. The competing theory suggests only that $p < .40$ and is therefore satisfied by the alternative hypotheses listed below. For each alternative calculate the power of the statistical test at the $\alpha = .01$ level for $N = 50$ observations.

(a) H_1: $p = .395$ (b) H_1: $p = .38$
(c) H_1: $p = .35$ (d) H_1: $p = .20$

7. Suppose the theorist in Exercise 9.1.5 (p. 342) can deduce only that if the theory is flawed, $P(A) \neq .80$. That is, if $p = P(A)$, then

H_1: $p = .80$

H_1: $p \neq .80$

Use the binomial distribution (Table 2 of Appendix VIII) to determine the rejection rule for each of the following:

(a) $\alpha \cong .01$, $N = 25$ (b) $\alpha \cong .035$, $N = 18$

Use the normal approximation to the binomial to determine the rejection rule for each of the following. Express critical values in terms of X.

(c) $\alpha \doteq .01$, $N = 25$ (d) $\alpha \doteq .035$, $N = 18$

8. Suppose the competing theory in Exercise 9.1.3 (p. 341) implied only that $p \neq .50$. That is,

H_0: $p = .50$

H_1: $p \neq .50$

Use the binomial distribution (Table 2 of Appendix VIII) to determine the rejection rule for each of the following:

(a) $\alpha \cong .025$, $N = 10$ (b) $\alpha \cong .0125$, $N = 20$

Use the normal approximation to the binomial to determine the rejection rule for each of the following. Express critical values in terms of X.

(c) $\alpha \doteq .025$, $N = 10$ (d) $\alpha \doteq .0125$, $N = 20$

9. A distributor of computer chips suspects that more than 10 percent of the chips received from her supplier are defective. She decides to draw a sample of 50 chips from the next shipment and test them.
   (a) What is her test hypothesis?
   (b) What is her alternative hypothesis?
   (c) If she uses the normal approximation to the distribution of X/N, what is the appropriate test statistic?
   (d) How should the test statistic be distributed if the test hypothesis is correct?
   (e) If the probability of a Type I error must not exceed .05, she should reject the test hypothesis if _____ of the 50 chips are defective.

10. An $AD \& D^{\circledR}$ player believes that his 6-sided die is not rolling 6s as often as it should. He decides to submit his suspicion to a statistical test. He rolls the die 60 times and records the number of times that 6 appears.
    (a) Define the random variable in this scenario and write the specification statement suggested by the player's "theory."
    (b) What is the test hypothesis?
    (c) What is the alternative hypothesis?
    (d) Define an appropriate test statistic, assuming that the player has available only the tables in Appendix VIII.
    (e) How should the test statistic be distributed if the test hypothesis is correct?
    (f) If the probability of a Type I error is not to exceed .025, he should reject the test hypothesis if he obtains _____ 6s in 60 rolls.

The manufacturer of Brand B disposable razor blades claims that more than two thirds of men who try both blades find Brand B as good or better than Gillette$^{\circledR}$. The claim is based on responses of men who use both razors under laboratory conditions and check ($\surd$) one of the following:

1. **Brand B gave me a better shave.** _____
2. **Gillette gave me a better shave.** _____
3. **I couldn't tell the difference.** _____

11. Suppose the manufacturer bases this claim on the number of respondents who choose #1 or #3.
    (a) Write the specification statement for the random variable.
    (b) What is the test hypothesis?
    (c) What is the alternative hypothesis?
    (d) If 50 responses are collected, what is the appropriate test statistic?
    (e) How should the test statistic be distributed if the test hypothesis is correct?
    (f) If α is not to exceed .01, the manufacturer's claim is supported if the value of the random variable is _____.

12. Suppose the manufacturer's claim is based on the number of men who check #2.
    (a) Write the specification statement for the random variable.
    (b) What is the test hypothesis?
    (c) What is the alternative hypothesis?
    (d) If 50 responses are collected, what is the appropriate test statistic?
    (e) How should the test statistic be distributed if the test hypothesis is correct?
    (f) If α is not to exceed .01, the manufacturer's claim is supported if the value of the random variable is _____.

An independent consumer protection testing laboratory subpoenas the data on which the Brand B manufacturer bases his claim. It is found that 6 preferred Brand B, 9 preferred Gillette®, and 35 couldn't tell the difference. The laboratory investigators repeat the experiment, but their questionnaire includes only 2 choices: Prefer B or Prefer Gillette®.

13. The laboratory tests the hypothesis "more men prefer Gillette" against the alternative "men can't tell the difference."
    (a) Based on the data provided by the manufacturer of Brand B, what is the test hypothesis?
    (b) What is the alternative hypothesis?
    (c) If $\alpha = .05$, what is the power of the test?
14. Suppose the laboratory tests the hypothesis "men can't tell the difference" against the alternative "men prefer Gillette."
    (a) What is the test hypothesis?
    (b) Based on the data provided by the manufacturer, what is the alternative hypothesis?
    (c) If $\alpha = .025$, what is the power of the test?

E. SUMMARY

The four major steps in testing a statistical hypothesis are *formulation*, *decisions*, *data collection*, and *conclusions*.

An hypothesis test begins with the formulation of a *theory*—some notion about how the world—or some little part of it—is put together.

The only thing that a statistician demands of a theory is that it have implications for the distribution of some measurement. The mathematical idealization of this distribution is called the *model*.

Hypotheses specify the property of the model that will be tested against a sample of real measurements. The most common hypotheses concern some parameter θ of the distribution specified in the model. The *test* hypothesis, denoted H_0, ordinarily takes the form

$$H_0: \theta = \theta_0$$

Since θ_0 is a single value, H_0 is said to be a *simple* hypothesis.

The *alternative hypothesis* may also be simple, in which case it takes the form

$$H_1: \theta = \theta_1$$

Sometimes the alternative hypothesis specifies a range of values for θ, rather than a single value. Such a *composite* alternative hypothesis may be *directional*, e.g.,

$$H_1: \theta > \theta_0$$

or

$$H_1: \theta < \theta_0$$

or it may be *nondirectional*, e.g.,

$$H_1: \theta \neq \theta_0$$

The first decision a scientist makes is the choice of a *test statistic*, W. The fundamental requirement for a test statistic is that its sampling distribution, if H_0 is correct, be different from its sampling distribution, if H_1 is correct. In addition, the experimenter must be able to specify ϕ_0, the distribution of W when H_0 is correct, and the expected value of W when H_0 is correct. It must also be true that

$$E(W|H_0 \text{ correct}) \neq E(W|H_1 \text{ correct})$$

Next, the experimenter must decide on a *rejection rule*, which specifies the values of W that will lead to rejection of H_0 and the values of W that will lead to rejection of H_1. The *critical region* is the set of w-values that will cause the experimenter to reject H_0. The critical region depends on two considerations:

(1) The alternative hypothesis
  - If H_1 specifies that $\theta > \theta_0$, the rejection region ordinarily lies in the *upper* tail of ϕ_0.
  - If H_1 specifies that $\theta < \theta_0$, the rejection region ordinarily lies in the *lower* tail of ϕ_0.
  - If H_1 is nondirectional, the rejection region lies in *both* tails of ϕ_0.

(2) The significance level specified by the experimenter

Significance level is denoted α and is equal to the probability that w falls in the critical region when H_0 is correct. The significance level of the test is therefore the probability of *incorrectly* rejecting the test hypothesis, which is called a *Type I error*.

$$\text{Significance} = P(\text{rejecting } H_0 | H_0 \text{ correct}) = P(\text{Type I error}) = \alpha$$

A *Type II error* occurs if the alternative hypothesis is rejected incorrectly. The probability of a Type II error is denoted β.

$$P(\text{rejecting } H_1 | H_1 \text{ correct}) = P(\text{Type II error}) = \beta$$

The probability of a Type I error is predicated on the condition that H_0 is correct, and the probability of a Type II error is predicated on the condition that H_1 is correct. Consequently, α is always calculated under the distribution of W implied by H_0, and β is calculated under the distribution of W implied by H_1. If H_1 is composite or if ϕ_1 is unspecified, β cannot be calculated.

All other things being equal, decreasing α increases β, and vice versa. Therefore, choosing a significance level, which defines the rejection region, depends on the relative importance of avoiding a Type I and Type II error.

The appropriateness of many probability distributions as models for the distributions of test statistics imposes certain requirements on data collection. These requirements generally include random sampling, independence of observations, and some minimal sample size. In addition, sample size influences the probability of a Type II error. As sample size increases, the variability of the test statistic decreases. All other things being equal (e.g., α), this has the effect of reducing β.

Statistical conclusions are always stated in terms of *rejecting* an hypothesis, never *accepting* an hypothesis. This is because confirmatory evidence is always equivocal, but disconfirmatory evidence is unequivocal.

The conclusion one draws from a statistical test is ordinarily accompanied by a statement indicating the *significance probability*, or p-value, of the result. The p-value of a test is the probability of obtaining a value of the test statistic that is *at least* as favorable to the alternative hypothesis as the value actually obtained.

The *power* of a test is the probability of rejecting H_0 when H_1 is correct. This means that power is defined under the assumption that H_1 is correct, and since a rejection rule permits only two possible conclusions (reject H_0 or reject H_1), power must equal $1 - \beta$. Consequently, anything that reduces β will increase power. Therefore, power can be increased by (1) increasing α, (2) increasing sample size (or otherwise reducing the variance of the test statistic), or (3) formulating a new alternative hypothesis that increases $|\theta_1 - \theta_0|$.

The greater the power of a statistical test, the *smaller* the difference it can detect between θ_0 and the true parameter value θ. Therefore, more powerful tests are generally preferred to less powerful tests, but this is not always a sound principle. A test that is too powerful can lead to rejection of the test hypothesis even though the difference between θ and θ_0 is so small that it really isn't important.

Summary table of statistical tests

Testing hypotheses about p

Distribution of X	Sample Size	Test Statistic	Distribution of Test Statistic*
Using the Binomial Distribution			
$B(N, p)$	Any	X	$B(N, p_0)$
Using the Normal Approximation to the Binomial			
$B(N, p)$	$Np_0 \geq 5$ $N(1 - p_0) \geq 5$	$\dfrac{X - Np_0}{\sqrt{Np_0(1 - p_0)}}$	$N(0, 1)$
$B(N, p)$	$Np_0 \geq 5$ $N(1 - p_0) \geq 5$	$\dfrac{\dfrac{X}{N} - p_0}{\sqrt{\dfrac{p_0(1 - p_0)}{N}}}$	$N(0, 1)$

*If H_0 is correct.

CHAPTER 10

Testing hypotheses about population means

CHAPTER OUTLINE

In the preceding chapter we tracked a single experiment designed to test an hypothesis about the proportion of wolves capable of learning a simple verbal cue. This example allowed us to introduce various general aspects of hypothesis testing—formulation, decisions, data collection, and conclusions—which we will find a convenient framework for developing hypothesis tests in this and in subsequent chapters.

Implicitly the wolf problem also served to introduce some of the specific considerations involved in testing hypotheses about population means, because as we saw in Chapter 8, a proportion can be thought of as a special case of the mean. The majority of statistical inferences found in the literature of the behavioral, biological, and social sciences involve one or another test of population means. The mathematics of such tests is exceptionally well developed, and the repertoire of available tests is more extensive than tests about other population values.

Broadly speaking, there are three circumstances in which an investigator tests hypotheses about means. The simplest case is when the experimenter draws a sample of observations from a single population and tests the hypothesis

$$H_0: \mu = \mu_0$$

where μ is the unknown mean of the population on which the observations are made and μ_0 is a specific value. The value μ_0 ordinarily derives either from some (more or less articulated) theory or from some body of prior observations. The principles and techniques involved in testing one mean can be generalized to problems involving two population means, where the test hypothesis takes the form

$$H_0: \mu_1 = \mu_2$$

In tests of this sort the experimenter is concerned only about the *difference* between the two population means and needn't specify values for either μ_1 or μ_2.[1] Finally, the experimenter may be concerned about the means of many populations:

$$H_0: \mu_1 = \mu_2 = \cdots = \mu_J$$

The most common techniques for conducting tests about one or two population means are presented in this chapter, and much of the material will be familiar from our discussion of population proportions. Testing hypotheses about more than two means is the province of analysis of variance, which requires material developed in Chapter 11.

A. A "TEXTBOOK" EXAMPLE: THE NEANDERTHAL PROBLEM

Contemporary anthropologists have a number of hypotheses concerning the relationship of Neanderthal man (*Homo sapiens neanderthalensis*) to modern humans (*Homo sapiens sapiens*).[2] The prevailing consensus is that Neanderthal man and modern man are "cousins" that diverged from a common ancestral line as much as 250,000 years ago.[3] A persistent minority, however, maintain the older position that Neanderthal man was the immediate ancestor of modern humans. This is called the

---

[1] The test hypothesis often takes the form

$$H_0: \mu_1 - \mu_2 = 0$$

That is, the experimenter hypothesizes that there is *no* difference between the two population means. This is the most common situation in which the test hypothesis is called the *null* hypothesis.

[2] For a highly readable discussion of the status of Neanderthal man see Lasker, G. W. and Tyzzer, R. N. (1982). *Physical Anthropology* (3rd ed.). New York: Holt, Rinehart and Winston.

[3] The current version of this hypothesis is based on DNA evidence that modern humans evolved in Africa, whereas "classic" Neanderthal remains have been found only in Europe and the Middle East.

Neanderthal-phase hypothesis. Its simplicity is attractive, but it is inconsistent with the fossil record in one very important particular: Neanderthal man disappeared about 40,000 years ago, and modern man did not appear until about 35,000 years ago.

Between 1930 and 1970, however, a number of interesting fossil skeletons were found in the Middle East. In some respects the skeletons appear to be anatomically modern, but in other respects they are more typical of Neanderthal anatomy. Conventional dating methods indicate these specimens to be about 40,000 years old, and a more recently developed (though controversial) technique called thermoluminescence dating suggests that they may be even older.

Let us suppose that a certain anthropologist favors the Neanderthal-phase hypothesis. If the Middle Eastern skeletons are those of modern humans, the 5,000-year gap between Neanderthal man and modern man is eliminated, and his hypothesis becomes more plausible. If the skeletons are Neanderthal, the gap remains and the orthodox hypothesis is retained. A Neanderthal-phase theorist would therefore like to conclude that the skulls are modern.

1. Formulation

a. Theory. The scenario presented above offers two competing theories concerning the relationship of Neanderthal man to modern man: The currently popular hypothesis is that modern and Neanderthal man evolved separately from a common ancestor, and the Neanderthal-phase hypothesis is that modern humans are direct descendents of Neanderthal man.

b. Model. The significance of the antiquity of the Middle Eastern skeletons depends on whether they are modern or Neanderthal. Physical anthropologists use a number of skeletal landmarks and measurements to define hominid populations; one of the more traditional measures is *cranial capacity*, the volume of the skull measured in cubic centimeters.[4]

Because cranial capacity is correlated with overall body size, the anthropologist decides to base his comparison on male adults. The mean cranial capacity for modern, adult males is approximately 1,345 cc, and the variance is 28,900. The mean capacity of adult, male Neanderthal skulls is 1,575 cc. (Yes, that's right. The average Neanderthal had a larger brain than the average modern human. This should come as no surprise to those of you who follow contemporary politics.) The variance

---

[4]Skull measurements are important for several reasons. First, skulls are more durable than other parts of the skeleton and are most easily identified as human (or hominid) even by untrained persons. Consequently, there are more fossil skulls available for comparison than, say, femurs or scapulas. The emphasis on cranial capacity is a vestige of the biological determinism and social prejudice that pervaded much of nineteenth- and early twentieth-century anthropology. For a more complete discussion see Gould, S. J. (1981). *The Mismeasure of Man*. New York: Norton.

is 1,600.[5] Furthermore, cranial capacity, like many anatomic measurements, is polygenically determined. From our discussion of the Central Limit Theorem in Box 8.1 we know that the normal curve should therefore closely approximate the distribution of cranial capacities in either population.[6]

If we let the random variable X represent cranial capacity of adult males in the archaic Middle Eastern population, the preceding paragraph suggests two possible models for the distribution of X: If the Middle Eastern skulls are modern,

$$X: N(1,345, \quad 28,900)$$

If the Middle Eastern skulls are Neanderthal,

$$X: N(1,575, \quad 1,600)$$

c. Hypotheses. Both models provide specific values for the population mean of Middle Eastern fossil skull capacities, and we therefore have two simple hypotheses, $\mu = 1,575$ and $\mu = 1,345$. Either one could be cast in the role of the test hypothesis, but we know from Chapter 9 that the object of a statistical test is to *reject* the test hypothesis (in favor of the alternative). A Neanderthal-phase theorist would therefore be disposed to test the hypothesis that the Middle Eastern remains are Neanderthal.

$$H_0: \mu = 1,575$$

$$H_1: \mu = 1,345$$

2. Decisions

a. Test statistic. The simplest test statistic would be the cranial capacity, or volume, of one of the Middle Eastern fossil skulls. Recall from Chapter 8, however, that the variance of $\overline{X}$ decreases as sample size increases. The volume of a single skull may be thought of as the mean of a sample of $N = 1$ observations. The distribution of means for samples of $N > 1$ observations will therefore exhibit less variability than the distribution of individual observations. Accordingly, the mean capacity of, say, $N = 4$ skulls is a more *powerful* test statistic than the capacity of a single skull.

Furthermore, the sampling distribution of the mean $\overline{X}$ can be derived directly from the models given above. In general we know that if $X: N(\mu, \sigma^2)$, then $\overline{X}: N(\mu, \sigma^2/N)$ for samples of size N. If the test hypothesis is correct, the skulls are Neanderthal and $X: N(1,575, 1,600)$,

[5] Data on modern man are from Tobias, P. (1971). *The Brain in Hominid Evolution*. New York: Columbia University Press, p. 54. Data on Neanderthal man are from Holloway, R. L. (1985). The poor brain of *Homo sapiens neanderthalensis*: See what you please . . . . In E. Delson (Ed.). *Ancestors: The Hard Evidence*. New York: Liss, p. 320.
[6] The distribution of modern human skull capacities is actually skewed negatively, but for our purposes the departure from the normal can be ignored.

so for $N = 4$ skulls,

$$\bar{X}: N(1{,}575, \quad 400)$$

If the alternative hypothesis is correct, the skulls are modern and $X: N(1{,}345, 28{,}900)$, so for samples of 4 skulls,

$$\bar{X}: N(1{,}345, \quad 7{,}225)$$

Under either hypothesis the statistic $\bar{X}$ is normally distributed. In addition, $E(\bar{X}|H_0 \text{ correct}) = 1{,}575$, and $E(\bar{X}|H_1 \text{ correct}) = 1{,}345$. The sample mean therefore satisfies two important requirements for a test statistic: The expected value of $\bar{X}$ is different under the two hypotheses, and the distribution of $\bar{X}$ under H_0 is specified.

There is no table of cumulative probabilities for $X: N(1{,}575, \quad 400)$, but we know that any normally distributed random variable can be standardized and submitted to the distribution for $Z: N(0, 1)$. From Chapter 8,

$$Z_{\bar{X}} = \frac{\bar{X} - \mu_{\bar{X}}}{\sigma_{\bar{X}}} = \frac{\bar{X} - \mu_{\bar{X}}}{\sqrt{\sigma^2/N}}$$

so under the test hypothesis that $\mu_{\bar{X}} = \mu_0$, the appropriate test statistic is

$$Z_{\bar{X}} = \frac{\bar{X} - \mu_0}{\sigma_{\bar{X}}} \qquad\qquad [10.1]$$

or,

$$Z_{\bar{X}} = \frac{\bar{X} - \mu_0}{\sqrt{\sigma^2/N}} \qquad\qquad [10.2]$$

where σ^2 is the variance of X.

If X is normally distributed with known variance σ^2, then the appropriate statistic to test the hypothesis that $\mu = \mu_0$ is

$$Z_{\bar{X}} = \frac{\bar{X} - \mu_0}{\sigma_{\bar{X}}}$$

which is distributed $N(0, 1)$

If, as the test hypothesis specifies, the skulls are Neanderthal, then $\mu_0 = 1{,}575$ and $\sigma_{\bar{X}} = \sqrt{1{,}600/4} = 20$, so

$$Z_{\bar{X}} = \frac{\bar{X} - 1{,}575}{20} \qquad\qquad [10.3]$$

and will be normally distributed with an expected value of 0 and a variance of 1. If H_1 is correct and $E(\overline{X})$ is *in fact* 1,345, the expected value of $Z_{\overline{X}}$ as given in [10.3] will be smaller than 0:

$$E(Z_{\overline{X}}) = E\left(\frac{\overline{X} - 1{,}575}{20}\right)$$

$$= \left(\frac{1}{20}\right)E(\overline{X} - 1{,}575)$$

$$= \left(\frac{1}{20}\right)\left[E(\overline{X}) - 1{,}575\right] = \left(\frac{1}{20}\right)(1{,}345 - 1{,}575)$$

$$= \left(\frac{1}{20}\right)(-230) = -11.5$$

And, for the general alternative hypothesis, H_1: $\mu = \mu_1$,

$$E(Z_{\overline{X}}|H_1 \text{ correct}) = \left(\sqrt{\frac{1}{\sigma_{\overline{X}}^2}}\right)(\mu_1 - \mu_0) \qquad [10.4]$$

From [10.4] it can be seen that $E(Z_{\overline{X}}|H_1$ correct) will be *negative* if $\mu_1 < \mu_0$ and will be *positive* if $\mu_1 > \mu_0$.

b. Rejection rule. Because $E(Z_{\overline{X}}|H_1$ correct) is negative, the rejection region should be in the *lower* tail of the standard normal distribution. That is, the test hypothesis will be rejected if the observed value $z_{\overline{X}}$ is less than some critical value z_α, which, as we know from Chapter 9, depends on the significance level selected by the experimenter. Suppose the experimenter chooses to let $\alpha = .05$. Then, z_α must be such that

$$P(z_{\overline{X}} \le z_\alpha|H_0 \text{ correct}) = .05$$

Since $P(z_{\overline{X}} \le z_\alpha) = \mathbb{P}(z_\alpha)$, the appropriate entry in Table 3 of Appendix VIII is $z = -1.64$. This gives us the following rejection rule: If $z_{\overline{X}} \le -1.64$ reject H_0, but if $z_{\overline{X}} > -1.64$ reject H_1.

3. Data collection

In the finest tradition of L. S. B. Leaky, Donald Johanson, and Indiana Jones, our intrepid researcher braves the rigors and perils of dust-infested museums, computes the mean cranial capacity of the skulls in question, and standardizes the obtained value, $\bar{x}$. The mean cranial capacity of the four Middle Eastern skulls turns out to be 1,561.75 cc, so

$$z_{\overline{X}} = \frac{\bar{x} - 1{,}575}{20} \qquad [10.5]$$

$$= \frac{1{,}561.75 - 1{,}575}{20} \doteq -0.66$$

4. Conclusions

The value of $z_{\bar{x}}$ is greater than -1.64, so the researcher cannot reject the hypothesis that the skulls are Neanderthal. He must reject the hypothesis that the skulls are modern, and the 40,000-year gap between the disappearance of Neanderthal man and the first appearance of modern man remains a flaw in the Neanderthal-phase hypothesis.

5. The power of the statistical test

We know from Chapter 9 that the power of a statistical test is $1 - \beta$, that is, $1 - P(\text{rejecting } H_1 | H_1 \text{ correct})$. We also learned in Chapter 9 that calculation of power (or β) requires that we first determine the critical value of the test statistic in *natural* units. The value of $\bar{x}$ corresponding to $z_\alpha = -1.64$ is easily obtained from equation [**10.5**]:

$$\frac{\bar{x} - 1,575}{20} = -1.64$$

Therefore,

$$\bar{x} = 1,575 - 1.64(20)$$

$$\doteq 1,542$$

According to the rejection rule, the alternative hypothesis will be rejected if the sample mean is greater than 1,542. The probability of a Type II error is therefore the probability that the observed value $\bar{x}$ is greater than 1,542 *given that* $\bar{X}$ is distributed $N(1,345, 7,225)$. To determine

$$P\left[\bar{x} > 1,542 | \bar{X} : N(1,345, 7,225)\right]$$

the critical value of $\bar{x}$ is standardized with parameter values $\mu_{\bar{X}} = 1,345$ and $\sigma_{\bar{X}}^2 = 7,225$ and submitted to the table of standard normal cumulative probabilities. As in Chapter 9, the critical value standardized under H_1 is denoted z_β.

$$z_\beta = \frac{1,542 - 1,345}{\sqrt{7,225}} = \frac{197}{85} \doteq 2.32$$

and in Table 3 of Appendix VIII we find that $P(z > 2.32) \doteq .01017$. The probability of a Type II error is therefore approximately .01, and the power of the test approximately .99. The density functions $\phi(Z_{\bar{x}})$ under the test hypothesis and the alternative hypothesis are illustrated in Figure 10.1.

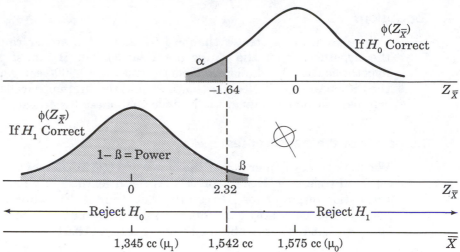

Figure 10.1. Distributions of mean cranial capacities of Middle Eastern skulls standardized under test hypothesis and under alternative hypothesis.

B. TESTING HYPOTHESES ABOUT ONE POPULATION MEAN μ

The example presented above was realistic (if a bit simplified), but it was unusual in several ways. First, the anthropologist had legitimate reason to assume that the distribution of X was closely approximated by the normal distribution. Consequently, $\overline{X}$ could likewise be assumed to be normally distributed. Second, the variance of X was known, so the variance of $\overline{X}$ could be computed. Third, an alternative model was available that gave a specific value for μ_1. This provided a simple alternative hypothesis and made it possible to calculate β.

Such ideal circumstances are very rare. Indeed, it is more commonly the case that the distribution of X is unspecified, the variance of X is unknown, and the alternative hypothesis is composite. In testing hypotheses about a population mean, the appropriate statistical test depends on what the experimenter knows or can legitimately assume about the distribution of $\overline{X}$. This in turn depends on (1) the distribution ϕ and variance σ^2 of the random variable X and (2) the size of the sample. The strategy for selecting a test statistic can therefore be laid out as a checklist or program of questions that the experimenter asks about the data and the distribution of the random variable. These questions are presented as a decision tree in Box. 10.1.[7]

Before we look at the various branches of the decision tree, there are three rather general points that Box 10.1 suggests about selection of a test statistic. First, the fundamental questions are whether the random variable X is normally distributed and whether the variance of X is known. Second, if the normality or variance of X is uncertain, one must rely on less powerful tests. Third, these uncertainties can be mitigated if

---

[7] Some of the decisions in Box 10.1 include unfamiliar terms (e.g., Use t). These will be explained as we proceed through the chapter.

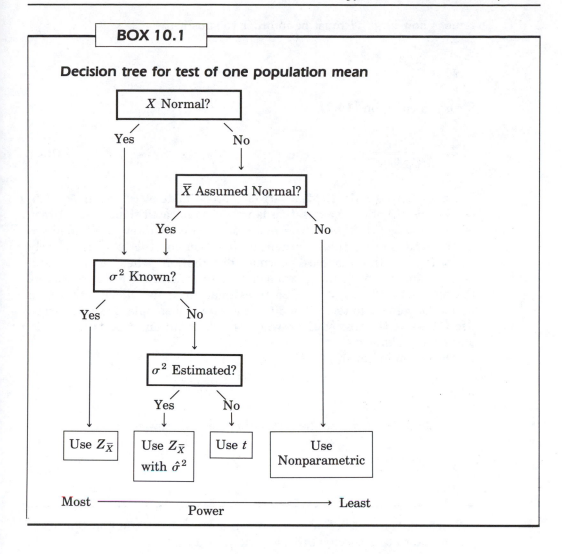

BOX 10.1

Decision tree for test of one population mean

X Normal?

Yes   No

$\overline{X}$ Assumed Normal?

Yes   No

σ^2 Known?

Yes   No

σ^2 Estimated?

Yes   No

Use $Z_{\overline{X}}$ Use $Z_{\overline{X}}$ with $\hat{\sigma}^2$ Use t Use Nonparametric

Most ⟶ Least
Power

the experimenter is willing to make certain assumptions. Since these assumptions are most often justified on the basis of sample size, the choice of a test statistic often depends ultimately on the size of one's sample.

1. Large samples ($N \geq 30$)

The mitigating effects of sample size on selection of a test statistic are most apparent in the situation when one has 30 or more observations. In most introductory statistics books, $N = 30$ is treated as the break point between "large" and "small" samples. This convention is based on two quite distinct considerations: the assumed normality of $\overline{X}$ and estimation of the variance σ^2. In Chapter 8 we said that if $N \geq 30$, one can usually invoke the Central Limit Theorem and assume that $\overline{X}$ is normally distributed. Consequently, one can ordinarily proceed *as if* X is normally distributed, even if the distribution of X is in fact unknown. The second issue is how large N must be in order to substitute

$$\hat{\sigma}^2 = \frac{\Sigma(x - \bar{x})^2}{N - 1}$$

for σ^2 in equation [10.2],

$$Z_{\bar{X}} = \frac{\bar{X} - \mu_0}{\sqrt{\hat{\sigma}^2/N}} \qquad\qquad \textbf{[10.6]}$$

For most purposes $\hat{\sigma}^2$ yields a satisfactory estimate of σ^2 when $N \geq 25$, but in this situation the question is more complicated than the accuracy of $\hat{\sigma}^2$, because [10.6] is *not* normally distributed. However, we shall see in the next section that the density function for this statistic closely *approximates* the standard normal distribution when $\mu = \mu_0$ and $N \geq 30$. Instead of defining two sample-size criteria ($N \geq 30$ to assume the normality of $\bar{X}$ and $N \geq 25$ to estimate σ^2), statisticians therefore find it convenient to take $N = 30$ as the minimal sample size required to use [10.6] as the test statistic when the distribution of X is uncertain and σ^2 is unknown.

The denominator in [10.6]

$$\sqrt{\hat{\sigma}^2/N}$$

is usually denoted $\hat{\sigma}_{\bar{X}}$, so the test statistic can be rewritten as

$$Z_{\bar{X}} = \frac{\bar{X} - \mu_0}{\hat{\sigma}_{\bar{X}}} \qquad\qquad \textbf{[10.7]}$$

When the variance of X is unknown and $N \geq 30$, the appropriate statistic to test the hypothesis that $\mu = \mu_0$ is

$$Z_{\bar{X}} = \frac{\bar{X} - \mu_0}{\hat{\sigma}_{\bar{X}}}$$

which is approximately distributed $N(0, 1)$

To see how this works in practice, we shall develop another research scenario.

a. Formulation

Theory: Example 10.1.1. Several years ago one of the authors proposed a model of cognitive evolution in canids, which predicted that wolves should be better able than domestic dogs to solve complex problems. The details of the theory are unimportant, except as they suggest a possible category of random variable (problem-solving tests) and a directional hypothesis (that wolves should perform better than dogs). So much for the theory.

Model: Example 10.1.2. To obtain experimental support for the theory we (the author and his co-workers) tested a group of wolf pups on a battery of problem-solving tasks originally developed for domestic dogs by J. P. Scott and J. L. Fuller.[8] On one such task a pup and a dish of food are placed on opposite sides of a wide, plywood barrier. The pup can see and smell the food through a wire-mesh window in the plywood, but it can't reach the food unless it leaves the window and runs around the end of the barrier. The score on this test is the number of errors–halts and reversals in direction–a pup makes in finding its way to the food. If the author's notions about evolution were correct, we should expect wolf pups to make fewer errors than domestic dog pups.

Performance data collected by Scott and Fuller for 204 dogs representing five breeds yielded a mean of 14.6 errors and a variance of 160.43. We know from Chapter 8 that the sample mean is a good estimator of μ and that $s^2[N/(N-1)]$ is an unbiased estimate of the population variance. From Scott and Fuller's data, then, the mean of the dog population μ_D is estimated to be 14.6, and the variance of the dog population σ_D^2 is estimated to be $(160.43)(204/203) \doteq 161.22$, with the distribution of X unspecified.

Hypotheses: Example 10.1.3. The theory implies that wolves should make fewer errors than dogs, so if μ_W is the mean number of errors for the wolf population, the partial model of dog performance provided by Scott and Fuller's data suggests the following hypotheses:

$$H_0: \mu_W = 14.6$$

$$H_1: \mu_W < 14.6$$

It will be noted that the *alternative* hypothesis is the result that would support our theory and that H_0 is the hypothesis we hoped to reject.

Example 10.2.1. Although the theory in Example 10.1.1 suggested a directional hypothesis, we know that this is not always the case. A different set of theoretical assumptions about the effects of domestication might have implied only that wolves and dogs should *differ* in problem-solving performance. In this case the alternative hypothesis is

$$H_1: \mu_W \neq 14.6$$

[8] Scott, J. P., and Fuller, J. L. (1965). *Genetics and the Social Behavior of the Dog*. Chicago: University of Chicago Press.

b. Decisions

Test statistic: Example 10.1.4. There is no reason to expect that the normal density function is an appropriate model for the distribution of X, but we are illustrating the large ($N \geq 30$) sample situation, so the Central Limit Theorem allows us to assume that $\overline{X}$ is (approximately) normally distributed no matter what the distribution of X might be. Nor do we know the variance of wolf errors σ_W^2, and the theory gives us no particular reason to assume that it is equal to the variance of dog errors, σ_D^2. However, $N \geq 30$ is large enough that we can estimate the value of σ_W^2 by testing our wolves and calculating $\hat{\sigma}_W^2$. Let us suppose that for a sample of 30 wolf pups $s^2 = 123.25$. Then,

$$\hat{\sigma}^2 = 123.25\left(\frac{30}{29}\right) = 127.5$$

Clearly, the availability of a large sample of observations confers a good bit of leverage. We began with only two pieces of information, the mean and variance for a large sample of dogs. But, thanks to the Central Limit Theorem and the unbiasedness of $\hat{\sigma}^2$, we can now make the following assertions: If H_0 is correct, then for a sample of 30 wolf pups $\overline{X}$ is (approximately) normally distributed with an expected value of 14.6 and a variance of $127.5/30 = 4.25$. If H_1 is correct, we can assume that $\overline{X}$ is normally distributed with a variance of 4.25 and an expected value of μ_W. We don't have a specific value for μ_W, but our theory tells us that it should be smaller than 14.6.

We have now bootstrapped ourselves into the same position we were in in our anthropology example. The appropriate test statistic is

$$Z_{\overline{X}} = \frac{\overline{X} - 14.6}{\sqrt{4.25}} \qquad\qquad [10.8]$$

which will be approximately distributed $N(0, 1)$ if the test hypothesis is correct.

Rejection rule: Example 10.1.5. Because the alternative hypothesis specifies that $\mu_W < 14.6$, the expected value of $Z_{\overline{X}}$ is negative if H_1 is correct, so the rejection region will lie in the lower tail of the standard normal distribution. A more exact statement of the rule requires that α be specified. If α is set equal to .025, the test hypothesis will be rejected in favor of H_1 if $z_{\overline{X}}$ is less than or equal to z_α where

$$P(z \leq z_\alpha | H_0 \text{ correct}) = .025$$

From Table 3 of Appendix VIII, the value for which $\mathbb{P} \doteq .025$ is $z = -1.96$.

Example 10.2.2. If we had begun with the nondirectional alternative hypothesis given in Example 10.2.1, we would use the same test statistic

$$Z_{\bar{X}} = \frac{\bar{X} - 14.6}{\sqrt{4.25}}$$

but we would need a rule that provides for rejection of the test hypothesis if $\bar{x}$ is *either* larger *or* smaller than 14.6. The convention is the same as we discussed in connection with two-tailed tests of hypotheses about proportions in Chapter 9 (pp. 369–370): Half of the critical region is in the lower tail of the standard normal curve, and half of the critical region is in the upper tail. For a specific level of significance α, we therefore have two critical values, $z_{\alpha/2}$ and $-z_{\alpha/2}$, where

$$P\left(z_{\bar{X}} \geq z_{\alpha/2} | H_0 \text{ correct}\right) = \frac{\alpha}{2}$$

and

$$P\left(z_{\bar{X}} \leq -z_{\alpha/2} | H_0 \text{ correct}\right) = \frac{\alpha}{2}$$

Thus, if we set $\alpha = .025$, then $z_{\alpha/2} = 2.24$ and $-z_{\alpha/2} = -2.24$. The rejection rule would therefore be, if $z_{\bar{X}}$ is *either* greater than (or equal to) 2.24 *or* less than (or equal to) -2.24, reject H_0; otherwise, reject H_1. It is important to note that the two-tailed test requires a *bigger* difference between $\bar{x}$ and μ_0 in order to reject H_0 than does the one-tailed test in the preceding example. That is, the two-tailed is *less powerful* than the one-tailed test. And, in general,

> For significance level α and sample size N, an upper-tailed test is more powerful than a two-tailed test if $\mu_1 > \mu_0$, and a lower-tailed test is more powerful than a two-tailed test if $\mu_1 < \mu_0$ (see Figure 10.2, p. 396).

c. Data collection and conclusions

Example 10.1.6. Returning to the one-tailed test in Example 10.1.5, let us suppose that the wolf pups averaged 9.5 errors, and that the value of our test statistic is therefore

$$\frac{9.5 - 14.6}{\sqrt{4.25}} \doteq -2.47$$

Since -2.47 is smaller than -1.96, we reject the test hypothesis in favor of the alternative, which supports the prediction that wolves are more adept at solving problems than are domestic dogs.

d. Postscript: significance, power, and sample size

Significance level and significance probability. In most basic research, the costs of a Type I error are so intangible that an experimenter has little basis for choosing a particular level of significance. For this reason it is no longer a common practice (if it ever was!) to establish a significance level or an exact rejection region before the test is conducted. As we mentioned in Chapter 9, most experimenters report instead the significance *probability* of their results, which is calculated after the test is conducted. The significance probability may be thought of as the smallest value of α for which the test hypothesis could be rejected with the obtained value of the test statistic.

Example 10.1.7. In Example 10.1.6, $z_{\bar{X}} \doteq -2.47$, and in Table 3 of Appendix VIII we find that $\mathbb{P}(-2.47) \doteq .006756$. If the experimenter had chosen *any* level of α equal to or greater than .006756 before the test was actually performed, this particular result would have allowed him to reject H_0.

Example 10.2.3. For a two-tailed test, the interpretation of significance probability is the same, but the calculation is slightly different. Remember that significance probability is the probability of obtaining a result that is at least as favorable to H_1 as the value actually obtained. With a nondirectional alternative hypothesis, a departure in *either* direction from the value expected under H_0 is favorable to H_1. The expected value of $Z_{\bar{X}}$ is zero under the test hypothesis, so the significance probability for a two-tailed test is the probability of obtaining a result *at least* as far from zero *in either direction* as the observed value, $z_{\bar{X}}$. Had $z_{\bar{X}} \doteq -2.47$ been reported for a two-tailed test, the significance probability of the result would be

$$P(z \leq -2.47 \text{ or } z \geq 2.47) = P(z \leq -2.47) + P(z \geq 2.47)$$

Since the normal curve is symmetrical about zero,

$$P(z \leq -2.47) + P(z \geq 2.47) = 2P(z \leq -2.47)$$

$$\doteq 2(.006756)$$

$$= .013152$$

Therefore, if $z_{\bar{X}} = -2.47$, the smallest α that would permit rejection of H_0 for a two-tailed test is .013152. This result would ordinarily be denoted ($p < .02$, two-tailed) or ($p < .015$, two-tailed).

Significance and power. Although the practice of reporting (acceptably low) *p*-values instead of tying oneself to an a priori α-value makes practical sense, it eliminates all advance consideration of β. In order to prevent problems that can arise from Type II errors, it is therefore a wise precaution to go through the formal routine of establishing a rejection rule before collecting data, even if the experimenter intends to ignore it and observe the current practice of reporting the p-value.

Determination of β in the anthropology example was a matter of routine calculation, because the experimenter had a specific value for μ_1. In Example 10.1 the alternative hypothesis is composite and, as indicated at the end of Chapter 9, determination of β therefore involves issues of *scientific* significance as well as *statistical* significance.

One approach is to decide in advance how small a difference between μ, the *true* value of the population mean, and μ_0 is scientifically important. In Chapter 9 we let δ denote the *smallest scientifically important, absolute difference* between the *true* population value θ and the value θ_0 given in the test hypothesis. In tests of a single population mean, the synthetic alternative hypothesis therefore takes one of three forms:

$$H_1: \mu = \mu_0 + \delta \ (\text{if } \mu_1 > \mu_0)$$

$$H_1: \mu = \mu_0 - \delta \ (\text{if } \mu_1 < \mu_0)$$

$$H_1: \mu = \mu_0 \pm \delta \ (\text{if } \mu_1 \neq \mu_0)$$

Example 10.1.8. Suppose it was decided that even a statistically significant result wouldn't really be very important unless wolves averaged at least 5 fewer errors than dogs. Since $\mu_0 = 14.6$, this gives us the following *synthetic* alternative hypothesis:

$$H_1: \mu = \mu_0 - 5 = 9.6$$

for which to calculate β.

If there is no scientific basis for establishing δ, the experimenter can at least get an idea of the relationship between β and all *possible* values of μ_1 by examining the power function of his test. Set α arbitrarily, calculate β for a number of possible values for μ_1 and plot the graph of $1 - \beta$ against these possible alternative mean values. Typical power curves for a one-tailed test (against the alternative $\mu > \mu_0$) and a two-tailed test are illustrated in Figure 10.2.

Power and sample size. We know that one way to control the probability of a Type II error (or, conversely, the *power* of a statistical test) is with sample size. When only a small pool of observations is available, one risks a test in which β is too large, that the test is not sensitive enough to detect real and scientifically important differences between μ_0 and the *real* population mean, μ. When the experimenter has access to large samples of potential observations, the risk is that the test might be overpowered, capable of detecting that μ is *statistically* significantly different from μ_0 even though the difference is *scientifically* insignificant.

In Chapter 9 we said that if one has (or can synthesize) a simple alternative hypothesis, it is sometimes possible to calculate the value of N that will yield specified values of α and β. For a test about one

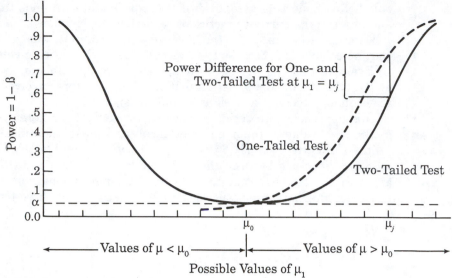

Figure 10.2. Power of directional (upper-tailed) and nondirectional (two-tailed) tests of hypothesis that $\mu = \mu_0$.

population mean it is shown in Box 10.2 that significance α and power $1 - \beta$ are obtained if

$$N \cong \left(\frac{\sigma(z_\beta - z_\alpha)}{\mu_0 - \mu_1} \right)^2 \qquad [10.9]$$

where

- σ is the standard deviation of X.
- z_α is the critical value of $\overline{X}$, standardized under the *test* hypothesis, that yields significance level α.
- z_β is the same value of $\overline{X}$, standardized under the *alternative* hypothesis, that yields $P(\text{Type II error}) = \beta$.

Example 10.1.9. In the preceding example, we decided that 5 errors was the smallest scientifically important difference between μ_D, the mean of the dog population, and μ_W, the mean of the wolf population. This gave us a synthetic value of $\mu_1 = 9.6$. In addition, let us suppose we want $\alpha = \beta = .025$. We have estimated σ^2 to be 127.5, so we may use $\sqrt{127.5} \doteq 11.29$ as an estimate of σ. Since $\mu_1 < \mu_0$, the critical value z_α lies in the lower tail of the standard normal curve and $\alpha = .025$,

$$P(z \leq z_\alpha | H_0 \text{ correct}) = .025$$

From Table 3 of Appendix VIII, $\mathbb{P}(-1.96) \doteq .025$, so $z_\alpha = -1.96$. Under the *alternative* hypothesis, the critical value lies in the *upper* tail of the

---

BOX 10.2

Calculating N for specific values of α and β

Let z_α be the critical value of Z for significance level α. Then, for $V(X) = \sigma^2$ and sample size N,

$$z_\alpha = \frac{\bar{x}_\alpha - \mu_0}{\sigma/\sqrt{N}}$$

where $\bar{x}_\alpha$ is the critical value of $\bar{X}$ in natural units. Therefore,

$$\bar{x}_\alpha = \mu_0 + z_\alpha(\sigma/\sqrt{N})$$

Let z_β be the z-value that gives $P(\text{Type II error}) = \beta$. Standardizing $\bar{x}_\alpha$ under the alternative hypothesis,

$$z_\beta = \frac{\bar{x}_\alpha - \mu_1}{\sigma/\sqrt{N}}$$

so

$$\bar{x}_\alpha = \mu_1 + z_\beta(\sigma/\sqrt{N})$$

Since $\bar{x}_\alpha$ is the *same* value whether it is expressed in terms of z_α or in terms of z_β,

$$\mu_0 + z_\alpha(\sigma/\sqrt{N}) = \mu_1 + z_\beta(\sigma/\sqrt{N})$$

Rearranging terms,

$$\mu_0 - \mu_1 = z_\beta(\sigma/\sqrt{N}) - z_\alpha(\sigma/\sqrt{N}) = \frac{\sigma}{\sqrt{N}}(z_\beta - z_\alpha)$$

so

$$(\mu_0 - \mu_1)(\sqrt{N}) = \sigma(z_\beta - z_\alpha)$$

and

$$\sqrt{N} = \frac{\sigma(z_\beta - z_\alpha)}{\mu_0 - \mu_1}$$

Squaring both sides,

$$N = \left(\frac{\sigma(z_\beta - z_\alpha)}{\mu_0 - \mu_1}\right)^2$$

normal curve. With $\beta = .025$, the critical value corresponds to z_β, where

$$P(z \geq z_\beta | H_1 \text{ correct}) = .025$$

which means that $\mathbb{P}(z_\beta) \doteq .975$. From Table 3, $z_\beta = 1.96$, so by equation [**10.9**],

$$N = \left(\frac{\sigma(z_\beta - z_\alpha)}{\mu_1 - \mu_2} \right)^2 \doteq \left(\frac{11.29[1.96 - (-1.96)]}{5} \right)^2 \cong 78$$

That is, we would need approximately 78 wolf pups to ensure that $\alpha = \beta = .025$ against the alternative hypothesis $\mu = 9.6$.

EXERCISES 10.1

In Exercises 1–4, the random variable X is known to be normally distributed. Test the hypothesis $\mu = \mu_0$ against the indicated alternative hypothesis. For each test state the appropriate rejection rule, the value of the test statistic, and the appropriate conclusion.

1.

	$\bar{x}$	μ_0	$\sigma_{\bar{X}}^2$	$H_1 : \mu$	α
(a)	92	100	225	= 85	.05
(b)	1,520	1,555	12,293	= 1,345	.01
(c)	1,520	1,345	28,900	= 1,555	.01
(d)	75	52	182.25	= 100	.05
(e)	16	24.5	14.5	= 19.75	.025

2.

	$\bar{x}$	μ_0	$\sigma_{\bar{X}}^2$	$H_1 : \mu$	α
(a)	.036	1.92	.53	= −2.50	.005
(b)	−26.3	−31.1	2.90	= −22.7	.0025
(c)	230	180	480	= 240	.01
(d)	240	150	2,100	= 100	.05
(e)	.00021	.0002	9.18×10^{-12}	= .00025	.0005

3.

	$\bar{x}$	σ^2	N	μ_0	$H_1 : \mu$	α
(a)	67	1,000	10	50	> 50	.05
(b)	81	6,750	30	100	< 100	.025
(c)	67	500	5	50	≠ 50	.05
(d)	100	25	100	98.6	> 98.6	.01
(e)	.0319	1	10,000	.001	> .001	.001

4.

	$\bar{x}$	σ^2	N	μ_0	$H_1 : \mu$	α
(a)	60.75	625	25	69	< 69	.05
(b)	405	17,430	15	480	< 480	.01
(c)	3.1	5	5	0	> 0	.001
(d)	.105	.139	100	.167	< .167	.05
(e)	.105	.139	100	.167	≠ .167	.05

5. A cross-country ski with a stiffness of 3 (on a scale of 1 to 5) should completely decamber (flatten out) if it is weighted with 150 lbs. A ski shop places an order for 40 skis of stiffness 3, but the 40 skis that are delivered have no stiffness rating marked on them. The shop owner tests the skis and finds that the mean weight to decamber the skis is 146.5 lbs with a variance of 132.34.
   (a) What are the hypotheses?
   (b) What is the appropriate test statistic?
   (c) Set α at any level you consider suitable and state the rejection rule appropriate to your hypotheses and the significance level you have chosen.
   (d) Calculate the value of your test statistic.
   (e) Report the appropriate conclusion.

6. The manufacturer of an inductively coupled plasma (IC-P) unit claims that the average downtime per operating week is less than 50 min. The manager of a laboratory that owns one of these instruments believes that it is out of service more than 50 min a week and records the number of minutes of downtime (X) over a period of 32 operating weeks. The manager finds that $\Sigma x = 1767$ and $\Sigma x^2 = 98{,}755$.
   (a) What are the hypotheses?
   (b) What is the appropriate test statistic?
   (c) Set α at any level you consider suitable and state the rejection rule appropriate to your hypotheses and the significance level you have chosen.
   (d) Calculate the value of your test statistic.
   (e) Report the appropriate conclusion.

7. When light is passed through a chemical substance, some wavelengths are absorbed more than other wavelengths. The purity of a substance can often be determined by measuring absorption at the wavelength that is known to be most readily absorbed by that substance. A chemical company is producing a particular solvent. To test for purity, a chemist draws a sample of 36 aliquots (observations of equal volume) from the production run and measures the absorbance of each aliquot at a wavelength of 575 nanometers. The expected absorbance of the solvent at this wavelength is .765, and the absorbance values obtained by the chemist are as follows:

.766	.765	.765	.760	.771	.755	.759	.772	.771
.769	.771	.773	.775	.770	.769	.772	.773	.768
.772	.766	.770	.768	.769	.770	.771	.774	.767
.766	.766	.772	.771	.770	.773	.774	.775	.768

   (a) What are the hypotheses?
   (b) What is the appropriate test statistic?
   (c) Set α at any level you consider suitable and state the rejection rule appropriate to your hypotheses and the significance level you have chosen.
   (d) Calculate the value of your test statistic.
   (e) Report the appropriate conclusion.

8. "Toe-in" is the degree to which the front wheels of a vehicle point inward. Specifications for a certain model of half-ton pickup truck call for .125 inches of toe-in. The manager of one assembly plant believes that the trucks produced on the day shift do not meet specifications. Toe-in is measured on a random sample of 30 trucks assembled by the day shift. In the following data,

"zero" means that the wheels point straight ahead; negative values mean that the wheels point outward.

.125	.241	.010	−.060	.185	.250
−.030	.000	.240	.130	.120	.122
.130	.450	.320	−.100	.010	.200
.160	.150	.123	.130	.220	−.005
.125	.063	.063	−.063	.130	.120

(a) What are the hypotheses?
(b) What is the appropriate test statistic?
(c) Set α at any level you consider suitable and state the rejection rule appropriate to your hypotheses and the significance level you have chosen.
(d) Calculate the value of your test statistic.
(e) Report the appropriate conclusion.

9. For every test in Exercise 10.1.1
(a) report the significance probability of the result
(b) assume that σ^2 is the same whether or not H_0 is correct, and calculate β and the power of the test.

10. For every test in Exercise 10.1.2
(a) report the significance probability of the result
(b) assume that σ^2 is the same whether or not H_0 is correct, and calculate β and the power of the test.

In Exercises 11 and 12 the random variable X is known to be normally distributed. Write the formula for the appropriate statistic to test the hypothesis that $\mu = \mu_0$, calculate the critical value of $\overline{X}$, and calculate the power of the test. *Note:* σ_0^2 is the variance of X if H_0 is correct, and σ_1^2 is the variance of X if H_1 is correct.

11.

	μ_0	μ_1	σ_0^2	σ_1^2	N	α
(a)	50	60	225	256	10	.01
(b)	50	60	225	256	25	.01
(c)	50	60	225	256	50	.01
(d)	50	60	225	256	50	.025
(e)	50	60	225	256	50	.05

12.

	μ_0	μ_1	σ_0^2	σ_1^2	N	α
(a)	80	70	625	900	15	.05
(b)	80	65	625	900	15	.05
(c)	80	60	625	900	15	.05
(d)	80	60	625	900	25	.05
(e)	80	60	625	900	35	.05

In Exercises 13 and 14, the random variable X is known to be normally distributed. Test the hypothesis that $\mu = \mu_0$. For each test, state the appropriate rejection rule, the value of the test statistic, the *p*-value of

the result, the appropriate conclusion, and the power of the test if a simple alternative hypothesis is indicated. In calculating power, assume σ^2 is the same under both hypotheses.

13.

	Σx	N	μ_0	μ_1	σ^2	α
(a)	920	10	100	85	225	.05
(b)	5	24	1/3	1/4	5/48	.025
(c)	9.45	15	.75	$\neq$ .75	.056	.05
(d)	14,711	27	500	> 500	10^4	.01

14.

	Σx	N	μ_0	μ_1	$\hat{\sigma}^2$	α
(a)	-402	31	-14	> -14	6.25	.01
(b)	19,140	120	160	< 160	12.25	.005
(c)	111	60	.031	= 2.52	20	.001
(d)	665	32	18	$\neq$ 18	50	.025

15. The experimenter in Exercise 10.1.13(d) feels that the smallest scientifically important difference between the *true* value of the population mean μ and the value μ_0 specified by the test hypothesis is one standard unit (i.e., the equivalent of one standard deviation of the random variable). The experimenter has also decided to set $\alpha = \beta = .01$.
(a) Reformulate the alternative hypothesis.
(b) Calculate the sample size the experimenter needs.

16. The experimenter in Exercise 10.1.14(b) calculates that the cost of a Type I error is 10 times greater than the cost of a Type II error. In addition, the experimenter feels that it is not important to detect a difference between the hypothesized value μ_0 and the true population value μ that is smaller than three units.
(a) Reformulate the alternative hypothesis.
(b) Calculate the sample size that the experimenter should use.

2. Small samples ($N < 30$): Student's t

When $N < 30$, one can neither estimate σ^2 from the data nor rely on the Central Limit Theorem to justify the assumption that $\bar{X}$ is normally distributed. The choice of an appropriate test therefore depends wholly on the normality of X and the experimenter's knowledge of σ^2, the variance of X. There are three possibilities:

(i) X is normally distributed and σ^2 is known. If X is normal, then $\bar{X}$ is also normally distributed for *any* sample size. If, in addition, the variance of X is known, then the experimenter has all of the information needed to specify the distribution of $\bar{X}$ under the test hypothesis:

$$\bar{X}: N(\mu_0, \sigma^2/N)$$

This is the case we developed in our Neanderthal example, and the appropriate test statistic is given in [**10.1**],

$$Z_{\overline{X}} = \frac{\overline{X} - \mu_0}{\sigma_{\overline{X}}}$$

(ii) X cannot be assumed to be normally distributed. If the number of observations is less than 30 and X cannot be assumed to be distributed normally, the experimenter's lot is an easy—but not necessarily happy—one. He or she must use what statisticians call a *nonparametric* statistic. Most nonparametric tests require minimal assumptions about one's random variable, place relatively simple demands on one's data (e.g., independent observations or equal numbers of observations in all groups or some minimal number of observations), and are uncommonly *robust*. That is, they are relatively insensitive to slight violations of these requirements. However, the trade-off is that nonparametric tests are not very powerful and often lead an experimenter to reject the alternative hypothesis incorrectly.

Since we do not cover nonparametric tests of means in this text, the reader's lot is happier than the experimenter's.

(iii) X is normally distributed, but σ^2 is unknown. Because the relative frequency distributions of so many of the variables studied by scientists are, in fact, approximated closely by the normal distribution (see Box 8.1) this case is very common in scientific research.

a. Decisions. *Test statistic.* In 1899 William Sealy Gosset (1876–1937), an Oxford-trained mathematician and chemist, was recruited by the Guinness brewery in Dublin. Brewers were just beginning to use quantitative methods to control the quality of beer and ale, and Gosset soon encountered a perplexing difficulty. Many of the statistical tests he needed to perform involve normal approximation methods. The variables of concern to Guinness were known to be normally distributed, but the chemistry of brewing makes it impossible to collect the large samples of independent observations that are needed to estimate variance. Gosset therefore derived the exact probability distribution for the quotient

$$\frac{\overline{X} - \mu}{S/\sqrt{N-1}} \qquad\qquad [\textbf{10.10}]$$

where $\overline{X}$ is assumed to be normally distributed. He denoted this statistic t and published it under the pen name "Student."[9]

---

[9] Guinness encouraged Gosset's mathematical research but, for reasons of industrial and commercial secrecy, required that his work appear under a pseudonym and that none of the company's data be used. (Plackett, R. L. (Ed.) (1990). '*Student' A Statistical Biography of William Sealy Gosset.* London: Oxford University Press, p. 17.) "Student" is the pen-name Gosset frequently used in correspondence with his "professor," Karl Pearson. (cf. Pearson, E. S. (1967). Studies in the history of probability and statistics. XVII. *Biometrika, 54,* 352.)

After calculating Student's t a few times, many students realize that

$$s/\sqrt{N-1} = \sqrt{\frac{\Sigma(x-\bar{x})^2}{N(N-1)}} = \sqrt{\frac{\hat{\sigma}^2}{N}} = \hat{\sigma}_{\bar{X}}$$

and conclude that [10.10] is just

$$Z_{\bar{X}} = \frac{\bar{X} - \mu}{\sigma_{\bar{X}}}$$

with $\sigma_{\bar{X}}$ estimated by $\hat{\sigma}_{\bar{X}}$. This is not the case. We know from Chapter 3 that Z is simply a numerical transformation of X and that the probability distribution of Z therefore depends entirely on the distribution of X. Similarly, the distribution of $Z_{\bar{X}}$ depends entirely on the distribution of $\bar{X}$. In expression [10.10] the numerator includes the random variable $\bar{X}$ and the denominator includes the random variable S. The distribution of Student's t therefore depends on the distributions of *two* random variables, $\bar{X}$ and S. It should therefore be apparent that the derivation of the t distribution proceeded along very different lines from the derivation of the standard normal distribution, which we sketched in Chapter 7.

Gosset demonstrated that for samples of N independent observations on a normally distributed random variable X, the density function of

$$t = \frac{\bar{X} - \mu}{S/\sqrt{N-1}}$$

is

$$\phi(t) = \frac{1}{K}\left(1 + \frac{t^2}{\nu}\right)^{-(1+\nu)/2} \qquad\qquad [10.11]$$

where $\nu = N - 1$ and K is a positive constant that depends on ν. It will be observed that neither μ nor $\sigma_{\bar{X}}^2$ appear in [10.11]. This means that $S/\sqrt{N-1}$ does not enter t as an estimator of $\sigma_{\bar{X}}$ even though $S/\sqrt{N-1}$ is numerically equal to $\hat{\sigma}_{\bar{X}}$ for any particular value of S. Indeed, the only parameter in the density function of Student's t is *degrees of freedom*, which is denoted by ν, the lowercase Greek letter *nu*. Accordingly, the notation $W : t(\nu)$ is a specification statement indicating that the random variable W is distributed as Student's t with ν degrees of freedom. It is customary to denote the ratio given in [10.10] or any other t-distributed variable as t, much as any standardized random variable is denoted Z. It is therefore important to distinguish t, the *random variable*, from $t(\nu)$, the statement specifying the *distribution* of a random variable.

The concept of degrees of freedom is elusive, and as we introduce other statistics for which degrees of freedom (often abbreviated *df*) is a parameter, it will seem as if *df* is defined differently for each such statistic. At base, however, degrees of freedom may be thought of as *the number of observations* that are free to assume any value of the random variable.

For example, let us suppose we define the function Sum of Squares as

$$Y = \sum_{}^{N} (X - \bar{X})^2$$

and let us suppose you are given the value of $\bar{x}$ and are asked to guess the values of the numbers $x_1, x_2, \ldots, x_N$. The first $N - 1$ numbers $x_1, x_2, \ldots, x_{N-1}$ are *free* to assume any values, but the last value x_N is *constrained* by the requirement that

$$\sum_{}^{N} x = N\bar{x}$$

That is, the last value *must* equal

$$N\bar{x} - \sum_{}^{N-1} x$$

The Sum of Squares therefore has $N - 1$ degrees of freedom. And, in general, if

1. one makes N observations $\{x_1, x_2, \ldots, x_N\}$ and
2. the observations are used to make independent estimates of J parameters and
3. the random variable Y is a function of the N observations and the J estimates,

then Y has $N - J$ degrees of freedom.

In Student's t,

$$\frac{\bar{X} - \mu}{S/\sqrt{N-1}}$$

one makes N observations and uses these observations to calculate the sample mean $\bar{X}$, which estimates the population mean μ in the calculation of S. Degrees of freedom for this ratio is therefore $N - 1$.

In Chapter 7 some important properties of the standard normal curve were revealed by an examination of its density function. This is also true for Student's t. First we note that [**10.11**] can be rewritten

$$\phi(t) = \frac{1}{K\left(1 + \dfrac{t^2}{\nu}\right)^{(1+\nu)/2}} \qquad [\textbf{10.12}]$$

just as we rewrote $\phi(z)$ in equation [**7.2**]. In this form, $\phi(t)$ can be seen to have many of the same properties as $\phi(z)$.

For any given sample size, ν and K are constants, so the value of ϕ depends entirely on t^2. Since t^2 is in the denominator, $\phi(t)$ *increases* as t^2 *decreases*, but t^2 can never be negative, so $\phi(t)$ must be greatest when $t^2 = 0$. Like the standard normal density function, $\phi(t)$ is therefore *unimodal* at zero, that is, when $\bar{x} = \mu$.

Another property that Student's t shares with the standard normal curve is that it is *symmetrical* about 0, its mode. This is because t enters ϕ squared and $\phi(t)$ is therefore equal to $\phi(-t)$. We know from Chapter 2 that if a frequency polygon is unimodal and symmetrical, the mode is equal to the mean. It should not be surprising, therefore, that $E(t) = 0$.

The same reasoning that let us deduce that the standard normal curve is *asymptotic to zero* can also be applied to Student's t: We can see from [10.12] that $\phi(t)$ tends to zero as t^2 increases, but we can also see that $\phi(t)$ is always positive. As t^2 increases (i.e., as $\bar{x}$ gets farther from μ), the density function of $t(\nu)$ therefore gets closer to the horizontal axis but never reaches it.

The last property of Student's t we wish to discuss is not based on its density function, but on its variance. We assert without proof that if $\nu > 2$,

$$ V(t) = \frac{\nu}{\nu - 2} $$

That is, $V(t) = (N - 1)/(N - 3)$. The smallest sample size for which variance is defined is $N = 4$, in which case $V(t) = 3$. As sample size increases, $V(t)$ decreases, approaching (but never quite reaching) 1.

From the properties we have presented, it should not be surprising that as N increases, $\phi(t)$ approaches $\phi(z)$, the density function for the standard normal random variable. If y is any real number, then for large samples, $P(t \leq y) \cong P(z \leq y)$. (This can be confirmed by taking a number between -3.09 and $+3.09$ and comparing its cumulative probability in Table 3 of Appendix VIII with its cumulative probability for various values of ν in Table 4.) This is, of course, a consequence of the Central Limit Theorem and the consistency of $\hat{\sigma}$ and is why for large samples

$$ \frac{\bar{X} - \mu}{\hat{\sigma}_{\bar{X}}} = \frac{\bar{X} - \mu}{s/\sqrt{N - 1}} $$

may be assumed approximately distributed $N(0, 1)$. The distributions of the standard normal random variable and Student's t for several values of ν are illustrated in Figure 10.3.

From Figure 10.3 we can see that Student's t is not *one* distribution, but a *family* of distributions, one for each possible sample size. It is also true, of course, that there is a different normal distribution for every possible combination of σ^2 and μ, but any normally distributed random variable can be standardized and probabilities determined by reference to a table of cumulative probabilities for the standard normal curve. There is no "standard" Student's t distribution, so in principle we need an "infinite" number of tables. This would require a very long appendix.

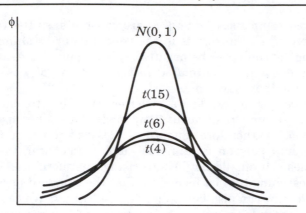

Figure 10.3. Probability density functions for standard normal random variable and for Student's *t* with $\nu = 15$, $\nu = 6$, and $\nu = 4$ (differences exaggerated to emphasize effects of *df*).

Fortunately, it isn't really necessary. In the first place, N is always an integer and is always greater than 1. Second,

$$\frac{\overline{X} - \mu}{s/\sqrt{N-1}} = \frac{\overline{X} - \mu}{\hat{\sigma}_{\overline{X}}}$$

is distributed approximately $N(0, 1)$ if $N \geq 30$. A tabulation of probabilities for Student's t therefore requires a maximum of only 29 tables, one each for $\nu = 1, 2, \ldots, 29$. Each table could be as extensive as Table 3 for the standard normal curve, but most t tables are very abbreviated and include t-values corresponding to only a small selection of cumulative probabilities.

In Table 4 of Appendix VIII each row represents a t distribution for one value of ν and includes tabled values of t for cumulative probabilities .60, .75, .80, .90, .95, .975, .99, .995, and .9995, as indicated by the column headings. The entry in row ν and column $\mathbb{P}$ is the value of t in the distribution of $t(\nu)$ with cumulative probability $\mathbb{P}$.

Example 10.3. What value in the distribution of t with 14 degrees of freedom has cumulative probability .995? Entering Table 4 in row 14, we find that the value in column .995 is $t = 2.977$.

Since $t(\nu)$ is symmetrical about zero, $\mathbb{P}(0) = .50$, and all of the entries in Table 4 must therefore be positive. One can (in principle) find the cumulative probability of a positive value t_i in the distribution $t(\nu)$ by entering Table 4 in the row corresponding to ν and moving across the row until the number is found.

Example 10.4.1. What is the cumulative probability of $t = 1.350$ if $t : t(13)$? In row 13 we find that 1.350 is in the column headed .90, so $\mathbb{P}(1.350) \doteq .90$.

Example 10.4.2. Likewise, one can (in principle) find the probability that a t-value exceeds some specific value, t_i. What is the probability that $t > 1.350$ if $t : t(13)$? In the preceding exercise we found that $\mathbb{P}(1.350) = P(t \le 1.350) \doteq .90$. Therefore, $\mathbb{P}(t > 1.350) \doteq 1 - .90 = .10$.

We have said that $P(t \le t_i)$ or $P(t > t_i)$ can "in principle" be found, because there are large gaps in the tabled cumulative probabilities, so the t-value in question usually falls between columns. If a particular t-value cannot be found (even if rounded to two places), and you need to find a cumulative probability, you can get an acceptable approximation by interpolation (see Appendix II).

Example 10.5.1. If a random variable is distributed as Student's t with 26 degrees of freedom, what is $\mathbb{P}(t = .470)$? In the row for $\nu = 26$, we do not find $t = .470$. The nearest entries are .256 (corresponding to $\mathbb{P} \doteq .60$) and .684 (corresponding to $\mathbb{P} \doteq .75$). The value $t = .470$ falls halfway between .256 and .684, so by interpolation

$$\mathbb{P}(.470) \doteq \frac{.60 + .75}{2} = .675$$

For the most part, however, you will be using Table 4 to determine significance probabilities for test statistics that are distributed as Student's t, so cumulative probabilities for t-values that "fall in the cracks" can be reported as inequalities (see Box 9.1, p. 358).

Example 10.5.2. Express the cumulative probability of $t = .470$ as given in the preceding example in terms of the cumulative probabilities of the nearest tabled value larger than .470 and the nearest tabled value smaller than .470. From Example 10.5.1,

$$.60 < \mathbb{P}(.470) < .75$$

Example 10.5.3. If $t : t(26)$, what is $P(t > .470)$? We know that $P(t \le .256) \doteq .60$, so $P(t > .256) \doteq 1 - .60 = .40$. We also know that $P(t \le .684) \doteq .75$, so $P(t > .684) \doteq .25$. Therefore, the probability that t will exceed any value between .256 and .684 is greater than .25 and less than .40:

$$.25 < P(t > .470) < .40$$

Although the t-values in Table 4 are positive, the symmetry of Student's t distributions makes it possible to use Table 4 to obtain cumulative probabilities of negative t-values. In any distribution that is symmetrical about zero, the positive half is the mirror image of the negative half. The probability that t is $greater$ than any positive value t_i must therefore equal the probability that t is less than the corresponding *negative* value, $-t_i$. That is,

$$P(t > t_i) = P(t < -t_i)$$

We also know that $P(t > t_i) = 1 - \mathbb{P}(t_i)$. Therefore,

$$\mathbb{P}(-t_i) = 1 - \mathbb{P}(t_i)$$

That is, the cumulative probability of any negative t-value, $-t_i$, can be obtained by taking 1 minus the cumulative probability of the corresponding positive value, t_i.

Example 10.6. If a random variable is distributed as $t(14)$, what is the probability that $t < -1.761$? The probability that t falls *below* -1.761 is equal to the probability that t falls *above* $+1.761$, and we know that $P(t > 1.761) = 1 - \mathbb{P}(1.761)$. In Table 4 we find that $\mathbb{P}(1.761) \doteq .95$ in $t(14)$, so

$$P(t < -1.761) \doteq 1 - .95 = .05$$

Example 10.7. Negative values also fall "in the cracks." If $t : t(5)$, what is $P(t < -3.800)$? Once again, we look to the corresponding *positive* value, but we find that $t = 3.800$ is not a tabled entry in $t(5)$. The value $t = 3.800$ lies between the columns for $\mathbb{P} = .99$ and $\mathbb{P} = .995$. Therefore, the probability that t is *greater* than 3.800 must be between $(1 - .99) = .01$ and $(1 - .995) = .005$. Because Student's t distributions are symmetrical about zero, it must likewise be true that

$$.005 < P(t < -3.800) < .01$$

───────── EXERCISES 10.2 ─────────────────────────

Find the cumulative probability for each of the t-values and degrees of freedom in Exercises 1–4. If the t-value given in the exercise does not appear in Table 4 of Appendix VIII, express the cumulative probability as an inequality. Hint: Draw pictures.

1. (a) $t = .718$, $\nu = 6$ (b) $t = 2.861$, $\nu = 19$
   (c) $t = .258$, $\nu = 14$ (d) $t = -3.012$, $\nu = 13$
   (e) $t = -3.551$, $\nu = 40$ (f) $t = -1.328$, $\nu = 19$
2. (a) $t = -1.812$, $\nu = 10$ (b) $t = 2.5$, $\nu = 23$
   (c) $t = -1$, $\nu = 1$ (d) $t = -.261$, $\nu = 9$
   (e) $t = -1.886$, $\nu = 2$ (f) $t = 3.499$, $\nu = 7$
3. (a) $t = .50$, $\nu = 10$ (b) $t = 2.8$, $\nu = 14$
   (c) $t = -1.00$, $\nu = 9$ (d) $t = 3$, $\nu = 15$
   (e) $t = 1.96$, $\nu = 22$ (f) $t = -2.83$, $\nu = 8$
4. (a) $t = 1.64$, $\nu = 1{,}000$ (b) $t = -3.0$, $\nu = 3$
   (c) $t = -.40$, $\nu = 19$ (d) $t = -1.64$, $\nu = 24$
   (e) $t = 2.23$, $\nu = 17$ (f) $t = -2.5$, $\nu = 1$

In Exercises 5 and 6 find the *t*-value corresponding to the indicated cumulative probability.

5. (a) .95 in $t(11)$ (b) .99 in $t(23)$
   (c) .75 in $t(17)$ (d) .05 in $t(15)$
   (e) .005 in $t(4)$ (f) .025 in $t(19)$
6. (a) .90 in $t(14)$ (b) .01 in $t(8)$
   (c) .40 in $t(18)$ (d) .0005 in $t(22)$
   (e) .975 in $t(13)$ (f) .95 in $t(2)$

In Exercises 6 and 7 find the indicated conditional probabilities. If the *t*-value given in the exercises does not appear in Table 4 of Appendix VIII, express your answer as an inequality. Hint: Draw pictures.

7. Find the following conditional probabilities.

   (a) $P[t \geq 2.160 | t : t(13)]$ (b) $P[t \leq -3.365 | t : t(5)]$
   (c) $P[t \geq 1.325 | t : t(20)]$ (d) $P[t < -1.761 | t : t(14)]$
   (e) $P[t \geq 2.9 | t : t(9)]$ (f) $P[t \leq -2.81 | t : t(3)]$
   (g) $P[-3.499 \leq t \leq 3.499 | t : t(7)]$ (h) $P[-2.16 \leq t \leq 2.16 | t : t(13)]$

8. Find the following conditional probabilities.
   (a) $P[t \leq -2.718 | t : t(11)]$
   (b) $P[t \geq 3.707 | t : t(6)]$
   (c) $P[-2.101 \leq t \leq 2.101 | t : t(18)]$
   (d) $P[-2.015 \leq t \leq 2.015 | t : t(5)]$
   (e) $P[t \leq -2.074$ or $t \geq 2.074 | t : t(22)]$
   (f) $P[t \leq -8.610$ or $t \geq 8.610 | t : t(4)]$
   (g) $P[t \leq -2.921$ or $t \geq 2.921 | t : t(16)]$
   (h) $P[t \leq -2.896$ or $t \geq 2.896 | t : t(8)]$

In Exercises 9–14 let X be normally distributed and let

$$t = \frac{\bar{x} - \mu}{s/\sqrt{N - 1}}$$

9. If $E(x) = \mu$, find the probability that
   (a) $t \geq 2.821$ for a sample of 10 observations.
   (b) $t \leq -1.761$ for a sample of 15 observations.
   (c) $-2.861 \leq t \leq 2.861$ for a sample of 20 observations.
   (d) $t \leq -2.132$ or $t \geq 2.132$ for a sample of 5 observations.
10. If $E(X) = \mu$, find the probability that
   (a) $t \geq 2.015$ for a sample of 6 observations.
   (b) $t \leq -2.5$ for a sample of 24 observations.
   (c) $-31.598 \leq t \leq 31.598$ for a sample of 3 observations.
   (d) $t \leq -2.567$ or $t \geq 2.567$ for a sample of 18 observations.
11. If $E(X) = 50$ and an experimenter makes 10 observations, what is the probability of obtaining a *t*-value
   (a) as large or larger than the experimenter's calculated value of t if $\bar{x} = 55$ and $s^2 = 21.30$?
   (b) as small or smaller than the experimenter's calculated value of t if $\bar{x} = 40.5$ and $s^2 = 158.75$?

12. If $E(X) = 50$ and an experimenter makes 10 observations, what is the probability of obtaining a t-value
    (a) as large or larger than the experimenter's calculated value of t if $\bar{x} = 55$ and $s^2 = 66.97$?
    (b) as small or smaller than the experimenter's calculated value of t if $\bar{x} = 40.5$ and $s^2 = 102.07$?
13. If $E(X) = 175$ and $s^2 = 1,560.25$, what is the probability of obtaining a t-value
    (a) as large or larger than the experimenter's calculated value of t if $\bar{x} = 211.7$ and $N = 5$?
    (b) as small or smaller than the experimenter's calculated value of t if $\bar{x} = 150$ and $N = 20$?
14. If $E(X) = 175$ and $s^2 = 1,560.25$, what is the probability of obtaining a t-value
    (a) as large or larger than the experimenter's calculated value of t if $\bar{x} = 200$ and $N = 10$?
    (b) as small or smaller than the experimenter's calculated value of t if $\bar{x} = 150$ and $N = 15$?

To see how Student's t ratio is used as a test statistic, let us return to the Neanderthal problem discussed at the beginning of the chapter.

Example 10.8.1. The value of 1,600 given earlier as the variance of Neanderthal skulls is, in fact, an estimate based on all skulls that have ever been catalogued. Let us suppose that our Neanderthal-phase theorist has examined all of these skulls and concluded that fewer than 25 of them can be unequivocally identified as adult males. He is therefore unwilling to rely on 1,600 as an estimate of σ^2. His own sample of $N = 4$ Middle Eastern skulls is too small for an accurate estimate of σ^2, so the appropriate test statistic is

$$t = \frac{\bar{X} - \mu_0}{s/\sqrt{N-1}} = \frac{\bar{X} - 1,575}{s/\sqrt{3}}$$

which will be distributed $t(3)$ if the test hypothesis is correct. And, in general (cf. Box 10.1),

---

If X is a normally distributed random variable with unknown variance and if $N < 30$, then the appropriate statistic to test the hypothesis that $\mu = \mu_0$ is

$$t = \frac{\bar{X} - \mu_0}{S/\sqrt{N-1}}$$

which is distributed $t(N - 1)$.

Rejection rule. Deciding on a rejection region for $Z_{\bar{X}}$ was predicated in part on the fact that if $\bar{X}$ is normally distributed, $Z_{\bar{X}}$ is normally distributed under *either* hypothesis. The situation is a bit different with

Student's t. If the alternative hypothesis is correct (i.e., if $\mu = \mu_1$), the probability model for

$$\frac{\bar{X} - \mu_0}{S/\sqrt{N-1}}$$

is a function called *noncentral* t. We won't discuss this distribution, except to say that $E(\text{noncentral } t)$ is positive if $\mu_1 > \mu_0$ and is negative if $\mu_1 < \mu_0$, so our basic strategy is the same as before: Since the expected value of the test statistic under H_0 is zero, the rejection region must fall in the upper tail of the appropriate t distribution when $\mu_1 > \mu_0$ and in the lower tail of the appropriate t distribution when $\mu_1 < \mu_0$.

Example 10.8.2. The anthropologist in Example 10.8.1 has a sample of 4 observations, and if H_0 is correct his statistic is distributed as $t(3)$. Let us suppose, as before, that he wants $\alpha = .05$. In the present example $\mu_1 < \mu_0$, so if H_1 is correct, the expected value of t is negative, and the critical value of t must lie in the lower tail of $t(3)$. This means that the critical value t_α is *negative* and must be such that

$$p[t \le t_\alpha | t : t(3)] = .05 \qquad\qquad [\textbf{10.13}]$$

We know from earlier discussion that if $\mathbb{P}(t_i) = 1 - \alpha$, then $\mathbb{P}(-t_i) = \alpha$. The critical value t_α that satisfies equation [**10.13**] is therefore the *negative* of the t-value for which $\mathbb{P} = (1 - .05) = .95$. From Table 4 of Appendix VIII, $\mathbb{P}(2.353) = .95$, so the anthropologist can reject the test hypothesis only if his calculated value of t is less than or equal to -2.353.

In the case of a nondirectional alternative hypothesis, the critical region is two intervals: (1) all values greater than or equal to $t_{\alpha/2}$ and (2) all values less than or equal to $-t_{\alpha/2}$, where

$$\mathbb{P}(-t_{\alpha/2}) = \left[1 - \mathbb{P}(t_{\alpha/2})\right] = \frac{\alpha}{2}$$

b. Data collection and conclusions

Example 10.8.3. The four skulls in the Middle Eastern sample have cranial capacities of 1,555, 1,520, 1,585, and 1,587 cc. The value of Student's t is therefore

$$\frac{\bar{x} - \mu_0}{s/\sqrt{N-1}} = \frac{1{,}561.75 - 1{,}575}{27.23/\sqrt{3}} \doteq -0.843$$

Since $t = -0.843$ is greater than -2.353, the anthropologist must again reject the alternative hypothesis.

c. The power of Student's t. Although Student's t approaches the standard normal as N increases, we know that for small samples sizes $t(\nu)$ exhibits more variability than the standard normal (cf. Figure 10.3).

From earlier discussion we also know that the power of a statistical test decreases as the variability of the test statistic increases. Accordingly, statistical tests based on samples that are small enough to require use of Student's t are less powerful than tests based on the standard normal distribution.

Example 10.9. Suppose that in Example 10.1 the experimenters had tested only 4 wolf pups instead of 30. Suppose further that the distribution of barrier-test scores for Scott and Fuller's dogs was approximately normal and that the experimenters therefore assume that wolf performance is normally distributed. Finally, let us suppose that the statistics calculated for these 4 pups are the same as given in Examples 10.1.4 and 10.1.6 for 30 pups, that is, $\bar{x} = 9.5$ and $\hat{\sigma}_{\bar{X}}^2 = 4.25$. The value of the test statistic therefore remains

$$\frac{\bar{x} - 14.6}{\hat{\sigma}_{\bar{X}}} = \frac{\bar{x} - 14.6}{s/\sqrt{N - 1}} \doteq \frac{9.5 - 14.6}{2.062} = -2.47$$

but with $N = 4$, it must be submitted to the cumulative probabilities for $t(3)$. In Table 4 of Appendix VIII we find that $\mathbb{P}(3.182) \doteq .975$. Since $\mu_1 < \mu_0$, the critical value of Student's t for $\alpha = .025$ is -3.182, and the test hypothesis cannot be rejected. Clearly, Student's t is not as sensitive as $Z_{\bar{X}}$ to differences between $\bar{X}$ and μ_0. That is, it is not as powerful.

In Figure 10.4 it can be seen that the greater variability of $t(\nu)$, in comparison with the standard normal curve, translates into less kurtosis.

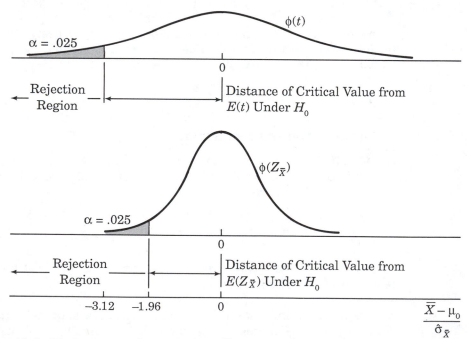

Figure 10.4. Distributions for Student's t and $Z_{\bar{X}}$ for same statistical test.

That is, the distribution of Student's t has less area in the peak and more in the tails. For any given value of α, the critical value of $(\overline{X} - \mu_0)/\hat{\sigma}_{\overline{X}}$ must therefore lie farther from the mean.

Calculation of power for Student's t requires tables for noncentral t, which are not widely available. This precludes the possibility of calculating N to achieve a desired β against a specific value of μ_1, but in general it may be safely assumed that when Student's t is indicated, the experimenter is unlikely to overpower his test and reject H_0 on the basis of scientifically insignificant differences.

BOX 10.3

Interval estimates of μ based on Student's t

Techniques for making interval estimates of μ for large samples were discussed in Chapter 8. When N is too small to estimate σ^2 but X is normally distributed, Student's t may be used to make an interval estimate of μ. For $N \geq 30$, we know that the limits of an interval estimate of confidence $\mathbb{C}$ for the population mean μ are $(\overline{x} - z_b\hat{\sigma}_{\overline{x}})$ and $(\overline{x} + z_b\hat{\sigma}_{\overline{x}})$ where z_b is that value for which $\mathbb{P} = (1 - \mathbb{C})/2$ under the standard normal curve and $\hat{\sigma}_{\overline{x}} = s/\sqrt{N-1}$.

It is similarly true for a sample of $N < 30$ observations that the limits of an interval estimate of μ with confidence $\mathbb{C}$ are $(\overline{x} - t_b\hat{\sigma}_{\overline{x}})$ and $(\overline{x} + t_b\hat{\sigma}_{\overline{x}})$ where $\mathbb{P}(t_b) = (1 + \mathbb{C})/2$ under the distribution of $t(N-1)$. For example, let $\overline{x}$ be the mean of a sample of 10 observations and let $\mathbb{C}$ be .98. In Table 4 of Appendix VIII we find that $\mathbb{P} = (1 + .98)/2 = .99$ corresponds to $t = 2.821$ in $t(9)$. The desired confidence interval is therefore $(\overline{x} - 2.821\hat{\sigma}_{\overline{x}})$ to $(\overline{x} + 2.821\hat{\sigma}_{\overline{x}})$.

Conversely, if $t_b > 0$ and $P(t \leq t_b) = \mathbb{P}$ in $t(N-1)$, then for a sample of N observations the confidence associated with the interval $(\overline{x} - t_b\hat{\sigma}_{\overline{x}})$ to $(\overline{x} + t_b\hat{\sigma}_{\overline{x}})$ is approximately $2\mathbb{P} - 1$.

EXERCISES 10.3

In Exercises 1 and 2 the random variable X is known to be normally distributed. Use Student's t to test the hypothesis that $\mu = \mu_0$ against the alternative hypothesis indicated. For each test specify the distribution of the test statistic, state the appropriate rejection rule, calculate the value of the test statistic, state the appropriate conclusion, and report your p-value.

1.

	$\overline{x}$	s^2	N	μ_0	$H_1 : \mu$	α
(a)	31	95	10	25	> 25	.05
(b)	185	1,380	5	250	< 250	.025
(c)	1,575	202,500	17	1,350	$\neq 1,350$	.01
(d)	14.7	1,255	4	50.5	< 50.5	.05
(e)	85	1,245	4	150	$\neq 150$	.05

2.

		$\bar{x}$	s^2	N	μ_0	$H_1 : \mu$	α
(a)		.026	.000036	20	.03	$= .02$	.005
(b)		-41	4267	7	-57	$= -32$	.05
(c)		3.87	25.19	24	0	$\neq 0$	.001
(d)		37.7	31.32	28	36.2	$\neq 36.2$	.05
(e)		1,360	100,450	15	1,000	$> 1,000$	.01

3. Let X be a normally distributed random variable. A sample of 9 observations yields a mean $\bar{x} = 25$ and a variance $s^2 = 24$. Calculate interval estimates for μ for the following levels of confidence:

    (a) $\mathbb{C} = .80$ (b) $\mathbb{C} = .90$ (c) $\mathbb{C} = .98$

4. The following scores were obtained for a sample of 10 individuals on a measure that is known to be normally distributed: 12, 14, 15, 15, 22, 17, 19, 18, 18, 23. Use these data to calculate interval estimates for μ for the following levels of confidence.

    (a) $\mathbb{C} = .95$ (b) $\mathbb{C} = .98$ (c) $\mathbb{C} = .99$

    (d) How do your answers to (a) and (c) differ from your answers to Exercises 8.3.9(a) and 8.3.9(c) on p. 320? Why?

5. A health physiologist measures the resting pulse rate of 27 men who run between 20 and 40 miles each week. The mean resting heart rate is 51 and the variance is 12.96. How confident can the physiologist be that the population mean for such recreational athletes is between

    (a) 49.25 and 52.75   (b) 49 and 53   (c) 50.5 and 51.5

6. The following concentrations of lead are found in 26 samples of hazardous waste (milligrams of lead per kilogram of waste):

    5.0   5.5   3.1   0.1   0.2   0.8   2.2   3.6   4.1
    6.2   0.0   0.6   4.6   4.8   0.3   0.4   1.3   1.1
    2.7   4.9   3.2   3.7   6.5   5.2   0.3   0.2

    (a) Calculate a 95 percent confidence interval for the mean concentration of lead in this source of waste.
    (b) How confident can the waste disposal company be that the concentration of lead from this source is between 1.5 and 3.9 mg per kg of waste?

In Exercises 7 and 8 test the hypothesis that $\mu = \mu_0$ against the alternative hypothesis indicated. State the test statistic for each test, specify the distribution of the test statistic, state the appropriate rejection rule, calculate the value of the test statistic, state the appropriate conclusion, and report your p-value. (The notation ? under the heading $\phi(X)$ indicates that the distribution of X is unknown.)

7.

	$\phi(X)$	$\bar{x}$	σ^2	s^2	N	μ_0	$H_1 : \mu$	α
(a)	Normal	52	20	—	7	50	> 50	.05
(b)	Normal	52	?	20	27	50	> 50	.05
(c)	Normal	52	?	20	22	50	> 50	.01
(d)	?	52	?	20	37	50	> 50	.01
(e)	?	60	20	—	14	50	> 50	.05

8.

	$\phi(X)$	$\bar{x}$	σ^2	s^2	N	μ_0	$H_1 : \mu$	α
(a)	?	94	196	—	18	100	< 100	.05
(b)	Normal	94	196	—	18	100	< 100	.05
(c)	Normal	95	?	256	25	100	< 100	.05
(d)	?	95	?	256	30	100	< 100	.05
(e)	Normal	95	?	256	29	100	< 100	.05

9. A comparative psychologist is studying the effects of food reward on problem solving in chimpanzees. Chimps are usually deprived of all food for 24 hr prior to administering this sort of a test in order to increase motivation, but the experimenter suspects that arousal of hunger may actually *interfere* with speed of learning and therefore tests the animals just after they eat. The mean for a sample of 25 chimps is 115 and the variance is 2,300. With food-deprived chimps, the average number of trials required to solve this problem is $\mu = 150$ and performance is approximately normally distributed.
(a) What are the experimenter's hypotheses?
(b) What is the appropriate test statistic?
(c) Set α at any level you consider suitable and state the rejection rule appropriate to your hypotheses and the significance level you have chosen.
(d) Calculate the value of your test statistic.
(e) Report the appropriate conclusion.

10. The mean score for a large sample of white, male college students was 93.69 on MACH IV, a test of interpersonal manipulativeness and similar qualities derived from Machiavelli's *The Prince* and *The Discourses*. An experimenter suspects because of their relative powerlessness, suppressed minority populations develop higher levels of interpersonal manipulativeness than do majority populations. For 62 nonwhite college males the mean was 97.25 with $s = 15.08$. [Data from Christie, R., and Geis, F. L. (1970). *Studies in Machiavellianism*. New York: Academic, p. 32.]
(a) What are the hypotheses?
(b) What is the appropriate test statistic?
(c) Set α at any level you consider suitable and state the rejection rule appropriate to your hypotheses and the significance level you have chosen.
(d) Calculate the value of your test statistic.
(e) Report the appropriate conclusion.

11. An airline advertises that its new flight schedules are more realistic than in the past and that delays are shorter. To test this claim, a consumer group samples 100 flights and records the number of minutes of delay between scheduled departure and actual departure. The sample average is 55 minutes with a standard deviation of 30 min. Under the old schedule the average delay was 58 min. Determine whether the sample data support the airline's claim.
(a) What are the hypotheses?
(b) What is the appropriate test statistic?
(c) Set α at any level you consider suitable and state the rejection rule appropriate to your hypotheses and the significance level you have chosen.
(d) Calculate the value of your test statistic.
(e) Report the appropriate conclusion.

12. A game designed to test interpersonal manipulation skills requires three players. A the end of the game the total points earned by the three participants is 100, and the expected value of any player's final score is 33 1/3. An experimenter predicts that players high in Machiavellianism will earn more points than expected on the basis of random chance. For 22 players with high MACH IV scores (see Exercise 10.3.10), the average game score was 47.32 points and $s^2 = 133$. [Source: Geis, F. L. (1970). The con game. In R. Christie and R. L. Geis (Eds.), *Studies in Machiavellianism*. New York: Academic, pp. 106–129.] Assume game scores are normally distributed.
    (a) What are the experimenters's hypotheses?
    (b) What is the value of the appropriate test statistic?
    (c) If $\alpha = .001$, what is the experimenter's conclusion?
    (d) What is the probability that the conclusion is wrong?

13. The airline in Exercise 10.3.11 has a run of misfortune–equipment failures, strikes, pilots grounded for substance abuse, and so on. They are taken over by another carrier, which enjoys a reputation for meeting its schedules. The consumer group plans to test the hypothesis that the new owner has made no improvement against the hypothesis that the airline's performance is now comparable to that of the parent company. They want $\alpha = .05$ and $\beta = .20$ and do not want to find a significant difference unless the airline's performance has improved by at least 10 min. How large a sample should the consumer group collect if the variance has not changed?

14. The population mean for white female college students on the MACH IV (see Exercise 10.3.10) is known to be 87.66, and the population standard deviation is known to be 13.45. (Geis, ibid.). The experimenter in Exercise 10.3.10 has two competing hypotheses concerning *nonwhite* college females:
    (1) The mean for *nonwhite* college females equals the mean for *white* college females.
    (2) The mean for nonwhite college *females* equals the mean for nonwhite college *males* (estimated to be 97.25).
    If variance for nonwhite females is the same as for white females and if the experimenter wants $\alpha = \beta = .01$, how many nonwhite females should be tested?

In Exercises 15 and 16, find the smallest sample size needed to claim significance at the level indicated for each test. (The notation ? under the heading $\phi(X)$ indicates that the distribution of X is unknown.)

15.

	$\phi(X)$	$\bar{x}$	σ^2	s^2	μ_0	$H_1 : \mu$	α
(a)	Normal	65	?	100	69	< 69	.05
(b)	?	21.0	?	670	33.33	< 33.33	.01
(c)	Normal	1,475	25,000	—	1,300	$> 1,300$	.005
(d)	Normal	28.5	?	2,055	45	$\neq 45$	.05

16.

	$\phi(X)$	$\bar{x}$	σ^2	s^2	μ_0	$H_1 : \mu$	α
(a)	?	85	?	1,383	100	< 100	.025
(b)	Normal	5.57	?	2.19	3.33	> 3.33	.005
(c)	Normal	87.66	181	—	93.69	< 93.69	.001
(d)	Normal	80	?	122	75	$\neq 75$	.05

17. Burrowing animals contribute to the nutrient supply of farmland soil by transporting plant matter underground to build nests, and the amount of material transported varies from crop to crop. On one test site where rye is the principal crop, common voles built 11 nests in a 1-year period and the average weight of each nest was 27 g. In alfalfa fields, nest weight is known to average 405 g. If nest weight is normally distributed, what variance s^2 would be required to reject the hypothesis that nests in rye fields weigh the same as nests in alfalfa fields in favor of the hypothesis that rye nests weigh less ($\alpha = .01$)? [*Source*: Ryszkowski, R. (1982). Structure and function of the mammal community in an agricultural landscape. *Acta Zoologica Fennica*, *169*, 45–59.]

18. The experimenter in Exercise 10.3.12 predicts that players low in Machiavellianism will earn fewer points than expected on the basis of random chance (33 1/3). For 22 players with low MACH IV scores, the average game score was 21.58 points, which yields $t = -5.58$. (From Geis, ibid.)
    (a) What was the variance for this group of scores?
    (b) If $\alpha = .005$, what sort of error might the experimenter have made?
    (c) What is the probability that the experimenter made this type of error?

C. TESTING HYPOTHESES ABOUT TWO POPULATION MEANS μ_1 AND μ_2

In all of the examples discussed in the preceding section, an experimenter tested the hypothesis that the mean μ of the population from which a sample was drawn was equal to the mean of a population for which μ (and possibly σ^2) was known, i.e.,

$$H_0: \mu = \mu_0$$

where μ_0 is a specified value.

In many cases, however, the experimenter does not have the luxury of testing sample data against a *known* population value. It was only by rare good fortune, for example, that the experimenters in Example 10.1 had access to large pool of data from which they could estimate the mean of the dog population. As we shall see below, comparison of wolf and dog performance on other tests required that we collect data on dogs as well as wolves. In these situations we were not especially concerned with the value of μ in *either* the wolf *or* the dog population, but only with whether or not they were different (in the expected direction!).

A concern with the *difference* between two population means, rather than their values per se, is typical of research designs in which the investigator manipulates an experimental variable and conducts a statistical test to determine its effects. In medical research, for example, one group of subjects (the control group) may be given an orthodox treatment and another group given an experimental treatment. The researcher's principal interest is the *difference* (in length of posttreatment survival, number of days to recovery, number of kilograms of weight lost, etc.) between the experimental population and the control population.

1. Formulation

a. Hypotheses. When an experimenter predicts that the mean of some measurement will be different for two populations, samples are drawn from *both* populations, and the experimenter tests the *null* hypothesis, the hypothesis that the parent populations have equal means:

$$H_0: \mu_1 - \mu_2 = 0 \quad (\text{or, } \mu_1 = \mu_2)$$

where μ_1 is the mean of X_1, the random variable X in Population 1, and μ_2 is the mean of X_2, the random variable X in Population 2.

In tests of two population means the alternative hypothesis may be directional

$$H_1: \mu_1 - \mu_2 > 0 \quad (\text{or, } \mu_1 > \mu_2)$$

or

$$H_1: \mu_1 - \mu_2 < 0 \quad (\text{or, } \mu_1 < \mu_2)$$

or it may be nondirectional

$$H_1: \mu_1 - \mu_2 \neq 0 \quad (\text{or, } \mu_1 \neq \mu_2)$$

Example 10.10.1. One of the tasks on which the author and his co-workers compared wolf and dog performance was to learn the concept of "oddity." On each trial a pup was presented three wooden blocks. Sometimes two blocks were white and one was black; sometimes two were black and one was white. Every time a pup chose the "odd" block, it was rewarded with food. Each pup received 15 trials a day until it reached a criterion of 13 correct choices out of 15 trials. The score recorded for each pup was total trials to criterion. Our theory predicted that on this type of task, dogs should perform better than wolves, that is, dogs were expected to require *fewer* trials. But, we had no prior data on which to base a numerical prediction of the average score in either group. Therefore, our hypotheses were

$$H_0: \mu_D - \mu_W = 0$$

$$H_1: \mu_D - \mu_W < 0$$

b. Model: The distribution of a difference $X_1 - X_2$. All of the hypotheses described above are formulated in terms of the difference $\mu_1 - \mu_2$. Accordingly, we will soon see that the test statistics we use to test these hypotheses revolve around the difference $\bar{X}_1 - \bar{X}_2$, where $\bar{X}_1$ is the mean of the sample drawn from Population 1, and $\bar{X}_2$ is the mean of the sample drawn from Population 2. Before we begin we must therefore consider some important statistical properties of the difference between two random variables, that is, properties of the random variable Y, where $Y = X_1 - X_2$.

Lemma 10.1. If X_1 and X_2 are independent, normally distributed random variables, then $Y = X_1 - X_2$ is also normally distributed. The proof of this lemma requires mathematics beyond the scope of this book.

Lemma 10.2. If X_1 and X_2 are random variables with $E(X_1) = \mu_1$ and $E(X_2) = \mu_2$, then $E(X_1 - X_2) = \mu_1 - \mu_2$.

This follows directly from the Algebra of Expectations:

$$X_1 - X_2 = X_1 + (-1)X_2$$

Therefore,

$$
\begin{aligned}
E(X_1 - X_2) &= E\big[X_1 + (-1)X_2\big] \\
&= E(X_1) + (-1)E(X_2) \\
&= \mu_1 - \mu_2
\end{aligned}
$$

Lemma 10.3. If X_1 and X_2 are independent random variables with $V(X_1) = \sigma_1^2$ and $V(X_2) = \sigma_2^2$, then $V(X_1 - X_2) = \sigma_1^2 + \sigma_2^2$.

Since X_1 and X_2 are independent,

$$
\begin{aligned}
V(X_1 - X_2) &= V(X_1) + V\big[(-1)X_2\big] \\
&= V(X_1) + (-1)^2 V(X_2) \\
&= \sigma_1^2 + \sigma_2^2
\end{aligned}
$$

2. Decisions

In testing hypotheses about one population mean, we know that the choice of a test statistic is driven by (1) whether or not the random variable X is normally distributed and whether or not the variance of X is known and (2) the assumptions that can be justified on the basis of sample size N. In testing hypotheses about two population means, the experimenter has two random variables X_1 and X_2 and draws samples of N_1 observations on X_1 and N_2 observations on X_2. The strategy for selecting an appropriate test statistic therefore depends on (1) the distributions $\phi(X_1)$ and $\phi(X_2)$ and variances σ_1^2 and σ_2^2 of the two variables and (2) the assumptions that can be justified on the basis of sample sizes N_1 and N_2. In Box 10.4 the questions an experimenter must ask are again laid out in the form of a decision tree.

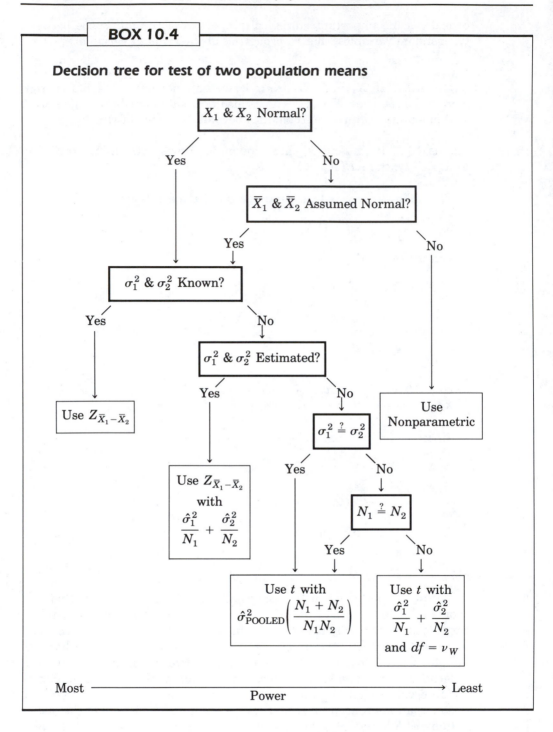

BOX 10.4

Decision tree for test of two population means

X_1 & X_2 Normal?

Yes

No

$\bar{X}_1$ & $\bar{X}_2$ Assumed Normal?

Yes

No

σ_1^2 & σ_2^2 Known?

Yes

No

Yes

Use $Z_{\bar{X}_1 - \bar{X}_2}$

σ_1^2 & σ_2^2 Estimated?

Yes

No

Use
Nonparametric

$\sigma_1^2 \overset{?}{=} \sigma_2^2$

Use $Z_{\bar{X}_1 - \bar{X}_2}$
with
$$\frac{\hat{\sigma}_1^2}{N_1} + \frac{\hat{\sigma}_2^2}{N_2}$$

Yes

No

$N_1 \overset{?}{=} N_2$

Yes

No

Use t with
$$\hat{\sigma}_{\text{POOLED}}^2 \left(\frac{N_1 + N_2}{N_1 N_2} \right)$$

Use t with
$$\frac{\hat{\sigma}_1^2}{N_1} + \frac{\hat{\sigma}_2^2}{N_2}$$
and $df = \nu_W$

Most ⟶ Least

Power

The first branch of the tree in Box 10.4 represents the ideal situation in which the two random variables are both known to be normally distributed and their variances σ_1^2 and σ_2^2 are known. This case practically never arises, because the values of μ_1 and μ_2 must be known in order to calculate σ_1^2 and σ_2^2. And, if μ_1 and μ_2 are known, one already knows whether or not they are equal. Nevertheless, this ideal case is logically possible and, like the Neanderthal problem discussed earlier, allows us to develop the mathematical bases for selecting an appropriate test statistic.

Let us suppose that X_1 is a normally distributed random variable with expected value μ_1 and variance σ_1^2. For samples of N_1 observations, we know that the sample means will be distributed normally with an expected value of $\mu_{\bar{X}_1} = \mu_1$ and a variance $\sigma_{\bar{X}_1}^2 = \sigma_1^2/N_1$. That is,

$$\bar{X}_1 : N\left(\mu_1, \sigma_1^2/N_1\right) \tag{10.14}$$

Likewise, if X_2 is a normally distributed random variable with expected value μ_2 and variance σ_2^2, then the sampling distribution of means for samples of N_2 observations will be normal with an expected value $\mu_{\bar{X}_2} = \mu_2$ and a variance $\sigma_{\bar{X}_2}^2 = \sigma_2^2/N_2$. That is,

$$\bar{X}_2 : N\left(\mu_2, \sigma_2^2/N_2\right) \tag{10.15}$$

Now, consider the random variable defined by taking the *difference* between the two sample means, $\bar{X}_1 - \bar{X}_2$. If $\bar{X}_1$ and $\bar{X}_2$ are independent of one another, then from Lemma 10.1 we know that the difference $\bar{X}_1 - \bar{X}_2$ is normally distributed. From Lemma 10.2 we know that

$$E\left(\bar{X}_1 - \bar{X}_2\right) = \mu_{\bar{X}_1} - \mu_{\bar{X}_2}$$

which from [10.14] and [10.15] becomes

$$E\left(\bar{X}_1 - \bar{X}_2\right) = \mu_1 - \mu_2 \tag{10.16}$$

From Lemma 10.3 we know that

$$V\left(\bar{X}_1 - \bar{X}_2\right) = \sigma_{\bar{X}_1}^2 + \sigma_{\bar{X}_2}^2$$

which from [10.14] and [10.15] becomes

$$V\left(\bar{X}_1 - \bar{X}_2\right) = \frac{\sigma_1^2}{N_1} + \frac{\sigma_2^2}{N_1} \tag{10.17}$$

Therefore,

$$\left(\bar{X}_1 - \bar{X}_2\right) : N\left(\mu_1 - \mu_2, \frac{\sigma_1^2}{N_1} + \frac{\sigma_2^2}{N_2}\right) \tag{10.18}$$

Expression **[10.18]** tells us that the *difference* between means of independent samples of observations on two normally distributed random variables is itself a normally distributed random variable with expected value equal to $\mu_1 - \mu_2$ and variance equal to $(\sigma_1^2/N_1) + (\sigma_2^2/N_2)$.

If we standardize the random variable $\bar{X}_1 - \bar{X}_2$, we obtain

$$Z_{\bar{X}_1 - \bar{X}_2} = \frac{(\bar{X}_1 - \bar{X}_2) - \mu_{\bar{X}_1 - \bar{X}_2}}{\sigma_{\bar{X}_1 - \bar{X}_2}} = \frac{(\bar{X}_1 - \bar{X}_2) - \mu_{\bar{X}_1 - \bar{X}_2}}{\sqrt{\sigma_{\bar{X}_1 - \bar{X}_2}^2}} \qquad [10.19]$$

$$= \frac{(\bar{X}_1 - \bar{X}_2) - (\mu_1 - \mu_2)}{\sqrt{\dfrac{\sigma_1^2}{N_1} + \dfrac{\sigma_2^2}{N_2}}}$$

which is distributed $N(0, 1)$.

If $\mu_1 = \mu_2$, the second term in the numerator of **[10.19]** is equal to zero. Therefore, if the test hypothesis in our ideal case is correct, then

$$Z_{\bar{X}_1 - \bar{X}_2} = \frac{\bar{X}_1 - \bar{X}_2}{\sigma_{\bar{X}_1 - \bar{X}_2}} = \frac{\bar{X}_1 - \bar{X}_2}{\sqrt{\dfrac{\sigma_1^2}{N_1} + \dfrac{\sigma_2^2}{N_2}}} \qquad [10.20]$$

is normally distributed with an expected value of 0 and a variance of 1. In the general case, the expected value of $Z_{\bar{X}_1 - \bar{X}_2}$ is equal to

$$E\left(\frac{\bar{X}_1 - \bar{X}_2}{\sqrt{\dfrac{\sigma_1^2}{N_1} + \dfrac{\sigma_2^2}{N_2}}}\right) = \frac{E(\bar{X}_1 - \bar{X}_2)}{\sqrt{\dfrac{\sigma_1^2}{N_1} + \dfrac{\sigma_2^2}{N_2}}} = \frac{\mu_1 - \mu_2}{\sqrt{\dfrac{\sigma_1^2}{N_1} + \dfrac{\sigma_2^2}{N_2}}}$$

and will therefore be greater than zero if $\mu_1 > \mu_2$ and will be less than zero if $\mu_1 < \mu_2$. For the case in which X_1 and X_2 are both normally distributed and σ_1^2 and σ_2^2 are known, **[10.20]** satisfies our requirements for a suitable test statistic.

If X_1 and X_2 are independent, normally distributed random variables and if $V(X_1) = \sigma_1^2$ and $V(X_2) = \sigma_2^2$, then for samples of N_1 observations on X_1 and N_2 observations on X_2, the appropriate statistic to test the hypothesis that $\mu_1 = \mu_2$ is

$$Z_{\bar{X}_1 - \bar{X}_2} = \frac{\bar{X}_1 - \bar{X}_2}{\sqrt{\dfrac{\sigma_1^2}{N_1} + \dfrac{\sigma_2^2}{N_2}}}$$

which is distributed $N(0, 1)$.

The rejection region for this statistic will lie in the upper tail of the standard normal distribution if the alternative hypothesis specifies that $(\mu_1 - \mu_2) > 0$ (i.e., $\mu_1 > \mu_2$), the lower tail if the alternative hypothesis specifies that $(\mu_1 - \mu_2) < 0$ (i.e., $\mu_1 < \mu_2$), and in both tails if the alternative hypothesis is nondirectional. The value of z_α (or $\pm z_{\alpha/2}$) will, of course, depend on the significance level chosen by the experimenter.

a. Large samples (N_1 and $N_2 \geq 30$). The ideal conditions discussed above (X_1 and X_2 normally distributed and known values of σ_1^2 and σ_2^2) are rare, but the case provides a suitable approximation to the more typical situation, where

- X_1 and X_2 cannot be assumed normally distributed, and
- The variances σ_1^2 and σ_2^2 are unknown, but
- Both samples include 30 or more observations.

In this situation both sample means can be assumed normally distributed under the Central Limit Theorem, and the experimenter can estimate σ_1^2 and σ_2^2. These estimates can be substituted into equation [10.20] to give

$$Z_{\bar{X}_1 - \bar{X}_2} = \frac{\bar{X}_1 - \bar{X}_2}{\sqrt{\dfrac{\hat{\sigma}_1^2}{N_1} + \dfrac{\hat{\sigma}_2^2}{N_2}}} \qquad\qquad [10.21]$$

$$= \frac{\bar{X}_1 - \bar{X}_2}{\sqrt{\hat{\sigma}_{\bar{X}_1 - \bar{X}_2}^2}}$$

This statistic will be approximately distributed $N(0, 1)$ if H_0 is correct, and the experimenter may proceed as described above for the ideal case.

If X_1 and X_2 are independent random variables with unknown variances, then for samples of $N_1 \geq 30$ observations on X_1 and $N_2 \geq 30$ observations on X_2, the appropriate statistic to test the hypothesis that $\mu_1 = \mu_2$ is

$$Z_{\bar{X}_1 - \bar{X}_2} = \frac{\bar{X}_1 - \bar{X}_2}{\sqrt{\dfrac{\hat{\sigma}_1^2}{N_1} + \dfrac{\hat{\sigma}_2^2}{N_2}}}$$

which is distributed approximately $N(0, 1)$.

———————— EXERCISES 10.4 ————————————————————————————

1. The random variable X_1 can take the values 15 and 29 and is uniformly distributed, i.e., $P(15) = P(29) = .50$. The random variable X_2 can take the values 4 and 8 and is likewise uniformly distributed.
   (a) Calculate $E(X_1)$ and $E(X_2)$.
   (b) Calculate $V(X_1)$ and $V(X_2)$.

   Let Y be $X_1 - X_2$. Calculate every possible y-value, and fill in the appropriate blanks with these y-values.

X_1:	15	29
X_2 : 4	_____	_____
8	_____	_____

   (c) Calculate the mean of the y-values entered in the table above and express this mean in terms of your answers to (a).
   (d) Calculate the variance of the y-values entered in the table above and express this variance in terms of your answers to (b).

2. (a) List all possible results of the experiment "make two observations on X_1 and record the values *in order*," where X_1 is defined in Exercise 10.4.1 above. Calculate the mean $\bar{x}_1$ of every possible sample and fill in the values of $\bar{X}_1$ in the top row of the table below.
   (b) List all possible results of the experiment "make two observations on X_2 and record the values *in order*," where X_2 is defined in Exercise 10.4.1 above. Calculate the mean $\bar{x}_2$ of every possible sample and fill in the values of $\bar{X}_2$ in the left margin of the table below.

   $\bar{X}_1$:

$\bar{X}_2$: _____	_____	_____	_____	_____
_____	_____	_____	_____	_____
_____	_____	_____	_____	_____
_____	_____	_____	_____	_____

   (c) Let $Y = \bar{X}_1 - \bar{X}_2$. Calculate every possible y-value, and fill in the appropriate blanks with these y-values.
   (d) Calculate the mean of the y-values entered in the table above and express this mean in terms of your answers to 10.4.1(a).
   (e) Calculate the variance of the y-values entered in the table above and express this variance in terms of your answers to 10.4.1(b).

In Exercises 3 and 4 the random variable X_1 is distributed $N(\mu_1, \sigma_1^2)$ and the random variable X_2 is distributed $N(\mu_2, \sigma_2^2)$. Find the probabilities indicated in the first column.

3.

	$P(x_1 - x_2)$	μ_1	σ_1^2	μ_2	σ_2^2
(a)	≥ 36.93	100	225	85	256
(b)	≤ -15.85	18.5	15.6	23.4	28.9
(c)	≥ 23	25	100	26	50
(d)	≥ 27.72 *or*	500	100	500	100
	≤ -27.72				
(e)	≤ 0	5.6×10^{-4}	π	5.6×10^{-4}	e

4.

	$P(x_1 - x_2)$	μ_1	σ_1^2	μ_2	σ_2^2
(a)	≥ 3.41	69	2.25	67	.5625
(b)	≥ 16.4	50	9	40	16
(c)	≤ -8.78	60	1.00	63	2.50
(d)	≤ -104.4	150	2,500	135	2,800
(e)	$\geq .595$	.75	.009375	.50	.0125

In Exercises 5 and 6 the random variable X_1 is distributed $N(\mu_1, \sigma_1^2)$ and the random variable X_2 is distributed $N(\mu_2, \sigma_2^2)$. Let N_1 observations be made on X_1 and N_2 observations be made on X_2. Find the probabilities indicated in the first column.

5.

	$P(\bar{x}_1 - \bar{x}_2)$	μ_1	σ_1^2	N_1	μ_2	σ_2^2	N_2
(a)	≥ 15.28	50	500	5	40	352	8
(b)	≤ -73	85	2,800	20	115	896	16
(c)	≤ -1.46	.33	39	300	.67	60	500
(d)	≤ 13	85	241	30	75	541	45
(e)	≤ -10.24	-5.5	15.5	5	1.05	11.75	6

6.

	$P(\bar{x}_1 - \bar{x}_2)$	μ_1	σ_1^2	N_1	μ_2	σ_2^2	N_2
(a)	≤ -319	1,200	2,050	55	1,500	2,350	40
(b)	$\geq .2238$	.80	.16	1,000	.60	.24	500
(c)	≥ -10	55	624	15	70	1017	20
(d)	≤ 17.5	45	234	288	25	310	205
(e)	≥ 13.24 or ≤ -13.24	100	361	15	100	225	100

In Exercises 7 and 8 the random variables X_1 and X_2 are known to be normally distributed. Test the hypothesis that $\mu_1 = \mu_2$ against the alternative hypothesis indicated. State the appropriate rejection rule, calculate the value of the test statistic, report the p-value, and state the appropriate conclusion.

7.

	$\bar{x}_1$	$\bar{x}_2$	σ_1^2	N_1	σ_2^2	N_2	H_1: $\mu_1 - \mu_2$	α
(a)	21.9	15.1	25	5	110	10	> 0	.05
(b)	171.95	178.50	100	35	150	30	< 0	.01
(c)	98.0	108.6	225	20	256	15	$\neq 0$	.05
(d)	12.5	14.0	45	300	20	200	< 0	.001
(e)	47	48.4	$41\frac{2}{3}$	300	$112\frac{1}{2}$	600	$\neq 0$	.01

8.

	$\bar{x}_1$	$\bar{x}_2$	σ_1^2	N_1	σ_2^2	N_2	H_1: $\mu_1 - \mu_2$	α
(a)	103.25	96.5	225	50	169	45	> 0	.01
(b)	142	145	190	95	220	115	< 0	.05
(c)	93.25	94.75	130.0	750	118.5	680	< 0	.005
(d)	5.6	72.45	7.74	20	14750	20	$\neq 0$	.01
(e)	.3625	.3721	.000065	12	.000055	9	$\neq 0$	.005

9. A university statistics instructor believes that transfer students from the regional agricultural college have better preparation in mathematics than transfer students from the local junior college. She administers the university's mathematics placement examination to all students enrolled in the statistics course and obtains the following results:

	Freshman and Sophomore Years Completed at	
	Agricultural College	**Junior College**
$\bar{x}$	33.1	28.5
N	29	26

From the records of past placement examinations it is known that variance for all transfer students from the agricultural college is 120 and that the variance for all junior college transfer students is 150. Placement test scores are approximately normally distributed.
(a) What are the hypotheses?
(b) What is the appropriate test statistic?
(c) If $\alpha = .05$, what is the rejection rule?
(d) Calculate the value of the test statistic.
(e) What is the appropriate conclusion?
(f) What is the p-value of the result?
(g) What is the smallest α that allows rejection of H_0?

10. A dairy farmer introduces two lines of sires into the gene pool of a dairy herd in order to determine whether the offspring of one line produce more milk than the offspring of the other. The herd standard deviation in annual (305-day) milk yield is 2,500 lbs, and the line breeding does not affect variability. The average annual yield (in pounds) for the first generation of offspring is as follows:

	Line 1	Line 2
$\bar{x}$	14,962	16,078
N	37	42

(a) What are the hypotheses?
(b) What is the appropriate test statistic?
(c) Set α at any level you consider suitable and state the rejection rule appropriate to your hypotheses and the significance level you have chosen.
(d) Calculate the value of your test statistic.
(e) Report the appropriate conclusion.

11. Through the 1960s and 1970s college administrators noted a nationwide decline in scores on the College Entrance Examination Board's Scholastic Aptitude Test (SAT). To test the hypothesis that trends at the University of Michigan–Flint were consistent with trends elsewhere, researchers compared SAT scores of 400 applicants to the university in 1964 with scores for 275 applicants in 1978 (Source: University of Michigan–Flint Office of Admissions):

$$\bar{x}_{1964} = 822; \ s^2_{1964} = 12{,}544 \quad (N_{1964} = 400)$$

$$\bar{x}_{1978} = 800; \ s^2_{1978} = 10{,}609 \quad (N_{1978} = 275)$$

(a) What are the hypotheses?
(b) What is the appropriate test statistic?
(c) Set α at any level you consider suitable and state the rejection rule appropriate to your hypotheses and the significance level you have chosen.
(d) Calculate the value of your test statistic.
(e) Report the appropriate conclusion.

12. In order to compare the effectiveness of two laboratory curricula, half of the 100 students enrolled in an introductory psychology course are assigned at random to Laboratory Section A, which uses one curriculum, and half to Laboratory Section B, which uses the other. All of the students are administered the same laboratory final examination. Students in Section A averaged 65 points with a standard deviation of 8 points. Students in Section B averaged 70 points with a standard deviation of 10 points.

(a) What are the researcher's hypotheses?
(b) What is the appropriate test statistic?
(c) Calculate the value of the test statistic.
(d) If the experimenter sets $\alpha = .01$, what is the appropriate conclusion?
(e) What is the smallest value of α that will permit rejection of the test hypothesis?

13. A large sample of college students was administered the D-Scale, a test designed to measure resistance to change in systems of belief (dogmatism). A social psychologist suspects that religious affiliation is related to such resistance. Use the following data to address this question.

	Religious Affiliation	
Dogmatism	**Catholics**	**Jews**
Mean	147.4	139.5
Standard deviation	30.0	24.5
Sample size	46	131

Source: Rokeach, M. (1960). *The Open and Closed Mind*. New York: Basic Books, p. 112.

(a) What are the hypotheses?
(b) What is the appropriate test statistic?
(c) Set α at any level you consider suitable and state the rejection rule appropriate to your hypotheses and the significance level you have chosen.
(d) Calculate the value of your test statistic.
(e) Report the appropriate conclusion.

14. University students in England were administered the California F-Scale, a test that measures authoritarian, or antidemocratic, personality traits. A social psychologist suspects that students who identify themselves with the Communist Party or the Liberal Party are less authoritarian than those who identify themselves with the Conservative Party. Use the following data to address this question.

Authoritarianism	Political party preference	
	Liberals & Communists	Conservatives
Mean	92.6	115.5
Standard deviation	16.6	25.0
Sample size	35	54

Source: Rokeach, M. (1960). *The Open and Closed Mind.* New York: Basic Books, p. 114.

(a) What are the hypotheses?
(b) What is the appropriate test statistic?
(c) Set α at any level you consider suitable and state the rejection rule appropriate to your hypotheses and the significance level you have chosen.
(d) Calculate the value of your test statistic.
(e) Report the appropriate conclusion.

b. Small samples (N_1 or $N_2 < 30$)

As before, there are three possibilities:

(i) X_1 and X_2 are *normally distributed and σ_1^2 and σ_2^2 are known.* If the variances are known or can be estimated from other, larger samples of data, one can proceed with the test statistic given in [10.20] or [10.21], since $\overline{X}_1$ and $\overline{X}_2$ can be assumed to be normally distributed under the Central Limit Theorem.

(ii) X_1 or X_2 *cannot be assumed normally distributed.* As before, this requires that the experimenter use a nonparametric test.

(iii) X_1 and X_2 are *normally distributed but σ_1^2 or σ_2^2 is unknown* (and cannot be estimated from some other source of data). In the first half of this chapter, we learned that when σ^2 is known, the statistic of choice for testing hypotheses about one population mean is

$$\frac{\overline{X} - \mu_0}{\sqrt{\dfrac{\sigma^2}{N}}} = \frac{\overline{X} - \mu_0}{\sqrt{\sigma_{\overline{X}}^2}}$$

which is normally distributed. We also learned that when σ^2 is not known, the appropriate test statistic is distributed as Student's t with

$N - 1$ degrees of freedom and can be written in the form

$$\frac{\overline{X} - \mu_0}{\sqrt{\dfrac{\hat{\sigma}^2}{N}}} = \frac{\overline{X} - \mu_0}{\sqrt{\hat{\sigma}_{\overline{X}}^2}}$$

Thus far in our discussion of hypotheses about two population means we have established that if σ_1^2 and σ_2^2 are both known, the test statistic is once again normally distributed:

$$\frac{\overline{X}_1 - \overline{X}_2}{\sqrt{\dfrac{\sigma_1^2}{N_1} + \dfrac{\sigma_2^2}{N_2}}} = \frac{\overline{X}_1 - \overline{X}_2}{\sqrt{\sigma_{\overline{X}_1 - \overline{X}_2}^2}}$$

As one might expect, the appropriate statistic for situations in which σ_1^2 and σ_2^2 are unknown is distributed as Student's t. Furthermore, the test statistic is a quotient in which the numerator is the difference between sample means $\overline{X}_1 - \overline{X}_2$ and the denominator is an expression that can be used to estimate the variance of $\overline{X}_1 - \overline{X}_2$. The situation is not quite as straightforward as in the case of one mean, however, because estimation of

$$\sigma_{\overline{X}_1 - \overline{X}_2}^2$$

depends on whether or not σ_1^2 and σ_2^2 can be assumed to be equal.

Equal variances. The pooled estimate of σ^2. If $\sigma_1^2 = \sigma_2^2 = \sigma^2$, then from [10.17] the variance of $\overline{X}_1 - \overline{X}_2$ is

$$\sigma_{\overline{X}_1 - \overline{X}_2}^2 = \frac{\sigma^2}{N_1} + \frac{\sigma^2}{N_2} = \sigma^2 \left(\frac{1}{N_1} + \frac{1}{N_2} \right) = \sigma^2 \left(\frac{N_1 + N_2}{N_1 N_2} \right)$$

Similarly, if it is *assumed* that $\sigma_1^2 = \sigma_2^2 = \sigma^2$, then the *estimated* variance of $\overline{X}_1 - \overline{X}_2$ is given by the equation

$$\hat{\sigma}_{\overline{X}_1 - \overline{X}_2}^2 = \hat{\sigma}^2 \left(\frac{N_1 + N_2}{N_1 N_2} \right) \qquad\qquad \textbf{[10.22]}$$

where $\hat{\sigma}^2$ is an estimate of the common population variance, σ^2.

When σ_1^2 and σ_2^2 are assumed to be equal, $\hat{\sigma}_1^2$ and $\hat{\sigma}_2^2$ are estimates of the same value, and either might in principle be substituted for $\hat{\sigma}^2$ in [10.22], but it can be shown that a better estimate is obtained if $\hat{\sigma}_1^2$ and $\hat{\sigma}_2^2$ are *pooled*. The pooled estimate of the variance is denoted $\hat{\sigma}_{\text{POOLED}}^2$, and in Box 10.5 we show that with two independent samples, the pooled

---

BOX 10.5

Pooled estimates and the pooled estimate of σ^2

A *pooled* estimate is one that combines estimates from two or more samples. If samples of size N_1 and N_2 are drawn from the same population, both sample means, $\bar{x}_1$ and $\bar{x}_2$, are estimates of the population mean μ. However, since the sample mean is a consistent estimator, the mean $\bar{x}$ of all $N_1 + N_2 = N$ observations is probably more accurate. The value of $\bar{x}$ is obtained by pooling the two sample means. The sum of observations in sample 1 is $N_1\bar{x}_1$, and the sum of observations in sample 2 is $N_2\bar{x}_2$. Therefore

$$\bar{x} = \frac{\sum\limits^{N} x}{N} = \frac{N_1\bar{x}_1 + N_2\bar{x}_2}{N_1 + N_2}$$

In this example, each mean $\bar{x}_j$ is *weighted*, or multiplied, by its sample size N_j and the sum of the weighted means is divided by the sum of the weights. Because calculation of a sample mean does not involve estimation of any other parameter, the sample size N_j is actually the *degrees of freedom* for the sample mean $\bar{x}_j$. In the general case of pooling, each estimate is weighted by its degrees of freedom, and the sum of the weighted estimates is divided by the total degrees of freedom.

The same principles apply to calculating a pooled estimate of the variance σ^2. However, in calculating $\hat{\sigma}^2$ one uses the sample mean $\bar{x}$ to estimate the population mean μ. Consequently, $\hat{\sigma}_1^2$ has $N_1 - 1$ degrees of freedom, and $\hat{\sigma}_2^2$ has $N_2 - 1$ degrees of freedom. If one has two independent estimates of σ^2 the pooled estimate of variance is therefore

$$\hat{\sigma}^2_{\text{POOLED}} = \frac{(N_1 - 1)\hat{\sigma}_1^2 + (N_2 - 1)\hat{\sigma}_2^2}{N_1 + N_2 - 2}$$

The formulae for $\hat{\sigma}_1^2$ and $\hat{\sigma}_2^2$ can be substituted into this equation to obtain a computational formula for $\hat{\sigma}^2_{\text{POOLED}}$:

$$\hat{\sigma}^2_{\text{POOLED}} = \frac{(N_1 - 1)\dfrac{\sum\limits^{N_1}(x - \bar{x}_1)^2}{N_1 - 1} + (N_2 - 1)\dfrac{\sum\limits^{N_2}(x - \bar{x}_2)^2}{N_2 - 1}}{N_1 + N_2 - 2}$$

$$= \frac{\sum\limits^{N_1}(x - \bar{x}_1)^2 + \sum\limits^{N_2}(x - \bar{x}_2)^2}{N_1 + N_2 - 2}$$

With $J > 2$ independent estimates of σ^2, the pooled estimate of variance generalizes to

$$\frac{(N_1 - 1)\hat{\sigma}_1^2 + (N_2 - 1)\hat{\sigma}_2^2 + \cdots + (N_J - 1)\hat{\sigma}_J^2}{N_1 + N_2 + \cdots + N_J - J}$$

estimate of the variance may be calculated from $\hat{\sigma}_1^2$ and $\hat{\sigma}_2^2$

$$\hat{\sigma}_{\text{POOLED}}^2 = \frac{(N_1 - 1)\hat{\sigma}_1^2 + (N_2 - 1)\hat{\sigma}_2^2}{N_1 + N_2 - 2} \qquad [10.23]$$

or by the computational formula

$$\hat{\sigma}_{\text{POOLED}}^2 = \frac{\sum\limits^{N_1}(x - \bar{x}_1)^2 + \sum\limits^{N_2}(x - \bar{x}_2)^2}{N_1 + N_2 - 2} \qquad [10.24]$$

When an estimate of the common variance is obtained by pooling, the estimated variance of $\bar{X}_1 - \bar{X}_2$ in [10.22] becomes

$$\hat{\sigma}_{\bar{X}_1 - \bar{X}_2}^2 = \hat{\sigma}_{\text{POOLED}}^2 \left(\frac{N_1 + N_2}{N_1 N_2} \right)$$

and, as we hinted at the beginning of this discussion,

$$\frac{\bar{X}_1 - \bar{X}_2}{\sqrt{\hat{\sigma}_{\text{POOLED}}^2 \left(\dfrac{N_1 + N_2}{N_1 N_2} \right)}} \qquad [10.25]$$

is distributed as Student's t and may be used to test the hypothesis that $\mu_1 = \mu_2$.

> If X_1 and X_2 are independent, normally distributed random variables with unknown variances that are assumed to be equal, the appropriate statistic to test the hypothesis that $\mu_1 = \mu_2$ is
>
> $$t = \frac{\bar{X}_1 - \bar{X}_2}{\sqrt{\hat{\sigma}_{\text{POOLED}}^2 \left(\dfrac{N_1 + N_2}{N_1 N_2} \right)}}$$
>
> which is distributed as $t(N_1 + N_2 - 2)$.

Example 10.10.2. To test the hypothesis formulated in Example 10.10.1 (p. 418), the oddity test was administered to 4 timber wolf pups and 4 dog (Alaskan Malamute) pups. Each pup's score was a sum based on at least 45 trials, each of which was scored 0 for an error and 1 for a correct choice. Under the Central Limit Theorem the random variables X_D and X_W were therefore assumed to be normally distributed. Assuming that the wolf and dog populations have the same variance, our test statistic is

$$\frac{\overline{X}_D - \overline{X}_W}{\sqrt{\hat{\sigma}^2_{\text{POOLED}}\left(\dfrac{4 + 4}{16}\right)}}$$

and is distributed as Student's t with $(4 + 4 - 2) = 6$ degrees of freedom. For the dogs, $\bar{x} = 93.75$ and $\hat{\sigma}^2 = 206.25$. For the wolf pups, $\bar{x} = 82.5$ and $\hat{\sigma}^2 = 2{,}775.0$. By equation [**10.23**],

$$\hat{\sigma}^2_{\text{POOLED}} = \frac{(3)(206.25) + (3)(2{,}775.0)}{6} = 1{,}490.625$$

Therefore, the value of our test statistic is

$$\frac{\bar{x}_D - \bar{x}_W}{\sqrt{\hat{\sigma}^2_{\text{POOLED}}\left(\dfrac{4 + 4}{16}\right)}} = \frac{93.75 - 82.5}{\sqrt{1{,}490.625\left(\dfrac{4 + 4}{16}\right)}} \doteq .412$$

If $N_1 = N_2$ it is easily demonstrated that

$$\hat{\sigma}^2_{\text{POOLED}}\left(\frac{N_1 + N_2}{N_1 N_2}\right) = \frac{\hat{\sigma}^2_1}{N_1} + \frac{\hat{\sigma}^2_2}{N_2}$$

so equation [**10.25**] can be rewritten,

$$t = \frac{\overline{X}_1 - \overline{X}_2}{\sqrt{\dfrac{\hat{\sigma}^2_1}{N_1} + \dfrac{\hat{\sigma}^2_2}{N_2}}} \qquad\qquad [\textbf{10.26}]$$

Since many calculators compute means and variance estimates in the same operation, [**10.26**] may be easier to calculate than [**10.25**].

Unequal variances. We shall see in later chapters that the assumption of equal variances is a requirement in many statistical procedures. Fortunately, the assumption can often be violated with relatively little risk of error if sample sizes are equal. When $N_1 = N_2$, the statistic given above in equation **[10.26]** remains distributed approximately as $t(N_1 + N_2 - 2)$ even if $\sigma_1^2 \neq \sigma_2^2$.

If X_1 and X_2 are independent, normally distributed random variables with unknown variances that cannot be assumed to be equal and if $N_1 = N_2$, the statistic

$$t = \frac{\overline{X}_1 - \overline{X}_2}{\sqrt{\dfrac{\hat{\sigma}_1^2}{N_1} + \dfrac{\hat{\sigma}_2^2}{N_2}}}$$

is distributed approximately as $t(N_1 + N_2 - 2)$ and may be used to test the hypothesis that $\mu_1 = \mu_2$.

Example 10.10.3. In the preceding example, the test statistic was calculated as if wolf and dog population variances were equal. The variance estimate for wolves ($\hat{\sigma}^2 = 2,775.0$) was more than 10 *times* the variance estimate for dogs ($\hat{\sigma}^2 = 206.25$), and the assumption of equal variances was therefore difficult to support, but the risk of error was minimized because sample sizes were equal.[10] It might also be noted that since $N_D = N_W$, we could have used equation **[10.26]** instead of **[10.25]** and obtained the same result:

$$t = \frac{\overline{X}_D - \overline{X}_W}{\sqrt{\dfrac{\hat{\sigma}_D^2}{N_D} + \dfrac{\hat{\sigma}_W^2}{N_W}}} \doteq \frac{93.75 - 82.5}{\sqrt{\dfrac{206.25}{4} + \dfrac{2,775.0}{4}}} \doteq \frac{11.25}{\sqrt{745.31}} \doteq .412$$

If variances are not equal and $N_1 \neq N_2$,

$$\frac{\overline{X}_1 - \overline{X}_2}{\sqrt{\dfrac{\hat{\sigma}_1^2}{N_1} + \dfrac{\hat{\sigma}_2^2}{N_2}}}$$

is still distributed approximately as Student's t, but ν is usually less than $N_1 + N_2 - 2$ and cannot be exactly determined. The most conservative approach to estimating ν is to let it equal the *smaller* of $N_1 - 1$ or

---

[10] Some texts suggest that Fisher's F test (developed in Chapter 11) be used to test the hypothesis that the two population variances are equal. For small samples, however, the test for equality of variances is sensitive to departures from normality that are small enough to ignore for purposes of using Student's t. Therefore, the experimenter is better advised to use samples of the same size if there is any doubt about equality of variances.

$N_2 - 1$.[11] It is conservative in that it minimizes the degrees of freedom and, therefore, the power of the test. This increases the likelihood that the alternative hypothesis will be rejected, i.e., that the test hypothesis will be "conserved." A less extreme correction recommended by Welch[12] adjusts degrees of freedom in accordance with differences between $\hat{\sigma}_1^2$ and $\hat{\sigma}_2^2$ and between N_1 and N_2. Degrees of freedom calculated by Welch's method is denoted ν_W in this text.

$$\nu_W = \frac{\left[(\hat{\sigma}_1^2/N_1) + (\hat{\sigma}_2^2/N_2)\right]^2}{\dfrac{\left(\hat{\sigma}_1^2/N_1\right)^2}{N_1 + 1} + \dfrac{\left(\hat{\sigma}_2^2/N_2\right)^2}{N_2 + 1}} - 2 \qquad [\,10.27\,]$$

If X_1 and X_2 are independent, normally distributed random variables with unknown variances that cannot be assumed to be equal and if $N_1 \neq N_2$, the appropriate statistic to test the hypothesis that $\mu_1 = \mu_2$ is

$$t = \frac{\overline{X}_1 - \overline{X}_2}{\sqrt{\dfrac{\hat{\sigma}_1^2}{N_1} + \dfrac{\hat{\sigma}_2^2}{N_2}}}$$

which is distributed approximately as Student's t with

$$\nu_W \cong \frac{\left[(\hat{\sigma}_1^2/N_1) + (\hat{\sigma}_2^2/N_2)\right]^2}{\dfrac{\left(\hat{\sigma}_1^2/N_1\right)^2}{N_1 + 1} + \dfrac{\left(\hat{\sigma}_2^2/N_2\right)^2}{N_2 + 1}} - 2$$

It should be understood that formula [10.27] seldom yields a whole number, and ν_W must therefore be rounded. The most conservative practice is to set *df* equal to the nearest integer *smaller* than ν_W. In addition, it sometimes happens that Welch's correction yields a value that is *larger* than $N_1 + N_2 - 2$, in which case many experimeters set $df = N_1 + N_2 - 2$.

In establishing a rejection rule for a t test of two population means, the experimenter must always use the t distribution appropriate to the test statistic. If sample sizes are equal or if population variances can otherwise be assumed equal, the distribution is $t(N_1 + N_2 - 2)$. If variances cannot be assumed equal and sample sizes are not equal, the distribution is $t(\nu)$, where ν depends on how conservatively one corrects

---

[11]When ν is determined in this way, [10.26] is sometimes called the Behrens-Fisher statistic.
[12]Welch, B. (1947). The generalization of 'Student''s problem when several different population variances are involved. *Biometrika*, *34*, 28–35.

degrees of freedom for unequal variances. Beyond that, the procedures are the same as in the t test of one population mean:

If the alternative states that $\mu_1 > \mu_2$, the critical region lies in the upper tail of $t(\nu)$ and includes all values greater than or equal to t_α, where $\mathbb{P}(t_\alpha) = 1 - \alpha$.

If H_1 states that $\mu_1 < \mu_2$, the critical region lies in the lower tail of $t(\nu)$ and includes all values less than or equal to t_α, where $\mathbb{P}(t_\alpha) = \alpha$.

If the alternative hypothesis is nondirectional, the test is two-tailed and the critical values are $t_{\alpha/2}$ and $-t_{\alpha/2}$.

Example 10.10.4. Suppose we want $\alpha = .025$ in our oddity-learning example. Since the alternative hypothesis (Example 10.10.1, p. 418) is $(\mu_D - \mu_W) < 0$, the critical region must include all values in $t(6)$ less than t_α, where $\mathbb{P}(t_\alpha) = .025$. In Table 4 of Appendix VIII we find that $\mathbb{P}(2.447) \doteq .975$. This means that $P(t \geq .975) \doteq (1 - .975) = .025$ and, because $\phi(t)$ is symmetrical, that $P(t \leq -2.447) \doteq .025$. Therefore, $t_\alpha = -2.447$ is the critical value for our test statistic. In Examples 10.10.2 and 10.10.3 we calculated Student's t to be approximately .412. This falls short of the critical value, so we were forced to reject the alternative hypothesis that dogs require fewer trials to reach criterion.

Example 10.10.5. The results in Example 10.10.4 were very perplexing for the experimenters. Not only did the wolves fail to perform more poorly than the dogs, they had the unpardonable effrontery to perform *better*, though the difference was not statistically significant. Another research team, familiar with the theory, suggested that we had misinterpreted the oddity task and that we should have predicted superior wolf performance. They therefore built an apparatus identical to ours, tested a group of 7 wolf pups, and compared their wolf pup performance with our dog pup performance to test the hypothesis

$$H_0: \mu_D - \mu_W = 0$$

against the alternative

$$H_1: \mu_D - \mu_W > 0$$

Please note that they were predicting that dogs would require *more* trials to learn the task.

There was no basis to assume that the wolf and dog distributions had the same variance (and, based on the variance estimates in Example 10.10.2, ample reason to suspect that they were *not* equal), and the wolf and dog samples were not of equal size ($N_D = 4$ and $N_W = 7$). Therefore, the appropriate test statistic is [10.26], and the experimenters estimated degrees of freedom by Welch's method. For the seven wolves in this study, $\hat{\sigma}^2 = 278.57$, and from Example 10.10.2, the variance estimate for dogs is 206.25, so by formula [10.27],

$$\nu_W = \frac{\left[(\hat{\sigma}_D^2/N_D) + (\hat{\sigma}_W^2/N_W)\right]^2}{\dfrac{(\hat{\sigma}_D^2/N_D)^2}{N_D + 1} + \dfrac{(\hat{\sigma}_W^2/N_W)^2}{N_W + 1}} - 2$$

$$= \frac{\left[(206.25/4) + (278.57)/7\right]^2}{\dfrac{(206.25/4)^2}{5} + \dfrac{(278.57/7)^2}{8}} - 2 \doteq 9.44 \cong 9$$

They set $\alpha = .01$, and since the alternative hypothesis is $(\mu_D - \mu_W) > 0$, the critical value corresponds to $\mathbb{P} \doteq .99$ in the distribution of t (9). In Table 4 of Appendix VIII this is found to be $t = 2.821$.

Their seven wolves averaged 55.71 trials to criterion, and the mean for the four dogs given in Example 10.10.2 was 93.75. Therefore,

$$\frac{\bar{x}_D - \bar{x}_W}{\sqrt{\dfrac{\hat{\sigma}_D^2}{N_D} + \dfrac{\hat{\sigma}_W^2}{N_W}}} = \frac{93.75 - 55.71}{\sqrt{\dfrac{206.25}{4} + \dfrac{278.57}{7}}} \doteq 3.98$$

The value of the test statistic is larger than 2.821, and the test hypothesis is rejected in favor of the alternative hypothesis. The result of this experiment is obviously more satisfying than the result reported in Example 10.10.4, but, then, the second research team had the advantage of hindsight. That's life in the wide world of research.

3. Data collection

a. Sample size and power. In our discussion of testing hypotheses about one population mean, it was pointed out that a test based on the standard normal distribution is always more powerful than a test based on Student's t (see Figure 10.4, p. 412). This is also true for tests about two population means. Consequently, the risk of overpowering one's test is ordinarily a realistic concern only when samples are large enough to use a test statistic that is approximately distributed $N(0, 1)$, for example, the statistic given in [10.20].

Calculation of sample sizes that yield a specific value of β against a simple alternative hypothesis is more complex than in the one-population test. Because the power of the test depends on the *sum* of two variances, (σ_1^2/N_1) and (σ_2^2/N_2), there are many pairs, N_1 and N_2, that will yield prescribed values of α and β. However, it can be shown that of all such pairs, the one that yields the *smallest* number of observations $N_1 + N_2$

occurs when

$$\frac{N_1}{N_2} = \frac{\sigma_1}{\sigma_2}$$

in which case

$$N_1 \cong \sigma_1(\sigma_1 + \sigma_2)\left(\frac{z_\beta - z_\alpha}{\mu_1 - \mu_2}\right)^2 \qquad\qquad\qquad [10.28]$$

and

$$N_2 \cong \sigma_2(\sigma_1 + \sigma_2)\left(\frac{z_\beta - z_\alpha}{\mu_1 - \mu_2}\right)^2 \qquad\qquad\qquad [10.29]$$

where σ_1 and σ_2 are estimated by $\sqrt{\hat\sigma_1^2}$ and $\sqrt{\hat\sigma_2^2}$. The values of μ_1 and μ_2 may in rare instances be dictated by the theoretical distributions of X_1 and X_2. It is more often the case that the difference $\mu_1 - \mu_2$ is a synthetic value δ equal to the smallest difference between the two population means that the experimenter considers scientifically significant.

b. Independence of observations. In testing hypotheses about *one* population, the principal *research design* consideration is that subjects be chosen at random from the parent population. Researchers have developed a wider variety of strategies for testing hypotheses about *two* (or more) populations (e.g., *matching* and *subjects-as-own-controls* designs discussed in Appendix VI), so there are additional issues that must be considered.

We began our discussion of testing hypotheses about two population means with three lemmas concerning the distribution of the difference $(X_1 - X_2)$. One of the fundamental assumptions embedded in these lemmas and, therefore, in the techniques we have developed is that X_1 and X_2 are *independent* random variables. When two variables represent a measurement performed on groups that are randomly selected from populations that are known to differ on a particular attribute (i.e., *quasi-experimental* research, Appendix VI) this assumption is ordinarily inherent in the research design. The assumption is also satisfied if subjects are assigned at random to undergo different experimental treatments (i.e., *randomized* designs, Appendix VI).

Other research designs, however, require what are called *correlated* groups, which means that X_1 and X_2 are *not* independent.

Groups are said to be *correlated* if the score of any particular member of Group 1 is expected to be closer to the score of some *particular* member of Group 2 than to the score of a *randomly* selected member of Group 2.

The most obvious such design is one in which the researcher performs the same measurement twice on the same individuals in order to assess the effect of some treatment that is administered between the first measurement and the second (i.e., subjects as own controls discussed in Appendix VI). This design is found in medical research and is also common in certain branches of psychology. Attitude change, for example, is often assessed by administering an attitude questionnaire, presenting the subjects some form of persuasive communication, and then administering the same attitude measure at a later date.

Groups are less obviously—but nonetheless—correlated when every individual in the first group is somehow associated with or related to an individual in the second group. For example, biological anthropologists have investigated the effects of introducing calcium into the diet of a population by comparing bone thickness in women of the generation before the calcium supplements were introduced (Group 1) with bone thickness of their daughters (Group 2). Clearly, the difference in bone thickness between a particular woman in Group 1 and *her own daughter* is expected to be less than the difference between a woman in Group 1 and *any woman chosen at random* from Group 2.

In designs of this sort the variable under investigation is *not* the difference of the averages, but the average of the differences.[13] When the two variables are *correlated*, the experimenter must treat the difference D between correlated *pairs* of scores as the random variable and test the null hypothesis using the methods appropriate to *one* population mean.

The basic strategy is very straightforward if we think of a subjects-as-own-controls experiment in which we obtain a pretreatment score (x) and a posttreatment score (y) on each of N individuals. The score of individual 1 on the first measurement is denoted x_1, and the score obtained by individual 1 on the second measurement is y_1. Likewise, x_2 and y_2 are the pair of scores obtained by individual 2, and so on. We calculate $d_i = x_i - y_i$ for each pair of scores and test the hypothesis

$$H_0: \mu_D = 0$$

against the appropriate alternative hypothesis. If it is expected that scores will increase after the intervening treatment, then

$$H_1: \mu_D < 0$$

If scores are expected to decrease, then

$$H_1: \mu_D > 0$$

If the direction of change is not predicted, then

$$H_1: \mu_D \neq 0$$

[13] If this distinction seems subtle, keep in mind that when X and Y are independent, the difference of the averages is equal to the mean of every x-score minus *every* y-score. When X and Y are correlated, every y-score is subtracted from only *one* x-score.

All decisions concerning the appropriate test statistic should be made in accordance with the assumptions governing tests about one population mean. (See Box 10.1.)

Example 10.11. In a test of leash training, experimenters recorded the number of "demerits" a pup earned for such faults as balking, biting the leash, dragging behind the trainer, etc. while walking a prescribed course of obstacles. Each pup was given 10 training sessions, and it was hypothesized that dog pups would show a significant reduction in demerits over the 10 days of training. Each pup's *difference* score d was the number of demerits earned on Day 1 (X) minus the number of demerits earned on Day 10 (Y). Therefore,

$$H_0: \mu_D = 0$$

$$H_1: \mu_D > 0$$

For a group of 4 Alaskan Malamute pups the average change $\bar{d}$ was 5.5 demerits and $s_D^2 = 2.75$, so

$$\frac{\bar{d} - \mu_0}{s/\sqrt{N-1}} \doteq \frac{5.5 - 0}{(1.66)/\sqrt{3}} \doteq 5.74$$

In Table 4 of Appendix VIII we find that for $t(3)$, $\mathbb{P}(5.84) = .995$, and $\mathbb{P}(4.54) = .99$, so $\mathbb{P}(5.74)$ is between .99 and .995. The result is therefore significant at the .01 level, but t is not quite big enough to be significant at the .005 level. This is expressed either by the annotation $.005 < p < .01$ or by the annotation $p < .01$, where p is understood to mean $P(t \geq 5.74 | H_0 \text{ correct})$.

The average difference for one group of correlated pairs can be compared with the average difference for a second group of correlated pairs using the methods developed for testing hypotheses about two means. For example, anthropologists might wish to compare mother-daughter differences among Aleuts (D_1) with mother-daughter differences among Inuits (D_2). If μ_{D1} is the population mean of Aleut differences and μ_{D2} is the population mean of Inuit differences, the test hypothesis is

$$H_0: \mu_{D1} - \mu_{D2} = 0$$

and the alternative hypothesis is formulated according to the anthropologists' theory. Selection of the test statistic is based on the same considerations that apply to any test of hypotheses about two population means (see Box 10.4).

D. TESTING HYPOTHESES ABOUT TWO POPULATION PROPORTIONS p_1 AND p_2

In Chapter 9 we developed methods for using the standard normal distribution to test hypotheses about the proportion X/N, where X is a binomially distributed random variable, and a hypothetical value p_0 is implied by a theoretical model. Questions about proportions also arise in comparisons of two populations. The usual test hypothesis is

$$H_0: p_1 - p_2 = 0 \quad (or, p_1 = p_2)$$

and no value is specified for the common population proportion, p.

We know from Chapter 7 that if $X: B(N, p)$ and if $Np \geq 5$ and $Nq \geq 5$, the proportion X/N is approximately normally distributed with an expected value of p and a variance of $(pq)/N$. Likewise, if $X_1: B(N_1, p_1)$ and if $X_2: B(N_2, p_2)$ and if N_1 and N_2 are large enough so that both (X_1/N_1) and (X_2/N_2) can be assumed to be approximately normally distributed, it follows from Lemmas 10.1, 10.2, and 10.3 that the *difference* $(X_1/N_1) - (X_2/N_2)$ is approximately normally distributed with an expected value of $p_1 - p_2$ and a variance of

$$\frac{p_1 q_1}{N_1} + \frac{p_2 q_2}{N_2}$$

However, if the test hypothesis is correct, then $p_1 = p_2 = p$, and $q_1 = q_2 = q = (1 - p)$, so the variance of $(X_1/N_1) - (X_2/N_2)$ is

$$\frac{p_1 q_1}{N_1} + \frac{p_2 q_2}{N_2} = \frac{pq}{N_1} + \frac{pq}{N_2} = pq \left(\frac{1}{N_1} + \frac{1}{N_2} \right) = pq \left(\frac{N_1 + N_2}{N_1 N_2} \right)$$

The common population proportion p is unknown but may be estimated by *pooling* the sample proportions (x_1/N_1) and (x_2/N_2), as discussed in Box 10.5:

$$\hat{p} = \frac{N_1 \left(\dfrac{x_1}{N_1} \right) + N_2 \left(\dfrac{x_2}{N_2} \right)}{N_1 + N_2} = \frac{x_1 + x_2}{N_1 + N_2} \qquad [10.30]$$

and the pooled estimate of q is $\hat{q} = 1 - \hat{p}$. Accordingly, if the population proportions p_1 and p_2 are equal, the following statistic is approximately distributed as the standard normal random variable and may therefore be used to test the hypothesis $H_0: p_1 - p_2 = 0$

$$Z_{\hat{p}_1 - \hat{p}_2} = \frac{\dfrac{X_1}{N_1} - \dfrac{X_2}{N_2}}{\sqrt{\hat{p}\hat{q} \left(\dfrac{N_1 + N_2}{N_1 N_2} \right)}} \qquad [10.31]$$

The normal approximation to this statistic is ordinarily quite satisfactory if the number of successes X and the number of failures $(N - X)$ are equal to or greater than 5 in both groups.

EXERCISES 10.5

Exercises 10.5 emphasize material covered after **Exercises 10.4** but also review techniques developed earlier in the chapter.

In Exercises 1 and 2 the random variables X_1 and X_2 are known to be normally distributed. Test the hypothesis that $\mu_1 = \mu_2$ against the alternative hypothesis indicated. State the test statistic for each test, specify the distribution of the test statistic, calculate the value of the test statistic, and report its *p*-value.

1.

	$\sigma_1^2 \overset{?}{=} \sigma_2^2$	$\bar{x}_1$	$\bar{x}_2$	s_1^2	N_1	s_2^2	N_2	H_1: $\mu_1 - \mu_2$
(a)	Yes	11.5	9.9	3.8	17	5.5	15	> 0
(b)	Yes	200	204.6	9.4	8	7.6	6	< 0
(c)	?	74.7	81	21	12	80	12	$\neq 0$
(d)	?	.013	.001	.01	6	.001	6	> 0
(e)	?	96	121	1,305	22	753	18	< 0
(f)	?	13.4	4.2	89	31	435	11	$\neq 0$

2.

	$\sigma_1^2 \overset{?}{=} \sigma_2^2$	$\bar{x}_1$	$\bar{x}_2$	s_1^2	N_1	s_2^2	N_2	H_1: $\mu_1 - \mu_2$
(a)	Yes	5,568	5,672	1,225	18	62,500	17	< 0
(b)	Yes	151	112	104	3	210	4	$\neq 0$
(c)	?	6.9	6.42	22	17	35	17	> 0
(d)	?	-22	-10	0.8	3	0.9	3	< 0
(e)	?	52.7	65.9	455	10	20	5	$\neq 0$
(f)	?	19.9	18.5	0.4	3	100	26	> 0

3. The dairy farmer in Exercise 10.4.10 (p. 426) receives $6.00 per hundred weight for milk. The first-year cost of artificially inseminating cows with semen from Line 2 will be approximately $75.00 per cow more than direct service with bulls from Line 1.
   (a) How much more milk will the farmer have to average using Line 2 than using Line 1 in order to make a profit in the first year of milk production? Let this value be δ.
   (b) Set $\alpha = \beta = .05$ and calculate the sample sizes the farmer requires to test the hypotheses

$$H_0: \mu_1 - \mu_2 = 0$$
$$H_1: \mu_1 - \mu_2 \leq -\delta$$

4. Scores for 1,596 Caucasian college students on the Mach IV test of Machiavellianism (see Exercise 10.3.10, p. 415) yielded the following:

Males ($N = 764$): $\bar{x} = 93.69$ $s = 14.37$
Females ($N = 832$): $\bar{x} = 87.77$ $s = 13.45$

Source: Christie, R., and Geis, F. L. (1970). *Studies in Machiavellianism.* New York: Academic, p. 32.

A psychologist is interested in gender differences and suspects that men are more Machiavellian than women, but he does not want to find a statistically significant difference that is smaller than the difference one expects to find for repeated testing of the *same* individual. Based on the reliability of the test, the psychologist estimates this to be 10 points. How large should the samples be if the psychologist wants $\alpha = \beta = .05$ against such a test?

5. Fifty graduates in psychology from a Midwestern, 4-year liberal arts college take the Graduate Record Examination Advanced Test in Psychology. Scores are reported below. The average score for graduates of the state's large universities on this examination is 505. The average score for graduates of state colleges in the region is 490. The Psychology Department chair believes that her department's curriculum is more nearly equivalent to those of the large universities. Test this belief.

460	420	569	355	390	442	410	534	320	472
540	601	528	480	580	609	502	462	545	497
620	515	508	655	545	445	485	455	518	558
619	435	408	350	510	430	380	398	468	660
571	359	635	478	425	350	585	599	370	519

(a) What are the hypotheses?
(b) What is the appropriate test statistic?
(c) Set α at any level you consider suitable and state the rejection rule appropriate to your hypotheses and the significance level you have chosen.
(d) Calculate the value of your test statistic.
(e) Report the appropriate conclusion.

6. The department chair in Exercise 5 is puzzled, because the average of the 50 scores is well below the national average. Then it occurs to her that some of the students completed the department's B.A. program, which is intended to meet a wide variety of student goals, and some completed the B.S. program, which is specifically designed to prepare students for graduate school. Use the test scores given below to test the hypotheses that follow from the department chair's flash of inspiration. You may assume that test scores are approximately normally distributed.

B.S. Scores						**B.A. Scores**			
460	534	569	320	390	472	355	410	420	442
540	462	528	545	580	497	480	502	601	609
620	455	508	518	545	558	655	485	515	445
619	398	408	468	510	660	350	380	435	430
571	599	635	370	425	519	478	585	359	350

(a) What are the hypotheses?
(b) If $\sigma^2_{BS} = \sigma^2_{BA}$, what is the appropriate test statistic?
(c) How is it distributed if the test hypothesis is correct?
(d) If $\alpha = .05$, what is the appropriate rejection rule?
(e) What is the value of the test statistic?
(f) What is the appropriate conclusion?
(g) If $\sigma^2_{BS} \neq \sigma^2_{BA}$ what is the appropriate test statistic?
(h) How is it distributed if the test hypothesis is correct?

7. "Log rolling" is a game used by psychologists to study social behavior. Players take the roles of legislators and bargain with one another to vote for or against various bills. In one study, every player was told that passage or defeat of each bill would have a particular payoff in votes received or lost by the player in the next election. Payoffs differed from player to player. In

Game 1 the "bills" concerned issues about which the players themselves held strong opinions. In Game 2 the "bills" concerned issues the players considered trivial. It was hypothesized that in Game 1, players who scored high on the Mach IV test of Machiavellianism (see Exercise 10.3.10) would be more successful than players who scored low on the Mach IV in persuading other "legislators" to take actions with favorable payoffs. It was hypothesized that in Game 2, without the distraction of their own strongly held positions, players who scored low on Mach IV would be as persuasive as high Mach players. Use the data presented below to test these hypotheses. Assume that population variances are equal and that scores are approximately normally distributed.

		Votes Earned ($\times 10,000$)	
		Game 1 (Emotional Issues)	Game 2 (Neutral Issues)
High Mach Players (N = 25)	$\bar{x}$	11.24	9.32
	s^2	19.9	15.12
Low Mach Players (N = 29)	$\bar{x}$	8.86	10.21
	s^2	16.9	11.60

Source: Geis, F., Weinheimer, S. and Berger, D. (1970). Playing legislature: Cool heads and hot issues. In R. Christie and F. L. Geis (Eds.), *Studies in Machiavellianism*, New York: Academic, 190–209. (Variances estimated.)

(a) What are the hypotheses?
(b) What is the appropriate test statistic?
(c) Calculate the value of your test statistic.
(d) What is the *p*-value of your test statistic?
(e) Report the appropriate conclusion.

The evolutionary theory discussed throughout this chapter suggests that wolves are not as easily trained as dogs. Wolves and Alaskan Malamutes were both administered the leash-training test described in Example 10.11. All subjects were led over the same course and were assigned demerits for such faults as balking, dragging, etc. The data reported below are demerits on Day 1 and demerits on Day 10 for the two groups. Use these data in Exercises 8 and 9. Similar data collected by Scott and Fuller (see footnote 8, p. 391) indicate that leash-training scores are normally distributed.

Wolves	Day 1	Day 10	Malamutes	Day 1	Day 10
1	8	6	1	6	1
2	5	5	2	7	0
3	5	7	3	5	2
4	9	8	4	6	1
5	0	1			
6	3	3			
7	4	5			

8. Test the hypothesis that wolves exhibit less improvement than Malamutes over 10 days of leash training.
   (a) What are the hypotheses?
   (b) What is the appropriate test statistic?
   (c) Calculate the value of your test statistic.
   (d) What is the p-value of your test statistic?
   (e) Report the appropriate conclusion.

9. Test the hypothesis that wolves exhibit *any* improvement over 10 days of leash training.
   (a) What are the hypotheses?
   (b) What is the appropriate test statistic?
   (c) Calculate the value of your test statistic.
   (d) What is the p-value of your test statistic?
   (e) Report the appropriate conclusion.

10. A farmer wants to increase the fat content of the milk produced by a dairy herd. A traditional measure for evaluating change in fat yield is daughter-dam comparisons. The farmer therefore test-breeds 10 cows to a new line of bulls and records fat yield for the mothers and their daughters over a 1-year (305-day) period. Test the hypothesis of improvement in milk-fat yield. (You may assume milk-fat production over a 1-year period to be normally distributed.)

Milk-fat Production for 1 Year (pounds)	
Dams	**Daughters**
387	402
423	460
495	505
533	580
520	510
610	590
530	512
522	541
399	442
460	475

(a) What are the hypotheses?
(b) What is the appropriate test statistic?
(c) Set α at .05 and state the rejection rule appropriate to your hypotheses.
(d) Calculate the value of your test statistic.
(e) Report the appropriate conclusion.

11. In Example 10.10.1 researchers predicted that dogs would perform better than wolves on a test of oddity learning, and the researchers who repeated the experiment in Example 10.10.5 predicted superior wolf performance. The two studies differed in one respect: The wolf pups in the first study were reared by a foster-mother wolf, and the wolf pups in the second were entirely hand-reared. Other research reported that hand-reared wolves were more interested in food reward than were mother-reared wolves. Since food reward was used in oddity-learning, it was predicted that the hand-reared wolves would perform better than the mother-reared wolves. The task was to select the odd-colored stimulus object in an array of three objects, and each pup received 15 trials each day. The measurement is the number of trials to reach

a criterion of 85 percent correct choices over two days (30 trials). Use the following data to test the prediction.

Trials to Criterion			
Hand-reared Wolves		**Mother-reared Wolves**	
90	150	210	
90	150	195	
75	120	150	
150		210	

Source: Frank, H. et al. (1989). Motivation and insight in wolf (*Canis lupus*) and Alaskan malamute (*Canis familiaris*). *Bulletin of the Psychonomic Society, 27(5)*, 455–458.

(a) What are the hypotheses?
(b) What is the appropriate test statistic?
(c) Set α at .05 and state the rejection rule appropriate to your hypotheses.
(d) Calculate the value of your test statistic.
(e) Report the appropriate conclusion.

Many predators (e.g., hawks and owls) disdain dead prey and drop food items that don't move. This preference may account for the fact that young animals of many wild species reflexively "freeze" when constricted around the neck. Since the leash-training test described in Exercises 8 and 9 was originally developed for domestic dogs, it employs a choke-chain around the neck. The researchers suspected that a "freeze" response in young wolves might therefore interfere with their leash-training performance, so some of the wolves were tested with a choke chain and some with a body harness. The data reported below are demerits recorded on Day 1 and on Day 10 for the two groups. Use these data in Exercises 12 and 13. As before, you may assume leash-training scores to be normally distributed.

	Choke Chain			Harness		
Wolf	**Day 1**	**Day 10**	Wolf	**Day 1**	**Day 10**
1	8	6	5	0	1
2	5	5	6	3	3
3	5	7	7	4	5
4	9	8			

12. The experimenters reasoned that if the hypothetical "freeze" reflex interferes with leash training, then wolves restrained by body harness should perform better than wolves restrained by choke chain at all stages of training. Assume that variances for the two groups are the same on Day 1 and that the variances for the two groups are the same on Day 10.
(a) What are the hypotheses?
(b) What is the appropriate test statistic?

(c) What is the value of your test statistic?
(d) What is the p-value of your result?
(e) Report the appropriate conclusion.

13. The experimenters reasoned that if the freeze response is innate, the choke-chain group should be more resistant to training than the harness group and should therefore exhibit less improvement over the 10 days. Do not assume that the relevant variances are equal.
(a) What are the hypotheses?
(b) What is the appropriate test statistic?
(c) What is the value of your test statistic?
(d) What is the p-value of your result?
(e) Report the appropriate conclusion.

Exercises 14 and 15 refer to the following results for samples of students in Texas and Utah who were administered a questionnaire designed to assess familial disposition to stroke and early coronary heart disease.

	Texas	Utah
Number of respondents	9,287	15,045
High-risk families:		
Stroke	399	471
Coronary disease	789	1,007

Source: Williams, R. R., et al. (1988). Health family trees: A tool for finding and helping young family members of coronary and cancer prone pedigrees in Texas and Utah. *American Journal of Public Health*, 78(10), 1283–1286.

14. Suppose an epidemiologist suspects that risk factors for early coronary disease differ from region to region and decides to use these data to test her suspicion.
(a) What are her hypotheses?
(b) What test statistic should she use?
(c) Set α at any level you consider suitable and state the rejection rule appropriate to her hypotheses and the significance level you have chosen.
(d) Calculate the value of the test statistic.
(e) Report the appropriate conclusion.

15. On the basis of a theory of environmental factors in stroke, an epidemiologist believes that familial disposition to stroke is more prevalent in Texas than in Utah. If he uses these data to get support for his theory,
(a) What are his hypotheses?
(b) What test statistic should he use?
(c) Set α at .0001 and state the appropriate rejection rule.

(d) Calculate the value of the test statistic.
(e) Report the appropriate conclusion.
(f) What is the smallest α that will permit rejection of H_0?

16. Biologists who study the social behavior of lions believe that one factor influencing the size of hunting groups is the size of the prey species the lions hunt. If this thinking is correct, lions should be observed hunting alone more often when hunting small prey, such as Thomson's gazelle, than when hunting larger prey, like wildebeest and zebra. Address this question using the following data:

Number of lions hunting:	Prey species	
	Thomson's gazelle	**Wildebeest & zebra**
1	185	33
2 or more	177	70

Source: Schaller, G. B. (1972). *The Serengeti lion.* Chicago: University of Chicago Press, p. 405.

(a) What are the hypotheses?
(b) What is the appropriate test statistic?
(c) Set α at any level you think suitable and state the appropriate rejection rule.
(d) Calculate the value of the test statistic.
(e) Report the appropriate conclusion.

17. The lion is popularly thought of as a hunter and the hyena as a scavenger. Test this bit of folk wisdom with the following data.

	Wildebeest	Zebra	Gazelle
Hyenas			
Killed	118	22	62
Scavenged	66	28	52
Lions*			
Killed	73	42	5
Scavenged	53	22	13

*Plains data only. Prey of unknown source not included.
Source: Schaller, G. B., ibid., pp. 434 and 451.

(a) What are the hypotheses?
(b) What is the appropriate test statistic?
(c) Set α at any level you think suitable and state the appropriate rejection rule.
(d) Calculate the value of the test statistic.
(e) Report the appropriate conclusion.

E. SUMMARY

The test hypothesis about a single population mean takes the form

$$H_0: \mu = \mu_0$$

The alternative hypothesis may be simple or composite, and if composite may be directional or nondirectional. That is, the alternative may take any of the following forms:

$$H_1: \mu = \mu_1$$

$$H_1: \mu > \mu_0$$

$$H_1: \mu < \mu_0$$

$$H_1: \mu \neq \mu_0$$

The appropriate test statistic depends on (1) whether or not X can be assumed to be normally distributed, (2) whether or not the population variance σ^2 is known, and (3) sample size.

If the test statistic is normally distributed, the probability of a Type II error and power are determined as follows: Calculate the value of $\bar{x}$ corresponding to z_α and calculate

$$z_\beta = \frac{\bar{x} - \mu_1}{\sigma_{\bar{X}}} \quad \text{or} \quad \frac{\bar{x} - \mu_1}{\hat{\sigma}_{\bar{X}}}$$

Then,

$$\beta = P(z_{\bar{X}} \leq z_\beta) \quad \text{if } \mu_1 > \mu_0$$

$$\beta = P(z_{\bar{X}} \geq z_\beta) \quad \text{if } \mu_1 < \mu_0$$

and the power of the statistical test is equal to $1 - \beta$.

To calculate β (and power) one must have a simple alternative hypothesis. When the natural alternative is composite, the experimenter can let δ be the smallest absolute difference between μ and μ_0 considered to be scientifically important and *synthesize* an alternative hypothesis of the form $\mu = \mu_0 + \delta$ or $\mu = \mu_0 - \delta$ or $\mu = \mu_0 \pm \delta$. In the case of the two-directional alternative ($\mu = \mu_0 \pm \delta$), the probability of a Type II error is the probability of rejecting H_1 when $\mu = (\mu_0 + \delta)$ *plus* the probability of rejecting H_1 when $\mu = (\mu_0 - \delta)$.

If α and β are disproportionate to the costs of a Type I and Type II error, β can be adjusted by (1) choosing a different value of α, (2) testing the hypothesis against a different value of μ_1, or (3) reducing or increasing the variance of the test statistic. Apart from changes in experimental design, the most direct way to change the variance of the test statistic is to reduce or increase sample size. To calculate the sample size required for a desired β, let z_β be the value of z that yields the desired value of β under the distribution specified in the alternative hypothesis. Then,

$$N \cong \left(\frac{\sigma(z_\beta - z_\alpha)}{\mu_0 - \mu_1} \right)^2 \quad \text{or} \quad \left(\frac{\hat{\sigma}(z_\beta - z_\alpha)}{\mu_0 - \mu_1} \right)^2$$

Student's t statistic is the ratio

$$t = \frac{\overline{X} - \mu}{S/\sqrt{N-1}}$$

The density function for Student's t is given in [**10.11**]:

$$\frac{1}{K} \left(1 + \frac{t^2}{\nu} \right)^{-(1+\nu)/2}$$

where K is a positive constant, and the parameter ν, called *degrees of freedom*, is equal to $N - 1$. Student's t is not a single distribution but rather a *family* of distributions, one for every possible sample size, N.

Like the standard normal distribution, all of the Student's t distributions (1) are unimodal at zero, (2) have an expected value of zero, (3) are symmetrical about zero, and (4) are asymptotic to zero. The variance of t is

$$\frac{\nu}{\nu - 2} = \frac{N-1}{N-3}$$

For small values of N, Student's t exhibits greater dispersion than the standard normal distribution, and statistical tests based on t are therefore less powerful than tests based on Z. But when $N \geq 30$, $V(t) \cong 1$, and the distribution of the t ratio is closely approximated by the standard normal distribution.

The test hypothesis for the two population means takes the form

$$H_0: \mu_1 - \mu_2 = 0 \quad (or, \ \mu_1 = \mu_2)$$

The alternative hypothesis is usually composite and will therefore take one of the following forms:

$$H_1: \mu_1 - \mu_2 > 0 \qquad (or, \mu_1 > \mu_2)$$

$$H_1: \mu_1 - \mu_2 < 0 \qquad (or, \mu_1 < \mu_2)$$

$$H_1: \mu_1 - \mu_2 \neq 0 \qquad (or, \mu_1 \neq \mu_2)$$

The appropriate test statistic depends on (1) whether or not X_1 and X_2 can be assumed to be normally distributed, (2) whether or not the population variances σ_1^2 and σ_2^2 are known, and (3) sample sizes. Statistics for testing hypotheses about μ_1 and μ_2 are based on the sampling distribution of $\bar{X}_1 - \bar{X}_2$, the difference between two sample means. If X_1 is a normally distributed random variable with expected value μ_1 and variance σ_1^2 and if X_2 is a normally distributed random variable with expected value μ_2 and variance σ_2^2, then $(\bar{X}_1 - \bar{X}_2)$ is normally distributed with an expected value of $\mu_1 - \mu_2$ and a variance of

$$\frac{\sigma_1^2}{N_1} + \frac{\sigma_2^2}{N_2}$$

If one's test statistic is normally distributed and hypothetical values for μ_1 and μ_2 are specified, the calculation of β and power are completely analogous to the calculation of β and power for tests of a single population mean. If the values of μ_1 and μ_2 are not specified but one has reliable estimates of σ_1^2 and σ_2^2, the following algorithms can be used to calculate the sample sizes required for a desired value of β for any specific *difference* between μ_1 and μ_2: For a one-tailed test,

$$N_1 \cong \hat{\sigma}_1 (\hat{\sigma}_1 + \hat{\sigma}_2) \left(\frac{z_\beta - z_\alpha}{\mu_1 - \mu_2} \right)^2$$

and

$$N_2 \cong \hat{\sigma}_2 (\hat{\sigma}_1 + \hat{\sigma}_2) \left(\frac{z_\beta - z_\alpha}{\mu_1 - \mu_2} \right)^2$$

where $\mu_1 - \mu_2$ will ordinarily be the synthetic value δ. For a two-tailed test, substitute $z_{\alpha/2}$ for z_α and $z_{\beta/2}$ for z_β.

If X and Y are scores for *correlated* groups, the random variable is $D = X - Y$, where d is the difference $x - y$ for each correlated pair. The appropriate test hypothesis is that the mean of difference is zero

$$H_0: \mu_D = 0$$

and is therefore treated as an hypothesis about one population mean. The most common correlated-groups designs are studies in which X and Y are repeated measurements on the same subjects or when X and Y are measurements on closely related pairs of individuals.

Summary table
of statistical tests

I. Testing hypotheses about μ

Distribution of X	σ^2	Sample size	Test statistic	Distribution of test statistic*
Normal	Known	Any	$\dfrac{\overline{X} - \mu_0}{\sigma_{\overline{X}}}$	$N(0, 1)$
Unknown	Known	$N \geq 30$		
Any	Unknown	$N \geq 30$	$\dfrac{\overline{X} - \mu_0}{\hat{\sigma}_{\overline{X}}}$	$N(0, 1)$
Normal	Unknown	$N < 30$	$\dfrac{\overline{X} - \mu_0}{S/\sqrt{N - 1}}$	$t(N - 1)$
Unknown	—	$N < 30$	Nonparametric	Various

II. Testing hypotheses about p_1 and p_2

Distribution of X_1 and X_2	Sample size	Test statistic	Distribution of test statistic*
$B(N_1, p_1)$ $B(N_2, p_2)$	$N_1\hat{p}_1 \geq 5$ $N_1(1 - \hat{p}_1) \geq 5$ $N_2\hat{p}_2 \geq 5$ $N_2(1 - \hat{p}_2) \geq 5$	$\dfrac{\dfrac{X_1}{N_1} - \dfrac{X_2}{N_2}}{\sqrt{\hat{p}\hat{q}\left(\dfrac{N_1 + N_2}{N_1 N_2}\right)}}$†	$N(0, 1)$

*If H_0 is correct.
†$\hat{p} = \dfrac{x_1 + x_2}{N_1 + N_2}$ and $\hat{q} = 1 - \hat{p}$.

III. Testing hypotheses about μ_1 and μ_2

Distribution of X_1 and X_2	σ_1^2 and σ_2^2	Sample size	Test statistic	Distribution of test statistic*
Normal Unknown	Known Known	Any $N_1 \geq 30$ $N_2 \geq 30$	$\dfrac{\overline{X}_1 - \overline{X}_2}{\sqrt{\dfrac{\sigma_1^2}{N_1} + \dfrac{\sigma_2^2}{N_2}}}$	$N(0, 1)$
Unknown	Unknown	$N_1 \geq 30$ $N_2 \geq 30$	$\dfrac{\overline{X}_1 - \overline{X}_2}{\sqrt{\dfrac{\hat{\sigma}_1^2}{N_1} + \dfrac{\hat{\sigma}_2^2}{N_2}}}$	$N(0, 1)$
Normal	Unknown assumed equal	$N_1 < 30$ or $N_2 < 30$	$\dfrac{\overline{X}_1 - \overline{X}_2}{\sqrt{\hat{\sigma}_{\text{POOLED}}^2 \left(\dfrac{N_1 + N_2}{N_1 N_2} \right)}}$†	$t(N_1 + N_2 - 2)$
Normal	Unknown	N_1 and $N_2 < 30$ $N_1 = N_2$	$\dfrac{\overline{X}_1 - \overline{X}_2}{\sqrt{\dfrac{\hat{\sigma}_1^2}{N_1} + \dfrac{\hat{\sigma}_2^2}{N_2}}}$	$t(N_1 + N_2 - 2)$
Normal	Unknown	N_1 or $N_2 < 30$ $N_1 \neq N_2$	$\dfrac{\overline{X}_1 - \overline{X}_2}{\sqrt{\dfrac{\hat{\sigma}_1^2}{N_1} + \dfrac{\hat{\sigma}_2^2}{N_2}}}$	$t(\nu_W)$‡
Unknown	—	N_1 or $N_2 < 30$	Nonparametric	Various

*If H_0 is correct.

†$\hat{\sigma}_{\text{POOLED}}^2 = \dfrac{(N_1 - 1)\hat{\sigma}_1^2 + (N_2 - 1)\hat{\sigma}_2^2}{N_1 + N_2 - 2}$

‡$\nu_W = \dfrac{\left[(\hat{\sigma}_1^2/N_1) + (\hat{\sigma}_2^2/N_2) \right]}{\dfrac{\left(\hat{\sigma}_1^2/N_1 \right)^2}{N_1 + 1} + \dfrac{\left(\hat{\sigma}_2^2/N_2 \right)^2}{N_2 + 1}}$

CHAPTER 11

Testing hypotheses about population variances

A. THE χ^2 FAMILY OF DISTRIBUTIONS

In the last chapter we found it necessary to introduce a new distribution, Student's t, in order to develop several tests of hypotheses about population means. Similarly, the methods involved in testing hypotheses about population variances require some understanding of the χ^2 (*chi*-square) family of distributions.[1] The χ^2 distribution is the probability model for the sum of squares of ν standard normal random variables, and the standard normal distribution therefore provides a springboard for the development of χ^2.

Let X be a normally distributed random variable in a population for which the mean is μ and the variance is σ^2. Then, $X_1 : N(\mu, \sigma^2)$ is the model for the experiment "make one observation on this population." And, if the observation is standardized, the random variable associated with our experiment is $Z_1 : N(0, 1)$, where

$$Z_1 = \frac{X_1 - \mu}{\sigma}$$

Now, let us suppose that we *square* our standardized normal random variable. Then,

$$Z_1^2 = \left(\frac{X_1 - \mu}{\sigma} \right)^2$$

and is said to be distributed as *chi*-square with *one* degree of freedom, denoted $\chi^2(1)$. Since X_1 represents a *single* observation that is free to assume *any* value of the random variable X, the notion of degrees of freedom has the same meaning here as in our discussion of Student's t.

Now, let us suppose we define the experiment "Make two independent observations on the population." Then, $X_1 : N(\mu, \sigma^2)$ and $X_2 : N(\mu, \sigma^2)$, and

$$Z_1^2 + Z_2^2 = \left(\frac{X_1 - \mu}{\sigma} \right)^2 + \left(\frac{X_2 - \mu}{\sigma} \right)^2$$

is the sum of squares of two independent standard normal random variables. Furthermore, X_1 and X_2 are *both* free to assume any value of the random variable X. The sum $Z_1^2 + Z_2^2$ is therefore said to be distributed as χ^2 with *two* degrees of freedom, denoted $\chi^2(2)$. And, in

[1] *Chi* is pronounced *k*ai, not *ch*ai. In ancient Rome, χriminals were χrucified, not *ch*rucified.

general,

If Y is the sum of squares of ν independent standard normal random variables,

$$Y = \sum_1^\nu Z_i^2 = Z_1^2 + Z_2^2 + \cdots + Z_\nu^2$$

then Y is distributed as χ^2 with ν degrees of freedom, denoted $Y : \chi^2(\nu)$.

We haven't given the formula for the density function of χ^2, but from the notation $\chi^2(\nu)$, the reader may deduce that the only parameter for *chi*-square is ν, degrees of freedom. Every possible value of ν defines a different χ^2 distribution, so χ^2—like Student's t—is characterized as a *family* of distributions.

1. Expected value and variance of $\chi^2(\nu)$

The expected value of a *chi*-square distributed random variable is easily derived from the Algebra of Expectations. If $Y = Z_1^2 + Z_2^2 + \cdots + Z_\nu^2$, then $Y : \chi^2(\nu)$, and

$$E(Y) = E\big(Z_1^2 + Z_2^2 + \cdots + Z_\nu^2\big)$$

From the Algebra of Expectations, then,

$$E(Y) = E\big(Z_1^2\big) + E\big(Z_2^2\big) + \cdots + E\big(Z_\nu^2\big)$$

From equation **[6.6]** we know that if $E(X) = \mu$ and $V(X) = \sigma^2$, then

$$\sigma^2 = E(X^2) - \mu^2$$

Therefore

$$E(X^2) = \mu^2 + \sigma^2$$

so, for any standard normal random variable squared,

$$E(Z^2) = \mu_Z^2 + \sigma_Z^2 = 0^2 + 1^2 = 1$$

and

$$E(Y) = \underbrace{1 + 1 + \cdots + 1}_{\nu \text{ times}}$$
$$= \nu$$

The expected value of any *chi*-square distributed random variable is equal to its degrees of freedom.

The variance of $Y : \chi^2(\nu)$ is not so easily derived, so we state without proof that the variance of any *chi*-square distributed random variable is equal to twice its degrees of freedom. That is, if $Y : \chi^2(\nu)$, then $V(Y) = 2\nu$.

2. Properties of χ^2 distributions

We have seen in earlier chapters that we can deduce many properties of a probability distribution by examining the formula for the density function of the random variable. Because χ^2 distributions are so intimately connected with the standard normal random variable, we can derive much of this information indirectly.

Let $Y : \chi^2(1)$. This means that the distribution of Y is the same as the distribution of a standardized normal random variable squared. We know that the standard normal random variable Z can assume any value, positive, negative, or zero, but z^2 can never be negative. Therefore, the random variable Y (or, for that matter, any χ^2 distributed random variable) can take on only nonnegative values.

If we examine Table 3 of Appendix VIII we see that the probability that z falls between -1 and $+1$ is about .68. That is,

$$P(-1 \leq z \leq 1) \doteq .68$$

If z is squared, then any z-value between -1 and 1 yields a z^2-value between 0 and 1. Therefore,

$$P(-1 \leq z \leq 1) = P(0 \leq z^2 \leq 1)$$

and since $y = z^2$,

$$P(0 \leq y \leq 1) = P(-1 \leq z \leq 1) \doteq .68$$

This tells us that approximately 68 percent of the area under $\chi^2(1)$ falls to the left of 1. We also know that the median of any distribution is the point at or below which 50 percent of the area falls, so the median for $\chi^2(1)$ must be less than 1. Furthermore, since Y has one degree of freedom (that is, $\nu = 1$), we know that $E(Y) = 1$. Therefore, the expected value is greater than the median. From our discussion of skew in Chapter 2 (see Figure 2.7, p. 63), this suggests that $\chi^2(1)$ is skewed to the right, which is evident in Figure 11.1.

Now, let Z_1 and Z_2 be independent, standard normal random variables and let

$$Y = Z_1^2 + Z_2^2$$

The distribution of Y is therefore $\chi^2(2)$. Once again, we can use what we know about the distribution of Z to make inferences about the distribution of Y. First,

$$y = z_1^2 + z_2^2$$

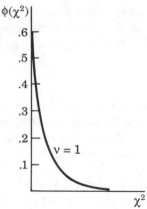

Figure 11.1. Distribution of χ^2 for $\nu = 1$.

so

$$P(y > 1) = P(z_1^2 + z_2^2 > 1)$$

Since y must always be greater than 1 if *either* z_1^2 or z_2^2 is greater than 1, it must be true that

$$P(y > 1) \geq P(z_1^2 > 1 \text{ or } z_2^2 > 1)$$

From the Addition Theorem [5.5],

$$P(z_1^2 > 1 \text{ or } z_2^2 > 1) \qquad\qquad\qquad \text{[11.1]}$$

$$= P(z_1^2 > 1) + P(z_2^2 > 1) - P(z_1^2 > 1 \text{ and } z_2^2 > 1)$$

so

$$P(y > 1) \geq P(z_1^2 > 1) + P(z_2^2 > 1) - P(z_1^2 > 1 \text{ and } z_2^2 > 1)$$

Now, z^2 will be greater than 1 when z is either greater than $+1$ or less than -1. Since the probability is approximately .68 that z is between -1 and $+1$, the probability that z is either greater than $+1$ or less than -1 must be approximately $(1 - .68) = .32$. This means that

$$P(z_1^2 > 1) = P(z_2^2 > 1) \doteq .32$$

Furthermore, since Z_1 and Z_2 are independent, we know from the Multiplication Theorem for Independent Events [5.4] that

$$P(z_1^2 > 1 \text{ and } z_2^2 > 1) = P(z_1^2 > 1)P(z_2^2 > 1)$$

$$\doteq (.32)(.32)$$

$$\doteq .10$$

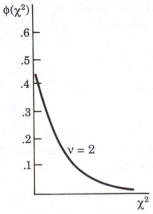

Figure 11.2. Distribution of χ^2 for $\nu = 2$.

Therefore, the inequality in expression **[11.1]** gives us

$$P(y > 1) \geq .32 + .32 - .10 = .54$$

This means that *at least* 54 percent of the area under $\chi^2(2)$ lies to the right of 1, which means that the median is greater than 1.

We have seen that the median of $\chi^2(1)$ is less than 1 and that the median of $\chi^2(2)$ is greater than 1. Accordingly, the distribution of $\chi^2(2)$ has more area in the upper tail than does the distribution of $\chi^2(1)$ and is somewhat less skewed. This is apparent in Figure 11.2. And in general, as ν increases, χ^2 becomes increasingly symmetrical. As we can see in Figure 11.3, the degree of skew in the distribution for χ^2 with $\nu = 10$ is barely noticeable.

The symmetry of χ^2 for large degrees of freedom should not be surprising. As discussed above, the distribution $\chi^2(\nu)$ is defined by a random variable Y that is the sum of ν independent and identically distributed random variables. These properties satisfy the conditions of the Central Limit Theorem, so for large values of ν, the distribution of χ^2 is approximately normal, with expected value equal to ν and variance equal to 2ν.

Cumulative probability distributions for 30 *chi*-square distributions appear in Table 5 of Appendix VIII. Table 5 is organized in the same way as Table 4. Each row represents a different χ^2 distribution, one for each

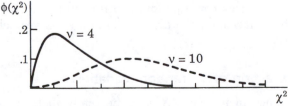

Figure 11.3. Distributions of χ^2 for $\nu = 4$ and for $\nu = 10$.

value of ν from $\nu = 1$ to $\nu = 30$, and the tabled entries are χ^2 values corresponding to the cumulative probabilities indicated by the column headings. For example, the entries in the row labeled $\nu = 16$ are the values of $\chi^2(16)$ corresponding to the cumulative probabilities listed at the top of the table. If we move across this row to the column headed .950, we find the value $\chi^2 = 26.3$, which tells us that if $Y : \chi^2(16)$, then $\mathbb{P}(26.3)$ is about .95.

Example 11.1. If $Y : \chi^2(3)$, what is the value of y for which $\mathbb{P} = .90$? To find this value we first find the row headed $\nu = 3$. Then we move across the table to the column headed .900. The value of y turns out to be 6.25.

Example 11.2. What is the median of the distribution of $\chi^2(9)$? Since the median is that value for which $\mathbb{P} = .50$, we want the entry in row 9 and column .50. This is seen to be $\chi^2 = 8.34$. Since the expected value of a *chi*-square distributed random variable with 9 degrees of freedom is 9, we see that the expected value is slightly larger than the median, indicating a slight positive skew.

Example 11.3. If $Y : \chi^2(11)$, for what value y_b is $P(y \geq y_b) = .05$? To find a value in the upper tail of $\chi^2(11)$, we first go to the row headed $\nu = 11$. Since $P(y \geq y_b) = 1 - \mathbb{P}(y_b)$, we find the y-value corresponding to $\mathbb{P} = 1 - .05 = .95$. This is found to be 19.7. Therefore, if $Y : \chi^2(11)$, then $P(y \geq 19.7) = .05$.

In principle, of course, there are an infinite number of χ^2 distributions, but–as we found with Student's t–we need table distributions for only a small number of ν values. Since χ^2 approaches the normal distribution as ν increases, the distribution of $X : N(\nu, 2\nu)$ ordinarily provides a satisfactory approximation to $\chi^2(\nu)$ for large values of ν.

Example 11.4. Let X be a normally distributed random variable and let $E(X) = 30$ and $V(X) = 60$. Use Table 3 of Appendix VIII to calculate $P(x \leq 34.8)$. Compare your answer with the $P(y \leq 34.8)$ if $Y : \chi^2(30)$.
For $x = 34.8$,

$$z = \frac{34.8 - 30}{\sqrt{60}} \doteq \frac{4.8}{7.75} \doteq .62$$

In Table 3 we find that $\mathbb{P}(.62) \doteq .73$. In Table 5 the cumulative probability or $y = 34.8$ in the row corresponding to $\nu = 30$ is .75. Even with as few as 30 degrees of freedom, the distribution of χ^2 is very close to the normal distribution.

3. The addition and subtraction of χ^2 random variables

Let us suppose once again that X is a normally distributed random variable and that X_1, X_2, X_3, X_4, X_5 represent five independent observations on X. If

$$Z_i = \frac{X_i - \mu}{\sigma}$$

then,

$$Y = Z_1^2 + Z_2^2 + Z_3^2 + Z_4^2 + Z_5^2$$

is distributed as $\chi^2(5)$. Let us also suppose that X' is another normally distributed random variable, which is independent of X, and that we make three independent observations on X', which we call X_1', X_2', and X_3'. Then,

$$Y' = Z_1'^2 + Z_2'^2 + Z_3'^2$$

is distributed as $\chi^2(3)$. Then,

$$W = Y + Y' = Z_1^2 + Z_2^2 + Z_3^2 + Z_4^2 + Z_5^2 + Z_1'^2 + Z_2'^2 + Z_3'^2$$

is the sum of squares of 8 independent standard normal random variables and is therefore distributed as χ^2 with eight degrees of freedom, that is, $W : \chi^2(5 + 3)$. And, in general,

---

Let Y and Y' be independent variables with Y distributed as $\chi^2(\nu_1)$ and Y' distributed as $\chi^2(\nu_2)$. If $W = Y + Y'$, then W is distributed as χ^2 with $\nu_1 + \nu_2$ degrees of freedom. That is, $W : \chi^2(\nu_1 + \nu_2)$.

It can also be shown that if W is defined as the *difference* between two independent *chi*-square random variables,

$$W = Y - Y'$$

where $Y : \chi^2(\nu_1)$ and $Y' : \chi^2(\nu_2)$ and $\nu_1 > \nu_2$, then W is distributed as $\chi^2(\nu_1 - \nu_2)$.

B. TESTING HYPOTHESES ABOUT ONE POPULATION VARIANCE σ^2

1. Formulation

a. Hypotheses. As with hypotheses about population means, the test hypothesis for a single population variance σ^2 must be simple. It therefore takes the form

$$H_0 : \sigma^2 = \sigma_0^2$$

where σ_0^2 is a specific value implied by the experimenter's theoretical model for the population distribution of the random variable, X.

The alternative hypothesis may be simple,

$$H_1: \sigma^2 = \sigma_1^2$$

or it may be composite, and if H_1 is composite it may be directional,

$$H_1: \sigma^2 > \sigma_0^2$$

or

$$H_1: \sigma^2 < \sigma_0^2$$

or it may be nondirectional,

$$H_1: \sigma^2 \neq \sigma_0^2$$

2. Decisions

a. Test statistic. In Chapter 10 we saw that the test statistic for hypotheses about μ was always some function of $\overline{X}$. Similarly, the appropriate test statistic for testing hypotheses about σ^2 is a function of $\hat{\sigma}^2$. For a sample of N observations, the statistic for testing the hypothesis that $\sigma^2 = \sigma_0^2$ is

$$\frac{\hat{\sigma}^2(N-1)}{\sigma_0^2} \qquad\qquad [11.2]$$

which is distributed as $\chi^2(N-1)$ if H_0 is correct.

> If X is a normally distributed random variable with variance σ^2, then the appropriate statistic to test the hypothesis that $\sigma^2 = \sigma_0^2$ is
>
> $$\frac{\hat{\sigma}^2(N-1)}{\sigma_0^2}$$
>
> which is distributed $\chi^2(N-1)$.

It is not immediately obvious that $\chi^2(N-1)$ is an appropriate model for the distribution of [11.2], so we shall derive it. First, it is shown in Box 11.1 that if $E(X) = \mu$ and $V(X) = \sigma^2$, then

$$\frac{\sum\limits_{N}^{N}\left(X_1 - \overline{X}\right)^2}{\sigma^2} = \sum\limits^{N} Z_i^2 - Z_{\overline{X}}^2$$

If the random variables $X_1, X_2, \ldots, X_N$ represent N independent observations on a normally distributed random variable, then we know from our discussion of χ^2 that

$$\sum_{i}^{N} Z_i^2$$

is distributed as $\chi^2(N)$. Likewise,

$$Z_{\bar{X}}^2$$

is a squared standard normal random variable and must therefore be distributed as $\chi^2(1)$. It can be shown that for normally distributed random variables, $\sum Z_i^2$ is independent of $Z_{\bar{X}}^2$, so

$$\frac{\sum_{i}^{N}(X_i - \bar{X})^2}{\sigma^2} = \sum_{i}^{N} Z_i^2 - Z_{\bar{X}}^2$$

is distributed as χ^2 with $(N - 1)$ degrees of freedom.

Furthermore, we know from **[8.9]** that

$$\hat{\sigma}^2 = \frac{\sum_{i}^{N}(x_i - \bar{x})^2}{N - 1}$$

so

$$\hat{\sigma}^2(N - 1) = \sum_{i}^{N}(x_i - \bar{x})^2$$

and

$$\frac{\hat{\sigma}^2(N - 1)}{\sigma^2} = \frac{\sum_{i}^{N}(x_i - \bar{x})^2}{\sigma^2}$$

Therefore, when H_0 is correct and $\sigma^2 = \sigma_0^2$,

$$\frac{\hat{\sigma}^2(N - 1)}{\sigma_0^2} : \chi^2(N - 1)$$

It follows that the expected value of the test statistic is $N - 1$ when H_0 is correct.

It is important to keep in mind that χ^2 is the probability model for the sum of squares of *independent* and *normally distributed* random variables. The claim that $\hat{\sigma}^2(N - 1)/\sigma_0^2$ is distributed as $\chi^2(N - 1)$ therefore requires that the N observations be independent of one another and the random variable under observation be normally distributed. In contrast to the rather liberal normality assumption underlying applications of Student's t, moreover, this assumption must be rigidly satisfied if the χ^2 distributions are to be used without serious inferential errors.

BOX 11.1

$\Sigma (X_i - \bar{X})^2 / \sigma^2$ **equals** $\Sigma z_i^2 - z_{\bar{x}}^2$

Consider the difference $x_i - \mu$. Clearly,

$$x_i - \mu = (x_i - \mu) + (0)$$

so

$$x_i - \mu = (x_i - \mu) + (\bar{x} - \bar{x})$$

$$= (x_i - \bar{x}) + (\bar{x} - \mu) \qquad [1]$$

Squaring both sides of equation [1]

$$(x_i - \mu)^2 = \left[(x_i - \bar{x}) + (\bar{x} - \mu) \right]^2$$

$$= (x_i - \bar{x})^2 + 2(x_i - \bar{x})(\bar{x} - \mu) + (\bar{x} - \mu)^2$$

and

$$\sum_{}^{N} (x_i - \mu)^2$$

$$= \sum_{}^{N} \left[(x_i - \bar{x})^2 + 2(x_i - \bar{x})(\bar{x} - \mu) + (\bar{x} - \mu)^2 \right]$$

$$= \sum_{}^{N} (x_i - \bar{x})^2 + \sum_{}^{N} 2(x_i - \bar{x})(\bar{x} - \mu) + \sum_{}^{N} (\bar{x} - \mu)^2 \qquad [2]$$

For any particular sample of N observations the difference $(\bar{x} - \mu)$ is a constant, so the last term in [2] becomes $N(\bar{x} - \mu)^2$, and the number 2 is a constant, so $2(\bar{x} - \mu)$ can be factored out of the middle term. Therefore,

$$\sum_{}^{N} (x_i - \mu)^2$$

$$= \sum_{}^{N} (x_i - \bar{x})^2 + 2(\bar{x} - \mu) \sum_{}^{N} (x_i - \bar{x}) + N(\bar{x} - \mu)^2 \qquad [3]$$

The middle term of equation [3] includes the factor

$$\sum_{}^{N} (x_i - \bar{x})$$

which is the sum of deviations about the mean and is therefore equal to zero. The middle term therefore drops out of equation [3], giving us

$$\sum_{}^{N} (x_i - \mu)^2 = \sum_{}^{N} (x_i - \bar{x})^2 + N(\bar{x} - \mu)^2 \qquad [4]$$

Dividing both sides of equation [4] by σ^2 we obtain

$$\frac{\sum\limits_{i}^{N}(x_i - \mu)^2}{\sigma^2} = \frac{\sum\limits_{i}^{N}(x_i - \bar{x})^2}{\sigma^2} + \frac{N(\bar{x} - \mu)^2}{\sigma^2}$$

and rearranging terms,

$$\frac{\sum\limits_{i}^{N}(x_i - \bar{x})^2}{\sigma^2} = \frac{\sum\limits_{i}^{N}(x_i - \mu)^2}{\sigma^2} - \frac{N(\bar{x} - \mu)^2}{\sigma^2} \qquad [5]$$

Now let us suppose that the values $x_1, \ldots, x_N$ are independent observations on the random variable X and that the observations have not yet been taken. Equation [5] therefore becomes

$$\frac{\sum\limits_{i}^{N}(X_i - \bar{X})^2}{\sigma^2} = \frac{\sum\limits_{i}^{N}(X_i - \mu)^2}{\sigma^2} - \frac{N(\bar{X} - \mu)^2}{\sigma^2}$$

$$= \sum\limits_{i}^{N}\left(\frac{X_i - \mu}{\sigma}\right)^2 - \frac{(\bar{X} - \mu)^2}{\sigma^2/N}$$

$$= \sum\limits_{i}^{N}\left(\frac{X_i - \mu}{\sigma}\right)^2 - \frac{(\bar{X} - \mu)^2}{\sigma_{\bar{X}}^2}$$

$$= \sum\limits_{i}^{N}\left(\frac{X_i - \mu}{\sigma}\right)^2 - \left(\frac{\bar{X} - \mu}{\sigma_{\bar{X}}}\right)^2$$

$$= \sum\limits_{i}^{N}Z_i^2 - Z_{\bar{X}}^2$$

b. Rejection rule. As always, we want a rejection region that is more likely to include the observed value of our test statistic if H_1 is correct than if H_0 is correct. This, of course, depends on the expected value of the test statistic under H_0 and under H_1.

If H_0 is correct, we know that our test statistic [11.2] is distributed $\chi^2(N - 1)$ and must therefore have expected value $N - 1$. If the alternative hypothesis is correct, $\sigma^2 = \sigma_1^2$, so,

$$E\left(\frac{\hat{\sigma}^2(N-1)}{\sigma_0^2}\right) = (N-1)E\left(\frac{\hat{\sigma}^2}{\sigma_0^2}\right)$$

$$= (N-1)\left(\frac{E(\hat{\sigma}^2)}{\sigma_0^2}\right) = (N-1)\left(\frac{\sigma_1^2}{\sigma_0^2}\right)$$

Consequently, if $\sigma_1^2 > \sigma_0^2$, the expected value of the test statistic is *larger* than $N - 1$. Therefore, if $\sigma_1^2 > \sigma_0^2$, the appropriate rejection region falls in the *upper* tail of $\chi^2(N-1)$. If $\sigma_1^2 < \sigma_0^2$, the expected value of the test statistic is *smaller* than $N-1$, so when $\sigma_1^2 < \sigma_0^2$, the rejection region falls in the *lower* tail of $\chi^2(N-1)$.

Example 11.5. Recall from the "Neanderthal problem" discussed in Chapter 10 that the variance in cranial capacity for modern man is $\sigma^2 = 28,900$ and that on the basis of present data, the variance of skull capacities in the Neanderthal population is assumed to be smaller than that of modern man. The four Middle Eastern fossil skulls of dubious ancestry had capacities of 1,555, 1,520, 1,585, and 1,587 cc. Let $\alpha = .025$ and test the hypothesis that the variance of the Middle Eastern fossil population is the same as the variance for the modern human population. The test hypothesis is

$$H_0: \sigma^2 = 28,900$$

The alternative hypothesis is guided by the assumption that if the Middle Eastern skulls are not modern they are probably Neanderthal, and that the variance is therefore less than 28,900. Therefore,

$$H_1: \sigma^2 < 28,900$$

For the Middle Eastern skulls,

$$\hat{\sigma}^2 = s^2 \frac{N}{N-1} \doteq 741\left(\frac{4}{3}\right) = 988$$

The value of our test statistic is

$$\frac{\hat{\sigma}^2(N-1)}{\sigma_0^2} = \frac{988(3)}{28,900} \doteq .103$$

Since $N - 1 = 3$ and the alternative hypothesis specifies that $\sigma_1^2 < \sigma_0^2$, our rejection region lies in the lower tail of $\chi^2(3)$. With $\alpha = .025$, the test hypothesis is rejected if the test statistic is less than or equal to χ_a^2 where $\mathbb{P}(\chi_a^2) = .025$. In Table 5 of Appendix VIII we find that $\mathbb{P}(.216) = .025$, and the test hypothesis is therefore rejected.

If we examine the distribution of $\chi^2(3)$ a little more closely, we find that the value of our test statistic falls between the entries corresponding to $\mathbb{P} = .005$ and $\mathbb{P} = .01$. The significance level of this result would, therefore, ordinarily be reported with the annotation $.005 < p < .01$ or $p < .01$.

Example 11.6. Test the hypothesis ($\alpha = .05$) that the variance of Middle Eastern skulls is the same as the variance for Neanderthal skulls against the alternative hypothesis that the variances of Middle Eastern and

Neanderthal skulls are different. That is,

$H_0: \sigma^2 = 1{,}600$

$H_1: \sigma^2 \neq 1{,}600$

As always, a nondirectional alternative calls for a two-tailed test. With $N - 1 = 3$ the rejection region for this test is all values in the distribution of $\chi^2(3)$ less than χ_a^2 or greater than χ_b^2, where $\mathbb{P}(\chi_a^2) = \alpha/2$ and $\mathbb{P}(\chi_b^2) = 1 - \alpha/2$. Since $\alpha = .05$, our critical values therefore correspond to $\mathbb{P} = .025$ and $\mathbb{P} = .975$. In Table 5 of Appendix VIII we find that these values are $\chi_a^2 = .216$ and $\chi_b^2 = 9.35$. The test hypothesis will therefore be rejected if the value of our test statistic is either (equal to or) less than .216 or (equal to or) greater than 9.35.

The value of our test statistic is

$$\frac{\hat{\sigma}^2(N-1)}{\sigma_0^2} = \frac{988(3)}{1{,}600} \doteq 1.85$$

and the alternative hypothesis is therefore rejected.

BOX 11.2

Interval estimates of σ^2

Since

$$\frac{\hat{\sigma}^2(N-1)}{\sigma^2}$$

is distributed as χ^2 with $N - 1$ degrees of freedom, this ratio can be used to calculate a confidence estimate of σ^2. If $X: N(\mu, \sigma^2)$, then for N independent observations on X, confidence $= \mathbb{C}$ that

$$\frac{(N-1)\hat{\sigma}^2}{\chi_b^2} \leq \sigma^2 \leq \frac{(N-1)\hat{\sigma}^2}{\chi_a^2}$$

where $\mathbb{P}(\chi_a^2) = \dfrac{1 - \mathbb{C}}{2}$ and $\mathbb{P}(\chi_b^2) = \dfrac{1 + \mathbb{C}}{2}$ in $\chi^2(N-1)$.

EXERCISES 11.1

In Exercises 1 and 2, the random variable Y is distributed as *chi*-square with v degrees of freedom. Find the cumulative probability for the given y-value in the distribution indicated.

1.

	y	ν
(a)	9.59	20
(b)	29.3	25
(c)	.554	5
(d)	.04	2
(e)	6.63	1

2.

	y	ν
(a)	7.26	15
(b)	8.34	9
(c)	21.9	11
(d)	53	28
(e)	1,104	1,000

In Exercises 3 and 4, the random variable Y is distributed as *chi*-square with ν degrees of freedom. Find the probability that y is greater than or equal to the given y-value in the distribution indicated.

3.

	y	ν
(a)	28.9	18
(b)	14.0	22
(c)	1.24	6
(d)	12.0	4
(e)	.500	1

4.

	y	ν
(a)	20.8	26
(b)	18.5	12
(c)	19.0	9
(d)	6.83	3
(e)	5.99	1

In Exercises 5 and 6, the random variable X is normally distributed. Test the hypothesis that $\sigma^2 = \sigma_0^2$. State your rejection rule, give the value of your test statistic, and state your conclusion.

5.

	s^2	N	σ_0^2	σ_1^2	α
(a)	24	15	15	30	.05
(b)	46	10	144	100	.05
(c)	113	30	225	< 225	.01
(d)	110	16	350	≠ 350	.01

6.

	s^2	N	σ_0^2	σ_1^2	α
(a)	.06	5	.35	.035	.05
(b)	60,000	12	29,000	> 29,000	.001
(c)	2.4	25	7.5	≠ 7.5	.01
(d)	300	15	150	175	.025

7. Test the hypothesis that $\sigma^2 = 45,000$ against the alternative hypothesis that $\sigma^2 < 45,000$, where $s^2 = 27,000$ for a sample of 20 independent observations on a normally distributed random variable. State the p-value of your result.

8. Test the hypothesis that $\sigma^2 = 10,000$ against the alternative hypothesis that $\sigma^2 > 10,000$, where $s^2 = 14,600$ for a sample of 15 independent observations on a normally distributed random variable. State the p-value of your result.

9. Use the following data to test the hypothesis that $\sigma^2 = 30$ against the hypothesis that $\sigma^2 > 30$.

$$5, 8, 12, 18, 26, 25, 11$$

(a) What is the appropriate test statistic?
(b) How is the test statistic distributed if H_0 is correct?
(c) What is the value of the test statistic?

(d) What is the p-value of the result?
(e) What is the appropriate conclusion?

10. Use the following data to test the hypothesis that $\sigma^2 = 60$ against the hypothesis that $\sigma^2 < 60$.

-8.3	2.6	-8.1
1.5	2.8	-1.6
-4.3	-3.1	5.0
-6.2	0.2	0.0

(a) What is the appropriate test statistic?
(b) How is the test statistic distributed if H_0 is correct?
(c) What is the value of the test statistic?
(d) What is the p-value of the result?
(e) What is the appropriate conclusion?

11. Scott and Fuller (p. 391n) found wide differences in the variability exhibited by five highly specialized breeds of domestic dogs in training and problem-solving tests. In order to compare dog performance with wolf performance, Frank and Frank[2] wanted a breed of dog that exhibited the same degree of variability found in the general dog population. They chose the Alaskan Malamute and compared the variability in Malamute performance with the variability observed in Scott and Fuller's entire five-breed sample. On a series of eight complex manipulation tasks the pooled variance for Scott and Fuller's 141 dogs was 6.55. Scores for a sample of 4 Alaskan Malamutes were 0, 1, 2, and 3.
(a) What are the hypotheses?
(b) If the scores are assumed to be normally distributed, what is the appropriate test statistic?
(c) Set α at .05 and state the rejection rule appropriate to your hypotheses and the significance level you have chosen.
(d) Calculate the value of your test statistic.
(e) What is the appropriate conclusion?

12. In Exercise 10.5.6 (p. 442) the Psychology Department chair in a 4-year liberal arts college found that a sample of 20 graduates with B.A. degrees in psychology obtained the following scores on the Graduate Record Examination Advanced Test in Psychology:

355	410	420	442
480	502	601	609
655	485	515	445
350	380	435	430
378	585	359	350

The variance for all persons taking this examination is 10,000. The department chair expects that persons with degrees in psychology should be less variable than the general student population.
(a) What are the hypotheses?
(b) What is the appropriate test statistic?

[2] Frank, H., and Frank, M. G. (1987). The University of Michigan canine information-processing project (1979–1981). In H. Frank (Ed.), *Man and Wolf: Advances, Issues and Problems in Captive Wolf Research* (143–167). Dr W. Junk Publishers, Dordrecht: The Netherlands.

      (c) Calculate the value of your test statistic.

      (d) What is the p-value of your test statistic?

      (e) What is the appropriate conclusion?

13. Calculate a 95 percent confidence interval for σ^2 using the data in Exercise 11.1.9.

14. Calculate a 99 percent confidence interval for σ^2 using the data in Exercise 11.1.10.

C. FISHER'S F DISTRIBUTION

The last distribution we will introduce in this text is Fisher's F distribution—or, more simply, the F distribution. The reader will undoubtedly be overjoyed to learn that we will have relatively little to say about it. (This is the *good* news. The *bad* news is that it takes a bit of practice to use the tables of cumulative probabilities for Fisher's F.) Fisher's F distribution is the probability model for the ratio of *two* independent *chi*-square random variables, each divided by its own degrees of freedom. That is,

If Y and W are independent random variables, and if Y is distributed as $\chi^2(\nu_1)$ and W is distributed as $\chi^2(\nu_2)$, then

$$\frac{Y/\nu_1}{W/\nu_2}$$

is said to be distributed as Fisher's F with ν_1 and ν_2 degrees of freedom, denoted $F(\nu_1, \nu_2)$.

The expected value of Fisher's F depends wholly upon ν_2. We state without proof that

$$E\left(\frac{Y/\nu_1}{W/\nu_2}\right) = \frac{\nu_2}{\nu_2 - 2}$$

so as ν_2 increases, the expected value approaches 1.

    The parameters for Fisher's F distribution are ν_1 (sometimes called *degrees of freedom for numerator*) and ν_2 (also called *degrees of freedom for denominator*). Every combination of ν_1 and ν_2 defines a different Fisher's F distribution. To use the cumulative probability tables for Fisher's F distribution in Table 6 of Appendix VIII, begin by finding the *block* of rows corresponding to ν_2. Each of the 24 columns in block ν_2 corresponds to the value of ν_1 indicated at the top of the page, and each row corresponds to the cumulative probability indicated by the row heading (labeled $\mathbb{P}$).

Example 11.7. Suppose the random variable X is distributed as Fisher's F with $\nu_1 = 5$ and $\nu_2 = 2$, that is, $X: F(5, 2)$. What is the cumulative probability of $x = 9.29$? The tables for $\nu_2 = 2$ are the middle block of rows on pages 730 and 731. Within this block, the column headed $\nu_1 = 5$ on page 730 includes F-values for the $\mathbb{P}$-values indicated in the margins of the page. If you run down the column of F values, you will find that $F = 9.29$ lies in the row labeled $\mathbb{P} = .90$. Therefore, the cumulative probability of $x = 9.29$ is .90.

Example 11.8. If the random variable X is distributed as $F(60, 20)$, what x-value cuts off the upper 1 percent of the distribution? The table for $F(60, 20)$ is the column headed 60 in the block for $\nu_2 = 20$ on page 739. The value that cuts off the upper 1 percent is the value for which $\mathbb{P} = .99$, which is found to be $F = 2.61$.

D. TESTING HYPOTHESES ABOUT TWO POPULATION VARIANCES σ_1^2 AND σ_2^2

1. Formulation

a. Hypotheses. In this section of our chapter we develop methods for testing that hypothesis of the form

$$H_0: \sigma_1^2 = \sigma_2^2$$

against any of the following alternatives:

$$H_1: \sigma_1^2 > \sigma_2^2$$

$$H_1: \sigma_1^2 < \sigma_2^2$$

$$H_1: \sigma_1^2 \neq \sigma_2^2$$

2. Decisions

a. Test statistic. Let us suppose that X_1 is a normally distributed random variable with variance σ_1^2. If we make N_1 independent observations and calculate

$$Y = \frac{\hat{\sigma}_1^2 (N_1 - 1)}{\sigma_1^2}$$

we know from earlier in the chapter that $Y: \chi^2(N_1 - 1)$.

Now, let us suppose that X_2 is a normally distributed random variable with variance σ_2^2 and that we make N_2 independent observations on X_2

and calculate

$$W = \frac{\hat{\sigma}_2^2(N_2 - 1)}{\sigma_2^2}$$

Then, $W : \chi^2(N_2 - 1)$. If X_1 and X_2 are independent,[3] then by the definition of Fisher's F given above,

$$\frac{Y/(N_1 - 1)}{W/(N_2 - 1)} \qquad\qquad [11.3]$$

is distributed as Fisher's F with $\nu_1 = N_1 - 1$ and $\nu_2 = N_2 - 1$.

Consider the numerator in expression [11.3]:

$$\frac{Y}{N_1 - 1} = \frac{\hat{\sigma}_1^2(N_1 - 1)/\sigma_1^2}{N_1 - 1} = \frac{\hat{\sigma}_1^2}{\sigma_1^2}$$

For the denominator in [11.3], it is similarly the case that

$$\frac{W}{N_2 - 1} = \frac{\hat{\sigma}_2^2(N_2 - 1)/\sigma_2^2}{N_2 - 1} = \frac{\hat{\sigma}_2^2}{\sigma_2^2}$$

Therefore,

$$\frac{Y/(N_1 - 1)}{W/(N_2 - 1)} = \frac{\hat{\sigma}_1^2/\sigma_1^2}{\hat{\sigma}_2^2/\sigma_2^2} \qquad\qquad [11.4]$$

It is noted, however, that under the test hypothesis, $\sigma_1^2 = \sigma_2^2$, so equation [11.4] becomes

$$\frac{Y/(N_1 - 1)}{W/(N_2 - 1)} = \frac{\hat{\sigma}_1^2}{\hat{\sigma}_2^2}$$

When the test hypothesis is correct,

$$\frac{\hat{\sigma}_1^2}{\hat{\sigma}_2^2}$$

is therefore distributed as Fisher's F with degrees of freedom for numerator equal to $N_1 - 1$ and degrees of freedom for denominator equal to $N_2 - 1$.

---

[3] Ordinarily, X_1 and X_2 are either observations made on independent samples drawn from the same population or measurements of the same attribute in two distinct populations.

Furthermore, if the test hypothesis is correct, and $\hat{\sigma}_1^2/\hat{\sigma}_2^2$ is therefore distributed as Fisher's F, then

$$E\left(\frac{\hat{\sigma}_1^2}{\hat{\sigma}_2^2}\right) = \frac{N_2 - 1}{N_2 - 3}$$

On the other hand, if $\sigma_1^2 \neq \sigma_2^2$, the probability model for the distribution of $\hat{\sigma}_1^2/\hat{\sigma}_2^2$ is a function called *noncentral F*. Under the noncentral F distribution, it can be shown that

$$E\left(\frac{\hat{\sigma}_1^2}{\hat{\sigma}_2^2}\right) > \frac{N_2 - 1}{N_2 - 3}$$

if $\sigma_1^2 > \sigma_2^2$ and that

$$E\left(\frac{\hat{\sigma}_1^2}{\hat{\sigma}_2^2}\right) < \frac{N_2 - 1}{N_2 - 3}$$

if $\sigma_1^2 < \sigma_2^2$. Therefore, the ratio

$$\frac{\hat{\sigma}_1^2}{\hat{\sigma}_2^2} \qquad\qquad [11.5]$$

satisfies all of our criteria for a good test statistic.

If X_1 and X_2 are independent, normally distributed random variables and if $V(X_1) = \sigma_1^2$ and $V(X_2) = \sigma_2^2$, then for samples of N_1 observations on X_1 and N_2 observations on X_2, the appropriate statistic to test the hypothesis that $\sigma_1^2 = \sigma_2^2$ is

$$\frac{\hat{\sigma}_1^2}{\hat{\sigma}_2^2}$$

which is distributed as $F(N_1 - 1, N_2 - 1)$ if the test hypothesis is correct.

---

b. Rejection rule. As always, the rejection region follows the alternative hypothesis. If H_1 is directional and specifies that $\sigma_1^2 > \sigma_2^2$, then the appropriate rejection region lies in the upper tail of $F(N_1 - 1, N_2 - 1)$. For significance level α, the critical value is F_b where $\mathbb{P}(F_b) = 1 - \alpha$, and H_0 is rejected if the test statistic is equal to or less than F_b.

If H_1 is directional and specifies that $\sigma_1^2 < \sigma_2^2$, then the appropriate rejection region lies in the lower tail of $F(N_1 - 1, N_2 - 1)$. For significance level α, the critical value is F_a where $\mathbb{P}(F_a) = \alpha$, and H_0 is rejected if the test statistic is equal to or less than F_a.

If H_1 is nondirectional, the experimenter must conduct a two-tailed test. For significance level α, the critical values are F_a and F_b where $\mathbb{P}(F_a) = \alpha/2$ and $\mathbb{P}(F_b) = 1 - \alpha/2$, and H_0 is rejected if the test statistic is either (equal to) or less than F_a or (equal to) or greater than F_b.

Example 11.9. A sample of 11 Classical Neanderthal European skulls has variance $s_1^2 = 11{,}175.72$, and a sample of 4 Middle Eastern skulls has variance $s_2^2 = 741.69$. Assuming that skull capacities are normally distributed, set $\alpha = .05$ and test the hypothesis that the two populations have the same variance. The test hypothesis in this example is

$$H_0: \sigma_1^2 = \sigma_2^2$$

Let us suppose that on the basis of Middle Eastern skulls found at other sites, the experimenter has independent evidence that Middle Eastern skulls may be less variable than Neanderthal skulls. This suggests a directional alternative hypothesis:

$$H_1: \sigma_1^2 > \sigma_2^2$$

If the test hypothesis is correct, the ratio

$$\frac{\hat{\sigma}_1^2}{\hat{\sigma}_2^2}$$

is distributed $F(10, 3)$, and the critical value is the value F_b for which the cumulative probability is $1 - .05 = .95$. In Table 6 of Appendix VIII we find in the distribution of $F(10, 3)$ that $\mathbb{P}(8.79) \doteq .95$.

The value of the test statistic is

$$\frac{s_1^2 \left(\dfrac{N_1}{N_1 - 1} \right)}{s_2^2 \left(\dfrac{N_2}{N_2 - 1} \right)} \doteq \frac{11{,}175.72 \left(\dfrac{11}{10} \right)}{741.69 \left(\dfrac{4}{3} \right)} \doteq 12.43$$

and the test hypothesis is rejected in favor of the alternative hypothesis, that Classical Neanderthal skulls are more variable than Middle Eastern Neanderthal(?) skulls.

Example 11.10. What is the p-value for the result in Example 11.9? In the table for $F(10, 3)$ we find that 12.43 falls between $F = 8.79$, for which $\mathbb{P} = .95$, and $F = 14.4$, for which $\mathbb{P} = .975$. Therefore, the p-value for this result would be reported as $.025 < p < .05$ or $p < .05$.

───────── EXERCISES 11.2 ─────────────────────────────

In Exercise 1 and 2 the random variable X is distributed as χ^2 with ν_1 degrees of freedom, and the random variable Y is distributed as χ^2 with ν_2 degrees of freedom. The random variable W is defined as

$(X/v_1)/(Y/v_2)$. **Find the probabilities indicated for the specified values of v_1 and v_2.**

1. (a) $P(w \geq .490)$ for $v_1 = 1$ and $v_2 = 10$
   (b) $P(w \leq 2.25)$ for $v_1 = 30$ and $v_2 = 15$
   (c) $P(w \geq 2.73)$ for $v_1 = 24$ and $v_2 = 30$
   (d) $P(w \leq .360)$ for $v_1 = 10$ and $v_2 = 20$
   (e) $P(w \geq 6.54)$ for $v_1 = 11$ and $v_2 = 7$
   (f) $P(.111 \leq w \leq 5.41)$ for $v_1 = 3$ and $v_2 = 5$
   (g) $P(w \geq 3.42$ or $w \leq .360)$ for $v_1 = 20$ and $v_2 = 10$
   (h) $P(w \geq 24{,}400$ or $w \leq .085)$ for $v_1 = 12$ and $v_2 = 1$
   (i) $P(.349 \leq w \leq 2.86)$ for $v_1 = 15$ and $v_2 = 15$
   (j) $P(w \geq 99.4$ or $w \leq .171)$ for $v_1 = 20$ and $v_2 = 2$
2. (a) $P(w \leq 2.32)$ for $v_1 = 10$ and $v_2 = 10$
   (b) $P(w \geq 1.38)$ for $v_1 = 60$ and $v_2 = 15$
   (c) $P(w \leq .040)$ for $v_1 = 5$ and $v_2 = 30$
   (d) $P(w \geq 3.46)$ for $v_1 = 9$ and $v_2 = 20$
   (e) $P(w \leq .433)$ for $v_1 = 20$ and $v_2 = 11$
   (f) $P(w \geq 4.43$ or $w \leq .292)$ for $v_1 = 24$ and $v_2 = 12$
   (g) $P(.046 \leq w \leq 999)$ for $v_1 = 7$ and $v_2 = 2$
   (h) $P(w \geq 7.57$ or $w \leq .001)$ for $v_1 = 1$ and $v_2 = 8$
   (i) $P(.143 \leq w \leq 4.16)$ for $v_1 = 12$ and $v_2 = 12$
   (j) $P(w \geq 22.4$ or $w \leq .123)$ for $v_1 = 15$ and $v_2 = 6$

In Exercises 3 and 4, $X: F(v_1, v_2)$. For each distribution, fill in the expected value of X, the median ($Mdn.$) of X (i.e., the value of X for which $\mathbb{P} = .50$), and calculate $E(X) - Mdn.$

3.

(v_1, v_2)	$(1, 5)$	$(10, 5)$	$(20, 5)$	$(50, 5)$	$(100, 5)$
$E(X)$	————	————	————	————	————
$Mdn.$	————	————	————	————	————
Difference	————	————	————	————	————

What properties of the F-distribution does this exercise suggest?

4.

(v_1, v_2)	$(10, 3)$	$(10, 5)$	$(10, 15)$	$(10, 20)$	$(10, 60)$
$E(X)$	————	————	————	————	————
$Mdn.$	————	————	————	————	————
Difference	————	————	————	————	————

What properties of the F distribution does this exercise suggest?

5. Fill in the blanks in the following table.

	x	v	$P(t \geq x)$	x^2	v_1	v_2	$P(F \geq x^2)$
(a)	———	9	.05	3.36	1	9	———
(b)	2.064	———	.025	4.26	———	24	.05
(c)	1.753	15	———	3.07	1	———	.10
(d)	———	3	.005	34.1	———	3	.01
(e)	———	40	.0005	———	1	40	.001

From answers to (a) through (e), we might imagine that if $t_i > 0$, then $P(F \geq \underline{\quad}) = 2P(t \geq \underline{\quad})$ for $t(\nu)$ and $F(\underline{\quad}, \underline{\quad})$.

6. Fill in the blanks in the following table.

	χ^2	ν	$\mathbb{P}$	χ^2/ν	F	ν_1	ν_2	$\mathbb{P}$
(a)	———	1	.90	———	2.71	1	∞	———
(b)	12.5	———	.75	———	1.25	———	∞	.75
(c)	———	24	.50	———	.972	24	———	.50
(d)	4.60	15	———	———	.307	———	∞	.005
(e)	———	30	.01	———	———	30	∞	.01

From your answers to (a) through (e) you might suspect that in general if $X: F(\nu, \underline{\quad})$ and $Y: \chi^2(\underline{\quad})$, then for any given cumulative probability, $x = \underline{\quad}$. This means that $\chi^2(1)$ is exactly the same distribution as $\underline{\quad\quad}$.

7. Let $X: \chi^2(\nu_1)$, let $Y: \chi^2(\nu_2)$, and let $W = (X/\nu_1)/(Y/\nu_2)$. Find the following probabilities.
   (a) $P(w \leq .194)$ for $\nu_1 = 6$ and $\nu_2 = 2$
   (b) $P(w \geq 1/.194)$ for $\nu_1 = 2$ and $\nu_2 = 6$
   (c) $P(w \leq .219)$ for $\nu_1 = 7$ and $\nu_2 = 15$
   (d) $P(w \geq 1/.219)$ for $\nu_1 = 15$ and $\nu_2 = 7$
   (e) From (a) through (d) it is apparent that if $P(w \leq w_i) = \alpha$ in $F(\nu_1, \nu_2)$, then $P(w \geq 1/w_i) = \alpha$ in $F(\underline{\quad}, \underline{\quad})$.

8. Let $X: \chi^2(\nu_1)$, let $Y: \chi^2(\nu_2)$, and let $W = (X/\nu_1)/(Y/\nu_2)$. Find the following probabilities.
   (a) $P(w \geq 2.61)$ for $\nu_1 = 5$ and $\nu_2 = 9$
   (b) $P(w \leq 1/2.61)$ for $\nu_1 = 9$ and $\nu_2 = 5$
   (c) $P(w \geq 6.61)$ for $\nu_1 = 20$ and $\nu_2 = 8$
   (d) $P(w \leq 1/6.61)$ for $\nu_1 = 8$ and $\nu_2 = 20$
   (e) From (a) through (d) it is apparent that if $P(w \geq w_i) = \alpha$ in $F(\nu_1, \nu_2)$, then $P(w \leq \underline{\quad}) = \alpha$ in $F(\nu_2, \nu_1)$.

In Exercises 9 and 10 test the hypothesis that $\sigma_1^2 = \sigma_2^2$ against the indicated alternative hypothesis for the given values of s_1^2, s_2^2, N_1, N_2, and α. Report the p-value of the result and the appropriate conclusion.

9.

	$H_1: \sigma_1^2$	s_1^2	N_1	s_2^2	N_2	α
(a)	$> \sigma_2^2$	5,638	6	1,625	9	.05
(b)	$< \sigma_2^2$	48.2	16	182.2	21	.01
(c)	$\neq \sigma_2^2$	29,500	10	113,100	25	.05
(d)	$\neq \sigma_2^2$	8,100	21	2,716	31	.01
(e)	$< \sigma_2^2$	3.39×10^{-5}	25	8.47×10^{-5}	10	.025

10.

	$H_1: \sigma_1^2$	s_1^2	N_1	s_2^2	N_2	α
(a)	$\neq \sigma_2^2$	2,450	13	1,200	41	.05
(b)	$< \sigma_2^2$	185	3	7,400	5	.025
(c)	$> \sigma_2^2$	.267	25	.134	25	.05
(d)	$> \sigma_2^2$	33,460	16	20,275	11	.01
(e)	$\neq \sigma_2^2$	9,751	7	609	7	.01

11. In Example 10.10.4 (p. 435) it was reported that the difference between oddity-learning performance by four wolf pups and four Alaskan Malamute pups was not significant. In Example 10.10.5 (p. 435) a group of seven wolves did perform significantly better than the dogs. The first group of wolves was mother-reared and the second group was hand-reared. It was suspected that the poor overall performance of mother-reared wolves was because they were more variable than the hand-reared wolves in their reaction to the "artificial" food used as reward. Variances for the two groups were compared.

Mother-reared wolves ($N = 4$): $s^2 = 2{,}081.25$

Hand-reared wolves ($N = 7$): $s^2 = 238.77$

(a) What are the hypotheses?
(b) What is the appropriate test statistic?
(c) Set α at any level you consider suitable and state the rejection rule appropriate to your hypotheses and the significance level you have chosen.
(d) Calculate the value of your test statistic.
(e) Report the appropriate conclusion.

12. A company that sells replicas of medieval weapons and armor establishes specifications and has the forgework and assembly performed by small smithies in various parts of the world. An Austrian smith and an Italian smith both produce replicas of the same eighth-century Viking sword. The contractor believes that the swords produced by the Italian smith are more variable in weight than those produced in Austria. Two shipments of weapons are carefully weighed, with the following results:

Italian swords		Austrian swords	
Pounds	**Ounces**	**Pounds**	**Ounces**
3	2	3	3
2	14	3	5
3	7	3	4
3	5	3	7
3	1	2	15
3	0	3	1
2	12	2	14
2	15	2	10
2	10	2	12
3	9	3	6

(a) What are the hypotheses?
(b) What is the appropriate test statistic?
(c) Set α at any level you consider suitable and state the rejection rule appropriate to your hypotheses and the significance level you have chosen.
(d) Calculate the value of your test statistic.
(e) Report the appropriate conclusion.

13. In Exercise 10.5.6 (p. 442) the chair of the Psychology Department separated scores on the Graduate Record Examination Advanced Test in Psychology

according to the degree program from which the students graduated. The data are as follows:

B.S. Scores						B.A. Scores			
460	534	569	320	390	472	355	410	420	442
540	462	528	545	580	497	480	502	601	609
620	455	508	518	545	558	655	485	515	445
619	398	408	468	510	660	350	380	435	430
571	599	635	370	425	519	478	585	359	350

Because the B.S. program has more prescribed courses and allows fewer electives than the B.A. program, she suspects that B.S. graduates will exhibit less variation in test scores.

(a) What are her hypotheses?

(b) What is the appropriate test statistic?

(c) Set α at any level you consider suitable and state the rejection rule appropriate to your hypotheses and the significance level you have chosen.

(d) Calculate the value of your test statistic.

(e) Report the appropriate conclusion.

14. Several years ago one of the authors taught a daytime section and a night section of the same course. Students in the night section were a more varied group (students in other disciplines satisfying cognate requirements, students with full-time jobs, retired persons, etc.) than the day students, and the author suspected that they might also be more variable in academic performance.

Day: 19, 19, 17, 17, 16, 16, 15, 15, 11, 11, 18, 6, 14,
        16, 12, 15, 13, 18,   0, 19, 14, 15, 12, 11, 13

Night: 22, 22, 20, 19, 17, 16, 15, 14, 12, 6, 2, 9, 21

(a) What are his hypotheses?

(b) What is the appropriate test statistic?

(c) Set α at any level you consider suitable and state the rejection rule appropriate to your hypotheses and the significance level you have chosen.

(d) Calculate the value of your test statistic.

(e) Report the appropriate conclusion.

E. SUMMARY

If Y is the sum of squares of ν independent squared standard normal random variables,

$$Y = Z_1^2 + Z_2^2 + \cdots + Z_\nu^2$$

then Y is said to be distributed as χ^2 with ν degrees of freedom. The parameter for χ^2 is ν, and every possible value of ν defines a different distribution. The following are true for all χ^2 distributions:

- A *chi*-square random variable can assume only nonnegative values.
- If $Y : \chi^2(\nu)$, then $E(Y) = \nu$ and $V(Y) = 2\nu$.
- If $Y : \chi^2(\nu_1)$ and $Y' : \chi^2(\nu_2)$, then if Y and Y' are independent, $Y + Y' : \chi^2(\nu_1 + \nu_2)$ and if $\nu_1 > \nu_2$, $Y - Y' : \chi^2(\nu_1 - \nu_2)$.

The distribution of χ^2 for small values of ν is positively skewed, but χ^2 becomes less skewed as ν increases. For very large values of ν the distribution is almost symmetrical. This is because χ^2 models the distribution of the sum of ν independent and identically distributed random variables and therefore satisfies the conditions under which the Central Limit Theorem applies. As ν increases, χ^2 therefore approaches the normal distribution, with $\mu = \nu$ and $\sigma^2 = 2\nu$.

For a sample of N independent observations on a random variable $X : N(\mu, \sigma_0^2)$, the test statistic given in **[11.2]**

$$\frac{\hat{\sigma}^2(N-1)}{\sigma_0^2}$$

is distributed $\chi^2(N-1)$ and therefore has an expected value of $N-1$.

If $\sigma^2 > \sigma_0^2$, the expected value of the test statistic **[11.2]** is larger than $N-1$, and the critical region for the alternative hypothesis $\sigma^2 > \sigma_0^2$ therefore falls in the upper tail of $\chi^2(N-1)$.

If $\sigma^2 < \sigma_0^2$, the expected value of the test statistic is smaller than $N-1$, and the critical region for the alternative hypothesis $\sigma^2 < \sigma_0^2$ therefore falls in the lower tail of $\chi^2(N-1)$.

An interval estimate for σ^2 can be calculated as follows: If $X : N(\mu, \sigma^2)$, then for N independent observations, confidence equals $\mathbb{C}$ that

$$\frac{(N-1)\hat{\sigma}^2}{\chi_b^2} \leq \sigma^2 \leq \frac{(N-1)\hat{\sigma}^2}{\chi_a^2}$$

where $\mathbb{P}(\chi_a^2) = (1 - \mathbb{C})/2$ and $\mathbb{P}(\chi_b^2) = (1 + \mathbb{C})/2$ in $\chi^2(N-1)$.

If $Y : \chi^2(\nu_1)$ and if $W : \chi^2(\nu_2)$, then if Y and W are independent,

$$\frac{Y/\nu_1}{W/\nu_2}$$

is distributed as Fisher's F with ν_1 and ν_2 degrees of freedom, denoted $F(N_1 - 1, N_2 - 1)$. The expected value of Fisher's F is $\nu_2/(\nu_2 - 2)$.

If $X_1 : N(\mu_1, \sigma_1^2)$, then for a sample of N_1 independent observations $\hat{\sigma}_1^2/\sigma_1^2$ is a *chi*-square random variable divided by its own degrees of freedom, $\nu_1 = N_1 - 1$. Likewise, if $X_2 : N(\mu_2, \sigma_2^2)$, then for a sample of N_2 independent observations $\hat{\sigma}_2^2/\sigma_2^2$ is a *chi*-square random variable divided by its own degrees of freedom, $\nu_2 = N_2 - 1$. Therefore, if X_1 and

X_2 are independent,

$$\frac{\hat{\sigma}_1^2/\sigma_1^2}{\hat{\sigma}_2^2/\sigma_2^2}$$

is distributed as Fisher's F with $N_1 - 1$ and $N_2 - 1$ degrees of freedom. If $\sigma_1^2 = \sigma_2^2$, it follows that the statistic given in [11.5]

$$\frac{\hat{\sigma}_1^2}{\hat{\sigma}_2^2}$$

is distributed as $F(N_1 - 1, N_2 - 1)$ and has an expected value of $(N_2 - 1)/(N_2 - 3)$.

If $\sigma_1^2 > \sigma_2^2$, the expected value of the test statistic [11.5] is larger than $(N_2 - 1)/(N_2 - 3)$, and the critical region for the alternative hypothesis $\sigma_1^2 > \sigma_2^2$ therefore falls in the upper tail of $F(N_1 - 1, N_2 - 1)$.

If $\sigma_1^2 < \sigma_2^2$, the expected value of the test statistic is less than $(N_2 - 1)/(N_2 - 3)$, and the critical region for the alternative hypothesis $\sigma_1^2 < \sigma_2^2$ therefore falls in the lower tail of $F(N_1 - 1, N_2 - 1)$.

Summary table of statistical tests

I. Testing hypotheses about σ^2

Distribution of X	H_0	H_1	Test statistic	Distribution of test statistic*
Normal	$\sigma^2 = \sigma_0^2$	Any	$\dfrac{\hat{\sigma}^2(N - 1)}{\sigma_0^2}$	$\chi^2(N - 1)$
Assumption: Observations of X independent				

II. Testing hypotheses about σ_1^2 and σ_2^2

Distribution of X_1 and X_2	H_0	H_1	Test statistic	Distribution of test statistic*
Normal	$\sigma_1^2 = \sigma_2^2$	Composite	$\dfrac{\hat{\sigma}_1^2}{\hat{\sigma}_2^2}$	$F(N_1 - 1, N_2 - 1)$
Assumptions: Observations of both X_1 and X_2 independent X_1 and X_2 independent of one another				

*If H_0 correct.

CHAPTER 12

Testing hypotheses about many population means,

$\mu_1, \mu_2, \ldots, \mu_J$: introduction to analysis of variance

CHAPTER OUTLINE

D. DECISIONS: TEST STATISTIC AND REJECTION RULE FOR ANOVA

Mean squares
- Expected value of MS_W
- Expected value of MS_A
- Distribution of MS_A/MS_W

Rejection rule for ANOVA

E. RESULTS: THE ANOVA SUMMARY TABLE

Exercises 12.3

F. EVALUATION OF THE ANALYSIS OF VARIANCE: POST HOC COMPARISONS

Formulation: Hypotheses about contrasts

Scheffé's test statistic

Exercises 12.4

G. SUMMARY

A. DOUBLE SUMMATION NOTATION

In Chapter 1 we introduced Σ-notation, which has allowed us to write expressions of the form

$$\sum_i^N x_i$$

to represent a sum of N numbers. The numbers in such a sum ordinarily constitute a single collection, such as the SAT scores for N applicants for admission to a university. In some situations, however, a collection of N numbers can be separated into J distinct groups, where Group 1 includes N_1 observations, Group 2 includes N_2 observations, ..., and Group J includes N_J observations.

Example 12.1. Scholastic Aptitude Test (SAT) scores of 15 college applicants in Table 12.1 have been separated according to the high school from which each applicant graduated.

The sum of SAT scores for high school 1 is

$$\sum_i^{N_1} x_i = \sum_i^5 x_i = 5{,}500;$$

Table 12.1. SAT scores of 15 college applicants
representing three high schools

High school 1	High school 2	High school 3
$x_1 = 1{,}026$	$x_1 = 1{,}100$	$x_1 = 916$
$x_2 = 995$	$x_2 = 1{,}180$	$x_2 = 1{,}120$
$x_3 = 1{,}255$	$x_3 = 860$	$x_3 = 1{,}077$
$x_4 = 1{,}121$	$x_4 = 1{,}031$	$x_4 = 928$
$x_5 = 1{,}103$	$x_5 = 1{,}079$	$x_5 = 959$
Σ 5,500	5,250	5,000

The sum for the second high school is

$$\sum_{i}^{N_2} x_i = \sum_{i}^{5} x_i = 5{,}250$$

and the sum for High School 3 is

$$\sum_{i}^{N_3} x_i = \sum_{i}^{5} x_i = 5{,}000$$

The sum of SAT scores for the entire pool of $N_1 + N_2 + N_3 = N$ applicants is therefore

$$\sum^{N} x = \sum_{i}^{N_1} x_i + \sum_{i}^{N_2} x_i + \sum_{i}^{N_3} x_i = 5{,}500 + 5{,}250 + 5{,}000 = 15{,}750$$

In situations like this, the notation x_i is ambiguous. For example, one can't tell if x_2 in Table 12.1 refers to the second score in Group 1 ($x_2 = 995$), the second score in Group 2 ($x_2 = 1{,}180$), or the second score in Group 3 ($x_2 = 1{,}120$). We therefore identify each score with *two* subscripts, x_{ij}, much as a person has two names. The second subscript, like a surname, indicates the group (or family) membership and therefore runs from 1 to J. The first subscript distinguishes observations within the same group, just as a given name distinguishes family members from one another. For Group j, the subscript i therefore runs from 1 to N_j. In the present example the score x_{23} refers unambiguously to the second score in the group of applicants from High School 3, which is equal to 1,120.

In general, therefore, if we have a total of N observations distributed among J groups with $N_1, N_2, \ldots, N_J$ observations, respectively, then

$$\sum_i^N x = \sum_i^{N_1} x_{i1} + \sum_i^{N_2} x_{i2} + \cdots + \sum_i^{N_J} x_{iJ} \qquad [\,12.1\,]$$

This tells us that the overall sum,

$$\sum_N x_i$$

also called the *grand* sum, can be thought of as a sum of J *group* sums. The sum of sums in expression [**12.1**] is made more compact with the use of *double* summation signs:

$$\underbrace{\sum_i^{N_1} x_{i1} + \sum_i^{N_2} x_{i2} + \cdots + \sum_i^{N_J} x_{iJ}}_{J \text{ sums}} = \sum_j^J \left(\sum_i^{N_j} x_{ij} \right) \qquad [\,12.2\,]$$

Double summations are ordinarily written without parentheses,

$$\sum_j^J \sum_i^{N_j} x_{ij}$$

so our grand sum of 15 SAT scores might be expressed as

$$\sum_j^3 \sum_i^5 x_{ij} = 15{,}750 \qquad [\,12.3\,]$$

To simplify notation, the upper limits of summation are often omitted, so when the reader encounters the expression

$$\sum_j \sum_i x_{ij}$$

it must be remembered that

$$\sum_i$$

(called the *within-groups sum*) runs from 1 to N_j, the number of observations in Group j, and

$$\sum_j$$

(called the *across-groups sum*) runs from 1 to J, the number of groups.

The order in which the subscripted Σ-signs are written indicates the order in which the summations are to be performed. If you look at the right-hand side of expression [12.2], it is clear that the summation inside the parentheses (the within-groups sum) must be calculated *before* the summation on the outside of the parentheses (the across-groups sum). If the data are laid out as in Table 12.1, one first computes every column total and then adds the J totals.

Even without parentheses, the order of operations is always the same: Start with the summation nearest the summand, x, and work to the left. The conventions for double summation generalize to triple summations, quadruple summations, and so on. For example, the expression

$$\sum_{k}^{K} \sum_{j}^{J} \sum_{i}^{N_{jk}} x_{ijk}$$

means that the reader should first add with respect to i and obtain JK subgroup sums, then add the J sums in each of the K groups, then add the K group sums. That is,

$$\sum_{k}^{K} \sum_{j}^{J} \sum_{i}^{N_{jk}} x_{ijk} = \sum_{k}^{K} \left(\sum_{j}^{J} \left(\sum_{i}^{} x_{ijk} \right) \right)$$

1. The algebra of double summations

The Algebra of Summations (Appendix I) introduced in Chapter 2 can be expanded to include rules governing multiple summations.

Example 12.2. Let us suppose that all of the SAT scores in Example 12.1 are multiplied by 10. Then,

$$\sum_{j}^{3} \sum_{i}^{N_{j}} 10x_{ij} = \sum_{i}^{N_{1}} 10x_{i1} + \sum_{i}^{N_{2}} 10x_{i2} + \sum_{i}^{N_{3}} 10x_{i3}$$

$$= 10 \sum_{i}^{N_{1}} x_{i1} + 10 \sum_{i}^{N_{2}} x_{i2} + 10 \sum_{i}^{N_{3}} x_{i3}$$

$$= 10 \left(\sum_{i}^{N_{1}} x_{i1} + \sum_{i}^{N_{2}} x_{i2} + \sum_{i}^{N_{3}} x_{i3} \right) = 10 \sum_{j}^{3} \sum_{i}^{N_{j}} x_{ij}$$

And, in general, if all N scores are multiplied by a constant, c, then, the grand sum of the products cx_{ij} is

$$\sum_{j}^{J} \sum_{i}^{N_{j}} cx_{ij} = c \sum_{j}^{J} \sum_{i}^{N_{j}} x_{ij}$$

Example 12.3. Let us suppose that all of the SAT scores for High School 1 are multiplied by 2, all the scores for High School 2 are multiplied by 3, and all of the scores for High School 3 are multiplied by 4. Then,

$$\sum_{i}^{N_1} 2x_{i1} + \sum_{i}^{N_2} 3x_{i2} + \sum_{i}^{N_3} 4x_{i3} = 2\sum_{i}^{N_1} x_{i1} + 3\sum_{i}^{N_2} x_{i2} + 4\sum_{i}^{N_3} x_{i3}$$

And, in general, if all of the scores in Group j are multiplied by a constant c_j, then the grand sum of products $c_j x_{ij}$ is

$$\sum_{j}^{J} \sum_{i}^{N_j} c_j x_{ij} = \sum_{j}^{J} c_j \sum_{i}^{N_j} x_{ij}$$

If we put parentheses around the within-group sum $\sum_{i}$, the preceding expression becomes

$$\sum_{j}^{J} \left(\sum_{i}^{N_j} c_j x_{ij} \right) = \sum_{j}^{J} \left(c_j \sum_{i}^{N_j} x_{ij} \right)$$

and we see that this is nothing more than an application of the old and by now familiar business of factoring a constant out of a sum. From Example 12.3 we can also see that c_j is a *variable* across the J groups, so it is obvious that c_j can not be factored out of the across-group sum $\sum_{j}$.

────── EXERCISES 12.1 ──────────────────────────────────

Exercises 1–14 refer to the following set of data.

Group 1	Group 2	Group 3	Group 4	Group 5
5	10	15	12	14
6	22	19	10	11
7	18	21	19	22
	51	24	3	26
			22	19
			24	

1. $N =$ _____   **2.** $J =$ _____

3. $N_4 =$ _____   **4.** $x_{42} =$

5. The observation with numerical value 12 can be denoted _____ .

6. $\sum^{N_3} x_{i3} =$ _____   **7.** $\sum_{j=1}^{3} \sum_{i=1}^{N_j} x_{ij} =$ _____

8. $\sum_{j=4}^{5} \sum_{i=1}^{N_j} x_{ij} = $ _____

9. $\bar{x}_3 = $ _____

10. $\bar{x} = $ _____

11. $\sum_{j}^{J} N_j \bar{x}_j = $ _____

12. $\sum_{j}^{J} \sum_{i}^{N_j} x_{ij} = $ _____

13. Write the formula for the mean of Group 4. Include numerical values of the limits of summation.

14. $\sum_{i}^{N} x_i = $ _____

Exercises 15–28 refer to the following set of data:

Group 1	Group 2	Group 3	Group 4
11	29	18	31
15	22	11	28
19	21	14	5
23	27	19	14
26	29	15	19
	32	20	16
			21

15. $J = $ _____

16. $N = $ _____

17. $x_{52} = $ _____

18. $N_3 = $ _____

19. $\sum_{i}^{N_2} x_{i2} = $ _____

20. The observation with numerical value 27 can be denoted _____.

21. $\bar{x} = $ _____

22. $\bar{x}_2 = $ _____

23. $\sum_{j}^{J} \sum_{i}^{N_j} x_{ij} = $ _____

24. $\sum_{j}^{J} N_j \bar{x}_j = $ _____

25. $\sum_{i}^{N} x_i = $ _____

26. Write the formula for the mean of Group 3. Include numerical values of the limits of summation.

27. Every observation is multiplied by 14.3. Calculate the sum of these products and express the sum in double summation notation.

28. Every observation in Group j is multiplied by N_j/N. Calculate the sum of these products over all N observations. Express the sum of products in summation notation.

B. LOGIC AND OVERVIEW OF ANALYSIS OF VARIANCE (ANOVA)

1. Formulation

Suppose that the director of admissions at a university wants to find out whether the three high schools in the immediate area provide comparable levels of college preparation. The director therefore takes SAT scores for large samples of applicants from these high schools and organizes them like the data in Table 12.1. The layout of the data, expressed in general notation, is given in Table 12.2:

Table 12.2. SAT scores of college applicants representing three high schools

High school 1	High school 2	High school 3	
x_{11}	x_{12}	x_{13}	
x_{21}	x_{22}	x_{23}	
. . .	. . .	. . .	
$x_{N_1 1}$	$x_{N_2 2}$	$x_{N_3 3}$	
$\displaystyle\sum^{N_1} x_{i1}$	$\displaystyle\sum^{N_2} x_{i2}$	$\displaystyle\sum^{N_3} x_{i3}$	$\displaystyle\sum^{N} x$
$\bar{x}_1$	$\bar{x}_2$	$\bar{x}_3$	$\bar{x}$

In Table 12.2 we have included the group means, $\bar{x}_j$, as well as the grand sum

$$\sum^{N} x = \sum^{N_1} x_{i1} + \sum^{N_2} x_{i2} + \sum^{N_3} x_{i3} = \sum\sum x$$

and grand mean

$$\bar{x} = \frac{\sum\sum x}{N} \qquad\qquad [12.4]$$

where $N = N_1 + N_2 + N_3$.

To simplify things and to satisfy certain assumptions discussed later, we shall assume for the moment that all three samples (i.e., groups) are the same size, which we denote N_g. That is,

$$N_1 = N_2 = N_3 = N_g$$

The grand mean $\bar{x}$ is therefore equal to the mean of the group means:

$$\frac{\bar{x}_1 + \bar{x}_2 + \bar{x}_3}{3} = \frac{\dfrac{\sum x_{i1}}{N_1} + \dfrac{\sum x_{i2}}{N_2} + \dfrac{\sum x_{i3}}{N_3}}{3} = \frac{\dfrac{\sum x_{i1}}{N_g} + \dfrac{\sum x_{i2}}{N_g} + \dfrac{\sum x_{i3}}{N_g}}{3}$$

$$= \frac{\dfrac{1}{N_g}\left(\sum x_{i1} + \sum x_{i2} + \sum x_{i3}\right)}{3} = \frac{\sum x_{i1} + \sum x_{i2} + \sum x_{i3}}{3N_g}$$

$$= \frac{\sum\sum x}{N}$$

a. Hypotheses. The basic question that the admissions officer wants to ask is whether the three high school populations have the same mean for SAT scores. Therefore, the test hypothesis in this example is

$$H_0: \mu_1 = \mu_2 = \mu_3$$

and in general, the test hypothesis for equality of J population means takes the form

$$H_0: \mu_1 = \mu_2 = \cdots = \mu_J$$

The test hypothesis is incorrect if two or more of the populations have different means, so the most general form of the alternative hypothesis is, simply,

$$H_1: H_0 \text{ incorrect}$$

2. Decisions

a. Test statistic. In Chapter 10 we discussed a number of ways to test hypotheses about means, and at first blush it might seem an obvious tactic to test for the equality of several population means by testing the hypothesis

$$H_0: \mu_i - \mu_j = 0$$

for every pair of populations. With three populations this would require three separate tests:

$$H_0: \mu_1 - \mu_2 = 0 \qquad H_0: \mu_1 - \mu_3 = 0 \qquad H_0: \mu_2 - \mu_3 = 0$$
$$H_1: \mu_1 - \mu_2 \neq 0 \qquad H_1: \mu_1 - \mu_3 \neq 0 \qquad H_1: \mu_2 - \mu_3 \neq 0$$

This approach creates a number of problems. First, if all tests are conducted at significance level α, the overall significance level is *greater* than α. Let us suppose for the moment that the three tests given above are independent of one another. If we let $\alpha = .05$ for each test, the probability of *avoiding* a Type I error on each test is $1 - .05 = .95$. By the Multiplication Theory for Independent Events, the probability of avoiding a Type I error on *all three* tests is therefore

$$(.95)(.95)(.95) = (.95)^3 \doteq .8574$$

which means that the probability of *making* a Type I error on *at least* one of the tests is approximately

$$1 - .8574 = .1426$$

That is, our *overall* α is not .05, but about .14. And, in general, if K independent tests are conducted at significance level α, the probability of at least one Type I error is

$$1 - (1 - \alpha)^K$$

The calculations in the preceding paragraph assumed that our three tests are independent. It is easily seen that they are not. Suppose we draw samples from all three populations. If $\bar{x}_1 - \bar{x}_2 = -10$ and $\bar{x}_1 - \bar{x}_3 = 5$, we immediately know that $\bar{x}_2 - \bar{x}_3 = 15$:

$$\bar{x}_2 - \bar{x}_3 = (\bar{x}_1 - \bar{x}_3) - (\bar{x}_1 - \bar{x}_2) = 5 - (-10) = 15$$

In this instance, only two of the differences are independent, and in the general case of J samples, the largest possible number of independent differences is $J - 1$. For three or more samples, the number of possible pairs $({}_JC_2)$ is always greater than $J - 1$, so if one tests every difference, some tests will necessarily be dependent. When all K tests are *not* independent, the overall probability of a Type I error may be impossible to calculate, but if all tests are conducted at significance level α, the probability can be as great as $K\alpha$, assuming, of course, that $K\alpha$ does not exceed 1.

One way to resolve the problem is to test every difference at a significance level of α/K. This ensures that the overall probability of a Type I error is no greater than α, but the dependency among tests makes multiple results difficult to interpret. In the case of the three sample means discussed above, the third difference, $\bar{x}_2 - \bar{x}_3 = 15$, yields redundant information, that is, information that is implicit in the first two differences. Nevertheless, the third difference might well be significant even if the first two (-10 and 5) are not. This is difficult to reconcile with the redundancy of the third difference and leaves the experimenter with results that support the hypotheses that $\mu_3 = \mu_1$ and that $\mu_2 = \mu_1$ as well as the contradictory hypothesis that $\mu_2 \neq \mu_3$.

A strategy that circumvents these difficulties was developed by Sir Ronald Fisher in the 1920s and hinges on the reasoning that differences among J *population* means $\mu_1, \mu_2, \ldots, \mu_J$ should be reflected in *variability* among the means of *samples* drawn from these populations, $\bar{x}_1, \bar{x}_2, \ldots, \bar{x}_J$.

Variability among sample means. In Chapter 8 we discovered that we could estimate the variance of the sample mean $\sigma_{\bar{X}}^2$ by calculating the unbiased estimate of variance $\hat{\sigma}^2$ from a sample of N observations on X and dividing by N. In the present circumstance we have J samples and may view their sample means $\bar{x}_1, \ldots, \bar{x}_J$ as J observations on the random variable $\bar{X}$. This suggests an obvious alternative strategy for estimating $\sigma_{\bar{X}}^2$. In general, we know from expression **[8.9]** that the variance σ^2 of *any* random variable X can be estimated by

$$\frac{\sum(x_i - \bar{x})^2}{N - 1}$$

where N is the number of observations. We have already established that for J samples of equal size, the mean of the sample means $\bar{x}_j$ is equal to the grand mean $\bar{x}$. For J observations on the random variable $\bar{X}$, the variance of $\bar{X}$ can therefore be estimated by

$$\frac{\sum\limits_{}^{J}\left(\bar{x}_j - \bar{x}\right)^2}{J - 1} = \frac{\left(\bar{x}_1 - \bar{x}\right)^2 + \cdots + \left(\bar{x}_J - \bar{x}\right)^2}{J - 1}$$

which we denote est.$\sigma_{\bar{X}}^2$:

$$\text{est.}\sigma_{\bar{X}}^2 = \frac{\sum\limits_{}^{J}\left(\bar{x}_j - \bar{x}\right)^2}{J - 1} \qquad \qquad [12.5]\,^1$$

From Chapter 8 we know that for samples of size N, the variance of the mean $\sigma_{\bar{X}}^2$ is σ^2/N. For samples of size N_g, therefore,

$$\sigma_{\bar{X}}^2 = \frac{\sigma^2}{N_g}$$

and multiplying both sides by the sample size N_g,

$$N_g\sigma_{\bar{X}}^2 = \sigma^2$$

Similarly, the product

$$N_g\left(\text{est.}\sigma_{\bar{X}}^2\right)$$

provides an *estimate* of σ^2.

In our SAT example the grand population variance σ^2 captures overall variability in test scores for applicants from all three schools, so the quantity $N_g(\text{est.}\sigma_{\bar{X}}^2)$ must likewise reflect *all* sources of test performance variability. Some of this variability will derive from individual differences in ability, motivation, test anxiety, and other factors that account for performance differences in any group of subjects. The variability attributable to such sources is called *error variance*.

In addition, some of the variability among test scores may be attributable to differences among the high schools from which the applicants graduated. Analysis of variance evolved in the context of experimental research, where subjects are assigned to groups that are administered different experimental treatments, so this source of variability is called *treatment variance*.

---

[1] The quantity $\hat{\sigma}^2/N$ is also an unbiased estimator of $\sigma_{\bar{X}}^2$, but it is not algebraically equivalent to expression [12.5] and seldom yields the same value. The two estimators must therefore be distinguished notationally, and although "est." is awkward, it is a common alternative to "^" for denoting "estimator of."

However, if the test hypothesis is correct, there is *no* treatment variance, and any variability observed among our three sample means must be the sort we would expect whenever we draw three independent samples from the same population, that is, error variance, which we denote σ_e^2. Therefore, if our test hypothesis is correct, $N_g(\text{est.}\sigma_{\bar{X}}^2)$ is an unbiased estimator of σ_e^2. That is, the estimate of variance among sample means multiplied by sample size is an unbiased estimate of error variance. On the other hand, if the test hypothesis is not correct, then $\text{est.}\sigma_{\bar{X}}^2$ taps treatment variance *in addition* to error variance, and the value of $N_g(\text{est.}\sigma_{\bar{X}}^2)$ is therefore expected to be *larger* than σ_e^2.

Variability within groups. Now, let us consider the variability among scores *within* any single group, say, High School j. For N_j observations the estimate of variance in SAT scores for all graduates of this high school is

$$\hat{\sigma}_j^2 = \frac{\displaystyle\sum^{N_j} (x_{ij} - \bar{x}_j)^2}{N_j - 1}$$

Since every member of this group graduated from the *same* high school, they all had the same *treatment*, so $\hat{\sigma}_j^2$ can reflect *no* treatment variance. That is, *all* of the variability in test performance among applicants who graduated from the same high school must be entirely attributable to error variance *whether or not* the test hypothesis is correct.

In our example this means that we have three independent estimates of error variance, $\hat{\sigma}_1^2$, $\hat{\sigma}_2^2$, and $\hat{\sigma}_3^2$. From Chapter 10 we know that the best possible estimate of error variance can therefore be obtained by pooling our three independent estimates (see Box 10.5, p. 430):

$$\hat{\sigma}_{\text{POOLED}}^2 = \frac{(N_1 - 1)\hat{\sigma}_1^2 + (N_2 - 1)\hat{\sigma}_2^2 + (N_3 - 1)\hat{\sigma}_3^2}{N_1 + N_2 + N_3 - 3} \qquad \textbf{[12.6]}$$

$$= \hat{\sigma}_e^2$$

We see, then, that expressions **[12.5]** and **[12.6]** are both estimators of σ_e^2, but with a difference that is central to our hypothesis of equal population means and that leads us to a test statistic:

- $N_g(\text{est.}\sigma_{\bar{X}}^2)$ estimates σ_e^2 *only if H_0 is correct.*
- $\hat{\sigma}_{\text{POOLED}}^2$ estimates σ_e^2 *whether or not H_0 is correct.*

When the test hypothesis is correct, **[12.5]** and **[12.6]** are therefore estimates of the *same* variance (σ_e^2), and from Chapter 11 this means that

$$\frac{N_g\left(\text{est.}\sigma_{\bar{X}}^2\right)}{\hat{\sigma}_{\text{POOLED}}^2} \qquad \textbf{[12.7]}$$

is distributed as Fisher's F with expected value $v_2/(v_2 - 2)$, where v_2 is degrees of freedom for the denominator, $\hat{\sigma}^2_{\text{POOLED}}$. We'll discuss degrees of of freedom later in the chapter. The point to note right now is that if H_1 is correct, the numerator $N_g(\text{est.}\sigma^2_{\bar{X}})$ captures treatment effects in addition to error, so

$$E\left[N_g\left(\text{est.}\sigma^2_{\bar{X}}\right)\right] > E\left(\hat{\sigma}^2_{\text{POOLED}}\right)$$

and the expected value of [**12.7**] must be greater than $v_2/(v_2 - 2)$.

If all samples are of the same size N_g (and assuming we can furnish parameter values v_1 and v_2), the ratio given in expression [**12.7**] therefore satisfies the requirements for a test statistic for the hypothesis.

$$H_0 \colon \mu_1 = \mu_2 = \cdots = \mu_J$$

Example 12.4. If we use the data in Table 12.1 to illustrate calculation of [**12.7**], we obtain the following: For the numerator of [**12.7**],

$$\text{est.}\sigma^2_{\bar{X}} = \frac{\sum\limits_{}^{J}\left(\bar{x}_j - \bar{x}\right)^2}{J - 1}$$

$$= \frac{\left(\dfrac{5{,}500}{5} - 1{,}050\right)^2 + \left(\dfrac{5{,}250}{5} - 1{,}050\right)^2 + \left(\dfrac{5{,}000}{5} - 1{,}050\right)^2}{2}$$

$$= 2{,}500$$

so

$$N_g\left(\text{est.}\sigma^2_{\bar{X}}\right) = 5(2{,}500) = 12{,}500$$

If we calculate the unbiased estimator of the variance for each of the three groups, we obtain

$$\hat{\sigma}^2_1 = 10{,}244$$

$$\hat{\sigma}^2_2 = 14{,}175.5$$

$$\hat{\sigma}^2_3 = 8{,}562.5$$

so the pooled estimate of the variance is

$$\hat{\sigma}^2_{\text{POOLED}} = \frac{4(10{,}244) + 4(14{,}175.5) + 4(8{,}562.5)}{12}$$

$$= 10{,}994$$

Therefore,

$$\frac{N_g\left(\text{est.}\sigma_{\bar{X}}^2\right)}{\hat{\sigma}_{\text{POOLED}}^2} = \frac{12{,}500}{10{,}994} \doteq 1.14$$

b. Rejection rule. If the test hypothesis is correct, we know that the test statistic [**12.7**] is distributed as $F(\nu_1, \nu_2)$. We can't define the rejection region until we have values for ν_1 and ν_2, but we do know that the ratio in expression [**12.7**] is expected to be *larger* under the alternative hypothesis than under the test hypothesis. The critical region must therefore lie in the *upper* tail of $F(\nu_1, \nu_2)$.

EXERCISES 12.2

1. Each of the following data sets lists the scores for three groups of experimental subjects.

Data Set 1

	Group	
1	2	3
1	6	9
1	6	9
1	6	9
1	6	9
1	6	9
1	6	9

Data Set 2

	Group	
1	2	3
1	6	9
1	6	9
6	9	1
6	9	1
9	1	6
9	1	6

Data Set 3

	Group	
1	2	3
1	6	9
1	6	9
1	6	9
9	1	6
9	1	6
9	1	6

For *each* data set,
(a) calculate the (grand) variance for all 18 subjects, s^2
(b) calculate the mean for each group ($\bar{x}_1$, $\bar{x}_2$, and $\bar{x}_3$) and the variance of these three means ($s_{\bar{X}}^2$)
(c) calculate the variance for each group (s_1^2, s_2^2, and s_3^2)
(d) indicate the source(s) of variability reflected in the grand variance.

2. *Experiment 1.* Five groups of college freshmen are administered the same problem-solving task. Each group is subjected to a different type of distraction while working the problems, and the experimenter records the number of minutes required by every subject to complete the task.

		Group		
1	2	3	4	5
3	5	8	11	15
3	5	8	11	15
3	5	8	11	15
3	5	8	11	15
3	5	8	11	15
3	5	8	11	15

(a) Calculate the variance for all 30 subjects.
(b) Calculate the mean $\bar{x}_j$ for each group, and calculate the variance of these five means, $s_{\bar{X}}^2$.
(c) Calculate the variance s_j^2 for each group.
(d) The variability among the 30 subjects is attributable to_____.

Experiment 2. Ten rats from each of six genetic strains are fed a high calorie diet for one week. The amount of weight (in grams) gained by each rat is recorded below.

		Strain			
1	2	3	4	5	6
10	21	5	15	9	24
12	22	7	18	12	26
15	23	15	22	15	23
18	24	23	24	15	25
25	25	28	27	16	26
30	25	30	31	28	27
30	26	35	43	35	23
35	27	40	18	42	22
30	28	37	30	38	26
45	29	30	22	40	28

(e) Calculate the variance for all 60 subjects.
(f) Calculate the mean $\bar{x}_j$ for each group, and calculate the variance of these six means, $s_{\bar{X}}^2$.
(g) Calculate the variance s_j^2 for each group.
(h) The variability among the 60 subjects is attributable to_____.
3. Consider the following scores obtained by subjects in three experimental treatment groups:

	Group	
1	2	3
2	3	4
4	6	8
6	9	12
8	12	16
10	15	20

(a) Calculate the mean $\bar{x}_j$ for each group.
(b) Use your answer to (a) to calculate an estimate of the grand population variance, σ^2.
(c) Calculate the unbiased estimate of the variance $\hat{\sigma}_j^2$ for each group.
(d) Use your answer to (c) to calculate the pooled estimate of the grand population variance, $\hat{\sigma}_{POOLED}^2$.
(e) Are your two estimates (b) and (d) of the population variance the same? Why do you suppose this is so?
Note: Save your answers for use in Exercise 12.3.1.

4. Consider the following scores obtained by subjects in four experimental treatment groups:

Group 1	Group 2	Group 3	Group 4
75	250	175	100
175	150	125	200
25	100	225	50
125	200	75	150

(a) Calculate the mean $\bar{x}_j$ for each group.
(b) Use your answer to (a) to calculate an estimate of the grand population variance, σ^2.
(c) Calculate the unbiased estimate of the variance $\hat{\sigma}_j^2$ for each group.
(d) Use your answer to (c) to calculate the pooled estimate of the grand population variance, $\hat{\sigma}_{POOLED}^2$.
(e) Are your two estimates (b) and (d) of the population variance the same? Why do you suppose this is so?

5. Consider the following scores obtained by subjects in three experimental treatment groups:

Group 1	Group 2	Group 3
7.5	12.5	6.5
9.5	14.5	14.5
3.5	8.5	10.5
5.5	10.5	16.5
11.5	4.5	8.5
13.5	6.5	12.5

(a) Calculate the mean $\bar{x}_j$ for each group.
(b) Use your answer to (a) to calculate an estimate of the grand population variance, σ^2.
(c) Calculate the unbiased estimate of the variance $\hat{\sigma}_j^2$ for each group.
(d) Use your answer to (c) to calculate the pooled estimate of the grand population variance, $\hat{\sigma}_{POOLED}^2$.
(e) Are your two estimates (b) and (d) of the population variance the same? Why do you suppose this is so?

6. Consider the following scores obtained by subjects in four experimental treatment groups:

Group 1	Group 2	Group 3	Group 4
10	9	15	16
9	10	13	13
8	8	10	14
7	6	14	17
6	5	12	12
5	7	11	15

(a) Calculate the mean $\bar{x}_j$ for each group.

(b) Use your answer to (a) to calculate an estimate of the grand population variance, σ^2.

(c) Calculate the unbiased estimate of the variance $\hat{\sigma}_j^2$ for each group.

(d) Use your answer to (c) to calculate the pooled estimate of the grand population variance, $\hat{\sigma}_{POOLED}^2$.

(e) Are your two estimates (b) and (d) of the population variance the same? Why do you suppose this is so?

Note: Save your answers for use in Exercise 12.3.2.

C. FOUNDATIONS OF ANALYSIS OF VARIANCE

1. Assumptions underlying analysis of variance

In Exercise 11.2.5 we saw that the distribution of $F(1, \nu)$ is closely related to $t(\nu)$: For $F(1, \nu)$ and $t(\nu)$,

$$P\left(F \geq t_i^2\right) = 2P\left(t \geq t_i\right)$$

where $t_i > 0$. Since the test statistic for analysis of variance is distributed as Fisher's F, it should not be surprising that analysis of variance and the t test for two population means are also closely related. An analysis of variance performed on observations from $J = 2$ groups will yield exactly the same p-value as a nondirectional t test performed on the same data. Accordingly, the assumptions underlying the analysis of variance are the same as the assumptions required for Student's t, but generalized from two samples to J samples.

a. Independence of observations. In Chapter 11 we pointed out that Fisher's F is derived from the joint distribution of two independent χ^2 random variables and that estimates $\hat{\sigma}_1^2$ and $\hat{\sigma}_2^2$ of variance σ^2 must therefore be based on independent samples, each of which is comprised of independent observations, in order to assume that the ratio

$$\frac{\hat{\sigma}_1^2}{\hat{\sigma}_2^2}$$

is distributed as $F(\nu_1, \nu_2)$. This requirement extends to the test statistic for analysis of variance. Even if the test hypothesis is correct, one can assume that

$$\frac{N_g\left(\text{est.}\sigma_{\bar{X}}^2\right)}{\hat{\sigma}_{POOLED}^2}$$

is distributed as Fisher's F only if all N observations are independent of one another, both within groups and across groups.

Independence of observations is the most important assumption underlying the F test in analysis of variance, and violation can lead to serious inferential errors. In experimental research, independence is achieved by random *assignment* of subjects to the J treatment groups. When subjects are *selected* from natural populations, the assumption of independence may require some justification.[2] Consider our SAT example. Most public school systems require students to attend the high school that serves the district where they live. Therefore, the director of admissions cannot assume that students were assigned to their high schools at random. The admissions officer must, instead, assume explicitly that the school district in which a student lives is independent of factors that contribute to SAT performance. If there are wide socioeconomic differences among the school districts or if closely related family members are more likely to live in the same district than in different districts, the assumption might be difficult to support.

b. Normal distribution of X. The second assumption underlying analysis of variance is also related to the requirement that $\hat{\sigma}_1^2$ and $\hat{\sigma}_2^2$ be independent of one another. We know that

$$N_g\left(\text{est.}\,\sigma_{\bar{X}}^2\right)$$

is calculated from sample *means* $\bar{x}_j$ and that

$$\hat{\sigma}_{\text{POOLED}}^2$$

is calculated from *variance* estimates $\hat{\sigma}_j^2$. Therefore, Fisher's F is an appropriate model for the distribution of [12.7] only when means and variances of X are independent. It turns out that this can be safely assumed only if X is normally distributed. The analysis of variance therefore assumes that X is normally distributed in all J populations, but this assumption is essentially a mathematical formality. In practice, the probability of a Type I error remains close to α even if the population distributions depart considerably from the normal.

c. Equality of J population variances. If the pooled estimate of variance is to estimate the error variance of the grand population, it must be assumed that the J samples are drawn from populations with equal variances. That is,

$$\sigma_1^2 = \sigma_2^2 = \cdots = \sigma_J^2$$

Because it is virtually impossible to identify all sources of error, this assumption (called *homogeneity of variance* or *homoscedasticity*) is often difficult to justify, and violation can have serious effects on the validity of one's inferences *if sample sizes differ markedly* from group to group. On

---

[2] See Appendix VI for a discussion of randomization strategies in experimental, quasi-experimental, and correlational research designs.

the other hand, the assumption may be violated without serious risk if the number of observations in each group is the same, that is, if

$$N_1 = N_2 = \cdots = N_J$$

This is, of course, precisely what we assumed in our development–which brings us to our last point. Expression [12.7] is a test statistic for analysis of variance only when all samples are of the same size, N_g. A more general expression (which reduces to [12.7] when sample sizes are equal) is developed later in the chapter.

2. The linear model for analysis of variance

As we have seen, the strategy underlying analysis of variance is based on the notion that all of the variability among N scores obtained from J distinct populations is attributable to two sources: variability that is associated with systematic differences among the J populations (treatment) and *all other sources* of variability (error), which are assumed to operate randomly throughout the grand population and therefore to contribute variability in equal measure to all J populations.

Implicitly, this means that if there were neither error nor treatment effects, all of the scores in the grand population would be identical. Thus, any score may be thought of as having three components: a "true" value that is common to all scores, a treatment component, and an error component. This notion is expressed mathematically in the *linear model for analysis of variance*. To understand this model, consider the following story.[3]

Grandma's birthday party. On her 84th birthday Grandma hosts a birthday party attended by her entire family. To celebrate the occasion she decides to distribute her collection of 84 silver dollars among her 12 grandchildren, giving $7 to each grandchild.

After the gifts are distributed Grandpa points out to the grandchildren that 4 of them are high school students living at home and receiving generous allowances, 4 are college students with considerably higher living expenses, and 4 are graduate students who are barely scraping by. The grandchildren therefore agree that each of the high school students should contribute $2 and each of the college students should contribute $1 to a pool divided equally among the graduate students.

When the party is over the grandchildren organize a poker game. The graduate students play at one table, the college students play at another table, and the high school students at yet another table. They agree to play only with the money they received at the party, so every graduate student begins with a stake of $7 + 3 = \$10$, every college student begins with a stake of $7 - 1 = \$6$, and every high school student begins with a stake of $7 - 2 = \$5$. The entire series of transactions, including each player's balance at the end of the game, is illustrated in Table 12.3.

[3] From Li, C. C. (1964). *Introduction to Experimental Statistics*, New York: McGraw-Hill, pp. 48–51.

Table 12.3. Gifts, contributions, and poker earnings

Group	Grandchild	Gift (1)	Contribution (2)	Stake (3)	Poker earnings (4)	Balance (5)
1	1	$7	+$3	$10	+$1	$11
Graduate	2	7	+3	10	−2	8
Students	3	7	+3	10	+2	12
	4	7	+3	10	−1	9
	1	7	−$1	$6	$0	$6
2	2	7	−1	6	−2	4
College	3	7	−1	6	+4	10
Students	4	7	−1	6	−2	4
	1	$7	−$2	$5	−$2	$3
3	2	7	−2	5	+4	9
High School	3	7	−2	5	−3	2
Students	4	7	−2	5	+1	6
Σ		$84	0	$84	0	$84

Each grandchild's balance in column 5 can be expressed as the sum of column 1 plus column 2 plus column 4. That is,

Balance = gift + contribution + earnings

Furthermore, if we define this balance as the random variable X, then the "score" for grandchild i in Group j becomes

x_{ij} = gift + contribution + earnings

Consider the gift component of the score. If we look at column 1 in Table 12.3 we see that this component is a constant for all of the grandchildren in our population. This constant is therefore equal to the mean of the 12 gifts. Moreover, since subsequent transactions involve only redistribution of the money disbursed by Grandma, the total at the end of the game is exactly equal to the total of the gifts, and the mean of the gifts (column 1) is exactly equal to μ, the mean of the balances in column 5. Therefore, we see that

$x_{ij} = \mu$ + contribution + earnings

Now, let us consider the contribution component of each score. The contribution varies *among* groups, but it is a constant *within* groups. Therefore, each individual's contribution depends entirely upon whether the contributor belongs to Group 1, 2, or 3. The component of any score x_{ij} that is entirely attributable to group membership or assignment is known as a *treatment effect* and is denoted by the lowercase Greek *tau*, τ. Since the value of τ differs from group to group, it is identified with the subscript j. Therefore, we can now rewrite the expression for the final balance of grandchild i in Group j as follows:

$$x_{ij} = \mu + \tau_j + \text{earnings}$$

We see in column 3 that all of the players in Group j begin the game with the same stake, which is equal to the gift plus contribution, that is, $\mu + \tau_j$. The poker game simply redistributes the total number of dollars with which the players begin, so at the end of the game the average balance for Group j must likewise be equal to $\mu + \tau_j$:

$$\mu_j = \mu + \tau_j$$

so

$$\tau_j = \mu_j - \mu$$

That is, treatment effect is equal to the difference between the grand mean and the group mean. The test hypothesis for analysis of variance can therefore be formulated in terms of treatment effects:

$$H_0: \mu_1 = \mu_2 = \cdots = \mu_J = \mu$$

can be rewritten

$$H_0: (\mu_1 - \mu) = (\mu_2 - \mu) = \cdots = (\mu_J - \mu) = 0$$

or,

$$H_0: \tau_1 = \tau_2 = \cdots = \tau_J = 0$$

Finally, let us consider the earnings component of each grandchild's score. We see in column 4 of Table 12.3 that this differs unpredictably from person to person. Besides individual differences in poker skill, this component depends on such qualities as attentiveness to the game, willingness to take risks, interpersonal perceptiveness, and–most important–the fall of the cards. These sorts of random influences are designated *error*, denoted e. And, since each person's error component is unique to that individual, it is identified with double subscripts. Therefore, the balance for grandchild i in Group j becomes

$$x_{ij} = \mu + \tau_j + e_{ij} \qquad [12.8]$$

We know that every grandchild in Group J begins the poker game with μ_j dollars. The amount that he or she wins or loses (e_{ij}) is therefore the difference between the group mean and the grandchild's balance at the end of the game. That is,

$$e_{ij} = x_{ij} - \mu_j$$

The model for the score of a single individual [12.8] suggests an obvious way to represent the random variable X that yielded all of the scores, x_{ij}:

$$X = \mu + \tau + \mathscr{E} \qquad\qquad\qquad [12.9]$$

where the grand population mean μ is a constant, τ is the random variable that takes on values

$$\tau_1, \tau_2, \ldots, \tau_J$$

and $\mathscr{E}$ is the random variable whose values are the error components

$$e_{11}, e_{12}, \ldots, e_{N_J J}$$

of our N observations x_{ij}.

If we write the expected value of X for the grand population in terms of expression [12.9] we obtain

$$E(X) = E(\mu + \tau + \mathscr{E})$$

$$= \mu + \mu_\tau + \mu_e$$

where μ_τ is the expected value of τ, and μ_e is the expected value of $\mathscr{E}$. We see from the bottom line of Table 12.3 that the average gift and the average poker earnings in our population of grandchildren is 0, and in general it is assumed that $E(\tau) = 0$ and $E(\mathscr{E}) = 0$.

The linear model for the random variable X is

$$X = \mu + \tau + \mathscr{E}$$

where μ is the grand population mean, $E(\tau) = 0$, and $E(\mathscr{E}) = 0$.

If we write the grand population variance σ^2 in terms of the linear model, it is easy to formalize our earlier, informal claim that the only sources of variability in X are treatment and error:

$$V(X) = V(\mu + \tau + \mathscr{E})$$

The grand mean μ is a constant, and it is assumed that the variables τ and $\mathscr{E}$ are independent of one another, so by the Algebra of Variances,

$$V(X) = \sigma_\tau^2 + \sigma_e^2$$

where σ_τ^2 is the variance of τ and σ_e^2 is the variance of $\mathscr{E}$.

Our earlier plausibility argument that error is the only source of variability *within* each population can likewise be seen to follow directly from the assumptions of the linear model. Let X_j denote the random variable X in Group j and let $\mathscr{E}_j$ denote the random variable that gave us the N_j error components

$$e_{1j}, \ldots, e_{N_j j}$$

of the scores in Group j. Then,

$$X_j = \mu + \tau_j + \mathscr{E}_j \qquad\qquad [12.10]$$

and the variance for population j is

$$V(X_j) = V(\mu + \tau_j + \mathscr{E}_j)$$

or

$$\sigma_j^2 = V(\mu + \tau_j + \mathscr{E}_j)$$

The grand mean μ is a constant, and since all of the members in Group j receive the same treatment, τ_j is also a constant. The variance of X_j therefore reduces to

$$\sigma_j^2 = V(\mathscr{E}_j)$$

Furthermore, it will be recalled that analysis of variance assumes all population variances to be equal. If error is the only source of variability within populations, this means that

$$V(\mathscr{E}_1) = \cdots = V(\mathscr{E}_J)$$

The common error variance may be denoted $V(\mathscr{E})$ or σ_e^2. The assumption of homogeneity of variance can therefore be stated more exactly: All population variances are assumed to be equal to the error variance, σ_e^2. That is,

$$\sigma_1^2 = \cdots = \sigma_J^2 = \sigma_e^2$$

We have also said that analysis of variance assumes X_j to be normally distributed. Since error is the only source of variability in X_j, this assumption means that $\mathscr{E}$ is normally distributed in every population.

3. Variability as sums of squares

We have argued that if group sizes $N_1, \ldots, N_J$ are equal, the hypothesis test for analysis of variance can be conducted using the statistic

$$\frac{N_g\left(\text{est.}\sigma_{\bar{X}}^2\right)}{\hat{\sigma}_{\text{POOLED}}^2}$$

and in the last section we saw that this argument rests implicitly on the assumptions of the linear model. In this section we use the linear model to develop a more general form of the test statistic, which can be used when group sizes are unequal and which allows us to get a grip on degrees of freedom for analysis of variance in a reasonably straightforward fashion.

Equation [12.8] represents the score of individual i in Group j as the sum

$$x_{ij} = \mu + \tau_j + e_{ij}$$

We know that

$$\tau_j = \mu_j - \mu$$

and

$$e_{ij} = x_{ij} - \mu_j$$

so [12.8] can be rewritten

$$x_{ij} = \mu + (\mu_j - \mu) + (x_{ij} - \mu_j)$$

For a sample of N observations among J groups, this becomes

$$x_{ij} = \bar{x} + (\bar{x}_j - \bar{x}) + (x_{ij} - \bar{x}_j) \tag{12.11}$$

Beginning in Chapter 2 we have used one or another function of the sum of squared deviations about the mean

$$\sum_{i=1}^{N} (x_i - \bar{x})^2$$

to capture the variability in a distribution. This *sum of squares* appears in the numerator of s^2 and in the numerator of $\hat{\sigma}^2$. Likewise, the total variability in any collection of N scores x_{ij} from J populations can be expressed in terms of the sum of squared deviations from the grand mean, $\bar{x}$. Since the scores are separated by groups, the total sums of squares (SS_{TOTAL}) is usually written using double summation notation:

$$SS_{\text{TOTAL}} = \sum_{j}^{J} \sum_{i}^{N_j} (x_{ij} - \bar{x})^2 \tag{12.12}$$

In Box 12.1 we show that if the scores x_{ij} are represented as in [12.11], then SS_{TOTAL} can be expressed as the sum of two quantities,

$$SS_{\text{TOTAL}} = \sum_{j} \sum_{i} (x_{ij} - \bar{x}_j)^2 + \sum_{j} N_j (\bar{x}_j - \bar{x})^2 \tag{12.13}$$

BOX 12.1

Partitioning the total sums of squares

From equation [12.11]

$$x_{ij} = \bar{x} + (\bar{x}_j - \bar{x}) + (x_{ij} - \bar{x}_j)$$

Subtracting $\bar{x}$ from both sides and rearranging terms,

$$x_{ij} - \bar{x} = (x_{ij} - \bar{x}_j) + (\bar{x}_j - \bar{x})$$

so

$$(x_{ij} - \bar{x})^2 = \left[(x_{ij} - \bar{x}_j) + (\bar{x}_j - \bar{x})\right]^2$$

and

$$\sum_j \sum_i (x_{ij} - \bar{x})^2 = \sum_j \sum_i \left[(x_{ij} - \bar{x}_j) + (\bar{x}_j - \bar{x})\right]^2 \tag{1}$$

The summation on the left side of equation [1] defines SS_{TOTAL}, so if we expand the square on the right side,

$$SS_{\text{TOTAL}} = \sum_j \sum_i \left[(x_{ij} - \bar{x}_j)^2 + 2(x_{ij} - \bar{x}_j)(\bar{x}_j - \bar{x}) + (\bar{x}_j - \bar{x})^2\right]$$

$$= \sum_j \sum_i (x_{ij} - \bar{x}_j)^2 - \sum_j \sum_i 2(x_{ij} - \bar{x}_j)(\bar{x}_j - \bar{x}) + \sum_j \sum_i (\bar{x}_j - \bar{x})^2$$

Consider the middle term in the preceding expression:

$$\sum_j \sum_i 2(x_{ij} - \bar{x}_j)(\bar{x}_j - \bar{x}) = 2 \sum_j \sum_i (x_{ij} - \bar{x}_j)(\bar{x}_j - \bar{x})$$

Since $(\bar{x}_j - \bar{x})$ is a constant with respect to the within-group sum it may be factored out of $\sum_i$, so

$$2 \sum_j \sum_i (x_{ij} - \bar{x}_j)(\bar{x}_j - \bar{x}) = 2 \sum_j (\bar{x}_j - \bar{x}) \sum_i (x_{ij} - \bar{x}_j)$$

In addition, it can be seen that within each group

$$\sum_i (x_{ij} - \bar{x}_j)$$

is the sum of deviations about the group mean and must therefore equal zero. Therefore,

$$\sum_j \sum_i 2(x_{ij} - \bar{x}_j)(\bar{x}_j - \bar{x}) = 0$$

and the total sum of squares [1] reduces to

$$\sum_j \sum_i (x_{ij} - \bar{x})^2 = \sum_j \sum_i (x_{ij} - \bar{x}_j)^2 + \sum_j \sum_i (\bar{x}_j - \bar{x})^2 \qquad [2]$$

As before, we note that for any group j, the quantity $(\bar{x}_j - \bar{x})^2$ is a constant, so

$$\sum_i^{N_j} (\bar{x}_j - \bar{x})^2 = N_j (\bar{x}_j - \bar{x})^2$$

and

$$\sum_j \sum_i (\bar{x}_j - \bar{x})^2 = \sum_j N_j (\bar{x}_j - \bar{x})^2$$

Therefore, equation [2] can be rewritten

$$\sum_j \sum_i (x_{ij} - \bar{x})^2 = \sum_j \sum_i (x_{ij} - \bar{x}_j)^2 + \sum_j N_j (\bar{x}_j - \bar{x})^2$$

The first term on the right-hand side of equation [12.13]

$$\sum_j \sum_i (x_{ij} - \bar{x}_j)^2 \qquad [12.14]$$

is the sum of squared differences between every score in Group j and the mean for Group j. This quantity can therefore reflect only variation *within* groups and is, accordingly, called the *sum of squares within groups*, denoted SS_W.

The second term on the right-hand side of equation [12.13]

$$\sum_j N_j (\bar{x}_j - \bar{x})^2 \qquad [12.15]$$

is based on differences between the J group means $\bar{x}_j$ and the grand mean $\bar{x}$. It therefore reflects variation *across* or *among* groups and is called the *sum of squares among groups*, denoted SS_A. Therefore,

$$SS_{\text{TOTAL}} = SS_W + SS_A$$

Decomposing the total sum of squares into the sums of squares within groups and the sum of squares among groups is known as *partitioning* the sum of squares.

The *total* sum of squares

$$SS_{TOTAL} = \sum_{j}^{J} \sum_{i}^{N_j} (x_{ij} - \bar{x})^2$$

captures all of the variability in a collection of

$$N_1 + N_2 + \cdots + N_J = N$$

scores and is equal to the sum of squares *within* groups,

$$SS_W = \sum_{j} \sum_{i} (x_{ij} - \bar{x}_j)^2$$

plus the sum of squares *among groups,*

$$SS_A = \sum_{j} N_j (\bar{x}_j - \bar{x})^2$$

That is,

$$SS_{TOTAL} = SS_W + SS_A$$

a. Computational formulae and degrees of freedom. In Appendix VII we derive the following computational formulae for sums of squares:

$$SS_{TOTAL} = \sum_{j} \sum_{i} x_{ij}^2 - N\bar{x}^2 \tag{12.16}$$

$$SS_W = \sum_{j} \sum_{i} x_{ij}^2 - \sum_{j} N_j \bar{x}_j^2 \tag{12.17}$$

$$SS_A = \sum_{j} N_j \bar{x}_j^2 - N\bar{x}^2 \tag{12.18}$$

It will be observed that these expressions permit computation of sums of squares from three basic quantities:

$$A = \sum_{j} \sum_{i} x_{ij}^2 \qquad B = \sum_{j} N_j \bar{x}_j^2 \qquad D = N\bar{x}^2$$

Recall from the discussion of degrees of freedom in Chapter 10 that if (1) an experimenter makes N observations and (2) the observations are used to estimate J parameters and (3) the random variable Y is a function of the N observations and the J estimates, then Y has $N - J$

degrees of freedom. Using these guidelines, we can easily determine the degrees of freedom for our sums of squares from the basic quantities A, B, and D.

From equation [**12.16**] we see that

$$SS_{\text{TOTAL}} = A - D$$

where A is the sum of N independent observations, x_{ij}, and the grand mean $\bar{x}$ in quantity D estimates the grand population mean μ. Therefore, SS_{TOTAL} is a function of N observations which are used to calculate one estimate, so SS_{TOTAL} has $N - 1$ degrees of freedom.

Equation [**12.17**] can be written

$$SS_W = A - B$$

Quantity A is the sum of N independent observations, and the J group means in quantity B are estimates of the J population means, $\mu_1, \ldots, \mu_J$. Degrees of freedom for SS_W is therefore $N - J$.

In expression [**12.18**] the random variable is $\bar{X}$, the sample mean. Therefore, quantity B is the sum of J independent observations $\bar{x}_j$, and the grand population mean $\bar{x}$ in quantity D is an estimate of $\mu_{\bar{X}}$. Thus,

$$SS_A = B - D$$

has $J - 1$ degrees of freedom.

In Chapter 11 we explained that degrees of freedom are additive for sums of squares of independent normally distributed random variables so it is not surprising that

$$df_{SSW} + df_{SSA} = df_{SST}$$

that is,

$$(N - J) + (J - 1) = N - 1$$

D. DECISIONS: TEST STATISTIC AND REJECTION RULE FOR ANOVA

1. Mean squares

A sum of squares divided by its degrees of freedom is defined as a *mean of squares*. Accordingly, we define the *mean squares* total (MS_{TOTAL}), the *mean squares* within groups (MS_W), and the *mean squares* among groups (MS_A) as follows:

BOX 12.2

Formulae for mean squares

		Definitional	Computational

$$MS_{\text{TOTAL}} = \frac{SS_{\text{TOTAL}}}{N-1} = \frac{\sum_j \sum_i (x_{ij} - \bar{x})^2}{N-1} = \frac{\sum_j \sum_i x_{ij}^2 - N\bar{x}^2}{N-1}$$

$$MS_W = \frac{SS_W}{N-J} = \frac{\sum_j \sum_i (x_{ij} - \bar{x}_j)^2}{N-J} = \frac{\sum_j \sum_i x_{ij}^2 - \sum_j N_j \bar{x}_j^2}{N-J}$$

$$MS_A = \frac{SS_A}{J-1} = \frac{\sum_j N_j (\bar{x}_j - \bar{x})^2}{J-1} = \frac{\sum_j N_j \bar{x}_j^2 - N\bar{x}^2}{J-1}$$

The definitions of the various mean squares allow us to complete our earlier plausibility argument for the distribution of

$$\frac{N_g \left(\text{est.} \sigma_{\bar{X}}^2 \right)}{\hat{\sigma}_{\text{POOLED}}^2}$$

and to generalize this statistic to tests in which sample sizes N_j are not equal.

a. Expected value of MS_W. First we note that

$$\sum (x - \bar{x})^2 = (N-1)\hat{\sigma}^2$$

Therefore

$$MS_W = \frac{\sum_j \sum_i (x_{ij} - \bar{x}_j)^2}{N-J} = \frac{(N_1 - 1)\hat{\sigma}_1^2 + \cdots + (N_J - 1)\hat{\sigma}_J^2}{N_1 + \cdots + N_J - J}$$

which is the general expression for the pooled estimate of variance $\hat{\sigma}_{\text{POOLED}}^2$ given in Box 10.5 (p. 430). Earlier in the chapter we argued that $E(\hat{\sigma}_{\text{POOLED}}^2) = \sigma_e^2$. Since $\hat{\sigma}_{\text{POOLED}}^2 = MS_W$, it follows that their expected values are equal.

From the definition of MS_W in Box 12.2

$$E(MS_W) = E\left(\frac{SS_W}{N-J} \right)$$

$$= \frac{1}{N-J} E(SS_W)$$

and from equation **[18]** in Appendix VII

$$E(SS_W) = \sigma_e^2(N - J)$$

Therefore,

$$E(MS_W) = \frac{1}{N - J}\sigma_e^2(N - J) \qquad \qquad [12.19]$$

$$= \sigma_e^2$$

The expected value of mean squares within groups is the error variance, σ_e^2.

$$E(MS_W) = \sigma_e^2$$

---

b. Expected value of MS_A. By the definitional formula of MS_A given in Box 12.2,

$$MS_A = \frac{\sum\limits_{j}^{J} N_j(\bar{x}_j - \bar{x})^2}{J - 1}$$

If $N_1 = N_2 = \cdots = N_J = N_g$, then

$$MS_A = \frac{\sum\limits_{j}^{J} N_g(\bar{x}_j - \bar{x})^2}{J - 1} = \frac{N_g\sum\limits_{j}^{J}(\bar{x}_j - \bar{x})^2}{J - 1}$$

It will be noted that for J sample means, $\bar{x}_1, \ldots, \bar{x}_J$,

$$\text{est.}\sigma_{\bar{X}}^2 = \frac{\sum\limits_{j}^{J}(\bar{x}_j - \bar{x})^2}{J - 1}$$

as given in equation **[12.5]**. When all J means are based on samples of equal size, we therefore see that

$$MS_A = N_g\left(\text{est.}\sigma_{\bar{X}}^2\right)$$

We argued earlier that

$$N_g\left(\text{est.}\sigma_{\bar{X}}^2\right)$$

estimates the grand population variance and therefore taps both error variance and treatment variance, if any. This may now be formally proved by taking the expected value of MS_A.

$$E(MS_A) = E\left(\frac{SS_A}{J-1}\right) = \frac{1}{J-1}E(SS_A)$$

From formula [20] in Appendix VII

$$E(SS_A) = \sum_j N_j \tau_j^2 + (J-1)\sigma_e^2$$

where $\tau_j = \mu_j - \mu$. Therefore,

$$E(MS_A) = \frac{1}{J-1}\left(\sum_j N_j \tau_j^2 + (J-1)\sigma_e^2\right)$$

$$= \frac{\sum_j N_j \tau_j^2}{J-1} + \sigma_e^2 \qquad\qquad [12.20]$$

The expected value of mean squares among groups is error variance plus a term that can be zero only if $\tau_j = 0$,

$$E(MS_A) = \frac{\sum_j N_j \tau_j^2}{J-1} + \sigma_e^2$$

---

c. Distribution for MS_A / MS_W. Equation [12.19] tells us that

$$E(MS_W) = \sigma_e^2$$

Therefore, MS_W is an unbiased estimator of σ_e^2 *whether or not* the test hypothesis is correct.

Equation [12.20] tells us that

$$E(MS_A) = \sigma_e^2 + \frac{\sum_j N_j \tau_j^2}{J-1}$$

If the test hypothesis is correct, however, $\mu_1 = \cdots = \mu_J = \mu$, so

$$\tau_j = \mu_j - \mu = 0, \qquad \text{and}$$

$$E(MS_A) = \sigma_e^2 + \frac{\sum_j N_j \tau_j^2}{J-1}$$

$$= \sigma_e^2 + 0$$

$$= \sigma_e^2$$

Therefore, MS_A is an unbiased estimator of σ_e^2 *if and only if the test hypothesis is correct*, in which case the ratio

$$\frac{MS_A}{MS_W}$$

is distributed as Fisher's $F(\nu_1, \nu_2)$, where

$$\nu_1 = df_{SSA} = J - 1$$

and

$$\nu_2 = df_{SSW} = N - J$$

2. Rejection rule for ANOVA

If the alternative hypothesis is correct, $E(MS_A) > E(MS_W)$. As we indicated earlier, therefore, the rejection region for an analysis of variance always lies in the *upper* tail of $F(J - 1, N - J)$. For significance level α the critical region is all values greater than or equal to F_b, where $\mathbb{P}(F_b) = 1 - \alpha$.

Example 12.5. In Example 12.4 (pp. 494–495),

$$J - 1 = 2$$

and

$$N - J = 15 - 3 = 12$$

If the admissions officer wants $\alpha = .05$, the critical value is the value in $F(2, 12)$ for which $\mathbb{P} = 1 - .05 = .95$. In Table 6 of Appendix VIII we find that $\mathbb{P}(3.89) = .95$, so the test hypothesis is rejected if $MS_A/MS_W \geq 3.89$. We calculated

$$\frac{N_g\left(\text{est.}\sigma_{\bar{X}}^2\right)}{\hat{\sigma}_{\text{POOLED}}^2} = \frac{MS_A}{MS_W} \doteq 1.14$$

so the alternative hypothesis must be rejected.

In the next section we will see that the results of an analysis of variance are ordinarily reported in terms of the *p*-value of the test statistic. In the distribution of $F(2, 12)$ we see that $.50 < \mathbb{P}(1.14) < .75$, so at least 25 percent of the area under $F(2, 12)$ lies above the calculated value of the test statistic. Therefore, $p > .25$, and even without a predetermined significance level the admissions officer would probably reject the alternative hypothesis.

E. RESULTS: THE ANOVA SUMMARY TABLE

To see how ANOVA results are typically reported, let us work through an entire example.

Example 12.6. Until recently animals were considered members of the same species if they produced fertile offspring. This definition breaks down when we realize that such apparently distinct species as the bobcat and the domestic cat produce viable offspring, as do matings of dogs and wolves. The problem becomes even more complex when a biologist is asked to determine whether two breeding populations are separate sub-species or simply local variants of the same species. In this regard, some biologists recognize as many as 26 subspecies of gray wolf (*Canis lupus*). These classifications are based on differences in size, coloration, body conformation, and so on. Whatever the physical characteristics, the crite-rion for recognizing different populations as distinct subspecies is that they display more variation across groups than within groups, and as such the matter of distinguishing among subspecies can be thought of as an analysis of variance problem.

Let us suppose that we have three mated pairs of gray wolves cap-tured, respectively, in Ontario, Minnesota, and Alaska. Let us further suppose that the three females have delivered litters of pups that were removed from their mothers at birth and reared on the same commercial diet for 17 days. To explore the possibility that the three pairs represent distinct subspecies, we propose to test the hypothesis that the popula-tions are not different in average 10-day body weight. Thus, our random variable is weight measured in ounces at 10 days of age, and our hypotheses are

$$H_0: \mu_1 = \mu_2 = \mu_3$$

$$H_1: H_0 \text{ incorrect}$$

The data for this problem appear in Table 12.4. If H_0 is correct, then

Table 12.4. Body weight at 10 days of age for three litters of gray wolf pups

	Group 1 (Ontario)	Group 2 (Minnesota)	Group 3 (Alaska)	
	22	25	29	
	24	26	31	
	24	29	32	
	27	32	33	
	28		35	
$\sum$	125	112	160	$\sum \sum = 397$
$\bar{x}_j$	25	28	32	$\bar{x} \doteq 28.36$

with a total of $N = 14$ observations among $J = 3$ groups, the ratio

$$\frac{MS_A}{MS_W}$$

is distributed as $F(2, 11)$. If we set $\alpha = .01$, the critical value of our test statistic is found in Table 6 of Appendix VIII to be $F_b = 7.21$

We calculate the value of our test statistic using the computational formulae in Box 12.2, so we first compute the basic quantities

$$A = \sum_j \sum_i x_{ij}^2 = 11,455$$

$$B = \sum N_j \bar{x}_j^2 = 5(25)^2 + 4(28)^2 + 5(32)^2$$

$$= 3,125 + 3,136 + 5,120 = 11,381$$

$$D = N\bar{x}^2 \doteq 14(28.36)^2 \doteq 11,260.05$$

Then,

$$SS_A = B - D \doteq 11,381 - 11,260.05 = 120.95$$

and

$$MS_A = \frac{SS_A}{J - 1} \doteq \frac{120.95}{2} \doteq 60.48$$

Similarly,

$$SS_W = A - B = 11,455 - 11,381 = 74$$

and

$$MS_W = \frac{SS_W}{N - J} = \frac{74}{11} \doteq 6.73$$

so

$$\frac{MS_A}{MS_W} \doteq \frac{60.48}{6.73} \doteq 8.99$$

This exceeds our critical F-value, and we conclude that the three populations of wolves from which we drew our samples do not have identical mean 10-day body weights. Insofar as weight is relevant to wolf taxonomy, the result would suggest that the three breeding populations constitute distinct subspecies.

To facilitate representation of data in analysis of variance, the important computations are usually, though by no means always, presented in an analysis of variance summary table, such as illustrated in Table 12.5.

Table 12.5. Analysis of variance (ANOVA) summary table for Example 12.6

Source	SS	df	MS	F
Treatments (across groups)	120.95	2	60.48	8.99*
Error (within groups)	74	11	6.73	
Totals	194.95	13		

*$p < .005$.

Total sums of squares and total degrees of freedom need not be calculated in analysis of variance, but they are often computed and displayed in the summary table as an independent check on the computation of sums of squares and degrees of freedom for treatments and error. It is seen in Table 12.5 that treatments and error account for all $N - 1 = 13$ degrees of freedom and that $SS_{\text{TOTAL}} = A - D$ is equal to $SS_A + SS_W$.

─────── EXERCISES 12.3 ───────

You may assume that all random variables in the following exercises are normally distributed and that group variances are equal.

1. In Exercises 12.2.3 you were asked to calculate $\bar{x}_j$ and $\hat{\sigma}^2_{\text{POOLED}}$ for the following groups of scores:

Group 1	Group 2	Group 3
2	3	4
4	6	8
6	9	12
8	12	16
10	15	20

(a) Calculate MS_A from either the definitional or computational formula. Compare your answer to the value you computed for $N_g(\text{est.}\sigma^2_{\bar{X}})$ in Exercise 12.2.3(b). What do you notice?
(b) Calculate MS_W from either the definitional or computational formula. Compare your answer to the value you computed for $\hat{\sigma}^2_{\text{POOLED}}$ in Exercise 12.2.3(d). What do you notice?
(c) How is the ratio MS_A/MS_W distributed if $\mu_1 = \mu_2 = \mu_3$?
(d) Calculate MS_A/MS_W.
(e) Find the cumulative probability of the value you calculated for MS_A/MS_W.

(f) What is the probability of making a Type I error if you reject the hypothesis that $\mu_1 = \mu_2 = \mu_3$?

2. In Exercise 12.2.6 you were asked to calculate $\bar{x}_j$ and $\hat{\sigma}^2_{POOLED}$ for the following groups of scores:

Group 1	Group 2	Group 3	Group 4
10	9	15	16
9	10	13	13
8	8	10	14
7	6	14	17
6	5	12	12
5	7	11	15

(a) Calculate MS_A from either the definitional or computational formula. Compare your answer to the value you computed for $N_g(\text{est.}\sigma_{\bar{X}}^2)$ in Exercise 12.2.6(b). What do you notice?

(b) Calculate MS_W from either the definitional or computational formula. Compare your answer to the value you computed for $\hat{\sigma}^2_{POOLED}$ in Exercise 12.2.6(d). What do you notice?

(c) How is MS_A/MS_W distributed if $\mu_1 = \mu_2 = \mu_3 = \mu_4$?

(d) Calculate MS_A/MS_W.

(e) Find the cumulative probability of the value you calculated for MS_A/MS_W.

(f) What is the probability of making a Type I error if you reject the hypothesis that $\mu_1 = \mu_2 = \mu_3 = \mu_4$?

3. Experimenters are investigating differences in the incentive value of various types of food reward for wolves. They administer a discrimination-learning task to three groups of mother-reared, unsocialized wolf pups. The scores given below are the number of trials to reach criterion. Group 1 received chunks of kibble, Group 2 received sardines, and Group 3 received Reese's Pieces®.

Group	1	2	3
	10	4	10
	0	0	0
	8	5	9
	1	0	1
	1	1	10

(a) What are the hypotheses?

(b) Give the *definitional* formula for the appropriate test statistic to test this hypothesis.

(c) What is the expected value of the test statistic if the test hypothesis is correct?

(d) Would you expect the observed value of the test statistic to be larger or smaller than your answer to 12.3.3(c) if the test hypothesis is not correct?

(e) Why?

(f) If $\alpha = .05$, what is your rejection rule?

(g) Calculate the value of the test statistic using the *definitional* formulae for SS_W and SS_A. Confirm your calculations with the *computational* formulae for SS_W and SS_A.

(h) What is your conclusion?

4. The experimenters in Exercise 12.3.3 repeat the experiment using wolf pups that were hand-reared and intensively socialized to humans. Their scores are as follows:

Group	1	2	3
	6	1	8
	5	3	7
	3	2	5
	3	1	4
	3	3	6

(a) What is the test hypothesis?
(b) What is the appropriate test statistic?
(c) How is the test statistic distributed if H_0 is correct?
(d) If the signfiicance level α is .05, what is the critical value of the test statistic?
(e) Calculate the value of the test statistic using the *definitional* formulae for SS_W and SS_A. Confirm your calculations with the *computational* formulae for SS_W and SS_A.
(f) What conclusion should the experimenters draw?
(g) If they reject the test hypothesis, _____ < p < _____ that they have made a Type I error. Calculate the three group means for the socialized pups used in this exercise and the three group means for the unsocialized pups in Exercise 12.3.3.
(h) Calculate the following differences: $\bar{x}$ for Group 1 socialized minus $\bar{x}$ for Group 1 unsocialized, $\bar{x}$ for Group 2 socialized minus $\bar{x}$ for Group 2 unsocialized, $\bar{x}$ for Group 3 socialized minus $\bar{x}$ for Group 3 unsocialized.
(i) What do you notice about these three differences?
(j) Given your answer to 12.3.4(i), how can you explain the difference between the conclusions you reached in Exercises 12.3.3(h) and 12.3.4(f)? (If you were not assigned Exercise 12.3.3, check the answer in the back of the book.)

5. Egbert owns an Italian sports car (Brand α). Egbert's brother, Ethelred, owns a German sports car (Brand $\mathcal{B}$). When they purchased their automobiles from F.F.I. (Foolish Foreign Investments), Ltd., Ethelred's brother-in-law, Ernie, bought a British *Trouble* (abbreviated *TR*). After 6 months of recording maintenance costs, the three stooges agree that the dealership is suitably named, but they still argue about whether their expenses are due to automotive design or piracy in F.F.I., Ltd.'s service department. If it is the latter, the average maintenance cost per month should be identical for all three, but if it is design, average maintenance costs should differ from car to car. They therefore compare maintenance bills for the 6 months:

	α	$\mathcal{B}$	*TR*
February	$73	$47	$58
March	55	30	47
April	90	70	73
May	70	45	55
June	85	55	65
July	77	53	62

Test the hypothesis that monthly maintenance costs are identical for the three makes of automobile.
(a) What are the hypotheses?
(b) What is the appropriate test statistic?
(c) How is the test statistic distributed if H_0 is correct?
(d) What is the value of the test statistic?
(e) What is the p-value of the test statistic?
(f) What is the appropriate conclusion?

6. A major problem in assigning grades in any course is that test performance is often confounded by test anxiety. Since anxiety seems higher among freshmen than among seniors, one might hypothesize that it is a function of age. Suppose the following data are scores on the Test Anxiety Questionnaire for students in an undergraduate psychology course.

	Age	
16–20	**26–30**	**36–40**
18	12	23
11	28	3
33	2	8
14	8	7
26	28	11
1	19	
	15	

(a) What is the test hypothesis?
(b) What is the appropriate test statistic?
(c) How is the test statistic distributed if H_0 is correct?
(d) What is the value of the test statistic?
(e) What is the appropriate conclusion?

7. A test of inhibition was administered to four Alaskan Malamutes and four wolves, all reared by the same mother wolf, and seven hand-reared wolves. Each animal was placed on a platform and punished if it attempted to leave before a prescribed number of seconds had elapsed. The *mother-reared* wolves and dogs received a sharp tug on a choke chain. The *hand-reared* wolves were divided into two subgroups: Four were punished by a tug on the chain and three were shot with felt pellets by a concealed sniper. The number of seconds that each subject remained on the platform was recorded for each trial, and each subject's highest score on each of 10 consecutive days was noted ("best" daily score). The sum of each subject's 10 best daily scores is given below.

Number of seconds on platform (sum of 10 "best" daily scores)			
Group 1 Hand-reared / choke chain (wolves)	Group 2 Mother-reared / choke chain (wolves)	Group 3 Hand-reared / pellets (wolves)	Group 4 Mother-reared / choke chain (Malamutes)
221	267	413	523
345	241	565	905
362	564	657	990
63	397		540

(a) What is the test hypothesis?
(b) What is the appropriate test statistic?
(c) How is the test statistic distributed if H_0 is correct?
(d) What is the value of the test statistic?
(e) What is the p-value of the test statistic?
(f) What is the appropriate conclusion?

8. Students in a comparative psychology course conducted an experiment to determine whether or not blowflies (*Sarcophaga*) can discriminate among different kinds of food. Each fly was placed in a dish containing one of three solutions. Group 1 received 5 percent sucrose. Group 2 received 5 percent dextrose. Group 3 received 5 percent saccharin. The following data are *durations* (in seconds) of proboscis extension.

Group 1	Group 2	Group 3
8.00	10.40	0.00
9.55	4.37	0.00
3.00	4.69	6.60
12.14	33.66	17.00
15.70	8.73	0.00
32.00	4.28	1.00

Test the hypothesis ($\alpha = .05$) that the insects could not detect differences in these chemicals.
(a) What is the test hypothesis?
(a) What is the appropriate test statistic?
(c) How is the test statistic distributed if H_0 is correct?
(d) Use the definitional or computational formula to calculate the value of the test statistic.
(e) What is the appropriate conclusion?
(f) What is the p-value of the test statistic?
(g) Calculate the mean and variance for each group. Using *only* this information and the sample sizes, compute the value of the test statistic.

9. Genetic information is transmitted by three principal mechanisms. Single-locus traits (such as ABO blood type) are carried on single genes. Karyotypic traits (such as sex) involve entire chromosomes. Polygenetic traits (such as height) involve the effects of many genes, which are often located on different chromosomes. Any of these three mechanisms can be implicated in mental retardation. An experimenter believes it may be possible to distinguish among genetic categories of mental retardation on the basic of standard mental tests. The experimenter administers a nonverbal intelligence test to three groups of mentally retarded subjects, all of whom have a mental age of 7 years (± 2 months), and obtains the following scores:

Single-locus (phenylketonuria)	Karyotypic (Down's Syndrome)	Polygenic
56	70	77
53	61	73
68	67	63
	62	74
		68

The theory is supported if test performance is influenced by the genetic mechanisms of retardation.
(a) What is the test hypothesis?
(b) What is the appropriate test statistic?
(c) How is the test statistic distributed if H_0 is correct?
(d) What is the value of the test statistic?
(e) What is the p-value of the test statistic?
(f) What is the appropriate conclusion?

10. A psychologist is trying to determine whether different temporal patterns of conditional and unconditional stimuli yield differences in resistance to extinction. Twenty-four subjects are assigned at random to four experimental groups: Group 1 is given *delayed* conditioning; Group 2 is given *simultaneous* conditioning; Group 3 is given *trace* conditioning; Group 4 is given *backward* conditioning. After all subjects have acquired the conditioned response, the experimenter begins extinction trials. The following data are the number of trials required to extinguish the response:

Group 1	Group 2	Group 3	Group 4
33	31	29	19
47	37	41	13
33	23	27	21
35	28	29	19
25	21	17	23
37	31	33	17

(a) What is the test hypothesis?
(b) What is the appropriate test statistic?
(c) How is the test statistic distributed if H_0 is correct?
(d) If $\alpha = .025$, what is the critical value of the test statistic?
(e) What is the value of the test statistic?
(f) What is the appropriate conclusion?
(g) Calculate the mean and variance for each group. Using *only* this information and the sample sizes, compute the value of the test statistic.

11. Five breeds of 10-week-old puppies were administered eight problem-solving tasks in which a food dish could be extracted from a puzzle box if the puppy performed certain manipulations, e.g., tugging on a wooden dowel tied to the food dish. Two experimental trials were administered every day for 4 days. The number of problems solved by a random sample of 21 puppies of each breed is given below. Use these data to determine whether or not the average number of problems solved differs from breed to breed.
(a) What are the hypotheses?
(b) What is the appropriate test statistic?
(c) How is the test statistic distributed if H_0 is correct?
(d) What is the value of the test statistic?
(e) What is the p-value of the test statistic?
(f) What is the appropriate conclusion?

Wirehaired Terriers	Beagles	Shelties	Basenjis	Cocker Spaniels
1	0	0	5	0
6	6	0	8	0
0	2	6	7	1
0	5	6	8	4
0	2	8	8	0
1	3	3	1	1
8	4	6	8	0
7	3	8	5	2
2	3	6	7	6
5	4	7	8	3
2	0	7	7	3
1	4	1	5	1
4	0	0	7	6
2	1	0	8	4
1	0	2	7	0
7	0	2	8	0
8	0	2	6	0
7	0	0	3	2
3	6	1	8	1
7	6	1	3	4
2	6	1	2	0

Source: Individual scores furnished by J. P. Scott. Results reported in Scott, J. P., and Fuller, J. L. (1965). *Genetics and the Social Behavior of the Dog*. Chicago: University of Chicago Press, p. 231.

12. Six batches of doughnuts were cooked in each of four types of fat. The number of grams of fat absorbed by each batch is recorded below. Use these data to test a claim that the amount of fat absorbed by doughnuts depends on the type of fat in which the doughnuts are cooked.

Fat 1	Fat 2	Fat 3	Fat 4
164	178	175	155
172	191	193	166
168	197	178	149
177	182	171	164
156	185	163	170
195	177	176	168

Source: Lowe, B (1935). Iowa Agricultural Experimental Station. Cited in Snedecor, G. W. and Cochran, W. G. (1980) *Statistical Methods* (Seventh ed.). Ames, Iowa: The Iowa State University Press, pp. 215–216.

(a) What are the hypotheses?
(b) What is the appropriate test statistic?
(c) How is the test statistic distributed if H_0 is correct?
(d) What is the value of the test statistic?
(e) What is the p-value of the test statistic?
(f) What is the appropriate conclusion?

13. Blood plasma concentration of high-density cholesterol (HDL) is believed to be associated with reduced risk of coronary artery disease. To investigate mechanisms that might account for the low incidence of heart disease in long-distance runners, HDL concentrations (milliliters HDL per 100 milliliters of blood plasma) were measured for 20 elite runners, 8 good runners, and 72 nonrunners. Use the data summarized below to test the possibility that running increases HDL concentration.

	Elite Runners $(N = 20)$	Good Runners $(N = 8)$	Nonrunners $(N = 72)$
$\bar{x}$	56	52	49
$\hat{\sigma}^2$	146.41	118.81	110.25

Source: Martin, R. P., et al. (1978). Blood chemistry and lipid profiles of elite distance runners. In P. Milvey (Ed.), *The Long Distance Runner*. New York: Urizen, p. 93.

(a) What are the hypotheses?
(b) What is the appropriate test statistic?
(c) How is the test statistic distributed if H_0 is correct?
(d) What is the value of the test statistic?
(e) What is the p-value of the test statistic?
(f) What is the appropriate conclusion?

14. Resting heart rate is one measure of cardiovascular fitness. The following statistics summarize resting heart rates for 20 elite runners, 8 good runners, and 10 untrained men matched for body-fat composition with the runners. Use these data to test the claim that running improves cardiovascular fitness.

	Elite runners $(N = 20)$	Good runners $(N = 8)$	Nonrunners $(N = 10)$
$\bar{x}$	47.1	52.4	65.0
$\hat{\sigma}^2$	28.31	42.25	94.09

Source: Pollock, M. L. (1978). Submaximal working capacity of elite distance runners. In Milvey, ibid. pp. 62–63.

(a) What are the hypotheses?
(b) What is the appropriate test statistic?
(c) How is the test statistic distributed if H_0 is correct?
(d) What is the value of the test statistic?
(e) What is the p-value of the test statistic?
(f) What is the appropriate conclusion?

F. EVALUATION OF THE ANALYSIS OF VARIANCE: POST HOC COMPARISONS

The analysis of variance is an omnibus test. A significant F ratio MS_A/MS_W implies that two or more of J population means are unequal, but it does not tell the experimenter *which* population means. When an F test for analysis of variance is significant, an experimenter frequently wants to "snoop" the data and find out which of the populations are different from the others. Investigations of this sort are called *post hoc comparisons*, because they are ordinarily conducted only after a significant analysis of variance is obtained.[4] The most widely used methods of post hoc comparison are Tukey's test, the Newman-Keuls test, and Scheffé's test.

If all samples are of the same size, Tukey's method and the Newman-Keuls method allows one to make all of the $J(J - 1)/2$ pairwise comparisons among population means, that is, to test for differences between μ_1 and μ_2, between μ_1 and μ_3, and so on. Unlike, say, multiple t tests for differences between means, Tukey's test preserves one's level of α and for this reason is sometimes called Tukey's *HSD* test, where *HSD* stands for Honestly Significant Difference. In most cases, this is also true for Newman-Keuls. Both tests are simple, but they require special tables that are ordinarily found only in more advanced texts.

Scheffé's method is more versatile than the others in three ways. First, it does not require equal sample sizes. Second, it permits one to test not only pairwise comparisons but also multiple comparisons. That is, one can test μ_1 against the average of μ_2 and μ_3, or the average of μ_2 and μ_3 against the average of μ_4 and μ_5 and μ_7, and so on. Third, critical values and significance probabilities for Scheffé's statistic are based on Fisher's F distributions, which are tabled in almost every statistics book. The only disadvantage to Scheffé's method is that for pairwise comparisons, it is the least powerful of the three.

1. Formulation: hypotheses about contrasts

A *contrast* among J population means μ_j is a weighted sum

$$c_1\mu_1 + c_2\mu_2 + \cdots + c_J\mu_J$$

where $c_j \neq 0$ for *some j* and $\sum_j^J c_j = 0$.

[4] In some cases, an experimenter begins an investigation with a hunch that only a particular treatment or treatments will differ from the others or that one or another pair of groups will be different. If an experimenter has such *planned* comparisons in mind before data are collected, there is a wide variety of techniques that can be used *instead* of analysis of variance. Alternatively, the researcher might wish to test the omnibus hypothesis that all population means are equal and then proceed with post hoc comparisons if the analysis of variance yields significant results. One can also proceed directly to post hoc comparisons after simple inspection of group means.

A contrast is denoted by the lowercase Greek letter *psi*, ψ, which may be subscripted if the experimenter considers more than one set of weights $c_1, \ldots, c_J$. Thus, some of the contrasts we might define for five populations are:

$$\psi_1 = (0)\mu_1 + (2)\mu_2 + (1)\mu_3 + (0)\mu_4 + (-3)\mu_5$$

$$\psi_2 = (1)\mu_1 + (2)\mu_2 + (3)\mu_3 + (6)\mu_4 + (-12)\mu_5$$

$$\psi_3 = (0)\mu_1 + (0)\mu_2 + (0)\mu_3 + (1)\mu_4 + (-1)\mu_5$$

$$\psi_4 = (1/2)\mu_1 + (1/2)\mu_2 + (0)\mu_3 + (0)\mu_4 + (-1)\mu_5$$

It will be noted that ψ_3 can be rewritten

$$\psi_3 = \mu_4 - \mu_5$$

Therefore, the hypothesis that $\psi_3 = 0$ is equivalent to the hypothesis that $\mu_4 - \mu_5 = 0$, i.e, that $\mu_4 = \mu_5$.

Likewise, ψ_4 can be rewritten

$$\psi_4 = \frac{\mu_1 + \mu_2}{2} - \mu_5$$

so the hypothesis that $\psi_4 = 0$ is equivalent to the hypothesis that

$$\frac{\mu_1 + \mu_2}{2} - \mu_5 = 0$$

i.e., the hypothesis that the *average* of μ_1 and μ_2 is equal to μ_5.

And in general, any such comparison among J population means can therefore be formulated in terms of a test hypothesis about a contrast:

$$H_0: c_1\mu_1 + c_2\mu_2 + \cdots + c_J\mu_J = 0$$

or

$$H_0: \psi = 0$$

The alternative hypothesis for a post hoc comparison is always nondirectional, so

$$H_1: \psi \neq 0$$

This is because directional predictions that are formulated in advance should be tested by *planned* comparisons, and directional hypotheses formulated *after* inspection of results should be tested with new data.

2. Scheffé's test statistic

The test statistic for Scheffé's Multiple Comparison Test is

$$\frac{\left(\sum\limits_{j}^{J} c_j \bar{x}_j\right)^2}{(J-1)MS_W \sum\limits_{j}^{J} \dfrac{c_j^2}{N_j}}$$

$$= \frac{\left(c_1\bar{x}_1 + c_2\bar{x}_2 + \cdots + c_J\bar{x}_J\right)^2}{(J-1)MS_W\left(\dfrac{c_1^2}{N_1} + \dfrac{c_2^2}{N_2} + \cdots + \dfrac{c_J^2}{N_J}\right)} \qquad [12.21]$$

Scheffé[5] has shown that critical values for this statistic can be obtained from the distribution of Fisher's F with $J-1$ degrees of freedom in the numerator and $N-J$ degrees of freedom in the denominator. Most nondirectional hypotheses call for *two*-tailed tests, but positive contrasts ($\psi > 0$) and negative contrasts ($\psi < 0$) *both* tend to increase the value of [12.21]. Scheffé's statistic therefore takes an *upper*-tailed test. For significance level α, the test hypothesis is rejected if [12.21] is greater than or equal to F_b, where $\mathbb{P}(F_b) = 1 - \alpha$ in $F(J-1, N-J)$.[6]

Example 12.7. In Example 12.6 (p. 515) we concluded that mean 10-day body weights are not identical in populations of wolf pups from Ontario, Minnesota, and Alaska. However, the analysis did not reveal whether all three of the means are different or whether one of the three is different from the other two. Gray wolves found in Minnesota and Ontario are usually Eastern Timberwolves (*C. lupus lycaon*), which is a medium-sized subspecies, but the Interior Alaskan wolf (*C. lupus pambeselius*) is the largest of all subspecies. We might therefore be interested in comparing the mean of Group 3 (Alaskan pups) with the average of the means for Groups 1 and 2. That is, we might wish to test the hypothesis that

$$\frac{\mu_1 + \mu_2}{2} = \mu_3$$

which can be expressed as the following contrast:

$$H_0\colon \psi = (1/2)\mu_1 + (1/2)\mu_2 + (-1)\mu_3 = 0$$

---

[5] Scheffé, H. A method for judging all contrasts in the analysis of variance. *Biometrika*, 40, 87–104.
[6] Scheffé does not include $(J-1)$ in the denominator of the test statistic. Instead, he sets the critical value equal to $(J-1)F_b$.

Scheffé's statistic is therefore

$$\frac{[(1/2)25 + (1/2)28 - 32]^2}{(2)6.73\left(\dfrac{1/4}{5} + \dfrac{1/4}{4} + \dfrac{1}{5}\right)} \doteq 7.19$$

With $J = 3$ populations, we have 2 degrees of freedom in the numerator, and with a total of 14 observations, $N - J = 11$ degrees of freedom in the denominator. If we set $\alpha = .05$, the critical value F_b has cumulative probability .95 in $F(2, 11)$. In Table 6 of Appendix VIII we find that $\mathbb{P}(3.98) = .95$ in $F(2, 11)$, so we can reject the hypothesis that the contrast is equal to zero and conclude that the mean 10-day body weight for Population 3 is greater than the average of the means for Populations 1 and 2.

The test for one particular contrast ψ seems little different from many of the tests we developed throughout this text. What makes Scheffé's and other such methods different is that they *control* the probability of a Type I error for *multiple* comparisons. The contrast tested in Example 12.7 is ψ_4 in the following list of several comparisons[7] that the analysis of variance in Example 12.6 might suggest.

$$
\begin{array}{lll}
\psi_1: & (1)\mu_1 + (-1)\mu_2 + (0)\mu_3 & (\mu_1 \; vs \; \mu_2) \\
\psi_2: & (1)\mu_1 + (0)\mu_2 + (-1)\mu_3 & (\mu_1 \; vs \; \mu_3) \\
\psi_3: & (0)\mu_1 + (1)\mu_2 + (-1)\mu_3 & (\mu_2 \; vs \; \mu_3) \\
\psi_4: & (1/2)\mu_1 + (1/2)\mu_2 + (-1)\mu_3 & (\mu_1 \text{ and } \mu_2 \; vs \; \mu_3) \\
\psi_5: & (1/2)\mu_1 + (-1)\mu_2 + (1/2)\mu_3 & (\mu_1 \text{ and } \mu_3 \; vs \; \mu_2) \\
\psi_6: & (-1)\mu_1 + (1/2)\mu_2 + (1/2)\mu_3 & (\mu_2 \text{ and } \mu_3 \; vs \; \mu_1)
\end{array}
$$

pairwise comparisons

If we tested each of these contrasts at $\alpha = .05$, as in Example 12.7, our *overall* probability of making one or more Type I errors would not exceed .05. In addition, the (omnibus) analysis of variance in Example 12.6 was significant at $\alpha = .005$, so we can be *certain* that at least one contrast (though not necessarily one of those listed) is significant at $\alpha = .005$.

The assumptions underlying Scheffé's test are essentially the same as for analysis of variance:

- Observations are independent, both within and across populations.
- The random variable is normally distributed in all J populations.
- The variances σ_j^2 of the J populations are equal.

And, like the F test for analysis of variance, Scheffé's test is relatively insensitive to departures from normality and homogeneity of variance.

---

[7] For any J population means, the number of contrasts that satisfy the requirements that some $c \neq 0$ and $\Sigma c = 0$ is, of course, unlimited. If sample sizes differ widely, for example, the experimenter may wish to assign (nonzero) weights that are proportional to sample size.

—————— EXERCISES 12.4 ——————————————————

1. In Exercise 12.3.7, three groups of wolf pups and one group of dog pups were administered a test of inhibition learning. The task required that subjects remain on a plywood platform, and data for the three groups are given below.

Number of seconds on platform (sum of 10 "best" daily scores)			
Group 1 Hand-reared / choke chain (wolves)	Group 2 Mother-reared / choke chain (wolves)	Group 3 Hand-reared / pellets (wolves)	Group 4 Mother-reared / choke chain (Malamutes)
221	267	413	523
345	241	565	905
362	564	657	990
63	397		540
$\sum$ 991	1,469	1,635	2,959
$\bar{x}$ 247.75	367.25	545	739.50

Use Scheffé's method to determine whether to not the form of punishment (choke chain versus pellets) exerts an influence on inhibition learning in wolves, whether or not socialization (mother rearing versus hand rearing) exerts an influence on inhibition learning in wolves, and whether or not wolves and Malamutes differ.

(a) Formulate the appropriate contrasts to compare the average of (1) the wolf pups' choke-chain means with the mean of the pellets group, (2) the means for hand-reared wolf pups with the mean for mother-reared wolf pups, and (3) the wolf pup means with the mean of the Malamute pups.

(b) What are the hypotheses?

(c) What is the appropriate test statistic?

(d) How is the test statistic distributed if the test hypotheses are correct?

(e) If the significance level α is .05, what is the critical value of the test statistic?

(f) What are the values of the test statistic?

(g) What are the appropriate conclusions?

2. Use Scheffé's method to test the hypothesis that the significance obtained in Exercise 12.3.4 is due to differences between sardines and other reinforcements.

Kibble	Sardines	Reese's Pieces [®]
6	1	8
5	3	7
3	2	5
3	1	4
3	3	6

(a) Formulate the appropriate contrast to compare the mean of the sardines group with the average of the means for the other groups.
(b) What are the hypotheses?
(c) What is the appropriate test statistic?
(d) How is the test statistic distributed if H_0 is correct?
(e) If the significance level α is .05, what is the critical value of the test statistic?
(f) What is the value of the test statistic?
(g) What is the appropriate conclusion?

In Exercise 12.3.10, subjects in a classical conditioning experiment were assigned to four groups: delayed conditioning (Group 1), simultaneous conditioning (Group 2), trace conditioning (Group 3), and backward conditioning (Group 4). The number of trials required to extinguish the conditioned response was recorded for every subject. The means and variances for the four groups were as follows:

	Delayed ($N = 5$)	Simultaneous ($N = 5$)	Trace ($N = 5$)	Backward ($N = 5$)
$\bar{x}$	35	28.5	29.33	18.67
s^2	42.67	28.58	51.22	9.89

3. It is known that *backward* conditioning yields the least effective resistance to extinction. Use Scheffé's method to test the hypothesis that any differences detected by analysis of variance are due to the performance of the group that received backward conditioning.
    (a) Formulate the appropriate contrast to compare the mean of the backward-conditioning group with the average of the means of the other groups.
    (b) What are the hypotheses?
    (c) What is the appropriate test statistic?
    (d) How is the test statistic distributed if H_0 is correct?
    (e) If the significance level α is .05, what is the critical value of the test statistic?
    (f) What is the value of the test statistic?
    (g) What is the appropriate conclusion?
4. It is known that *delayed* conditioning yields the most effective resistance to extinction. Use Scheffé's method to test the hypothesis that any differences detected by analysis of variance are due to the performance of the group that received delayed conditioning.
    (a) Formulate the appropriate contrast to compare the mean of the delayed-conditioning group with the average of the means of the other groups.
    (b) What are the hypotheses?
    (c) What is the appropriate test statistic?
    (d) How is the test statistic distributed if H_0 is correct?
    (e) If the significance level α is .01, what is the critical value of the test statistic?
    (f) What is the value of the test statistic?
    (g) What is the appropriate conclusion?

5. In Exercise 12.3.11, five groups of puppies were administered eight puzzle-solving tasks.

Wirehaired Terriers	Beagles	Shelties	Basenjis	Cocker Spaniels
1	0	0	5	0
6	6	0	8	0
0	2	6	7	1
0	5	6	8	4
0	2	8	8	0
1	3	3	1	1
8	4	6	8	0
7	3	8	5	2
2	3	6	7	6
5	4	7	8	3
2	0	7	7	3
1	4	1	5	1
4	0	0	7	6
2	1	0	8	4
1	0	2	7	0
7	0	2	8	0
8	0	2	6	0
7	0	0	3	2
3	6	1	8	1
7	6	1	3	4
2	6	1	2	0
$\sum$ 74	55	67	129	38
$\bar{x}$ 3.52	2.62	3.19	6.14	1.81

A researcher observes that basenjis appear to score higher than the other breeds, cocker spaniels appear to score lower than the other breeds, and the remaining three breeds all score about midway between basenjis and cockers. Use Scheffé's method to test this interpretation.

(a) Formulate the appropriate contrasts to compare the mean of the basenjis with the average of the means of the other groups, the mean of the cocker spaniels with the average of the means of the other groups, and the average of the means of basenjis and cockers with the average of the means of the other groups.

(b) What are the hypotheses?

(c) What is the appropriate test statistic?

(d) How is the test statistic distributed if the test hypotheses are correct?

(e) If the significance level α is .05, what is the critical value of the test statistic?

(f) What are the values of the test statistic?

(g) What are the appropriate conclusions?

6. Use Scheffé's method to determine whether there are *any* pairwise differences among the groups of blowflies in Exercise 12.3.8.

	Sucrose $(N = 6)$	Dextrose $(N = 6)$	Saccharine $(N = 6)$
$\bar{x}$	13.398	11.022	4.100
s^2	84.163	107.986	38.783

(a) Write the expressions for the appropriate contrasts.
(b) What are the hypotheses?
(c) What is the appropriate test statistic for each test?
(d) How is the test statistic distributed if the test hypotheses are correct?
(e) Calculate the values of the test statistic.
(f) If the critical value is set at 4.76 for each test, what is the *overall* probability of a Type I error, i.e., the probability that at least one of the tests will lead to an incorrect rejection of a test hypothesis?
(g) What are the appropriate conclusions?

7. In Exercise 12.3.13 summary statistics for blood plasma concentration of high-density cholesterol (HDL) were reported for 20 elite runners, 8 good runners, and 72 nonrunners.

	Elite runners $(N = 20)$	Good runners $(N = 8)$	Nonrunners $(N = 72)$
$\bar{x}$	56	52	49
$\hat{\sigma}^2$	146.41	118.81	110.25

Use Scheffé's method to test the hypothesis that elite runners differ from the other groups and to test the hypothesis that runners differ from nonrunners.
(a) Formulate the appropriate contrast to compare the mean of elite runners with the average of the means of the other groups and to compare the mean of nonrunners with the average of the means of the two groups of runners.
(b) What are the hypotheses?
(c) What is the appropriate test statistic?
(d) How is the test statistic distributed if the test hypotheses are correct?
(e) If the significance level α is .05, what is the critical value of the test statistic?
(f) What is the value of the test statistic?
(g) What is the appropriate conclusion?

8. In Exercise 12.3.14 summary statistics for resting heart rates were reported for 20 elite runners, 8 good runners, and 10 untrained men matched for body-fat composition with the runners.

	Elite runners $(N = 20)$	Good runners $(N = 8)$	Nonrunners $(N = 10)$
$\bar{x}$	47.1	52.4	65.0
$\hat{\sigma}^2$	28.31	42.25	94.09

Use Scheffé's method to test the hypothesis that elite runners differ from the other groups and to test the hypothesis that runners differ from nonrunners.

(a) Formulate the appropriate contrast to compare the mean of elite runners with the average of the means of the other groups and to compare the mean of nonrunners with the average of the means of the two groups of runners.
(b) What are the hypotheses?
(c) What is the appropriate test statistic?
(d) How is the test statistic distributed if the test hypotheses are correct?
(e) If the significance level α is .01, what is the critical value of the test statistic?
(f) What is the value of the test statistic?
(g) What is the appropriate conclusion?

G. SUMMARY

The most basic analysis of variance is a technique for testing the hypothesis that J population means are equal:

$$H_0: \mu_1 = \mu_2 = \cdots = \mu_J = \mu$$

When $J = 2$, the analysis of variance gives the same result as a nondirectional t test for two population means, and the assumptions underlying analysis of variance are seen to be a generalization from two groups to J groups of the assumptions required for Student's t:

- The J groups are independent.
- X is normally distributed in all J populations.
- All J population variances are equal.

The t test for two means and the analysis of variance for J means are mathematically related, but the two techniques take somewhat different approaches. The t test focuses on the *location* of the difference $\bar{x}_1 - \bar{x}_2$; the analysis of variance focuses on the *dispersion* of the sample means $\bar{x}_j$.

Analysis of variance is based on a *linear model* for the score of individual i in Population j:

$$x_{ij} = \mu + \tau_j + e_{ij}$$

where μ is the mean of the grand population, τ_j represents the effect of treatment j (or membership in Population j), and e_{ij} is called error, the contribution to x of everything *other than* the treatment.

The treatment effect τ_j is equal to the difference between the grand mean and the mean of Population j,

$$\tau_j = \mu - \mu_j$$

and is therefore a constant for all scores in Group j, but a variable across groups. The hypothesis H_0: $\mu_j = \mu$ is equivalent to the hypothesis H_0: $\tau_j = 0$.

The error component e_{ij} is the difference between the score of individual ij and the mean of Population j,

$$e_{ij} = x_{ij} - \mu_j$$

and therefore varies from individual to individual within each group. Since all subjects in Group j receive the same treatment, the only source of variability within each group is error, and the assumption of equal population variances means that the common population variance σ_j^2 is equal to error variance σ_e^2.

The total variability among all N observations reflects both sources of variability represented in the linear model, error and treatment, and is captured by the *total sum of squares*,

$$SS_{\text{TOTAL}} = \sum \sum (x_{ij} - \bar{x})^2$$

which can be partitioned into the *sum of squares within groups* and the *sum of squares among groups*:

$$SS_{\text{TOTAL}} = SS_W + SS_A$$

$$= \sum_j \sum_i (x_{ij} - \bar{x}_j)^2 + \sum_j N_j (\bar{x}_j - \bar{x})^2$$

A sum of squares divided by its degrees of freedom is called a *mean squares* (*MS*). The degrees of freedom for SS_W is $N - J$, and the degrees of freedom for SS_A is $J - 1$. Therefore,

$$MS_W = \frac{\sum_j \sum_i (x_{ij} - \bar{x}_j)^2}{N - J}$$

and

$$MS_A = \frac{\sum_j N_j (\bar{x}_j - \bar{x})^2}{J - 1}$$

It is seen that the formula for MS_W is identical to the formula for the pooled estimate of variance, and the common variance estimated by MS_W must therefore be error variance, σ_e^2. That is,

$$E(MS_W) = \sigma_e^2$$

The mean squares among groups is based on differences between the J sample means $\bar{x}_j$ and the grand mean $\bar{x}$ and estimates the grand population variance σ^2. The mean squares among groups therefore taps treatment variance as well as error variance. It is shown that

$$E(MS_A) = \frac{\sum N_j \tau_j^2}{J - 1} + \sigma_e^2$$

Under the test hypothesis, however, all τ_j equal zero, so $E(MS_A) = \sigma_e^2$.

Sample means and variances are independent for normally distributed random variables, so when the test hypothesis is correct, MS_A and MS_W are independent estimates of the same variance σ_e^2 and the ratio

$$\frac{MS_A}{MS_W}$$

is therefore distributed as Fisher's F with $J - 1$ degrees of freedom in the numerator and $N - J$ degrees of freedom in the denominator.

If MS_A/MS_W is statistically significant, the experimenter can conclude that at least two of the J population means are unequal, but the F test does not indicate *which* means. This question is addressed in post hoc tests, which permit comparisons among specific means. The most versatile of these is Scheffé's Multiple Comparison Test. Any comparison among J means can be expressed as *contrast*,

$$\psi = c_1 \mu_1 + c_2 \mu_2 + \cdots + c_J \mu_J$$

where $c_j \neq 0$ for some j and $\Sigma c_j = 0$. Scheffé's statistic is

$$\frac{\left(\sum_j^J c_j \bar{x}_j \right)^2}{(J - 1) MS_W \sum_j^J \frac{c_j^2}{N_j}}$$

Critical values for Scheffé's statistic can be obtained from the distribution of $F(J - 1, N - J)$.

Summary table of statistical tests

Testing hypotheses about $\mu_1, \mu_2, \ldots,$ and μ_J

H_0	H_1	Test statistic	Distribution of test statistic*
Analysis of variance			
$\mu_j = \mu$ (or $\tau_j = 0$) for all j	H_0 incorrect	$\dfrac{MS_A}{MS_W}$	$F(J-1, N-J)$
Contrasts			
$\psi = 0$	$\psi \neq 0$	$\dfrac{\left(\sum\limits_{j}^{J} c_j \bar{x}_j\right)^2}{(J-1)MS_W \sum\limits_{j}^{J} \dfrac{c_j^2}{N_j}}$	$F(J-1, N-J)$

*If H_0 is correct.

Assumptions: Observations independent
X_j normally distributed
$\sigma_1^2 = \cdots = \sigma_J^2$ (can be violated if $N_1 = \cdots = N_J$)

Sums and means of squares for analysis of variance

Source	Sums of squares		Mean squares
	Definitional	Computational	
Within groups	$\sum\limits_{j}\sum\limits_{i}(x_{ij} - \bar{x}_j)^2$	$A - B$	$\dfrac{SS_W}{N-J}$
Among groups	$\sum\limits_{j} N_j(\bar{x}_j - \bar{x})^2$	$B - D$	$\dfrac{SS_A}{J-1}$

Basic quantities for computational formulae:
$$A = \sum_{j}\sum_{i} x_{ij}^2 \qquad B = \sum_{j} N_j \bar{x}_j^2 \qquad D = N\bar{x}^2$$

PART IV

Hypothesis testing: intermediate techniques

CHAPTER 13

More complex analyses of variance

CHAPTER OUTLINE

We introduced analysis of variance in Chapter 12 as a method of generalizing tests of population means to J populations. In fact, however, the linear model on which analysis of variance is based is extremely versatile. It can be adapted to so wide a variety of experiments that graduate courses and advanced texts devoted entirely to analysis of variance are frequently titled "Experimental Design." The present chapter introduces two of the most common extensions of analysis of variance. Our development of these techniques is deliberately terse and much more algorithmic than in earlier chapters. This is because the mathematical and conceptual underpinnings are, except as noted, essentially variations of the linear model and the partition of sums of squares developed in Chapter 12.[1]

We showed in Chapter 12 that the analysis of variance as a test of J population means is closely related to the t-test for two independent group means. Accordingly, all of the ANOVA problems and examples in Chapter 12 are characterized by two properties that are also found in tests of two means. First, the J populations represent different variations of a *single* factor, e.g., locations of capture of wolves, high schools attended by college applicants, etc. Such analyses are described as *one-way*. In this chapter, we extend ANOVA to problems involving *two* experimental factors, the *two-way* analysis of variance. In addition, it was assumed throughout Chapter 12 that the J groups were comprised of *different* individuals. One-way analyses of this sort are additionally characterized as *simple*. The present chapter introduces techniques that may be used when the J treatments are repeated measurements on the *same* individuals.

A. TWO-WAY ANALYSIS OF VARIANCE

Consider once again the concerns of a university admissions director about differences among three high schools (pp. 488–489, 513). Let us suppose that the director is now persuaded that in general they provide comparable college preparation but suspects that they may not all provide comparable training to both minority and nonminority students. It may therefore be useful to classify the data in a two-way scheme that separates Scholastic Aptitude Test (SAT) scores both by high school and by ethnicity. In a two-way data layout, one experimental factor is represented by rows and the second factor is represented by columns. In Table 13.1, we represent the two ethnic classifications as rows. Row 1 will include the scores of all minority applicants, and row 2 will include the scores of all nonminority applicants. The second factor, high school, is represented as columns. Scores of all students from High School 1 will be entered in column 1, scores for students from High School 2 in column 2, and High School 3 scores in column 3.

The scores of subjects who receive any particular *combination* of treatments is called a *cell* and is denoted by *two* subscripts, j and k, indicating the row and the column, respectively. In Table 13.1, the scores of all minority applicants (row 1) from High School 2 (column 2) will therefore appear in cell 1, 2.

Half the battle in handling two-way analysis of variance is keeping the notation straight. In Table 13.2 we have filled in the cells represented in Table 13.1 with scores expressed in general notation. Each score is identified with three subscripts, i, j, and k. As indicated above, the subscripts j and k specify the cell in which the score falls (row j and

---

[1] Some of the variations, however, are elegant and fascinating in their own right. The interested student is referred to Hays, W. L. (1988). *Statistics* (4th ed.). Fort Worth TX: Holt, Rinehart and Winston.

Table 13.1. Data layout for tabulating SAT scores of college applicants representing two ethnic classifications and three high schools

	High School 1	High School 2	High School 3	
Minority applicants	cell 1, 1	cell 1, 2	cell 1, 3	row 1
Nonminority applicants	cell 2, 1	cell 2, 2	cell 2, 3	row 2
	column 1	column 2	column 3	

Table 13.2. SAT scores of college applicants representing two ethnic classifications and three high schools

	High School 1	High School 2	High School 3	Row Totals
Minority applicants	x_{111} x_{211} $\ldots$ $x_{N_{11}11}$	x_{112} x_{212} $\ldots$ $x_{N_{12}12}$	x_{113} x_{213} $\ldots$ $x_{N_{13}13}$	
Cell totals for row 1	$\displaystyle\sum_{i}^{N_{11}} x_{i11}$	$\displaystyle\sum_{i}^{N_{12}} x_{i12}$	$\displaystyle\sum_{i}^{N_{13}} x_{i13}$	$\displaystyle\sum_{k}\sum_{i} x_{i1k}$
Nonminority applicants	x_{121} x_{221} $\ldots$ $x_{N_{21}21}$	x_{122} x_{222} $\ldots$ $x_{N_{22}22}$	x_{123} x_{223} $\ldots$ $x_{N_{23}23}$	
Cell totals for row 2	$\displaystyle\sum_{i}^{N_{21}} x_{i21}$	$\displaystyle\sum_{i}^{N_{22}} x_{i22}$	$\displaystyle\sum_{i}^{N_{23}} x_{i23}$	$\displaystyle\sum_{k}\sum_{i} x_{i2k}$
Column totals	$\displaystyle\sum_{j}\sum_{i} x_{ij1}$	$\displaystyle\sum_{j}\sum_{i} x_{ij2}$	$\displaystyle\sum_{j}\sum_{i} x_{ij3}$	$\displaystyle\sum\sum\sum x_{ijk}$

column k). The first subscript i identifies the *individual* in that cell. Thus,

$$x_{ijk}$$

is the score of individual i in cell jk. (If R or C is greater than 10, it may be necessary to separate *numerical* subscripts with commas to avoid confusion, for example, to distinguish $x_{21,1,3}$ from $x_{2,11,3}$. Commas are ordinarily omitted if subscripts are in general notation, e.g., x_{ijk}.) The number of observations in cell jk is denoted N_{jk}, so the scores in cell jk are

$$x_{1jk}$$

$$x_{2jk}$$

$$\cdots$$

$$x_{N_{jk}jk}$$

as seen in Table 13.2, and the sum of the scores in cell jk is

$$\sum_i^{N_{jk}} x_{ijk}$$

The scores for nonminority students from High School 3 appear in cell $2,3$ (that is, row 2 and column 3) of Table 13.2, and the total for this group is therefore represented as

$$\sum_i^{N_{23}} x_{i23}$$

The number of columns is denoted C, so k runs from 1 to C. The total for any particular row j is calculated by adding the C cell totals in row j and is therefore denoted

$$\sum_k^C \sum_i x_{ijk}$$

In Table 13.2, row 1 represents the scores of minority students, so the total of minority students' scores is the sum of the $C = 3$ cell totals in row 1 and is represented by the first sum on the right-hand side of the table,

$$\sum_k^3 \sum_i x_{i1k}$$

The row subscript j is replaced by the number 1, because we are adding only the cell totals in row 1.

The number of rows is denoted R, so j runs from 1 to R. The total for any particular column k is calculated by adding the R cell totals in column k and is therefore denoted

$$\sum_{j}^{R} \sum_{i} x_{ijk}$$

Scores of applicants from High School 3 are given in the third column of Table 13.2, and the total for this group is calculated by adding the $R = 2$ cell totals in column 3, which is represented by the third sum at the bottom of the table,

$$\sum_{j}^{2} \sum_{i} x_{ij3}$$

The column subscript k is replaced by the number 3, because we are adding only the cell totals in column 3.

As the reader might expect, the mean for any particular cell is identified by row and column subscripts, so $\bar{x}_{23}$ denotes the mean of nonminority applicants (row 2) from High School 3 (column 3). Denoting row and column means requires a new convention. In general notation, the mean for row j is $\bar{x}_j$ and the mean for column k is $\bar{x}_k$, but if a particular row or column is specified by number, it is not always possible to tell whether the number indicates a column or a row. For example, $\bar{x}_2$ might represent the mean of row 2 or the mean of column 2. To avoid this sort of ambiguity, we let $\bar{x}_{.2}$ denote the mean of row 2 and $\bar{x}_{.2}$ denote the mean of column 2. The dot ($\cdot$) is a "phantom" subscript or placeholder that tells the reader whether the number (e.g., 2) stands for j or for k.

1. Interaction and the linear model for two-way ANOVA

The feature of two-way ANOVA that most distinguishes it from one-way ANOVA is the possibility that the two sets of treatments *interact*.

Example 13.1. Let us suppose that our admissions officer obtains the following means for his two-way analysis.

Table 13.3. SAT means for minority and nonminority students from three high schools

	High School			Row means
	1	2	3	
Minority	$\bar{x}_{11} = 900$	$\bar{x}_{12} = 1,100$	$\bar{x}_{13} = 1,300$	$\bar{x}_{1.} = 1,100$
Nonminority	$\bar{x}_{21} = 1,300$	$\bar{x}_{22} = 1,100$	$\bar{x}_{23} = 900$	$\bar{x}_{2.} = 1,100$
Column means	$\bar{x}_{.1} = 1,100$	$\bar{x}_{.2} = 1,100$	$\bar{x}_{.3} = 1,100$	

Treatment effects of the row factor and treatment effects of the column factor are called *main* effects. It is apparent from the row means that there is no main effect attributable to ethnicity, and it is apparent from the column means that there is no main effect on SAT score attributable to high school. But, we see that the *combination* of high school and ethnicity does produce systematic differences in SAT performance. That is, the high school factor has *one* effect on minority students and a *different* effect on nonminority students. The two factors are therefore said to exhibit *interaction*. In general,

> Interaction is present when the effects of one factor are not the same across all treatments of the second factor.

One can have main effects with or without interaction, and interaction can occur in the absence of main effects, as in Example 13.1. In Figure 13.1(a) the cell means for each high school in Example 13.1 are plotted against ethnicity. Other possible patterns of main and interaction effects are illustrated in Figures 13.1(b) and 13.1(c). When interaction is present, the graphs for at least two of the groups in this sort of plot will *not* be parallel.

In principle, then, there are *four* sources of variability in any two-way classification of scores: (1) the effects of the treatments represented by the R rows, (2) the effects of the treatments represented by the C columns, (3) the interaction of the row and column treatments, and (4) error. These are represented in the linear model for a single score, x_{ijk}:

$$x_{ijk} = \mu + \tau_j + \gamma_k + \tau\gamma_{jk} + e_{ijk}$$

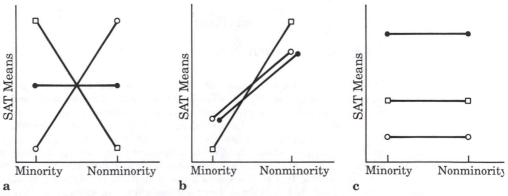

Figure 13.1. SAT means for minority and nonminority students from High School 1 (○), High School 2 (●) and High School 3 (□). (a) Interaction but no main effects, (b) ethnicity effects and interaction, (c) high school effects only.

2. Fixed effects and random effects

In some experiments the R row treatments and the C column treatments exhaust all of the populations of interest to the experimenter. In our SAT example, for instance, the two ethnic classifications (minority and nonminority) are logically exhaustive, and we have assumed that the admissions officer is concerned only about three particular high schools.

In designs of this sort, the treatment effects are said to be *fixed*. In other experiments the experimenter may intend to generalize beyond the treatments actually observed to *all* possible variations of the experimental factor or factors. For example, the admissions officer might have selected three high schools at random from among all of the high schools that furnish large numbers of applicants. This is typically the situation when treatments represent quantitatively different levels of the *same* factor, such as drug dosages in medical research, number of reinforcements in a learning experiment, or hours of weekly aerobic training in a study of cardiovascular fitness. When the particular treatments employed in an analysis of variance are viewed as a *sample* chosen from a wider population of similar treatments, treatment effects are said to be *random*.[2]

In two-way (and higher-order) ANOVA, it is also possible that one factor (represented by either rows or columns) has fixed effects and the other has random effects. Such a design is called a *mixed*-effects analysis of variance.

3. Hypotheses and test statistics for two-way ANOVA

The variability associated with row effects (τ), column effects (γ, the lowercase Greek *gamma*), interaction of rows and columns ($\tau\gamma$), and error (e) can be expressed in terms of sums and means of squares, just as in one-way analysis of variance. Likewise, ratios of means squares may be submitted to Fisher's F distribution to test the three hypotheses associated with any two-way analysis of variance:

(1) H_0:   No row effects

(2) H_0:   No column effects

(3) H_0:   No interaction

---

[2] The distinction between fixed and random effects also applies to one-way analysis of variance. In one-way ANOVA fixed- and random-effects analyses use the same test statistic, but somewhat different test hypotheses must be formulated. In any one-way ANOVA $E(\tau)$ is assumed to be 0. In the fixed-effect experiment $\tau_1 = \cdots = \tau_J$ therefore implies that $\tau_j = 0$ for all j, and the test hypothesis can therefore be formulated as H_0: $\tau_j = 0$. In the random-effects design the J treatments in the experiment are only a sample of all *possible* treatments, so $\tau_1 = \cdots = \tau_J = 0$ does not necessarily imply that all $\tau = 0$. Since $E(\tau) = 0$, one can conclude that all τ are zero if and only if the population *variance* of τ is zero. Therefore, the test hypothesis for a random-effects experiment is formulated as H_0: $V(\tau) = 0$. There are also some subtle differences in the assumptions underlying random- and fixed-effects one-way ANOVA. The interested reader is referred to Hays, W. L. (1988). *Statistics* (4th ed.). Forth Worth TX: Holt, Rinehart and Winston, pp. 489–490.

a. Sums and means of squares. For a two-way analysis of variance, the total sum of squares SS_{TOTAL} is

$$\sum_{k}^{C} \sum_{j}^{R} \sum_{i}^{N_{jk}} (x_{ijk} - \bar{x})^2$$

where $\bar{x}$ is the grand mean calculated across all N observations. The difference between any particular score and the grand mean $(x_{ijk} - \bar{x})$ is unchanged if we add *and* subtract the cell mean $\bar{x}_{jk}$, the row mean $\bar{x}_j$, and the column mean $\bar{x}_k$, so

$$x_{ijk} - \bar{x} = (x_{ijk} - \bar{x}_{jk}) + (\bar{x}_j - \bar{x}) + (\bar{x}_k - \bar{x}) + (\bar{x}_{jk} - \bar{x}_j - \bar{x}_k + \bar{x})$$

where N_j is the number of observations in row j and N_k is the number of observations in column k. And, it can be shown by extension of the same sort of algebraic operations as in Box 12.1 that the total sum of squares can be partitioned as follows:

$$SS_{\text{TOTAL}} = \sum \sum \sum (x_{ijk} - \bar{x}_{jk})^2 + \sum_{j}^{R} N_j (\bar{x}_j - \bar{x})^2 \qquad \textbf{[13.1]}$$

$$+ \sum_{k}^{C} N_k (\bar{x}_k - \bar{x})^2 + \sum_{j}^{R} \sum_{k}^{C} (\bar{x}_{jk} - \bar{x}_j - \bar{x}_k + \bar{x})^2$$

The first term on the right-hand side of equation **[13.1]**

$$\sum \sum \sum (x_{ijk} - \bar{x}_{jk})^2$$

reflects differences of every score from its cell mean, $\bar{x}_{jk}$. Since every individual in cell jk receives the same combination of column and row treatments, these differences can reflect only error. Accordingly this quantity is called the *sum of squares error* (abbreviated SS_{ERROR}) and is analogous to SS_W in one-way analysis of variance. The sum of squares error is a function of N observations that are used to calculate RC cell means $\bar{x}_{jk}$, which estimate the population cell means, μ_{jk}. Not surprisingly, SS_{ERROR} has $N - RC$ degrees of freedom. We can therefore define *mean* squares error as follows:

$$MS_{\text{ERROR}} = \frac{\sum \sum \sum (x_{ijk} - \bar{x}_{jk})^2}{N - RC} \qquad \textbf{[13.2]}$$

The second term in [**13.1**]

$$\sum_{j}^{R} N_j(\bar{x}_j - \bar{x})^2$$

is called the *sum of squares for rows* (SS_{ROW}) because it is based on differences between the R row means and the grand mean. This sum of squares, like SS_A in one-way analysis of variance, taps the effects of both row treatments and error. The random variable in SS_{ROW} is the R row means $\bar{X}_j$, and the grand mean estimates the expected value of the row means, so degrees of freedom for SS_{ROW} is $R - 1$, and the mean squares for rows is

$$MS_{\text{ROW}} = \frac{\sum_{j}^{R} N_j(\bar{x}_j - \bar{x})^2}{R - 1} \qquad [\textbf{13.3}]$$

The third term

$$\sum_{k}^{C} N_k(\bar{x}_k - \bar{x})^2$$

is called the *sum of squares for columns* (SS_{COL}) because it is based on differences between the C column means and the grand mean. This sum of squares therefore reflects both the effects of column treatments and error. As the reader might now suspect, degrees of freedom for SS_{COL} is $C - 1$, so the mean squares for columns is

$$MS_{\text{COL}} = \frac{\sum_{k}^{C} N_k(\bar{x}_k - \bar{x})^2}{C - 1} \qquad [\textbf{13.4}]$$

The last term in equation [**13.1**]

$$\sum_{j}^{R} \sum_{k}^{C} \left(\bar{x}_{jk} - \bar{x}_j - \bar{x}_k + \bar{x}\right)^2$$

is a residual. That is, it is equal to

$$SS_{\text{TOTAL}} - SS_{\text{ERROR}} - SS_{\text{ROW}} - SS_{\text{COL}}$$

and must therefore reflect the only other source of variability represented in the linear model, the *interaction* between row and column treatments. Accordingly, it is called the *sum of squares for interaction*, abbreviated SS_{INT}. Degrees of freedom for SS_{INT} is equal to

$$df_{\text{TOTAL}} - df_{\text{ERROR}} - df_{\text{ROW}} - df_{\text{COL}}$$

Since SS_{TOTAL} is calculated from N independent observations and the N values are used to calculate only one estimate, $\bar{x}$, degrees of freedom for SS_{TOTAL} is $N - 1$. Therefore,

$$df_{\text{INT}} = (N - 1) - (N - RC) - (R - 1) - (C - 1)$$

$$= RC - R - C + 1$$

$$= (R - 1)(C - 1)$$

and

$$MS_{\text{INT}} = \frac{\displaystyle\sum_{j}^{R}\sum_{k}^{C}\left(\bar{x}_{jk} - \bar{x}_{j} - \bar{x}_{k} + \bar{x}\right)^2}{(R - 1)(C - 1)} \qquad [13.5]$$

In one-way analysis of variance, we derived computational formulae to calculate sums and means of squares using three basic quantities, A, B, and D. If every cell in a two-way analysis of variance has the same number of observations $N_{jk} = N_g$, the computational formulae in Box 13.1 are similarly faster and involve fewer opportunities for error than the definitional formulae discussed above.

b. Fisher's _F_ for two-way ANOVA. For a fixed-effects analysis of variance it can be shown that

$$E(MS_{\text{ERROR}}) = \sigma_e^2$$

whether or not there are either main effects or interaction. If there are *no* row effects, it can also be shown that

$$E(MS_{\text{ROW}}) = \sigma_e^2$$

However, if there *are* differences among the row means μ_j, then

$$E(MS_{\text{ROW}}) > \sigma_e^2$$

Therefore, the test statistic for row effects is

$$\frac{MS_{\text{ROW}}}{MS_{\text{ERROR}}}$$

which is distributed as Fisher's F with $df_{\text{ROW}} = R - 1$ in the numerator and $df_{\text{ERROR}} = N - RC$ in the denominator.

Similarly, the test statistic for column effects is

$$\frac{MS_{\text{COL}}}{MS_{\text{ERROR}}}$$

BOX 13.1

Computational formulae for two-way ANOVA

If every cell has the same number of observations $N_{jk} = N_g$, it can be shown that the sums of squares for two-way analysis of variance can be expressed in terms of five basic quantities:

$$A = \sum_k \sum_j \sum_i x_{ijk}^2 \qquad B_R = \frac{\sum_j^R \left(\sum_k \sum_i x_{ijk} \right)^2}{CN_g} \qquad B_C = \frac{\sum_k^C \left(\sum_j \sum_i x_{ijk} \right)^2}{RN_g}$$

$$D = N\bar{x}^2 = \frac{\left(\sum_k \sum_j \sum_i x_{ijk} \right)^2}{N} \qquad G = \frac{\sum_j^R \sum_k^C \left(\sum_i x_{ijk} \right)^2}{N_g}$$

For purposes of computation,

$$SS_{\text{ERROR}} = A - G \text{ and}$$

$$MS_{\text{ERROR}} = \frac{A - G}{N - RC}$$

$$SS_{\text{ROW}} = B_R - D \text{ and}$$

$$MS_{\text{ROW}} = \frac{B_R - D}{R - 1}$$

$$SS_{\text{COL}} = B_C - D \text{ and}$$

$$MS_{\text{COL}} = \frac{B_C - D}{C - 1}$$

$$SS_{\text{INT}} = G - B_R - B_C + D \text{ and}$$

$$MS_{\text{INT}} = \frac{G - B_R - B_C + D}{(R - 1)(C - 1)}$$

which is distributed as Fisher's F with $df_{\text{COL}} = C - 1$ in the numerator and $df_{\text{ERROR}} = N - RC$ in the denominator.

It can also be shown that if there is no interaction between row and column treatments, then

$$E(MS_{\text{INT}}) = \sigma_e^2$$

but that

$$E(MS_{\text{INT}}) > \sigma_e^2$$

if intersection is present. The test statistic for interaction is therefore

$$\frac{MS_{\text{INT}}}{MS_{\text{ERROR}}}$$

which is distributed as Fisher's F with $df_{\text{INT}} = (R - 1)(C - 1)$ in the numerator and $df_{\text{ERROR}} = N - RC$ in the denominator.

Example 13.2. An experimenter is investigating the effects of different cues on reaction times for men and women. The cue for one group is an audible tone, the cue for the second group is a flash of light, and the cue for the third group is a mild electrical pulse from a metal band around one wrist. All subjects are instructed to press the [ENTER] key on a computer keyboard as soon as the cue is detected, and the time between cue and response is automatically recorded. Each subject receives 12 trials, and the scores given in Table 13.4 are total times (in seconds) for all participants. Since the research is preliminary, the experimenter is concerned only with the particular cues used in the study.

Table 13.4. Reaction time totals for 15 men and 15 women

	Tone	Light	Pulse
Men	10.0	6.0	9.1
	7.2	3.7	5.8
	6.8	5.1	6.0
	6.0	4.0	4.0
	5.0	3.2	5.1
Women	10.5	6.6	7.3
	8.8	4.9	6.1
	9.2	2.5	5.2
	8.1	4.2	2.5
	13.4	1.8	3.9

Since all six cells have the same number of observations, we can use the computational algorithms in Box 13.1. The basic quantity A is the sum of the 30 values x_{ijk}^2, and quantity D is $(1/30)$ times the squared sum of the 30 values x_{ijk}. Most calculators will compute sums and sums of squares at the same time, so A and D can be calculated in the same operation.

$$A = \sum_k \sum_j \sum_i x_{ijk}^2 = (10)^2 + (7.2)^2 + \cdots + (3.9)^2 = 1{,}306.68$$

$$D = \frac{\left(\sum_k \sum_j \sum_i x_{ijk}\right)^2}{N} = \frac{(10 + 7.2 + \cdots + 3.9)^2}{30} \doteq 1{,}104.13$$

Next, calculate the sums of all six cells and use these values to calculate the two row sums and the three column sums.

	Cell totals		Row totals
35	22	30	87
50	20	25	95
Column Totals 85	42	55	

The quantity B_R is the sum of the squared row totals divided by CN_g, the number of columns times the number of observations in each cell:

$$B_R = \frac{\sum\limits_{j}^{R}\left(\sum\limits_{k}\sum\limits_{i} x_{ijk}\right)^2}{CN_g} = \frac{(87)^2 + (95)^2}{(3)(5)} \doteq 1,106.27$$

The quantity B_C is the sum of the squared column totals divided by RN_g, the number of rows times the number of observations in each cell:

$$B_C = \frac{\sum\limits_{k}^{C}\left(\sum\limits_{j}\sum\limits_{i} x_{ijk}\right)^2}{RN_g} = \frac{(85)^2 + (42)^2 + (55)^2}{(2)(5)} = 1,201.40$$

The quantity G is the sum of the squared cell totals divided by the number of observations in each cell:

$$G = \frac{\sum\limits_{j}^{R}\sum\limits_{k}^{C}\left(\sum\limits_{i} x_{ijk}\right)^2}{N_g} = \frac{(35)^2 + (22)^2 + \cdots + (25)^2}{5} = 1,226.80$$

We can now calculate the means of squares needed to conduct our three F tests:

$$MS_{ERROR} = \frac{A - G}{N - RC} \doteq \frac{1,306.68 - 1,226.80}{30 - (2)(3)} \doteq 3.33$$

$$MS_{ROW} = \frac{B_R - D}{R - 1} \doteq \frac{1,106.27 - 1,104.13}{1} = 2.14$$

$$MS_{COL} = \frac{B_C - D}{C - 1} \doteq \frac{1,201.4 - 1,104.13}{2} \doteq 48.64$$

$$MS_{INT} = \frac{G - B_R - B_C + D}{(R - 1)(C - 1)} \doteq \frac{1,226.80 - 1,106.27 - 1,201.40 + 1,10}{(1)(2)}$$

$$= 11.63$$

The F ratio for row effects is

$$\frac{MS_{\text{ROW}}}{MS_{\text{ERROR}}} \doteq \frac{2.14}{3.33} \doteq .64$$

In $F(1, 24)$ we see that $\mathbb{P}(.469) = .50$ and $\mathbb{P}(1.39) = .75$, which means that the probability of observing a value as large or larger than .64 is at least .25. Therefore, $p > .25$, and the experimenter should conclude that gender exerts no main effect on reaction time.

The F ratio for column effects is

$$\frac{MS_{\text{COL}}}{MS_{\text{ERROR}}} \doteq \frac{48.64}{3.33} \doteq 14.61$$

In $F(2, 24)$ we see that $\mathbb{P}(14.0) > .9995$, so the main effects for columns is significant at the .0005 level (i.e., $p < .0005$), and the experimenter should conclude that subjects have different reaction times to the three different cues.

The F ratio for interaction is

$$\frac{MS_{\text{INT}}}{MS_{\text{ERROR}}} \doteq \frac{11.63}{3.33} \doteq 3.49$$

In $F(2, 24)$ we see that $\mathbb{P}(3.40) = .95$, so interaction is significant at the .05 level (i.e., $p < .05$), and the experimenter should conclude that men and women have different patterns or profiles of differences across the three cues.

The results of our two-way analysis of variance are summarized in Table 13.5. Totals for sums of squares and degrees of freedom are included, as in the one-way ANOVA summary table (p. 516).

Table 13.5. Analysis of variance (ANOVA) summary table for Example 13.2

Source	SS	df	MS	F
Rows	2.14	1	2.14	.64
Columns	97.27	2	48.63	14.61[†]
Interaction	23.26	2	11.63	3.49*
Error	79.88	24	3.33	
Totals	202.55	29		

$*p < .05.$
$[†]p < .0005.$

Tests for random-effects and mixed-effects designs. For a random-effects analysis of variance the situation is different. The expected values for both MS_{ROW} and MS_{COL} include an interaction component, and it can be shown that the appropriate test statistics are as follows:

- Row treatments:

$$\frac{MS_{\text{ROW}}}{MS_{\text{INT}}} : F(df_{\text{ROW}}, df_{\text{INT}})$$

- Column treatments:

$$\frac{MS_{\text{COL}}}{MS_{\text{INT}}} : F(df_{\text{COL}}, df_{\text{INT}})$$

- Interaction:

$$\frac{MS_{\text{INT}}}{MS_{\text{ERROR}}} : F(df_{\text{INT}}, df_{\text{ERROR}})$$

In a random-effects analysis of variance, the F test for interaction is therefore seen to be identical to the test in fixed-effects analysis and should be conducted before the tests for main effects. If interaction effects are significant, the F tests for rows and columns are as given above. If interaction is not significant, one must decide whether the evidence justifies the stronger conclusion that interaction is, in fact, absent altogether. When interaction is *not* present, MS_{ROW} and MS_{COL} tap only treatment effects and error, so the appropriate denominator in the F ratios for main effects is an estimator of σ_e^2. Furthermore, we know that MS_{INT} estimates error variance when there is no interaction, in which case the best possible estimate of σ_e^2 is therefore obtained by *pooling* MS_{INT} and MS_{ERROR}, as given in equation **[13.6]** below. One is clearly justified in concluding that interaction is absent when $MS_{\text{INT}} \leq MS_{\text{ERROR}}$. If MS_{INT} is greater than MS_{ERROR}, Paull[3] suggests that the pooled estimate of error can be used when

(1) df for error > 6,
(2) df for interaction > 6,
(3) F ratio for interaction < 2.00.

If the experimenter decides that interaction is absent, the tests for main effects therefore become

- Row treatments:

$$\frac{MS_{\text{ROW}}}{MS_{\text{POOLED ERROR}}} : F(df_{\text{ROW}}, df_{\text{POOLED ERROR}})$$

---

[3] Paull, A. E. (1950). On a preliminary test for pooling mean squares in the analysis of variance. *Annals of Mathematical Statistics, 21,* 539–556.

- Column treatments:

$$\frac{MS_{\text{ROW}}}{MS_{\text{POOLED ERROR}}} : F(df_{\text{ROW}}, df_{\text{POOLED ERROR}})$$

where $MS_{\text{POOLED ERROR}}$ is

$$\frac{SS_{\text{POOLED ERROR}}}{df_{\text{POOLED ERROR}}} = \frac{SS_{\text{INT}} + SS_{\text{ERROR}}}{df_{\text{INT}} + df_{\text{ERROR}}} \qquad [13.6]$$

and

$$df_{\text{INT}} + df_{\text{ERROR}} = (R - 1)(C - 1) + (N - RC)$$

For a mixed-effects analysis of variance, Fisher's F test for interaction is, as one might expect, the same as in the fixed-effects and in the random-effects analyses. To test for main effects, the mean squares for one factor is divided by MS_{INT} and the mean squares for the other is divided by MS_{ERROR} and these ratios are submitted to Fisher's F in the usual way. *However*, there is a counterintuitive twist: The denominator in the F ratio for the *random*-effects factor is MS_{ERROR}, and the denominator in the F ratio for the *fixed*-effects factor is MS_{INT}. If the F test for interaction does not yield significant results, one must, as in the random effects analysis, decide whether there is sufficient evidence to conclude that interaction is absent. If $MS_{\text{INT}} \leq MS_{\text{ERROR}}$ or if Paull's criteria are satisfied, then the mean squares for the fixed-effects factor should be tested against the pooled mean squares for error.

Example 13.3. Suppose the experimenter in Example 13.2 wanted to generalize his conclusion to cues involving other sense modalities or to visual, auditory, and tactile cues in general. The column treatments would then be a random sample of all possible cues, but there are only two possible genders, so the row treatments are fixed. The experiment would therefore be a *mixed*-effects design. The F ratios for the *mixed* factor (cues) and for interaction are the same as in the fixed-effects analysis, but interaction was significant, so the mean squares for the *fixed* factor (gender) is divided by MS_{INT} and the ratio submitted to the F table for $df_{\text{ROW}} = 1$ degree of freedom in the numerator and $df_{\text{INT}} = (1)(2) = 2$ degrees of freedom in the denominator:

$$\frac{MS_{\text{ROW}}}{MS_{\text{INT}}} \doteq \frac{2.14}{11.63} \doteq .184$$

for which $\mathbb{P}$ in $F(1, 2)$ is less than .10, making the p-value greater than .90. As in Example 13.2, therefore, gender exerts no main effect on reaction time.

Results of random- and mixed-effects analysis are ordinarily summarized in the same format as results of fixed-effects ANOVA, illustrated in

Example 13.2. Only the quantities used to calculate the F ratios in the last column are different.

B. ANALYSIS OF VARIANCE FOR ONE EXPERIMENTAL TREATMENT WITH REPEATED MEASURES

Recall from Chapter 10 (pp. 437–439) that if X_1 and X_2 are measurements taken on the same group of individuals, the two variables are not independent, and one cannot test the hypothesis that $\mu_1 = \mu_2$ with techniques developed to test hypotheses about two means. This problem generalizes to situations in which each of N subjects is administered J different treatments. Individual factors such as ability, personality, etc., are likely to make two scores obtained by the *same* individual more similar than scores obtained by *different* individuals. That is, the J scores of the same individual, i

$$x_{i1}, x_{i2}, \ldots, x_{iJ}$$

can be expected to *covary*, and the assumption of independence across the J groups is therefore violated.

This problem is addressed by a *two-way* analysis of variance procedure, in which the J measures are treated as one factor, and the N subjects are treated as the second factor. Technically, a repeated measures analysis is always either a random-effects or a mixed-effects analysis, because the subjects are a random sample drawn from the population of all possible subjects.

The total sum of squares is calculated exactly as in simple one-way analysis of variance. The notation is slightly different, because each group includes the same N subjects and the within-group sum therefore runs from 1 to N, rather than 1 to N_j:

$$SS_{\text{TOTAL}} = \sum_{j}^{J} \sum_{i}^{N} \left(x_{ij} - \bar{x} \right)^2$$

The total number of observations is therefore NJ, and the degrees of freedom for SS_{TOTAL} is $NJ - 1$.

The partitioning of the total sum of squares likewise proceeds very much as in one-way analysis of variance (see Box 12.1, pp. 506–507), but with one important difference: In simple one-way ANOVA we add and subtract the *group* (i.e., treatment) mean $\bar{x}_j$ to every difference $(x_{ij} - \bar{x})$. In repeated-measures ANOVA, we add and subtract the *individual* mean $\bar{x}_i$, that is, the mean for individual i calculated across the J repeated measures:

$$x_{ij} - \bar{x} = x_{ij} - \bar{x} + \bar{x}_i - \bar{x}_i$$

When differences are expressed in this form, the total sum of squares can be partitioned as follows:

$$\sum_{j}^{J} \sum_{i}^{N} \left(x_{ij} - \bar{x} \right)^2 = J \sum_{i}^{N} \left(\bar{x}_i - \bar{x} \right)^2 + \sum_{j}^{J} \sum_{i}^{N} \left(x_{ij} - \bar{x}_i \right)^2$$

The first term,

$$J \sum_{i}^{N} \left(\bar{x}_i - \bar{x} \right)^2$$

captures differences between individual means and the grand mean. Accordingly, this quantity is called the *sum of squares across subjects*, denoted SS_{AS}, and has $N - 1$ degrees of freedom.

The second term

$$\sum_{j}^{J} \sum_{i}^{N} \left(x_{ij} - \bar{x}_i \right)^2$$

reflects differences between the J scores obtained by individual i and the mean for individual i and is therefore called the *sum of squares within subjects*. This quantity is abbreviated SS_{WS} and has $N(J - 1)$ degrees of freedom.

The most obvious source of variability within each subject's J scores is the effects of the J treatments. As in one-way analysis of variance, treatment effects are captured by the sum of squared differences between the means of the J treatment groups and the grand mean:

$$\sum_{j} N \left(\bar{x}_j - \bar{x} \right)^2$$

In the context of repeated-measures analysis of variance, this quantity is called the *sum of squares across treatments* and is denoted SS_{AT}. Like the sum of squares among groups in one-way ANOVA, SS_{AT} has $J - 1$ degrees of freedom.

If the sum of squares across treatments is subtracted from the sum of squares within subjects, the difference

$$\sum_{j}^{J} \sum_{i}^{N} \left(x_{ij} - \bar{x}_i \right)^2 - \sum_{j} N \left(\bar{x}_j - \bar{x} \right)^2$$

captures all *residual* sources of within-subjects variation, including interaction between subjects and treatments, and is therefore denoted SS_{RESIDUAL}. Degrees of freedom for SS_{RESIDUAL} is $(J - 1)(N - 1)$, and it can be shown that

$$E(MS_{\text{RESIDUAL}}) = E\left(\frac{SS_{\text{RESIDUAL}}}{(J - 1)(N - 1)} \right) = \sigma_e^2$$

The hypothesis of interest in a single-factor repeated-measures experiment is the same as in one-way ANOVA for J independent groups:

$$H_0: \mu_1 = \mu_2 = \cdots = \mu_J$$

Likewise, the numerator is the mean squares across treatment groups, and the denominator is the unbiased estimate of error variance, which in this case is the mean squares for residual variability within subjects. Therefore, the appropriate test statistic is

$$\frac{MS_{AT}}{MS_{RESIDUAL}}$$

[13.7]

which is distributed as Fisher's F with $J - 1$ degrees of freedom in the numerator and $(N - 1)(J - 1)$ degrees of freedom in the denominator. The various sums of squares required for the F test can be expressed in terms of four basic quantities, which are given in Box 13.2.

Example 13.4. Schroeder[4] had 11 archers shoot one flight of six arrows at each of three distances, 30, 40, and 50 yards. The experiment was repeated on six consecutive days so that every archer shot the three distances in every possible order. Scores for every archer at each distance are given in Table 13.6.

Table 13.6. Total scores for 11 archers shooting at three distances

Archer	30 yards	40 yards	50 yards	$\sum_{j}^{3} x_{ij}$
1	423	306	158	887
2	406	335	264	1,005
3	541	418	173	1,132
4	549	388	294	1,231
5	611	380	240	1,231
6	623	450	186	1,259
7	616	414	339	1,369
8	619	559	331	1,509
9	742	590	462	1,794
10	708	697	460	1,865
11	846	646	419	1,911
$\sum_{i}^{11} x_{ij}$	6,684	5,183	3,326	$\sum \sum = 15,193$

[4] Schroeder, E. M. (1945). *On measurement of motor skills.* New York: King's Crown Press.

| BOX 13.2 |

Basic quantities and computational formulae for repeated-measures ANOVA with J treatments on N subjects

$$A = \sum_{j}^{J} \sum_{i}^{N} x_{ij}^2 \qquad B_S = \frac{\sum_{i}^{N} \left(\sum_{j}^{J} x_{ij} \right)^2}{J}$$

$$B_\tau = \frac{\sum_{j}^{J} \left(\sum_{i}^{N} x_{ij} \right)^2}{N} \qquad D = \frac{\left(\sum_{j}^{J} \sum_{i}^{N} x_{ij} \right)^2}{NJ}$$

For purposes of computation,

$SS_{AT} = B_\tau - D$ and

$$MS_{AT} = \frac{B_\tau - D}{J - 1}$$

$SS_{WS} = A - B_S$, so

$SS_{RESIDUAL} = (A - B_S) - (B_\tau - D) = A - B_S - B_\tau + D$, and

$$MS_{RESIDUAL} = \frac{A - B_S - B_\tau + D}{(N - 1)(J - 1)}$$

Therefore,

$$\frac{MS_{AT}}{MS_{RESIDUAL}} = \frac{(B_\tau - D)(N - 1)}{A - B_S - B_\tau + D}$$

which is distributed as Fisher's F with $\nu_1 = J - 1$ and $\nu_2 = (N - 1)(J - 1)$.

Target faces are larger at longer distances. Assuming that scores are normally distributed at every distance, test the hypothesis that target size compensates for distance. That is, test the hypothesis that mean scores for the three distances are equal:

$$H_0: \mu_1 = \mu_2 = \mu_3$$

Since every archer shot at each distance, this is a repeated-measures design with one experimental factor, distance. The basic computational

quantities in Box 13.2 are as follows:
Quantity A is the sum of the 33 squared scores,

$$A = \sum_{j}^{3} \sum_{i}^{11} x_{ij}^2 = 7{,}970{,}897$$

Quantity B_S is the sum of the squared subject totals divided by the number of treatments,

$$B_S = \frac{\sum_{i}^{11} \left(\sum_{j}^{3} x_{ij} \right)^2}{3} = \frac{(887)^2 + (1{,}005)^2 + (1{,}132)^2 + \cdots + (1{,}911)^2}{3}$$

$$= \frac{22{,}193{,}845}{3} \doteq 7{,}397{,}948.33$$

Quantity B_τ is the sum of the squared treatment group totals divided by the number of subjects

$$B_\tau = \frac{\sum_{j}^{3} \left(\sum_{i}^{11} x_{ij} \right)^2}{11} = \frac{(6{,}684)^2 + (5{,}183)^2 + (3{,}326)^2}{11}$$

$$= \frac{82{,}601{,}621}{11} \doteq 7{,}509{,}238.27$$

Quantity D is the square of the grand sum divided by the total number of observations, $NJ = (11)(3) = 33$.

$$D = \frac{\left(\sum_{j}^{3} \sum_{i}^{11} x_{ij} \right)^2}{33} = \frac{(15{,}193)^2}{33} = \frac{230{,}827{,}249}{33} \doteq 6{,}994{,}765.12$$

By the computational formula for the F ratio in Box 13.2,

$$\frac{MS_{AT}}{MS_{\text{RESIDUAL}}} = \frac{(B_\tau - D)(N - 1)}{A - B_S - B_\tau + D}$$

$$\doteq \frac{(7{,}509{,}238.27 - 6{,}994{,}765.12)(10)}{7{,}970{,}897 - 7{,}397{,}948.33 - 7{,}509{,}238.27 + 6{,}994{,}765.12}$$

$$\doteq 87.98$$

Mean squares across treatments has

$$J - 1 = 2$$

degrees of freedom, and mean squares residual has

$$(N - 1)(J - 1) = (10)(2) = 20$$

degrees of freedom. From Table 6 of Appendix VIII, $\mathbb{P}(87.98) > .9995$ in the distribution of $F(2, 20)$, so $p < .0005$. We reject the hypothesis that scores are the same at different ranges, which suggests that target size does not, in fact, compensate for distance.

Suppose the data in Table 13.6 were mistakenly treated as scores obtained by *three different* groups of archers and MS_A and MS_W calculated for simple one-way ANOVA (see Box 12.2, p. 510). We would find that

$$MS_A = MS_{AT}$$

but that

$$MS_W > MS_{\text{RESIDUAL}}$$

resulting in a smaller, and therefore less significant, F-ratio. This is because the sum of squares within groups as calculated for simple one-way ANOVA includes (in addition to error variance) the *covariance* of scores within subjects. In the repeated-measures partition, this source of variability is included in the sum of squares across subjects (SS_{AS}). This underscores why the assumption of independence is so important in the simple one-way ANOVA: If the assumption is violated, the covariance inflates MS_W, so $E(MS_W) > \sigma_e^2$.

—————— EXERCISES 13.1 ——————

Perform the analysis of variance appropriate to each of the exercises given below. You may assume that data are normally distributed and that cell variances in each experiment are equal.

1. The following scores were obtained in a two-way, fixed-effects experiment.

Factor 1	Experimental factor 2 $(C = 3)$		
$(R = 4)$	Treatment 1	Treatment 2	Treatment 3
Treatment 1	33 30 27	25 26 26	35 30 34
Treatment 2	27 31 33	28 30 32	32 30 28
Treatment 3	33 34 28	33 30 29	30 28 31
Treatment 4	26 28 34	24 26 28	35 34 33

(a) What are the test and alternative hypotheses for this analysis?
(b) Write the definitional formula of the test statistic appropriate to each test and state how it is distributed if H_0 is correct.
(c) Calculate the test statistic for each test and report its p-value.

2. The following scores were obtained in a two-way, fixed-effects experiment.

Factor 1	Experimental factor 2 $(C = 4)$			
$(R = 3)$	Treatment 1	Treatment 2	Treatment 3	Treatment 4
Treatment 1	11 9 −13 −7	32 24 13 11	29 22 18 11	61 56 39 44
Treatment 2	−27 −13 −24 −16	9 −8 5 −6	36 24 38 22	39 36 24 21
Treatment 3	−7 −13 −2 −18	−21 −39 −26 −34	−29 −35 −45 −51	−44 −56 −52 −40

(a) What are the test and alternative hypotheses for this analysis?
(b) Write the definitional formula of the test statistic appropriate to each test and state how it is distributed if H_0 is correct.

(c) Calculate the test statistic for each test and report its p-value.

3. Assume that the data in Exercise 13.1.1 are taken from a random-effects experiment.

(a) What are the test and alternative hypotheses for this analysis? Which hypothesis should be tested first and why?

(b) What is the test statistic for the first hypothesis, and how is it distributed if H_0 is correct?

(c) Calculate the test statistic for the first hypothesis and report its p-value.

(d) What are the test statistics for the remaining hypotheses, and how are they distributed if H_0 is correct?

(e) Calculate the test statistics given in your answer to (d) and report their p-values. How do your results differ from the results obtained in Exercise 13.1.1? Why?

4. Assume that the data in Exercise 13.1.2 are taken from a random-effects experiment.

(a) What are the test and alternative hypotheses for this analysis? Which hypothesis should be tested first and why?

(b) What is the test statistic for the first hypothesis, and how is it distributed if H_0 is correct?

(c) Calculate the test statistic for the first hypothesis and report its p-value.

(d) What are the test statistics for the remaining hypotheses, and how are they distributed if H_0 is correct?

(e) Calculate the test statistics given in your answer to (d) and report their p-values. How do your results differ from the results obtained in Exercise 13.1.2? Why?

5. The following scores are measures of reading improvement for four types of remedial reading program. Scores are classified according to the method of instruction originally received by the child at school and by the type of method used in the remedial program.

Remedial method	Original instruction			
	Alphabetic	Kinesthetic	Phonic	Visual
Alphabetic	.987	1.142	1.028	1.032
	1.094	1.016	.961	1.115
	1.002	1.130	1.102	1.064
Kinesthetic	1.186	1.025	1.135	1.039
	1.060	1.067	1.174	1.191
	1.161	1.104	1.168	1.162
Phonic	1.075	1.064	1.112	1.016
	1.126	.984	1.008	1.055
	1.053	.936	1.096	.947
Visual	1.281	1.134	1.198	1.012
	1.190	1.192	1.066	1.080
	1.265	1.113	1.079	1.076

Source: Burt, C. and Lewis, R. B. (1946). Teaching backward readers. *British Journal of Educational Psychology, 16,* 116–132.

Assume that the experimenters are interested only in these four instructional methods.

(a) What are the test and alternative hypotheses for this analysis?

(b) Write the definitional formula of the test statistic appropriate to each test and state how it is distributed if H_0 is correct.

(c) Calculate the test statistic for each test and report its p-value.

(d) State your conclusions.

6. A student database published by Professor Roger E. Kirk (Department of Psychology, Baylor University) includes the sex, major, GPA, score on a mathematical skills test, and final course grade for 461 introductory statistics students. The first eight men and eight women listed in the database in each of three majors obtained the following mathematical skills scores.

	Major		
Sex	Psychology	Physical therapy	Pre-medicine
Men	29	34	39
	17	30	37
	37	34	35
	45	25	41
	25	24	43
	16	31	35
	34	34	42
	39	40	40
Women	26	40	33
	39	34	38
	42	38	38
	17	34	41
	45	28	35
	35	28	33
	42	37	39
	30	44	38

Source: Kirk, R. E. (1990). *Statistics: An introduction* (3rd ed.). Forth Worth, TX: Holt, Rinehart and Winston, pp. 685–697.

Assume that all of the students in Professor Kirk's classes major in psychology, physical therapy, or pre-medicine.

(a) What are the test and alternative hypotheses for this analysis?

(b) Write the definitional formula of the test statistic appropriate to each test and state how it is distributed if H_0 is correct.

(c) Calculate the test statistic for each test and report its p-value.

(d) State your conclusions.

7. Many anticancer drugs have toxic effects, and two of the variables related to toxicity are *cumulative dosage* (the total amount of the drug administered over a series of treatments) and *dosage intensity* (the amount administered in each treatment). Suppose that cyclophosphamide is administered by injection to 45 postoperative cancer patients in 2, 3, or 4 treatments and that the cumulative dosage is 2,400, 3,600, or 4,800 milligrams per square meter of body surface (mg/m^2). To measure toxicity, white blood cell count (WBC) is

recorded for every patient 1 week after each treatment and averaged over the entire course of therapy. The values tabled below are each patient's average WBC × 1,000.

Intensity	Cumulative dosage of cyclophosphamide (mg / m²)		
	2,400	3,600	4,800
Low (4 treatments)	1.9	3.0	2.3
	3.3	1.8	2.9
	2.8	2.4	2.4
	2.4	2.6	1.7
	2.6	2.2	2.2
Medium (3 treatments)	2.4	1.9	2.4
	2.7	1.6	2.1
	2.6	2.7	1.6
	2.0	2.0	1.2
	1.8	2.3	2.0
High (2 treatments)	1.7	1.4	1.6
	2.4	1.7	1.2
	1.4	0.6	1.2
	1.8	1.8	2.0
	1.2	2.0	1.0

(a) Is this a fixed-effects experiment, a random-effects experiment, or a mixed-effects experiment?
(b) State the test hypotheses for this experiment. Which hypothesis should be tested first? Why?
(c) What is the test statistic for the first hypothesis, and how is it distributed if H_0 is correct?
(d) Calculate the test statistic for the first hypothesis and report its p-value.
(e) What are the test statistics for the remaining hypotheses, and how are they distributed if H_0 is correct?
(f) Calculate the test statistics given in your answer to (e) and report their p-values.
(g) State your conclusions.

8. Corn planted on irrigated plots and on nonirrigated plots was fertilized with 60, 120, or 180 lbs of nitrogen. The yield in bushels per acre for each plot is given below.
(a) Is this a fixed-effects experiment, a random-effects experiment, or a mixed-effects experiment?
(b) State the test hypotheses for this experiment. Which hypothesis should be tested first? Why?
(c) What is the test statistic for the first hypothesis, and how is it distributed if H_0 is correct?
(d) Calculate the test statistic for the first hypothesis and report its p-value.
(e) What are the test statistics for the remaining hypotheses, and how are they distributed if H_0 is correct?
(f) Calculate the test statistics given in your answer to (e) and report their p-values.
(g) State your conclusions.

	Nonirrigated plots										
				Level of fertilization							
60				**120**				**180**			
90	83	85	86	95	80	88	78	107	95	88	89
92	98	112	79	89	98	104	86	92	106	91	87
81	74	82	85	92	81	78	89	93	74	94	83

	Irrigated plots										
				Level of fertilization							
60				**120**				**180**			
80	102	60	73	87	109	104	114	100	105	114	114
121	99	90	109	110	94	118	131	119	123	113	126
78	136	119	116	98	133	122	136	122	132	136	133

Source: Adapted from Snedecor, G. W., and Cochran, W. G. (1980). *Statistical methods*. Ames, IA: The Iowa State University Press, p. 329.

9. Attitude change is typically studied by taking preexperimental attitude measurements, presenting information that is counter to subjects' attitudes (the experimental treatment), and then measuring (postexperimental) attitudes. Sometimes, postexperimental attitudes are measured immediately after treatment and again several weeks later in order to assess the stability of change and to detect changes that occur only after the information is thoroughly integrated into the subject's belief system. The following data are scores for 10 white subjects on a measure of antiblack attitudes.

Subject	Preexperimental attitude	Postexperimental attitude	
		1 hour	**6 weeks**
1	51	46	49
2	45	40	41
3	44	39	40
4	38	39	34
5	46	40	45
6	40	39	36
7	45	40	41
8	44	41	40
9	43	43	40
10	48	43	46

(a) What is the test hypothesis?
(b) Write the definitional formula of test statistic and state how it is distributed if H_0 is correct.
(c) Calculate the test statistic and report its *p*-value.
(d) What is the appropriate conclusion?

10. The amount of litter (branches, bark, etc.) dropped by a tree is one measure of its health. A one-week accumulation of litter within the drip line of six red oak trees was collected and weighed. The same measurement was taken after deep-root fertilization and after injection of soluble insecticide pellets.

	Litter (pounds)		
Tree	Before treatment	After fertilization	After pesticide
1	12	6	2
2	3	0	1
3	14	8	3
4	6	3	4
5	9	6	7
6	11	4	6

(a) What is the test hypothesis?
(b) Write the definitional formula of the test statistic appropriate to each test and state how it is distributed if H_0 is correct.
(c) Calculate the test statistic for each test and report its p-value.
(d) What is the appropriate conclusion?

C. SUMMARY

In this chapter we extend the basic logic of ANOVA to

- experiments involving two factors, and
- single-factor experiments in which the experimenter takes J repeated measurements on the same subjects.

If data are arranged so that treatments of one factor are represented by R rows and treatments of the second factor are represented by C columns, then the linear model for the score of individual i in row j and column k is

$$x_{ijk} = \mu + \tau_j + \gamma_k + \tau\gamma_{jk} + e_{ijk}$$

where

- μ is the mean of the grand population,
- τ_j is the effect of row treatment j,
- γ_k is the effect of column treatment k,
- $\tau\gamma_{jk}$ is the *interaction* of treatments j and k, and
- e_{ijk} is the error component of the score.

The *main* effects τ and γ are analogous to effects in the one-way analysis, as is error. Interaction is said to occur when the effects of one factor are not the same across all treatments of the second factor.

In two-way analysis of variance the total sum of squares

$$SS_{\text{TOTAL}} = \sum \sum \sum (x_{ijk} - \bar{x})$$

is partitioned into components corresponding to the terms of the linear model

$$SS_{\text{TOTAL}} = SS_{\text{ERROR}} + SS_{\text{ROW}} + SS_{\text{COL}} + SS_{\text{INT}}$$

where

$$SS_{\text{ERROR}} = \sum \sum \sum (x_{ijk} - \bar{x}_{jk})^2$$

and has $N - RC$ degrees of freedom,

$$SS_{\text{ROW}} = \sum_{j}^{R} N_j (\bar{x}_j - \bar{x})^2$$

and has $R - 1$ degrees of freedom,

$$SS_{\text{COL}} = \sum_{k}^{C} N_k (\bar{x}_k - \bar{x})^2$$

and has $C - 1$ degrees of freedom, and

$$SS_{\text{INT}} = \sum_{j}^{R} \sum_{k}^{C} \left(\bar{x}_{jk} - \bar{x}_j - \bar{x}_k + \bar{x} \right)^2$$

and has $(R - 1)(C - 1)$ degrees of freedom.

Mean squares for each component is SS divided by df, but the appropriate F ratio depends upon whether the treatment effects are *fixed*, *random*, or *mixed*. If an experimenter's inferences apply *only* to those treatments actually represented in the experiment, the effects are said to be fixed. If the experimenter considers the treatments to be a *sample* of a larger population of potential treatments, the effects are said to be random. A mixed-effect analysis of variance is one in which one factor is fixed and the other random.

In a one-way analysis of variance with repeated measures, the total sum of squares is partitioned into the *sum of squares across subjects* and the *sum of squares within subjects*:

$$SS_{\text{TOTAL}} = SS_{AS} + SS_{WS}$$

$$= J \sum_{i}^{N} (\bar{x}_i - \bar{x})^2 + \sum_{j}^{J} \sum_{i}^{N} (x_{ij} - \bar{x}_i)^2$$

The most obvious source of variation with a subject's J scores is treatment effects, which is captured exactly as in simple one-way analysis of variance: The *sum of squares across treatments* is

$$SS_{AT} = \sum_j N(\bar{x}_j - \bar{x})^2$$

All other variation within subjects is *residual*:

$$SS_{\text{RESIDUAL}} = SS_{WS} - SS_{AT}$$

Degrees of freedom for SS_{AT} is $J - 1$, so

$$MS_{AT} = \frac{SS_{AT}}{J - 1}$$

and degrees of freedom for SS_{RESIDUAL} is $(N - 1)(J - 1)$. Therefore,

$$MS_{\text{RESIDUAL}} = \frac{SS_{\text{RESIDUAL}}}{(N - 1)(J - 1)}$$

and it can be shown that

$$E(MS_{\text{RESIDUAL}}) = \sigma_e^2$$

For J repeated measures taken on the same group of N individuals, the test statistic appropriate to the hypothesis

$$H_0: \mu_1 = \cdots = \mu_J$$

is

$$\frac{MS_{AT}}{MS_{\text{RESIDUAL}}}$$

This statistic is distributed as Fisher's F with $\nu_1 = J - 1$ and $\nu_2 = (N - 1)(J - 1)$.

Summary table of statistical tests

Two-way analysis of variance

H_0	H_1	Test statistic	Distribution of test statistic*
Fixed effects			
$\tau_j = 0$ for all j	H_0 incorrect	$\dfrac{MS_{\text{ROW}}}{MS_{\text{ERROR}}}$	$F(R - 1, N - RC)$
$\gamma_k = 0$ for all k	H_0 incorrect	$\dfrac{MS_{\text{COL}}}{MS_{\text{ERROR}}}$	$F(C - 1, N - RC)$
$\tau\gamma_{jk} = 0$ for all jk	H_0 incorrect	$\dfrac{MS_{\text{INT}}}{MS_{\text{ERROR}}}$	$F[(R - 1)(C - 1), (N - RC)]$
Random effects			
$V(\tau) = 0$	H_0 incorrect	$\dfrac{MS_{\text{ROW}}}{MS_{\text{INT}^\dagger}}$	$F[(R - 1), (R - 1)(C - 1)]$
$V(\gamma) = 0$	H_0 incorrect	$\dfrac{MS_{\text{COL}}}{MS_{\text{INT}^\dagger}}$	$F[(C - 1), (R - 1)(C - 1)]$
$V(\tau\gamma) = 0$	H_0 incorrect	$\dfrac{MS_{\text{INT}}}{MS_{\text{ERROR}}}$	$F[(R - 1)(C - 1), (N - RC)]$

Note: For mixed-effects ANOVA, see p. 553.
For repeated-measures ANOVA see Box 13.2, p. 557.
*If H_0 is correct.
†If there is no interaction, denominator is

$$MS_{\text{POOLED ERROR}} = \frac{SS_{\text{INT}} + SS_{\text{ERROR}}}{(R - 1)(C - 1) + (N - RC)}$$

and $\nu_2 = (R - 1)(C - 1) + (N - RC)$.

Sums and means of squares for two-way analysis of variance

| Source | Sums of squares | | Mean squares |
	Definitional	Computational*	
Error	$\displaystyle\sum\sum\sum(x_{ijk} - \bar{x}_{jk})^2$	$A - G$	$\dfrac{SS_{\text{ERROR}}}{N - RC}$
Rows	$\displaystyle\sum_{j}^{R} N_j(\bar{x}_j - \bar{x})^2$	$B_R - D$	$\dfrac{SS_{\text{ROW}}}{R - 1}$
Columns	$\displaystyle\sum_{k}^{C} N_k(\bar{x}_k - \bar{x})^2$	$B_C - D$	$\dfrac{SS_{\text{COL}}}{C - 1}$
Inter-action	$\displaystyle\sum_{k}^{R}\sum_{j}^{C}(\bar{x}_{jk} - \bar{x}_j - \bar{x}_k + \bar{x})^2$	$G - B_R - B_C + D$	$\dfrac{SS_{\text{INT}}}{(R-1)(C-1)}$
Pooled error	$SS_{\text{INT}} + SS_{\text{ERROR}}$	$A - B_R - B_C + D$	$\dfrac{SS_{\text{INT}} + SS_{\text{ERROR}}}{(R-1)(C-1) + (N-RC)}$

Basic quantities for computational formulae:

$$A = \sum_{k}\sum_{j}\sum_{i} x_{ijk}^2 \qquad B_R = \frac{\displaystyle\sum_{j}^{R}\left(\sum_{k}\sum_{i} x_{ijk}\right)^2}{CN_g} \qquad B_C = \frac{\displaystyle\sum_{k}^{C}\left(\sum_{j}\sum_{i} x_{ijk}\right)^2}{RN_g}$$

$$D = \frac{\left(\displaystyle\sum_{k}\sum_{j}\sum_{i} x_{ijk}\right)^2}{N} \qquad G = \frac{\displaystyle\sum_{k}^{C}\sum_{j}^{R}\left(\sum_{i} x_{ijk}\right)^2}{N_g}$$

*May be used only if the number of observations is the same in every cell. The (common) number of observations in each cell is denoted N_g in the basic quantities given above.

CHAPTER 14

Testing hypotheses about covariation:

correlation and regression

CHAPTER OUTLINE

Our treatment of statistical description in Part I first developed the techniques by which statisticians describe relative frequency distributions of one measurement (say, height) and then discussed the problem of describing the strength and magnitude of the relationship between *two* measurements (e.g., height and weight) taken on the same group of N individuals. And, just as we have subsequently discussed techniques for making inferences about population distributions of a single variable X, we now take up the problem of making inferences about joint distributions of *two* variables, X and Y.

A. THE JOINT DISTRIBUTION OF TWO VARIABLES

Joint probability distributions are a mathematical model for the joint distributions of data introduced in Chapter 4. In Chapter 4 we developed the notion of correlation with an example that was based on the joint distribution of two *continuous* random variables, weight and height. The hypothesis tests discussed in this chapter are likewise concerned with continuous variables, but as we discovered in Chapters 6 and 7, continuous probability distributions are more easily understood if fundamental concepts are first introduced for the discrete case.

1. Joint discrete probability distributions

Example 14.1. Suppose we flip a coin three times and we let X be the number of heads and let Y be the longest *run* (i.e., the longest unbroken string) of either heads or tails. For the result HTT, the number of heads is one, so $x = 1$, and the run of two tails is longer than the run of one head, so $y = 2$. The sample space and value sets of X and Y are given below:

Result	X	Y
HHH	3	3
HHT	2	2
HTH	2	1
HTT	1	2
THH	2	2
THT	1	1
TTH	1	2
TTT	0	3

The eight possible results of this experiment are equally likely, so

$$P(HHH) = P(HHT) = \cdots = P(TTT) = \tfrac{1}{8}$$

We can therefore calculate the *joint* probability of every *pair* of values, $P(x_j, y_k)$. For example, the probability that $x = 1$ *and* $y = 2$, which we denote $P(1, 2)$, is

$$P(HTT) + P(TTH) = \tfrac{2}{8}$$

The entries in the body of Table 14.1 are the joint probabilities, $P(x, y)$.

Table 14.1. Joint and marginal probability distributions of X and Y

		X				
		0	1	2	3	$P(y)$
Y	1	0	$\frac{1}{8}$	$\frac{1}{8}$	0	$\frac{1}{4}$
	2	0	$\frac{2}{8}$	$\frac{2}{8}$	0	$\frac{1}{2}$
	3	$\frac{1}{8}$	0	0	$\frac{1}{8}$	$\frac{1}{4}$
$P(x)$		$\frac{1}{8}$	$\frac{3}{8}$	$\frac{3}{8}$	$\frac{1}{8}$	

The row and column totals in the margins of Table 14.1 are labeled $P(y)$ and $P(x)$ because they are the probabilities associated with the individual variables X and Y. For example, three of the outcomes include one and only one head, so $P(x = 1) = 3/8$, which is the sum of the entries in the second column,

$$P(x = 1) = P(1, 1) + P(1, 2) + P(1, 3)$$

$$= \tfrac{1}{8} + \tfrac{2}{8} + 0 = \tfrac{3}{8}$$

Table 14.1 therefore includes both the *joint* probability distribution of (X, Y) and the *marginal* probability distributions of X and Y and is therefore completely analogous to the joint and marginal frequency distributions displayed in Table 4.2 (p. 100).

Because joint probabilities are associated with values of *two* random *variables*, they are sometimes called *bivariate* probabilities and the distribution of $P(x, y)$ a bivariate probability distribution. In Chapter 6 we let $P(x)$ denote the function that assigns a probability $P(x_i)$ to every value x_i of a discrete random variable X. For two discrete random variables X and Y, we let $P(x, y)$ denote the *joint* probability function that assigns a probability to every *pair* of values, ($x = x_j$ and $y = y_k$). That is,

---

If X and Y are discrete random variables and if

$$P(x_j, y_k) = P(x = x_j \text{ and } y = y_k),$$

then the function P is the *joint probability function* of X and Y.

The reader will observe that in discussion of jointly distributed random variables, we use the subscript j instead of the subscript i to identify a particular x-value. The subscripting convention for handling bivariate distributions follows the conventions established for two-way analysis of variance (Chapter 13): The subscript j indicates a particular x-value, and the subscript k indicates a particular y-value. The subscript i is reserved (as in Chapter 4) to denote the *individual* who obtained a particular score or pair of scores.

Conditional distributions. In Chapter 5 we introduced the notion of *conditional* probability, asking the question, "What is the probability of Event 2 *given* Event 1?" That is, what is $P(E_2|E_1)$ where E_1 and E_2 are events in the same sample space? There is a natural extension of the notion of conditional probability of two events to conditional probability distributions of random variables. We may ask the question, "What is the probability of every possible y-value *given that $x = x_j$?*"

Example 14.2. Let us consider the *conditional* probabilities of Y given that $x = 1$ in Example 14.1. We note in Table 14.1 that if $x = 1$, then Y takes on only two values, $y = 1$ or $y = 2$. To calculate the conditional probabilities $P(y = 1|x = 1)$ and $P(y = 2|x = 1)$ recall from Chapter 5 (equation [5.2]) that

$$P(E_2|E_1) = \frac{P(E_1 \text{ and } E_2)}{P(E_1)}$$

Similarly, to obtain the conditional probability of y_k given x_j, we divide the *joint* probability $P(x = x_j \text{ and } y = y_k)$ by the probability of x_j. From the marginal distribution of X in Table 14.1 we find that $P(x = 1) = 3/8$, and from joint distribution we see that $P(x = 1 \text{ and } y = 1)$ is $1/8$. Therefore,

$$P(y = 1|x = 1) = \frac{P(1,1)}{P(x = 1)} = \frac{\frac{1}{8}}{\frac{3}{8}} = \frac{1}{3}$$

We calculate $P(y = 2|x = 1)$ in the same way. The joint probability $P(x = 1 \text{ and } y = 2)$ is $2/8$, so

$$P(y = 2|x = 1) = \frac{P(1,2)}{P(x = 1)} = \frac{\frac{2}{8}}{\frac{3}{8}} = \frac{2}{3}$$

We use the notation $Y|(x = x_j)$ or $Y|x_j$ to distinguish the conditional distribution of Y from the overall, or marginal, distribution of Y.

Therefore,

$Y\|(x = 1)$	$P(y\|x = 1)$
1	$\frac{1}{3}$
2	$\frac{2}{3}$

Of course, one can also deal with rows and consider the probability that x takes on a particular value x_j given that y is equal to y_k. That is, we can calculate $P(x = x_j | y = y_k)$.

Example 14.3. What is the conditional distribution of X given that $y = 2$? If $y = 2$ we see from Table 14.1 that X takes on the value 1 with probability 2/8 and that X takes on the value 2 with probability 2/8. Since these probabilities are the same, the random variable $X|(y = 2)$ is uniformly distributed, so

$X\|(y = 2)$	$P(x\|y = 2)$
1	$\frac{1}{2}$
2	$\frac{1}{2}$

These conditional probabilities can be calculated from the joint distribution of (X, Y) and the marginal distribution of Y exactly as we calculated the probabilities for $Y|(x = 1)$:

$$P(x = 1 | y = 2) = \frac{P(1,2)}{P(y = 2)} = \frac{\frac{2}{8}}{\frac{1}{2}} = \frac{2}{4} = \frac{1}{2}$$

and

$$P(x = 2 | y = 2) = \frac{P(2,2)}{P(y = 2)} = \frac{\frac{2}{8}}{\frac{1}{2}} = \frac{2}{4} = \frac{1}{2}$$

In general, therefore,

If X and Y are discrete random variables, then

$$P(y = y_k | x = x_j) = \frac{P(x_j, y_k)}{P(x_j)}$$

and

$$P(x = x_j | y = y_k) = \frac{P(x_j, y_k)}{P(y_k)}$$

Independence. Recall from Chapter 5 that two events E_1 and E_2 are independent if and only if $P(E_2|E_1) = P(E_2)$. This same notion can be extended to the independence of random variables:

> The discrete random variables X and Y are independent if $P(y = y_k | x = x_j) = P(y = y_k)$ for all values k and j.

We see that the random variables in Example 14.1 are *not* independent since, for example,

$$\tfrac{1}{2} = P(x = 1|y = 2) \neq P(x = 1) = \tfrac{3}{8}$$

In Chapter 5 we pointed out that independence of events is symmetrical, i.e., that if E_2 is independent of E_1, then E_1 must be independent of E_2. It is similarly true that independence of discrete random variables is symmetrical. If X is independent of Y, then Y is independent of X.

We also have a multiplication theorem for independent random variables:

> If discrete random variables X and Y are independent, then $P(x_j, y_k) = P(x_j)P(y_k)$ for all pairs (x_j, y_k).

That is, if X and Y are independent, then every value $P(x_j, y_k)$ in a table like Table 14.1 is always the product of the marginal probabilities at the ends of its row and column. We know that the random variables in Example 14.1 are not independent, so we must find at least one instance in which the joint probability is not the product of the corresponding marginal probabilities. In Table 14.1, this happens to be true for all joint probabilities. For example, $P(x = 1$ and $y = 2)$ is

$$P(1, 2) = \tfrac{2}{8} = \tfrac{1}{4}$$

but from the marginal probability distributions,

$$P(x = 1)P(y = 2) = \left(\tfrac{3}{8}\right)\left(\tfrac{1}{2}\right) = \tfrac{3}{16}$$

a. Histograms for jointly distributed discrete random variables. Graphic representations of probability distributions have served us well throughout our study of statistics. The histogram not only gives us a "picture" of the distribution, but has helped us to understand the notion of probability as "area." Recall as well that we used the histogram to bridge the gap between discrete and continuous distributions by thinking of continuous distributions as histograms with infinitely many bars.

The simplest graphic representation of a joint discrete distribution is the sort of line graph we introduced in Chapter 1 (see p. 11). To construct a line graph for a bivariate probability distribution, we represent X on one axis, Y on another axis, and joint probability $P(x, y)$ on a third axis. Figure 14.1 is a line graph for the joint distribution in Table 14.1:

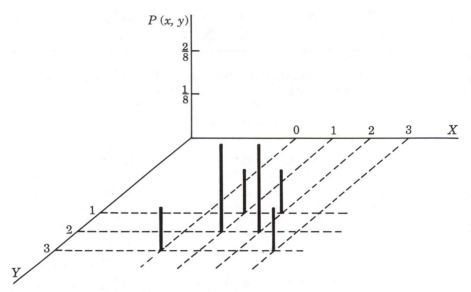

Figure 14.1. Line graph for joint probability distribution in Table 14.1.

In Chapter 1 we made histograms out of line graphs by fattening the lines into bars by extending them half a measurement unit to the right and to the left. To make the line graph for a joint distribution into a histogram, we need to make solid bars by extending the line representing

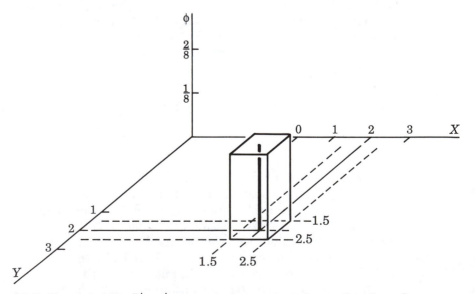

Figure 14.2. The probability $P(x, y)$ represented as a volume for $x = 2$ and $y = 2$.

every pair (x_j, y_k) half a unit in both directions on *both* the X-axis *and* the Y-axis. This is illustrated for the pair of values ($x = 2$, $y = 2$) in Figure 14.2. The base of each bar is one unit by one unit and so has area 1. It follows that the *volume* of the histogram bar equals its height and thus represents probability. The probability $P(2, 2)$ is represented as the volume of the solid constructed over the square $x = 1.5$ to $x = 2.5$ and $y = 1.5$ to $y = 2.5$.

The entire joint distribution in Table 14.1 is represented as volume in Figure 14.3.

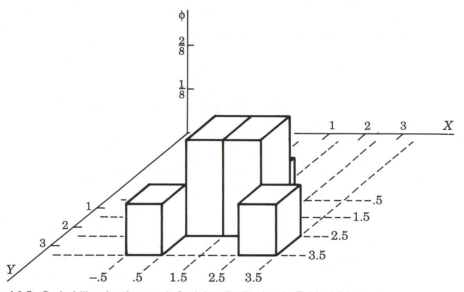

Figure 14.3. Probability density graph for joint distribution in Table 14.1.

We know from Chapter 6 that when the probability of an x-value is represented as area of a histogram bar, the height of the bar is probability density. When the volume of histogram solid represents the joint probability of x_j and y_k, height of the solid likewise represents probability density. It will be noted that the vertical axis is therefore labeled ϕ in both Figure 14.2 and Figure 14.3.

2. Jointly distributed continuous random variables

In our development of the normal distribution in Chapter 7 we standardized the binomial random variable and let N get very large. By so doing, the number of histogram bars increased to the point that their flat tops blended into a smooth curve, the standard normal density function. If we increase the number of histogram bars for a discrete bivariate distribution, the flat tops of the solid histogram bars will similarly blend together to form a smooth surface, the density function for a *continuous* bivariate distribution.

Just as the *area* under the *curve* $\phi(X)$ of a continuously distributed random variable X represents probability, so does the *volume* under the *surface* $\phi(X, Y)$ of a continuous bivariate distribution represent probability. For example, the probability $P(a \leq x \leq b, c \leq y \leq d)$ is the volume under this surface over the rectangle that extends from a to b on the X-axis and from c to d on the Y-axis.

Population parameters for conditional probability distributions are subscripted to identify the value of the *given* variable. That is, the population mean of the conditional distribution of Y given x_j is denoted μ_j or, if there is any possibility of ambiguity,

$$\mu_{Y|x_j}$$

and the variance of $Y|x_j$ is σ_j^2 or

$$\sigma_{Y|x_j}^2$$

Likewise, the population mean of X given y_k is denoted μ_k or

$$\mu_{X|y_k}$$

and the conditional variance of $X|y_k$ is σ_k^2 or

$$\sigma_{X|y_k}^2$$

Independence. The formal definition of independence for continuous random variables is analogous to the definition of independence for discrete random variables: Continuous random variables X and Y are independent if the probability density of y_k in the conditional distribution $Y|x_j$ is equal to the probability density of y_k in the marginal distribution of Y. That is, X and Y are independent if $\phi(y = y_k | x = x_j) = \phi(y_k)$ for all values k and j. And, as before, X is independent of Y if and only if Y is independent of X. Finally, if X and Y are independent, then the joint density of x_j and y_k is the product of their marginal densities, that is, $\phi(x_j, y_k) = \phi(x_j)\phi(y_k)$.

a. The bivariate normal distribution. A particularly important distribution is the bivariate normal distribution, illustrated in Figure 14.4.

The surface of a bivariate normal surface has "normal" cross sections. That is, the *conditional* distributions obtained by slicing the surface perpendicular to either the X- or Y-axis are shaped like a normal curve. This is seen in Figure 14.5, which shows the cross section of the bivariate normal distribution at $x = x_j$.

The height of the cross-sectional slice at $y = y_a$ is the probability density of y_a when $x = x_j$. Likewise, the area of the cross section between y_a and y_b is the *conditional* probability that y falls between y_a and y_b, that is,

$$P\left(y_a \leq y \leq y_b | x = x_j\right)$$

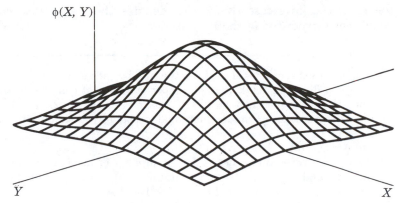

Figure 14.4. Bivariate normal distribution.

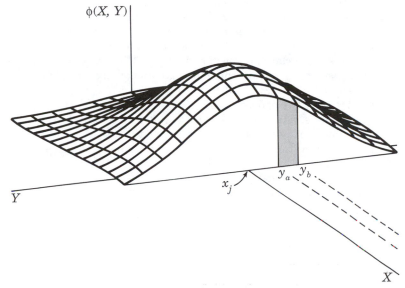

Figure 14.5. Conditional distribution of $Y|x_j$ shown as cross-section of bivariate distribution. Shaded area is $P(y_a \leq y \leq y_b|x = x_j)$.

This curve, then, represents the probability distribution of $Y|x_j$, and "slicing" into the surface is exactly analogous to looking at row or column distributions for jointly distributed discrete variables.

If X and Y are bivariate normally distributed random variables with marginal means μ_X and μ_Y and marginal variances σ_X^2 and σ_Y^2, it can be shown that the mean $\mu_{Y|x_j}$ of the conditional distribution of $Y|(x = x_j)$ is equal to

$$\mu_Y + \rho\frac{\sigma_Y}{\sigma_X}(x_j - \mu_X)$$

and variance $\sigma_{Y|x_j}^2$ of the conditional distribution is equal to

$$\sigma_Y^2(1 - \rho^2)$$

where ρ, the lowercase Greek *rho*, denotes the *population* correlation coefficient introduced in the next section.

─────── EXERCISES 14.1 ───────────────────────────────

1. Flip a coin three times. Let X be the number of heads that come up in the first two flips, and let Y be the number of tails that come up in the last two flips. (For example, the result *HTT* has $x = 1$ and $y = 2$.)
   (a) Construct the joint and marginal probability distribution table for X and Y.
   (b) Are X and Y independent?

2. Flip a coin three times. Let X be the number of heads that come up in the first flip and let Y be the number of tails that come up in the last two flips. (For example, the result *HTT* has $x = 1$ and $y = 2$.)
   (a) Construct the joint and marginal probability distribution table for X and Y.
   (b) Are X and Y independent?

3. The 6 sides of a fair die are labeled 1, 1, 2, 2, 3, 3. Roll the die twice. Let $x = 2$ if the rolls come up the same; let $x = 4$ if the sum is 3 or if the first die comes up exactly 2 larger than the second; let $x = 6$ otherwise. Let $y = 0$ if the second die comes up even and let $y = 1$ if the second die comes up odd.
   (a) Construct the joint and marginal probability distribution table for X and Y.
   (b) Are X and Y independent?

4. The 6 sides of a fair die are labeled 1, 1, 2, 2, 3, 3. Roll the die twice. Let X be the number of 2s that come up and Y be the largest number that comes up.
   (a) Construct the joint and marginal probability distribution table for X and Y.
   (b) Are X and Y independent?

5. For the joint distribution:

		X	
Y	2	5	6
5	$\frac{4}{48}$	$\frac{1}{48}$	$\frac{1}{48}$
9	$\frac{8}{48}$	$\frac{2}{48}$	$\frac{2}{48}$
11	$\frac{13}{48}$	$\frac{3}{48}$	$\frac{2}{48}$
17	$\frac{7}{48}$	$\frac{2}{48}$	$\frac{3}{48}$

   (a) Calculate the marginal distributions of X and Y.
   (b) Are X and Y independent?
   (c) Write out the distributions $X|(y = 9)$ and $Y|(x = 5)$.

6. For the joint distribution:

		X		
Y	3	4	7	11
0	$\frac{3}{32}$	$\frac{1}{16}$	$\frac{1}{32}$	$\frac{1}{16}$
3	$\frac{1}{8}$	$\frac{1}{12}$	$\frac{1}{24}$	$\frac{1}{12}$
4	$\frac{3}{32}$	$\frac{1}{16}$	$\frac{1}{48}$	$\frac{7}{96}$
7	$\frac{1}{16}$	$\frac{1}{24}$	$\frac{1}{32}$	$\frac{1}{32}$

(a) Calculate the marginal distributions of X and Y.

(b) Are X and Y independent?

(c) Write out the distributions $X|(y = 7)$ and $Y|(x = 11)$.

7. Draw the histogram for the joint distribution

		X
Y	1	2
1	0	$\frac{2}{3}$
2	$\frac{1}{3}$	0
3	0	0

8. Draw the histogram for the joint distribution

		X
Y	1	2
1	0	$\frac{1}{4}$
2	$\frac{1}{2}$	0
3	0	$\frac{1}{4}$

In Exercises 9 and 10, random variable X takes on only the values 2, 4, and 6, and random variable Y takes on only the values 1, 2, 3, and 5.

9. Suppose the joint distribution of X, Y is given by

$$P(x, y) = \frac{k}{xy}$$

(a) Write the joint distribution with every probability expressed in terms of k.

(b) Find the number k.

(c) Calculate the joint and marginal distributions.

(d) Write the conditional distributions for $X|(y = 3)$ and $Y|(x = 6)$.

10. Suppose the joint distribution of X, Y is given by

$$P(x, y) = \frac{kx}{y}$$

(a) Write the joint distribution with every probability expressed in terms of k.

(b) Find the number k.

(c) Calculate the joint and marginal distributions.

(d) Write the conditional distributions for $X|(y = 3)$ and $Y|(x = 6)$.

Cognitive psychologists devise games in which two players, **A and B**, make independent decisions. Players then report their choices, and the joint results of their decisions yield specified payoffs or costs to each. Probabilities of players' choices illuminate the strategies that players evolve in the course of play. The marginal probabilities shown in Exer-

cises 11 and 12 indicate that players have evolved strategies that minimize their maximal losses.[1] Determine how much Player A should
expect to win (or lose) with this strategy.

11. Player A picks a number X that is 1, 2, 3, or 4. Player B picks a number Y
    that is 1, 2, or 3. If the numbers chosen are the same, the game is a draw.
    Otherwise the person picking the larger number wins $2, unless it is *exactly*
    2 larger, in which case the person choosing the larger number loses $3.

X	$P(x)$	Y	$P(y)$
1	0	1	$\frac{14}{49}$
2	$\frac{15}{49}$	2	$\frac{20}{49}$
3	$\frac{20}{49}$	3	$\frac{15}{49}$
4	$\frac{14}{49}$		

12. Player A picks a number X that is 1, 2, 3, or 4. Player B picks a number Y
    that is 1, 2, or 3. If A and B choose the same number, the game is a draw.
    Otherwise the person picking the larger number wins $2, unless the sum is 5,
    in which case the person picking the larger number loses $3.

X	$P(x)$	Y	$P(y)$
1	$\frac{5}{7}$	1	$\frac{4}{7}$
2	0	2	$\frac{3}{7}$
3	0	3	0
4	$\frac{2}{7}$		

13. A person flips six coins. Let X be the number of heads that come up and let
    Y be the longest run. Write out the distribution of $Y|(x = 4)$.

14. A person rolls three ordinary 6-sided dice. Let X be the smallest number
    that shows up and let Y be the sum of the spots that come up on all three
    dice. Write out the distribution of $X|(y = 15)$.

**In Exercises 15 and 16 the distribution of (X, Y) is bivariate normal. Use
the parameter values given for μ_X, μ_Y, σ_X^2, σ_Y^2, and ρ to obtain the
indicated conditional probabilities from Table 3 of Appendix VIII.**

15. $\mu_X = 12$; $\mu_Y = 25$; $\sigma_X^2 = 4$; $\sigma_Y^2 = 9$; $\rho = 0.5$
    (a) $P(18 \leq y \leq 22 | x = 8)$
    (b) $P(18 \leq y \leq 22 | x = 4)$
    (c) $P(23 \leq y \leq 27 | x = 12)$

16. $\mu_X = 16$; $\mu_Y = 18$; $\sigma_X^2 = 25$; $\sigma_Y^2 = 16$; $\rho = -0.25$
    (a) $P(20 \leq y \leq 22 | x = 5)$
    (b) $P(18 \leq y \leq 22 | x = 19)$
    (c) $P(15 \leq y \leq 21 | x = 16)$

[1] The *minimax* strategy of decision making is discussed in Chapter 9 (pp. 350–352) in
connection with choosing α and β.

B. THE POPULATION CORRELATION COEFFICIENT ρ

In Chapter 4 we developed the Pearson Product Moment Coefficient of Correlation

$$r_{XY} = \frac{\sum\limits_{i}^{N}(x_i - \bar{x})(y_i - \bar{y})}{Ns_X s_Y} \qquad [14.1]$$

as an index that describes the direction and strength of linear relationship between two measurements, X and Y, in a collection of data. It is easily seen (cf. p. 111) that equation [14.1] is equivalent to

$$r_{XY} = \frac{\sum\limits^{N} z_X z_Y}{N}$$

That is, the correlation r is the *average* product of standardized scores

$$z_X = \frac{x_i - \bar{x}}{s_X} \quad \text{and} \quad z_Y = \frac{y_i - \bar{y}}{s_Y}$$

for all N individuals in the data collection.

Likewise, the *population* correlation of the variables X and Y is equal to the *expected value* of the product of the standardized random variables Z_X and Z_Y:

$$\rho_{XY} = E(Z_X Z_Y) \qquad [14.2]$$

Like r_{XY}, the population correlation coefficient ρ_{XY} can assume values from -1 to $+1$.

1. Testing hypotheses about ρ

Descriptive statistics like the mean $\bar{x}$ and variance s^2 and correlation coefficient r summarize a good bit of information, and as long as we confine our interpretations of these indices to the collection of observations from which they are calculated, we can get away with very few assumptions about our measurements.

We have seen that the situation is quite different when we move from description of data to inferences about populations. Statistical inference requires that we use probability distributions as models for population distributions, and if our inferences are to be meaningful, our data must therefore satisfy certain assumptions to ensure that the models are appropriate.

In testing hypotheses about ρ, the joint distribution of X and Y is assumed to be *bivariate normal*. We will not present the bivariate normal density function, but for our purposes it has three important properties. First, if $\phi(X, Y)$ bivariate normal, the marginal distributions of X and Y

are normal, i.e., $X: N(\mu_X, \sigma_X^2)$ and $Y: N(\mu_Y, \sigma_Y^2)$. Second, if $\phi(X, Y)$ is bivariate normal, all conditional distributions of Y on x and X on y are normal. Finally, if $\phi(X, Y)$ is bivariate normal, then X and Y are independent *if and only if* ρ_{XY} is zero. The property that $\rho_{XY} = 0$ implies independence of X and Y for bivariate normal distributions is the basis for the following tests.

a. Fisher's Z_r. If the joint distribution of X and Y is bivariate normal and the number of pairs x_i, y_i is greater than 10, the most versatile test statistic is based on Sir Ronald Fisher's transformation of the correlation r to a normally distributed random variable, which we denote Z_r:

$$z_r = \frac{1}{2} \, ln \left(\frac{1 + r_{XY}}{1 - r_{XY}} \right)$$

[14.3]

where ln denotes the natural logarithm of the value in parentheses and is a standard single-key feature on most hand calculators.[2] For moderate sample sizes (i.e., $N > 10$), the distribution of Fisher's Z_r is approximately normal with

$$E(Z_r) = \frac{1}{2} \, ln \left(\frac{1 + \rho_{XY}}{1 - \rho_{XY}} \right)$$

which is ordinarily denoted ζ (the lowercase Greek *zeta*)[3] and

$$V(Z_r) = \frac{1}{N - 3}$$

Fisher's r-to-z transformation allows us to use the techniques developed for testing hypotheses about μ to test any hypothesis of the form

$$H_0: \rho = \rho_0$$

where $-1 \le \rho_0 \le +1$, against either a simple or composite alternative hypothesis. If the test hypothesis is correct, then the expected value of Z_r

---

[2] In general, if $b^y = x$, then y is said to be the logarithm to the base b of x, denoted $y = \log_b x$. The base of the *natural* logarithm system is the irrational number $e = 2.718281828\ldots$. Therefore, if $e^y = x$, then $\log_e x = y$. In modern texts, this is denoted $ln(x) = y$. Before the development of hand calculators, tables of logarithms (most commonly to the base 10) were widely used even in middle school, because they simplify computations of products and powers.

[3] Actually, $E(Z_r) = \zeta + r/[2(N - 1)]$. If r is calculated from one sample of N observations, the bias is a negligible fraction of

$$\sqrt{1/(N - 3)}$$

and is ordinarily ignored. If r is a pooled estimate of ρ calculated from several independent samples, the bias can be important and should be subtracted from z_r.

equals

$$\frac{1}{2} \ln \left(\frac{1 + \rho_0}{1 - \rho_0} \right)$$

which we denote ζ_0. We know that if a normally distributed random variable X has expected value μ_0 and variance σ^2, then

$$\frac{X - \mu_0}{\sigma}$$

is distributed $N(0, 1)$. Thus, if $\rho = \rho_0$, and the expected value of Z_r therefore equals ζ_0, then the statistic

$$\frac{Z_r - \zeta_0}{\sqrt{1/(N - 3)}} \qquad\qquad [14.4]$$

is approximately distributed $N(0, 1)$.

To test the hypothesis that $\rho = \rho_0$, then, calculate [14.4] and submit the value to the table of cumulative probabilities for the standard normal random variable (Table 3 of Appendix VIII). If ρ is greater than ρ_0, then the expected value of the test statistic is positive. When the alternative hypothesis specifies $\rho > \rho_0$, the rejection region is therefore in the upper tail of the standard normal curve. If $\rho < \rho_0$ under H_1, the rejection region is in the lower tail.

If ρ is the correlation of bivariate normally distributed random variables, then for $N > 10$ the statistic

$$\frac{Z_r - \zeta_0}{\sqrt{1/(N - 3)}}$$

is distributed approximately $N(0, 1)$ and may be used to test the hypothesis that $\rho = \rho_0$.

Example 14.4. A particular model of assortative mating predicts that the correlation between heights of husbands and wives should be no greater than .70, which implies the following hypotheses:[4]

$H_0: \rho = .70$

$H_1: \rho > .70$

---

[4] If ρ_0 is positive, the alternative hypothesis typically specifies that ρ is *less than* ρ_0, and if ρ_0 is negative, the alternative hypothesis typically specifies that ρ is *greater than* ρ_0. We have chosen this example to illustrate that this is not always the case.

In the course of reading this book, a geneticist encounters the data given in Example 4.2, in which we calculated the correlation for heights of 15 married couples to be .89. The geneticist is willing to assume that the joint distribution of husbands' and wives' heights is bivariate normal[5] and uses these data to calculate [14.4], which will be distributed $N(0, 1)$. The direction of the alternative calls for a rejection region in the upper tail of the normal curve, so if we set $\alpha = .01$, the critical value z_α is 2.33.

To calculate [14.4] we first normalize $r = .89$ using Fisher's r-to-z transformation as given in equation [14.3]:

$$z_r = \frac{1}{2} \ln\left(\frac{1 + .89}{1 - .89}\right) \doteq \frac{1}{2}(2.84) = 1.42$$

We then perform the same computation setting $\rho_{XY} = \rho_0$ in order to obtain ζ_0:

$$\zeta_0 = \frac{1}{2} \ln\left(\frac{1 + .70}{1 - .70}\right) \doteq \frac{1}{2}(1.73) \doteq .87$$

The value of our test statistic is therefore

$$\frac{1.42 - .87}{\sqrt{1/12}} \doteq 1.91$$

This is less than the critical value, 2.33, so the alternative hypothesis is rejected, and the assortative mating model remains uncontested.

In most correlational research, it is rare that either theoretical considerations or prior observations permit a researcher to specify a nonzero value for ρ_0. In most instances a researcher simply suspects either that ρ is positive or that ρ is negative and formulates the directionally appropriate alternative against the test hypothesis

$$H_0: \rho = 0$$

When one speaks of finding a "significant" correlation, it is therefore understood to mean "significantly different from zero." If $\rho_{XY} = 0$, then the expected value of Z_r is

$$\zeta_0 = \frac{1}{2} \ln\left(\frac{1 + 0}{1 - 0}\right) = \frac{1}{2} \ln(1) = 0$$

and [14.4] becomes

$$\frac{Z_r}{\sqrt{1/(N - 3)}} \qquad\qquad [14.5]$$

---

[5] For reasons discussed in Box 8.1, the assumption of bivariate normality is very reasonable for many biological variables. This is, in part, why much of the pioneering work in correlation was conducted by persons investigating genetic inheritance.

which is distributed approximately $N(0, 1)$. When the alternative hypothesis specifies $\rho > 0$, the rejection region is therefore in the upper tail of the standard normal curve. If $\rho < 0$ under H_1, the rejection region is in the lower tail.

If ρ is the correlation of bivariate normally distributed random variables, then for $N > 10$ the statistic

$$\frac{Z_r}{\sqrt{1/(N-3)}}$$

is distributed approximately $N(0, 1)$ and may be used to test the hypothesis that $\rho = 0$.

Example 14.5. In Exercise 4.3.15 the correlation r between number of cubs born and number of dominance encounters won by 12 female hyenas was found to be .54. Assuming that the joint distribution of birth rate and dominance success is bivariate normal, test the hypothesis ($\alpha = .05$) that the correlation is significant.

By equation [14.3],

$$z_r = \frac{1}{2} \, ln \left(\frac{1 + .54}{1 - .54} \right) \doteq \frac{1}{2}(1.21) \doteq .60$$

The standard deviation of Z_r is

$$\sqrt{1/(12-3)} \doteq .333$$

so by expression [14.5] the value of the test statistic is

$$\frac{z_r}{\sqrt{1/(N-3)}} \doteq \frac{.60}{.333} \doteq 1.80$$

The conventional wisdom among ethologists is that dominant animals are more reproductively successful than subordinate animals, so the logical alternative hypothesis is that the correlation is positive. The rejection region therefore lies in the upper tail of the normal curve, and with $\alpha = .05$, the critical value z_α is 1.64. Since $1.80 > 1.64$, the test hypothesis is rejected in favor of the alternative. Tough mommies have more babies than wimpy mommies.

b. Student's transformation. We have just seen that if $N > 10$, we can use [14.5] to test the hypothesis that $\rho = 0$. For smaller samples, this hypothesis can be tested using the test statistic

$$t_r = \frac{r_{XY}\sqrt{N - 2}}{\sqrt{1 - r_{XY}^2}} \qquad\qquad [14.6]$$

which is distributed approximately as Student's t with $N - 2$ degrees of freedom if the test hypothesis is correct and if $N > 4$.

If $N > 4$, the statistic

$$t_r = \frac{r_{XY}\sqrt{N - 2}}{\sqrt{1 - r_{XY}^2}}$$

is distributed approximately as $t(N - 2)$ and may be used to test the hypothesis that $\rho = 0$.

Like Fisher's r-to-z, Student's transformation is used only if $\phi(X, Y)$ can be assumed to be bivariate normal.

Example 14.6. In Exercise 4.3.16 the correlation r between number of cubs surviving to 1 year and number of dominance encounters won by 9 female hyenas was found to be .82. Assuming that the joint distribution of survivorship and dominance success is bivariate normal, test the hypothesis ($\alpha = .05$) that $\rho = 0$ against the alternative that the correlation is positive. By equation [**14.6**],

$$t_r = \frac{.82\sqrt{9 - 2}}{\sqrt{1 - .672}} \doteq 3.79$$

In Table 4 of Appendix VIII, we find $.995 < \mathbb{P}(3.79) < .9995$ in $t(7)$. Therefore, $p < .005 < \alpha$, so the test hypothesis is rejected.

2. Testing hypotheses about ρ_1 and ρ_2

In Chapter 10 we developed methods based on the normal distribution for testing hypotheses about the difference between two population means, $\mu_1 - \mu_2$. Fisher's r-to-z transformation permits us to extend these methods to hypotheses about the difference between two correlations, $\rho_1 - \rho_2$. Let r_1 be the correlation of X and Y calculated from a sample of N_1 observations on Population 1 and let r_2 be the correlation of X and Y calculated from a sample of N_2 observations on Population 2. Then, from the Algebra of Expectations,

$$E\left(Z_{r_1} - Z_{r_2}\right) = \zeta_1 - \zeta_2$$

And since samples drawn from distinct populations are ordinarily assumed to be independent,

$$V\left(Z_{r_1} - Z_{r_2}\right) = \frac{1}{N_1 - 3} + \frac{1}{N_2 - 3}$$

Therefore, if the joint distribution of X and Y is bivariate normal in both populations and if N_1 and N_2 are both greater than 10, then

$$\frac{(Z_{r_1} - Z_{r_2}) - (\zeta_1 - \zeta_2)}{\sqrt{\dfrac{1}{N_1 - 3} + \dfrac{1}{N_2 - 3}}}$$

is distributed approximately $N(0, 1)$. If $\rho_1 = \rho_2$, then $\zeta_1 - \zeta_2 = 0$, and

$$Z_{r_1 - r_2} = \frac{Z_{r_1} - Z_{r_2}}{\sqrt{\dfrac{1}{N_1 - 3} + \dfrac{1}{N_2 - 3}}} \qquad [14.7]$$

may therefore be used to test the hypothesis

$$H_0: \rho_1 - \rho_2 = 0$$

against a directional or nondirectional composite hypothesis or against a simple hypothesis.

If $N_1 > 10$ and $N_2 > 10$, then

$$Z_{r_1 - r_2} = \frac{Z_{r_1} - Z_{r_2}}{\sqrt{\dfrac{1}{N_1 - 3} + \dfrac{1}{N_2 - 3}}}$$

is distributed approximately $N(0, 1)$ and may be used to test the hypothesis that $\rho_1 = \rho_2$.

Example 14.7. A psychologist suspects that there is a higher correlation between citizenship grades and academic grades assigned by teachers with "authoritarian" personalities (see p. 428) than by teachers who do not possess this trait. Suppose that $r_1 = .56$ is the correlation of citizenship grades (X) and academic grades (Y) received by 12 students from a teacher who is high on a measure of authoritarianism. Suppose further that $r_2 = .42$ is the correlation of citizenship and academic grades received by 14 students from a teacher who obtains a low score on the same test of authoritarianism. Test the hypotheses

$$H_0: \rho_1 - \rho_2 = 0$$

$$H_1: \rho_1 - \rho_2 > 0$$

From equation [**14.3**]

$$z_{r_1} = \frac{1}{2} \ln \left(\frac{1 + r_1}{1 - r_1} \right) \doteq .633$$

and

$$z_{r_2} = \frac{1}{2} \ln \left(\frac{1 + r_2}{1 - r_2} \right) \doteq .448$$

From equation [**14.7**], therefore,

$$z_{r_1 - r_2} = \frac{z_{r_1} - z_{r_2}}{\sqrt{\dfrac{1}{N_1 - 3} + \dfrac{1}{N_2 - 3}}} \doteq \frac{.633 - .448}{\sqrt{\frac{1}{9} + \frac{1}{11}}} \doteq \frac{.185}{.450} \doteq .411$$

In Table 3 of Appendix VIII we find that $\mathbb{P}(.41) \doteq .6591$, so $p > .34$, and the alternative hypothesis is rejected.

3. Interpreting tests about ρ: linearity, reduction of uncertainty, and independence

In Chapter 4 we said that two measurements X and Y are statistically related to the degree that uncertainty about Y is reduced by knowing that the value of X is x_i, and reduction of uncertainty was expressed in terms of the correlation ratio

$$\frac{s_Y^2 - s_{Y|x}^2}{s_Y^2} \qquad\qquad [14.8]$$

In addition, we said that if the relationship between X and Y is *linear*, the correlation ratio is equal to the coefficient of determination, r^2, but that if the relationship is not linear, the correlation ratio is greater than r^2.

The connection between reduction of uncertainty, linearity, and correlation also holds true at the population level: The population coefficient of determination is the square of the population correlation coefficient ρ_{XY}^2, and the population correlation ratio is

$$\eta^2 = \frac{\sigma_Y^2 - \sigma_{Y|x}^2}{\sigma_Y^2} \qquad\qquad [14.9]$$

where η is the lowercase Greek *eta*.

The essence of a statistical relationship is that changes in one variable X are accompanied by systematic changes in another variable Y. The variables X and Y are therefore said to *covary*, that is, *vary together*. The conditional variance $\sigma_{Y|x}^2$ is the variance of Y when X is held constant, so the correlation ratio η^2 indicates the proportion of variance of Y that is eliminated if X does not vary. Therefore,

> The correlation ratio η^2 is interpreted as the proportion of variance of Y *attributable to* covariation with X.

In Chapter 4 we said that if two sets of measurements X and Y are linearly related, then the correlation ratio is equal to the coefficient of determination, and we illustrated (pp. 120–121) that the coefficient of determination is less than the correlation ratio if the relationship between X and Y is nonlinear. This likewise holds true for the population values η^2 and ρ^2_{XY}. Therefore,

> The coefficient of determination ρ^2 is interpreted as the proportion of variance of Y attributable to *linear* covariation with X.

To say that covariation of X and Y is linear means that the rule by which X and Y are related is described by the general equation for a straight line. In Chapter 4 we denoted the general linear equation for N pairs of scores

$$y = bx + a$$

The population parameter corresponding to a is denoted α (not to be confused with the probability of making a Type I error) and the parameter corresponding to b is denoted β (not to be confused with the probability of making a Type II error). The linear function that models the relationship of Y to X is therefore

$$y = \beta x + \alpha$$

If $\rho = 1$ or -1, the model is a perfect representation of the relationship of X to Y; all of the pairs (x, y) lie on the line $y = \beta x + \alpha$. To the extent that ρ departs from zero, the experimenter may therefore conclude that the relationship of X to Y is linear. Furthermore, the algebraic sign of ρ indicates whether the slope β is positive or negative. However, concluding that $\rho \neq 0$ does not necessarily imply a *strong* linear relationship between X and Y. In our discussions of power in Chapters 9 and 10 we pointed out that a difference can be statistically significant without having any scientific or practical importance. It is likewise true that if r is significantly different from zero, linear estimates of Y calculated from known values of X will be characterized by significantly less uncertainty than estimates made in ignorance of X. However, unless ρ is *exactly* $+1$ or -1, the variability of Y is also influenced by factors independent of X, so the linear function $y = \beta x + \alpha$ is an imperfect model. This means that the y-values associated with any particular value of X exhibit some degree of *residual* uncertainty, which is reflected in the conditional variance, $\sigma^2_{Y|x}$.

In this regard, it will be noted that

$$\eta^2 = \frac{\sigma^2_Y - \sigma^2_{Y|x}}{\sigma^2_Y}$$

$$= \frac{\sigma^2_Y}{\sigma^2_Y} - \frac{\sigma^2_{Y|x}}{\sigma^2_Y} = 1 - \frac{\sigma^2_{Y|x}}{\sigma^2_Y} \qquad [14.10]$$

The last term in [14.10],

$$\frac{\sigma^2_{Y|x}}{\sigma^2_Y}$$

indicates the proportion of Y variability that remains when X is held constant and which must therefore be due to sources that are independent of X, which we call error. Unfortunately, the proportion of variability in Y that is due to error can be very large, even when r is significantly different from zero. This is because low to moderate values of r, even if statistically significant, yield very small values of r^2. A correlation of .30, for example, is significantly greater than 0 at the .05 level for any sample of more than 31 observations. Nevertheless, X accounts for only $(.30)^2 = 9$ percent of the variance of Y.

A more dramatic demonstration was recently published by Standing et al.[6] The authors calculated correlations for 135 educational and biographical variables (e.g., paternal income, nutritional habits, sports activity, magazine reading, and second-grade mathematics achievement) for 2,058 Montreal grade school students. On average, each variable correlated significantly ($\alpha = .05$) with 41 percent of the other variables with an average (absolute) correlation of .07.

Clearly, the reduction of uncertainty about Y occasioned by knowledge of X may really be very small, and—just as one should consider the magnitude of a statistically significant difference between μ and μ_0—one should therefore consider the value of r^2 when interpreting significant results of an hypothesis test about ρ.

In Chapter 4 we explained that a high (positive) correlation does not imply causation. The same warning applies to interpreting a statistically significant correlation. The conclusion that X and Y share common sources of variation does not necessarily mean that X *causes* Y or that Y *causes* X.

We also know from Chapter 4 that one must be cautious in interpreting low (i.e., near zero) values of r. Failure to reject the hypothesis that $\rho = 0$ must likewise be interpreted judiciously. If X and Y are indepen-

[6]Standing, L., Sproule, R., and Khouzam, N. (1991). Empirical statistics. IV. Illustrating Meehl's sixth law of soft psychology: Everything correlates with everything. *Psychological Reports, 69,* 123–126.

dent, then $\rho = 0$ and an experimenter will probably obtain a value of r that is near zero. Recall, however, that $\rho = 0$ ordinarily implies independence of X and Y *only* if $\phi(X, Y)$ is bivariate normal.[7] This means that *all* covariation in a bivariate normal distribution must be linear, which is not true for other joint distributions. Therefore, unless an experimenter is confident that the bivariate normal distribution is an appropriate model for the joint distribution of X and Y, a low correlation cannot be taken to imply an absence of covariation. As we know from Chapter 4, absolutely small values of r will also be obtained when the relationship of X and Y is nonlinear. One approach to teasing apart independence and nonlinearity is taken up in the next section.

EXERCISES 14.2

In the following exercises, assume that $\phi(X, Y)$ is bivariate normal. In Exercises 1 and 2 calculate Student's transformation for every correlation. If the number of observations (N) is greater than 10, also calculate Fisher's *r-to-z*.

1.

	r	N
(a)	.42	35
(b)	−.78	15
(c)	.73	10
(d)	.73	11
(e)	.91	8

2.

	r	N
(a)	.81	12
(b)	−.46	8
(c)	−.85	8
(d)	.23	10
(e)	.23	75

In Exercises 3 and 4, test the hypothesis that $\rho = \rho_0$ against the indicated alternative hypothesis and at the indicated significance level. Report the value of the test statistic, the *p*-value of your result, and your conclusion. If $\rho_0 = 0$, use Student's transformation only if sample size does not permit use of Fisher's *r-to-z*.

3.

	r	N	ρ_0	$H_1 : \rho$	α
(a)	.31	32	0	> 0	.05
(b)	.43	15	0	> 0	.05
(c)	−.56	103	−.70	$> -.70$	.01
(d)	.26	51	.50	$\neq .50$	.05
(e)	.53	11	0	> 0	.05
(f)	.53	10	0	> 0	.05
(g)	−.23	20	0	< 0	.05
(h)	.76	6	0	$\neq 0$	.05

[7] If X, Y is bivariate normal, the marginal distributions of X and Y are normal, but *bivariate* normality of the joint distribution is by no means guaranteed by normality of the marginal distributions.

4.

	r	N	ρ_0	$H_1 : \rho$	α
(a)	$-.38$	20	0	< 0	.05
(b)	.10	543	0	> 0	.01
(c)	.41	200	.50	$< .50$	.05
(d)	.81	35	.70	$> .70$	.05
(e)	.29	9	0	> 0	.05
(f)	$-.81$	5	0	< 0	.05
(g)	$-.60$	11	0	$\neq 0$	.05
(h)	$-.76$	10	0	$\neq 0$	.05

For each set of X- and Y-values in Exercises 5 and 6, calculate r, test the indicated hypotheses using Fisher's r-to-z or Student's transformation as appropriate to the hypotheses and sample sizes. Report the p-value of your result and your conclusion.

5.

(a)		(b)		(c)	
X	Y	X	Y	X	Y
14	10	148	151	131	207
10	11	152	159	126	200
13	12	121	167	130	202
13	11	136	150	132	209
11	9	141	164	144	216
9	10	149	152	129	203
11	11	137	160	140	203
7	9	135	170	136	211
12	10	129	160	128	216
13	12	128	163	138	201
		155	154	149	221
				129	210
$H_0: \rho = 0$		$H_0: \rho = -.70$		$H_0: \rho = .80$	
$H_1: \rho > 0$		$H_1: \rho > -.70$		$H_1: \rho < .80$	

6.

(a)		(b)		(c)	
X	Y	X	Y	X	Y
11	24	90	87	58	139
9	17	100	39	54	127
10	17	120	75	57	130
12	18	150	75	59	146
8	16	170	72	60	137
10	20	180	54	53	132
7	12	200	50	51	112
9	16	200	45	58	147
		220	45	62	147
		220	45	56	139
		220	84	55	136
		230	39		
$H_0: \rho = 0$		$H_0: \rho = 0$		$H_0: \rho = .85$	
$H_1: \rho > 0$		$H_1: \rho < 0$		$H_1: \rho \neq .85$	

7. The following scores on a mathematics skills test (X) and final grade ($A = 4$, $B + = 3.5$, etc.) in a statistics course (Y) were selected at random from the student data reported by Kirk (op. cit.).

X	Y	X	Y	X	Y	X	Y
19	1	29	3	39	4	14	1
38	4	35	1	30	2	33	3.5
35	2	17	2	30	3	42	4
28	3	8	3	40	2.5	41	4
36	2	21	1	39	3.5	34	3
42	3	40	4	38	3	39	4
44	4	34	3.5	23	3.5	31	2
43	4	28	1	41	4	31	4
33	2.5	36	4	31	2	32	4
18	2	42	2.5	42	4	28	2

Test the hypothesis ($\alpha = .05$) that at least 50 percent of the variation in statistics grades is attributable to linear covariation with the mathematical skills test. (*Hint:* The proportion of Y-variance attributable to X is the *square* of the correlation ρ_{XY} when the relationship of X and Y is linear.)

8. Data Set 1 is drawn from Population 1 and Data Set 2 is drawn from Population 2.

DATA SET 1

x, y	x, y	x, y	x, y	x, y
4, 4	7, 8	10, 12	13, 16	16, 20
4, 5	7, 9	10, 13	13, 17	16, 21
4, 6	7, 10	10, 14	13, 18	16, 22
4, 7	7, 11	10, 15	13, 19	16, 23

DATA SET 2

x, y	x, y	x, y	x, y	x, y
4, 4	7, 6	15, 10	13, 6	16, 4
4, 5	7, 7	15, 11	13, 7	16, 5
4, 6	7, 8	15, 12	13, 8	16, 6
4, 7	7, 9	15, 13	13, 9	16, 7

Test the hypotheses

$$H_0: \rho_1 = \rho_2$$

$$H_1: \rho_1 \neq \rho_2$$

(a) Calculate r_1 the correlation of X and Y in data set 1, and calculate r_2 the correlation of X and Y in data set 2.
(b) Calculate the value of your test statistic.
(c) Report the p-value of the test statistic.

9. (a) Calculate the correlation ratio

$$\frac{s_Y^2 - s_{Y|X}^2}{s_Y^2}$$

for each data set in Exercise 8. What does your answer imply about the strength of the relationship between X and Y in the two populations?
(b) Calculate the coefficient of determination r^2 for each data set.
(c) Reconcile your answers to (a) and (b).

C. REGRESSION

1. The population regression equation

We know from Chapter 4 that the linear equation

$$y = bx + a$$

that best summarizes a scatter plot of jointly distributed scores is obtained when

$$b = r\frac{s_Y}{s_X}$$

and

$$a = \bar{y} + b\bar{x}$$

Our presentation in Chapter 4 was confined to the descriptive task of predicting the value of Y from a specified value of X by the least squares linear regression equation **[4.9]**

$$\hat{y} = b(x - \bar{x}) + \bar{y}$$

It was also explained in Chapter 4 that if X and Y are linearly related, the least squares regression line may be interpreted as the locus of conditional means of $Y|x$. It is similarly true that if random variables X and Y are linearly related, the locus of conditional population means $\mu_{Y|x}$ is the *population* regression equation

$$y = \beta x + \alpha$$

where

$$\beta = \rho\frac{\sigma_Y}{\sigma_X}$$

and

$$\alpha = \mu_Y + \beta\mu_X$$

In this chapter we consider the N pairs of values (x_i, y_i) not merely as a collection of observations but as a *sample* drawn from a larger population, and we develop methods to test hypotheses about linearity and about departure from linearity.

2. Formulating hypotheses about regression

From earlier discussion we know that the population coefficient of determination ρ_{XY}^2 is interpreted as the proportion of variance of Y attributable to *linear* covariation with X. Accordingly, the hypothesis that X and Y are linearly related is

$$H_0: \rho^2 = 0$$

and the alternative hypothesis is

$$H_1: \rho^2 > 0$$

We also know that the population correlation ratio η^2 indicates the proportion of variability in Y attributable to covariation with X. That is, the correlation ratio captures *all* sources of covariation, linear as well as nonlinear. The difference $\eta^2 - \rho^2$ therefore indicates the extent to which the relationship of X to Y departs from linearity, and the hypotheses for nonlinearity are

$$H_0: \eta^2 - \rho^2 = 0$$

and

$$H_1: \eta^2 - \rho^2 > 0$$

The reader may note that the hypothesis for linear regression seems redundant with the test hypothesis formulated earlier to test for significance of correlation. After all, if ρ is *different* from zero, it follows that ρ^2 should be *greater* than zero. The fact of the matter is that hypotheses about correlation and regression do not always arise out of the same research. In part, this is because correlation focuses only on the degree of linear covariation, and regression considers the additional question of prediction. In part, it is because questions about correlation and regression are in many ways appropriate to different research designs.

In Appendix VI we explain that correlational designs are *observational*, rather than experimental. That is, correlational research is concerned about attributes that subjects possess before they are selected for study. Such research may be predictive or not. If subjects bring *both* X- and Y-attributes to the research setting, it is likely that the researcher is interested only in the covariation of X and Y and that neither variable is a sensible predictor of the other. For example, a developmental biologist might formulate hypotheses about the correlation of height and weight, but height and weight are contemporaneous and equally accessible, so it would seem just as meaningful—or just as meaningless—to predict height from weight as to predict weight from height. In other correlational studies, attribute Y might not be present or measurable until some time after attribute X is measured, in which case X might be a perfectly sensible predictor of Y. For example, aptitude tests are explicitly designed to predict future performance, so it seems more reasonable to predict first-year college grades from SAT scores than to "anticipate the past"[8] by "predicting" SAT scores from college grades.

In experimental research, such as we discussed in Chapter 12, subjects are typically assigned at random to J groups. The members of each group are administered some experimental treatment X, and the experimenter then obtains scores on variable Y, which are expected to reflect the *effect* of the treatments. In experimental research, the independent variable X is therefore the logical predictor not only because it precedes Y, but also because experimental research (as opposed to observational research) is often conducted to test a theoretical *causal* connection between X and Y. If experimental treatments are *qualitative* (e.g., different drugs in medical research or different reinforcers in a learning experiment) the experimenter may only predict that the y-scores will differ from group to group. But if the experimental treatments are *quantitative*, e.g., different dosages of the *same* drug or different magnitudes of the *same* reinforcer, the experimenter may formulate hypotheses about regression.

The most fundamental characteristic of experimental research is that the experimenter determines each subject's "score" on the independent variable (X). The experimenter can therefore ensure that every x-value is *replicated* (that is, occurs more than once) and thus be certain of observing two or more y-scores for every value of X. This is by no means assured in observational research, and we will see below that it is essential to distinguishing error from nonlinearity.

3. Test statistics: analysis of variance for regression

The sort of experiment described in the preceding paragraph is similar in many ways to those discussed in Chapter 12. Not surprisingly, the test for significance of regression is a form of one-way analysis of variance. The analysis of variance for regression differs in one important respect from the analysis of variance for differences among J group means: The regression analysis partitions SS_A into *two* components, one of which reflects linear regression and the other of which reflects departure of the conditional means from linearity.

a. Components of variability: sums of squares. To begin, we observe that y_{ij}, the y-score obtained by individual i with an x-score of x_j, can be expressed as the sum

$$y_{ij} = \bar{y} + (y_{ij} - \bar{y}_j) + (\bar{y}_j - \hat{y}_j) + (\hat{y}_j - \bar{y}) \qquad [14.11]$$

[8] Sheridan, R. B. *The Rivals*. Act IV; Scene ii. Mrs. Malaprop to Sir Anthony.

where $\bar{y}$ is the mean of the marginal distribution of Y, and $\bar{y}_j$ is the mean of the conditional distribution of $Y|x_j$, and $\hat{y}_j$ is the predicted y-value for individuals with $x = x_j$. Subtracting $\bar{y}$ from both sides of [14.11] gives us

$$y_{ij} - \bar{y} = (y_{ij} - \bar{y}_j) + (\bar{y}_j - \hat{y}_j) + (\hat{y}_j - \bar{y}) \qquad [14.12]$$

If we square both sides of [14.12] and sum over all N individuals, we obtain

$$\sum_j \sum_i (y_{ij} - \bar{y})^2 = \sum \sum [(y_{ij} - \bar{y}_j) + (\bar{y}_j - \hat{y}_j) + (\hat{y}_j - \bar{y})]^2 \quad [14.13]$$

The left-hand side of [14.13] is familiar to use from Chapter 12 as the total sum of squares (SS_{TOTAL}). By a proof that is very similar to the one in Box 12.1 (pp. 506–507), the expression on the right-hand side of [14.13] can be partitioned into three components:

$$\sum_j \sum_i (y_{ij} - \bar{y})^2$$

$$= \sum_j \sum_i (y_{ij} - \bar{y}_j)^2 + \sum_j N_j (\bar{y}_j - \hat{y}_j)^2 + \sum_j N_j (\hat{y}_j - \bar{y})^2 \quad [14.14]$$

The first component of [14.14]

$$\sum_j \sum_i (y_{ij} - \bar{y}_j)^2$$

is simply the sum of squared deviations of every y-value from its conditional mean. If we think of all persons with a score of x_j as a treatment group (e.g., all persons who receive x_j amount of the experimental treatment), then this component is familiar as the *within-group* sum of squares (SS_W) and must therefore reflect variability due only to error. It is therefore abbreviated

$$SS_{\text{ERROR}}$$

The second component in equation [14.14]

$$\sum_j N_j (\bar{y}_j - \hat{y}_j)^2$$

reflects another source of variability within groups. Each deviation is the difference between the conditional mean and the predicted y-score for all persons with $x = x_j$. Two points are important here:

- It is shown in Box 14.1 that in any distribution of data, the mean is the value about which the sum of squared deviations is smallest.
- We also know that we have chosen a and b so that the sum of squared deviations about $\hat{y}$ is smaller than the sum of squared deviations about any *linear* function of X.

BOX 14.1

Proof that $\bar{x}$ is the value about which sum of squared deviations is smallest

1. Let $k = \bar{x} + c$, where c is any constant.

2. $\dfrac{\displaystyle\sum_{}^{N}(x_i - k)^2}{N} = \dfrac{\sum x_i^2 - 2\sum x_i k + \sum k^2}{N}$

3. $= \dfrac{\sum x_i^2 - 2k\sum x_i + Nk^2}{N} = \dfrac{\sum x_i^2}{N} - 2k\dfrac{\sum x_i}{N} + k^2$

4. $= \dfrac{\sum x_i^2}{N} - 2k\bar{x} + k^2 = \dfrac{\sum x_i^2}{N} - 2(\bar{x} + c)\bar{x} + (\bar{x} + c)^2$

5. $= \dfrac{\sum x_i^2}{N} - 2\bar{x}^2 - 2c\bar{x} + \bar{x}^2 + 2c\bar{x} + c^2$

6. $= \dfrac{\sum x_i^2}{N} - \bar{x}^2 + c^2$

7. Therefore,

$$\dfrac{\displaystyle\sum_{}^{N}(x_i - k)^2}{N}$$

assumes the smallest possible value when $c = 0$, i.e., when $k = \bar{x}$.

Therefore, if means $\bar{y}_j$ lie in a straight line, then $\bar{y}_j$ must equal $\hat{y}_j$, so $\sum N_j(\bar{y}_j - \hat{y}_j)^2 = 0$. If the conditional means lie on a curve (as in Figure 14.6, p. 606) or otherwise depart from linearity, this sum of squares will be greater than zero. Since conditional sample means are estimates of the conditional population means μ_j, this component must reflect deviations of the population means from linearity and is therefore abbreviated

$$SS_{\text{DEV. FROM LINEAR}}$$

The last component in equation **[14.14]**

$$\sum_{j} N_j(\hat{y}_j - \bar{y})^2$$

reflects the amount of variability due to linear regression. We know from equations **[4.9]** and **[4.7]** that

$$\hat{y} = b(x - \bar{x}) + \bar{y}$$

$$= \left(r\frac{s_Y}{s_X}\right)(x - \bar{x}) + \bar{y}$$

Therefore, if the correlation is zero, $\hat{y} = \bar{y}$ for all values of X, so $(\hat{y}_j - \bar{y})^2 = 0$. Consequently, small values of $\sum N_j(\hat{y}_j - \bar{y})^2$ indicate an absence of linear correlation, and large values indicate the presence of linear correlation. This component is therefore abbreviated

$$SS_{\text{LINEAR}}$$

Equation [**14.14**] may therefore be rewritten as

$$SS_{\text{TOTAL}} = SS_{\text{ERROR}} + SS_{\text{DEV. FROM LINEAR}} + SS_{\text{LINEAR}}$$

We said earlier that the coefficient of determination r^2 is interpreted as the proportion of variance in Y that may be attributed to linear covariation with X. This may now be expressed more formally:

$$r^2 = \frac{SS_{\text{LINEAR}}}{SS_{\text{TOTAL}}}$$

We have also said that the correlation ratio reflects all sources of covariation. In this connection, it can be shown that

$$\frac{s_Y^2 - s_{Y|x}^2}{s_Y^2} = \frac{SS_{\text{LINEAR}} + SS_{\text{DEV. FROM LINEAR}}}{SS_{\text{TOTAL}}}$$

It is easily seen that if $SS_{\text{DEV. FROM LINEAR}}$ is zero, then linear correlation accounts for all variability other than error, and as indicated earlier,

$$\frac{s_Y^2 - s_{Y|x}^2}{s_Y^2} = r^2$$

Finally, it might be noted that if every value x_j occurs only once, then there is only one value in the distribution of $Y|x_j$. Therefore, $y_{ij} = \bar{y}_j$, and equation [**14.14**] becomes

$$\sum_j \sum_i (y_{ij} - \bar{y})^2 = \sum_j \sum_i (y_{ij} - \hat{y}_j)^2 + \sum_j N_j(\hat{y}_j - \bar{y})^2$$

or, since $N_j = 1$,

$$\sum_j (y_j - \bar{y})^2 = \sum_j (y_j - \hat{y}_j)^2 + \sum_j (\hat{y}_j - \bar{y})^2$$

The second term $\sum(\hat{y}_j - \bar{y})^2$ captures linear regression, but error and deviations from linear regression are pooled in the first term $\sum(y_j - \hat{y}_j)^2$. As noted on page 598, it is therefore possible to distinguish error from nonlinearity only if x-values are replicated.

b. Mean squares and the F ratio

Degrees of freedom. Since SS_{TOTAL} and SS_{ERROR} are defined exactly as SS_{TOTAL} and SS_W in Chapter 12, they have $N - 1$ and $N - J$ degrees of freedom, respectively. In addition, it can be shown that SS_{LINEAR} has only one degree of freedom. By the additive property of degrees of freedom, $SS_{\text{DEV. FROM LINEAR}}$ must therefore have

$$(N - 1) - (N - J) - 1 = J - 2$$

degrees of freedom.

By the definition of mean squares given in Chapter 12, then,

$$MS_{\text{ERROR}} = \frac{SS_{\text{ERROR}}}{N - J}$$

$$MS_{\text{DEV. FROM LINEAR}} = \frac{SS_{\text{DEV. FROM LINEAR}}}{J - 2}$$

and

$$MS_{\text{LINEAR}} = \frac{SS_{\text{LINEAR}}}{1} = SS_{\text{LINEAR}}$$

Expected values of mean squares. From our discussion of MS_W in Chapter 12,

$$E(MS_{\text{ERROR}}) = \sigma_e^2$$

We have already explained that $SS_{\text{DEV. FROM LINEAR}}$ captures departure of the conditional population means μ_j from linearity. However, the conditional mean $\overline{Y}_j$ is a random variable, so any observed value $\bar{y}_j$ is likely to differ from $E(\overline{Y}_j) = \mu_j$. The quantity $\Sigma N_j(\hat{y}_j - \bar{y}_j)^2$ must therefore tap error as well as nonlinearity. Accordingly, if nonlinearity is *absent*, the only source of variability in $\Sigma N_j(\hat{y}_j - \bar{y}_j)^2$ is error, and

$$E(MS_{\text{DEV. FROM LINEAR}}) = \sigma_e^2$$

Finally, it can be shown that

$$E(MS_{\text{LINEAR}}) = \sigma_e^2 + N\rho^2\sigma_Y^2$$

Fisher's F test. If ρ^2 is zero, MS_{LINEAR} is an unbiased estimate of σ_e^2, and MS_{ERROR} is an unbiased estimator of σ_e^2 whether ρ^2 is zero or not. Therefore, to test the hypothesis

$$H_0: \rho^2 = 0$$

or the equivalent hypothesis

$$H_0: \beta = 0$$

submit the ratio

$$\frac{MS_{\text{LINEAR}}}{MS_{\text{ERROR}}}$$

[14.15]

to Fisher's F for 1 and $(N - J)$ degrees of freedom. If $\rho^2 = 0$, and MS_{ERROR} and MS_{LINEAR} are therefore independent estimates of the same variance, the expected value of [14.11] is

$$\frac{df_{\text{ERROR}}}{df_{\text{ERROR}} - 2} = \frac{N - J}{N - J - 2}$$

If $\rho^2 > 0$, the expected value is larger than $(N - J)/(N - J - 2)$, so the rejection region lies in the upper tail of $F(1, N - J)$.

To test the nonlinearity hypothesis

$$H_0: \eta^2 - \rho^2 = 0$$

submit the ratio

$$\frac{MS_{\text{DEV. FROM LINEAR}}}{MS_{\text{ERROR}}}$$

[14.16]

to Fisher's F for $(J - 2)$ and $(N - J)$ degrees of freedom.

Computational shortcuts for calculating MS. The Fisher's F test can be conducted by calculating the various mean squares directly from the formulae for SS in equation [14.14]. This requires many intermediate calculations and affords many opportunities for error. Fortunately, we can exploit several of the computational formulae we have already developed and express all necessary quantities in terms of Σx, Σx^2, Σy, Σy^2, and Σxy.

First, SS_{TOTAL} is defined exactly as in Chapter 12, so

$$SS_{\text{TOTAL}} = \sum^{N} y^2 - N\bar{y}^2$$

Likewise, we know that SS_{ERROR} is identical to SS_W, and the same computational formula presented in Chapter 12 may be used here.

$$SS_{\text{ERROR}} = SS_W = \sum^{N} y^2 - \sum_{j} N_j \bar{y}_j^2$$

so

$$MS_{\text{ERROR}} = \frac{\sum^{N} y^2 - \sum_{j} N_j \bar{y}_j^2}{N - J}$$

To derive a computational form of MS_{LINEAR} we take advantage of [4.9] and substitute $b(x_j - \bar{x}) + \bar{y}$ for $\hat{y}_j$ in

$$SS_{\text{LINEAR}} = \sum_j N_j(\hat{y}_j - \bar{y})^2$$

A little judicious algebraic rearrangement gives us

$$SS_{\text{LINEAR}} = Nb^2 s_X^2$$

Since df for $SS_{\text{LINEAR}} = 1$, it must also be true that

$$Nb^2 s_X^2 = MS_{\text{LINEAR}}$$

From the computational formula for b given in equation [4.10],

$$b^2 = \left(\frac{N\sum xy - \sum x \sum y}{N\sum x^2 - (\sum x)^2}\right)^2$$

and from the computational formula for variance, [2.11]

$$s_X^2 = \frac{\sum x^2}{N} - \bar{x}^2$$

The sum of squares for deviations from linear regression requires no special formula. It is obtained from the other sums of squares by subtraction:

$$SS_{\text{DEV. FROM LINEAR}} = SS_{\text{TOTAL}} - SS_{\text{ERROR}} - SS_{\text{LINEAR}}$$

so

$$MS_{\text{DEV. FROM LINEAR}} = \frac{SS_{\text{TOTAL}} - SS_{\text{ERROR}} - SS_{\text{LINEAR}}}{J - 2}$$

Example 14.8. Suppose that the following data represent recreational athletes' resting heart rate and the number of days each month that they perform vigorous aerobic exercise for 20 minutes or more (see Table 14.3). The basic computational quantities are as follows:

$$\sum x = 6(8) + 6(12) + 6(20) + 6(24) = 384$$

$$\sum x^2 = 6(64) + 6(144) + 6(400) + 6(576) = 7,104$$

$$\sum y = 440 + 333 + 315 + 310 = 1,398$$

$$\sum y^2 = 84,908$$

$$\sum xy = 21,256$$

Table 14.2. Resting heart rate (per minute) by number of days of exercise

	Days of exercise per month (x)			
	8	12	20	24
Heart Rate (y)	82	65	60	52
	78	50	47	41
	87	62	44	57
	58	55	66	50
	70	52	41	47
	65	49	57	63
Σy	440	333	315	310

Therefore,

$$SS_{\text{TOTAL}} = \sum y^2 - N\bar{y}^2 = 84{,}908 - 24\left(\frac{1{,}398}{24}\right)^2 = 3{,}474.5$$

$$SS_{\text{ERROR}} = \sum y^2 - \sum_j N_j \bar{y}_j^2$$

$$= 84{,}908 - \left[6\left(\frac{440}{6}\right)^2 + 6\left(\frac{333}{6}\right)^2 + 6\left(\frac{315}{6}\right)^2 + 6\left(\frac{310}{6}\right)^2\right]$$

$$\doteq 84{,}908 - 83{,}302.33 = 1{,}605.67$$

$$SS_{\text{LINEAR}} = Nb^2 s_X^2$$

$$= 24\left(\frac{24(21{,}256) - (384)(1{,}398)}{24(7{,}104) - (384)^2}\right)^2 \left(\frac{7{,}104}{24} - (16)^2\right) \doteq 1{,}288.07$$

$$SS_{\text{DEV. FROM LINEAR}} \doteq 3{,}474.5 - 1{,}605.67 - 1{,}288.07 = 580.76$$

Fisher's F test for linear regression therefore yields a value of

$$\frac{SS_{\text{LINEAR}}/df_{\text{LINEAR}}}{SS_{\text{ERROR}}/df_{\text{ERROR}}} \doteq \frac{1{,}288.07/(1)}{1{,}605.67/(24-4)} \doteq 16.04$$

which from Table 6 of Appendix VIII is seen to be significant in $F(1, 20)$

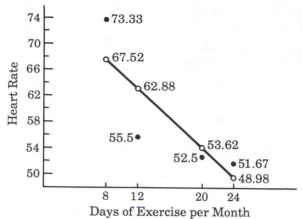

Figure 14.6. Observed mean (•) and predicted heart rate (○) for various levels of exercise.

at .001. The *F* ratio for deviations from linear regression is

$$\frac{SS_{\text{DEV. FROM LINEAR}}/df_{\text{DEV. FROM LINEAR}}}{SS_{\text{ERROR}}/df_{\text{ERROR}}} \doteq \frac{580.77/(4-2)}{1{,}605.67/(24-4)}$$

$$\doteq 3.62$$

In the distribution of $F(2, 20)$ we find $.95 < \mathbb{P}(3.62) < .975$, so $p < .05$. The data given in Table 14.2 clearly indicate both linear and nonlinear covariation. This is seen in Figure 14.6, where we have plotted both the mean heart rate $\bar{y}_j$ for each group and the predicted heart rate $\hat{y}_j$ calculated by the least squares regression equation **[4.9]**,

$$\hat{y}_j = b(x_j - \bar{x}) + \bar{y}$$

As exercise physiologists have observed, heart rate exhibits a steady decrease with increased aerobic exercise, but the rate of decrease levels off beyond 3 aerobic workouts per week (12 per month). The steep drop between 8 and 12 days of exercise per month accounts for most of the deviation from linear regression, which is seen in the differences between the means $\bar{y}_j$ (•) and the predicted values $\hat{y}_j$ (○).

Assumptions underlying the analysis of variance for regression. The assumptions that must be met in order to test hypotheses about regression using analysis of variance are considerably less demanding than those required for testing hypotheses about correlation. In experimental research, the distribution of X is determined by the experimenter, because the frequency of x_j is the number of persons assigned to treatment group j. Furthermore, we know that the analysis of variance for regression can reveal significant nonlinear covariation. It should therefore be obvious that bivariate normality, which is essential to testing hypotheses about correlation, is not assumed in the test for linear regression.

Basically, the assumptions to be satisfied are the same as those discussed for analysis of variance in Chapter 12.

First, the conditional distributions of Y on x must be normal. This embraces the assumption common to all one-way ANOVA that error be normally distributed within each of the J populations.

Second, it is assumed that the conditional variances $\sigma_{Y|x}^2$ are equal. That is, the conditional distributions of Y are homoscedastic. As we have discovered elsewhere, this assumption can be nominally satisfied if all J sample sizes are equal.

Finally, it is assumed that all observations y_{ij} are independent of one another.

EXERCISES 14.3

In the following exercises, assume that the conditional variances $\sigma_{Y|x}^2$ in each data set are equal. In Exercises 1–4, (a) test the hypothesis that X and Y are linearly related and (b) test the hypothesis that X and Y are nonlinearly related.

1.

	X			
	10	40	90	140
Y	6	20	24	32
	10	8	22	20
	2	14	20	26

2.

	X				
	1.7	1.9	2.4	2.9	3.4
Y	41	29	20	17	7
	33	31	23	13	9
	37	33	17	9	11

3.

	X				
	1.5	1.7	1.9	2.5	2.8
Y	29	19	25	17	15
	33	23	17	11	17
	37	27	21	14	13

4.

	X				
	10	50	100	160	190
Y	8	26	26	20	12
	12	20	28	25	8
	16	14	24	15	16

5. Use the data below from Exercises 4.3.11 (p. 122) and 4.4.3 (p. 135) to
   (a) test the hypothesis that reinforcement frequency and habit strength are linearly related
   (b) test the hypothesis that reinforcement frequency and habit strength are nonlinearly related.

	Frequency of reinforcement Number of times response must be performed to receive reinforcer				
	5	**10**	**15**	**20**	**25**
	118	107	87	141	106
	87	83	100	130	151
	113	75	126	109	110
Habit strength	80	78	122	103	143
Trials to extinction	109	111	96	136	155
	68	82	88	99	112
	110	112	123	103	148
	75	111	130	137	115

6. Use the data below from Exercises 4.3.12 (p. 122) and 4.4.4 (p. 135) to
   (a) test the hypothesis that reinforcement frequency and habit strength are linearly related, and
   (b) test the hypothesis that reinforcement frequency and habit strength are nonlinearly related.

	Frequency of reinforcement Minutes between reinforcements				
	.50	**1**	**5**	**10**	**20**
	109	106	87	141	107
	75	151	100	130	83
	113	148	126	109	75
Habit strength	80	143	122	103	78
Trials to extinction	118	155	96	136	111
	68	112	88	99	82
	110	110	123	103	112
	87	115	130	137	111

7. The data below are averages of white blood cell counts ($\times 1,000$) recorded weekly during chemotherapy for 45 postoperative cancer patients. Each patient received a total dosage of 2,400, 3,600, or 4,800 mg/m^2 of cyclophosphamide (see Exercise 13.1.7).
   (a) Test the hypothesis that total dosage of cyclophosphamide and white blood cell count are linearly related.
   (b) Test the hypothesis that total dosage of cyclophosphamide and white blood cell are nonlinearly related.

Total dosage of cyclophosphamide (mg/m^2)					
2400		3600		4800	
2.3	2.4	3.4	2.4	2.7	1.8
3.7	2.2	2.2	2.7	3.3	2.4
3.2	2.1	2.8	1.8	2.8	2.0
2.8	2.8	3.0	2.1	2.1	1.6
3.0	1.8	2.6	1.0	2.6	1.6
2.8	2.2	2.3	2.2	2.8	2.4
3.1	1.6	2.0	2.4	2.5	1.4
3.0		3.1		2.0	

8. The data below are atomic absorption readings for 15 samples of a particular element of known concentration. Five of the samples are 3 weeks old, five are 5 weeks old, and five are 7 weeks old. The atomic absorption readings for the 15 samples are given below. (See Exercise 4.4.8, p. 137.)
(a) Test the hypothesis that age and atomic absorption are linearly related.
(b) Test the hypothesis that age and atomic absorption are nonlinearly related.

	Age of sample		
	3 weeks	5 weeks	7 weeks
	58	40	25
Atomic	60	37	24
absorption	55	42	26
	56	40	23
	61	39	25

D. SUMMARY

In Chapter 4 we introduced the notion of a joint distribution of data and developed several indices for describing the strength and direction of a statistical relation. In this chapter we take up methods for testing hypotheses about joint distributions and statistical relations.

The mathematical model for a joint distribution of data is the joint *probability* distribution. If X and Y are *discrete* random variables, the joint probability $P(x_j, y_k)$ is the probability that $x = x_j$ and that $y = y_k$, and the function $P(x, y)$ that assigns a probability to every pair of values (x_j, y_k) is called the *joint probability function* of X and Y. Because joint probabilities are associated with values of two random variables, they are sometimes called *bivariate* probabilities.

The notions of conditional probability and independence of two events E_1 and E_2 can be extended to conditional probability and independence of two discrete random variables, X and Y. The conditional probability that

$P(y = y_k)$ given that $P(x = x_j)$ is denoted $P(y = y_k | x = x_j)$ and is equal to

$$\frac{P(x_j, y_k)}{P(x_j)}$$

Likewise, the conditional probability that $x = x_j$ given that $y = y_k$ is

$$P\big(x = x_j | y = y_k\big) = \frac{P(x_j, y_k)}{P(y_k)}$$

If discrete random variables X and Y are independent, then

$$P(x_j, y_k) = P(x_j)P(y_k)$$

In Chapter 7 we developed the notion of graphically representing probability of a random variable X as *area*. Similarly, bivariate probability can be thought of as *volume*. In a joint probability histogram, X is represented on one axis, Y is represented on another axis, and probability density ϕ is represented on a third axis. Any pair of values x, y is represented as the rectangle with sides $x - .5$ to $x + .5$ and $y - .5$ to $y + .5$, and the probability of x, y is represented as the volume of the solid with height $\phi(x, y)$ over the rectangle. For a continuous random variable X, the tops of the histogram bars become a smooth curve, describing the density function $\phi(X)$; for a continuous bivariate distribution, the tops of the rectangular solids become a smooth surface, defining the bivariate density function $\phi(X, Y)$.

If one "slices" into a bivariate surface perpendicular to the X-axis at x_j, the cross section is the *conditional* distribution of Y given that $x = x_j$. Similarly, the cross section obtained by slicing into a bivariate surface perpendicular to the Y-axis at y_k is the conditional distribution of X given that $y = y_k$. The height of the conditional distribution $Y | x_j$ at y_k is the probability density of y_k when $x = x_j$, and the height of the conditional distribution $X | y_k$ at x_j is the probability density of x_j when $y = y_k$.

The mean of the conditional distribution of $Y | x_j$ may be denoted μ_j or

$$\mu_{Y | x_j}$$

and the variance of the conditional distribution of $Y | x_j$ may be denoted σ_j^2 or

$$\sigma_{Y | x_j}^2$$

Similarly, the mean of the conditional distribution of $X | y_k$ may be denoted μ_k or

$$\mu_{X | y_k}$$

and the variance of the conditional distribution of $X|y_k$ may be denoted σ_k^2 or

$$\sigma_{X|y_k}^2$$

Continuous random variables X and Y are independent if the probability density of y_k in the conditional distribution $Y|x_j$ is equal to the probability density of y_k in the marginal distribution of Y, that is, if $\phi(y = y_k|x = x_j) = \phi(y)$. And if X and Y are independent, then the joint density of x_j and y_k is the product of their marginal densities, that is,

$$\phi(x_j, y_k) = \phi(x_j)\phi(y_k)$$

A joint density function is said to be *bivariate normal* if the marginal distributions of X and Y are both normal and if the bivariate distribution has "normal" cross sections, that is, if $\phi(Y|x_j)$ is normal for all x_j and if $\phi(X|y_k)$ is normal for all y_k.

The population correlation of the variables X and Y is denoted ρ_{XY} and is equal to the expected value of the product $Z_X Z_Y$. If the joint distribution of X and Y is bivariate normal and $N > 10$, one can test the hypothesis

$$H_0: \rho_{XY} = \rho_0$$

using Fisher's r-to-z transformation

$$z_r = \frac{1}{2} \ln\left(\frac{1 + r_{XY}}{1 - r_{XY}}\right)$$

which is distributed approximately normally with expected value ζ equal to

$$\frac{1}{2} \ln\left(\frac{1 + \rho_{XY}}{1 - \rho_{XY}}\right)$$

and variance equal to $1/(N - 3)$.

If the test hypothesis is

$$H_0: \rho_{XY} = 0$$

one may instead use Student's transformation

$$\frac{r_{XY}\sqrt{N - 2}}{\sqrt{1 - r_{XY}^2}}$$

If $\phi(X, Y)$ is bivariate normal and $N > 4$, this statistic is distributed as $t(N - 2)$ when the test hypothesis is correct.

If $\phi(X, Y)$ is bivariate normal in two populations, one can test the hypothesis

$$H_0: \rho_1 - \rho_2 = 0$$

using Fisher's r-to-z transformation of the difference, $r_1 - r_2$:

$$Z_{r_1 - r_2} = \frac{Z_{r_1} - Z_{r_2}}{\sqrt{\dfrac{1}{N_1 - 3} + \dfrac{1}{N_2 - 3}}}$$

which is distributed approximately $N(0, 1)$ if both sample sizes, N_1 and N_2, are greater than 10.

The extent to which two random variables X and Y are statistically related can be thought of as the degree to which the variability of Y is reduced if x is known and is quantitatively expressed as the *population correlation ratio*, denoted η^2:

$$\eta^2 = \frac{\sigma_Y^2 - \sigma_{Y|x}^2}{\sigma_Y^2}$$

The correlation ratio may be thought of as the proportion of variability in Y that is attributable to factors that also contribute to the variability of X. If X and Y covary and the relationship is *linear*, then η^2 is equal to the *population* coefficient of determination, ρ_{XY}^2. The coefficient of determination is therefore interpreted as the proportion of variance of Y attributable to *linear* correlation with X.

To the extent that ρ departs from zero, the relationship of X to Y is approximated by the equation for a straight line,

$$y = \beta x + \alpha$$

In addition, the algebraic sign of ρ indicates whether the slope (β) of the line is positive or negative. However, unless ρ is *exactly* 1, the variability of Y is also influenced by factors that are independent of X, and the linear function $y = \beta x + \alpha$ is an imperfect model. This means that the y-values associated with any particular x-value exhibit some degree of *residual* uncertainty, or *error*, which is reflected in the conditional variance, $\sigma_{Y|x}^2$.

Significant values of r must be interpreted cautiously. The proportion of variability in Y due to error can be substantial, even if r is significantly different from zero. Therefore, one should consider the value of r^2 when interpreting significant results of an hypothesis test about ρ. Furthermore, one should never interpret a significant correlation as implying causation.

Nonsignificant values of r must also be cautiously interpreted. If X and Y are independent, then $\rho_{XY} = 0$. But, in general, $\rho_{XY} = 0$ implies that X and Y are independent only if $\phi(X, Y)$ is bivariate normal. If the joint distribution of X and Y is *not* bivariate normal, then one can obtain values of r near zero even when η^2 is large. This is most commonly found if the relationship of X to Y is curvilinear.

Correlation problems typically involve situations where X and Y represent attributes that occur "naturally" in the same individuals and neither variable, therefore, can be thought of as a "predictor" of the other. In other situations, one variable X is clearly the independent, or predictor, variable and the other variable Y is the dependent variable. This is true for experiments in which subjects in J groups receive quantities $x_1, x_2, \ldots, x_J$ of some treatment, and the experimenter records subjects' subsequent performance on variable Y. In this situation, the experimenter ordinarily assigns equal numbers of subjects to each group, so the marginal distribution of X is uniform, rather than normal, and the strength of relationship between X and Y cannot be assessed by correlation.

From Chapters 12 and 13 we know that the *effect* of treatment X on variable Y in such an experiment can be tested by a (random-effects) one-way analysis of variance. Analysis of variance is also used to test the more explicit hypothesis that X and Y are *linearly* related and the hypothesis that X and Y are curvilinearly related.

The total variability among N scores in J experimental groups defined by x-values $x_1, x_2, \ldots, x_J$ is expressed as the total sum of squares (SS_{TOTAL})

$$\sum \sum (y_{ij} - \bar{y})^2$$

which can be partitioned into three components:

$$SS_{\text{TOTAL}} = \sum \sum (y_{ij} - \bar{y}_j)^2 + \sum_j N_j(\bar{y}_j - \hat{y}_j)^2 + \sum_j N_j(\hat{y}_j - \bar{y})^2$$

where y_{ij} is the score of individual i in group j, and $\bar{y}_j$ is the mean of y-values for individuals in group j, and $\hat{y}_j$ is the predicted y-value for persons in group j, and $\bar{y}$ is the grand mean of all y-values.

The first component

$$\sum (y_{ij} - \bar{y}_j)^2$$

is the sum of squares within groups and, as in one-way analysis of variance, reflects only random error. It is therefore denoted SS_{ERROR}. Degrees of freedom for SS_{ERROR} is $N - J$.

The second component

$$\sum_j N_j(\bar{y}_j - \hat{y}_j)^2$$

reflects deviation of predicted y-scores from the conditional mean of Y for individuals with $x = x_j$. If the relationship of X and Y is linear, it can be shown that $\bar{y}_j = \hat{y}_j$, so this component reflects deviations from linearity and is denoted $SS_{\text{DEV. FROM LINEAR}}$. Degrees of freedom for $SS_{\text{DEV. FROM LINEAR}}$ is $J - 2$.

If correlation is zero, the regression line is parallel to the X-axis, so all predicted values $\hat{y}_j$ are equal to the grand mean, $\bar{y}$. Therefore, the last component

$$\sum_j N_j(\hat{y}_j - \bar{y})^2$$

must reflect the degree of linear relationship between X and Y and is therefore denoted SS_{LINEAR}, which has one degree of freedom.

As in analysis of variance, the mean squares (MS) for each component is sum of squares (SS) divided by degrees of freedom, and ratios of mean squares are distributed as Fisher's F. These ratios can be used to test the hypothesis of significant linearity, which may be formulated as either

$$H_0 : \beta = 0$$

or

$$H_0 : \rho^2 = 0$$

and the hypothesis of significant curvilinearity (deviations from linearity)

$$H_0 : \eta^2 - \rho^2 = 0$$

as described in the Summary Table of Statistical Tests.

Summary table of statistical tests

I. Testing hypotheses about correlation

H_0	H_1	Sample size	Test statistic	Distribution of test statistic*
$\rho = \rho_0$	Any	$N > 10$	$\dfrac{Z_r - \zeta_0}{\sqrt{1/(N-3)}}$ [†]	$N(0,1)$
$\rho = 0$	Any	$N > 10$	$\dfrac{Z_r}{\sqrt{1/(N-3)}}$	$N(0,1)$
$\rho = 0$	Any	$N > 4$	$\dfrac{r_{XY}\sqrt{N-2}}{\sqrt{1 - r_{XY}^2}}$	$t(N-2)$
$\rho_1 - \rho_2 = 0$	Any	$N_1 > 10$ $N_2 > 10$	$\dfrac{Z_{r_1} - Z_{r_2}}{\sqrt{\dfrac{1}{N_1 - 3} + \dfrac{1}{N_2 - 3}}}$	$N(0,1)$

Assumptions: $\phi(X, Y)$ bivariate normal.
Samples independent for $H_0 : \rho_1 - \rho_2 = 0$

---

*If H_0 correct.

[†]$z_r = \dfrac{1}{2} \ln\left(\dfrac{1+r}{1-r}\right)$ and $\zeta_0 = \dfrac{1}{2} \ln\left(\dfrac{1 + \rho_0}{1 - \rho_0}\right)$

II. Testing hypotheses about regression

H_0	H_1	Test statistic	Distribution of test statistic*
$\beta = 0$ or $\rho^2 = 0$	$\beta \neq 0$ or $\rho^2 \neq 0$	$\dfrac{MS_{\text{LINEAR}}}{MS_{\text{ERROR}}}$	$F(1, N - J)$
$\eta^2 - \rho^2 = 0$	$\eta^2 - \rho^2 \neq 0$	$\dfrac{MS_{\text{DEV. FROM LINEAR}}}{MS_{\text{ERROR}}}$	$F(J - 2, N - J)$

Assumptions: Observations independent.

$Y|x_j$ normally distributed.

$\sigma_1^2 = \cdots = \sigma_J^2$ (can be assumed if $N_1 = \cdots = N_J$).

*If H_0 correct.

Sums and means of squares for analysis of variance for regression

Source	Sums of Squares		Mean Squares
	Definitional	Computational	
Error	$\displaystyle\sum_j \sum_i (y_{ij} - \bar{y}_j)^2$	$\displaystyle\sum^N y^2 - \sum_j N_j \bar{y}_j^2$	$\dfrac{SS_{\text{ERROR}}}{N - J}$
Linearity	$\displaystyle\sum_j N_j (\hat{y}_j - \bar{y})^2$	$Nb^2 s_X^{2*}$	$\dfrac{SS_{\text{LINEAR}}}{1}$
Deviations from linearity	$\displaystyle\sum_j N_j (\bar{y}_j - \hat{y}_j)^2$	$\displaystyle\sum^N y^2 - N\bar{y}^2 -$ $SS_{\text{ERROR}} - SS_{\text{LINEAR}}$	$\dfrac{SS_{\text{DEV. FROM LINEAR}}}{J - 2}$

$$*b^2 = \left(r \frac{s_Y}{s_X} \right)^2 = \left(\frac{N \sum xy - \sum x \sum y}{N \sum x^2 - \left(\sum x \right)^2} \right)^2.$$

CHAPTER 15

Testing hypotheses about entire distributions: Pearson's *chi*-square

Fourfold point correlation: The *phi* coefficient
A test for correlated proportions
A test of homogeneity when expected frequencies are small
Exercises 15.3

E. SUMMARY

The methods developed in this chapter differ in several important and interesting ways from the hypothesis tests developed in Chapters 9 through 14.

- Every technique developed in earlier chapters allows us to test hypotheses about *one* parameter that captures *one* property of a distribution. For example, in Chapter 10 we developed tests about the population mean μ, which describes the location of a distribution, and in Chapter 11 we developed tests about the population variance σ^2, which describes dispersion. Pearson's *chi*-square techniques permit us to test hypotheses about *entire* distributions, which necessarily comprehend *all* properties of distributions (location, dispersion, skew, kurtosis, etc.).
- From Chapter 5 we know that a random variable is a rule that assigns a number to every possible result of a random experiment. When we encounter values of random variables, we have become accustomed to performing arithmetic operations on them—adding them, subtracting them, and so forth. We assume that our operations have meaning, but whether they do, in fact, have meaning depends on certain properties that define the *level of measurement* of the random variable. Measurements can yield *nominal* scales, *ordinal* scales, *interval* scales, or *ratio* scales. The properties of these scales are discussed in Appendix IV and summarized in Box 15.1. It will be noted that ordinary arithmetic operations cannot be sensibly performed on the values produced by ordinal and nominal measurements. It should therefore be apparent that all the hypothesis-testing techniques we have developed thus far are strictly applicable only to interval- or ratio-scaled random variables. The data used to calculate Pearson's *chi*-square statistics are not the *measurements* themselves, but the *frequencies* associated with the measurements. Pearson's *chi*-square techniques therefore permit inferences about observations at *any* level of measurement and are the basis for the most widely used methods for testing hypotheses about distributions of nominal measurements.
- Barring application of the Central Limit Theorem, all the tests developed in Chapters 9 through 14 require that the statistician be able to model the distribution of the random variable. That is, the

---

| **BOX 15.1** |

Levels of measurement: summary of properties

Ratio scales

Properties: units of equal size at all points along scale, true zero point

Examples: length, number of correct answers on examination, weight

Meaningful arithmetic operations: addition, subtraction, multiplication, division

Interval scales

Properties: units of equal size at all points along scale, arbitrary zero point

Examples: temperature (F or C), IQ, z-scores

Meaningful arithmetic operations: addition, subtraction

Ordinal scales

Properties: numbers indicate only the ordering of observations with respect to some criterion.

Examples: rankings, place in a race, percentile ranks

Meaningful arithmetic operations: none

Nominal scales

Properties: set of discrete, qualitative categories or attributes; numbers used as shorthand names with no quantitative meaning

Examples: sex, political party, experimental treatment group, species

Meaningful arithmetic operations: none

---

statistician must be able to assume that the population distribution of the random variable is closely approximated by some *known* probability function or probability density function. For example, we know from Chapter 10 that in order to submit the test statistic

$$\frac{\overline{X} - \mu_0}{S/\sqrt{N-1}}$$

to the distribution of Student's t, it must be assumed that X is normally distributed. Similarly, the assumption that

$$\frac{\hat{\sigma}_1^2}{\hat{\sigma}_2^2}$$

is distributed as Fisher's F requires that X_1 and X_2 both be normally distributed. If the experimenter cannot formulate a model for the distribution of the random variable, we have from time to time said that one must rely on *nonparametric* techniques. Pearson's *chi*-square techniques do not require the experimenter to specify a probability model for the distribution of the random variable. Not surprisingly, the techniques developed in this chapter are the basis for many statistical methods that are frequently characterized as "nonparametric."

A. TESTING HYPOTHESES ABOUT ONE DISTRIBUTION: GOODNESS-OF-FIT TESTS

We began Chapter 10 with techniques for testing hypotheses about *one* population mean μ, and we began Chapter 11 by developing techniques for testing hypotheses about *one* population variance σ^2. We know that these tests are used when an experimenter has theoretical reasons to hypothesize that the population parameter equals some *specific* value, e.g., that the population mean equals μ_0 or that the population variance equals σ_0^2. In the present chapter we begin with a similar situation, in which the experimenter is concerned about *one* population distribution and hypothesizes that the distribution of frequencies in this population fits some *specific* theoretical distribution of frequencies.

a. Hypotheses. When measurements are quantitative, the theoretical, or *expected*, frequencies are sometimes based on a known probability distribution, and the hypotheses can be formulated accordingly:

H_0: $X : \phi_0$

H_1: X not distributed as ϕ_0

where ϕ_0 may be either a density function or a probability function. For example, an experimenter who is testing the fairness of an ordinary six-sided die might formulate the hypotheses

H_0: X uniformly distributed

H_1: X not uniformly distributed

where X is the number of spots observed on a single roll of the die. If the experimenter intends to roll the die, say, 120 times, the test hypothesis implies that the number 1 should appear about $\frac{1}{6}(120) = 20$ times, the number 2 should appear about $\frac{1}{6}(120) = 20$ times, and so on. Under the test hypothesis, the *expected* frequencies are therefore as shown in Table 15.1.

In this text, an expected, or theoretical, frequency is denoted by a boldface capital **F**, and in tests about one population, the subscript j

Table 15.1. Expected frequencies for 120 rolls of a fair die

Value (x)	1	2	3	4	5	6
Expected Frequency (**F**)	20	20	20	20	20	20

identifies the column, or *cell*, to which it belongs. Cell j may correspond to an interval of values or to a discrete numerical value. In the present example, for instance, F_1 is the number of observations expected to fall in the first cell, that is, the number of times that the value $x = 1$ is expected to appear; the number of observations expected to fall in the second cell is denoted F_2, and so on. If cells correspond to nominal categories or attributes, the subscript j may be an alphabetic abbreviation for a noun that indicates a group or an adjective indicating some property or characteristic, rather than a number.

When observations are nominal, rather than quantitative, expected frequencies are seldom based on formal probability models like the uniform distribution. It is more typically the case that the expected frequencies derive from earlier observations of some particular population. In such instances, the test hypothesis simply states that the C cell frequencies obtained from sample data are expected to equal the corresponding theoretical frequencies. The cell frequencies that are actually *observed* when the experimenter collects data are denoted f_j, so the test hypothesis takes the form[1]

$$H_0: f_1 = F_1, \text{ and } f_2 = F_2, \text{ and } \ldots, \text{ and } f_C = F_C,$$

or

$$H_0: f_j = F_j$$

and the alternative hypothesis is, simply,

$$H_1: H_0 \text{ incorrect}$$

b. Test statistic. In earlier chapters we used "theories" from animal behavior and paleontology to develop statistical techniques. As unlikely as it may seem, one also can find theories that generate statistically testable hypotheses in biblical history.

Example 15.1. Historians and theologians have speculated for years about why the Egyptians invited the Hebrews to Goshen in the time of Joseph

---

[1] It should be understood that the test hypothesis does not really imply that f_j is *equal* to F_j. The hypothesis should be interpreted to mean that F_j is the *expected value* of f_j in the same sense the μ is the expected value of X. Some texts therefore denote observed and expected frequencies O_j and E_j, respectively, or use similar notation to emphasize this interpretation, e.g., f_{oj} and f_{ej}.

(ca. 1700 B.C.) and later, certainly by the time of Moses (ca. 1200 B.C.), enslaved the Hebrews. According to one theory, the "Egyptians" of Joseph's time were actually *Hyksos*, "foreign kings" from beyond the Caucasus Mountains who, by virtue of military high technology (the horse-drawn chariot), conquered Egypt in about 1780 B.C. After 200 years of equine proliferation, the Hyksos were driven out by the Egyptian baron, Ahmose of Thebes, who had mastered the art of chariot warfare. Ahmose eliminated his rival barons and enslaved the Hebrews and other Semitic tribes who had supported the Hyksos regime.[2]

In principle, this historical explanation is testable, because

1. The Hyksos were Caucasoid, and many scholars believe that the Egyptians of the Middle Kingdom were Negroid.
2. Racially distinct populations have characteristic profiles of A, B, and O blood types.
3. It is possible to rehydrate mummified cells and test for blood type.

Suppose an experimenter therefore reasons that if Joseph's "Egyptians" were really Hyksos, the frequencies of O, A, and B blood types found in mummies of that era should be typical of Caucasian populations. Among Caucasian people, the proportion with type O blood is approximately .484, the proportion with type A blood is approximately .423, and the proportion with type B blood is approximately .093.[3] Suppose further that the experimenter obtains permission to take blood scrapings from 60 Egyptian mummies of Joseph's era. He can then calculate the expected frequencies (summarized below in Table 15.2) by multiplying sample size ($N = 60$) by the blood-type proportions for modern Caucasian people.

Type O: $F_O = (60)(.484) \doteq 29.0$

Type A: $F_A = (60)(.423) \doteq 25.4$

Type B: $F_B = (60)(.093) \doteq 5.6$

Table 15.2. Expected frequencies of blood types for 60 Egyptian mummies (circa 1700 B.C.)

Blood Type	O	A	B	Σ
Expected Frequency (**F**)	29.0	25.4	5.6	60

[2] For a delightful treatment of this scenario (and just about everything else of consequence since the Big Bang), see Gonick, L. (1990). *The Cartoon History of the Universe*. New York: Doubleday.

[3] If blood type AB is included, the (approximate) percentages are 47 percent type O, 41 percent type A, 9 percent type B, and 3 percent type AB. Source: Guyton, A. C. (1971). *Textbook of Medical Physiology* (4th ed.). Philadelphia: Saunders, p. 127.

and in general,

$$\mathbf{F}_j = Np_j \qquad\qquad\qquad\qquad [15.1]$$

where N is the sample size, and p_j is the *population proportion* in Category j.

Now, let us imagine that blood samples from the 60 mummies are collected and typed, and yield the frequencies in Table 15.3.

Table 15.3. Observed frequencies of blood types for 60 Egyptian mummies (circa 1700 B.C.)

Blood Type	O	A	B	Σ
Observed Frequency (f)	20	23	17	60

If the test hypothesis is correct, any differences between the observed frequencies f_j and their corresponding expected frequencies $\mathbf{F}_j$ must be entirely attributable to random error, and such differences should therefore be relatively small. If the alternative hypothesis is correct, there should be substantial differences between the observed and expected frequencies.

One way to determine how well the observed distribution "fits" the expected distribution might therefore be to calculate the difference between the expected and observed frequencies for every cell and sum the differences:

$$\sum_{j}^{c} (f_j - \mathbf{F}_j)$$

In the present example, however, we find that this sum is zero:

$$\sum_{j}^{c} (f_j - \mathbf{F}_j) = (20 - 29.0) + (23 - 25.4) + (17 - 5.6)$$

$$= (-9) + (-2.4) + (11.4)$$

$$= 0$$

Indeed, the sum of differences $(f_j - \mathbf{F}_j)$ will *always* be zero, because (as seen in Tables 15.2 and 15.3) the sum of the observed frequencies and the sum of the expected frequencies are *both* equal to sample size, that is,

$$\sum f_j = \sum \mathbf{F}_j = N$$

If the observed frequency in one cell (e.g., the frequency of type A) is smaller than the expected frequency for that cell, this deficiency must therefore be distributed as a surplus among the other observed frequencies. Deficiencies are *negative* differences and surpluses are *positive* differences, so the net difference is zero.

We encountered much the same problem in Chapter 2 when we contemplated using the sum of deviations $(x_i - \bar{x})$ as an index of dispersion, and the problem was overcome by *squaring* each difference. If we employ the same tactic here, we obtain

$$\sum_{j}^{C} (f_j - \mathbf{F}_j)^2 = (20 - 29.0)^2 + (23 - 25.4)^2 + (17 - 5.6)^2$$

$$= (9)^2 + (-2.4)^2 + (11.4)^2$$

$$= 81 + 5.76 + 129.96$$

$$= 216.72$$

This gets rid of the minus signs, but unless the sum of squared differences is exactly zero (which occurs only when $f_j = \mathbf{F}_j$ in all C cells), it is difficult to interpret. The ambiguity is apparent if we consider the squared difference between observed and expected frequencies for just one cell. For type A blood,

$$(f_A - \mathbf{F}_A)^2 = (23 - 25.4)^2 = (-2.4)^2 = 5.76$$

This same squared difference would be obtained if $f_A = 5$ and $\mathbf{F}_A = 2.6$. Were these indeed our values, the observed frequency would be almost twice as large as the expected frequency. On the other hand, $(f_A - \mathbf{F}_A)^2$ also equals 5.76 if $f_A = 1{,}003$ and $\mathbf{F}_A = 1{,}000.6$, but in this case the observed frequency is only about 2 one-thousandths larger than expected. Clearly, the squared difference between an observed frequency and the corresponding expected frequency is meaningful only *in comparison with* the expected frequency. Therefore, the difference between f_j and $\mathbf{F}_j$ for any cell is captured by the ratio

$$\frac{(f_j - \mathbf{F}_j)^2}{\mathbf{F}_j}$$

and the sum of these ratios across all C cells

$$\sum_{j}^{C} \frac{(f_j - \mathbf{F}_j)^2}{\mathbf{F}_j} \qquad\qquad \textbf{[15.2]}$$

reflects the departure from perfect fit across the entire distribution.

In 1900 Karl Pearson (1857–1936) demonstrated that under certain experimental conditions, this statistic is distributed approximately as χ^2, and it is therefore known as Pearson's *chi*-square or Pearson's approximate *chi*-square. In mathematics "approximate" is often indicated by a tilde ($\sim$), so we denote Pearson's *chi*-square $\tilde{\chi}^2$. In the present example,

$$\tilde{\chi}^2 = \sum_{j}^{C} \frac{(f_j - \mathbf{F}_j)^2}{\mathbf{F}_j} = \frac{(20 - 29.0)^2}{29.0} + \frac{(23 - 25.4)^2}{25.4} + \frac{(17 - 5.6)^2}{5.6}$$

$$\doteq 2.793 + .227 + 23.207$$

$$\doteq 26.23$$

Degrees of freedom for Pearson's chi-square. We know from Chapter 11 that a *chi*-square distribution is specified by degrees of freedom, ν. We therefore need degrees of freedom for $\tilde{\chi}^2$. As in the case of Student's t, χ^2, and Fisher's F, degrees of freedom for Pearson's *chi*-square may be thought of as the number of observations that are "free" to assume *any* value. With Pearson's *chi*-square, however, the reasoning is slightly different in two respects:

First, *the observations are frequencies associated with the C cells.* The observations are not the values of the random variable, which may define the cells. In this respect, Pearson's *chi*-square is like the binomial random variable.

Second, even if no parameters are estimated, *the number of cell frequencies that are "free" is constrained by sample size.* In calculating expected frequencies in a goodness-of-fit test, the experimenter can in principle assign *any* positive values to the first $C - 1$ cells, but since

$$\sum f_j = \sum \mathbf{F}_j = N$$

the value of the *last* cell must always equal sample size minus the cumulative expected frequency of the first $C - 1$ cells. That is,

$$\mathbf{F}_C = N - \sum_{1}^{C-1} \mathbf{F}_j$$

In Example 15.1, we calculated expected frequencies by multiplying blood-type proportions in the Caucasian population by the number of mummies in the sample. The expected frequency of type B specimens was therefore $(60)(.093) \doteq 5.6$. We obtain the same result if we subtract from 60 the sum of expected frequencies for blood types O and A:

$$\mathbf{F}_B = 60 - (29.0 + 25.4) = 5.6$$

Of the C observations (cell frequencies) only one is constrained, so number of degrees of freedom is $\nu = C - 1$. In Example 15.1, the number of cells C is 3, so Pearson's *chi*-square is distributed as χ^2 with

$$\nu = 3 - 1 = 2$$

degrees of freedom.

The hypothesis that a population distribution across C mutually exclusive categories is identical to a theoretical, or expected, distribution can be tested using Pearson's *chi*-square, denoted $\tilde{\chi}^2$.

$$\tilde{\chi}^2 = \sum_j^C \frac{\left(f_j - \mathbf{F}_j\right)^2}{\mathbf{F}_j}$$

where f_j is the *observed* frequency in Category j, $\mathbf{F}_j$ is the *expected* frequency in Category j, and $\tilde{\chi}^2$ is distributed approximately as $\chi^2(C - 1)$.

c. Rejection rule. We know that if the fit between the observed and expected distributions is perfect (i.e., if $f_j = \mathbf{F}_j$ for all C cells), then all of the squared differences $(f_j - \mathbf{F}_j)^2$ are zero. When corresponding observed and expected frequencies are not identical, these *squared* differences must be positive. Consequently the sum of the ratios $(f_j - \mathbf{F}_j)^2/\mathbf{F}_j$ increases as the differences between observed and expected frequencies increase, so the rejection region for Pearson's *chi*-square is always in the *upper* tail of $\chi^2(C - 1)$.

If we set α in advance, the test hypothesis is rejected if $\tilde{\chi}^2 \geq \chi^2_\alpha$, where $\mathbb{P}(\chi^2_\alpha) = 1 - \alpha$ in $\chi^2(C - 1)$. If the experimenter in Example 15.1 wants $\alpha = .05$, the critical value has a cumulative probability of .95 in the distribution of $\chi^2(2)$. In Table 6 of Appendix VIII we find that $\mathbb{P}(5.99) \doteq .95$, and since

$$\tilde{\chi}^2 = 26.23 > 5.99$$

the test hypothesis is rejected. Whether or not the Egyptians of Joseph's time were Hyksos, the distribution of blood types for mummies of Joseph's era does *not* "fit" the distribution for modern European Caucasians. As always, one can simply report the significance probability of a result instead of establishing the significance level of the test. In the present example we find that $\mathbb{P}(26.23) > .999$, so the experimenter would probably annotate his result $p < .001$.

Example 15.2. Earlier, we pointed out that one might test the fairness of a 6-sided die by rolling the die 120 times and testing the hypotheses

H_0: X uniformly distributed

H_1: X not uniformly distributed

The expected frequencies for this experiment are $\mathbf{F}_j = 20$ (see Table 15.1), and let us suppose that the experimenter rolls the die and obtains

Table 15.4. Observed frequencies for 120 rolls of a fair die

Value (x)	1	2	3	4	5	6
Observed Frequency (f)	16	15	29	28	18	14

the following observed frequencies:

Calculating Pearson's *chi*-square from formula **[15.2]**,

$$\tilde{\chi}^2 = \frac{(16-20)^2}{20} + \frac{(15-20)^2}{20} + \frac{(29-20)^2}{20}$$

$$+ \frac{(28-20)^2}{20} + \frac{(18-20)^2}{20} + \frac{(14-20)^2}{20}$$

$$= \frac{16}{20} + \frac{25}{20} + \frac{81}{20} + \frac{64}{20} + \frac{4}{20} + \frac{36}{20} = 11.3$$

There are six cells, so $\nu = 6 - 1 = 5$, and we find in the distribution of $\chi^2(5)$ that $\mathbb{P}(11.1) = .95$. Therefore, the significance probability of the result is $p < .05$, and the hypothesis is rejected for any $\alpha \geq .05$. Incidentally, the sample mean for these data is about 3.49. For a fair die the expected value is 3.50, so a test of means would not reveal the bias.

1. Goodness of fit as a test of proportions

In Chapter 9 we developed methods to test hypotheses about the binomial proportion. Since a binomial experiment can have only two results, the test hypothesis

$$H_0: p = p_0$$

implicitly specifies values for *two* proportions, the proportion of successes *and* the proportion of failures. The proportion of successes in a *sample* of N observations is X/N and the proportion of failures is $(N - X)/N$. If we denote the corresponding *population* proportions p_S and p_F, the binomial test hypothesis can be written

$$H_0: p_S = p_0 \quad \text{and} \quad p_F = 1 - p_0$$

Similarly, the experimenter in Example 15.1 might have formulated his hypotheses in terms of the *proportions* of blood types in the population of Egypt, circa 1700 B.C.

$$H_0: p_O = .484, \text{ and } p_A = .423, \text{ and } p_B = .093$$

The goodness-of-fit test can therefore be thought of as a generalization of the test of proportions from the binomial case to the *multinomial* case, that is, to the case where each observation can be assigned to any of $C > 2$ mutually exclusive categories. If a frequency distribution has only two categories, it should not be surprising that Pearson's *chi*-square yields the same result as a test of binomial proportions. More formally, for $C = 2$ categories,

$$\tilde{\chi}^2 = Z^2_{X/N}$$

where

$$Z_{X/N} = \frac{\dfrac{X}{N} - p_0}{\sqrt{\dfrac{p_0(1 - p_0)}{N}}}$$

is the statistic for testing hypotheses about the binomial proportion p using the normal approximation to the binomial (see Chapter 9).

Example 15.3. Consider the binomial hypothesis test embedded in Exercise 9.1.20(f) (p. 346): An experiment can have two results, A or B, and is to be repeated 50 times. The number of times A occurs is defined as X, and p denotes $P(A$ occurs$)$. The experimenter formulates the hypotheses

$$H_0\colon p = .50$$

$$H_1\colon p > .50$$

Suppose Event A occurs 31 times. Then,

$$z_{X/N} = \frac{\dfrac{31}{50} - .50}{\sqrt{\dfrac{(.50)(.50)}{50}}} = \frac{.12}{\sqrt{.005}}$$

and

$$z^2_{X/N} = \frac{(.12)^2}{.005} = 2.88$$

The same hypotheses can be tested using Pearson's *chi*-square. We know from Chapter 6 that if X is a binomial random variable, then $E(X) = Np$ and we also know that a binomial random variable is the number, or *frequency*, of successes. Therefore, $E(X)$ is the *expected* frequency of successes, that is, $\mathbf{F} = Np$. For $N = 50$ observations, the test hypothesis implies that the expected frequencies of Events 1 and 2 are

$$F_1 = Np_0 = (50)(.50) = 25$$

$$F_1 = N(1 - p_0) = (50)(.50) = 25$$

And since x is the number of times Event 1 is actually observed in N trials, $f_1 = x$ and $f_2 = N - x$:

$$f_1 = x = 31$$

$$f_2 = 50 - 31 = 19$$

Therefore,

$$\tilde{\chi}^2 = \sum_j^2 \frac{(f_j - F_j)^2}{F_j} = \frac{(31 - 25)^2}{25} + \frac{(19 - 25)^2}{25}$$

$$= 1.44 + 1.44$$

$$= 2.88$$

———— EXERCISES 15.1 ————

In Exercises 1 and 2, test the hypothesis that the expected frequency distribution is an appropriate model for the distribution of observed frequencies. (a) State the hypotheses, (b) write the expression for the appropriate test statistic, (c) indicate how the test statistic is distributed if the test hypothesis is correct, (d) calculate the value of the test statistic, and (e) report the *p*-value.

1.

	Category		
	1	**2**	**3**
Observed frequency (*f*)	5	15	20
Expected frequency (**F**)	10	20	10

2.

	Category				
	1	**2**	**3**	**4**	**5**
Observed frequency (*f*)	30	26	29	27	28
Expected frequency (**F**)	40	35	30	25	20

3. It is believed that the random variable X is distributed as follows:

x	1	2	3	4	5
$P(x)$	.178	.384	.311	.112	.015

An experimenter makes 600 observations and obtains the following distribution of scores:

x	1	2	3	4	5
f	94	209	199	86	12

Test the hypothesis that the probability distribution is an appropriate model for the distribution of X.

(a) What are the hypotheses?
(b) What is the appropriate test statistic?
(c) How is the test statistic distributed if the test hypothesis is correct?
(d) What are the expected frequencies?
(e) What is the value of the test statistic?
(f) What is the p-value of the test statistic?

4. It is believed that the random variable X is distributed as follows:

x	x_1	x_2	x_3	x_4	x_5	x_6	x_7
$P(x)$	.016	.094	.234	.312	.234	.094	.016

An experimenter makes 350 observations and obtains the following distribution of scores:

x	x_1	x_2	x_3	x_4	x_5	x_6	x_7
f	4	26	68	103	112	31	6

Test the hypothesis that the probability distribution is an appropriate model for the distribution of X.

(a) What are the hypotheses?
(b) What is the appropriate test statistic?
(c) How is the test statistic distributed if the test hypothesis is correct?
(d) What are the expected frequencies?
(e) What is the value of the test statistic?
(f) What is the p-value of the test statistic?

5. The following are the first 100 numbers in Table 1 of Appendix VIII. Test the hypothesis that Table 1 is a random sample of uniformly distributed random numbers from 0 to 99,999. For purposes of this exercise use 10 intervals of equal width.

10480	15011	01536	02011	81647
91646	69179	14194	62590	36207
20969	99570	91291	90700	22368
46573	25595	85393	30995	89198
27982	53402	93965	34095	52666
19174	39615	99505	24130	48360
22527	97265	76393	64809	15179
23830	49340	32081	30680	19655
63348	58629	42167	93093	06243
61680	07856	16376	39440	53537
71341	57004	00849	74917	97758
16379	37570	39975	81837	16656
06121	91782	60468	81305	49684
60672	14110	06927	01263	54613
77921	06907	11008	42751	27756
53498	18602	70659	90655	15053
21916	81825	44394	42880	99562
72905	56420	69994	98872	31016
71194	18738	44013	48840	63213
21069	10634	12952	96301	91997

(a) What are the hypotheses?
(b) What is the appropriate test statistic?
(c) How is the test statistic distributed if the test hypothesis is correct?
(d) Tabulate the observed frequencies.
(e) What are the expected frequencies?
(f) What is the value of the test statistic?
(g) What is the conclusion?

6. The random variable X can assume only the values 1, 2, 3, and 4, and an experimenter believes that the function $P(x) = .10x$ is a model for the distribution of X. A random sample of 50 observations yields the following results:

x	1	2	3	4
f	3	7	13	27

(a) What test hypothesis would the experimenter formulate?
(b) What is the alternative hypothesis?
(c) What is the appropriate test statistic?
(d) How is the test statistic distributed if the test hypothesis is correct?
(e) What are the expected frequencies?
(f) What is the value of the test statistic?
(g) What is the conclusion?

Many parts of Chrysler, Ford, and General Motors vehicles are manufactured by independent suppliers. Labor advocates claim that workers

in supplier plants are at greater risk of injury than workers in the Big Three plants, because suppliers are not unionized and almost half the supplier plants are small enough to be exempt from federal and state safety and health regulation. This claim was addressed by J. Daugherty, B. Shellum, and J. May in a *Detroit Free Press* feature series ("Workers at Risk," July 7–9, 1990) that was based in part on accident records obtained from the Michigan Labor Department's Bureau of Safety and Regulation. The exercises in this chapter that refer to the "Workers at Risk" study cite data that appeared in this series.

7. One claim discussed in the "Workers at Risk" study is that regulations are enforced less vigorously in smaller plants and that workers in smaller plants are therefore at greater risk than workers in larger plants. The study grouped 6,550 supplier plants into five size categories, with each category including approximately 64,000 workers. Since each category has roughly the same number of workers, frequency of injury should be approximately uniformly distributed across the five categories if risk is unrelated to plant size. Test this hypothesis using the following data.

Number of Workers	1–44 Group 1	45–124 Group 2	125–273 Group 3	274–672 Group 4	673–4,547 Group 5
Number of injuries (1986–1989*)	6,297	7,739	8,186	7,896	1,618

*Injuries for 1989 are seasonally adjusted estimates based on first 6 months.

(a) What are the hypotheses?
(b) What is the appropriate test statistic?
(c) How is the test statistic distributed if the test hypothesis is correct?
(d) Calculate the value of the test statistic.
(e) State the p-value of the test statistic.
(f) State the appropriate conclusion.
(g) Do the data support the claim that injury rates are higher in smaller plants than in larger plants?

8. Group 5 (673 to 4,547 workers) plants include a disproportionate number of white-collar workers who are not involved in high-risk activities. Test the hypothesis in Exercise 15.1.7 excluding plants in Group 5.
(a) What are the hypotheses?
(b) What is the appropriate test statistic?
(c) How is the test statistic distributed if the test hypothesis is correct?
(d) Calculate the value of the test statistic.
(e) State the p-value of the test statistic.
(f) State the appropriate conclusion.
(g) Do the data support the claim that injury rates are higher in smaller plants than larger plants?

9. The following exercise is quoted from Crow, J. F. (1960) *Genetics Notes* (4th ed.). Minneapolis: Burgess Publishing Co., p. 137.

In some matings involving different color phases of the mink, Shackelford [*Am. Naturalist* 83:49–67; 1949] found 53 dark, 17 platinum, 24 aleutian, and 8 sapphire offspring from a series of matings. He had theoretical reasons to suspect that there were two genes involved and that a $9:3:3:1$ ratio was to be expected. Were the results in agreement with this hypothesis?

(a) What are the hypotheses?
(b) What is the appropriate test statistic?
(c) How is the test statistic distributed if the test hypothesis is correct?
(d) Tabulate the observed and expected frequencies.
(e) What is the value of the test statistic?
(f) What is the conclusion?

10. A geneticist believes that gene frequencies for blood antigens M and N in a certain population yield proportions .314, .194, and .492 for blood types M, N, and MN, respectively. In a sample of 375 persons, 101 were blood type M, 86 were type N, and 188 were type MN. Test the hypothesis that the geneticist is correct.

(a) What are the hypotheses?
(b) What is the appropriate test statistic?
(c) How is the test statistic distributed if the test hypothesis is correct?
(d) What are the expected frequencies?
(e) What is the value of the test statistic?
(f) What is the conclusion?

11. The "Workers at Risk" study claims that industrial injuries are increasing in supplier plants. The data presented below are the number of injuries for supplier plants in Groups 1, 2, 3, and 4. Group 5 plants (673 to 4,547 workers) are excluded from the analysis because they include a disproportionate number of workers not involved in high-risk production jobs.

Number of injuries in supplier plants with 45 – 672 workers

Year	1986	1987	1988	1989*
Number of injuries	7,149	7,225	7,703	8,131

*Seasonally adjusted estimates based on first 6 months of 1989.

The number of workers was not reported by year, but if the number remained constant from year to year, the number of injuries should likewise remain constant if the injury rate did not change from year to year. Use the data to address the conclusion drawn by the "Workers at Risk" study.

(a) What are the hypotheses?
(b) What is the appropriate test statistic?
(c) How is the test statistic distributed if the test hypothesis is correct?
(d) Calculate the value of the test statistic.
(e) State the p-value of the test statistic.
(f) State the appropriate conclusion.
(g) Do the data support the claim that injury rates are increasing over time?

12. Auto industry critics claim that the Big Three are increasingly subcontracting their most dangerous jobs in order to reduce the cost of worker compensation. If the claim is correct, the number of injuries in Big Three plants should be declining over time. The data presented below are the number of injuries for Big Three plants.

Year	1986	1987	1988	1989*
Number of injuries	3,745	3,727	3,881	3,385

*Seasonally adjusted estimates based on first 6 months of 1989.

The number of workers was not reported by year, but if it is assumed that the number remained constant from year to year,[4] the number of injuries should likewise remain constant if the injury rate did not change from year to year. Use the data to address the concerns of auto industry critics.
(a) What are the hypotheses?
(b) What is the appropriate test statistic?
(c) How is the test statistic distributed if the test hypothesis is correct?
(d) Calculate the value of the test statistic.
(e) State the p-value of the test statistic.
(f) State the appropriate conclusion.
(g) Do the data support the claim that injury rates are decreasing over time?

13. Game animals taken by human hunters are generally considered to represent a random sample of the game species. The age distribution of zebra given below is estimated from a sample of 313 zebra shot by human hunters. Age categories range from 0–9 months (Category 1) to 12–20+ years (Category 8) and are based on tooth development and wear.

Age Category	1	2	3	4	5	6	7	8
$\hat{p}$*	.029	.077	.083	.073	.109	.291	.227	.112

*$\Sigma\hat{p} = 1.001$ due to rounding.

Wildlife biologists claim that predators eliminate the weakest specimens so, unlike human hunters, they preferentially take the very young and the very old. Test this hypothesis using the following distribution of zebra taken by lions.

Age Category	1	2	3	4	5	6	7	8
f	15	9	12	18	11	34	20	55

Source: Schaller, G. B. (1972). *The Serengeti Lion*. Chicago: University of Chicago Press, p. 440.

(a) What are the hypotheses?
(b) What is the appropriate test statistic?
(c) How is the test statistic distributed if the test hypothesis is correct?
(d) What are the expected frequencies?
(e) What is the value of the test statistic?
(f) What is the conclusion?

[4] There is good reason to suspect that the number of employees did not remain constant in the Big Three. Chrysler and GM sales declined, which was reflected in plant closings and layoffs. Ford sales increased during this period.

14. Recall from Chapter 6 that attributes (strength, wisdom, etc.) of *Advanced Dungeons & Dungeons*® characters are determined by rolling a 6-sided die three times, so that each attribute can have any integer value from 3 to 18. It is therefore important that dice be fair. One of the authors invested in a pair of professional casino dice advertised as approved by the Nevada Gaming Commission. After rolling up several characters with disappointingly weak attributes, he began to suspect that one of the dice might be biased toward low values, so he tested it with a run of 300 rolls. The results are given below.

Score	1	2	3	4	5	6
f	42	36	62	56	65	39

Test the hypothesis that the expected value for this particular die is equal to the value expected of a *fair die*.
(a) What is the test hypothesis?
(b) What is the alternative hypothesis?
(c) What is the value of the appropriate test statistic?
(d) What is the conclusion?

Now test the hypothesis that the die is evenly balanced.

(e) What are the hypotheses?
(f) What is the appropriate test statistic?
(g) How is the test statistic distributed if the test hypothesis is correct?
(h) What are the expected frequencies?
(i) What is the value of the test statistic?
(j) What is your conclusion?
(k) What do your answers to (d) and (j) tell you about the definition of a "fair" die?

15. The "Workers at Risk" study reports that in 1988, the last year for which complete records were available, the Big Three employed 308,696 workers and recorded 3,881 injuries. Injury figures for 1988 are given below for supplier plants grouped by size (number of workers).

Plant Size	1–44 (Group 1)	45–124 (Group 2)	125–273 (Group 3)	274–672 (Group 4)	673–4547 (Group 5)
Number of Injuries	1,581	1,947	2,022	2,153	381
Number* of Workers	64,161	64,261	64,155	64,032	64,482

*Annual average, 1986–1989.

Assume that 1988 figures constitute a representative sample of observations on the supplier plants and use the normal approximation to the binomial to test the claim that injuries are more frequent in supplier plants than in Big Three plants. Set α equal to any value for which there is a tabled cumulative probability in Appendix VIII.
(a) What are the hypotheses?
(b) What is the appropriate test statistic?
(c) How is the test statistic distributed if the test hypothesis is correct?

(d) Calculate the value of the test statistic.
(e) What is the conclusion?

Perform a goodness-of-fit test to test the same claim. Set α equal to any value for which there is a tabled cumulative probability in Appendix VIII.

(f) What are the hypotheses?
(g) What is the appropriate test statistic?
(h) How is the test statistic distributed if the test hypothesis is correct?
(i) Calculate the value of the test statistic.
(j) What is the conclusion?
(k) What do you note about (d) and (i)?

16. Group 5 plants (673 to 4,547 workers) include a disproportionate number of white-collar workers who are not involved in high-risk activities. Test the hypothesis in Exercise 15.15 excluding plants in Group 5.
(a) What are the hypotheses?
(b) What is the appropriate test statistic?
(c) How is the test statistic distributed if the test hypothesis is correct?
(d) Calculate the value of the test statistic.
(e) What is the conclusion?

Perform a goodness-of-fit test to test the same claim. Set α equal to any value for which there is a tabled cumulative probability in Appendix VIII.

(f) What are the hypotheses?
(g) What is the appropriate test statistic?
(h) How is the test statistic distributed if the test hypothesis is correct?
(i) Calculate the value of the test statistic.
(j) What is the conclusion?
(k) What do you note about (d) and (i)?

B. TESTING HYPOTHESES ABOUT TWO OR MORE DISTRIBUTIONS: TESTS OF ASSOCIATION

In the last section we saw that Pearson's *chi*-square may be used to determine whether or not a distribution of observed frequencies "fits" a distribution of expected frequencies that the experimenter can specify *before* data are collected. We pointed out that this is very similar to a test of one population mean, where the investigator specifies an hypothesized mean value μ_0, or a test of one population variance, where the investigator specifies an hypothesized value σ_0^2. We know, however, that a scientist will often formulate the test hypothesis that two population means are equal or that two population variances are equal *without* specifying values for μ_1 and μ_2 or for σ_1^2 and σ_2^2. In this section we introduce Pearson's *chi*-square methods for testing hypotheses about two or more frequency distributions, for which the experimenter has no a priori expected frequencies.

As in the case of testing the hypothesis that two parameters (e.g., μ_1 and μ_2) are equal, the experimenter tests the hypothesis that the two (or more) distributions exhibit the same profile of frequencies across two or more categories, which may represent numerical values, numerical intervals, or nominal classes. The frequencies in tests of association are arranged in R rows and C columns. In the discussion that follows, each row represents one distribution and, just as in goodness-of-fit tests, each column represents one category. The number of cells in a test of association is therefore RC, and each cell is identified by two subscripts. To avoid confusion we use the same subscript notation as in two-way analysis of variance. That is, the subscript j indicates the row ($j = 1, \ldots, R$), and the subscript k indicates the column ($k = 1, \ldots, C$). Accordingly, f_{jk} denotes the observed frequency in cell jk and the corresponding expected frequency is denoted $\mathbf{F}_{jk}$.

1. Tests of homogeneity

We have seen that a goodness-of-fit test can be thought of as testing the hypothesis that C proportions in *one* particular population assume the values,

$$p_1, p_2, \ldots, p_C$$

If an experimenter hypothesizes that the profile of proportions across C categories is the same in *two or more* well-defined populations, the hypothesis is said to test for *homogeneity* of proportions, or *homogeneity* of probability.

a. Hypotheses. In tests of homogeneity, each proportion is identified with double subscripts that identify its row and column. The distribution of proportions for R populations across C categories is given in general notation in Table 15.5.[5] The marginal row proportion $p_{j.}$ is the proportion in Population j (across all categories), and the marginal column proportion $p_{.k}$ is the proportion in Category k (across all populations).

Table 15.5. Proportions across C categories in R populations

	Categories						
Populations	1	2	$\cdots$	k	$\cdots$	C	Rows
1	p_{11}	p_{12}	$\cdots$	p_{1k}	$\cdots$	p_{1C}	$p_{1.}$
2	p_{21}	p_{22}	$\cdots$	p_{2k}	$\cdots$	p_{2C}	$p_{2.}$
$\vdots$				$\vdots$			$\vdots$
j	p_{j1}	p_{j2}	$\cdots$	p_{jk}	$\cdots$	p_{jC}	$p_{j.}$
$\vdots$				$\vdots$			$\vdots$
R	p_{R1}	p_{R2}	$\cdots$	p_{Rk}	$\cdots$	p_{RC}	$p_{R.}$
Columns	$p_{.1}$	$p_{.2}$	$\cdots$	$p_{.k}$	$\cdots$	$p_{.C}$	

[5] In this layout, each row represents the distribution for one population. This is consistent with the customary practice of displaying the values of a random variable as the columns of a frequency table or as points on the horizontal axis of a graph. However, if there are more populations than categories, the populations may be represented as columns and the categories as rows.

The test hypothesis for homogeneity of proportions states that proportions in Category 1 are identical across all R populations

$$p_{11} = p_{21} = p_{31} = \cdots = p_{R1}$$

and that the proportions in Category 2 are identical across all R populations

$$p_{12} = p_{22} = p_{32} = \cdots = p_{R2}$$

and that proportions are likewise identical across all R populations in Categories 3, 4, ..., and C:

$$p_{13} = p_{23} = p_{33} = \cdots = p_{R3}$$

$$p_{14} = p_{24} = p_{34} = \cdots = p_{R4}$$

$$\cdots$$

$$p_{1C} = p_{2C} = p_{3C} = \cdots = p_{RC}$$

Since all of the proportions for Category 1 are hypothesized to be equal, we use $p_{\cdot 1}$ to denote the *common* proportion for Category 1. Likewise, all of the proportions in column 2 are hypothesized to equal the same value, $p_{\cdot 2}$, and so on.[6] The test hypothesis can therefore be written

$$H_0: p_{1k} = p_{2k} = p_{3k} = \cdots = p_{Rk} = p_{\cdot k}$$

for $k = 1, 2, \ldots, C$, and the alternative hypothesis is

$$H_1: H_0 \text{ incorrect}$$

Example 15.4. In Example 15.1 we rejected the hypothesis that the proportions of blood types O, A, and B in mummies of the time of Joseph are the same as modern European Caucasians. Let's suppose that the experimenter realizes a little belatedly that the assumption on which he based his expected frequencies is flawed. The Hyksos of 1700 B.C. were a reproductively isolated people, and it is quite likely that their profile of O, A, and B blood types was quite different from that of any contemporary population. A more appropriate test, he reasons, would compare blood-type frequencies for mummies of Joseph's time with blood-type frequencies for mummies of Moses's time. If we use the subscripts O, A, and B as before to indicate blood type and, in addition, let the Egyptians of Joseph's era be Population 1 and Egyptians of Moses's era be Population 2, the experimenter's test hypothesis becomes:

$$H_0: p_{1O} = p_{2O} = p_O$$

$$p_{1A} = p_{2A} = p_A$$

$$p_{1B} = p_{2B} = p_B$$

---

[6] If subscripts are such that row and column indices cannot be confused (e.g., when the C categories are identified by alphabetic abbreviations), the row "placeholder" ($\cdot$) may be unnecessary.

where p_O, p_A, and p_B are the *unspecified* proportions of type O, A, and B blood common to both populations.

b. Test statistic. The basic strategy for testing hypotheses about homogeneity of proportions in R populations is the same as in testing hypotheses about proportions in one population. We want to compare observed frequencies with expected frequencies. In the test of homogeneity, however, we do not have a theoretical distribution of proportions

$$p_{.1}, p_{.2}, \ldots, p_{.C}$$

from which to calculate expected frequencies, so we use the observed frequencies to obtain *estimates* of these proportions,

$$\hat{p}_{.1}, \hat{p}_{.2}, \ldots, \hat{p}_{.C}$$

To see how this works, let us suppose our intrepid scientist obtains blood specimens from mummies of Moses's time (circa 1200 B.C.) and finds that 45 of them can be typed as O, A, or B as follows: $f_O = 26$, $f_A = 12$, $f_B = 7$. If we lay out these frequencies along with those from his earlier sample, we obtain the joint distribution in Table 15.6.

Table 15.6. Observed frequencies of blood types for 60 Egyptian mummies (circa 1700 B.C.) and 45 Egyptian mummies (circa 1200 B.C.)

Blood Type	O	A	B	Σ
Population 1 (Joseph)	20	23	17	60
Population 2 (Moses)	26	12	7	45
Σ		46	35	24	105

In addition to the observed frequencies and sample sizes (row totals), it will be noted that Table 15.6 includes the grand total ($N = 105$) and the C column totals. The total for the first column is the frequency of mummies with type O blood tallied across both samples and may therefore be denoted f_O:

$$\sum_{j}^{2} f_{jO} = f_O = 46$$

Likewise, for blood types A and B

$$\sum_{j}^{2} f_{jA} = f_A = 35$$

$$\sum_{j}^{2} f_{jB} = f_B = 24$$

If we divide each column total f_k by the grand total, we therefore obtain the *relative* frequency of every blood type.

$$\frac{f_O}{N} = \frac{46}{105} \doteq .4381$$

$$\frac{f_A}{N} = \frac{35}{105} \doteq .3333$$

$$\frac{f_B}{N} = \frac{24}{105} \doteq .2286$$

We know from Chapter 8 that relative frequency is an estimator of p, and the values f_k/N therefore estimate the proportions, p_O, p_A, and p_B that we should expect in both groups, if the two populations have identical blood-type profiles. That is,

$$\frac{f_O}{N} \doteq .4381 = \hat{p}_O$$

$$\frac{f_A}{N} \doteq .3333 = \hat{p}_A$$

$$\frac{f_B}{N} \doteq .2286 = \hat{p}_B$$

In the goodness-of-fit test we calculated expected frequency for Category k by multiplying the number of observations N by the theoretical proportion p_k,

$$\mathbf{F}_k = Np_k$$

For the test of homogeneity, we may similarly obtain the expected frequency for Category k in Population j by multiplying the number of observations from Population j by the *estimated* common proportion for Category k:

$$\mathbf{F}_{jk} = N_j \hat{p}_k \qquad\qquad [15.3]$$

Or,

$$\mathbf{F}_{jk} = N_j \left(\frac{f_k}{N} \right) \qquad\qquad [15.4]$$

For $N_1 = 60$ mummies of Joseph's time,

Type O: $\mathbf{F}_{1O} = N_1 \hat{p}_O = 60(.4381) \doteq 26.3$

Type A: $\mathbf{F}_{1A} = N_1 \hat{p}_A = 60(.3333) \doteq 20.0$

Type B: $\mathbf{F}_{1B} = N_1 \hat{p}_B = 60(.2286) \doteq 13.7$

and for $N_2 = 45$ mummies of Moses's time,

Type O: $\mathbf{F}_{O2} = N_2 \hat{p}_O = 45(.4381) \doteq 19.7$

Type A: $\mathbf{F}_{2A} = N_2 \hat{p}_A = 45(.3333) \doteq 15.0$

Type B: $\mathbf{F}_{2B} = N_2 \hat{p}_B = 45(.2286) \doteq 10.3$

These expected frequencies appear in Table 15.7.

Table 15.7. Expected frequencies of blood types for 60 Egyptian mummies (circa 1700 B.C.) and 45 Egyptian mummies (circa 1200 B.C.)

Blood Type	O	A	B	Σ
Population 1 (Joseph)	26.3	20.0	13.7	60
Population 2 (Moses)	19.7	15.0	10.3	45
Σ		46	35	24	105

As in the goodness-of-fit test, the test statistic is the sum of the ratios $(f - \mathbf{F})^2/\mathbf{F}$ calculated over all cells. In the test of homogeneity, however, we have R rows and C columns, so the formula for Pearson's *chi*-square is ordinarily written using double summation notation, which we introduced in Chapter 12.

$$\sum_{j}^{R} \sum_{k}^{C} \frac{\left(f_{jk} - \mathbf{F}_{jk}\right)^2}{\mathbf{F}_{jk}} \qquad [15.5]$$

For Example 15.4,

$$\tilde{\chi}^2 = \sum_{j}^{2} \sum_{k}^{3} \frac{\left(f_{jk} - \mathbf{F}_{jk}\right)^2}{\mathbf{F}_{jk}}$$

$$= \frac{(20 - 26.3)^2}{26.3} + \frac{(23 - 20.0)^2}{20.0} + \frac{(17 - 13.7)^2}{13.7}$$

$$+ \frac{(26 - 19.7)^2}{19.7} + \frac{(12 - 15.0)^2}{15.0} + \frac{(7 - 10.3)^2}{10.3}$$

$$= \frac{39.69}{26.3} + \frac{9}{20} + \frac{10.89}{13.7} + \frac{39.69}{19.7} + \frac{9}{15} + \frac{10.89}{10.3} \doteq 6.43$$

Practical shortcuts for calculating expected frequencies. All six of the expected frequencies in Table 15.7 were calculated by formula [15.3], but in this example only two of the expected frequencies had to be calculated in this way. The number of mummies from Joseph's time is 60, so the number with type B blood must be 60 minus the number with type O and type A blood:

$$\mathbf{F}_{1B} \doteq 60 - (26.3 + 20) = 13.7$$

which is exactly what we obtained by formula **[15.3]**,

$$N_1 \hat{p}_B \doteq 60(.2286) = 13.7$$

Similarly, we used formula **[15.3]** to calculate the expected frequencies for Moses's time, but these three values are also determined. Our estimate

$$\hat{p}_O \doteq .4381$$

of the common proportion of type O mummies was obtained by dividing the type O total $f_O = 46$ by the grand total 105. No matter how we elect to shuffle these 46 between our two populations, the total must consequently remain equal to 46. Therefore, the number of type O mummies from Moses's time must equal 46 minus the number of type O mummies from Joseph's time. That is,

$$\mathbf{F}_{2O} = 46 - \mathbf{F}_{1O} \doteq 46 - 26.3 = 19.7$$

Likewise, there are 35 mummies with type A blood, so

$$\mathbf{F}_{2A} = 35 - \mathbf{F}_{1A} \doteq 35 - 20 = 15$$

and there are 24 type B mummies, so

$$\mathbf{F}_{2B} = 24 - \mathbf{F}_{1B} \doteq 24 - 13.7 = 10.3$$

In all of the examples we have discussed, populations are represented as rows and categories as columns. The logic of the homogeneity test therefore led us to calculate expected frequencies by multiplying the number of individuals in row j by the proportion of observations in column k, as shown in equation **[15.4]**:

$$\mathbf{F}_{jk} = N_j \left(\frac{f_k}{N} \right)$$

We pointed out earlier, however, that when there are more populations than categories, data are often tabulated with populations represented as columns and categories as rows. Logically, the expected frequency $\mathbf{F}_{jk}$ is therefore obtained by multiplying the number of individuals in column k

by the proportion of observations in row j:

$$\mathbf{F}_{jk} = N_k \left(\frac{f_j}{N} \right)$$

Happily, it is not necessary to remember both of these formulae; they are algebraically identical. Since N_j is the number, or *frequency*, of observations in row j, it can be denoted f_j. Therefore,

$$N_j \left(\frac{f_k}{N} \right) = f_j \left(\frac{f_k}{N} \right) = \frac{f_j f_k}{N}$$

Similarly, N_k can be denoted f_k, so

$$N_k \left(\frac{f_j}{N} \right) = f_k \left(\frac{f_j}{N} \right) = \frac{f_j f_k}{N}$$

Whether populations are represented as rows or as columns, the expected frequency $\mathbf{F}_{jk}$ equals

$$\frac{f_j f_k}{N}$$

BOX 15.2

Calculating expected frequencies for tests of association

The expected frequency in cell jk is the product of the observed frequency in row j multiplied by the observed frequency in column k divided by the total number of observations.

$$\mathbf{F}_{jk} = \frac{f_j f_k}{N}$$

Degrees of freedom. Degrees of freedom for Pearson's *chi*-square test of homogeneity has exactly the same meaning as in the goodness-of-fit test: Degrees of freedom is the number of "free" cells, i.e., the number of cells to which the experimenter can assign any expected frequency.

In the last section we discovered that it was unnecessary to compute all six expected frequencies in Example 15.4 by formula [15.3]. Once the expected frequencies were calculated for two blood types in mummies from Joseph's era, all of the remaining expected frequencies were determined. The last cell in the first row was *constrained* by the observed row total (the number of mummies from Joseph's time), and all of the expected frequencies in the second row were *constrained* by the observed column totals (the total number of mummies in each blood group). In this example, then, two of the six cells were "free" to assume values assigned by the experimenter.

In the general case of testing homogeneity of proportions across R populations, it is similarly true that every row *except the last row* has $C - 1$ free cells and that the last row has *no* free cells. Therefore, the number of free cells is $(R - 1)(C - 1)$, and Pearson's *chi*-square statistic for testing homogeneity of proportions is (approximately) distributed as χ^2 with $(R - 1)(C - 1)$ degrees of freedom. In Example 15.4, $R = 2$ and $C = 3$, so

$$\nu = (R - 1)(C - 1) = (1)(2) = 2$$

which agrees with the number of free cells we found.

The hypothesis that R population distributions are identically distributed across C categories can be tested using Pearson's *chi*-square,

$$\tilde{\chi}^2 = \sum_j^R \sum_k^C \frac{\left(f_{jk} - \mathbf{F}_{jk}\right)^2}{\mathbf{F}_{jk}}$$

where f_{jk} is the *observed* frequency in Population j and Category k and $\mathbf{F}_{jk}$ is the *expected* frequency in Population j and Category k, and $\tilde{\chi}^2$ is distributed approximately as χ^2 with $(R - 1)(C - 1)$ degrees of freedom.

---

c. Rejection rule. As before, $\tilde{\chi}^2$ is zero if $f_{jk} = \mathbf{F}_{jk}$ in all RC cells and increases as the difference between corresponding observed and expected frequencies increases. The rejection region for a Pearson's *chi*-square test of homogeneity is, therefore, always in the upper tail of $\chi^2([R - 1][C - 1])$. In Example 14.4 we calculated $\tilde{\chi}^2$ to be 6.43, and in Table 6 of Appendix VIII we find that the cumulative probability of 6.43 falls between .950 and .975 in $\chi^2(2)$. The hypothesis of homogeneity is therefore rejected for any $\alpha \geq .05$, which supports the argument that the Egyptians of Joseph's era were a different people from Egyptians of Moses's era.[7]

---

[7]Ascertaining blood-type profiles for ancient populations is an uncertain enterprise, made all the more difficult by the possible instability of type A antigens. In our examples, the profiles for mummies of Joseph's era were based on blood-type gene frequencies for modern Egypt, especially the Nile Valley, and northern Transcaucasia. These profiles are quite distinct from profiles for immediately adjacent regions and surprisingly similar to one another. Profiles for mummies of Moses's era were based on blood-type gene frequencies for north African populations west of Egypt. Source: Mourant, A. E., et al. (1983). *Blood Relations*. New York: Oxford University Press, 133–135.

2. Tests of independence

In tests of homogeneity, the experimenter draws samples of $N_1, N_2, \ldots, N_R$ observations from $R \geq 2$ *well-defined populations*. In other situations, the experimenter begins with a sample of N observations drawn from *one* population and then classifies the N subjects according to their status on *two* variables, X and Y, where X has R mutually exclusive categories, and Y has C mutually exclusive categories. The experimenter in this situation is typically concerned with whether or not the two variables are *independent*.

In Chapter 14 we discussed independence of random variables in the context of Pearson's correlation, ρ, but techniques for testing hypotheses about ρ can be used only with bivariate normally distributed random variables. These techniques are necessarily limited to ratio- or interval-scaled random variables. The statistic we used to test hypotheses about homogeneity [15.5] can be used to test hypotheses about independence for variables at any level of measurement, so long as the variables can be partitioned into mutually exclusive (and therefore discrete) categories.

Example 15.5. We began this chapter with the simple example of an experimenter who hypothesizes that a particular six-sided die is fair and that rolls of the die should therefore be uniformly distributed. Let us suppose that the experimenter is a player of *Advanced Dungeons & Dragons*® and that he believes that he exerts psychokinetic influence on his dice. More particularly, he believes that as he becomes more fatigued, his rolls become worse and worse—that is, he obtains lower and lower scores on his dice. He decides to test his hypothesis by recording the scores he rolls during his first hour of play, during his third hour of play, and during his fifth hour of play.[8] If X represents the period in which the scores were recorded (first hour, third hour, fifth hour) and Y represents the scores on the die, the joint frequency distribution of his data might look like Table 15.8.[9]

Table 15.8. Joint and marginal frequencies of dice rolls in three periods of *AD & D* play

Period of play (X)	Score on die (Y)						$\Sigma = f_j$
	1	2	3	4	5	6	
1 (first hour)	5	9	14	12	11	16	67
2 (third hour)	16	14	19	21	9	14	93
3 (fifth hour)	18	12	19	15	14	8	86
$\Sigma = f_k$	39	35	52	48	34	38	$\Sigma\Sigma = 246$

[8] A good *AD & D* session can run 8 or 10 hours, and players become very, very tired.
[9] In the context of Pearson's *chi*-square, a table representing the classification of observations in terms of two categorical variables is often called a *contingency table*.

a. Test hypothesis. Recall from Chapter 14 that if two discrete random variables X and Y are independent, then the joint probability $P(x_j, x_k)$ equals the product of the corresponding marginal probabilities

$$P(x_j, y_k) = P(x_j)P(y_k)$$

for all pairs (x_j, y_k). Accordingly, the hypotheses for Pearson's chi-square test of independence are

$$H_0: P(x_j, y_k) = P(x_j)P(y_k)$$

for all (x_j, y_k) and

$$H_1: P(x_j, y_k) \neq P(x_j)P(y_k)$$

for some (x_j, y_k).

b. Test statistic. We have already said that the test statistic for Pearson's chi-square test of independence is the same as for Pearson's chi-square test of homogeneity:

If X is partitioned into R mutually exclusive categories and Y is partitioned into C mutually exclusive categories, the independence of X and Y can be tested using Pearson's chi-square,

$$\tilde{\chi}^2 = \sum_j^R \sum_k^C \frac{\left(f_{jk} - \mathbf{F}_{jk}\right)^2}{\mathbf{F}_{jk}}$$

where f_{jk} and $\mathbf{F}_{jk}$ are, respectively, the observed and expected joint frequencies in Category j of X and Category k of Y, and $\tilde{\chi}^2$ is distributed approximately as χ^2 with $(R - 1)(C - 1)$ degrees of freedom.

However, the logic by which we obtained expected frequencies is somewhat different.

The probability that any observation falls in cell jk of the joint frequency distribution is $P(x_j, y_k)$, and if the test hypothesis is correct $P(x_j, y_k)$ equals $P(x_j)P(y_k)$. We do not know the values $P(x_j)$ and $P(y_k)$, but we know that relative frequency is a consistent and unbiased estimator of probability. Therefore, the marginal frequency for row j divided by the grand total estimates $P(x_j)$:

$$\frac{\sum_k^C f_{jk}}{N} = \frac{f_j}{N} = \hat{P}(x_j)$$

In Example 15.5, the estimated probability of rolling a die during the second period of play is

$$\hat{P}(x = 2) = \frac{93}{246} \doteq .378$$

Similarly, the marginal frequency for column k divided by the grand total (N) estimates $P(y_k)$:

$$\frac{\sum\limits_{j}^{R} f_{jk}}{N} = \frac{f_k}{N} = \hat{P}(y_k)$$

From the data in Table 15.8, we can estimate the probability of rolling a 3 by dividing the number of times that 3 was rolled during all three gaming periods by the total number of rolls:

$$\hat{P}(y = 3) = \frac{52}{246} \doteq .211$$

Since X and Y are hypothesized to be independent, we *estimate* $P(x_j, y_k)$, the probability that an observation falls in cell jk, as

$$\hat{P}(x_j, y_k) = \hat{P}(x_j)\hat{P}(y_k) = \left(\frac{f_j}{N}\right)\left(\frac{f_k}{N}\right)$$

If die-rolling results are independent of the length of time our experimenter plays, the probability of rolling a 3 during the second period of play is therefore estimated to be

$$(.378)(.211) \doteq .080$$

In the test for homogeneity, we obtained the expected frequency for cell jk by multiplying each row total N_j by the estimated proportion $\hat{p}_k$. In tests of independence, we multiply the grand total N by the estimated joint probability $\hat{P}(x_j, y_k)$:

$$\mathbf{F}_{jk} = N\hat{P}(x_j, y_k) \qquad\qquad [15.6]$$

that is,

$$\mathbf{F}_{jk} = N\left(\frac{f_j}{N}\right)\left(\frac{f_k}{N}\right) \qquad\qquad [15.7]$$

From equation [15.6], the frequency with which our *AD & D*® player expects to roll a 3 in the second period of play is

$$246(.080) \doteq 19.7$$

(If the remaining expected frequencies are calculated, the reader can confirm that the value of Pearson's *chi*-square from [15.5] is approximately 12.48. From Table 6 in Appendix VIII, $\mathbb{P}(12.48) < .75$ in the distribution of $\chi^2(10)$, so $p > .25$ and the alternative hypothesis should probably be rejected. The performance of the dice appears to be independent of the player's fatigue.)

Although the logic by which we obtained the expected frequencies for our test of independence led us through calculations that differ somewhat from those used in tests of homogeneity, it is easily seen they are computationally identical. In our discussion of computational shortcuts (pp. 640–642) we showed that for the test of homogeneity, the expected frequency is

$$\mathbf{F}_{jk} = N_k\left(\frac{f_j}{N}\right) = \frac{f_j f_k}{N}$$

Likewise, from equation [15.7],

$$\mathbf{F}_{jk} = N\left(\frac{f_j}{N}\right)\left(\frac{f_k}{N}\right) = \frac{f_j f_k}{N}$$

3. The fourfold table

Many variables of interest to scientists are binomial, so the test of association for two binomial random variables is very common. This may be a test of homogeneity, as in testing the hypothesis that the proportions of registered Democrats and registered Republicans are the same for men and for women. Or it may be a test of independence, as in testing the hypothesis that passing a course in statistics is independent of passing a concurrent course in English. Whether the test hypothesis is formulated in terms of homogeneity or in terms of independence, each variable is binomial and each is therefore represented by two categories (success and failure). Thus, $R = C = 2$, and the observed frequencies can be arranged in a 2×2 table, or a *fourfold* table, as illustrated in Table 15.9:

Table 15.9. Fourfold table of observed frequencies

	Variable (Population) 1		
Variable (Population) 2	$f_{11} = a$	$f_{12} = b$	$a+b$
	$f_{21} = c$	$f_{22} = d$	$c+d$
	$a+c$	$b+d$	

With only four cells, it is more convenient to dispense with subscripts, so observed frequencies f_{11}, f_{12}, f_{21}, and f_{22} are respectively denoted a, b, c, and d, as illustrated in Table 15.9. Using this notational convention, it can be shown by routine, but tedious, algebra that formula [15.5] is equivalent to

$$\tilde{\chi}^2 = \frac{N(ad - bc)^2}{(a+b)(c+d)(a+c)(b+d)} \tag{15.8}$$

where $N = a + b + c + d$, the total number of observations.[10]

Example 15.6. Several years ago, Frank and Wilcox[11] developed an entry-level test for firefighters. The examination was administered to 29 white applicants and 37 minority applicants. Seven of the white applicants failed, and 17 of the minority applicants failed. The disproportionate number of minority applicants who failed the test led to a claim in federal court that the examination constituted an "artificial barrier" to minority employment. Test the hypothesis that proportions of applicants who passed and failed the examination were the same for minority and for white applicants. The alternative hypothesis, which supports the legal claim, is that the proportions are different in the two populations.

	White	Minority	Σ
Pass	$a = 22$	$b = 20$	42
Fail	$c = 7$	$d = 17$	24
Σ	29	37	

By formula [15.8]

$$\tilde{\chi}^2 = \frac{N(ad - bc)^2}{(a + b)(c + d)(a + c)(b + d)}$$

$$= \frac{66(374 - 140)^2}{(42)(24)(29)(37)} = \frac{66(234)^2}{1{,}081{,}584}$$

$$\doteq 3.34$$

In Table 6 of Appendix VIII we find $.90 < \mathbb{P}(3.34) < .95$ in $\chi^2(1)$, so $.05 < p < .10$, and the (alternative) hypothesis that the test is biased against minority applicants is rejected for any $\alpha \le .05$.

---

[10] Like the normal approximation to the binomial, Pearson's *chi*-square approximates discrete multinomial distributions with the continuous *chi*-square distributions. Many older texts therefore recommend using the formula

$$\tilde{\chi}^2 = \frac{N(|ad - bc| - N/2)^2}{(a + b)(c + d)(a + c)(b + d)}$$

instead of [15.8] with fourfold tables, in order to correct for continuity. However, several authors have determined *exact* distributions of Pearson's *chi*-square using both algorithms and have concluded that formula [15.8] usually yields a better approximation to $\chi^2(1)$, especially for small samples. For example, see Grizzle, J. E. (1967). Continuity correction in the $\tilde{\chi}^2$-test for 2 × 2 tables. *The American Statistician*, October, 28–31.

[11] Frank, H., and Wilcox, C. (1978). Development and preliminary cross-validation of a two-step procedure for firefighter selection, *Psychological Reports*, 43, 27–36.

—————— EXERCISES 15.2 ——————————————————————————

1. Test the hypothesis that the proportions of individual in Categories 1, 2, 3, and 4 are the same in Populations 1 and 2.

Population	Category 1	2	3	4
1	31	40	17	3
2	16	35	22	8

   (a) What are the hypotheses?
   (b) What is the appropriate test statistic?
   (c) How is the test statistic distributed if the test hypothesis is correct?
   (d) What are the expected frequencies?
   (e) What is the value of the test statistic?
   (f) What is the p-value of the test statistic?
   (g) What is the appropriate conclusion?

2. Test the hypothesis that the proportions of individuals in Categories 1, 2, and 3 are the same in Populations 1, 2, and 3.

Population	Category 1	2	3
1	8	66	76
2	7	31	20
3	10	50	20

   (a) What are the hypotheses?
   (b) What is the appropriate test statistic?
   (c) How is the test statistic distributed if the test hypothesis is correct?
   (d) What are the expected frequencies?
   (e) What is the value of the test statistic?
   (f) What is the p-value of the test statistic?
   (g) What is the appropriate conclusion?

3. Three hundred and twenty persons are classified according to their status on variable X and on variable Y. Use the observed frequencies given below to test the hypothesis that X and Y are independent.

	y_1	y_2	y_3	y_4
x_1	2	28	34	16
x_2	15	39	76	30
x_3	3	25	30	22

   (a) What are the hypotheses?
   (b) What is the appropriate test statistic?
   (c) How is the test statistic distributed if the test hypothesis is correct?
   (d) What are the expected frequencies?

    (e) What is the value of the test statistic?

    (f) What is the *p*-value of the test statistic?

    (g) What is the appropriate conclusion?

4. One hundred and fifty persons are classified according to their status on variable X and on variable Y. Use the observed frequencies given below to test the hypothesis that X and Y are independent.

	y_1	y_2	y_3
x_1	2	4	24
x_2	7	13	25
x_3	21	13	41

    (a) What are the hypotheses?

    (b) What is the appropriate test statistic?

    (c) How is the test statistic distributed if the test hypothesis is correct?

    (d) What are the expected frequencies?

    (e) What is the value of the test statistic?

    (f) What is the *p*-value of the test statistic?

    (g) What is the appropriate conclusion?

5. Scott and Fuller's (1965) *Genetics and the Social Behavior of the Dog* reports results for a series of manipulation tasks administered to large samples of several breeds of domestic dog. The frequency distribution for scores obtained by 28 cocker spaniels and 28 basenjis are given below.

Number correct	0	1	2	3	4	5	6	7	8
Cocker spaniels	11	6	3	2	4	0	2	0	0
Basenjis	1	1	1	2	1	3	1	6	12

Source: Data furnished by J. P. Scott.

Because they are behavioral geneticists, Scott and Fuller were not especially concerned about breed differences in the *number* of correct solutions. They were concerned about differences in *form* or shape of the distributions, because this can provide important clues about the underlying genetic mechanism of a trait. In the next section we will see that use of Pearson's *chi*-square when $v > 1$ assumes that $\mathbf{F}_{jk} \geq 5$ in 80 percent of the cells. To satisfy this assumption, collapse the data into three intervals: poor (0 to 2 correct), moderate (3 to 5 correct), and good (6 to 8 correct), and test the hypothesis that ability to perform complex manipulation tasks is distributed no differently in the gene pools of cockers and basenjis.

    (a) What is the test hypothesis?

    (b) What is the appropriate test statistic?

    (c) How is the test statistic distributed if the test hypothesis is correct?

    (d) What are the observed frequencies?

    (e) What are the expected frequencies?

    (f) What is the value of the test statistic?

    (g) What is the *p*-value of the test statistic?

(h) Frank and Frank administered the same manipulation tasks to four wolf pups and four Alaskan Malamute pups to test the hypothesis that wolves perform complex manipulation tasks better than dogs. Performance differences between the two groups were tested using Student's t, and the editor of *Animal Behaviour* rejected the paper on the grounds that there was no evidence that scores were normally distributed. Would the editor have turned down a similar paper by Scott and Fuller based on their data? Why or why not?

6. Frequencies of good (6 to 8 correct) and poor (0 to 2 correct) performance for all six breeds of dog tested by Scott and Fuller (see Exercise 15.2.5) are given below.

	Breed				
Performance	**Basenjis**	**Beagles**	**Cockers**	**Shelties**	**Wirehaired Terriers**
Good	19	8	2	8	7
Poor	3	11	20	13	11

Test the hypothesis that "good" and "poor" performance are homogeneously distributed across all five breeds.
(a) What is the test hypothesis?
(b) Write the formula for the appropriate test statistic.
(c) How is the test statistic distributed if the test hypothesis is correct?
(d) What are the expected frequencies?
(e) What is the value of the test statistic?
(f) What is the p-value of the test statistic?

7. A poll conducted by a nationally known research organization asked 1,500 Americans whether or not they approved of increasing taxes to reduce the national deficit. Each respondent was also asked to indicate his or her political affiliation. The entries in the table below are *proportions* of the entire sample of 1,500:

	Response	
Political affiliation	**Approve**	**Disapprove**
Democrats	.26	.09
Republicans	.16	.24
Independents	.15	.10

Test the hypothesis that approval and disapproval are homogeneously distributed across political preference. Hint: sample proportion = f/N, so $f = N$(sample proportion).
(a) What is the test hypothesis?
(b) Write the formula for the appropriate test statistic.
(c) How is the test statistic distributed if the test hypothesis is correct?
(d) What are the observed frequencies?
(e) What are the expected frequencies?
(f) What is the value of the test statistic?
(g) What is the p-value of the test statistic?

8. The following distributions are grades that were assigned in fall and spring sections of the same developmental psychology course using the same grading scale.

	Grade				
Semester	*A*	*B*	*C*	*D*	*E*
Fall	9	9	18	9	6
Spring	6	10	13	4	4

Test the hypothesis that grades are distributed the same way in fall and spring.

(a) What is the test hypothesis?
(b) Write the formula for the appropriate test statistic.
(c) How is the test statistic distributed if the test hypothesis is correct?
(d) What are the expected frequencies?
(e) What is the value of the test statistic?
(f) What is the *p*-value of the test statistic?

9. A former student of ours conducted an experiment studying the effects of population density on alcohol consumption in rats. Her "inner-city" group lived in overcrowded conditions, and her "suburban" group had a relatively comfortable amount of living space. Each animal had two bottles in its cage. One contained pure water and the other was infused with 80-proof vodka. Each day she recorded the amount drunk from each bottle, and after three weeks she computed the percentage of alcohol in each animal's total intake. Her distribution for the two groups is given below.

	Percent of alcohol in liquid intake					
	1.5 or less	1.5–4.5	4.5–6.5	6.5–8.5	8.5–10.5	10.5 or more
Inner city	4	10	16	12	6	2
Suburban	9	13	8	6	2	2

Test the hypothesis ($\alpha = .05$) that alcohol consumption is related to crowding.
(a) What are the hypotheses?
(b) What is the appropriate test statistic?
(c) How is the test statistic distributed if the test hypothesis is correct?
(d) What are the expected frequencies?
(e) What is the value of the test statistic?
(f) What is the *p*-value of the test statistic?
(g) What is the appropriate conclusion?

10. Prey animals killed by lions in the Serengeti were classified according to species and principal habitat where the kill was recorded.

Type of habitat	Prey species		
	Wildebeest	**Zebra**	**Thomson's gazelle**
Plains	73	42	5
Woodland	99	71	21
Mixed/riverine (Seronera region)	112	83	232

Source: Schaller, G. B. (1972) *The Serengeti Lion*. Chicago: University of Chicago Press, p. 434.

Test the hypothesis that habitat type is independent of prey species.

(a) What are the hypotheses?
(b) What is the appropriate test statistic?
(c) How is the test statistic distributed if the test hypothesis is correct?
(d) What are the expected frequencies?
(e) What is the value of the test statistic?
(f) What is the p-value of the test statistic?
(g) What is the appropriate conclusion?

11. As mentioned in Exercise 15.1.7, the "Workers at Risk" study claims that smaller plants receive less aggressive enforcement of occupational safety and health requirements than larger plants. Accordingly, it is also claimed that the injury rate in smaller supplier plants is increasing faster than in larger supplier plants. Use the data presented below to test the hypothesis that the number of injuries in smaller plants and larger plants exhibited the same profile of changes in years 1986 to 1989. For purposes of this analysis, "small" supplier plants are those with 45 to 124 workers (Groups 1 and 2) and "large" supplier plants are those with 125 to 672 workers (Groups 3 and 4).

Plant Size	Year			
	1986	**1987**	**1988**	**1989***
Small	3,088	3,313	3,528	4,107
Large	4,061	3,912	4,175	4,024

*Seasonally adjusted estimates based on first 6 months of 1989.

(a) What are the hypotheses?
(b) What is the appropriate test statistic?
(c) How is the test statistic distributed if the test hypothesis is correct?
(d) Calculate the value of the test statistic.
(e) State the p-value of the test statistic.
(f) State the appropriate conclusion.
(g) Does the conclusion address the claim that injury rates are increasing faster in small plants than in large plants?

12. In Exercise 15.1.12 the hypothesis was tested that injury rates remain consistent over time in Big Three plants. The test assumed that number of employees remained approximately equal to the 1986–1989 average reported in the "Workers at Risk" study over the four years. However, sales of

Chrysler and General Motors products declined during this period, and both companies closed a number of plants and laid off thousands of workers. Ford sales increased, which is typically accompanied by an increase in the work force. Changes in the numbers of workers should be reflected in changes in numbers of injuries. Use the following data to test the hypothesis that injury rates differed from year to year among the Big Three automobile companies.

Company	Year			
	1986	**1987**	**1988**	**1989***
Chrysler	635	762	721	548
Ford	465	562	832	893
GM	2,645	2,403	2,328	1,944

*Seasonally adjusted estimates based on first 6 months of 1989.

(a) What are the hypotheses?
(b) What is the appropriate test statistic?
(c) How is the test statistic distributed if the test hypothesis is correct?
(d) Calculate the value of the test statistic.
(e) State the *p*-value of the test statistic.
(f) State the appropriate conclusion.
(g) What is the most plausible interpretation of (f)?
(h) What bearing does your answer to (g) have on the test conducted in Exercise 15.1.12?

13. Male lions are much larger than females, and males make more frequent solitary kills than females. For some predators the size of preferred prey depends on the typical size of the predator; for other predators the size of the preferred prey depends on the number of individuals in the hunting group. In the table below, 231 kills are grouped according to the sex of the lion(s) observed feeding and the size of the prey species. Topi-hartebeest, wildebeest, and zebra are classified as medium-sized prey, eland, buffalo, and giraffe are classified as large.

Sex of lion(s)	Prey species size	
	Medium	**Large**
Females only	152	6
Males only	62	11

Source: Schaller, G. B. (1972), ibid., p. 44.

Test the hypothesis that prey size is associated with predator size.

(a) What is the appropriate test statistic?
(b) How is the test statistic distributed if the test hypothesis is correct?
(c) Calculate the value of the test statistic.
(d) State the *p*-value of the test statistic.
(e) State the appropriate conclusion.

14. Many animal species have evolved forms of ritualized aggressive display that permit resolution of conflict without physical contact that can reduce the likelihood of survival. However, if the evolutionary concept of "fitness' is interpreted as reproductive success rather than individual survival, the likelihood of serious (contact) combat should be greater in situations where the reproductive payoff is large than when the reproductive payoff is small. Watson (1991) observed combat between male sierra dome spiders on webs of inseminated females and on webs of nearly mature virgin females. Females that are already inseminated will bear fewer offspring from a second insemination, so we might predict that males will fight more vigorously for access to a virgin female than for access to a mated female.

	Female status	
Fight intensity	**Virgin**	**Mated**
Display only	2	9
Contact fighting	14	7

Source: Watson, P. J. (1991). Multiple paternity as genetic bet-hedging in female sierra dome spiders, *Linyphia litigiosa* (Linyphiidae). *Animal Behaviour*, *41*, 343–360.

Test the hypothesis that fights between males are more intense in the presence of virgin females than mated females.

(a) What is the appropriate test statistic?
(b) How is the test statistic distributed if the test hypothesis is correct?
(c) Calculate the value of the test statistic.
(d) State the *p*-value of the test statistic.
(e) State the appropriate conclusion.

15. One of the high-risk occupations highlighted in the "Workers at Risk" study is stamping press operator. The data given below show a downward trend in stamping press injuries in the Big Three and an upward trend in supplier plants. Test the appropriate hypothesis to ascertain whether or not these trends are statistically significant.

Year	1986	1987	1988	1989*
Big Three	165	134	119	91
Suppliers	684	685	718	722

*Seasonally adjusted estimates based on first 6 months of 1989.

(a) What are the hypotheses?
(b) What is the appropriate test statistic?
(c) How is the test statistic distributed if the test hypothesis is correct?
(d) Calculate the value of the test statistic.
(e) State the *p*-value of the test statistic.
(f) State the appropriate conclusion.
(g) Does the conclusion (f) address the claim that injury rates are decreasing in Big Three plants and increasing in supplier plants?

C. ASSUMPTIONS UNDERLYING PEARSON'S *CHI*-SQUARE

At the beginning of the chapter we pointed out that test statistics developed in earlier chapters are appropriate to hypotheses that focus on particular properties of a distribution. In this connection, the reader may have noted that every test statistic developed in Chapter 10 was a function of the estimator $\bar{X}$ and every test statistic developed in Chapter 11 involved the estimator $\hat{\sigma}^2$. In contrast, goodness-of-fit tests and tests of association are omnibus techniques that detect differences in kurtosis, location, skew, dispersion, and so on. Concomitantly, the various formulae for Pearson's *chi*-square include *no* obvious estimator of any particular parameter. Consequently, these statistics are often characterized as "estimation-free" or "nonparametric." At least in the case of Pearson's *chi*-square, these labels are manifestly inappropriate, because the marginal relative frequencies f_j/N or f_k/N are used to estimate, respectively, the probabilities $P(x_j)$ and $P(y_k)$.

Perhaps the feature of Pearson's *chi*-square statistics that most clearly distinguishes them from our other test statistics is in the *assumptions* that must be met if the tests are to yield meaningful results. In order to assume that the distribution of any Pearson's *chi*-square statistic is approximated by $\chi^2(\nu)$, it must be true that:

1. Every observation falls into one and only one cell. In goodness-of-fit tests this means that the C columns define mutually exclusive categories. In tests of association, the R rows likewise define mutually exclusive categories.
2. The N observations are independent of one another.
3. The sample size N is large.

The last requirement is the most difficult to pin down. There is almost universal agreement that $\tilde{\chi}^2$ approximates the *chi*-square distribution if no expected frequency is less than 10. For expected frequencies smaller than 10, "large enough" depends on the form of the distribution, the number of degrees of freedom, and the significance level of the test. To be conservative, we recommend the following:

- If $\nu = 1$, then $N \geq 20$ and $\mathbf{F}_{jk} \geq 5$ in all cells.
- If $\nu > 1$, then $\mathbf{F}_{jk} \geq 5$ in at least 80 percent of the cells. In addition, all $\mathbf{F}_{jk} \geq 2$ if $\nu < 30$ and all $\mathbf{F}_{jk} \geq 1$ if $\nu > 30$.[12]

Unlike the other tests we have developed, the assumptions underlying Pearson's *chi*-square are primarily concerned with data collection rather than the probability distribution of the random variable. Some writers have therefore suggested that test statistics like $\tilde{\chi}^2$ should be called

---

[12] See Cochrane, W. G. (1952). The χ^2 test of goodness of fit, *Annals of Mathematical Statistics*, *23*, 315–345; and Cochrane, W. G. (1954). Some methods for strengthening the common χ^2 tests, *Biometrics*, *10*, 417–451.

"distribution-free." This label is also misleading, however, because the investigator must always be able to specify the probability distribution of the *test statistic*, even if the distribution of the *random variable* cannot be specified. In the long run, it may be more useful to ignore existing nomenclature and think of tests like Pearson's *chi*-square as *minimal assumption* techniques. Although this feature makes Pearson's *chi*-square statistics extremely versatile, it should be understood that they do not have the power of other test statistics, which demand more exacting assumptions.

D. SPECIAL APPLICATIONS OF PEARSON'S *CHI*-SQUARE

1. Goodness-of-fit test for normality

Earlier, we reminded the reader that many of the tests developed in Chapters 10 through 13 require that random variables be normally distributed and that one must often resort to nonparametric methods if normality cannot be assumed. Another option is to use sample data to test the hypothesis that one's random variable is normally distributed.

Normally distributed random variables are continuous, so one must first group the values of the random variable into C intervals. We have become rather accustomed to grouping data into intervals of equal width, but we know that normal distributions have only a small proportion of area under the tails of the curve. If one defines intervals of equal width, the expected frequencies in the extreme cells can therefore be very small. Indeed, unless one has a very large sample, these values may even be less than 1. One way to avoid this problem is to define intervals associated with equal *proportions* of area under the normal curve. That is, if one wants C cells, they should be defined in such a way that each interval has a probability of $1/C$ under the normal curve. Since $\mathbf{F} = Np$, this practice yields the same expected frequency in every cell,

$$\mathbf{F}_j = N\left(\frac{1}{C}\right)$$

Furthermore, calculations are simplified if expected frequencies are whole numbers, so C should be a divisor of N if possible.

Example 15.7. Suppose we make 75 observations on a random variable X from a population that is known to have mean $\mu = 65$ and variance $\sigma^2 = 169$. Suppose further that we want to test the hypothesis that X is normally distributed. If we want the same expected frequency in each interval, we might choose 3 intervals in such a way that 25 observations are expected to fall in each cell, or we might choose 5 intervals, each with an expected frequency of 15, or 15 intervals with $\mathbf{F}_j = 5$. For purposes of illustration, let's use 5 intervals.

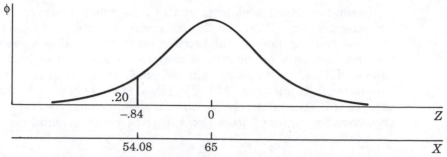

Figure 15.1. Proportion of area under normal curve below $x = 54.08$.

Since $C = 5$ and we want to expect the same proportion of observations in each interval, then each cell must obviously include 1/5, or 20 percent, of the observations. The upper limit of the first interval therefore corresponds to the z-value for which $\mathbb{P}(z) = .20$. In Table 3 of Appendix VIII, the cumulative probability nearest .20 is .2005, for which $z = -.84$. Therefore,

$$-.84 = \frac{x - \mu}{\sigma} = \frac{x - 65}{13}$$

and the corresponding x-value is

$$x \doteq 13(-.84) + 65 = 54.08$$

Consequently, one should expect about 20 percent of the observed x-values to fall at or below about 54.08 (see Figure 15.1).

We know from Chapter 7 that the proportion of area under the normal curve between two values x_a and x_b is equal to $\mathbb{P}(x_b) - \mathbb{P}(x_a)$, and we know that the cumulative probability of $x = 54.08$ is .20. We want the second interval to include 20 percent of the area under the normal curve, so the upper limit of the interval must therefore have a cumulative probability of .40. In Table 3 of Appendix VIII, the nearest entry to .40 is .4013 and corresponds to $z = -.25$. Therefore,

$$-.25 = \frac{x - 65}{13}$$

and

$$x \doteq 13(-.25) + 65 = 61.75$$

One should therefore expect about 20 percent of the observed x-values to fall between 54.08 and 61.75 (see Figure 15.2).

We use the same strategy to calculate the upper limits of the remaining intervals. In Table 3 we find $\mathbb{P}(.25) = .60$, $\mathbb{P}(.84) = .80$, and z-values greater than .80 account for the upper 20 percent of the standard normal

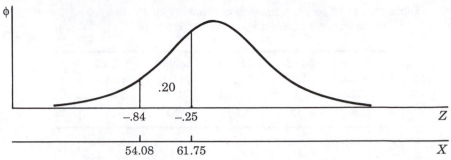

Figure 15.2. Proportion of area under normal curve between $x = 54.08$ and $x = 61.75$.

curve. Solving as before for the corresponding x-values, we find that for $z = .25$,

$$x \doteq 13(.25) + 65 = 68.25$$

for $z = .84$,

$$x \doteq 13(.84) + 65 \doteq 75.92$$

and the fifth interval (like the first) is open-ended (see Figure 15.3).

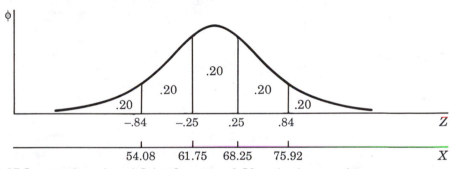

Figure 15.3. x- and z-values defining five areas of .20 under the normal curve.

Finally, we calculate the expected frequency in each cell. Since every cell is expected to include 20 percent of the 75 observations,

$$\mathbf{F}_j = (.20)(75) = 15$$

which gives us the expected frequencies shown in Table 15.10.

Table 15.10. Expected frequencies for 75 observations on a normally distributed random variable X with $\mu = 65$ and $\sigma^2 = 169$

X	54.08 or less	54.08– 61.75	61.75– 68.25	68.25– 75.92	75.92 or more	Σ
F	15	15	15	15	15	75

Table 15.11. Observed frequency distribution for 75 measurements of X with $\mu = 65$ and $\sigma^2 = 169$

X	54.08 or less	54.08– 61.75	61.75– 68.25	68.25– 75.92	75.92 or more	Σ
f	5	11	14	20	25	75

Now, let us suppose that the experimenter observes the frequencies of x-values shown in Table 15.11.

From the formula for Pearson's *chi*-square given in [15.2],

$$\tilde{\chi}^2 = \sum_j^5 \frac{(f_j - \mathbf{F}_j)^2}{\mathbf{F}_j} = \frac{(5-15)^2}{15} + \frac{(11-15)^2}{15} + \frac{(14-15)^2}{15}$$

$$+ \frac{(20-15)^2}{15} + \frac{(25-15)^2}{15}$$

$$= \frac{100}{15} + \frac{16}{15} + \frac{1}{15} + \frac{25}{15} + \frac{100}{15} \doteq 16.13$$

For $C = 5$, degrees of freedom ν is $5 - 1 = 4$, and in Table 6 of Appendix VIII we find that the cumulative probability of 16.13 falls between .995 and .999 in $\chi^2(4)$. The signficance probability of the result is $.001 < p < .005$, and the hypothesis that X is normally distributed is rejected for any $\alpha \geq .005$. A visual inspection of Table 15.11 reveals that the departure from normality is apparent in the negative skew of the observations.

In Example 15.7 we knew (or had reason to assume) that $\mu = 65$ and $\sigma^2 = 169$. If we had *estimated* these parameters from the 75 observations, i.e., if $\bar{x} = 65$ and $\hat{\sigma}^2 = 169$, these estimates would have reduced degrees of freedom by two. That is,

$$\nu = (5 - 1 - 2) = 2$$

so the value of Pearson's *chi*-square would be submitted to the distribution of $\chi^2(2)$.

The goodness-of-fit test for normality is outlined in Box 15.3.

BOX 15.3

Goodness-of-fit test for normality

1. Choose C, the number of intervals.

2. Find the z-value corresponding to the upper limit of each interval. For the first interval, $\mathbb{P}(z) = 1/C$; for the second interval, $\mathbb{P}(z) = 2/C$; for the third interval, $\mathbb{P}(z) = 3/C$; and so on. The last interval (like the first interval) is open-ended.

3. Calculate the x-values of the class limits,
$$x = \mu + z\sigma$$

4. Calculate the expected frequencies. Every interval is associated with the same proportion $(1/C)$, so the expected frequency in every interval is
$$\mathbf{F} = N\left(\frac{1}{C}\right)$$
where n is the total number of observations.

5. Pearson's $\tilde{\chi}^2$ is calculated by formula [15.1], but if μ and σ are estimated, $\nu = C - 3$.

---

2. The median test

If the C categories are values of intervals of an ordinal, ratio, or interval scale one can use Pearson's *chi*-square to test the hypothesis that R populations have the same median. The test hypothesis is formulated in terms of homogeneity

$$H_0: p_{j1} = p_{j2} = .50$$

for $j = 1, \ldots, R$, but the method for calculating expected frequencies is more akin to goodness-of-fit tests, because the proportions falling in each category (.50) are specified in advance, rather than estimated.

To test the hypothesis of equal medians, one first combines all of the data into a single distribution of C categories and calculates the combined, or *grand*, median $\tilde{x}$ (see Chapter 2, pp. 28–37). Then data are arranged in an $R \times 2$ table, where Category 1 is all values of the random variable below $\tilde{x}$ and Category 2 is all values of the random variable above $\tilde{x}$. The observed frequency f_{j1} is therefore the number of observations in Population j falling below $\tilde{x}$, and f_{j2} is the number of observations in Population j that lie above $\tilde{x}$.

Under the test hypothesis, all of the R populations have the same median (equal to the grand median, $\tilde{x}$), so half the observations in each population are expected to fall above $\tilde{x}$ and half are expected to fall below $\tilde{x}$. Therefore,

$$\mathbf{F}_{j1} = \mathbf{F}_{j2} = N_j(.50)$$

for all $j = 1, 2, \ldots, R$ populations.

Example 15.8. One of the earliest behavioral genetics studies involved selection of rats for the ability to learn mazes. Suppose that "maze-bright" rats are crossed with three standard strains of laboratory rats and their offspring are given 10 trials in a 15-unit maze, that is, a maze with 15 choice points. The number of errors over the 10 trials is recorded for each rat, as given in Table 15.12. The total number of rats is even ($N = 64$), so from Chapter 2 we know that the grand median is the average of the middle two values when the scores are arranged in numerical order. We find that the values in places 32 and 33 are 65 and 68, respectively, so the grand median $\tilde{x}$ is 66.5 errors.

Table 15.12. Number of maze-running errors for three groups of hybrid rats

Group 1	70, 36, 42, 40, 80, 52, 29, 51, 39, 81, 28, 39, 45, 69, 30, 60, 37, 75, 33, 83
Group 2	68, 128, 22, 11, 42, 126, 86, 79, 20, 120, 41, 31, 82, 88, 19, 64, 30, 45, 83, 28, 26, 84, 90, 89, 87, 85
Group 3	62, 111, 54, 115, 91, 56, 65, 86, 91, 130, 76, 71, 60, 53, 84, 73, 81, 85

There are 20 rats in Group 1, so if the median for Group 1 is 66.5, then half the rats, or 10, should fall below 66.5, and 10 should fall above 66.5. Likewise half of the 26 rats in Group 2 should fall above 66.5 and half should fall below 66.5, and half of the 18 rats in Group 3 should fall above and half should fall below 66.5. These expected frequencies are given in Table 15.13, along with the corresponding observed frequencies tallied from Table 15.12. From formula [**15.5**]

$$\tilde{\chi}^2 = \sum_j^3 \sum_k^2 \frac{(f_{jk} - F_{jk})^2}{F_{jk}} = \frac{(14 - 10)^2}{10} + \frac{(6 - 10)^2}{10}$$

$$+ \frac{(12 - 13)^2}{13} + \frac{(14 - 13)^2}{13}$$

$$+ \frac{(5 - 9)^2}{9} + \frac{(13 - 9)^2}{9}$$

$$= \frac{16}{10} + \frac{16}{10} + \frac{1}{13} + \frac{1}{13} + \frac{16}{9} + \frac{16}{9} \doteq 6.91$$

The frequency distribution for a median test has R rows and $C = 2$ columns, so degrees of freedom is always

$$(R - 1)(2 - 1) = R - 1$$

There are three populations in Example 15.8, so $\nu = 3 - 1 = 2$. In Table 6 of Appendix VIII we find that $.950 < \mathbb{P}(6.91) < .975$ in $\chi^2(2)$, so the hypothesis of equal medians is rejected for $\alpha \geq .05$.

Table 15.13. Observed and expected frequencies of rats in three hybrid groups falling above and falling below the grand median

	Expected frequencies		Observed frequencies	
	Below	Above	Below	Above
Group	$\tilde{x} = 66.5$	$\tilde{x} = 66.5$	$\tilde{x} = 66.5$	$\tilde{x} = 66.5$
1	10	10	14	6
2	13	13	12	14
3	9	9	5	13

As presented above, the *chi*-square test for equality of medians is strictly appropriate only when N is even and the middle two values are unequal. In other cases it frequently happens that one or more observed values is exactly equal to the grand median. In this circumstance, one can either drop these scores from the analysis or define Category 1 as all values falling *at or* below $\tilde{x}$. The latter strategy is preferred[13] and is consistent with the conception of the median for grouped data as discussed in Chapter 2 (p. 29). The *chi*-square test for equality of medians is ordinarily performed on ungrouped data, but can also be used with grouped data if one has access to the individual scores in the median interval.

If $R = 2$, the observed frequencies for the median test are arranged in a fourfold table, so the test statistic is most easily calculated by formula **[15.8]**. Since a median test for two populations has one degree of freedom, the test should be performed only if $N \geq 20$ (i.e., when $\mathbf{F}_{jk} = .50N \geq 10$).

BOX 15.4

The median test

1. Combine all data into a single distribution of C categories and compute the grand median $\tilde{x}$.
2. Arrange data in an $R \times 2$ table, where Category 1 is all values of the random variable (at or) below $\tilde{x}$ and Category 2 is all values of the random variable above $\tilde{x}$.
3. Observed frequencies: In Population j, the number of observations (at or) below $\tilde{x}$ is f_{j1} and the number of observations above $\tilde{x}$ is f_{j2}.
4. Expected frequencies: If H_0 is correct, half the observations in every population fall (at or) below the common median $\tilde{x}$ and half fall above $\tilde{x}$, so

$$\mathbf{F}_{j1} = \mathbf{F}_{j2} = N_j(.50)$$

for all $j = 1, 2, \ldots, R$ populations.
5. Pearson's $\tilde{\chi}^2$ is calculated by formula **[15.8]** and has $R - 1$ degrees of freedom.

---

[13] Siegel, S., and Castellan, N. J. (1988). *Nonparametric statistics for the behavioral sciences* (2nd ed.). New York: McGraw-Hill, pp. 200–201.

3. Fourfold point correlation: the *phi* coefficient

In general, the ordering of the rows and columns in a fourfold test of independence is arbitrary. In Example 15.6, for instance, it makes no difference whether white applicants are represented in the first column or in the second. There are circumstances, however, in which rows and columns have *quantitative* meaning, that is, where classification of an observation in the second row implies more of some attribute than classification in the first row and where the second column implies more of some attribute than the first column. When this is true, it is meaningful to ask about the *strength* of the relationship between X and Y. The value $\tilde{\chi}^2$ is not itself an easily interpretable index of the strength of association, because Pearson's *chi*-square (like the covariance) increases with sample size. And we know from Chapter 14 that significance level is not a dependable guide to strength of association. However, the strength of association between X and Y is captured by a statistic called the *fourfold point correlation*, which is analogous to the Pearson correlation coefficient r.

The fourfold point correlation is denoted φ, the lowercase Greek *phi* (not to be confused with ϕ, which denotes a density function), and is sometimes called the *phi* coefficient:

$$\varphi = \frac{ad - bc}{\sqrt{(a + b)(c + d)(a + c)(b + d)}}$$ [15.9]

It is easily seen that φ is a simple transformation of the fourfold $\tilde{\chi}^2$. From formulae [15.8] and [15.9],

$$\tilde{\chi}^2 = \frac{N(ad - bc)^2}{(a + b)(c + d)(a + c)(b + d)}$$

$$= N\varphi^2$$

Accordingly, $N\varphi^2$ is distributed approximately as $\chi^2(1)$ and the significance test of the fourfold $\tilde{\chi}^2$ implicitly tests the hypothesis that the population fourfold correlation is zero.

Example 15.9. All students who wish to enroll in a particular statistics course must first take a diagnostic mathematics skills test. At the end of the semester, the instructor categorizes each student according to whether the student's diagnostic test score was in the top or bottom half of the class and whether the student's score on the final examination was in the top or bottom half of the class.

		Diagnostic test	
		Bottom half	Top half
Final	Bottom half	$a = 15$	$b = 8$
examination	Top half	$c = 8$	$d = 15$

By formula **[15.9]**

$$\varphi = \frac{ad - bc}{\sqrt{(a + b)(c + d)(a + c)(b + d)}}$$

$$= \frac{225 - 64}{\sqrt{(23)(23)(23)(23)}} \doteq .304$$

The fourfold $\tilde{\chi}^2$ can be calculated directly from φ to test the hypothesis that the population point correlation is zero, i.e., that X and Y are independent:

$$\tilde{\chi}^2 = N\varphi^2 \doteq 46(.304)^2 \doteq 4.25$$

In Table 6 of Appendix VIII we find $.95 < \mathbb{P}(4.25) < .975$ in $\chi^2(1)$. The hypothesis that the diagnostic test and the final examination are unrelated is therefore rejected for $\alpha \geq .05$.

The connection between φ and the Pearson correlation r is seen if we assign the values 0 and 1 to our categories.

Example 15.10. For the data in Example 15.9 let

$x = 0$ for students in the bottom half of the diagnostic test distribution

$x = 1$ for students in the top half of the diagnostic test distribution

$y = 0$ for students in the bottom half of the final examination distribution

$y = 1$ for students in the top half of the final examination distribution.

By the computational formula for Pearson's correlation coefficient **[4.6]**,

$$r_{XY} = \frac{N\sum xy - \sum x \sum y}{\sqrt{N\sum x^2 - (\sum x)^2}\sqrt{N\sum y^2 - (\sum y)^2}}$$

For each of the 15 students in the top half of *both* tests, $xy = (1)(1) = 1$.

For all other students, $x = 0$ or $y = 0$, so xy must equal 0. Therefore

$$\sum xy = 15(0)(0) + 8(0)(1) + 8(1)(0) + 15(1)(1)$$

$$= 15$$

Since half of the 46 students received scores $x = 1$ and half received scores $x = 0$,

$$\sum x = 23(0) + 23(1) = 23$$

and

$$\left(\sum x\right)^2 = 23^2 = 529$$

Likewise, half the students received scores $y = 1$ and half received scores $y = 0$, so

$$\sum y = 23(0) + 23(1) = 23$$

and

$$\left(\sum y\right)^2 = 23^2 = 529$$

Since $1^2 = 1$ and $0^2 = 0$,

$$\sum x^2 = \underbrace{1^2 + \cdots + 1^2}_{23 \text{ times}} + \underbrace{0^2 + \cdots + 0^2}_{23 \text{ times}} = 23$$

Likewise,

$$\sum y^2 = 23$$

Therefore,

$$r_{XY} = \frac{46(15) - (23)(23)}{\sqrt{46(23) - 529}\,\sqrt{46(23) - 529}}$$

$$= \frac{690 - 529}{\sqrt{529}\,\sqrt{529}} \doteq .304$$

4. A test for correlated proportions

We know that one of the assumptions underlying use of Pearson's *chi*-square is that observations are independent. Violation of this assumption can be a matter of careless research design, but sometimes correlation among observations is really a necessary and central feature of the researcher's data. For example, testing procedures sometimes have R judges assigning one of C possible scores to the performance of each of N individuals. This is the situation in Olympic figure skating competition where nine judges each assign a number between 0.0 and 6.0 to the program performed by every competitor. Even if the judges make their ratings independently, their nine distributions across scoring categories will not be wholly independent, because they are all based on the *same* sample of behaviors. Superficially, this situation seems much like Example 15.9, but in that example, the diagnostic test and the final examination were two *different* samples of behavior.

When $R = C = 2$, the problem of testing correlated proportions is therefore analogous to testing mean differences for correlated groups discussed in Chapter 10 (pp. 437–439).

Example 15.11. The ski skills test for National Ski Patrol candidates is ordinarily conducted by at least two examiners. Let us suppose that Examiner 1 and Examiner 2 test a group of 20 candidates, with the following results:

		Examiner 1	
		Pass	Fail
Examiner 2	Pass	$a = 8$	$b = 3$
	Fail	$c = 7$	$d = 2$

The 20 candidates can be thought of as a sample from the population of *all* candidates to be tested by this team of examiners. If p_1 is the population proportion passed by Examiner 1 and p_2 is the population proportion passed by Examiner 2, the hypothesis

$$H_0: p_1 = p_2$$

can be tested by calculating

$$\tilde{\chi}^2 = \frac{(|b - c| - 1)^2}{b + c} \tag{15.10}$$

which is distributed as χ^2 with one degree of freedom. In this example, formula [15.10] gives a value of .9, and we should probably conclude that the two examiners do not pass significantly different proportions of candidates ($p > .75$).

5. A test of homogeneity when expected frequencies are small

We have mentioned elsewhere that current scientific practices favor the use of large samples, but in branches of research that depend on naturalistic observation rather than experimental manipulation, a researcher may have little control over sample sizes. For example, geneticists are often concerned with the distribution of rare mutations, and both ethologists and wildlife biologists can spend years collecting observations on one or a few small groups of animals. If the species is threatened or endangered or occupies a remote wilderness habitat, the problem of sample size can be very troublesome. In many instances, such research involves qualitative data (categories of behavior, species of prey consumed, etc.) that are ideally suited to Pearson's *chi*-square techniques, except for the requirement of large expected frequencies.

The British geneticist J. B. S. Haldane[14] developed the following statistic for comparisons of $R = 2$ distributions of frequencies across C categories when expected frequencies fail to satisfy the requirements for a conventional Pearson's *chi*-square test. The statistic is distributed (approximately) as $\chi^2(C - 1)$ and can be used even if expected frequencies are smaller than 1.

$$N\left(1 - \frac{N\sum\left(\dfrac{a_j b_j}{N_j}\right)}{AB}\right)$$

[15.11]

Haldane's statistic assumes the notation in the following layout of observed frequencies:

	Categories				Σ
Populations	a_1 b_1	a_2 b_2	$\cdots$ $\cdots$	a_C b_C	A B
Σ	N_1	N_2	$\cdots$	N_C	$\Sigma\Sigma = N$

Thus, A and B are the sums of rows 1 and 2, respectively, N is the grand total, and $N_1, N_2, \ldots, N_C$ are the column sums.

[14] Haldane, J. B. S. (1945). The use of χ^2 as a test of homogeneity in a ($n \times 2$)-fold table when expectations are small. *Biometrika*, *33*, 234–238. (Our notation differs slightly from Haldane's.)

For the special case where $C = 2$, expression [15.11] reduces to

$$N\left(1 - \frac{N\left(\dfrac{a_1 b_1}{N_1} + \dfrac{a_2 b_2}{N_2}\right)}{AB}\right) \qquad\qquad [15.12]$$

which is distributed (approximately) as $\chi^2(1)$.

Example 15.12. An entry-level test for police officers was administered to 13 minority applicants and 45 white applicants. Six of the minority applicants failed, and 15 of the white applicants failed. The disproportionate number of minority applicants who failed the test led to a claim in federal court[15] that the examination constituted an "artificial barrier" to minority employment. The claim is supported by rejection of the hypothesis that the distribution of passes and failures is the same in both populations.

	Pass	Fail	Σ
White	$a_1 = 30$	$a_2 = 15$	$A = 45$
Minority	$b_1 = 7$	$b_2 = 6$	$B = 13$
Σ	$N_1 = 37$	$N_2 = 21$	$N = 58$

The proportion of applicants who failed was $21/58 \doteq .362$, and the number of minority applicants was 13. By formula [15.3], the frequency of minority applicants expected to fail the examination is

$$13(.362) \doteq 4.7$$

which is too small to use Pearson's *chi*-square. However, by formula [15.12]

$$N\left(1 - \frac{N\left(\dfrac{a_1 b_1}{N_1} + \dfrac{a_2 b_2}{N_2}\right)}{AB}\right) = 58\left(1 - \frac{58\left(\dfrac{(30)(7)}{37} + \dfrac{(15)(6)}{21}\right)}{(45)(13)}\right)$$

$$\doteq .718$$

In Table 6 of Appendix VIII we find that $\mathbb{P}(.718)$ falls between .50 and .75 in $\chi^2(1)$, so $p > .25$. If $\alpha \le .25$, the test hypothesis cannot be rejected.

[15] *Holliman v. Price, et al.* U.S. District Court, Eastern District of Michigan, January 3, 1973.

———————— EXERCISES 15.3 ————————————————

1. In Exercises 12.3.11 and 12.4.5 it was assumed that puzzle-box performance for five breeds of domestic dog was normally distributed. Use the following data to test this hypothesis. For purposes of this exercise, use six intervals.

Puzzle-box scores for 126 dogs representing 5 breeds

Score	0	1	2	3	4	5	6	7	8
Frequency	28	16	14	9	9	6	14	14	16

(a) What are the hypotheses?
(b) What are the intervals?
(c) What are the expected frequencies?
(d) What are the observed frequencies?
(e) What is the value of Pearson's *chi*-square?
(f) What is the *p*-value of the test statistic?
(g) What is the appropriate conclusion?

2. Damage inflicted by the spell *Fireball* in *Advanced Dungeons & Dragons*® is calculated by throwing one 6-sided die for every proficiency level of the spell-caster. A player whose character is a sixth-level wizard records the damage inflicted by 70 *Fireball* spells.

22	25	24	20	21
22	18	22	21	14
19	26	23	24	21
25	22	15	31	19
21	18	23	26	28
15	18	23	25	12
22	20	15	22	23
24	21	18	23	20
23	16	24	19	15
16	30	17	25	19
18	14	22	17	13
21	32	19	19	20
21	17	28	21	20
16	19	12	27	10

(a) How should the damage be distributed if all of the dice are evenly balanced? *Hint*: See Exercise 8.2.14 (p. 295) and recall the Algebra of Expectations.

Test the hypothesis that the distribution of scores given above is distributed as indicated in your answer to (a).

(b) What are the hypotheses?
(c) Write the formula for the appropriate test statistic.
(d) How is the test statistic distributed if the test hypothesis is correct?
(e) Tabulate the observed and expected frequencies.
(f) What is the value of the test statistic?
(g) What is your conclusion?

Siamese fighting fish exhibit more persistent aggression if they are "primed" by prior visual exposure to another male. To investigate the effects of priming, a stimulus fish and subject fish are placed in adjacent tanks separated by a removable opaque partition, which permits the experimenter to vary exposure time.

3. In a typical priming demonstration, experimental subjects are placed in the tank for 15 minutes with the partition removed, and control subjects are placed in the tank for 15 minutes with the partition in place so that the stimulus fish cannot be seen. Fish in both conditions are then exposed to the stimulus fish for 15 minutes, during which period the experimenter records the number of seconds that subject fish display threat behavior. The following scores are aggression display times for 13 unprimed (control) Siamese fighting fish and for 13 primed (experimental) fish. If prior exposure to another male increases the tendency to threat behavior, median display time should be greater for the primed group.

Total display time (in seconds)	
Unprimed subjects	**Primed subjects**
40	900
859	900
252	900
818	900
872	900
890	900
112	700
54	898
875	856
236	793
614	900
359	900
728	900

(a) What is the test hypothesis?
(b) What is the appropriate test statistic?
(c) How is the test statistic distributed if H_0 is correct?
(d) What is the value of the test statistic?
(e) What is the p-value of the test statistic?
(f) What is the conclusion?

4. Siamese fighting fish often abandon even fierce combat if a visual obstacle (e.g., a rock or piece of coral) happens to come between opponents. To investigate the effects on priming of this apparently short "memory," Bronstein[16] inserted the partition between tanks for periods of 0, 1, and 5 minutes after priming. Aggression display times for subjects in these three conditions are recorded below. If persistence of aggression is affected by post-priming delay, the medians for the three groups should be different.

[16] Bronstein, P. (1988) The priming and retention of agonistic motivation in male Siamese fighting fish, *Betta splendens*. *Animal Behaviour*, *37*, 165–166.

Total display time (in seconds)		
Delay following priming (in minutes)		
0	1	5
725	438	367
900	900	285
900	711	318
900	900	867
900	900	278
900	880	885
900	451	853
170	892	896
700	883	862
898	796	432
856	791	884
492	749	
793		
378		
900		
900		
900		
734		

Source: Data furnished by P. Bronstein.

(a) What is the test hypothesis?
(b) What is the appropriate test statistic?
(c) How is the test statistic distributed if H_0 is correct?
(d) What is the value of the test statistic?
(e) What is the p-value of the test statistic?
(f) What is the conclusion?

5. The following data were used in Exercise 10.5.16 (p. 447) to test the hypothesis that solitary lions hunt small prey (Thomson's gazelle) more frequently than larger prey (Wildebeest and zebra). Calculate the fourfold point correlation between number of lions in the hunting group and prey size.

Number of lions	Prey species	
hunting	**Thomson's gazelle**	**Wildebeest & zebra**
1	185	33
2 or more	177	70

Source: Schaller, G. B. (1972). *The Serengeti lion*. Chicago: University of Chicago Press, p. 405.

6. Exercise 15.2.14 (p. 655) tested the association between intensity of fighting between male sierra dome spiders and the mating status of females in close proximity. Insemination of a virgin female produces more offspring than insemination of a female that is already inseminated, so access to a virgin female represents a larger reproductive payoff than access to a mated female.

Obtain a quantitative measure of the statistical relationship between fighting intensity and reproductive payoff by calculating the fourfold point correlation.

	Female status	
Fight intensity	**Mated**	**Virgin**
Display only	9	2
Contact fighting	7	14

7. In the table below, 108 kills by lions are classified both by prey species and by the type of cover in which the prey was taken: 1 = little or no cover, 2 = no cover but near some, 3 = scattered shrubs, 4 = thickets or tall grass, 5 = thickets along river.

Prey species	Type of cover				
	1	**2**	**3**	**4**	**5**
Thomson's gazelle	3	5	3	46	22
Wildebeest & zebra	1	14	0	1	13

Source: Schaller, G. B. (1972). *The Serengeti Lion*. Chicago: University of Chicago Press, p. 444.

Test the hypothesis that the two distributions are the same.

(a) What are the hypotheses?
(b) What are the expected frequencies?
(c) What is the appropriate test statistic?
(d) How is the test statistic distributed if the test hypothesis is correct?
(e) Calculate the value of the test statistic.
(f) What is the p-value of the test statistic?
(g) What is the appropriate conclusion?

8. The table below records the species of prey taken in 123 kills by lions and whether the kill occurred at night or during the day.

Time of kill	Prey species		
	Thomson's gazelle	**Wildebeest & zebra**	**Other**
Day	78	17	7
Night	7	11	3

Source: Schaller, G. B. (1972). *The Serengeti Lion*. Chicago: University of Chicago Press, p. 444.

Test the hypothesis that the distribution of prey species killed by lions is the same during the day as during the night.

(a) What are the hypotheses?
(b) What are the expected frequencies?

(c) What is the appropriate test statistic?
(d) How is the test statistic distributed if the test hypothesis is correct?
(e) Calculate the value of the test statistic.
(f) What is the *p*-value of the test statistic?
(g) What is the appropriate conclusion?

E. SUMMARY

Pearson's *chi*-square techniques differ in the following ways from other methods we have developed for testing hypotheses:

- Pearson's *chi*-square statistics can be used to test omnibus hypotheses that *simultaneously* address such properties as location, dispersion, skew, and kurtosis.
- Pearson's *chi*-square techniques can be used with nominal and ordinal data, as well as with interval and ratio data.
- Pearson's *chi*-square techniques can be used when one cannot specify a probability model for the distribution of the random variable.

A *goodness-of-fit test* is used to test the hypothesis that the distribution of frequencies in *one* particular population across C categories fits some *specified* theoretical distribution of frequencies. The most general test hypothesis for a goodness-of-fit test is

$$H_0 \colon f_j = \mathbf{F}_j$$

which is understood to mean that for sample size N, the expected number of observations falling in Category j is $\mathbf{F}_j$, where $j = 1, 2, \ldots, C$. The test statistic for goodness of fit is given in [15.1],

$$\tilde{\chi}^2 = \sum_j^C \frac{(f_j - \mathbf{F}_j)^2}{\mathbf{F}_j}$$

where f_j is the *observed* frequency in Category j and $\mathbf{F}_j$ is the *expected* frequency implied by the theoretical distribution. Degrees of freedom for this statistic is $C - 1$.

The goodness-of-fit test can be thought of as a generalization of the test of proportions from the binomial case to the *multinomial* case, that is, to the case where each observation can be assigned to any of $C > 2$ mutually exclusive categories. If $C = 2$, Pearson's *chi*-square yields the same result as a test of binomial proportions. That is,

$$\tilde{\chi}^2 = Z_{X/N}^2$$

where $Z_{X/N}$ is the statistic for testing hypotheses about the binomial proportion p using the normal approximation to the binomial.

A Pearson's *chi*-square *test of association* is used to test hypotheses about $R \geq 2$ frequency distributions, in which the experimenter has no a priori expected frequencies.

The hypothesis that the profile of proportions across C categories is the same in *two or more* well-defined populations is said to test for *homogeneity* of proportions, or *homogeneity* of probability. The test hypothesis can be written most compactly as

$$H_0: p_{1k} = p_{2k} = p_{3k} = \cdots = p_{Rk} = p_{\cdot k}$$

where p_{jk} is the proportion of cases in Category k (where $k = 1, 2, \ldots, C$) for Population j. The notation $p_{\cdot k}$ (or p_k) stands for the common proportion in Category k, and the test hypothesis therefore states that the proportions in Category 1 are equal to $p_{\cdot 1}$ across all R populations, that the proportions in Category 2 are equal to $p_{\cdot 2}$ across all R populations, and that proportions are likewise identical across all R populations in Categories $3, 4, \ldots$, and C:

$$p_{11} = p_{21} = p_{31} = \cdots = p_{R1} = p_{\cdot 1}$$

$$p_{12} = p_{22} = p_{32} = \cdots = p_{R2} = p_{\cdot 2}$$

$$p_{13} = p_{23} = p_{33} = \cdots = p_{R3} = p_{\cdot 3}$$

$$p_{14} = p_{24} = p_{34} = \cdots = p_{R4} = p_{\cdot 4}$$

$$\cdots$$

$$p_{1C} = p_{2C} = p_{3C} = \cdots = p_{RC} = p_{\cdot C}$$

This notation is decidedly easier to keep straight if the Categories are represented as Columns and the populations, therefore, as rows.

The expected frequency of observations in the *cell* corresponding to Category k in Population j is obtained by multiplying the number of observations drawn from Population j (denoted N_j) by the *estimated* common proportion in Category k. That is,

$$\mathbf{F}_{jk} = N_j \hat{p}_k = N_j \left(\frac{f_k}{N} \right)$$

where N is the total number of observations and f_k is the number of observations in Category k totaled across all R samples. If the number of observations from Population j is denoted f_j, then

$$\mathbf{F}_{jk} = \frac{f_j f_k}{N}$$

The appropriate test statistic is given in **[15.5]**:

$$\tilde{\chi}^2 = \sum_j^R \sum_k^C \frac{(f_{jk} - \mathbf{F}_{jk})^2}{\mathbf{F}_{jk}}$$

and is approximately distributed as χ^2 with $(R - 1)(C - 1)$ degrees of freedom.

In tests of homogeneity the experimenter draws samples of $N_1, N_2, \ldots, N_R$ observations from $R \geq 2$ *well-defined populations*. In other situations, the experimenter begins with a sample of N observations drawn from *one* population and then classifies the subjects according to their status on *two* variables, X and Y, where X defines R mutually exclusive categories and Y defines C mutually exclusive categories. In this situation, the experimenter is concerned with whether or not the two variables are *independent*.

From the definition of independent random variables, the hypotheses for Pearson's *chi*-square test of independence are

$$H_0: P(x_j, y_k) = P(x_j)P(y_k)$$

for all (x_j, y_k) and

$$H_1: P(x_j, y_k) \neq P(x_j)P(y_k)$$

for at least one pair (x_j, y_k), or,

$$H_1: H_0 \text{ incorrect}$$

The expected frequency of observations in cell jk follows directly from the test hypothesis,

$$\mathbf{F}_{jk} = NP(x_j, y_k) = NP(x_j)P(y_k)$$

and may be calculated by estimating the marginal probabilities $P(x_j)$ and $P(y_k)$:

$$\hat{P}(x_j) = \frac{f_j}{N} \quad \text{and} \quad \hat{P}(y_k) = \frac{f_k}{N}$$

where f_j is the marginal frequency of x_j and f_k is the marginal frequency of y_k. However, it is easily seen that

$$N\left(\frac{f_j}{N}\right)\left(\frac{f_k}{N}\right) = \frac{f_j f_k}{N}$$

so the calculation of expected frequencies $\mathbf{F}_{jk}$ for a test of independence is the same as in the test for homogeneity. The statistic is likewise the same as in the test for homogeneity.

When $R = C = 2$, the observed frequencies in a test of association are arranged in a 2×2 table, or a *fourfold* table:

a	b	$a + b$
c	d	$c + d$
$a + c$	$b + d$	

Observed frequencies f_{11}, f_{12}, f_{21}, and f_{22} are respectively denoted a, b, c, and d. Using this notational convention, it can be shown that formula **[15.5]** is equivalent to formula **[15.8]**

$$\tilde{\chi}^2 = \frac{N(ad - bc)^2}{(a + b)(c + d)(a + c)(b + d)}$$

where $N = a + b + c + d$, the total number of observations. Since $(R - 1)(C - 1) = 1$, the fourfold Pearson's *chi*-square is distributed (approximately) as $\chi^2(1)$.

The approximation of Pearson's *chi*-square statistics by $\chi^2(\nu)$ assumes the following:

1. Every observation falls into one and only one cell.
2. The N observations are independent of one another.
3. The sample size N is large. For $\nu = 1$ this assumption is ordinarily satisfied if $N \geq 20$ and $\mathbf{F}_{jk} \geq 5$. If $\nu > 1$, the approximation is usually satisfactory if $\mathbf{F}_{jk} \geq 5$ in at least 80 percent of the cells and never smaller than 2 unless $\nu > 30$.

There are a number of specific applications of Pearson's *chi*-square that are particularly useful.

Many inferential techniques require that one's random variable(s) be normally distributed. If normality cannot be assumed, Pearson's *chi*-square can sometimes be used to test the hypothesis of normality. Although the number of observations expected in the tails of a normal curve is very small, the test can be performed with moderate-sized samples if C intervals are defined so that every interval is associated with the *same* proportion $(1/C)$ of area under the normal curve.

A variant of the test for homogeneity can be used to test for equality of medians in $R \geq 2$ populations: For Population j, the number of observations at or below the grand median for *all* observations $\tilde{x}$ is f_{j1} and the number of observations above the grand median is f_{j2}. The expected frequencies (at or) below and above the grand median for Population j are

$$\mathbf{F}_{j1} = \mathbf{F}_{j2} = N_j(.50)$$

for $j = 1, 2, \ldots, R$.

If the rows (X) and columns (Y) in a fourfold test of independence have *quantitative* meaning, the *strength* of association between X and Y

is indicated by the *fourfold point correlation* given in equation [**15.9**]:

$$\varphi = \frac{ad - bc}{\sqrt{(a + b)(c + d)(a + c)(b + d)}}$$

The point correlation is also called the *phi* coefficient and is analogous to the Pearson correlation coefficient r. The significance of φ is determined by the significance of the fourfold $\tilde{\chi}^2$ calculated by [**15.8**].

If a test of proportions concerns R *correlated* groups, the assumption of independence is necessarily violated, and conventional Pearson's *chi*-square tests are inappropriate. However, if $R = C = 2$, the hypothesis

$$H_0: p_1 = p_2$$

when proportions are assumed to be correlated can be tested by calculating

$$\tilde{\chi}^2 = \frac{(|b - c| - 1)^2}{b + c}$$

given in [**15.10**], which is distributed as $\chi^2(1)$.

If expected frequencies are too small to satisfy the assumptions for Pearson's *chi*-square, homogeneity of proportions across C categories for $R = 2$ populations can be tested using a statistic developed by J. B. S. Haldane [**15.11**]:

$$N\left(1 - \frac{N\Sigma\left(\dfrac{a_j b_j}{N_j}\right)}{AB}\right)$$

where a_j is the frequency observed in Column j of the first row, b_j is the frequency observed in Column j of the second row, A is the first row total, B is the second row total, N_j is the total for Column j, and N is the total number of observations. Haldane's statistic is distributed (approximately) as $\chi^2(C - 1)$.

If $C = 2$, expression [**15.11**] reduces to [**15.12**]:

$$N\left(1 - \frac{N\left(\dfrac{a_1 b_1}{N_1} + \dfrac{a_2 b_2}{N_2}\right)}{AB}\right)$$

which is distributed (approximately) as $\chi^2(1)$.

Summary table
of statistical tests

H_0	Test statistic	Distribution of test statistic*
Goodness-of-Fit Tests		
Any of the following: $f_j = \mathbf{F}_j$ $X: \phi_0$ $\dfrac{f_j}{N} = p_j$	$\displaystyle\sum_j^C \frac{(f_j - \mathbf{F}_j)^2}{\mathbf{F}_j}$	$\chi^2(C-1)$
Tests of Association		
$p_{1k} = p_{2k} = \cdots = p_{Rk}$	$\displaystyle\sum_j^R \sum_k^C \frac{(f_{jk} - \mathbf{F}_{jk})^{2\dagger}}{\mathbf{F}_{jk}}$	$\chi^2[(R-1)(C-1)]$
$P(x_j, y_k) = P(x_j)P(y_k)$	Same	Same
$p_{1k} = p_{2k}, k = 1,2$	$\dfrac{N(ad - bc)^{2\ddagger}}{(a+b)(c+d)(a+c)(b+d)}$	$\chi^2(1)$
$P(x_j, y_k) = P(x_j)P(y_k)$, $j = 1,2$ and $k = 1,2$	Same	Same

*if H_0 is correct.

$\dagger \mathbf{F}_{jk} = \dfrac{f_j f_k}{N}$.

$\ddagger$ Data layout:

$a = f_{11}$	$b = f_{12}$
$c = f_{21}$	$d = f_{22}$

Special applications

Goodness of fit for normality with C intervals of equal probability under the normal curve

$$\sum_{j}^{C} \frac{(f_j - \mathbf{F}_j)^2}{\mathbf{F}_j} : \chi^2(\nu) \qquad \text{where } \mathbf{F}_j = N(1/C) \text{ and}$$
$$\nu = C - 1 - \text{number of estimated parameters}$$

The median test. Category 1 is observations at or below the grand median, and Category 2 is observations above the grand median.

For $R > 2$ populations · **For $R = 2$ populations**

$$\sum_{j}^{R} \sum_{k}^{2} \frac{(f_{jk} - \mathbf{F}_{jk})^2}{\mathbf{F}_{jk}} : \chi^2(R - 1)$$

where $\mathbf{F}_{j1} = \mathbf{F}_{j2} = N_j(.50)$

$$\frac{N(ad - bc)^2}{(a + b)(c + d)(a + c)(b + d)} : \chi^2(1)^*$$

Fourfold point correlation

$$\varphi = \frac{ad - bc}{\sqrt{(a + b)(c + d)(a + c)(b + d)}} = \sqrt{\frac{\chi^2}{N}} \qquad N\varphi^2 : \chi^2(1)$$

Test of equality of correlated proportions where $R = C = 2$

$$\tilde{\chi}^2 = \frac{(|b - c| - 1)^2}{b + c} : \chi^2(1)^*$$

Test of homogeneity for small $\mathbf{F}$

For $C > 2$ categories · **For $C = 2$ categories**

$$N\left(1 - \frac{N\sum\left(\dfrac{a_j b_j}{N_j}\right)}{AB}\right) : \chi^2(C - 1)^{\dagger} \qquad N\left(1 - \frac{N\left(\dfrac{a_1 b_1}{N_1} + \dfrac{a_2 b_2}{N_2}\right)}{AB}\right) : \chi^2(1)^{\dagger}$$

---

*Data layout:

$a = f_{11}$	$b = f_{12}$
$c = f_{21}$	$d = f_{22}$

†Data layout:

a_1	$a_2 \cdots$	a_C	A
b_1	$b_2 \cdots$	b_C	B
N_1	$N_2 \cdots$	N_C	N

APPENDICES

The algebra of summations

Rule 1: If c is a constant, then

$$\sum_{}^{N} c = Nc$$

Proof:

Think of $\sum^{N} c$ as $\sum^{N} x$ where every x-value equals c. Then,

$$\sum_{}^{N} x = \underbrace{c + c + \cdots + c}_{N \text{ times}} = Nc$$

Rule 2: If c is a constant and $x_1, x_2, \ldots, x_N$ is any collection of N numbers, then

$$\sum_{}^{N} cx = c \sum_{}^{N} x$$

Proof:

$$\sum_{}^{N} cx = cx_1 + cx_2 + \cdots + cx_N$$

$$= c(x_1 + x_2 + \cdots + x_N) = c \sum_{}^{N} x$$

Corollaries of Rule 2 for two-way data arrays. A collection of N numbers is sometimes arranged into J distinct groups of size $N_1, N_2, \ldots,$ and N_J, respectively, where $N_1 + N_2 + \cdots + N_J = N$. Then,

$$\sum_{}^{N} x = \sum_{j}^{J} \sum_{i}^{N_j} x_{ij}$$

Corollary 1: If c is a constant, then

$$\sum_{j}^{J}\sum_{i}^{N_j} cx_{ij} = c\sum_{j}^{J}\sum_{i}^{N_j} x_{ij}$$

Proof:

$$\sum_{j}^{J}\sum_{i}^{N_j} cx_{ij} = \sum^{N} cx$$

$$= c\sum x$$

$$= c\sum_{j}^{J}\sum_{i}^{N_j} x_{ij}$$

Corollary 2: If c_1 is a constant for Group 1, and c_2 is a constant for Group 2,..., and c_J is a constant for Group J, then

$$\sum_{j}^{J}\sum_{i}^{N_j} c_j x_{ij} = \sum_{j}^{J} c_j \sum_{i}^{N_j} x_{ij}$$

Proof:

$$\sum_{j}^{J}\sum_{i}^{N_j} c_j x_{ij} = \sum_{i}^{N_1} c_1 x_{i1} + \sum_{i}^{N_2} c_2 x_{i2} + \cdots + \sum_{i}^{N_J} c_J x_{iJ}$$

$$= c_1 \sum_{i}^{N_1} x_{i1} + c_2 \sum_{i}^{N_2} x_{i2} + \cdots + c_J \sum_{i}^{N_J} x_{iJ}$$

$$= \sum_{j}^{J} c_j \sum_{i}^{N_j} x_{ij}$$

Rule 3: If $x_1, x_2, \ldots, x_N$ is any set of N numbers and $y_1, y_2, \ldots, y_N$ is another set of N numbers, then

$$\sum^{N} x_i y_i = x_1 y_1 + x_2 y_2 + \cdots + x_N y_N$$

This is true by definition.

Rule 4: If $x_1, x_2, \ldots, x_N$ is any set of N numbers and $y_1, y_2, \ldots, y_N$ is another set of N numbers, then

$$\sum_{}^{N}(x_i + y_i) = \sum_{}^{N}x_i + \sum_{}^{N}y_i$$

Proof:

$$\sum_{}^{N}(x_i + y_i) = (x_1 + y_1) + (x_2 + y_2) + \cdots + (x_N + y_N)$$

$$= (x_1 + x_2 + \cdots + x_N) + (y_1 + y_2 + \cdots + y_N)$$

$$= \sum_{}^{N}x_i + \sum_{}^{N}y_i$$

Linear interpolation

Linear interpolation is a method that gives a good guess of an unknown value between two known values. For example, suppose you are told that at 8:00 a.m. the temperature was 50°F and at 11:00 a.m. the same day it was 56°F. Based on this information alone, what is a good guess of the temperature at 10:00 a.m.? We assume that since 56 is greater than 50, the temperature gradually increased during the morning. Of course, it may actually have risen sharply to, say, 60°, and then fallen off to 56° because of clouds or rain. But we are not privy to such meteorological data; we must use only the given information. The safest guess (that is, the guess that would be closest to correct most often) is that the 6° rise from 50° to 56° was spread out evenly across the 3-hr time interval. So, it was 52° at 9:00 a.m. and 54° at 10:00 a.m. In other words, since 10:00 is two thirds of the way from 8:00 a.m. to 11:00 a.m., we will assume that temperature likewise climbed two thirds of the way from 50° to 56°.

Of course, linear interpolation doesn't generally involve such simple whole numbers. Nevertheless, the method always works in precisely the way illustrated above. We can write a fancy formula to express what we did: First, we calculated the number of hours from 8:00 a.m. to 10:00 a.m. and divided by the number of hours from 8:00 a.m. to 11:00 a.m. This gives us the *proportion* that 8:00 to 10:00 is of the entire time interval, 8:00 to 11:00. We then multiplied this proportion by the number of degrees between 50 and 56 and added this number to 50:

$$50° + \frac{10:00 - 8:00}{3}(56° - 50°) = 54°$$

This looks exactly like expressions [2.1] and [3.2], for example.

In general, interpolation is called for when we have three known quantities, a, b, and c, such that c lies between a and b; we have two other quantities A and B, such that A corresponds to a, and B corresponds to b, and we wish to guess or estimate the quantity C that corresponds to c:

$a \rightarrow A$

$c \rightarrow C(?)$

$b \rightarrow B$

To find C we first calculate the proportion of the distance from a to b that c is

$$\frac{c - a}{b - a}$$

we multiply this proportion by the distance from A to B,

$$\frac{c - a}{b - a}(B - A)$$

and then add the result to A:

$$C = A + \frac{c - a}{b - a}(B - A)$$

Graphically, we are picking C as the y-coordinate of c on the line between points (a, A) and (b, B):

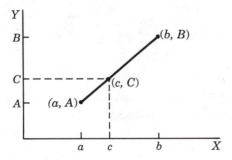

Since we are selecting a *line* as the basis of our guess, this process is called *linear* interpolation.

Regression coefficients

A. REGRESSION COEFFICIENTS FOR STANDARDIZED RANDOM VARIABLES

Let Z_X and Z_Y be standardized random variables, where

$$z_X = \frac{x - \bar{x}}{s_X} \quad \text{and} \quad z_Y = \frac{y - \bar{y}}{s_Y}$$

and let

$$\hat{z}_Y = \frac{\hat{y} - \bar{y}}{s_Y}$$

be the value of Z_Y predicted from z_X by the equation $\hat{z}_Y = bz_X + a$. The values of a and b that minimize the variance of estimate

$$\frac{\sum\limits_{}^{N}(\hat{z}_Y - z_Y)^2}{N} \qquad [1]$$

are $a = 0$ and $b = r_{XY}$, and the least squares regression equation for standardized random variables is therefore

$$\hat{z}_Y = r_{XY}z_X \qquad [2]$$

Proof:

1. Substituting $bz_X + a$ for $\hat{z}_Y$ in expression [1],

$$\frac{\sum\limits^{N}(\hat{z}_Y - z_Y)^2}{N} = \frac{\sum\limits^{N}[(bz_X + a) - z_Y]^2}{N}$$

$$= \frac{\sum\limits^{N}[(bz_X - z_Y) + a]^2}{N}$$

$$= \frac{\sum\limits^{N}[(bz_X - z_Y)^2 + 2a(bz_X - z_Y) + a]^2}{N}$$

$$= \frac{\sum\limits^{N}(bz_X - z_Y)^2}{N} + 2a\frac{\sum\limits^{N}(bz_X - z_Y)}{N} + \frac{\sum a^2}{N} \quad [3]$$

2. Consider the middle term of expression [3]:

$$2a \frac{\sum\limits^{N}(bz_X - z_Y)}{N} = 2a\left(\frac{\sum\limits^{N} bz_X}{N} - \frac{\sum\limits^{N} z_Y}{N}\right)$$

$$= 2a\left(\frac{b\sum\limits^{N} z_X}{N} - \frac{\sum\limits^{N} z_Y}{N}\right)$$

$$= 2a(b\bar{z}_X - \bar{z}_Y) = 2a[(b \cdot 0) - 0] = 0$$

3. Now consider the last term of expression [3]: Since a is a constant,

$$\frac{\sum a^2}{N} = \frac{Na^2}{N} = a^2$$

4. The variance of estimate [3] therefore reduces to

$$\frac{\sum\limits^{N}(bz_X - z_Y)^2}{N} + a^2 \qquad\qquad [4]$$

5. Since the variance of estimate is as small as possible, a must equal 0.

6. When $a = 0$, expression [4] becomes

$$\frac{\sum\limits^{N}(bz_X - z_Y)^2}{N} = \frac{\sum\limits^{N}(b^2 z_X^2 - 2bz_X z_Y + z_Y^2)}{N}$$

$$= \frac{b^2 \sum\limits^{N} z_X^2}{N} - \frac{2b \sum\limits^{N} z_X z_Y}{N} + \frac{\sum\limits^{N} z_Y^2}{N} \qquad\qquad [5]$$

7. From Chapter 3 we know that

$$\frac{\sum z^2}{N} = s_Z^2 = 1$$

and from Chapter 4 we know that

$$\frac{\sum z_X z_Y}{N} = r_{XY}$$

8. Substituting these into [5], we obtain

$$\frac{\sum\limits^{N}(bz_X - z_Y)^2}{N} = b^2 - 2br_{XY} + 1 \qquad\qquad [6]$$

9. Set $c = b - r_{XY}$, so $b = r_{XY} + c$. Then from [6], the variance of estimate is

$$(r_{XY} + c)^2 - 2(r_{XY} + c)r_{XY} + 1$$

$$= r_{XY}^2 + 2r_{XY}c + c^2 - 2r_{XY}^2 - 2r_{XY}c + 1$$

$$= 1 - r_{XY}^2 + c^2$$

10. Once again, since error is assumed to be minimum, c must be zero, and since $b = r_{XY} + c$ (see Step 9), it follows that $b = r_{XY}$. **[Q.E.D.]**

B. REGRESSION COEFFICIENTS FOR RANDOM VARIABLES IN NATURAL UNITS

Let X and Y be random variables and let $\hat{y}$ be the value of Y predicted from x by the equation $\hat{y} = a + bx$. The values of a and b that minimize the variance of estimate

$$\frac{\sum\limits_{}^{N} (\hat{y}_i - y_i)^2}{N}$$

are

$$b = r\frac{s_Y}{s_X} \quad \text{and} \quad a = \bar{y} - b\bar{x}$$

Proof:

We know from Chapter 3 that standardization merely moves the zero point and changes the units of measurement. It follows that we can obtain regression coefficients a and b by substituting

$$\frac{x - \bar{x}}{s_X}$$

for z_X and substituting

$$\frac{\hat{y} - \bar{y}}{s_Y}$$

for $\hat{z}_Y$ into the least squares regression equation for standardized random variables (equation [2], p. 688)

$$\hat{z}_Y = r_{XY}z_X$$

1. $\hat{z}_Y = \dfrac{\hat{y} - \bar{y}}{s_Y} = r_{XY}\left(\dfrac{x - \bar{x}}{s_X}\right)$

2. $\hat{y} - \bar{y} = r_{XY}s_Y\left(\dfrac{x - \bar{x}}{s_X}\right)$

3. $\hat{y} = r_{XY}s_Y\left(\dfrac{x - \bar{x}}{s_X}\right) + \bar{y}$

4. From Chapter 4 we know that

$$r_{XY} = \frac{C_{XY}}{s_X s_Y}$$

5. Therefore,

$$\hat{y} = \left(\frac{C_{XY}}{s_X s_Y}\right)s_Y\left(\frac{x - \bar{x}}{s_X}\right) + \bar{y} = \frac{C_{XY}}{s_X}\left(\frac{x - \bar{x}}{s_X}\right) + \bar{y}$$

$$= \frac{C_{XY}}{s_X^2}(x - \bar{x}) + \bar{y} = \left(\frac{C_{XY}}{s_X^2}\right)x + \bar{y} - \left(\frac{C_{XY}}{s_X^2}\right)\bar{x}$$

6. This means that

$$b = \left(\frac{C_{XY}}{s_X^2}\right) = r_{XY}\frac{s_Y}{s_X}$$

and that

$$a = \bar{y} - \left(\frac{C_{XY}}{s_X^2}\right)\bar{x} = \bar{y} - b\bar{x} \qquad\qquad \textbf{[Q.E.D.]}$$

APPENDIX IV

Scales of measurement

The numbers that a random variable assigns to the results of an experiment may constitute a *nominal* scale, an *ordinal* scale, an *interval* scale, or a *ratio* scale. These scales are thought of as *levels* of measurement because they constitute a hierarchy. The numbers assigned at each level have more *quantitative* meaning than do numbers at the next lower level and are therefore appropriate to more algebraic operations. And, scales at each level possess all the properties of lower-level scales.

Nominal scales. Consider a grand prix in which five cars have qualified. The starting lineup includes one Ferrari, one Eagle, one Honda, one Cooper-Maserati, and one Lotus. To facilitate identification of the cars for officials and spectators, the cars are assigned numbers as follows:

Car	Number
Cooper-Maserati	1
Ferrari	2
Honda	3
Eagle	4
Lotus	5

In this example, the numbers assigned to our units of observation (cars) are simply arbitrary symbols that serve to distinguish them from one another. It should therefore be apparent that numbers used in this fashion have no quantitative meaning. Designating the Honda as entry number 3, for example, in no way implies that it possesses less of some measurable attribute than entries 4 and 5 or that it possesses two more units of some attribute than entry number 1. Indeed, the number serves merely as a shorthand *name*, and any such ordered array of numbers is therefore called a *nominal* scale.

Random variables yield nominally scaled variables when observations are mapped into a set of numbers that represent *qualitative* categories. Bib numbers assigned to runners or to cross-country skiers sometimes represent categorical variables. In a ski marathon, for example, the first digit often distinguishes touring class skiers from racing class skiers, and the second digit may indicate gender. When the public address system announces, "The first woman racer is approaching the finish line," it need not be assumed that the official can tell touring skis from racing skis at 200 yards or even that a man is distinguishable from a woman at that distance. The announcer need only be able to read the bib number and know what categories the number represents.

Examples of nominally scaled variables in the behavioral and biological sciences include such qualitative characteristics as sex, political party affiliation, blood type, agreement or disagreement with a statement on an opinion questionnaire, presence or absence of a particular biochemical condition, and so forth. Such nominal categories can be identified either by names (or other words, e.g., male and female; Democrat, Republican, and independent) or by numbers, but numbers used in this way have no quantitative meaning. Typically, therefore, statisticians define another random variable, such as "let X be the *number of individuals* in Category 1." The *frequencies* thereby assigned to qualitative observations are the basis for much of probability theory.

Ordinal scales. Now let us suppose that the race is over and that the cars finished in the following order:

Car	Position
Ferrari	1
Lotus	2
Eagle	3
Honda	4
Cooper-Maserati	5

In this instance, the numbers *do* represent some quantitative attribute, for example, time. Assuming that all of the cars finished, we know that car number 1 spent less total time on the course than any of the others, that car number 2 spent less time than cars 3, 4, and 5, but more time than car number 1, and so forth. However, this scale does not tell us how *many* more minutes car number 2 spent on the course than car number 1. We cannot make quantitative statements about *differences* between the amount of this attribute possessed by the two observations. So, even though the difference in *position* between the Eagle and the Maserati

$$5 - 3 = 2$$

is the same as the difference between the Eagle and the Ferrari

$$3 - 1 = 2$$

we cannot infer that the corresponding time differences were equal. The scale in this example provides information only about the *order* in which the observations may be arrayed and is therefore an example of an *ordinal* scale.

Even with well-developed measurement techniques, ordinal scaling is often the highest level of measurement that can be legitimately assumed in the behavioral sciences. In an intelligence study, for example, a psychologist may be reasonably certain that an IQ of 160 indicates more of some attribute (e.g., aptitude for academic achievement) than an IQ of 150. But the psychologist also knows that measurement of this attribute breaks down at the extremes. It would therefore be difficult to justify the assumption that the difference in aptitude between persons with IQ scores of 150 and 160 is the same as the difference between persons scoring, say, 95 and 105.

Interval scales. A scale is said to have interval properties if units are the same size everywhere on the scale. What this means is that a unit of measurement at *one* place on the scale represents the *same* difference in the *measured characteristic* as one unit anywhere else on the scale. From the last paragraph, we see that the IQ scale does *not* possess interval properties, because an IQ point might represent a greater increment of aptitude at one place on the IQ scale than at another place. In contrast, 1°F represents 1/180 of the heat required to raise water under standard atmospheric conditions from the melting point to the boiling point. With this in mind, let us return to our motor racing scenario.

Suppose that all of the cars are equipped with telemetry devices that record cockpit temperature, and let us imagine that the highest recorded temperature for each car was as follows:

Car	Highest Recorded Cockpit Temperature (°F)
Eagle	144
Lotus	142
Ferrari	138
Cooper-Maserati	124
Honda	120

Since a difference of 1° at one place on the scale represents the same difference in heat as a difference of 1° anywhere else on the scale, equal intervals in Fahrenheit temperature reflect equal differences in heat. It is therefore perfectly reasonable to say that the difference between the Honda and the Maserati (124 − 120 = 4) is the same as the difference between the Ferrari and the Lotus (142 − 138 = 4). That is, 4°F represents an interval of the same magnitude whether it begins at 120 or 138 (or anyplace else). Had temperature been measured in degrees Celsius, the difference between the Honda and the Maserati would have been only about 2.2°, but it would still be the same as the difference between the Ferrari and the Lotus. It follows that an increase or decrease of 4°F (or 2.2°C) represents the same change in heat whether one begins at 120 or 138 (or anyplace else). Therefore, addition and subtraction can be performed meaningfully on measurements of temperature or, for that matter, on any interval-scaled measurements.

It is also meaningful to perform multiplication and division on *differences* between values of interval-scaled variables. For example, the difference between the Ferrari and the Lotus is *twice* as large as the difference between the Lotus and the Eagle, and this holds true whether differences are expressed in Fahrenheit degrees or in degrees Celsius. Although ratios of *differences* are meaningful for interval-scaled variables, ratios of *values* are not meaningful. The peak temperature in the

Eagle was 144°F and that of the Honda was 120°F, but we cannot calculate $144 \div 120 = 1.2$ and say that at peak temperature the Eagle was 20 percent warmer than the Honda. Neither 0°C nor 0°F represents a true zero point. (Absolute zero, the complete absence of heat, is -273°C, or approximately -460°F.) Ratios of values measured from arbitrary zero points (such as temperatures in the Fahrenheit and Celsius scales) are meaningless. This becomes obvious if we convert the temperatures of the Eagle and Honda to degrees Celsius. In the Celsius scale, the two temperatures are, respectively, 62.2 and 48.9. Their ratio is therefore $62.2 \div 48.9 \doteq 1.27$, which is different from the ratio 1.2 we obtained in degrees Fahrenheit. And, in general, multiplication and division of interval-scaled values are not meaningful operations.

Ratio scales. Finally, let us suppose that when the timekeeper posts the official times, recorded to the nearest minute, the results are as follows:

Car	Time in Minutes
Ferrari	68
Lotus	72
Eagle	75
Honda	81
Cooper-Maserati	84

Measurements of time have interval properties, so we can array the observations in a meaningful quantitative order, and we can make meaningful comparisons among differences between automobiles (e.g., claim that the difference between the Lotus and the Eagle is the same as the difference between the Honda and the Maserati). In addition, measured time has a true zero point; *zero* minutes represents a passage of *no* time. *Measurements* of time are therefore in the same *ratio* as the *times* they represent, so it makes perfect sense to say that the Cooper-Maserati spent 20 percent longer on the course than did the Ferrari. Whether time is measured in minutes, hours, or nanoseconds, the Ferrari time multiplied by 1.2 equals the Cooper-Maserati time, and in general, operations of multiplication and division are meaningful for ratio-scaled measurements. In Chapter 3 we stated that two measurements can be compared only if they are expressed in the same units of measurement and taken from the same reference point. These requirements can be satisfied only if the measurements are random variables with interval or ratio properties.

The algebra of expectations

Rule 1: If c is a constant, then

$$E(c) = c$$

If the same value c is assigned to every element in a sample space, then the value set has only one value, c. Therefore, $P(c) = 1$, and

$$E(c) = cP(c) = c(1) = c$$

Rule 2: If c is a constant and X is a random variable with expectation $E(X) = \mu$, then

$$E(X + c) = \mu + c$$

Proof:

$$E(X + c) = \sum_{i}^{n} (x_i + c) P(x_i + c)$$

$$= \sum_{i}^{n} x_i P(x_i + c) + \sum_{i}^{n} cP(x_i + c)$$

However, if c is to be added to every value, the probability of any value $(x_i + c)$ must be exactly equal to the probability of the corresponding x-value. That is, $P(x_i + c) = P(x_i)$, so

$$E(X + c) = \sum_{i}^{n} x_i P(x_i) + \sum_{i}^{n} cP(x_i)$$

$$= \mu + c\sum_{i}^{n} P(x_i) = \mu + c(1)$$

$$= \mu + c$$

Rule 3: If c is a constant and X is a random variable with expectation $E(X) = \mu$, then

$$E(cX) = c\mu$$

Proof:

If $c = 0$, the rule follows from Rule 1:

$$E(0X) = E(0) = 0 = 0\mu$$

Now suppose $c \neq 0$:

$$E(cX) = \sum_i^n (cx_i)P(cx_i)$$

However, if c is to be multiplied by every value, the probability of any value cx_i must be exactly equal to the probability of the corresponding x-value. That is, $P(cx_i) = P(x_i)$, so

$$E(cX) = \sum_i^n cx_i P(x_i)$$

$$= c \sum_i^n x_i P(x_i)$$

$$= c\mu$$

Rule 4: If X is a random variable with expectation $E(X) = \mu_X$ and Y is a random variable with expectation $E(Y) = \mu_Y$, then $E(X + Y) = \mu_X + \mu_Y$.

Proof:

$$E(X + Y) = \sum_i^n \sum_j^m (x_i + y_j)P(x_i, y_j)$$

$$= \sum_i^n \sum_j^m x_i P(x_i, y_j) + \sum_i^n \sum_j^m y_j P(x_i, y_j)$$

Consider the first term:

$$\sum_i^n \sum_j^m x_i P(x_i, y_j)$$

$$= \sum_j^m x_1 P(x_1, y_j) + \sum_j^m x_2 P(x_2, y_j) + \cdots + \sum_j^m x_n P(x_n, y_j)$$

The value x_1 is a constant in the first term, x_2 is a constant in the second term, etc. By Rule 2 of the Algebra of Summations $\sum\sum x_i P(x_i, y_j)$ therefore becomes

$$x_1 \sum_j^m P(x_1, y_j) + x_2 \sum_j^m P(x_2, y_j) + \cdots + x_n \sum_j^m P(x_n, y_j)$$

Note that

$$\sum_{j}^{m} P(x_i, y_j)$$

is the sum of the probabilities of x_1 across *all possible* values of y, which must be, simply, equal to the probability of x_1. That is,

$$\sum_{j}^{m} P(x_1, y_j) = P(x_1)$$

Therefore,

$$x_1 \sum_{j}^{m} P(x_1, y_j) = x_1 P(x_1)$$

Similarly,

$$x_2 \sum_{j}^{m} P(x_2, y_j) = x_2 P(x_2)$$

$$x_3 \sum_{j}^{m} P(x_3, y_j) = x_3 P(x_3)$$

$$\cdots$$

$$x_n \sum_{j}^{m} P(x_n, y_j) = x_n P(x_n)$$

Therefore,

$$\sum_{i}^{n} \sum_{j}^{m} x_i P(x_i, y_j) = x_1 P(x_1) + x_2 P(x_2) + \cdots + x_n P(x_n)$$

$$= E(X) = \mu_X$$

Similarly,

$$\sum_{i}^{n} \sum_{j}^{m} y_j P(x_i, y_j) = \mu_Y$$

so

$$E(X + Y) = \mu_X + \mu_Y$$

Rule 5: If X and Y are independent random variables with $E(X) = \mu_X$ and $E(Y) = \mu_Y$, then $E(XY) = \mu_X \mu_Y$.

Proof:

$$E(XY) = \sum_i^n \sum_j^m (x_i y_j) P(x_i, y_j)$$

Since X and Y are independent, $P(x_i, y_j) = P(x_i)P(y_j)$, so

$$E(XY) = \sum_i^n \sum_j^m (x_i y_j) P(x_i) P(y_j)$$

$$= \sum_j^m (x_1 y_j) P(x_1) P(y_j) + \sum_j^m (x_2 y_j) P(x_2) P(y_j)$$

$$+ \cdots + \sum_j^m (x_n y_j) P(x_n) P(y_j)$$

$$= x_1 P(x_1) \sum_j^m y_j P(y_j) + x_2 P(x_2) \sum_j^m y_j P(y_j)$$

$$+ \cdots + x_n P(x_n) \sum_j^m y_j P(y_j)$$

$$= x_1 P(x_1) E(Y) + x_2 P(x_2) E(Y) + \cdots + x_n P(x_n) E(Y)$$

$$= [x_1 P(x_1) + x_2 P(x_2) + \cdots + x_n P(x_n)] E(Y)$$

$$= E(X) E(Y) = \mu_X \mu_Y$$

Corollary: If X and Y are independent random variables, the covariance of X and Y is zero.

Proof:
The covariance of random variables X and Y is denoted $cov(X, Y)$, and by definition,

$$cov(X, Y) = E(XY) - E(X)E(Y)$$

Let $E(X) = \mu_X$ and $E(Y) = \mu_Y$. If X and Y are independent, then $E(XY) = \mu_X \mu_Y$ by Rule 5. Therefore,

$$cov(X, Y) = \mu_X \mu_Y - \mu_X \mu_Y = 0$$

Rule 6: If c is a constant, then

$$V(c) = 0$$

By definition, a constant does not vary and must therefore have zero variance. The formal proof is trivial:

Proof:

It is shown in Chapter 6 that if $E(X) = \mu$, then $V(X) = E(X^2) - \mu^2$. This can be rewritten

$$V(X) = [E(X)]^2$$

Likewise,

$$V(c) = E(c^2) - [E(c)]^2$$

By Rule 1 of the Algebra of Expectations, $E(c) = c$, so

$$[E(c)]^2 = c^2$$

Since c^2 is also a constant,

$$E(c^2) = c^2$$

Therefore,

$$V(c) = E(c^2) - [E(c)]^2 = c^2 - c^2 = 0$$

Rule 7: If c is a constant and X is a random variable with variance $V(X) = \sigma^2$, then

$$V(X + c) = \sigma^2$$

Proof:

$$V(X + c) = E[(X + c)^2] - [E(X + c)]^2$$

$$= E(X^2 + 2cX + c^2) - [E(X + c)]^2$$

By Rule 2 of the Algebra of Expectations,

$$E(X + c) = E(X) + c$$

Therefore,

$$V(X + c) = E(X^2 + 2cX + c^2) - [E(X) + c]^2$$

$$= E(X^2) + 2cE(X) + c^2 - [E(X)]^2 - 2cE(X) - c^2$$

$$= E(X^2) - [E(X)]^2$$

$$= V(X) = \sigma^2.$$

Rule 8: If c is a constant and X is a random variable with variance $V(X) = \sigma^2$, then

$$V(cX) = c^2\sigma^2$$

Proof:

$$V(cX) = E\left[(cX)^2\right] - \left[E(cX)\right]^2$$

$$= E(c^2X^2) - \left[E(cX)\right]^2$$

By Rule 3 of the Algebra of Expectations,

$$E(cX) = cE(X)$$

Therefore,

$$V(cX) = c^2E(X^2) - \left[cE(X)\right]^2$$

$$= c^2\left\{E(X^2) - \left[E(X)\right]^2\right\}$$

$$= c^2V(X) = c^2\sigma^2$$

Rule 9: If X and Y are independent random variables with variances $V(X) = \sigma_X^2$ and $V(Y) = \sigma_Y^2$, then

$$V(X + Y) = \sigma_X^2 + \sigma_Y^2$$

Proof:

$$V(X + Y) = E(X + Y)^2 - \left[E(X + Y)\right]^2$$

Expanding the first term, $E(X + Y)^2$,

$$V(X + Y) = E(X^2 + 2XY + Y^2) - \left[E(X + Y)\right]^2$$

By Rules 3 and 4 of the Algebra of Expectations,

$$E(X^2 + 2XY + Y^2) = E(X^2) + 2E(XY) + E(Y^2)$$

and by Rule 4 of the Algebra of Expectations,

$$\left[E(X + Y)\right]^2 = \left[E(X) + E(Y)\right]^2$$

Expanding the square, this becomes

$$\left[E(X)\right]^2 + 2E(X)E(Y) + \left[E(Y)\right]^2$$

so

$$V(X + Y) = E(X^2) + 2E(XY) + E(Y^2)$$

$$-\{[E(X)]^2 + 2E(X)E(Y) + [E(Y)]^2\}$$

Removing braces and collecting terms,

$$V(X + Y) = E(X^2) - [E(X)]^2 + E(Y^2)$$

$$-[E(Y)]^2 + 2[E(XY) - E(X)E(Y)]$$

$E(XY) - E(X)E(Y) = cov(X, Y)$ and by the corollary to Rule 5 of the Algebra of Expectations is equal to zero if X and Y are independent. Therefore,

$$V(X + Y) = \{E(X^2) - [E(X)]^2\} + \{E(Y^2) - [E(Y)]^2\}$$

$$= V(X) + V(Y) = \sigma_X^2 + \sigma_Y^2$$

Fundamentals of research design

A. EXPERIMENTAL AND OBSERVATIONAL RESEARCH

An *experiment* is a procedure in which the investigator manipulates one variable (called the *independent* variable) and observes the effects of these manipulations on another variable (called the *dependent* variable). For example, suppose that a particular food preservative is suspected of causing cancer. To test this hypothesis, a researcher might add the preservative to the diets of 100 rats and then compare their subsequent incidence of cancer with that of the untreated population from which the rats were selected. In this example the experimenter wants to draw an inference about *all rats* that might conceivably be administered the same dosage of food preservative. The target population is therefore defined by a variable that is manipulated by the experimenter.

In *observational* research the investigator is concerned about some attribute that subjects possess *before* they are selected for study. For example, in the comparisons of wolf and dog behavior discussed in Chapter 9, we wanted to make inferences about the population of wolves. The target population was therefore defined by species, an attribute that is determined at conception.

In experimental research, subjects that receive the experimental treatment (e.g., doses of a suspected cancer-causing chemical) are typically selected *at random* from a general population. In observational research, subjects may instead be selected *at random* from the population that possesses a particular attribute.

Observational research in which the "independent variable" is the *criterion* that defines the target population is sometimes called *quasi-experimental*. Comparisons between individuals representing naturally occurring populations (e.g., men and women) or between a general population and a well-defined subpopulation (e.g., persons who live near a toxic waste site and the general population) may be thought of as quasi-experimental studies.

In other observational research, every member of a general population may possess *some degree* of an attribute, for example, height or intelligence. The researcher therefore selects a *random sample* of individuals from the general population and measures the attribute. This method of collecting data is typical of correlation and regression studies, which were introduced in Chapter 4, and is sometimes called *correlational*. Research strategy considerations involved in testing hypotheses with correlational data are discussed in Chapter 14.

B. THE DESIGN OF EXPERIMENTS

The following discussion examines strategies for assigning subjects to two or more experimental treatment groups (one of which may be "no treatment"). Many of these strategies are therefore inapplicable to quasi-experimental designs. Likewise, many of these considerations are inapplicable to one-sample experiments, where a researcher tests the hypothesis that the parameter θ of the target population is equal to a specified value θ_0. Indeed, the considerations that call for these strategies most often arise *because* the experimenter does not have sufficient information to conduct a one-sample experiment.

Suppose a psychiatrist believes that many memory problems associated with aging are the result of progressive inability to metabolize calcium properly and that memory loss might be reduced by a type of drug called a calcium channel blocker. She therefore proposes to administer a calcium channel blocker to a group of hospitalized patients over 60 years of age and then have the subjects perform a series of memory tasks.

Experimental and control groups. If the experimenter has performance data that permit her to model the distribution of test performance for *untreated* persons, she can perform a one-sample experiment. She can administer the drug to a single group of subjects, measure their performance on the memory task, and test the hypothesis

$$H_0 : \theta = \theta_0$$

where θ is the parameter of the treated population and θ_0 is the known (or estimated) parameter value for the untreated population.

If data on the untreated population are not available she would have to test at least *two* groups, one of which received the drug and the other of which received either a placebo or nothing at all. In all other respects the experiment would be conducted under identical conditions for both groups. The group that is subjected to the experimental treatment is called the *experimental* group, and the group that receives no treatment (or no treatment that is relevant to the experimental task) is called the *control* group.

Randomization. There is also reason to believe that calcium channel blockers have no remedial effect on memory deficits caused by physical changes in brain cells, such as those associated with Alzheimer's disease. Many forms of cell damage are difficult, expensive, or impossible to diagnose reliably, so it may not be feasible to eliminate persons with these conditions from the experiment or to assign equal numbers of such persons to both groups. The experimenter should therefore assign subjects to her two groups completely at random—for example, by assigning persons born on odd-numbered dates to one group and persons born on even-numbered dates to the other, by flipping a coin, etc. As discussed in Chapter 8, two groups selected in this way would constitute *random samples*. The experimenter could therefore assume that the proportion of individuals with medical conditions that might affect responsiveness to the treatment would be essentially the same in both groups. These conditions would still contribute to variability in memory, but they would contribute only to performance differences *within* the groups, not to differences *between* the groups.

Matching. Cell pathology is not the only factor that might influence performance on a memory task. If the ability to metabolize calcium decreases with age, then age itself might affect either a subject's response to the drug or performance on the task irrespective of the drug. The experimenter has access to this variable, however, and she can balance its effects by matching rather than by relying on randomization. Following her initial random assignments, for example, she might group her subjects into age brackets—60 to 64, 65 to 69, 70 to 74, etc. If she found that one bracket was disproportionately represented in the control group and another bracket disproportionately represented in the experimental group, she could select individuals at random from the overrepresented brackets and drop them from the study or reassign them.

Like randomization, matching serves to eliminate the influence of variables other than the experimental treatment that might introduce systematic differences between experimental subjects and control subjects. The essential distinction is that matching depends on systematic group assignment rather than the laws of probability to ensure that extraneous variables are identically distributed in the two groups.

Subjects as own controls. A more sophisticated method of avoiding spurious group differences is to use the *same* subjects in both the control condition and in the experimental condition. This use of subjects as their *own* controls assures that all variables are identical in both experimental and control subjects and therefore eliminates possible sources of systematic variability that the experimenter might not anticipate or be able to control by matching.

Unfortunately, this design introduces a new set of problems. Performance on any experimental task is likely to improve with practice. If the experimental trials follow the control trials and if the performance is significantly better in the experimental condition, this result may as easily be attributed to practice as to the effects of the drug treatment. To eliminate the possibility of such practice effects, the subjects should be randomly separated into two groups. One group performs the task first in the experimental condition and then in the control condition. For the second group, this order is reversed. A *counterbalanced* design of this sort, however, requires that the entire experiment be conducted twice. The investigator must therefore be certain that both sets of experimental trials and both sets of control trials are administered under identical conditions. This always includes such obvious precautions as giving the same instructions, following the same procedures, and using tasks of equivalent difficulty. In some experiments it might also include control of more subtle variables, such as the time of day that the experiment is administered, the gender of the tester, and even the temperature of the room.

It should also be emphasized that observations on the *same* subjects under two (or more) conditions are *not* independent, so one must use statistical techniques appropriate to *correlated groups* (cf. Chapter 10, pp. 437–439) or *repeated measures* (cf. Chapter 13, pp. 554–559).

Sums of squares

A. COMPUTATIONAL FORMULAE

By definition

$$SS_{\text{TOTAL}} = \sum_j \sum_i (x_{ij} - \bar{x})^2$$

Expanding the square and distributing the sum we obtain

$$SS_{\text{TOTAL}} = \sum_j \sum_i (x_{ij} - 2x_{ij}\bar{x} + \bar{x})^2$$

$$= \sum_j \sum_i x_{ij} - 2\bar{x} \sum_j \sum_i x_{ij} + N\bar{x}^2 \qquad [1]$$

By definition, however,

$$\bar{x} = \frac{\overset{N}{\sum} x}{N} = \frac{\overset{J}{\sum} \overset{N_j}{\sum} x}{N}$$

so

$$\sum_j \sum_i x_{ij} = N\bar{x}$$

and [1] becomes

$$SS_{\text{TOTAL}} = \sum_j \sum_i x_{ij} - 2\bar{x}N\bar{x} + N\bar{x}^2 = \sum_j \sum_i x_{ij} - 2N\bar{x}^2 + N\bar{x}^2$$

$$= \sum_j \sum_i x_{ij} - N\bar{x}^2 \qquad [2]$$

The sum of squares within groups can also be expressed in a form that is more computationally convenient than the definitional formula:

$$SS_W = \sum_j \sum_i (x_{ij} - \bar{x}_j)^2 = \sum_j \sum_i \left(x_{ij}^2 - 2x_{ij}\bar{x}_j + \bar{x}_j^2\right)$$

In distributing the sum, note that $\bar{x}_j^2$ is a constant with respect to the within-group sum $\sum\limits_i$. Therefore,

$$SS_W = \sum_j \sum_i x_{ij}^2 - 2\sum_j \bar{x}_j \sum_i x_{ij} + \sum_j N_j \bar{x}_j^2 \qquad [3]$$

Consider the middle term on the right-hand side of [3]:

$$\sum_i x_{ij} = N_j \bar{x}_j$$

so

$$2\sum_j \bar{x}_j \sum_i x_{ij} = 2\sum_j \bar{x}_j N_j \bar{x}_j = 2\sum_j N_j \bar{x}_j^2$$

and equation [3] can be rewritten

$$SS_W = \sum_j \sum_i x_{ij}^2 - 2\sum_j N_j \bar{x}_j^2 + \sum_j N_j \bar{x}_j^2$$

$$= \sum_j \sum_i x_{ij}^2 - \sum_j N_j \bar{x}_j^2 \qquad [4]$$

To obtain the computational formula for SS_A, we simply subtract formula [4] from formula [2], that is,

$$SS_A = S_{\text{TOTAL}} - SS_W$$

$$= \sum_j \sum_i x_{ij}^2 - N\bar{x}^2 - \left(\sum_j \sum_i x_{ij}^2 - \sum_j N_j \bar{x}_j^2 \right)$$

$$= \sum_j N_j \bar{x}_j^2 - N\bar{x}^2 \qquad [5]$$

B. EXPECTED VALUES OF SUMS OF SQUARES

1. Basic quantities

If we examine the computational formulae for sums of squares, [2], [4], and [5], we note that they are comprised of three basic quantities:

$$A = \sum_j \sum_i x_{ij}^2$$

$$B = \sum_j N_j \bar{x}_j^2$$

$$D = N\bar{x}^2$$

and

$$SS_{\text{TOTAL}} = A - D$$

$$SS_W = A - B$$

$$SS_A = B - D$$

The expected values of A, B, and D are derived below. It is assumed that the N observations x_{ij} are yet to be taken and may therefore be thought of as N independent random variables, X_{ij}.

a. E(A)

$$E(A) = E\left(\sum_j \sum_i X_{ij}^2\right)$$

Since the expected value of a sum is the sum of expected values,

$$E(A) = \sum_j E\left(\sum_i X_{ij}^2\right) = \sum_j E\left(X_{1j}^2 + X_{2j}^2 + \cdots + X_{N_Jj}^2\right)$$

$$= \sum_j \left[E\left(X_{1j}^2\right) + E\left(X_{2j}^2\right) + \cdots + E\left(X_{N_Jj}^2\right)\right] \qquad [6]$$

From equation [8.8], we know that $E(X^2) = \sigma^2 + \mu^2$. Therefore, $E(X_{ij}^2) = \sigma_j^2 + \mu_j^2$, and equation [6] can be rewritten

$$E(A) = \sum_j \left[\underbrace{\left(\sigma_j^2 + \mu_j^2\right) + \left(\sigma_j^2 + \mu_j^2\right) + \cdots + \left(\sigma_j^2 + \mu_j^2\right)}_{N_j \text{ times}}\right] \qquad [7]$$

$$= \sum_j N_j\left(\sigma_j^2 + \mu_j^2\right)$$

$$= \sum_j N_j\sigma_j^2 + \sum_j N_j\mu_j^2$$

Since we are assuming that all of the population variances σ_j^2 are equal and that the only source of variability within each population is error variance, $\sigma_j^2 = \sigma_e^2$, equation [7] becomes

$$E(A) = \sum_j N_j\sigma_e^2 + \sum_j N_j\mu_j^2$$

$$= \sigma_e^2 \sum_j N_j + \sum_j N_j\mu_j^2$$

Since $\Sigma_j N_j = N$

$$E(A) = N\sigma_e^2 + \sum_j N_j \mu_j^2 \qquad [8]$$

b. $E(B)$

$$E(B) = E\left(\sum_j N_j \bar{X}_j^2\right)$$

$$= E\left(N_1 \bar{X}_1^2 + N_2 \bar{X}_2^2 + \cdots + N_J \bar{X}_J^2\right)$$

$$= N_1 E\left(\bar{X}_1^2\right) + N_2 E\left(\bar{X}_2^2\right) + \cdots + N_J E\left(\bar{X}_J^2\right) \qquad [9]$$

By equation [8.12]

$$E(\bar{X}^2) = \frac{\sigma^2}{N} + \mu^2$$

so

$$E\left(\bar{X}_j^2\right) = \frac{\sigma_j^2}{N} + \mu_j^2$$

and equation [9] can be rewritten

$$E(B) = N_1\left(\frac{\sigma_1^2}{N_1} + \mu_1^2\right) + N_2\left(\frac{\sigma_2^2}{N_2} + \mu_2^2\right) + \cdots + N_J\left(\frac{\sigma_J^2}{N_J} + \mu_J^2\right)$$

$$= \sum_j N_j\left(\frac{\sigma_j^2}{N_j} + \mu_j^2\right) = \sum_j \left(\sigma_j^2 + N_j \mu_j^2\right) = \sum_j \sigma_j^2 + \sum_j N_j \mu_j^2$$

Since all J population variances σ_j^2 are assumed equal to σ_e^2,

$$E(B) = J\sigma_e^2 + \sum_j N_j \mu_j^2 \qquad [10]$$

c. $E(D)$

$$E(D) = E(N\bar{X}^2)$$

$$= NE(\bar{X}^2)$$

The mean $\bar{X}$ is the grand population mean,

$$\frac{\sum X}{N}$$

where the sum includes all $N_1 + N_2 + \cdots + N_J = N$ independent observations:

$$X_{11}, X_{21}, \ldots, X_{N_1 1}$$
$$X_{12}, X_{22}, \ldots, X_{N_2 2}$$
$$\cdots$$
$$X_{1J}, X_{2J}, \ldots, X_{N_J J}$$

To simplify the following derivation, we will for the moment renumber these observations with consecutive single subscripts and call them

$$X_1, X_2, \ldots, X_N$$

Therefore,

$$E(D) = NE\left(\frac{\sum\limits_{}^{N} X}{N}\right)^2 = \frac{N}{N^2} E\left(\sum^N X\right)^2$$

$$= \frac{1}{N} E(X_1 + X_2 + \cdots + X_N)^2$$

$$= \frac{1}{N} E(X_1^2 + \cdots + X_N^2 + 2X_1 X_2 + \cdots + 2X_{N-1} X_N)$$

By the Algebra of Expectations, then,

$$E(D) = \frac{1}{N}\Big[E(X_1^2 + \cdots + E(X_N^2)$$

$$+ 2E(X_1 X_2) + \cdots + 2E(X_{N-1} X_N)\Big] \qquad [11]$$

By equation [8.12] $E(X^2) = \sigma^2 + \mu^2$, so

$$E(X_i^2) = \sigma_i^2 + \mu_i^2$$

and since our N random variables represent independent observations, $E(X_i X_j) = \mu_i \mu_j$ by Rule 5 of the Algebra of Expectations. Therefore, equation [11] becomes

$$E(D) = \frac{1}{N}\Big[(\sigma_1^2 + \mu_1^2) + \cdots + (\sigma_N^2 + \mu_N^2) \qquad [12]$$

$$+ 2\mu_1 \mu_2 + \cdots + 2\mu_{N-1}\mu_N \Big]$$

$$= \frac{1}{N}\Big[(\sigma_1^2 + \cdots + \sigma_N^2) + (\mu_1^2 + \cdots + \mu_N^2$$

$$+ 2\mu_1 \mu_2 + \cdots + 2\mu_{N-1}\mu_N)\Big]$$

$$= \frac{1}{N}\Big[(\sigma_1^2 + \cdots + \sigma_N^2) + (\mu_1 + \cdots + \mu_N)^2\Big]$$

Equation [12] can be rewritten in double subscripts:

$$E(D) = \frac{1}{N}\left[(\sigma_{11}^2 + \cdots + \sigma_{N_J J}^2) + (\mu_{11} + \cdots + \mu_{N_J J})^2\right] \qquad [13]$$

Since N_1 of the N variances in equation [13] are equal to σ_1, N_2 of them are equal to $\sigma_2^2, \ldots$, and N_J of them are equal to σ_J^2,

$$\sigma_{11}^2 + \cdots + \sigma_{N_J J}^2 = N_1 \sigma_1^2 + \cdots + N_J \sigma_J^2$$

Likewise,

$$\mu_{11} + \cdots + \mu_{N_J J} = N_1 \mu_1^2 + \cdots + N_J \mu_J^2 \qquad [14]$$

so

$$E(D) = \frac{1}{N}\left[(N_1 \sigma_1^2 + \cdots + N_J \sigma_J^2) + (N_1 \mu_1 + \cdots + N_J \mu_J)^2\right] \qquad [15]$$

However, we are assuming that all of the population variances σ_j^2 are equal to σ_e^2. Furthermore,

$$N_1 \mu_1 + \cdots + N_J \mu_J = \overset{N_1}{\sum} X_{i1} + \cdots + \overset{N_J}{\sum} X_{iJ} = \sum\sum X = N\mu \qquad [16]$$

Therefore, equation [15] becomes

$$E(D) = \frac{1}{N}\left[(N_1 \sigma_e^2 + \cdots + N_J \sigma_e^2) + (N\mu)^2\right] \qquad [17]$$

$$= \frac{1}{N}\left[\sigma_e^2(N_1 + \cdots + N_J) + N^2\mu^2\right]$$

$$= \frac{1}{N}(\sigma_e^2 N + N^2\mu^2)$$

$$= \sigma_e^2 + N\mu^2$$

2. Expected value of SS_W

By equation [4]

$$E(SS_W) = E\left(\sum_j \sum_i X_{ij}^2 - \sum_j N_j \bar{X}_j^2\right)$$

$$= E\left(\sum_j \sum_i X_{ij}^2\right) - E\left(\sum_j N_j \bar{X}_j^2\right)$$

From equation [8]

$$E\left(\sum_j \sum_i X_{ij}^2\right) = N\sigma_e^2 + \sum_j N_j \mu_j^2$$

and from equation [10]

$$E\left(\sum_j N_j \bar{X}_j^2\right) = J\sigma_e^2 + \sum_j N_j \mu_j^2$$

Therefore,

$$E(SS_W) = \left(N\sigma_e^2 + \sum_j N_j \mu_j^2\right) - \left(J\sigma_e^2 - \sum_j N_j \mu_j^2\right)$$

$$= N\sigma_e^2 - J\sigma_e^2$$

$$= \sigma_e^2(N - J) \qquad\qquad\qquad [18]$$

3. Expected value of SS_A

By equation [5]

$$E(SS_A) = E\left(\sum_j N_j \bar{X}_j^2 - N\bar{X}^2\right)$$

$$= E\left(\sum_j N_j \bar{X}_j^2\right) - E(N\bar{X}^2)$$

From equation [10]

$$E\left(\sum_j N_j \bar{X}_j^2\right) = J\sigma_e^2 + \sum_j N_j \mu_j^2$$

and from equation [17]

$$E(N\bar{X}^2) = \sigma_e^2 + N\mu^2$$

Therefore,

$$E(SS_A) = \left(J\sigma_e^2 + \sum_j N_j \mu_j^2\right) - (\sigma_e^2 + N\mu^2)$$

$$= \left(J\sigma_e^2 + \sum_j N_j \mu_j^2 - \sigma_e^2 - N\mu^2\right)$$

$$= \sum_j N_j \mu_j^2 - N\mu^2 + (J - 1)\sigma_e^2 \qquad\qquad [19]$$

In this regard,

$$\sum_j N_j(\mu_j - \mu)^2 = \sum_j N_j(\mu_j^2 - 2\mu_j\mu + \mu^2)$$

$$= \sum_j N_j\mu_j^2 - 2\sum N_j\mu_j\mu + \sum N_j\mu^2$$

$$= \sum_j N_j\mu_j^2 - 2\mu\sum N_j\mu_j + N\mu^2$$

From equation [16]

$$\sum N_j\mu_j = \sum\sum X = N\mu$$

so

$$\sum_j N_j\mu_j^2 - 2\mu\sum N_j\mu_j + N\mu^2 = \sum_j N_j\mu_j^2 - 2\mu N\mu + N\mu^2$$

$$= \sum_j N_j\mu_j^2 - 2N\mu^2 + N\mu^2$$

$$= \sum_j N_j\mu_j^2 - N\mu^2$$

Therefore, equation [19] can be rewritten,

$$E(SS_A) = \sum_j N_j(\mu_j - \mu)^2 + (J - 1)\sigma_e^2$$

From our discussion of the linear model in Chapter 12, we know that $\mu_j - \mu$ is defined as the treatment effect τ_j, so

$$E(SS_A) = \sum_j N_j\tau_j^2 + (J - 1)\sigma_e^2 \qquad [20]$$

Tables

Table 1. Uniformly distributed random numbers

Line/Col.	1	2	3	4	5	6	7	8	9	10	11	12	13	14
1	10480	15011	01536	02011	81647	91646	69179	14194	62590	36207	20969	99570	91291	90700
2	22368	46573	25595	85393	30995	89198	27982	53402	93965	34095	52666	19174	39615	99505
3	24130	48360	22527	97265	76393	64809	15179	23830	49340	32081	30680	19655	63348	58629
4	42167	93093	06243	61680	07856	16376	39440	53537	71341	57004	00849	74917	97758	16379
5	37570	39975	81837	16656	06121	91782	60468	81305	49684	60672	14110	06927	01263	54613
6	77921	06907	11008	42751	27756	53498	18602	70659	90655	15053	21916	81825	44394	42880
7	99562	72905	56420	69994	98872	31016	71194	18738	44013	48840	63213	21069	10634	12952
8	96301	91977	05463	07972	18876	20922	94595	56869	69014	60045	18425	84903	42508	32307
9	89579	14342	63661	10281	17453	18103	57740	84378	25331	12566	58678	44947	05585	56941
10	85475	36857	43342	53988	53060	59533	38867	62300	08158	17983	16439	11458	18593	64952
11	28918	69578	88231	33276	70997	79936	56865	05859	90106	31595	01547	85590	91610	78188
12	63553	40961	48235	03427	49626	69445	18663	72695	52180	29847	12234	90511	33703	90322
13	09429	93969	52636	92737	88974	33488	36320	17617	30015	08272	84115	27156	30613	74952
14	10365	61129	87529	85689	48237	52267	67689	93394	01511	26358	85104	20285	29975	89868
15	07119	97336	71048	08178	77233	13916	47564	81056	97735	85977	29372	74461	28551	90707
16	51085	12765	51821	51259	77452	16308	60756	92144	49442	53900	70960	63990	75601	40719
17	02368	21382	52404	60268	89368	19885	55322	44819	01188	65255	68435	44919	05944	55157
18	01011	54092	33362	94904	31273	04146	18594	29852	71585	85030	51132	01915	92747	64951
19	52162	53916	46369	58586	23216	14513	83149	98736	23495	64350	94738	17752	35156	35749
20	07056	97628	33787	09998	42698	06691	76988	13602	51851	46104	88916	19509	25625	58104
21	48663	91245	85828	14346	09272	30168	90229	04734	59193	22178	30421	61666	99904	32812
22	54164	58492	22421	74103	47070	25306	76468	26384	58151	06646	21524	15227	96909	44592
23	23639	32363	05597	24200	13363	38005	94342	28728	35806	06912	17012	64161	18296	22851
24	29334	27001	87637	87308	58731	00256	45834	15398	46557	41135	10367	07684	36188	18510
25	02488	33062	28834	07351	19731	92420	60952	61280	50001	67658	32586	86679	50720	94953
26	81525	72295	04839	96423	24788	82651	66566	14778	76797	14780	13300	87074	79666	95725
27	29676	20591	68086	26432	46901	20849	89768	81536	86645	12659	92259	57102	80428	25280
28	00742	57392	39064	66432	84673	40027	32832	61362	98947	96067	64760	64584	96096	98253
29	05366	04213	25669	26422	44407	44048	37937	63904	45766	66134	75470	66520	34693	90449
30	91921	26418	64117	94305	26766	25940	39972	22209	71500	64568	91402	42416	07844	69618
31	00582	04711	87917	77341	42206	35136	74087	99547	81817	42607	43808	76655	62028	76630
32	00725	69884	62797	56170	86324	88072	76222	36086	84637	93161	76038	65855	77919	88006
33	69011	65797	95876	55293	18988	27354	26575	08625	40801	59920	29841	80150	12777	48501
34	25976	57948	29888	88604	67917	48708	18912	82271	65424	69774	33611	54262	85963	03547
35	09763	83473	73577	12908	30883	18317	28290	35797	05998	41688	34952	37888	38917	88050
36	91567	42595	27958	30134	04024	86385	29880	99730	55536	84855	29080	09250	79656	73211
37	17955	56349	90999	49127	20044	59931	06115	29542	18059	02008	73708	83517	31603	42791
38	46503	18584	18845	49618	02304	51038	20655	58727	28168	15475	56942	53389	20562	87338
39	92157	89634	94824	78171	84610	82834	09922	25417	44137	48413	25555	21246	35509	20468
40	14577	62765	35605	81263	39667	47358	56873	56307	61607	49518	89656	20103	77490	18062

Table 1. Random numbers **717**

Table 1. (continued)

Line/Col.	1	2	3	4	5	6	7	8	9	10	11	12	13	14
41	98427	07523	33362	64270	01638	92477	66969	98420	04480	45585	46565	04102	46880	45709
42	34914	63976	88720	82765	34476	17032	87589	40836	32427	70002	70663	88863	77775	69348
43	70060	28277	39475	46473	23219	53416	94970	25832	69975	94884	19661	72828	00102	66794
44	53976	54914	06990	67245	68350	82948	11398	42878	80287	88267	47363	46634	06541	97809
45	76072	28515	40980	07391	58745	25774	22987	80059	39911	96189	41151	14222	60697	59583
46	90725	52210	83974	29992	65831	38857	50490	83765	55657	14361	31720	57375	56228	41546
47	64364	67412	33339	31926	14883	24413	59744	92351	97473	89286	35931	04110	23726	51900
48	08962	00358	31662	25388	61642	34072	81249	35648	56891	69352	48373	45578	78547	81788
49	95012	68379	93526	70765	10593	04542	76463	54328	02349	17247	28865	14777	62730	92277
50	15664	10493	20492	38391	91132	21999	59516	81652	27195	48223	46751	22923	32261	85653
51	16408	81899	04153	53381	79401	21438	83035	92350	36693	31238	59649	91754	72772	02338
52	18629	81953	05520	91962	04739	13092	97662	24822	94730	06496	35090	04822	86772	98299
53	73115	35101	47498	87637	99016	71060	88824	71013	18735	20286	23153	72924	35165	43040
54	57491	16703	23167	49323	45021	33132	12544	41035	80780	45393	44812	12515	98931	91202
55	30405	83946	23792	14422	15059	45799	22716	19792	09983	74353	68668	30429	70735	25499
56	16631	35006	85900	98275	32388	52390	16815	69298	82732	38480	73817	32523	41961	44437
57	96773	20206	42559	78985	05300	22164	24369	54224	35983	19687	11052	91491	60383	19746
58	38935	64202	14349	82674	66523	44133	00697	35552	35970	19124	63318	29686	03387	59846
59	31624	76384	17403	53363	44167	64486	64758	75366	76554	31601	12614	33072	60332	92325
60	78919	19474	23632	27889	47914	02584	37680	20801	72152	39339	34806	08930	85001	87820
61	03931	33309	57047	74211	63445	17361	62825	39908	05607	91284	68833	25570	38818	46920
62	74426	33278	43972	10119	89917	15665	52872	73823	73144	88662	88970	74492	51805	99378
63	09066	00903	20795	95452	92648	45454	09552	88815	16553	51125	79375	97596	16296	66092
64	42238	12426	87025	14267	20979	04508	64535	31355	86064	29472	47689	05974	52468	16834
65	16153	08002	26504	41744	81959	65642	74240	56302	00033	67107	77510	70625	28725	34191
66	21457	40742	29820	96783	29400	21840	15035	34537	33310	06116	95240	15957	16572	06004
67	21581	57802	02050	89728	17937	37621	47075	42080	96403	46826	68995	43805	33386	21597
68	55612	78095	83197	33732	05810	24813	86902	60397	16489	03264	88525	42786	05269	92532
69	44657	66999	99324	51281	84463	60563	79312	93454	68876	24571	93911	25650	12682	73572
70	91340	84979	46949	81973	37949	61023	43997	15263	80644	43942	89203	71795	99533	50501
71	91227	21199	31935	27022	84067	05462	35216	14486	29891	68607	41867	14951	91696	85065
72	50001	38140	66321	19924	72163	09538	12151	06878	91903	18749	34405	56087	82790	70925
73	65390	05224	72958	28609	81406	39147	25549	48542	42627	45233	57202	94617	23772	07896
74	27504	96131	83944	41575	10573	08619	64482	73923	36152	05184	94142	25299	84387	34925
75	37169	94851	39117	89632	00959	16487	65536	49071	39782	17095	02330	74301	00275	48280
76	11508	70225	51111	38351	19444	66499	71945	05422	13442	78675	84081	66938	93654	59894
77	37449	30362	06694	54690	04052	53115	62757	95348	78662	11163	81651	50245	34971	52924
78	46515	70331	85922	38329	57015	15765	97161	17869	45349	61796	66345	81073	49106	79860
79	30986	81223	42416	58353	21532	30502	32305	86482	05174	07901	54339	58861	74818	46942
80	63798	64995	46583	09765	44160	78218	83991	42865	92520	83531	80377	35909	81250	54238
81	82486	84846	99254	67632	43218	50076	21361	64816	51202	88124	41870	52689	51275	83556
82	21885	32906	92431	09060	64297	51674	64126	62570	26123	05155	59194	52799	28225	85762
83	60336	98782	07408	53458	13564	59089	26445	29789	85205	41001	12535	12133	14645	23541
84	43937	46891	24010	25560	86355	33941	35786	54990	71899	15475	95434	98227	21824	19585
85	97656	63175	89303	16275	07100	92063	21942	18611	47348	20203	18534	03862	78095	50136
86	03299	01221	05418	38982	55758	92237	26759	86367	21216	98442	08303	56613	91511	75928
87	79626	06486	03574	17668	07785	76020	79924	25651	83325	88428	85076	72811	22717	50585
88	85636	68335	47539	03129	65651	11977	02510	26113	99447	68645	34327	15152	55230	93448
89	18039	14367	61337	06177	12143	46609	32989	74014	64708	00533	35398	58408	13261	47908
90	08362	15656	60627	36478	65648	16764	53412	09013	07832	41574	17639	82163	60859	75567

Table 1. (continued)

Line/Col.	1	2	3	4	5	6	7	8	9	10	11	12	13	14
91	79556	29068	04142	16268	15387	12856	66227	38358	22478	73373	88732	09443	82558	05250
92	92608	82674	27072	32534	17075	27698	98204	63863	11951	34648	88022	56148	34925	57031
93	23982	25835	40055	67006	12293	02753	14827	22235	35071	99704	37543	11601	35503	85171
94	09915	96306	05908	97901	28395	14186	00821	80703	70426	74647	76310	88717	37890	40129
95	50937	33300	26695	62247	69927	76123	50842	43834	86654	70959	79725	93872	28117	19233
96	42488	78077	69882	61657	34136	79180	97526	43992	04098	73571	80799	76536	71255	64239
97	46764	86273	63003	93017	31204	36692	40202	35275	57306	55543	53203	18098	47635	88684
98	03237	45430	55417	63282	90816	17349	88298	90183	36600	78406	06216	95787	42579	90730
99	86591	81482	52667	61583	14972	90053	89534	76036	49199	43716	97548	04379	46370	28672
100	38534	01715	94964	87288	65680	43772	39560	12918	86537	62738	19636	51132	25739	56947

Source: Table 1 is reprinted with permission from *Handbook of Tables for Probability and Statistics*, 2nd edition, 1968, The Chemical Rubber Co., Cleveland, Ohio.

Table 2. Binomial distribution   **719**

Table 2. The binomial probability distribution

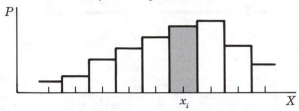

Table entry is height of shaded bar, $P(x = x_i)$.

N	x	.01	.05	.10	.15	.20	.25	.30	.35	.40	.45	.50	.55	.60	.65	.70	.75	.80	.85	.90	.95	.99
2	0	980	903	810	723	640	563	490	423	360	303	250	203	160	123	090	063	040	023	010	003	0+
	1	020	095	180	255	320	375	420	455	480	495	500	495	480	455	420	375	320	255	180	095	020
	2	0+	003	010	023	040	063	090	123	160	203	250	303	360	423	490	563	640	723	810	903	980
3	0	970	857	729	614	512	422	343	275	216	166	125	091	064	043	027	016	008	003	001	0+	0+
	1	029	135	243	325	384	422	441	444	432	408	375	334	288	239	189	141	096	057	027	007	0+
	2	0+	007	027	057	096	141	189	239	288	334	375	408	432	444	441	422	384	325	243	135	029
	3	0+	0+	001	003	008	016	027	043	064	091	125	166	216	275	343	422	512	614	729	857	970
4	0	961	815	656	522	410	316	240	179	130	092	063	041	026	015	008	004	002	001	0+	0+	0+
	1	039	171	292	368	410	422	412	384	346	299	250	200	154	111	076	047	026	011	004	0+	0+
	2	001	014	049	098	154	211	265	311	346	368	375	368	346	311	265	211	154	098	049	014	001
	3	0+	0+	004	011	026	047	076	111	154	200	250	299	346	384	412	422	410	368	292	171	039
	4	0+	0+	0+	001	002	004	008	015	026	041	063	092	130	179	240	316	410	522	656	815	961
5	0	951	774	590	444	328	237	168	116	078	050	031	018	010	005	002	001	0+	0+	0+	0+	0+
	1	048	204	328	392	410	396	360	312	259	206	156	113	077	049	028	015	006	002	0+	0+	0+
	2	001	021	073	138	205	264	309	336	346	337	313	276	230	181	132	088	051	024	008	001	0+
	3	0+	001	008	024	051	088	132	181	230	276	313	337	346	336	309	264	205	138	073	021	001
	4	0+	0+	0+	002	006	015	028	049	077	113	156	206	259	312	360	396	410	392	328	204	048
	5	0+	0+	0+	0+	0+	001	002	005	010	018	031	050	078	116	168	237	328	444	590	774	951
6	0	941	735	531	377	262	178	118	075	047	028	016	008	004	002	001	0+	0+	0+	0+	0+	0+
	1	057	232	354	399	393	356	303	244	187	136	094	061	037	020	010	004	002	0+	0+	0+	0+
	2	001	031	098	176	246	297	324	328	311	278	234	186	138	095	060	033	015	005	001	0+	0+
	3	0+	002	015	041	082	132	185	235	276	303	313	303	276	235	185	132	082	041	015	002	0+
	4	0+	0+	001	005	015	033	060	095	138	186	234	278	311	328	324	297	246	176	098	031	001
	5	0+	0+	0+	0+	002	004	010	020	037	061	094	136	187	244	303	356	393	399	354	232	057
	6	0+	0+	0+	0+	0+	0+	001	002	004	008	016	028	047	075	118	178	262	377	531	735	941
7	0	932	698	478	321	210	133	082	049	028	015	008	004	002	001	0+	0+	0+	0+	0+	0+	0+
	1	066	257	372	396	367	311	247	185	131	087	055	032	017	008	004	001	0+	0+	0+	0+	0+
	2	002	041	124	210	275	311	318	298	261	214	164	117	077	047	025	012	004	001	0+	0+	0+
	3	0+	004	023	062	115	173	227	268	290	292	273	239	194	144	097	058	029	011	003	0+	0+
	4	0+	0+	003	011	029	058	097	144	194	239	273	292	290	268	227	173	115	062	023	004	0+
	5	0+	0+	0+	001	004	012	025	047	077	117	164	214	261	298	318	311	275	210	124	041	002
	6	0+	0+	0+	0+	0+	001	004	008	017	032	055	087	131	185	247	311	367	396	372	257	066
	7	0+	0+	0+	0+	0+	0+	0+	001	002	004	008	015	028	049	082	133	210	321	478	698	932
8	0	923	663	430	272	168	100	058	032	017	008	004	002	001	0+	0+	0+	0+	0+	0+	0+	0+
	1	075	279	383	385	336	267	198	137	090	055	031	016	008	003	001	0+	0+	0+	0+	0+	0+
	2	003	051	149	238	294	311	296	259	209	157	109	070	041	022	010	004	001	0+	0+	0+	0+
	3	0+	005	033	084	147	208	254	279	279	257	219	172	124	081	047	023	009	003	0+	0+	0+
	4	0+	0+	005	018	046	087	136	188	232	263	273	263	232	188	136	087	046	018	005	0+	0+
	5	0+	0+	0+	003	009	023	047	081	124	172	219	257	279	279	254	208	147	084	033	005	0+
	6	0+	0+	0+	0+	001	004	010	022	041	070	109	157	209	259	296	311	294	238	149	051	003
	7	0+	0+	0+	0+	0+	0+	001	003	008	016	031	055	090	137	198	267	336	385	383	279	075
	8	0+	0+	0+	0+	0+	0+	0+	0+	001	002	004	008	017	032	058	100	168	272	430	663	923
9	0	914	630	387	232	134	075	040	021	010	005	002	001	0+	0+	0+	0+	0+	0+	0+	0+	0+
	1	083	299	387	368	302	225	156	100	060	034	018	008	004	001	0+	0+	0+	0+	0+	0+	0+
	2	003	063	172	260	302	300	267	216	161	111	070	041	021	010	004	001	0+	0+	0+	0+	0+
	3	0+	008	045	107	176	234	267	272	251	212	164	116	074	042	021	009	003	001	0+	0+	0+
	4	0+	001	007	028	066	117	172	219	251	260	246	213	167	118	074	039	017	005	001	0+	0+
	5	0+	0+	001	005	017	039	074	118	167	213	246	260	251	219	172	117	066	028	007	001	0+
	6	0+	0+	0+	001	003	009	021	042	074	116	164	212	251	272	267	234	176	107	045	008	0+
	7	0+	0+	0+	0+	0+	001	004	010	021	041	070	111	161	216	267	300	302	260	172	063	003
	8	0+	0+	0+	0+	0+	0+	0+	001	004	008	018	034	060	100	156	225	302	368	387	299	083
	9	0+	0+	0+	0+	0+	0+	0+	0+	0+	001	002	005	010	021	040	075	134	232	387	630	914
10	0	904	599	349	197	107	056	028	013	006	003	001	0+	0+	0+	0+	0+	0+	0+	0+	0+	0+
	1	091	315	387	347	268	188	121	072	040	021	010	004	002	001	0+	0+	0+	0+	0+	0+	0+
	2	004	075	194	276	302	282	233	176	121	076	044	023	011	004	001	0+	0+	0+	0+	0+	0+
	3	0+	010	057	130	201	250	267	252	215	166	117	075	042	021	009	003	001	0+	0+	0+	0+
	4	0+	001	011	040	088	146	200	238	251	238	205	160	111	069	037	016	006	001	0+	0+	0+
	5	0+	0+	001	008	026	058	103	154	201	234	246	234	201	154	103	058	026	008	001	0+	0+
	6	0+	0+	0+	001	006	016	037	069	111	160	205	238	251	238	200	146	088	040	011	001	0+
	7	0+	0+	0+	0+	001	003	009	021	042	075	117	166	215	252	267	250	201	130	057	010	0+
	8	0+	0+	0+	0+	0+	0+	001	004	011	023	044	076	121	176	233	282	302	276	194	075	004
	9	0+	0+	0+	0+	0+	0+	0+	001	002	004	010	021	040	072	121	188	268	347	387	315	091
	10	0+	0+	0+	0+	0+	0+	0+	0+	0+	001	003	006	013	028	056	107	197	349	599	904	

Table 2. (continued)

N	x	.01	.05	.10	.15	.20	.25	.30	.35	.40	.45	.50	.55	.60	.65	.70	.75	.80	.85	.90	.95	.99
11	0	895	569	314	167	086	042	020	009	004	001	0+	0+	0+	0+	0+	0+	0+	0+	0+	0+	0+
	1	099	329	384	325	236	155	093	052	027	013	005	002	001	0+	0+	0+	0+	0+	0+	0+	0+
	2	005	087	213	287	295	258	200	140	089	051	027	013	005	002	001	0+	0+	0+	0+	0+	0+
	3	0+	014	071	152	221	258	257	225	177	126	081	046	023	010	004	001	0+	0+	0+	0+	0+
	4	0+	001	016	054	111	172	220	243	236	206	161	113	070	038	017	006	002	0+	0+	0+	0+
	5	0+	0+	002	013	039	080	132	183	221	236	226	193	147	099	057	027	010	002	0+	0+	0+
	6	0+	0+	0+	002	010	027	057	099	147	193	226	236	221	183	132	080	039	013	002	0+	0+
	7	0+	0+	0+	0+	002	006	017	038	070	113	161	206	236	243	220	172	111	054	016	001	0+
	8	0+	0+	0+	0+	0+	001	004	010	023	046	081	126	177	225	257	258	221	152	071	014	0+
	9	0+	0+	0+	0+	0+	0+	001	002	005	013	027	051	089	140	200	258	295	287	213	087	005
	10	0+	0+	0+	0+	0+	0+	0+	0+	001	002	005	013	027	052	093	155	236	325	384	329	099
	11	0+	0+	0+	0+	0+	0+	0+	0+	0+	0+	0+	001	004	009	020	042	086	167	314	569	895
12	0	886	540	282	142	069	032	014	006	002	001	0+	0+	0+	0+	0+	0+	0+	0+	0+	0+	0+
	1	107	341	377	301	206	127	071	037	017	008	003	001	0+	0+	0+	0+	0+	0+	0+	0+	0+
	2	006	099	230	292	283	232	168	109	064	034	016	007	002	001	0+	0+	0+	0+	0+	0+	0+
	3	0+	017	085	172	236	258	240	195	142	092	054	028	012	005	001	0+	0+	0+	0+	0+	0+
	4	0+	002	021	068	133	194	231	237	213	170	121	076	042	020	008	002	001	0+	0+	0+	0+
	5	0+	0+	004	019	053	103	158	204	227	222	193	149	101	059	029	011	003	001	0+	0+	0+
	6	0+	0+	0+	004	016	040	079	128	177	212	226	212	177	128	079	040	016	004	0+	0+	0+
	7	0+	0+	0+	001	003	011	029	059	101	149	193	222	227	204	158	103	053	019	004	0+	0+
	8	0+	0+	0+	0+	001	002	008	020	042	076	121	179	213	237	231	194	133	068	021	002	0+
	9	0+	0+	0+	0+	0+	0+	001	005	012	028	054	092	142	195	240	258	236	172	085	017	0+
	10	0+	0+	0+	0+	0+	0+	0+	001	002	007	016	034	064	109	168	232	283	292	230	099	006
	11	0+	0+	0+	0+	0+	0+	0+	0+	0+	001	003	008	017	037	071	127	206	301	377	341	107
	12	0+	0+	0+	0+	0+	0+	0+	0+	0+	0+	0+	001	002	006	014	032	069	142	282	540	886
13	0	878	513	254	121	055	024	010	004	001	0+	0+	0+	0+	0+	0+	0+	0+	0+	0+	0+	0+
	1	115	351	367	277	179	103	054	026	011	004	002	0+	0+	0+	0+	0+	0+	0+	0+	0+	0+
	2	007	111	245	294	268	206	139	084	045	022	010	004	001	0+	0+	0+	0+	0+	0+	0+	0+
	3	0+	021	100	190	246	252	218	165	111	066	035	016	006	002	001	0+	0+	0+	0+	0+	0+
	4	0+	003	028	084	154	210	234	222	184	135	087	050	024	010	003	001	0+	0+	0+	0+	0+
	5	0+	0+	006	027	069	126	180	215	221	199	157	109	066	034	014	005	001	0+	0+	0+	0+
	6	0+	0+	001	006	023	056	103	155	197	217	209	177	131	083	044	019	006	001	0+	0+	0+
	7	0+	0+	0+	001	006	019	044	083	131	177	209	217	197	155	103	056	023	006	001	0+	0+
	8	0+	0+	0+	0+	001	005	014	034	066	109	157	199	221	215	180	126	069	027	006	0+	0+
	9	0+	0+	0+	0+	0+	001	003	010	024	050	087	135	184	222	234	210	154	084	028	003	0+
	10	0+	0+	0+	0+	0+	0+	001	002	006	016	035	066	111	165	218	252	246	190	100	021	0+
	11	0+	0+	0+	0+	0+	0+	0+	0+	001	004	010	022	045	084	139	206	268	294	245	111	007
	12	0+	0+	0+	0+	0+	0+	0+	0+	0+	0+	002	004	011	026	054	103	179	277	367	351	115
	13	0+	0+	0+	0+	0+	0+	0+	0+	0+	0+	0+	0+	001	004	010	024	055	121	254	513	878
14	0	869	488	229	103	044	018	007	002	001	0+	0+	0+	0+	0+	0+	0+	0+	0+	0+	0+	0+
	1	123	359	356	254	154	083	041	018	007	003	001	0+	0+	0+	0+	0+	0+	0+	0+	0+	0+
	2	008	123	257	291	250	180	113	063	032	014	006	002	001	0+	0+	0+	0+	0+	0+	0+	0+
	3	0+	026	114	206	250	240	194	137	085	046	022	009	003	001	0+	0+	0+	0+	0+	0+	0+
	4	0+	004	035	100	172	220	229	202	155	104	061	031	014	005	001	0+	0+	0+	0+	0+	0+
	5	0+	0+	008	035	086	147	196	218	207	170	122	076	041	018	007	002	0+	0+	0+	0+	0+
	6	0+	0+	001	009	032	073	126	176	207	209	183	140	092	051	023	008	002	0+	0+	0+	0+
	7	0+	0+	0+	002	009	028	062	108	157	195	209	195	157	108	062	028	009	002	0+	0+	0+
	8	0+	0+	0+	0+	002	008	023	051	092	140	183	209	207	176	126	073	032	009	001	0+	0+
	9	0+	0+	0+	0+	0+	002	007	018	041	076	122	170	207	218	196	147	086	035	008	0+	0+
	10	0+	0+	0+	0+	0+	0+	001	005	014	031	061	104	155	202	229	220	172	100	035	004	0+
	11	0+	0+	0+	0+	0+	0+	0+	001	003	009	022	046	085	137	194	240	250	206	114	026	0+
	12	0+	0+	0+	0+	0+	0+	0+	0+	001	002	006	014	032	063	113	180	250	291	257	123	008
	13	0+	0+	0+	0+	0+	0+	0+	0+	0+	0+	001	003	007	018	041	083	154	254	356	359	123
	14	0+	0+	0+	0+	0+	0+	0+	0+	0+	0+	0+	0+	001	002	007	018	044	103	229	488	869
15	0	860	463	206	087	035	013	005	002	0+	0+	0+	0+	0+	0+	0+	0+	0+	0+	0+	0+	0+
	1	130	366	343	231	132	067	031	013	005	002	0+	0+	0+	0+	0+	0+	0+	0+	0+	0+	0+
	2	009	135	267	286	231	156	092	048	022	009	003	001	0+	0+	0+	0+	0+	0+	0+	0+	0+
	3	0+	031	129	218	250	225	170	111	063	032	014	005	002	0+	0+	0+	0+	0+	0+	0+	0+
	4	0+	005	043	116	188	225	219	179	127	078	042	019	007	002	001	0+	0+	0+	0+	0+	0+
	5	0+	001	010	045	103	165	206	212	186	140	092	051	024	010	003	001	0+	0+	0+	0+	0+
	6	0+	0+	002	013	043	092	147	191	207	191	153	105	061	030	012	003	001	0+	0+	0+	0+
	7	0+	0+	0+	003	014	039	081	132	177	201	196	165	118	071	035	013	003	001	0+	0+	0+
	8	0+	0+	0+	001	003	013	035	071	118	165	196	201	177	132	081	039	014	003	0+	0+	0+
	9	0+	0+	0+	0+	001	003	012	030	061	105	153	191	207	191	147	092	043	013	002	0+	0+
	10	0+	0+	0+	0+	0+	001	003	010	024	051	092	140	186	212	206	165	103	045	010	001	0+
	11	0+	0+	0+	0+	0+	0+	001	002	007	019	042	078	127	179	219	225	188	116	043	005	0+
	12	0+	0+	0+	0+	0+	0+	0+	0+	002	005	014	032	063	111	170	225	250	218	129	031	0+
	13	0+	0+	0+	0+	0+	0+	0+	0+	0+	001	003	009	022	048	092	156	231	286	267	135	009
	14	0+	0+	0+	0+	0+	0+	0+	0+	0+	0+	0+	002	005	013	031	067	132	231	343	366	130
	15	0+	0+	0+	0+	0+	0+	0+	0+	0+	0+	0+	0+	0+	002	005	013	035	087	206	463	860

Table 2. Binomial distribution 721

Table 2. (continued)

p

N	x	.01	.05	.10	.15	.20	.25	.30	.35	.40	.45	.50	.55	.60	.65	.70	.75	.80	.85	.90	.95	.99
16	0	851	440	185	074	028	010	003	001	0+	0+	0+	0+	0+	0+	0+	0+	0+	0+	0+	0+	0+
	1	138	371	329	210	113	053	023	009	003	001	0+	0+	0+	0+	0+	0+	0+	0+	0+	0+	0+
	2	010	146	275	277	211	134	073	035	015	006	002	001	0+	0+	0+	0+	0+	0+	0+	0+	0+
	3	0+	036	142	229	246	208	146	089	047	022	009	003	001	0+	0+	0+	0+	0+	0+	0+	0+
	4	0+	006	051	131	200	225	204	155	101	057	028	011	004	001	0+	0+	0+	0+	0+	0+	0+
	5	0+	001	014	056	120	180	210	201	162	112	067	034	014	005	001	0+	0+	0+	0+	0+	0+
	6	0+	0+	003	018	055	110	165	198	198	168	122	075	039	017	006	001	0+	0+	0+	0+	0+
	7	0+	0+	0+	005	020	052	101	152	189	197	175	132	084	044	019	006	001	0+	0+	0+	0+
	8	0+	0+	0+	001	006	020	049	092	142	181	196	181	142	092	049	020	006	001	0+	0+	0+
	9	0+	0+	0+	0+	001	006	019	044	084	132	175	197	189	152	101	052	020	005	0+	0+	0+
	10	0+	0+	0+	0+	0+	001	006	017	039	075	122	168	198	198	165	110	055	018	003	0+	0+
	11	0+	0+	0+	0+	0+	0+	001	005	014	034	067	112	162	201	210	180	120	056	014	001	0+
	12	0+	0+	0+	0+	0+	0+	0+	001	004	011	028	057	101	155	204	225	200	131	051	006	0+
	13	0+	0+	0+	0+	0+	0+	0+	0+	001	003	009	022	047	089	146	208	246	229	142	036	0+
	14	0+	0+	0+	0+	0+	0+	0+	0+	0+	001	002	006	015	035	073	134	211	277	275	146	010
	15	0+	0+	0+	0+	0+	0+	0+	0+	0+	0+	0+	001	003	009	023	053	113	210	329	371	138
	16	0+	0+	0+	0+	0+	0+	0+	0+	0+	0+	0+	0+	0+	001	003	010	028	074	185	440	851
17	0	843	418	167	063	023	008	002	001	0+	0+	0+	0+	0+	0+	0+	0+	0+	0+	0+	0+	0+
	1	145	374	315	189	096	043	017	006	002	001	0+	0+	0+	0+	0+	0+	0+	0+	0+	0+	0+
	2	012	158	280	267	191	114	058	026	010	004	001	0+	0+	0+	0+	0+	0+	0+	0+	0+	0+
	3	001	041	156	236	239	189	125	070	034	014	005	002	0+	0+	0+	0+	0+	0+	0+	0+	0+
	4	0+	008	060	146	209	221	187	132	080	041	018	007	002	001	0+	0+	0+	0+	0+	0+	0+
	5	0+	001	017	067	136	191	208	185	138	087	047	021	008	002	001	0+	0+	0+	0+	0+	0+
	6	0+	0+	004	024	068	128	178	199	184	143	094	052	024	009	003	001	0+	0+	0+	0+	0+
	7	0+	0+	001	007	027	067	120	168	193	184	148	101	057	026	009	002	0+	0+	0+	0+	0+
	8	0+	0+	0+	001	008	028	064	113	161	188	185	154	107	061	028	009	002	0+	0+	0+	0+
	9	0+	0+	0+	0+	002	009	028	061	107	154	185	188	161	113	064	028	008	001	0+	0+	0+
	10	0+	0+	0+	0+	0+	002	009	026	057	101	148	184	193	168	120	067	027	007	001	0+	0+
	11	0+	0+	0+	0+	0+	001	003	009	024	052	094	143	184	199	178	128	068	024	004	0+	0+
	12	0+	0+	0+	0+	0+	0+	001	002	008	021	047	087	138	185	208	191	136	067	017	001	0+
	13	0+	0+	0+	0+	0+	0+	0+	001	002	007	018	041	080	132	187	221	209	146	060	008	0+
	14	0+	0+	0+	0+	0+	0+	0+	0+	0+	002	005	014	034	070	125	189	239	236	156	041	001
	15	0+	0+	0+	0+	0+	0+	0+	0+	0+	0+	001	004	010	026	058	114	191	267	280	158	012
	16	0+	0+	0+	0+	0+	0+	0+	0+	0+	0+	0+	001	002	006	017	043	096	189	315	374	145
	17	0+	0+	0+	0+	0+	0+	0+	0+	0+	0+	0+	0+	0+	001	002	008	023	063	167	418	843
18	0	835	397	150	054	018	006	002	0+	0+	0+	0+	0+	0+	0+	0+	0+	0+	0+	0+	0+	0+
	1	152	376	300	170	081	034	013	004	001	0+	0+	0+	0+	0+	0+	0+	0+	0+	0+	0+	0+
	2	013	168	284	256	172	096	046	019	007	002	001	0+	0+	0+	0+	0+	0+	0+	0+	0+	0+
	3	001	047	168	241	230	170	105	055	025	009	003	001	0+	0+	0+	0+	0+	0+	0+	0+	0+
	4	0+	009	070	159	215	213	168	110	061	029	012	004	001	0+	0+	0+	0+	0+	0+	0+	0+
	5	0+	001	022	079	151	199	202	166	115	067	033	013	004	001	0+	0+	0+	0+	0+	0+	0+
	6	0+	0+	005	030	082	144	187	194	166	118	071	035	015	005	001	0+	0+	0+	0+	0+	0+
	7	0+	0+	001	009	035	082	138	179	189	166	121	074	037	015	005	001	0+	0+	0+	0+	0+
	8	0+	0+	0+	002	012	038	081	133	173	186	167	125	077	038	015	004	001	0+	0+	0+	0+
	9	0+	0+	0+	0+	003	014	039	079	128	169	185	169	128	079	039	014	003	0+	0+	0+	0+
	10	0+	0+	0+	0+	001	004	015	038	077	125	167	186	173	133	081	038	012	002	0+	0+	0+
	11	0+	0+	0+	0+	0+	001	005	015	037	074	121	166	189	179	138	082	035	009	001	0+	0+
	12	0+	0+	0+	0+	0+	0+	001	005	015	035	071	118	166	194	187	144	082	030	005	0+	0+
	13	0+	0+	0+	0+	0+	0+	0+	001	004	013	033	067	115	166	202	199	151	079	022	001	0+
	14	0+	0+	0+	0+	0+	0+	0+	0+	001	004	012	029	061	110	168	213	215	159	070	009	0+
	15	0+	0+	0+	0+	0+	0+	0+	0+	0+	001	003	009	025	055	105	170	230	241	168	047	001
	16	0+	0+	0+	0+	0+	0+	0+	0+	0+	0+	001	002	007	019	046	096	172	256	284	168	013
	17	0+	0+	0+	0+	0+	0+	0+	0+	0+	0+	0+	0+	001	004	013	034	081	170	300	376	152
	18	0+	0+	0+	0+	0+	0+	0+	0+	0+	0+	0+	0+	0+	0+	002	006	018	054	150	397	835
19	0	826	377	135	046	014	004	001	0+	0+	0+	0+	0+	0+	0+	0+	0+	0+	0+	0+	0+	0+
	1	159	377	285	153	068	027	009	003	001	0+	0+	0+	0+	0+	0+	0+	0+	0+	0+	0+	0+
	2	014	179	285	243	154	080	036	014	005	001	0+	0+	0+	0+	0+	0+	0+	0+	0+	0+	0+
	3	001	053	180	243	218	152	087	042	017	006	002	0+	0+	0+	0+	0+	0+	0+	0+	0+	0+
	4	0+	011	080	171	218	202	149	091	047	020	007	002	001	0+	0+	0+	0+	0+	0+	0+	0+
	5	0+	002	027	091	164	202	192	147	093	050	022	008	002	001	0+	0+	0+	0+	0+	0+	0+
	6	0+	0+	007	037	095	157	192	184	145	095	052	023	008	002	001	0+	0+	0+	0+	0+	0+
	7	0+	0+	001	012	044	097	153	184	180	144	096	053	024	008	002	0+	0+	0+	0+	0+	0+
	8	0+	0+	0+	003	017	049	098	149	180	177	144	098	053	023	008	002	0+	0+	0+	0+	0+
	9	0+	0+	0+	001	005	020	051	098	146	177	176	145	098	053	022	007	001	0+	0+	0+	0+
	10	0+	0+	0+	0+	001	007	022	053	098	145	176	177	146	098	051	020	005	001	0+	0+	0+
	11	0+	0+	0+	0+	0+	002	008	023	053	098	144	177	180	149	098	049	017	003	0+	0+	0+
	12	0+	0+	0+	0+	0+	0+	002	008	024	053	096	144	180	184	153	097	044	012	001	0+	0+
	13	0+	0+	0+	0+	0+	0+	001	002	008	023	052	095	145	184	192	157	095	037	007	0+	0+
	14	0+	0+	0+	0+	0+	0+	0+	001	002	008	022	050	093	147	192	202	164	091	027	002	0+
	15	0+	0+	0+	0+	0+	0+	0+	0+	001	002	007	020	047	091	149	202	218	171	080	011	0+
	16	0+	0+	0+	0+	0+	0+	0+	0+	0+	0+	002	006	017	042	087	152	218	243	180	053	001
	17	0+	0+	0+	0+	0+	0+	0+	0+	0+	0+	0+	001	005	014	036	080	154	243	285	179	014
	18	0+	0+	0+	0+	0+	0+	0+	0+	0+	0+	0+	0+	001	003	009	027	068	153	285	377	159
	19	0+	0+	0+	0+	0+	0+	0+	0+	0+	0+	0+	0+	0+	0+	001	004	014	046	135	377	826

Table 2. (continued)

N	x	.01	.05	.10	.15	.20	.25	.30	.35	.40	.45	.50	.55	.60	.65	.70	.75	.80	.85	.90	.95	.99
20	0	818	358	122	039	012	003	001	0+	0+	0+	0+	0+	0+	0+	0+	0+	0+	0+	0+	0+	0+
	1	165	377	270	137	058	021	007	002	0+	0+	0+	0+	0+	0+	0+	0+	0+	0+	0+	0+	0+
	2	016	189	285	229	137	067	028	010	003	001	0+	0+	0+	0+	0+	0+	0+	0+	0+	0+	0+
	3	001	060	190	243	205	134	072	032	012	004	001	0+	0+	0+	0+	0+	0+	0+	0+	0+	0+
	4	0+	013	090	182	218	190	130	074	035	014	005	001	0+	0+	0+	0+	0+	0+	0+	0+	0+
	5	0+	002	032	103	175	202	179	127	075	036	015	005	001	0+	0+	0+	0+	0+	0+	0+	0+
	6	0+	0+	009	045	109	169	192	171	124	075	037	015	005	001	0+	0+	0+	0+	0+	0+	0+
	7	0+	0+	002	016	055	112	164	184	166	122	074	037	015	004	001	0+	0+	0+	0+	0+	0+
	8	0+	0+	0+	005	022	061	114	161	180	162	120	073	035	014	004	001	0+	0+	0+	0+	0+
	9	0+	0+	0+	001	007	027	065	116	160	177	160	119	071	034	012	003	0+	0+	0+	0+	0+
	10	0+	0+	0+	0+	002	010	031	069	117	159	176	159	117	069	031	010	002	0+	0+	0+	0+
	11	0+	0+	0+	0+	0+	003	012	034	071	119	160	177	160	116	065	027	007	001	0+	0+	0+
	12	0+	0+	0+	0+	0+	001	004	014	035	073	120	162	180	161	114	061	022	005	0+	0+	0+
	13	0+	0+	0+	0+	0+	0+	001	004	015	037	074	122	166	184	164	112	055	016	002	0+	0+
	14	0+	0+	0+	0+	0+	0+	0+	001	005	015	037	075	124	171	192	169	109	045	009	0+	0+
	15	0+	0+	0+	0+	0+	0+	0+	0+	001	005	015	036	075	127	179	202	175	103	032	002	0+
	16	0+	0+	0+	0+	0+	0+	0+	0+	0+	001	005	014	035	074	130	190	218	182	090	013	0+
	17	0+	0+	0+	0+	0+	0+	0+	0+	0+	0+	001	004	012	032	072	134	205	243	190	060	001
	18	0+	0+	0+	0+	0+	0+	0+	0+	0+	0+	0+	001	003	010	028	067	137	229	285	189	016
	19	0+	0+	0+	0+	0+	0+	0+	0+	0+	0+	0+	0+	0+	002	007	021	058	137	270	377	165
	20	0+	0+	0+	0+	0+	0+	0+	0+	0+	0+	0+	0+	0+	0+	001	003	012	039	122	358	818
21	0	810	341	109	033	009	002	001	0+	0+	0+	0+	0+	0+	0+	0+	0+	0+	0+	0+	0+	0+
	1	172	376	255	122	048	017	005	001	0+	0+	0+	0+	0+	0+	0+	0+	0+	0+	0+	0+	0+
	2	017	198	284	215	121	055	022	007	002	0+	0+	0+	0+	0+	0+	0+	0+	0+	0+	0+	0+
	3	001	066	200	241	192	117	058	024	009	003	001	0+	0+	0+	0+	0+	0+	0+	0+	0+	0+
	4	0+	016	100	191	216	176	113	059	026	009	003	001	0+	0+	0+	0+	0+	0+	0+	0+	0+
	5	0+	003	038	115	183	199	164	109	059	026	010	003	001	0+	0+	0+	0+	0+	0+	0+	0+
	6	0+	0+	011	054	122	177	188	156	105	057	026	009	003	001	0+	0+	0+	0+	0+	0+	0+
	7	0+	0+	003	020	065	126	172	180	149	101	055	025	009	002	0+	0+	0+	0+	0+	0+	0+
	8	0+	0+	001	006	029	074	129	169	174	144	097	053	023	008	002	0+	0+	0+	0+	0+	0+
	9	0+	0+	0+	002	010	036	080	132	168	170	140	093	050	021	006	001	0+	0+	0+	0+	0+
	10	0+	0+	0+	0+	003	014	041	085	134	167	168	137	089	046	018	005	001	0+	0+	0+	0+
	11	0+	0+	0+	0+	001	005	018	046	089	137	168	167	134	085	041	014	003	0+	0+	0+	0+
	12	0+	0+	0+	0+	0+	001	006	021	050	093	140	170	168	132	080	036	010	002	0+	0+	0+
	13	0+	0+	0+	0+	0+	0+	002	008	023	053	097	144	174	169	129	074	029	006	001	0+	0+
	14	0+	0+	0+	0+	0+	0+	0+	002	009	025	055	101	149	180	172	126	065	020	003	0+	0+
	15	0+	0+	0+	0+	0+	0+	0+	001	003	009	026	057	105	156	188	177	122	054	011	0+	0+
	16	0+	0+	0+	0+	0+	0+	0+	0+	001	003	010	026	059	109	164	199	183	115	038	003	0+
	17	0+	0+	0+	0+	0+	0+	0+	0+	0+	001	003	009	026	059	113	176	216	191	100	016	0+
	18	0+	0+	0+	0+	0+	0+	0+	0+	0+	0+	001	003	009	024	058	117	192	241	200	066	001
	19	0+	0+	0+	0+	0+	0+	0+	0+	0+	0+	0+	0+	002	007	022	055	121	215	284	198	017
	20	0+	0+	0+	0+	0+	0+	0+	0+	0+	0+	0+	0+	0+	001	005	017	048	122	255	376	172
	21	0+	0+	0+	0+	0+	0+	0+	0+	0+	0+	0+	0+	0+	0+	001	002	009	033	109	341	810
22	0	802	324	098	028	007	002	0+	0+	0+	0+	0+	0+	0+	0+	0+	0+	0+	0+	0+	0+	0+
	1	178	375	241	109	041	013	004	001	0+	0+	0+	0+	0+	0+	0+	0+	0+	0+	0+	0+	0+
	2	019	207	281	201	107	046	017	005	001	0+	0+	0+	0+	0+	0+	0+	0+	0+	0+	0+	0+
	3	001	073	208	237	178	102	047	018	006	002	0+	0+	0+	0+	0+	0+	0+	0+	0+	0+	0+
	4	0+	018	110	199	211	161	096	047	019	006	002	0+	0+	0+	0+	0+	0+	0+	0+	0+	0+
	5	0+	003	044	126	190	193	149	091	046	019	006	002	0+	0+	0+	0+	0+	0+	0+	0+	0+
	6	0+	001	014	063	134	183	181	139	086	043	018	006	001	0+	0+	0+	0+	0+	0+	0+	0+
	7	0+	0+	004	025	077	139	177	171	131	081	041	016	005	001	0+	0+	0+	0+	0+	0+	0+
	8	0+	0+	001	008	036	087	142	173	164	125	076	037	014	004	001	0+	0+	0+	0+	0+	0+
	9	0+	0+	0+	002	014	045	095	145	170	159	119	071	034	012	003	001	0+	0+	0+	0+	0+
	10	0+	0+	0+	001	005	020	053	101	148	169	154	113	066	029	010	002	0+	0+	0+	0+	0+
	11	0+	0+	0+	0+	001	007	025	060	107	151	168	151	107	060	025	007	001	0+	0+	0+	0+
	12	0+	0+	0+	0+	0+	002	010	029	066	113	154	169	148	101	053	020	005	001	0+	0+	0+
	13	0+	0+	0+	0+	0+	001	003	012	034	071	119	159	170	145	095	045	014	002	0+	0+	0+
	14	0+	0+	0+	0+	0+	0+	001	004	014	037	076	125	164	173	142	087	036	008	001	0+	0+
	15	0+	0+	0+	0+	0+	0+	0+	001	005	016	041	081	131	171	177	139	077	025	004	0+	0+
	16	0+	0+	0+	0+	0+	0+	0+	0+	001	006	018	043	086	139	181	183	134	063	014	001	0+
	17	0+	0+	0+	0+	0+	0+	0+	0+	0+	002	006	019	046	091	149	193	190	126	044	003	0+
	18	0+	0+	0+	0+	0+	0+	0+	0+	0+	0+	002	006	019	047	096	161	211	199	110	018	0+
	19	0+	0+	0+	0+	0+	0+	0+	0+	0+	0+	0+	002	006	018	047	102	178	237	208	073	001
	20	0+	0+	0+	0+	0+	0+	0+	0+	0+	0+	0+	0+	001	005	017	046	107	201	281	207	019
	21	0+	0+	0+	0+	0+	0+	0+	0+	0+	0+	0+	0+	0+	001	004	013	041	109	241	375	178
	22	0+	0+	0+	0+	0+	0+	0+	0+	0+	0+	0+	0+	0+	0+	002	007	028	098	324	802

Table 2. Binomial distribution 723

Table 2. (continued)

N	x										p											
		.01	.05	.10	.15	.20	.25	.30	.35	.40	.45	.50	.55	.60	.65	.70	.75	.80	.85	.90	.95	.99
23	0	794	307	089	024	006	001	0+	0+	0+	0+	0+	0+	0+	0+	0+	0+	0+	0+	0+	0+	0+
	1	184	372	226	097	034	010	003	001	0+	0+	0+	0+	0+	0+	0+	0+	0+	0+	0+	0+	0+
	2	020	215	277	188	093	038	013	004	001	0+	0+	0+	0+	0+	0+	0+	0+	0+	0+	0+	0+
	3	001	079	215	232	163	088	038	014	004	001	0+	0+	0+	0+	0+	0+	0+	0+	0+	0+	0+
	4	0+	021	120	204	204	146	082	037	014	004	001	0+	0+	0+	0+	0+	0+	0+	0+	0+	0+
	5	0+	004	051	137	194	185	133	076	035	013	004	001	0+	0+	0+	0+	0+	0+	0+	0+	0+
	6	0+	001	017	073	145	185	171	122	070	032	012	004	001	0+	0+	0+	0+	0+	0+	0+	0+
	7	0+	0+	005	031	088	150	178	160	113	064	029	011	003	001	0+	0+	0+	0+	0+	0+	0+
	8	0+	0+	001	011	044	100	153	172	151	105	058	026	009	002	0+	0+	0+	0+	0+	0+	0+
	9	0+	0+	0+	003	018	056	109	155	168	143	097	053	022	007	002	0+	0+	0+	0+	0+	0+
	10	0+	0+	0+	0+	006	026	065	117	157	164	136	090	046	018	005	001	0+	0+	0+	0+	0+
	11	0+	0+	0+	0+	002	010	033	074	123	159	161	130	082	040	014	003	0+	0+	0+	0+	0+
	12	0+	0+	0+	0+	0+	003	014	040	082	130	161	159	123	074	033	010	002	0+	0+	0+	0+
	13	0+	0+	0+	0+	0+	001	005	018	046	090	136	164	157	117	065	026	006	001	0+	0+	0+
	14	0+	0+	0+	0+	0+	0+	002	007	022	053	097	143	168	155	109	056	018	003	0+	0+	0+
	15	0+	0+	0+	0+	0+	0+	0+	002	009	026	058	105	151	172	153	100	044	011	001	0+	0+
	16	0+	0+	0+	0+	0+	0+	0+	001	003	011	029	064	113	160	178	150	088	031	005	0+	0+
	17	0+	0+	0+	0+	0+	0+	0+	0+	001	004	012	032	070	122	171	185	145	073	017	001	0+
	18	0+	0+	0+	0+	0+	0+	0+	0+	0+	001	004	013	035	076	133	185	194	137	051	004	0+
	19	0+	0+	0+	0+	0+	0+	0+	0+	0+	0+	001	004	014	037	082	146	204	204	120	021	0+
	20	0+	0+	0+	0+	0+	0+	0+	0+	0+	0+	0+	001	004	014	038	088	163	232	215	079	001
	21	0+	0+	0+	0+	0+	0+	0+	0+	0+	0+	0+	0+	001	004	013	038	093	188	277	215	020
	22	0+	0+	0+	0+	0+	0+	0+	0+	0+	0+	0+	0+	0+	001	003	010	034	097	226	372	184
	23	0+	0+	0+	0+	0+	0+	0+	0+	0+	0+	0+	0+	0+	0+	0+	001	006	024	089	307	794
24	0	786	292	080	020	005	001	0+	0+	0+	0+	0+	0+	0+	0+	0+	0+	0+	0+	0+	0+	0+
	1	190	369	213	086	028	008	002	0+	0+	0+	0+	0+	0+	0+	0+	0+	0+	0+	0+	0+	0+
	2	022	223	272	174	081	031	010	003	001	0+	0+	0+	0+	0+	0+	0+	0+	0+	0+	0+	0+
	3	002	086	221	225	149	075	031	010	003	001	0+	0+	0+	0+	0+	0+	0+	0+	0+	0+	0+
	4	0+	024	129	209	196	132	069	029	010	003	001	0+	0+	0+	0+	0+	0+	0+	0+	0+	0+
	5	0+	005	057	147	196	176	118	062	027	009	003	001	0+	0+	0+	0+	0+	0+	0+	0+	0+
	6	0+	001	020	082	155	185	160	106	056	024	008	002	0+	0+	0+	0+	0+	0+	0+	0+	0+
	7	0+	0+	006	037	100	159	176	147	096	050	021	007	002	0+	0+	0+	0+	0+	0+	0+	0+
	8	0+	0+	001	014	053	112	160	168	136	087	044	017	005	001	0+	0+	0+	0+	0+	0+	0+
	9	0+	0+	0+	004	024	067	122	161	161	126	078	038	014	004	001	0+	0+	0+	0+	0+	0+
	10	0+	0+	0+	001	009	033	079	130	161	155	117	069	032	011	003	0+	0+	0+	0+	0+	0+
	11	0+	0+	0+	0+	003	014	043	089	137	161	149	108	061	026	008	002	0+	0+	0+	0+	0+
	12	0+	0+	0+	0+	001	005	020	052	099	143	161	143	099	052	020	005	001	0+	0+	0+	0+
	13	0+	0+	0+	0+	0+	002	008	026	061	108	149	161	137	089	043	014	003	0+	0+	0+	0+
	14	0+	0+	0+	0+	0+	0+	003	011	032	069	117	155	161	130	079	033	009	001	0+	0+	0+
	15	0+	0+	0+	0+	0+	0+	001	004	014	038	078	126	161	161	122	067	024	004	0+	0+	0+
	16	0+	0+	0+	0+	0+	0+	0+	001	005	017	044	087	136	168	160	112	053	014	001	0+	0+
	17	0+	0+	0+	0+	0+	0+	0+	0+	002	007	021	050	096	147	176	159	100	037	006	0+	0+
	18	0+	0+	0+	0+	0+	0+	0+	0+	0+	002	008	024	056	106	160	185	155	082	020	001	0+
	19	0+	0+	0+	0+	0+	0+	0+	0+	0+	001	003	009	027	062	118	176	196	147	057	005	0+
	20	0+	0+	0+	0+	0+	0+	0+	0+	0+	0+	001	003	010	029	069	132	196	209	129	024	0+
	21	0+	0+	0+	0+	0+	0+	0+	0+	0+	0+	0+	001	003	010	031	075	149	225	221	086	002
	22	0+	0+	0+	0+	0+	0+	0+	0+	0+	0+	0+	0+	001	003	010	031	081	174	272	223	022
	23	0+	0+	0+	0+	0+	0+	0+	0+	0+	0+	0+	0+	0+	0+	002	008	028	086	213	369	190
	24	0+	0+	0+	0+	0+	0+	0+	0+	0+	0+	0+	0+	0+	0+	0+	001	005	020	080	292	786
25	0	778	277	072	017	004	001	0+	0+	0+	0+	0+	0+	0+	0+	0+	0+	0+	0+	0+	0+	0+
	1	196	365	199	076	024	006	001	0+	0+	0+	0+	0+	0+	0+	0+	0+	0+	0+	0+	0+	0+
	2	024	231	266	161	071	025	007	002	0+	0+	0+	0+	0+	0+	0+	0+	0+	0+	0+	0+	0+
	3	002	093	226	217	136	064	024	008	002	0+	0+	0+	0+	0+	0+	0+	0+	0+	0+	0+	0+
	4	0+	027	138	211	187	118	057	022	007	002	0+	0+	0+	0+	0+	0+	0+	0+	0+	0+	0+
	5	0+	006	065	156	196	165	103	051	020	006	002	0+	0+	0+	0+	0+	0+	0+	0+	0+	0+
	6	0+	001	024	092	163	183	147	091	044	017	005	001	0+	0+	0+	0+	0+	0+	0+	0+	0+
	7	0+	0+	007	044	111	165	171	133	080	038	014	004	001	0+	0+	0+	0+	0+	0+	0+	0+
	8	0+	0+	002	017	062	124	165	161	120	070	032	012	003	001	0+	0+	0+	0+	0+	0+	0+
	9	0+	0+	0+	006	029	078	134	163	151	108	061	027	009	002	0+	0+	0+	0+	0+	0+	0+
	10	0+	0+	0+	002	012	042	092	141	161	142	097	052	021	006	001	0+	0+	0+	0+	0+	0+
	11	0+	0+	0+	0+	004	019	054	103	147	158	133	087	043	016	004	001	0+	0+	0+	0+	0+
	12	0+	0+	0+	0+	001	007	027	065	114	151	155	124	076	035	011	002	0+	0+	0+	0+	0+
	13	0+	0+	0+	0+	0+	002	011	035	076	124	155	151	114	065	027	007	001	0+	0+	0+	0+
	14	0+	0+	0+	0+	0+	001	004	016	043	087	133	158	147	103	054	019	004	0+	0+	0+	0+
	15	0+	0+	0+	0+	0+	0+	001	006	021	052	097	142	161	141	092	042	012	002	0+	0+	0+
	16	0+	0+	0+	0+	0+	0+	0+	002	009	027	061	108	151	163	134	078	029	006	0+	0+	0+
	17	0+	0+	0+	0+	0+	0+	0+	001	003	012	032	070	120	161	165	124	062	017	002	0+	0+
	18	0+	0+	0+	0+	0+	0+	0+	0+	001	004	014	038	080	133	171	165	111	044	007	0+	0+
	19	0+	0+	0+	0+	0+	0+	0+	0+	0+	001	005	017	044	091	147	183	163	092	024	001	0+
	20	0+	0+	0+	0+	0+	0+	0+	0+	0+	0+	002	006	020	051	103	165	196	156	065	006	0+
	21	0+	0+	0+	0+	0+	0+	0+	0+	0+	0+	0+	002	007	022	057	118	187	211	138	027	0+
	22	0+	0+	0+	0+	0+	0+	0+	0+	0+	0+	0+	0+	002	008	024	064	136	217	226	093	002
	23	0+	0+	0+	0+	0+	0+	0+	0+	0+	0+	0+	0+	0+	002	007	025	071	161	266	231	024
	24	0+	0+	0+	0+	0+	0+	0+	0+	0+	0+	0+	0+	0+	0+	001	006	024	076	199	365	196
	25	0+	0+	0+	0+	0+	0+	0+	0+	0+	0+	0+	0+	0+	0+	0+	001	004	017	072	277	778

Probabilities calculated by authors using ASP® Version 2.10.

Table 3. Cumulative probabilities for the standard normal random variable

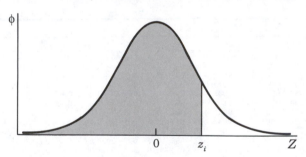

Table entry is shaded area, $\mathbb{P}(z_i)$.

z_i	.00	.01	.02	.03	.04	.05	.06	.07	.08	.09
− 3.0	$.0^2 1350$	$.0^2 1306$	$.0^2 1264$	$.0^2 1223$	$.0^2 1183$	$.0^2 1114$	$.0^2 1107$	$.0^2 1070$	$.0^2 1035$	$.0^2 1001$
− 2.9	$.0^2 1866$	$.0^2 1807$	$.0^2 1750$	$.0^2 1695$	$.0^2 1641$	$.0^2 1589$	$.0^2 1538$	$.0^2 1489$	$.0^2 1441$	$.0^2 1395$
− 2.8	$.0^2 2555$	$.0^2 2477$	$.0^2 2401$	$.0^2 2327$	$.0^2 2256$	$.0^2 2186$	$.0^2 2118$	$.0^2 2052$	$.0^2 1988$	$.0^2 1926$
− 2.7	$.0^2 3467$	$.0^2 3364$	$.0^2 3264$	$.0^2 3167$	$.0^2 3072$	$.0^2 2980$	$.0^2 2890$	$.0^2 2803$	$.0^2 2718$	$.0^2 2635$
− 2.6	$.0^2 4661$	$.0^2 4527$	$.0^2 4396$	$.0^2 4269$	$.0^2 4145$	$.0^2 4025$	$.0^2 3907$	$.0^2 3793$	$.0^2 3681$	$.0^2 3573$
− 2.5	$.0^2 6210$	$.0^2 6037$	$.0^2 5868$	$.0^2 5703$	$.0^2 5543$	$.0^2 5386$	$.0^2 5234$	$.0^2 5085$	$.0^2 4940$	$.0^2 4799$
− 2.4	$.0^2 8198$	$.0^2 7976$	$.0^2 7760$	$.0^2 7549$	$.0^2 7344$	$.0^2 7143$	$.0^2 6947$	$.0^2 6756$	$.0^2 6569$	$.0^2 6387$
− 2.3	.01072	.01044	.01017	$.0^2 9903$	$.0^2 9642$	$.0^2 9387$	$.0^2 9137$	$.0^2 8894$	$.0^2 8656$	$.0^2 8424$
− 2.2	.01390	.01355	.01321	.01287	.01255	.01222	.01191	.01160	.01130	.01101
− 2.1	.01786	.01743	.01700	.01659	.01616	.01578	.01539	.01500	.01463	.01426
− 2.0	.02275	.02222	.02169	.02118	.02068	.02018	.01970	.01923	.01876	.01831
− 1.9	.02872	.02807	.02743	.02680	.02619	.02559	.02500	.02442	.02385	.02330
− 1.8	.03593	.03515	.03438	.03362	.03288	.03216	.03144	.03074	.03005	.02938
− 1.7	.04457	.04363	.04272	.04182	.04093	.04006	.03920	.03836	.03754	.03673
− 1.6	.05480	.05370	.05262	.05155	.05050	.04947	.04846	.04746	.04648	.04551
− 1.5	.06681	.06552	.06426	.06301	.06178	.06057	.05938	.05821	.05705	.05592
− 1.4	.08076	.07927	.07780	.07636	.07493	.07353	.07215	.07078	.06944	.06811
− 1.3	.09680	.09510	.09342	.09176	.09012	.08851	.08691	.08534	.08379	.08226
− 1.2	.1151	.1131	.1112	.1093	.1075	.1056	.1038	.1020	.1003	.09853
− 1.1	.1375	.1335	.1314	.1292	.1271	.1251	.1230	.1210	.1190	.1170
− 1.0	.1587	.1562	.1539	.1515	.1492	.1469	.1446	.1423	.1401	.1379
− .9	.1841	.1814	.1788	.1762	.1736	.1711	.1685	.1660	.1635	.1611
− .8	.2119	.2090	.2061	.2033	.2005	.1977	.1949	.1922	.1894	.1867
− .7	.2420	.2389	.2358	.2327	.2297	.2266	.2236	.2206	.2177	.2148
− .6	.2743	.2709	.2676	.2643	.2611	.2578	.2546	.2514	.2483	.2451
− .5	.3085	.3050	.3015	.2981	.2946	.2912	.2877	.2843	.2810	.2776
− .4	.3446	.3409	.3372	.3336	.3300	.3264	.3228	.3192	.3156	.3121
− .3	.3821	.3783	.3745	.3707	.3669	.3632	.3594	.3557	.3520	.3483
− .2	.4207	.4168	.4129	.4090	.4052	.4013	.3974	.3936	.3897	.3859
− .1	.4602	.4562	.4522	.4483	.4443	.4404	.4364	.4325	.4286	.4247
− .0	.5000	.4960	.4920	.4880	.4840	.4801	.4761	.4721	.4681	.4641

Table 3. Standard normal distribution **725**

Table 3. (continued)

z_i	.00	.01	.02	.03	.04	.05	.06	.07	.08	.09
.0	.5000	.5040	.5080	.5120	.5160	.5199	.5239	.5279	.5319	.5359
.1	.5398	.5438	.5478	.5517	.5557	.5596	.5636	.5675	.5714	.5753
.2	.5793	.5832	.5871	.5910	.5948	.5987	.6026	.6064	.6103	.6141
.3	.6179	.6217	.6255	.6293	.6331	.6368	.6406	.6443	.6480	.6517
.4	.6554	.6591	.6628	.6664	.6700	.6736	.6772	.6808	.6844	.6879
.5	.6915	.6950	.6985	.7019	.7054	.7088	.7123	.7157	.7190	.7224
.6	.7257	.7291	.7324	.7357	.7389	.7422	.7454	.7486	.7517	.7549
.7	.7580	.7611	.7642	.7673	.7703	.7734	.7764	.7794	.7823	.7852
.8	.7881	.7910	.7939	.7967	.7995	.8023	.8051	.8078	.8106	.8113
.9	.8159	.8186	.8212	.8238	.8264	.8289	.8315	.8340	.8365	.8389
1.0	.8413	.8438	.8461	.8485	.8508	.8531	.8554	.8577	.8599	.8661
1.1	.8643	.8665	.8686	.8708	.8729	.8749	.8770	.8790	.8810	.8830
1.2	.8849	.8869	.8888	.8907	.8925	.8944	.8962	.8980	.8997	.90147
1.3	.90320	.90490	.90658	.90824	.90988	.91149	.91309	.91466	.91621	.91774
1.4	.91924	.92073	.92220	.92364	.92507	.92647	.92785	.92922	.93056	.93189
1.5	.93319	.93448	.93574	.93669	.93822	.93943	.94062	.94179	.94295	.94408
1.6	.94520	.94630	.94738	.94845	.94950	.95053	.95154	.95254	.95352	.95449
1.7	.95543	.95637	.95728	.95818	.95907	.95994	.96080	.96164	.96246	.96327
1.8	.96407	.96485	.96562	.96638	.96712	.96784	.96856	.96926	.96995	.97062
1.9	.97128	.97193	.97257	.97320	.97381	.97441	.97500	.97558	.97615	.97670
2.0	.97725	.97778	.97831	.97882	.97932	.97982	.98030	.98077	.98124	.98169
2.1	.98214	.98257	.98300	.98341	.98382	.98422	.98461	.98500	.98537	.98574
2.2	.98610	.98645	.98679	.98713	.98745	.98778	.98809	.98840	.98870	.98899
2.3	.98928	.98956	.98983	$.9^20097$	$.9^20358$	$.9^20613$	$.9^20863$	$.9^21106$	$.9^21344$	$.9^21576$
2.4	$.9^21802$	$.9^22024$	$.9^22240$	$.9^22451$	$.9^22656$	$.9^22857$	$.9^23053$	$.9^23244$	$.9^23431$	$.9^23613$
2.5	$.9^23790$	$.9^23963$	$.9^24132$	$.9^24297$	$.9^24457$	$.9^24614$	$.9^24766$	$.9^24915$	$.9^25060$	$.9^25201$
2.6	$.9^25339$	$.9^25473$	$.9^25604$	$.9^25731$	$.9^25855$	$.9^25975$	$.9^26093$	$.9^26207$	$.9^26319$	$.9^26427$
2.7	$.9^26533$	$.9^26636$	$.9^26736$	$.9^26833$	$.9^26928$	$.9^27020$	$.9^27110$	$.9^27197$	$.9^27282$	$.9^27365$
2.8	$.9^27445$	$.9^27523$	$.9^27599$	$.9^27673$	$.9^27744$	$.9^27814$	$.9^27882$	$.9^27948$	$.9^28012$	$.9^28074$
2.9	$.9^28134$	$.9^28193$	$.9^28250$	$.9^28305$	$.9^28359$	$.9^28411$	$.9^28462$	$.9^28511$	$.9^28559$	$.9^28605$
3.0	$.9^28650$	$.9^28694$	$.9^28736$	$.9^28777$	$.9^28817$	$.9^28856$	$.9^28893$	$.9^28930$	$.9^28965$	$.9^28999$

Source: Table 3 is taken with permission of the author from Hald, A. (1952). *Statistical Tables and Formulas*, John Wiley & Sons, New York.
Note: $.0^21350 = .001350$. $.9^28650 = .998650$.

Table 3A. Cumulative probabilities for selected extreme values of the standard normal random variable

z_i	.00	.10	.20	.30	.40	.50	.60	.70
3.0	$.9^287$	$.9^30$	$.9^33$	$.9^35$	$.9^37$	$.9^38$	$.9^38$	$.9^40$
4.0	$.9^47$					$.9^57$		
5.0	$.9^67$							

Probabilities calculated by authors using ASP ® Version 2.10.

Table 4. Student's *t*-values for selected cumulative probabilities

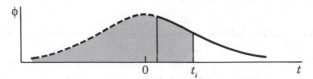

Table entry is value t_i corresponding to shaded area, $\mathbb{P}(t_i)$. Table includes only *t*-values under delineated portion of curve, for which $\mathbb{P}(t) \geq .60$.

ν	$\mathbb{P}$ .60	.75	.80	.90	.95	.975	.99	.995	.9995
1	.325	1.000	1.376	3.078	6.314	12.706	31.821	63.657	636.557
2	.289	.816	1.061	1.886	2.920	4.303	6.965	9.925	31.599
3	.277	.765	.978	1.638	2.353	3.182	4.541	5.841	12.924
4	.271	.741	.941	1.533	2.132	2.776	3.747	4.604	8.610
5	.267	.727	.920	1.476	2.015	2.571	3.365	4.032	6.869
6	.265	.718	.906	1.440	1.943	2.447	3.143	3.707	5.959
7	.263	.711	.896	1.415	1.895	2.365	2.998	3.499	5.408
8	.262	.706	.889	1.397	1.860	2.306	2.896	3.355	5.041
9	.261	.703	.883	1.383	1.833	2.262	2.821	3.250	4.781
10	.260	.700	.879	1.372	1.812	2.228	2.764	3.169	4.587
11	.260	.697	.876	1.363	1.796	2.201	2.718	3.106	4.437
12	.259	.695	.873	1.356	1.782	2.179	2.681	3.055	4.318
13	.259	.694	.870	1.350	1.771	2.160	2.650	3.012	4.221
14	.258	.692	.868	1.345	1.761	2.145	2.624	2.977	4.140
15	.258	.691	.866	1.341	1.753	2.131	2.602	2.947	4.073
16	.258	.690	.865	1.337	1.746	2.120	2.583	2.921	4.015
17	.257	.689	.863	1.333	1.740	2.110	2.567	2.898	3.965
18	.257	.688	.862	1.330	1.734	2.101	2.552	2.878	3.922
19	.257	.688	.861	1.328	1.729	2.093	2.539	2.861	3.883
20	.257	.687	.860	1.325	1.725	2.086	2.528	2.845	3.850
21	.257	.686	.859	1.323	1.721	2.080	2.518	2.831	3.819
22	.256	.686	.858	1.321	1.717	2.074	2.508	2.819	3.792
23	.256	.685	.858	1.319	1.714	2.069	2.500	2.807	3.768
24	.256	.685	.857	1.318	1.711	2.064	2.492	2.797	3.745
25	.256	.684	.856	1.316	1.708	2.060	2.485	2.787	3.725
26	.256	.684	.856	1.315	1.706	2.056	2.479	2.779	3.707
27	.256	.684	.855	1.314	1.703	2.052	2.473	2.771	3.690
28	.256	.683	.855	1.313	1.701	2.048	2.467	2.763	3.674
29	.256	.683	.854	1.311	1.699	2.045	2.462	2.756	3.659
30	.256	.683	.854	1.310	1.697	2.042	2.457	2.750	3.646
35	.255	.682	.852	1.306	1.690	2.030	2.438	2.724	3.591
40	.255	.681	.851	1.303	1.684	2.021	2.423	2.704	3.551
60	.254	.679	.848	1.296	1.671	2.000	2.390	2.660	3.460
120	.254	.677	.845	1.289	1.658	1.980	2.358	2.617	3.373
∞	.253	.674	.842	1.282	1.645	1.960	2.326	2.576	3.291

Values of Student's *t* calculated by authors using ASP® Version 2.10.

Table 5. *Chi*-square values for selected cumulative probabilities

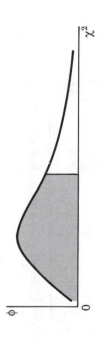

χ^2

Table entry is value of χ^2 corresponding to shaded area, $\mathbb{P}(\chi^2)$.

ν \ $\mathbb{P}$	.005	.010	.025	.050	.100	.250	.500	.750	.900	.950	.975	.990	.995	.999
1	.0000393	.000157	.000982	.00393	.0158	.102	.455	1.32	2.71	3.84	5.02	6.63	7.88	10.8
2	.0100	.0201	.0506	.103	.211	.575	1.39	2.77	4.61	5.99	7.38	9.21	10.6	13.8
3	.0717	.115	.216	.352	.584	1.21	2.37	4.11	6.25	7.81	9.35	11.3	12.8	16.3
4	.207	.297	.484	.711	1.06	1.92	3.36	5.39	7.78	9.49	11.1	13.3	14.9	18.5
5	.412	.554	.831	1.15	1.61	2.67	4.35	6.63	9.24	11.1	12.8	15.1	16.7	20.5
6	.676	.872	1.24	1.64	2.20	3.45	5.35	7.84	10.6	12.6	14.4	16.8	18.5	22.5
7	.989	1.24	1.69	2.17	2.83	4.25	6.35	9.04	12.0	14.1	16.0	18.5	20.3	24.3
8	1.34	1.65	2.18	2.73	3.49	5.07	7.34	10.2	13.4	15.5	17.5	20.1	22.0	26.1
9	1.73	2.09	2.70	3.33	4.17	5.90	8.34	11.4	14.7	16.9	19.0	21.7	23.6	27.9
10	2.16	2.56	3.25	3.94	4.87	6.74	9.34	12.5	16.0	18.3	20.5	23.2	25.2	29.6
11	2.60	3.05	3.82	4.57	5.58	7.58	10.3	13.7	17.3	19.7	21.9	24.7	26.8	31.3
12	3.07	3.57	4.40	5.23	6.30	8.44	11.3	14.8	18.5	21.0	23.3	26.2	28.3	32.9
13	3.57	4.11	5.01	5.89	7.04	9.30	12.3	16.0	19.8	22.4	24.7	27.7	29.8	34.5
14	4.07	4.66	5.63	6.57	7.79	10.2	13.3	17.1	21.1	23.7	26.1	29.1	31.3	36.1
15	4.60	5.23	6.26	7.26	8.55	11.0	14.3	18.2	22.3	25.0	27.5	30.6	32.8	37.7

Table 5. (continued)

ν	$\mathbb{P}$ .005	.010	.025	.050	.100	.250	.500	.750	.900	.950	.975	.990	.995	.999
16	5.14	5.81	6.91	7.96	9.31	11.9	15.3	19.4	23.5	26.3	28.8	32.0	34.3	39.3
17	5.70	6.41	7.56	8.67	10.1	12.8	16.3	20.5	24.8	27.6	30.2	33.4	35.7	40.8
18	6.26	7.01	8.23	9.39	10.9	13.7	17.3	21.6	26.0	28.9	31.5	34.8	37.2	42.3
19	6.84	7.63	8.91	10.1	11.7	14.6	18.3	22.7	27.2	30.1	32.9	36.2	38.6	43.8
20	7.43	8.26	9.59	10.9	12.4	15.5	19.3	23.8	28.4	31.4	34.2	37.6	40.0	45.3
21	8.03	8.90	10.3	11.6	13.2	16.3	20.3	24.9	29.6	32.7	35.5	38.9	41.4	46.8
22	8.64	9.54	11.0	12.3	14.0	17.2	21.3	26.0	30.8	33.9	36.8	40.3	42.8	48.3
23	9.26	10.2	11.7	13.1	14.8	18.1	22.3	27.1	32.0	35.2	38.1	41.6	44.2	49.7
24	9.89	10.9	12.4	13.8	15.7	19.0	23.3	28.2	33.2	36.4	39.4	43.0	45.6	51.2
25	10.5	11.5	13.1	14.6	16.5	19.9	24.3	29.3	34.4	37.7	40.6	44.3	46.9	52.6
26	11.2	12.2	13.8	15.4	17.3	20.8	25.3	30.4	35.6	38.9	41.9	45.6	48.3	54.1
27	11.8	12.9	14.6	16.2	18.1	21.7	26.3	31.5	36.7	40.1	43.2	47.0	49.6	55.5
28	12.5	13.6	15.3	16.9	18.9	22.7	27.3	32.6	37.9	41.3	44.5	48.3	51.0	56.9
29	13.1	14.3	16.0	17.7	19.8	23.6	28.3	33.7	39.1	42.6	45.7	49.6	52.3	58.3
30	13.8	15.0	16.8	18.5	20.6	24.5	29.3	34.8	40.3	43.8	47.0	50.9	53.7	59.7

Values of χ^2 calculated by authors using ASP® Version 2.10.

Table 6. Fisher's F distribution   **729**

Table 6. Fisher's F – values for selected cumulative probabilities

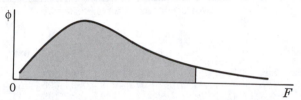

Table entry is value of F corresponding to shaded area, $\mathbb{P}(F)$.

Source: Table 6 is taken with permission from W. J. Dixon and F. J. Massey, Jr. (1969). *Introduction to Statistical Analysis* (3rd ed.), McGraw-Hill Book Co, New York. Tabled values are abstracted from the following sources:

1. All values for ν_1, ν_2 equal to 50, 100, 200, 500 are from A. Hald, *Statistical Tables and Formulas*, John Wiley & Sons, Inc., New York, 1952.

2. For cumulative proportions .5, .75, .9, .95, .975, .99, .995 most of the values are from M. Merrington and C. M. Thompson, *Biometrika*, vol. 33 (1943), p. 73.

3. For cumulative proportions .999 the values are from C. Colcord and L. S. Deming, *Sankhyā*, vol. 2 (1936), p. 423.

4. For cum. prop. $= \alpha < .5$ the values are the reciprocals of values for $1 - \alpha$ (with ν_1 and ν_2 interchanged). The values in Merrington and Thompson and in Colcord and Deming are to five significant figures, and it is hoped (but not expected) that the reciprocals are correct as givenl The values in Hald are to three significant figures, and the reciprocals are probably accurate within one to two digits in the third significant figure except for those values very close to unity, where they may be off four to five digits in the third significant figure.

5. Gaps remaining in the table after using the above sources were filled in by interpolation. See Dixon & Massey (1969), p. 485.

For sample sizes larger than, say, 30, a fairly good approximation to cumulative probabilities in the F distribution can be obtained from

$$\log_{10} F_\alpha(\nu_1, \nu_2) \cong \left(\frac{a}{\sqrt{h - b}} \right) - cg$$

where $h = 2\nu_1\nu_2/(\nu_1 + \nu_2)$, $g = (\nu_2 - \nu_1)/\nu_1\nu_2$, and a, b, c are functions of α given below:

				VALUES OF $\mathbb{P}$					
	.50	.75	.90	.95	.975	.99	.995	.999	.9995
a	0	.5859	1.1131	1.4287	1.7023	2.0206	2.2373	2.6841	2.8580
b	0	.58	.77	.95	1.14	1.40	1.61	2.09	2.30
c	.290	.355	.527	.681	.846	1.073	1.250	1.672	1.857

Table 6. (continued)

ν₁, DEGREES OF FREEDOM FOR NUMERATOR

ν₂, DEGREES OF FREEDOM FOR DENOMINATOR

	$\mathbb{P}$	1	2	3	4	5	6	7	8	9	10	11	12	$\mathbb{P}$
1	.0005	$.0^6 62$	$.0^3 50$	$.0^2 38$	$.0^2 94$	.016	.022	.027	.032	.036	.039	.042	.045	.0005
	.001	$.0^5 25$	$.0^2 10$	$.0^2 60$	.013	.021	.028	.034	.039	.044	.048	.051	.054	.001
	.005	$.0^4 62$	$.0^2 51$	.018	.032	.044	.054	.062	.068	.073	.078	.082	.085	.005
	.010	$.0^3 25$	.010	.029	.047	.062	.073	.082	.089	.095	.100	.104	.107	.010
	.025	$.0^2 15$	.026	.057	.082	.100	.113	.124	.132	.139	.144	.149	.153	.025
	.05	$.0^2 62$	.054	.099	.130	.151	.167	.179	.188	.195	.201	.207	.211	.05
	.10	.025	.117	.181	.220	.246	.265	.279	.289	.298	.304	.310	.315	.10
	.25	.172	.389	.494	.553	.591	.617	.637	.650	.661	.670	.680	.684	.25
	.50	1.00	1.50	1.71	1.82	1.89	1.94	1.98	2.00	2.03	2.04	2.05	2.07	.50
	.75	5.83	7.50	8.20	8.58	8.82	8.98	9.10	9.19	9.26	9.32	9.36	9.41	.75
	.90	39.9	49.5	53.6	55.8	57.2	58.2	58.9	59.4	59.9	60.2	60.5	60.7	.90
	.95	161	200	216	225	230	234	237	239	241	242	243	244	.95
	.975	648	800	864	900	922	937	948	957	963	969	973	977	.975
	.99	405^1	500^1	540^1	562^1	576^1	586^1	593^1	598^1	602^1	606^1	608^1	611^1	.99
	.995	162^2	200^2	216^2	225^2	231^2	234^2	237^2	239^2	241^2	242^2	243^2	244^2	.995
	.999	406^3	500^3	540^3	562^3	576^3	586^3	593^3	598^3	602^3	606^3	609^3	611^3	.999
	.9995	162^4	200^4	216^4	225^4	231^4	234^4	237^4	239^4	241^4	242^4	243^4	244^4	.9995
2	.0005	$.0^6 50$	$.0^3 50$	$.0^2 42$	.011	.020	.029	.037	.044	.050	.056	.061	.065	.0005
	.001	$.0^5 20$	$.0^2 10$	$.0^2 68$	.016	.027	.037	.046	.054	.061	.067	.072	.077	.001
	.005	$.0^4 50$	$.0^2 50$	.020	.038	.055	.069	.081	.091	.099	.106	.112	.118	.005
	.01	$.0^3 20$	.010	.032	.056	.075	.092	.105	.116	.125	.132	.139	.144	.01
	.025	$.0^2 13$	.026	.062	.094	.119	.138	.153	.165	.175	.183	.190	.196	.025
	.05	$.0^2 50$	.053	.105	.144	.173	.194	.211	.224	.235	.244	.251	.257	.05
	.10	.020	.111	.183	.231	.265	.289	.307	.321	.333	.342	.350	.356	.10
	.25	.133	.333	.439	.500	.540	.568	.588	.604	.616	.626	.633	.641	.25
	.50	.667	1.00	1.13	1.21	1.25	1.28	1.30	1.32	1.33	1.34	1.35	1.36	.50
	.75	2.57	3.00	3.15	3.23	3.28	3.31	3.34	3.35	3.37	3.38	3.39	3.39	.75
	.90	8.53	9.00	9.16	9.24	9.29	9.33	9.35	9.37	9.38	9.39	9.40	9.41	.90
	.95	18.5	19.0	19.2	19.2	19.3	19.3	19.4	19.4	19.4	19.4	19.4	19.4	.95
	.975	38.5	39.0	39.2	39.2	39.3	39.3	39.4	39.4	39.4	39.4	39.4	39.4	.975
	.99	98.5	99.0	99.2	99.2	99.3	99.3	99.4	99.4	99.4	99.4	99.4	99.4	.99
	.995	198	199	199	199	199	199	199	199	199	199	199	199	.995
	.999	998	999	999	999	999	999	999	999	999	999	999	999	.999
	.9995	200^1	200^1	200^1	200^1	200^1	200^1	200^1	200^1	200^1	200^1	200^1	200^1	.9995
3	.0005	$.0^5 46$	$.0^3 50$	$.0^2 44$	.012	.023	.033	.043	.052	.060	.067	.074	.079	.0005
	.001	$.0^5 19$	$.0^2 10$	$.0^2 71$	.018	.030	.042	.053	.063	.072	.079	.086	.093	.001
	.005	$.0^4 46$	$.0^2 50$	.021	.041	.060	.077	.092	.104	.115	.124	.132	.138	.005
	.01	$.0^3 19$	.010	.034	.060	.083	.102	.118	.132	.143	.153	.161	.168	.01
	.025	$.0^2 12$	.026	.065	.100	.129	.152	.170	.185	.197	.207	.216	.224	.025
	.05	$.0^2 46$	.052	.108	.152	.185	.210	.230	.246	.259	.270	.279	.287	.05
	.10	.019	.109	.185	.239	.276	.304	.325	.342	.356	.367	.376	.384	.10
	.25	.122	.317	.424	.489	.531	.561	.582	.600	.613	.624	.633	.641	.25
	.50	.585	.881	1.00	1.06	1.10	1.13	1.15	1.16	1.17	1.18	1.19	1.20	.50
	.75	2.02	2.28	2.36	2.39	2.41	2.42	2.43	2.44	2.44	2.44	2.45	2.45	.75
	.90	5.54	5.46	5.39	5.34	5.31	5.28	5.27	5.25	5.24	5.23	5.22	5.22	.90
	.95	10.1	9.55	9.28	9.12	9.01	8.94	8.89	8.85	8.81	8.79	8.76	8.74	.95
	.975	17.4	16.0	15.4	15.1	14.9	14.7	14.6	14.5	14.5	14.4	14.4	14.3	.975
	.99	34.1	30.8	29.5	28.7	28.2	27.9	27.7	27.5	27.3	27.2	27.1	27.1	.99
	.995	55.6	49.8	47.5	46.2	45.4	44.8	44.4	44.1	43.9	43.7	43.5	43.4	.995
	.999	167	149	141	137	135	133	132	131	130	129	129	128	.999
	.9995	266	237	225	218	214	211	209	208	207	206	204	204	.9995

Read $.0^3 56$ as .00056, 200^1 as 2,000, 162^4 as 1,620,000, and so on.

Table 6. Fisher's F distribution **731**

Table 6. (continued)

ν_1, DEGREES OF FREEDOM FOR NUMERATOR

P	15	20	24	30	40	50	60	100	120	200	500	∞	P	
.0005	.051	.058	062	.066	.069	.072	.074	.077	.078	.080	.081	.083	.0005	**1**
.001	.060	.067	.071	.075	.079	.082	.084	.087	.088	.089	.091	.092	.001	
.005	.093	.101	.105	.109	.113	.116	.118	.121	.122	.124	.126	.127	.005	
.01	.115	.124	.128	.132	.137	.139	.141	.145	.146	.148	.150	.151	.01	
.025	.161	.170	.175	.180	.184	.187	.189	.193	.194	.196	.198	.199	.025	
.05	.220	.230	.235	.240	.245	.248	.250	.254	.255	.257	.259	.261	.05	
.10	.325	.336	.342	.347	.353	.356	.358	.362	.364	.366	.368	.370	.10	
.25	.698	.712	.719	.727	.734	.738	.741	.747	.749	.752	.754	.756	.25	
.50	2.09	2.12	2.13	2.15	2.16	2.17	2.17	2.18	2.18	2.19	2.19	2.20	.50	
.75	9.49	9.58	9.63	9.67	9.71	9.74	9.76	9.78	9.80	9.82	9.84	9.85	.75	
.90	61.2	61.7	62.0	62.3	62.5	62.7	62.8	63.0	63.1	63.2	63.3	63.3	.90	
.95	246	248	249	250	251	252	252	253	253	254	254	254	.95	
.975	985	993	997	100^1	101^1	101^1	101^1	101^1	101^1	102^1	102^1	102^1	.975	
.99	616^1	621^1	623^1	626^1	629^1	630^1	631^1	633^1	634^1	635^1	636^1	637^1	.99	
.995	246^2	248^2	249^2	250^2	251^2	252^2	253^2	253^2	254^2	254^2	254^2	255^2	.995	
.999	616^3	621^3	623^3	626^3	629^3	630^3	631^3	633^3	634^3	635^3	636^3	637^3	.999	
.9995	246^4	248^4	249^4	250^4	251^4	252^4	252^4	253^4	253^4	253^4	254^4	254^4	.9995	
.0005	.076	.088	.094	.101	.108	.113	.116	.122	.124	.127	.130	.132	.0005	**2**
.001	.088	.100	.107	.114	.121	.126	.129	.135	.137	.140	.143	.145	.001	
.005	.130	.143	.150	.157	.165	.169	.173	.179	.181	.184	.187	.189	.005	
.01	.157	.171	.178	.186	.193	.198	.201	.207	.209	.212	.215	.217	.01	
.025	.210	.224	.232	.239	.247	.251	.255	.261	.263	.266	.269	.271	.025	
.05	.272	.286	.294	.302	.309	.314	.317	.324	.326	.329	.332	.334	.05	
.10	.371	.386	.394	.402	.410	.415	.418	.424	.426	.429	.433	.434	.10	
.25	.657	.672	.680	.689	.697	.702	.705	.711	.713	.716	.719	.721	.25	
.50	1.38	1.39	1.40	1.41	1.42	1.42	1.43	1.43	1.43	1.44	1.44	1.44	.50	
.75	3.41	3.43	3.43	3.44	3.45	3.45	3.46	3.47	3.47	3.48	3.48	3.48	.75	
.90	9.42	9.44	9.45	9.46	9.47	9.47	9.47	9.48	9.48	9.49	9.49	9.49	.90	
.95	19.4	19.4	19.5	19.5	19.5	19.5	19.5	19.5	19.5	19.5	19.5	19.5	.95	
.975	39.4	39.4	39.5	39.5	39.5	39.5	39.5	39.5	39.5	39.5	39.5	39.5	.975	
.99	99.4	99.4	99.5	99.5	99.5	99.5	99.5	99.5	99.5	99.5	99.5	99.5	.99	
.995	199	199	199	199	199	199	199	199	199	199	199	200	.995	
.999	999	999	999	999	999	999	999	999	999	999	999	999	.999	
.9995	200^1	200^1	200^1	200^1	200^1	200^1	200^1	200^1	200^1	200^1	200^1	200^1	.9995	
.0005	.093	.109	.117	.127	.136	.143	.147	.156	.158	.162	.166	.169	.0005	**3**
.001	.107	.123	.132	.142	.152	.158	.162	.171	.173	.177	.181	.184	.001	
.005	.154	.172	.181	.191	.201	.207	.211	.220	.222	.227	.231	.234	.005	
.01	.185	.203	.212	.222	.232	.238	.242	.251	.253	.258	.262	.264	.01	
.025	.241	.259	.269	.279	.289	.295	.299	.308	.310	.314	.318	.321	.025	
.05	.304	.323	.332	.342	.352	.358	.363	.370	.373	.377	.382	.384	.05	
.10	.402	.420	.430	.439	.449	.455	.459	.467	.469	.474	.476	.480	.10	
.25	.658	.675	.684	.693	.702	.708	.711	.719	.721	.724	.728	.730	.25	
.50	1.21	1.23	1.23	1.24	1.25	1.25	1.25	1.26	1.26	1.26	1.27	1.27	.50	
.75	2.46	2.46	2.46	2.47	2.47	2.47	2.47	2.47	2.47	2.47	2.47	2.47	.75	
.90	5.20	5.18	5.18	5.17	5.16	5.15	5.15	5.14	5.14	5.14	5.14	5.13	.90	
.95	8.70	8.66	8.63	8.62	8.59	8.58	8.57	8.55	8.55	8.54	8.53	8.53	.95	
.975	14.3	14.2	14.1	14.1	14.0	14.0	14.0	14.0	13.9	13.9	13.9	13.9	.975	
.99	26.9	26.7	26.6	26.5	26.4	26.4	26.3	26.2	26.2	26.2	26.1	26.1	.99	
.995	43.1	42.8	42.6	42.5	42.3	42.2	42.1	42.0	42.0	41.9	41.9	41.8	.995	
.999	127	126	126	125	125	125	124	124	124	124	124	123	.999	
.9995	203	201	200	199	199	198	198	197	197	197	196	196	.9995	

ν_2, DEGREES OF FREEDOM FOR DENOMINATOR

Table 6. (continued)

ν_1, DEGREES OF FREEDOM FOR NUMERATOR

ν_2, DEGREES OF FREEDOM FOR DENOMINATOR

	$\mathbb{P}$	1	2	3	4	5	6	7	8	9	10	11	12	$\mathbb{P}$
4	.0005	$.0^6 44$	$.0^3 50$	$.0^2 46$	.013	.024	.036	.047	.057	.066	.075	.082	.089	.0005
	.001	$.0^5 18$	$.0^2 10$	$.0^2 73$	.019	.032	.046	.058	.069	.079	.089	.097	.104	.001
	.005	$.0^4 44$	$.0^2 50$	.022	.043	.064	.083	.100	.114	.126	.137	.145	.153	.005
	.01	$.0^3 18$	.010	.035	.063	.088	.109	.127	.143	.156	.167	.176	.185	.01
	.025	$.0^2 11$	.026	.066	.104	.135	.161	.181	.198	.212	.224	.234	.243	.025
	.05	$.0^2 44$	.052	.110	.157	.193	.221	.243	.261	.275	.288	.298	.307	.05
	.10	.018	.108	.187	.243	.284	.314	.338	.356	.371	.384	.394	.403	.10
	.25	.117	.309	.418	.484	.528	.560	.583	.601	.615	.627	.637	.645	.25
	.50	.549	.828	.941	1.00	1.04	1.06	1.08	1.09	1.10	1.11	1.12	1.13	.50
	.75	1.81	2.00	2.05	2.06	2.07	2.08	2.08	2.08	2.08	2.08	2.08	2.08	.75
	.90	4.54	4.32	4.19	4.11	4.05	4.01	3.98	3.95	3.94	3.92	3.91	3.90	.90
	.95	7.71	6.94	6.59	6.39	6.26	6.16	6.09	6.04	6.00	5.96	5.94	5.91	.95
	.975	12.2	10.6	9.98	9.60	9.36	9.20	9.07	8.98	8.90	8.84	8.79	8.75	.975
	.99	21.2	18.0	16.7	16.0	15.5	15.2	15.0	14.8	14.7	14.5	14.4	14.4	.99
	.995	31.3	26.3	24.3	23.2	22.5	22.0	21.6	21.4	21.1	21.0	20.8	20.7	.995
	.999	74.1	61.2	56.2	53.4	51.7	50.5	49.7	49.0	48.5	48.0	47.7	47.4	.999
	.9995	106	87.4	80.1	76.1	73.6	71.9	70.6	69.7	68.9	68.3	67.8	67.4	.9995
5	.0005	$.0^6 43$	$.0^3 50$	$.0^2 47$	.014	.025	.038	.050	.061	.070	.081	.089	.096	.0005
	.001	$.0^5 17$	$.0^2 10$	$.0^2 75$	.019	.034	.048	.062	.074	.085	.095	.104	.112	.001
	.005	$.0^4 43$	$.0^2 50$	.022	.045	.067	.087	.105	.120	.134	.146	.156	.165	.005
	.01	$.0^3 17$	.010	.035	.064	.091	.114	.134	.151	.165	.177	.188	.197	.01
	.025	$.0^2 11$	.025	.067	.107	.140	.167	.189	.208	.223	.236	.248	.257	.025
	.05	$.0^2 43$	.052	.111	.160	.198	.228	.252	.271	.287	.301	.313	.322	.05
	.10	.017	.108	.188	.247	.290	.322	.347	.367	.383	.397	.408	.418	.10
	.25	.113	.305	.415	.483	.528	.560	.584	.604	.618	.631	.641	.650	.25
	.50	.528	.799	.907	.965	1.00	1.02	1.04	1.05	1.06	1.07	1.08	1.09	.50
	.75	1.69	1.85	1.88	1.89	1.89	1.89	1.89	1.89	1.89	1.89	1.89	1.89	.75
	.90	4.06	3.78	3.62	3.52	3.45	3.40	3.37	3.34	3.32	3.30	3.28	3.27	.90
	.95	6.61	5.79	5.41	5.19	5.05	4.95	4.88	4.82	4.77	4.74	4.71	4.68	.95
	.975	10.0	8.43	7.76	7.39	7.15	6.98	6.85	6.76	6.68	6.62	6.57	6.52	.975
	.99	16.3	13.3	12.1	11.4	11.0	10.7	10.5	10.3	10.2	10.1	9.96	9.89	.99
	.995	22.8	18.3	16.5	15.6	14.9	14.5	14.2	14.0	13.8	13.6	13.5	13.4	.995
	.999	47.2	37.1	33.2	31.1	29.7	28.8	28.2	27.6	27.2	26.9	26.6	26.4	.999
	.9995	63.6	49.8	44.4	41.5	39.7	38.5	37.6	36.9	36.4	35.9	35.6	35.2	.9995
6	.0005	$.0^6 43$	$.0^3 50$	$.0^2 47$	.014	.026	.039	.052	.064	.075	.085	.094	.103	.0005
	.001	$.0^5 17$	$.0^2 10$	$.0^2 75$	.020	.035	.050	.064	.078	.090	.101	.111	.119	.001
	.005	$.0^4 43$	$.0^2 50$	.022	.045	.069	.090	.109	.126	.140	.153	.164	.174	.005
	.01	$.0^3 17$	.010	.036	.066	.094	.118	.139	.157	.172	.186	.197	.207	.01
	.025	$.0^2 11$	.025	.068	.109	.143	.172	.195	.215	.231	.246	.258	.268	.025
	.05	$.0^2 43$	.052	.112	.162	.202	.233	.259	.279	.296	.311	.324	.334	.05
	.10	.017	.107	.189	.249	.294	.327	.354	.375	.392	.406	.418	.429	.10
	.25	.111	.302	.413	.481	.524	.561	.586	.606	.622	.635	.645	.654	.25
	.50	.515	.780	.886	.942	.977	1.00	1.02	1.03	1.04	1.05	1.05	1.06	.50
	.75	1.62	1.76	1.78	1.79	1.79	1.78	1.78	1.78	1.77	1.77	1.77	1.77	.75
	.90	3.78	3.46	3.29	3.18	3.11	3.05	3.01	2.98	2.96	2.94	2.92	2.90	.90
	.95	5.99	5.14	4.76	4.53	4.39	4.28	4.21	4.15	4.10	4.06	4.03	4.00	.95
	.975	8.81	7.26	6.60	6.23	5.99	5.82	5.70	5.60	5.52	5.46	5.41	5.37	.975
	.99	13.7	10.9	9.78	9.15	8.75	8.47	8.26	8.10	7.98	7.87	7.79	7.72	.99
	.995	18.6	14.5	12.9	12.0	11.5	11.1	10.8	10.6	10.4	10.2	10.1	10.0	.995
	.999	35.5	27.0	23.7	21.9	20.8	20.0	19.5	19.0	18.7	18.4	18.2	18.0	.999
	.9995	46.1	34.8	30.4	28.1	26.6	25.6	24.9	24.3	23.9	23.5	23.2	23.0	.9995

Table 6. Fisher's *F* distribution **733**

Table 6. (continued)

ν_1, DEGREES OF FREEDOM FOR NUMERATOR

ℙ	15	20	24	30	40	50	60	100	120	200	500	∞	ℙ	
.0005	.105	.125	.135	.147	.159	.166	.172	.183	.186	.191	.196	.200	.0005	**4**
.001	.121	.141	.152	.163	.176	.183	.188	.200	.202	.208	.213	.217	.001	
.005	.172	.193	.204	.216	.229	.237	.242	.253	.255	.260	.266	.269	.005	
.01	.204	.226	.237	.249	.261	.269	.274	.285	.287	.293	.298	.301	.01	
.025	.263	.284	.296	.308	.320	.327	.332	.342	.346	.351	.356	.359	.025	
.05	.327	.349	.360	.372	.384	.391	.396	.407	.409	.413	.418	.422	.05	
.10	.424	.445	.456	.467	.478	.485	.490	.500	.502	.508	.510	.514	.10	
.25	.664	.683	.692	.702	.712	.718	.722	.731	.733	.737	.740	.743	.25	
.50	1.14	1.15	1.16	1.16	1.17	1.18	1.18	1.18	1.18	1.19	1.19	1.19	.50	
.75	2.08	2.08	2.08	2.08	2.08	2.08	2.08	2.08	2.08	2.08	2.08	2.08	.75	
.90	3.87	3.84	3.83	3.82	3.80	3.80	3.79	3.78	3.78	3.77	3.76	3.76	.90	
.95	5.86	5.80	5.77	5.75	5.72	5.70	5.69	5.66	5.66	5.65	5.64	5.63	.95	
.975	8.66	8.56	8.51	8.46	8.41	8.38	8.36	8.32	8.31	8.29	8.27	8.26	.975	
.99	14.2	14.0	13.9	13.8	13.7	13.7	13.7	13.6	13.6	13.5	13.5	13.5	.99	
.995	20.4	20.2	20.0	19.9	19.8	19.7	19.6	19.5	19.5	19.4	19.4	19.3	.995	
.999	46.8	46.1	45.8	45.4	45.1	44.9	44.7	44.5	44.4	44.3	44.1	44.0	.999	
.9995	66.5	65.5	65.1	64.6	64.1	63.8	63.6	63.2	63.1	62.9	62.7	62.6	.9995	
.0005	.115	.137	.150	.163	.177	.186	.192	.205	.209	.216	.222	.226	.0005	**5**
.001	.132	.155	.167	.181	.195	.204	.210	.223	.227	.233	.239	.244	.001	
.005	.186	.210	.223	.237	.251	.260	.266	.279	.282	.288	.294	.299	.005	
.01	.219	.244	.257	.270	.285	.293	.299	.312	.315	.322	.328	.331	.01	
.025	.280	.304	.317	.330	.344	.353	.359	.370	.374	.380	.386	.390	.025	
.05	.345	.369	.382	.395	.408	.417	.422	.432	.437	.442	.448	.452	.05	
.10	.440	.463	.476	.488	.501	.508	.514	.524	.527	.532	.538	.541	.10	
.25	.669	.690	.700	.711	.722	.728	.732	.741	.743	.748	.752	.755	.25	
.50	1.10	1.11	1.12	1.12	1.13	1.13	1.14	1.14	1.14	1.15	1.15	1.15	.50	
.75	1.89	1.88	1.88	1.88	1.88	1.88	1.87	1.87	1.87	1.87	1.87	1.87	.75	
.90	3.24	3.21	3.19	3.17	3.16	3.15	3.14	3.13	3.12	3.12	3.11	3.10	.90	
.95	4.62	4.56	4.53	4.50	4.46	4.44	4.43	4.41	4.40	4.39	4.37	4.36	.95	
.975	6.43	6.33	6.28	6.23	6.18	6.14	6.12	6.08	6.07	6.05	6.03	6.02	.975	
.99	9.72	9.55	9.47	9.38	9.29	9.24	9.20	9.13	9.11	9.08	9.04	9.02	.99	
.995	13.1	12.9	12.8	12.7	12.5	12.5	12.4	12.3	12.3	12.2	12.2	12.1	.995	
.999	25.9	25.4	25.1	24.9	24.6	24.4	24.3	24.1	24.1	23.9	23.8	23.8	.999	
.9995	34.6	33.9	33.5	33.1	32.7	32.5	32.3	32.1	32.0	31.8	31.7	31.6	.9995	
.0005	.123	.148	.162	.177	.193	.203	.210	.225	.229	.236	.244	.249	.0005	**6**
.001	.141	.166	.180	.195	.211	.222	.229	.243	.247	.255	.262	.267	.001	
.005	.197	.224	.238	.253	.269	.279	.286	.301	.304	.312	.318	.324	.005	
.01	.232	.258	.273	.288	.304	.313	.321	.334	.338	.346	.352	.357	.01	
.025	.293	.320	.334	.349	.364	.375	.381	.394	.398	.405	.412	.415	.025	
.05	.358	.385	.399	.413	.428	.437	.444	.457	.460	.467	.472	.476	.05	
.10	.453	.478	.491	.505	.519	.526	.533	.546	.548	.556	.559	.564	.10	
.25	.675	.696	.707	.718	.729	.736	.741	.751	.753	.758	.762	.765	.25	
.50	1.07	1.08	1.09	1.10	1.10	1.11	1.11	1.11	1.12	1.12	1.12	1.12	.50	
.75	1.76	1.76	1.75	1.75	1.75	1.75	1.74	1.74	1.74	1.74	1.74	1.74	.75	
.90	2.87	2.84	2.82	2.80	2.78	2.77	2.76	2.75	2.74	2.73	2.73	2.72	.90	
.95	3.94	3.87	3.84	3.81	3.77	3.75	3.74	3.71	3.70	3.69	3.68	3.67	.95	
.975	5.27	5.17	5.12	5.07	5.01	4.98	4.96	4.92	4.90	4.88	4.86	4.85	.975	
.99	7.56	7.40	7.31	7.23	7.14	7.09	7.06	6.99	6.97	6.93	6.90	6.88	.99	
.995	9.81	9.59	9.47	9.36	9.24	9.17	9.12	9.03	9.00	8.95	8.91	8.88	.995	
.999	17.6	17.1	16.9	16.7	16.4	16.3	16.2	16.0	16.0	15.9	15.8	15.7	.999	
.9995	22.4	21.9	21.7	21.4	21.1	20.9	20.7	20.5	20.4	20.3	20.2	20.1	.9995	

ν_2, DEGREES OF FREEDOM FOR DENOMINATOR

Table 6. (continued)

ν_1, DEGREES OF FREEDOM FOR NUMERATOR

ν_2, DEGREES OF FREEDOM FOR DENOMINATOR

$\nu_2 = 7$

$\mathbb{P}$	1	2	3	4	5	6	7	8	9	10	11	12	$\mathbb{P}$
.0005	$.0^642$	$.0^350$	$.0^248$	.014	.027	.040	.053	.066	.078	.088	.099	.108	.0005
.001	$.0^517$	$.0^210$	$.0^276$	.020	.035	.051	.067	.081	.093	.105	.115	.125	.001
.005	$.0^442$	$.0^250$	.023	.046	.070	.093	.113	.130	.145	.159	.171	.181	.005
.01	$.0^317$	.010	.036	.067	.096	.121	.143	.162	.178	.192	.205	.216	.01
.025	$.0^210$	.025	.068	.110	.146	.176	.200	.221	.238	.253	.266	.277	.025
.05	$.0^242$	.052	.113	.164	.205	.238	.264	.286	.304	.319	.332	.343	.05
.10	.017	.107	.190	.251	.297	.332	.359	.381	.399	.414	.427	.438	.10
.25	.110	.300	.412	.481	.528	.562	.588	.608	.624	.637	.649	.658	.25
.50	.506	.767	.871	.926	.960	.983	1.00	1.01	1.02	1.03	1.04	1.04	.50
.75	1.57	1.70	1.72	1.72	1.71	1.71	1.70	1.70	1.69	1.69	1.69	1.68	.75
.90	3.59	3.26	3.07	2.96	2.88	2.83	2.78	2.75	2.72	2.70	2.68	2.67	.90
.95	5.59	4.74	4.35	4.12	3.97	3.87	3.79	3.73	3.68	3.64	3.60	3.57	.95
.975	8.07	6.54	5.89	5.52	5.29	5.12	4.99	4.90	4.82	4.76	4.71	4.67	.975
.99	12.2	9.55	8.45	7.85	7.46	7.19	6.99	6.84	6.72	6.62	6.54	6.47	.99
.995	16.2	12.4	10.9	10.0	9.52	9.16	8.89	8.68	8.51	8.38	8.27	8.18	.995
.999	29.2	21.7	18.8	17.2	16.2	15.5	15.0	14.6	14.3	14.1	13.9	13.7	.999
.9995	37.0	27.2	23.5	21.4	20.2	19.3	18.7	18.2	17.8	17.5	17.2	17.0	.9995

$\nu_2 = 8$

$\mathbb{P}$	1	2	3	4	5	6	7	8	9	10	11	12	$\mathbb{P}$
.0005	$.0^642$	$.0^350$	$.0^248$	.014	.027	.041	.055	.068	.081	.092	.102	.112	.0005
.001	$.0^517$	$.0^210$	$.0^276$	.020	.036	.053	.068	.083	.096	.109	.120	.130	.001
.005	$.0^442$	$.0^250$	.027	.047	.072	.095	.115	.133	.149	.164	.176	.187	.005
.01	$.0^317$	.010	.036	.068	.097	.123	.146	.166	.183	.198	.211	.222	.01
.025	$.0^210$	.025	.069	.111	.148	.179	.204	.226	.244	.259	.273	.285	.025
.05	$.0^242$	.052	.113	.166	.208	.241	.268	.291	.310	.326	.339	.351	.05
.10	.017	.107	.190	.253	.299	.335	.363	.386	.405	.421	.435	.445	.10
.25	.109	.298	.411	.481	.529	.563	.589	.610	.627	.640	.654	.661	.25
.50	.499	.757	.860	.915	.948	.971	.988	1.00	1.01	1.02	1.02	1.03	.50
.75	1.54	1.66	1.67	1.66	1.66	1.65	1.64	1.64	1.64	1.63	1.63	1.62	.75
.90	3.46	3.11	2.92	2.81	2.73	2.67	2.62	2.59	2.56	2.54	2.52	2.50	.90
.95	5.32	4.46	4.07	3.84	3.69	3.58	3.50	3.44	3.39	3.35	3.31	3.28	.95
.975	7.57	6.06	5.42	5.05	4.82	4.65	4.53	4.43	4.36	4.30	4.24	4.20	.975
.99	11.3	8.65	7.59	7.01	6.63	6.37	6.18	6.03	5.91	5.81	5.73	5.67	.99
.995	14.7	11.0	9.60	8.81	8.30	7.95	7.69	7.50	7.34	7.21	7.10	7.01	.995
.999	25.4	18.5	15.8	14.4	13.5	12.9	12.4	12.0	11.8	11.5	11.4	11.2	.999
.9995	31.6	22.8	19.4	17.6	16.4	15.7	15.1	14.6	14.3	14.0	13.8	13.6	.9995

$\nu_2 = 9$

$\mathbb{P}$	1	2	3	4	5	6	7	8	9	10	11	12	$\mathbb{P}$
.0005	$.0^641$	$.0^350$	$.0^248$	.015	.027	.042	.056	.070	.083	.094	.105	.115	.0005
.001	$.0^517$	$.0^210$	$.0^277$	.021	.037	.054	.070	.085	.099	.112	.123	.134	.001
.005	$.0^442$	$.0^250$	.023	.047	.073	.096	.117	.136	.153	.168	.181	.192	.005
.01	$.0^317$	.010	.037	.068	.098	.125	.149	.169	.187	.202	.216	.228	.01
.025	$.0^210$	.025	.069	.112	.150	.181	.207	.230	.248	.265	.279	.291	.025
.05	$.0^240$	.052	.113	.167	.210	.244	.272	.296	.315	.331	.345	.358	.05
.10	.017	.107	.191	.254	.302	.338	.367	.390	.410	.426	.441	.452	.10
.25	.108	.297	.410	.480	.529	.564	.591	.612	.629	.643	.654	.664	.25
.50	.494	.749	.852	.906	.939	.962	.978	.990	1.00	1.01	1.01	1.02	.50
.75	1.51	1.62	1.63	1.63	1.62	1.61	1.60	1.60	1.59	1.59	1.58	1.58	.75
.90	3.36	3.01	2.81	2.69	2.61	2.55	2.51	2.47	2.44	2.42	2.40	2.38	.90
.95	5.12	4.26	3.86	3.63	3.48	3.37	3.29	3.23	3.18	3.14	3.10	3.07	.95
.975	7.21	5.71	5.08	4.72	4.48	4.32	4.20	4.10	4.03	3.96	3.91	3.87	.975
.99	10.6	8.02	6.99	6.42	6.06	5.80	5.61	5.47	5.35	5.26	5.18	5.11	.99
.995	13.6	10.1	8.72	7.96	7.47	7.13	6.88	6.69	6.54	6.42	6.31	6.23	.995
.999	22.9	16.4	13.9	12.6	11.7	11.1	10.7	10.4	10.1	9.89	9.71	9.57	.999
.9995	28.0	19.9	16.8	15.1	14.1	13.3	12.8	12.4	12.1	11.8	11.6	11.4	.9995

Table 6. Fisher's *F* distribution   **735**

Table 6. (continued)

ν_1, DEGREES OF FREEDOM FOR NUMERATOR

$\mathbb{P}$	15	20	24	30	40	50	60	100	120	200	500	∞	$\mathbb{P}$	
.0005	.130	.157	.172	.188	.206	.217	.225	.242	.246	.255	.263	.268	.0005	**7**
.001	.148	.176	.191	.208	.225	.237	.245	.261	.266	.274	.282	.288	.001	
.005	.206	.235	.251	.267	.285	.296	.304	.319	.324	.332	.340	.345	.005	
.01	.241	.270	.286	.303	.320	.331	.339	.355	.358	.366	.373	.379	.01	
.025	.304	.333	.348	.364	.381	.392	.399	.413	.418	.426	.433	.437	.025	
.05	.369	.398	.413	.428	.445	.455	.461	.476	.479	.485	.493	.498	.05	
.10	.463	.491	.504	.519	.534	.543	.550	.562	.566	.571	.578	.582	.10	
.25	.679	.702	.713	.725	.737	.745	.749	.760	.762	.767	.772	.775	.25	
.50	1.05	1.07	1.07	1.08	1.08	1.09	1.09	1.10	1.10	1.10	1.10	1.10	.50	
.75	1.68	1.67	1.67	1.66	1.66	1.66	1.65	1.65	1.65	1.65	1.65	1.65	.75	
.90	2.63	2.59	2.58	2.56	2.54	2.52	2.51	2.50	2.49	2.48	2.48	2.47	.90	
.95	3.51	3.44	3.41	3.38	3.34	3.32	3.30	3.27	3.27	3.25	3.24	3.23	.95	
.975	4.57	4.47	4.42	4.36	4.31	4.28	4.25	4.21	4.20	4.18	4.16	4.14	.975	
.99	6.31	6.16	6.07	5.99	5.91	5.86	5.82	5.75	5.74	5.70	5.67	5.65	.99	
.995	7.97	7.75	7.65	7.53	7.42	7.35	7.31	7.22	7.19	7.15	7.10	7.08	.995	
.999	13.3	12.9	12.7	12.5	12.3	12.2	12.1	11.9	11.9	11.8	11.7	11.7	.999	
.9995	16.5	16.0	15.7	15.5	15.2	15.1	15.0	14.7	14.7	14.6	14.5	14.4	.9995	
.0005	.136	.164	.181	.198	.218	.230	.239	.257	.262	.271	.281	.287	.0005	**8**
.001	.155	.184	.200	.218	.238	.250	.259	.277	.282	.292	.300	.306	.001	
.005	.214	.244	.261	.279	.299	.311	.319	.337	.341	.351	.358	.364	.005	
.01	.250	.281	.297	.315	.334	.346	.354	.372	.376	.385	.392	.398	.01	
.025	.313	.343	.360	.377	.395	.407	.415	.431	.435	.442	.450	.456	.025	
.05	.379	.409	.425	.441	.459	.469	.477	.493	.496	.505	.510	.516	.05	
.10	.472	.500	.515	.531	.547	.556	.563	.578	.581	.588	.595	.599	.10	
.25	.684	.707	.718	.730	.743	.751	.756	.767	.769	.775	.780	.783	.25	
.50	1.04	1.05	1.06	1.07	1.07	1.07	1.08	1.08	1.08	1.09	1.09	1.09	.50	
.75	1.62	1.61	1.60	1.60	1.59	1.59	1.59	1.58	1.58	1.58	1.58	1.58	.75	
.90	2.46	2.42	2.40	2.38	2.36	2.35	2.34	2.32	2.32	2.31	2.30	2.29	.90	
.95	3.22	3.15	3.12	3.08	3.04	3.02	3.01	2.97	2.97	2.95	2.94	2.93	.95	
.975	4.10	4.00	3.95	3.89	3.84	3.81	3.78	3.74	3.73	3.70	3.68	3.67	.975	
.99	5.52	5.36	5.28	5.20	5.12	5.07	5.03	4.96	4.95	4.91	4.88	4.86	.99	
.995	6.81	6.61	6.50	6.40	6.29	6.22	6.18	6.09	6.06	6.02	5.98	5.95	.995	
.999	10.8	10.5	10.3	10.1	9.92	9.80	9.73	9.57	9.54	9.46	9.39	9.34	.999	
.9995	13.1	12.7	12.5	12.2	12.0	11.8	11.8	11.6	11.5	11.4	11.4	11.3	.9995	
.0005	.141	.171	.188	.207	.228	.242	.251	.270	.276	.287	.297	.303	.0005	**9**
.001	.160	.191	.208	.228	.249	.262	.271	.291	.296	.307	.316	.323	.001	
.005	.220	.253	.271	.290	.310	.324	.332	.351	.356	.366	.376	.382	.005	
.01	.257	.289	.307	.326	.346	.358	.368	.386	.391	.400	.410	.415	.01	
.025	.320	.352	.370	.388	.408	.420	.428	.446	.450	.459	.467	.473	.025	
.05	.386	.418	.435	.452	.471	.483	.490	.508	.510	.518	.526	.532	.05	
.10	.479	.509	.525	.541	.558	.568	.575	.588	.594	.602	.610	.613	.10	
.25	.687	.711	.723	.736	.749	.757	.762	.773	.776	.782	.787	.791	.25	
.50	1.03	1.04	1.05	1.05	1.06	1.06	1.07	1.07	1.07	1.08	1.08	1.08	.50	
.75	1.57	1.56	1.56	1.55	1.55	1.54	1.54	1.53	1.53	1.53	1.53	1.53	.75	
.90	2.34	2.30	2.28	2.25	2.23	2.22	2.21	2.19	2.18	2.17	2.17	2.16	.90	
.95	3.01	2.94	2.90	2.86	2.83	2.80	2.79	2.76	2.75	2.73	2.72	2.71	.95	
.975	3.77	3.67	3.61	3.56	3.51	3.47	3.45	3.40	3.39	3.37	3.35	3.33	.975	
.99	4.96	4.81	4.73	4.65	4.57	4.52	4.48	4.42	4.40	4.36	4.33	4.31	.99	
.995	6.03	5.83	5.73	5.62	5.52	5.45	5.41	5.32	5.30	5.26	5.21	5.19	.995	
.999	9.24	8.90	8.72	8.55	8.37	8.26	8.19	8.04	8.00	7.93	7.86	7.81	.999	
.9995	11.0	10.6	10.4	10.2	9.94	9.80	9.71	9.53	9.49	9.40	9.32	9.26	.9995	

ν_2, DEGREES OF FREEDOM FOR DENOMINATOR

Table 6. (continued)

ν_1, DEGREES OF FREEDOM FOR NUMERATOR

ν_2, DEGREES OF FREEDOM FOR DENOMINATOR

	ℙ	1	2	3	4	5	6	7	8	9	10	11	12	ℙ
10	.0005	$.0^641$	$.0^350$	$.0^249$	.015	.028	.043	.057	.071	.085	.097	.108	.119	.0005
	.001	$.0^517$	$.0^210$	$.0^277$	.021	.037	.054	.071	.087	.101	.114	.126	.137	.001
	.005	$.0^441$	$.0^250$	.023	.048	.073	.098	.119	.139	.156	.171	.185	.197	.005
	.01	$.0^317$	.010	.037	.069	.100	.127	.151	.172	.190	.206	.220	.233	.01
	.025	$.0^210$	.025	.069	.113	.151	.183	.210	.233	.252	.269	.283	.296	.025
	.05	$.0^241$	.052	.114	.168	.211	.246	.275	.299	.319	.336	.351	.363	.05
	.10	.017	.106	.191	.255	.303	.340	.370	.394	.414	.430	.444	.457	.10
	.25	.107	.296	.409	.480	.529	.565	.592	.613	.631	.645	.657	.667	.25
	.50	.490	.743	.845	.899	.932	.954	.971	.983	.992	1.00	1.01	1.01	.50
	.75	1.49	1.60	1.60	1.59	1.59	1.58	1.57	1.56	1.56	1.55	1.55	1.54	.75
	.90	3.28	2.92	2.73	2.61	2.52	2.46	2.41	2.38	2.35	2.32	2.30	2.28	.90
	.95	4.96	4.10	3.71	3.48	3.33	3.22	3.14	3.07	3.02	2.98	2.94	2.91	.95
	.975	6.94	5.46	4.83	4.47	4.24	4.07	3.95	3.85	3.78	3.72	3.66	3.62	.975
	.99	10.0	7.56	6.55	5.99	5.64	5.39	5.20	5.06	4.94	4.85	4.77	4.71	.99
	.995	12.8	9.43	8.08	7.34	6.87	6.54	6.30	6.12	5.97	5.85	5.75	5.66	.995
	.999	21.0	14.9	12.6	11.3	10.5	9.92	9.52	9.20	8.96	8.75	8.58	8.44	.999
	.9995	25.5	17.9	15.0	13.4	12.4	11.8	11.3	10.9	10.6	10.3	10.1	9.93	.9995
11	.0005	$.0^641$	$.0^350$	$.0^249$	.015	.028	.043	.058	.072	.086	.099	.111	.121	.0005
	.001	$.0^516$	$.0^210$	$.0^278$	.021	.038	.055	.072	.088	.103	.116	.129	.140	.001
	.005	$.0^440$	$.0^250$	.023	.048	.074	.099	.121	.141	.158	.174	.188	.200	.005
	.01	$.0^316$	.010	.037	.069	.100	.128	.153	.175	.193	.210	.224	.237	.01
	.025	$.0^210$	.025	.069	.114	.152	.185	.212	.236	.256	.273	.288	.301	.025
	.05	$.0^241$	.052	.114	.168	.212	.248	.278	.302	.323	.340	.355	.368	.05
	.10	.017	.106	.192	.256	.305	.342	.373	.397	.417	.435	.448	.461	.10
	.25	.107	.295	.408	.481	.529	.565	.592	.614	.633	.645	.658	.667	.25
	.50	.486	.739	.840	.893	.926	.948	.964	.977	.986	.994	1.00	1.01	.50
	.75	1.47	1.58	1.58	1.57	1.56	1.55	1.54	1.53	1.53	1.52	1.52	1.51	.75
	.90	3.23	2.86	2.66	2.54	2.45	2.39	2.34	2.30	2.27	2.25	2.23	2.21	.90
	.95	4.84	3.98	3.59	3.36	3.20	3.09	3.01	2.95	2.90	2.85	2.82	2.79	.95
	.975	6.72	5.26	4.63	4.28	4.04	3.88	3.76	3.66	3.59	3.53	3.47	3.43	.975
	.99	9.65	7.21	6.22	5.67	5.32	5.07	4.89	4.74	4.63	4.54	4.46	4.40	.99
	.995	12.2	8.91	7.60	6.88	6.42	6.10	5.86	5.68	5.54	5.42	5.32	5.24	.995
	.999	19.7	13.8	11.6	10.3	9.58	9.05	8.66	8.35	8.12	7.92	7.76	7.62	.999
	.9995	23.6	16.4	13.6	12.2	11.2	10.6	10.1	9.76	9.48	9.24	9.04	8.88	.9995
12	.0005	$.0^641$	$.0^350$	$.0^249$	.015	.028	.044	.058	.073	.087	.101	.113	.124	.0005
	.001	$.0^516$	$.0^210$	$.0^278$	.021	.038	.056	.073	.089	.104	.118	.131	.143	.001
	.005	$.0^439$	$.0^250$	.023	.048	.075	.100	.122	.143	.161	.177	.191	.204	.005
	.01	$.0^316$	.010	.037	.070	.101	.130	.155	.176	.196	.212	.227	.241	.01
	.025	$.0^210$	.025	.070	.114	.153	.186	.214	.238	.259	.276	.292	.305	.025
	.05	$.0^241$	.052	.114	.169	.214	.250	.280	.305	.325	.343	.358	.372	.05
	.10	.016	.106	.192	.257	.306	.344	.375	.400	.420	.438	.452	.466	.10
	.25	.106	.295	.408	.480	.530	.566	.594	.616	.633	.649	.662	.671	.25
	.50	.484	.735	.835	.888	.921	.943	.959	.972	.981	.989	.995	1.00	.50
	.75	1.46	1.56	1.56	1.55	1.54	1.53	1.52	1.51	1.51	1.50	1.50	1.49	.75
	.90	3.18	2.81	2.61	2.48	2.39	2.33	2.28	2.24	2.21	2.19	2.17	2.15	.90
	.95	4.75	3.89	3.49	3.26	3.11	3.00	2.91	2.85	2.80	2.75	2.72	2.69	.95
	.975	6.55	5.10	4.47	4.12	3.89	3.73	3.61	3.51	3.44	3.37	3.32	3.28	.975
	.99	9.33	6.93	5.95	5.41	5.06	4.82	4.64	4.50	4.39	4.30	4.22	4.16	.99
	.995	11.8	8.51	7.23	6.52	6.07	5.76	5.52	5.35	5.20	5.09	4.99	4.91	.995
	.999	18.6	13.0	10.8	9.63	8.89	8.38	8.00	7.71	7.48	7.29	7.14	7.01	.999
	.9995	22.2	15.3	12.7	11.2	10.4	9.74	9.28	8.94	8.66	8.43	8.24	8.08	.9995

Table 6. Fisher's F distribution **737**

Table 6. (continued)

ν_1, DEGREES OF FREEDOM FOR NUMERATOR

$\mathbb{P}$	15	20	24	30	40	50	60	100	120	200	500	∞	$\mathbb{P}$	
.0005	.145	.177	.195	.215	.238	.251	.262	.282	.288	.299	.311	.319	.0005	**10**
.001	.164	.197	.216	.236	.258	.272	.282	.303	.309	.321	.331	.338	.001	
.005	.226	.260	.279	.299	.321	.334	.344	.365	.370	.380	.391	.397	.005	
.01	.263	.297	.316	.336	.357	.370	.380	.400	.405	.415	.424	.431	.01	
.025	.327	.360	.379	.398	.419	.431	.441	.459	.464	.474	.483	.488	.025	
.05	.393	.426	.444	.462	.481	.493	.502	.518	.523	.532	.541	.546	.05	
.10	.486	.516	.532	.549	.567	.578	.586	.602	.605	.614	.621	.625	.10	
.25	.691	.714	.727	.740	.754	.762	.767	.779	.782	.788	.793	.797	.25	
.50	1.02	1.03	1.04	1.05	1.05	1.06	1.06	1.06	1.06	1.07	1.07	1.07	.50	
.75	1.53	1.52	1.52	1.51	1.51	1.50	1.50	1.49	1.49	1.49	1.48	1.48	.75	
.90	2.24	2.20	2.18	2.16	2.13	2.12	2.11	2.09	2.08	2.07	2.06	2.06	.90	
.95	2.85	2.77	2.74	2.70	2.66	2.64	2.62	2.59	2.58	2.56	2.55	2.54	.95	
.975	3.52	3.42	3.37	3.31	3.26	3.22	3.20	3.15	3.14	3.12	3.09	3.08	.975	
.99	4.56	4.41	4.33	4.25	4.17	4.12	4.08	4.01	4.00	3.96	3.93	3.91	.99	
.995	5.47	5.27	5.17	5.07	4.97	4.90	4.86	4.77	4.75	4.71	4.67	4.64	.995	
.999	8.13	7.80	7.64	7.47	7.30	7.19	7.12	6.98	6.94	6.87	6.81	6.76	.999	
.9995	9.56	9.16	8.96	8.75	8.54	8.42	8.33	8.16	8.12	8.04	7.96	7.90	.9995	
.0005	.148	.182	.201	.222	.246	.261	.271	.293	.299	.312	.324	.331	.0005	**11**
.001	.168	.202	.222	.243	.266	.282	.292	.313	.320	.332	.343	.353	.001	
.005	.231	.266	.286	.308	.330	.345	.355	.376	.382	.394	.403	.412	.005	
.01	.268	.304	.324	.344	.366	.380	.391	.412	.417	.427	.439	.444	.01	
.025	.332	.368	.386	.407	.429	.442	.450	.472	.476	.485	.495	.503	.025	
.05	.398	.433	.452	.469	.490	.503	.513	.529	.535	.543	.552	.559	.05	
.10	.490	.524	.541	.559	.578	.588	.595	.614	.617	.625	.633	.637	.10	
.25	.694	.719	.730	.744	.758	.767	.773	.780	.788	.794	.799	.803	.25	
.50	1.02	1.03	1.03	1.04	1.05	1.05	1.05	1.06	1.06	1.06	1.06	1.06	.50	
.75	1.50	1.49	1.49	1.48	1.47	1.47	1.47	1.46	1.46	1.46	1.45	1.45	.75	
.90	2.17	2.12	2.10	2.08	2.05	2.04	2.03	2.00	2.00	1.99	1.98	1.97	.90	
.95	2.72	2.65	2.61	2.57	2.53	2.51	2.49	2.46	2.45	2.43	2.42	2.40	.95	
.975	3.33	3.23	3.17	3.12	3.06	3.03	3.00	2.96	2.94	2.92	2.90	2.88	.975	
.99	4.25	4.10	4.02	3.94	3.86	3.81	3.78	3.71	3.69	3.66	3.62	3.60	.99	
.995	5.05	4.86	4.76	4.65	4.55	4.49	4.45	4.36	4.34	4.29	4.25	4.23	.995	
.999	7.32	7.01	6.85	6.68	6.52	6.41	6.35	6.21	6.17	6.10	6.04	6.00	.999	
.9995	8.52	8.14	7.94	7.75	7.55	7.43	7.35	7.18	7.14	7.06	6.98	6.93	.9995	
.0005	.152	.186	.206	.228	.253	.269	.280	.305	.311	.323	.337	.345	.0005	**12**
.001	.172	.207	.228	.250	.275	.291	.302	.326	.332	.344	.357	.365	.001	
.005	.235	.272	.292	.315	.339	.355	.365	.388	.393	.405	.417	.424	.005	
.01	.273	.310	.330	.352	.375	.391	.401	.422	.428	.441	.450	.458	.01	
.025	.337	.374	.394	.416	.437	.450	.461	.481	.487	.498	.508	.514	.025	
.05	.404	.439	.458	.478	.499	.513	.522	.541	.545	.556	.565	.571	.05	
.10	.496	.528	.546	.564	.583	.595	.604	.621	.625	.633	.641	.647	.10	
.25	.695	.721	.734	.748	.762	.771	.777	.789	.792	.799	.804	.808	.25	
.50	1.01	1.02	1.03	1.03	1.04	1.04	1.05	1.05	1.05	1.05	1.06	1.06	.50	
.75	1.48	1.47	1.46	1.45	1.45	1.44	1.44	1.43	1.43	1.43	1.42	1.42	.75	
.90	2.11	2.06	2.04	2.01	1.99	1.97	1.96	1.94	1.93	1.92	1.91	1.90	.90	
.95	2.62	2.54	2.51	2.47	2.43	2.40	2.38	2.35	2.34	2.32	2.31	2.30	.95	
.975	3.18	3.07	3.02	2.96	2.91	2.87	2.85	2.80	2.79	2.76	2.74	2.72	.975	
.99	4.01	3.86	3.78	3.70	3.62	3.57	3.54	3.47	3.45	3.41	3.38	3.36	.99	
.995	4.72	4.53	4.43	4.33	4.23	4.17	4.12	4.04	4.01	3.97	3.93	3.90	.995	
.999	6.71	6.40	6.25	6.09	5.93	5.83	5.76	5.63	5.59	5.52	5.46	5.42	.999	
.9995	7.74	7.37	7.18	7.00	6.80	6.68	6.61	6.45	6.41	6.33	6.25	6.20	.9995	

ν_2, DEGREES OF FREEDOM FOR DENOMINATOR

Table 6. (continued)

ν_1, DEGREES OF FREEDOM FOR NUMERATOR

ν_2, DEGREES OF FREEDOM FOR DENOMINATOR

ν_2	$\mathbb{P}$	1	2	3	4	5	6	7	8	9	10	11	12	$\mathbb{P}$
15	.0005	$.0^6 41$	$.0^3 50$	$.0^2 49$	.015	.029	.045	.061	.076	.091	.105	.117	.129	.0005
	.001	$.0^5 16$	$.0^2 10$	$.0^2 79$	.021	.039	.057	.075	.092	.108	.123	.137	.149	.001
	.005	$.0^4 39$	$.0^2 50$	.023	.049	.076	.102	.125	.147	.166	.183	.198	.212	.005
	.01	$.0^3 16$	.010	.037	.070	.103	.132	.158	.181	.202	.219	.235	.249	.01
	.025	$.0^2 10$	.025	.070	.116	.156	.190	.219	.244	.265	.284	.300	.315	.025
	.05	$.0^2 41$	.051	.115	.170	.216	.254	.285	.311	.333	.351	.368	.382	.05
	.10	.016	.106	.192	.258	.309	.348	.380	.406	.427	.446	.461	.475	.10
	.25	.105	.293	.407	.480	.531	.568	.596	.618	.637	.652	.667	.676	.25
	.50	.478	.726	.826	.878	.911	.933	.948	.960	.970	.977	.984	.989	.50
	.75	1.43	1.52	1.52	1.51	1.49	1.48	1.47	1.46	1.46	1.45	1.44	1.44	.75
	.90	3.07	2.70	2.49	2.36	2.27	2.21	2.16	2.12	2.09	2.06	2.04	2.02	.90
	.95	4.54	3.68	3.29	3.06	2.90	2.79	2.71	2.64	2.59	2.54	2.51	2.48	.95
	.975	6.20	4.76	4.15	3.80	3.58	3.41	3.29	3.20	3.12	3.06	3.01	2.96	.975
	.99	8.68	6.36	5.42	4.89	4.56	4.32	4.14	4.00	3.89	3.80	3.73	3.67	.99
	.995	10.8	7.70	6.48	5.80	5.37	5.07	4.85	4.67	4.54	4.42	4.33	4.25	.995
	.999	16.6	11.3	9.34	8.25	7.57	7.09	6.74	6.47	6.26	6.08	5.93	5.81	.999
	.9995	19.5	13.2	10.8	9.48	8.66	8.10	7.68	7.36	7.11	6.91	6.75	6.60	.9995
20	.0005	$.0^6 40$	$.0^3 50$	$.0^2 50$	.015	.029	.046	.063	.079	.094	.109	.123	.136	.0005
	.001	$.0^5 16$	$.0^2 10$	$.0^2 79$	.022	.039	.058	.077	.095	.112	.128	.143	.156	.001
	.005	$.0^4 39$	$.0^2 50$	.023	.050	.077	.104	.129	.151	.171	.190	.206	.221	.005
	.01	$.0^3 16$	.010	.037	.071	.105	.135	.162	.187	.208	.227	.244	.259	.01
	.025	$.0^2 10$	.025	.071	.117	.158	.193	.224	.250	.273	.292	.310	.325	.025
	.05	$.0^2 40$	.051	.115	.172	.219	.258	.290	.318	.340	.360	.377	.393	.05
	.10	.016	.106	.193	.260	.312	.353	.385	.412	.435	.454	.472	.485	.10
	.25	.104	.292	.407	.480	.531	.569	.598	.622	.641	.656	.671	.681	.25
	.50	.472	.718	.816	.868	.900	.922	.938	.950	.959	.966	.972	.977	.50
	.75	1.40	1.49	1.48	1.47	1.45	1.44	1.43	1.42	1.41	1.40	1.39	1.39	.75
	.90	2.97	2.59	2.38	2.25	2.16	2.09	2.04	2.00	1.96	1.94	1.91	1.89	.90
	.95	4.35	3.49	3.10	2.87	2.71	2.60	2.51	2.45	2.39	2.35	2.31	2.28	.95
	.975	5.87	4.46	3.86	3.51	3.29	3.13	3.01	2.91	2.84	2.77	2.72	2.68	.975
	.99	8.10	5.85	4.94	4.43	4.10	3.87	3.70	3.56	3.46	3.37	3.29	3.23	.99
	.995	9.94	6.99	5.82	5.17	4.76	4.47	4.26	4.09	3.96	3.85	3.76	3.68	.995
	.999	14.8	9.95	8.10	7.10	6.46	6.02	5.69	5.44	5.24	5.08	4.94	4.82	.999
	.9995	17.2	11.4	9.20	8.02	7.28	6.76	6.38	6.08	5.85	5.66	5.51	5.38	.9995
24	.0005	$.0^6 40$	$.0^3 50$	$.0^2 50$	.015	.030	.046	.064	.080	.096	.112	.126	.139	.0005
	.001	$.0^5 16$	$.0^2 10$	$.0^2 79$	.022	.040	.059	.079	.097	.115	.131	.146	.160	.001
	.005	$.0^4 40$	$.0^2 50$	.023	.050	.078	.106	.131	.154	.175	.193	.210	.226	.005
	.01	$.0^3 16$	.010	.038	.072	.106	.137	.165	.189	.211	.231	.249	.264	.01
	.025	$.0^2 10$	.025	.071	.117	.159	.195	.227	.253	.277	.297	.315	.331	.025
	.05	$.0^2 40$	.051	.116	.173	.221	.260	.293	.321	.345	.365	.383	.399	.05
	.10	.016	.106	.193	.261	.313	.355	.388	.416	.439	.459	.476	.491	.10
	.25	.104	.291	.406	.480	.532	.570	.600	.623	.643	.659	.671	.684	.25
	.50	.469	.714	.812	.863	.895	.917	.932	.944	.953	.961	.967	.972	.50
	.75	1.39	1.47	1.46	1.44	1.43	1.41	1.40	1.39	1.38	1.38	1.37	1.36	.75
	.90	2.93	2.54	2.33	2.19	2.10	2.04	1.98	1.94	1.91	1.88	1.85	1.83	.90
	.95	4.26	3.40	3.01	2.78	2.62	2.51	2.42	2.36	2.30	2.25	2.21	2.18	.95
	.975	5.72	4.32	3.72	3.38	3.15	2.99	2.87	2.78	2.70	2.64	2.59	2.54	.975
	.99	7.82	5.61	4.72	4.22	3.90	3.67	3.50	3.36	3.26	3.17	3.09	3.03	.99
	.995	9.55	6.66	5.52	4.89	4.49	4.20	3.99	3.83	3.69	3.59	3.50	3.42	.995
	.999	14.0	9.34	7.55	6.59	5.98	5.55	5.23	4.99	4.80	4.64	4.50	4.39	.999
	.9995	16.2	10.6	8.52	7.39	6.68	6.18	5.82	5.54	5.31	5.13	4.98	4.85	.9995

Table 6. Fisher's F distribution **739**

Table 6. (continued)

ν_1, DEGREES OF FREEDOM FOR NUMERATOR

$\mathbb{P}$	15	20	24	30	40	50	60	100	120	200	500	∞	$\mathbb{P}$	
.0005	.159	.197	.220	.244	.272	.290	.303	.330	.339	.353	.368	.377	.0005	**15**
.001	.181	.219	.242	.266	.294	.313	.325	.352	.360	.375	.388	.398	.001	
.005	.246	.286	.308	.333	.360	.377	.389	.415	.422	.435	.448	.457	.005	
.01	.284	.324	.346	.370	.397	.413	.425	.450	.456	.469	.483	.490	.01	
.025	.349	.389	.410	.433	.458	.474	.485	.508	.514	.526	.538	.546	.025	
.05	.416	.454	.474	.496	.519	.535	.545	.565	.571	.581	.592	.600	.05	
.10	.507	.542	.561	.581	.602	.614	.624	.641	.647	.658	.667	.672	.10	
.25	.701	.728	.742	.757	.772	.782	.788	.802	.805	.812	.818	.822	.25	
.50	1.00	1.01	1.02	1.02	1.03	1.03	1.03	1.04	1.04	1.04	1.04	1.05	.50	
.75	1.43	1.41	1.41	1.40	1.39	1.39	1.38	1.38	1.37	1.37	1.36	1.36	.75	
.90	1.97	1.92	1.90	1.87	1.85	1.83	1.82	1.79	1.79	1.77	1.76	1.76	.90	
.95	2.40	2.33	2.39	2.25	2.20	2.18	2.16	2.12	2.11	2.10	2.08	2.07	.95	
.975	2.86	2.76	2.70	2.64	2.59	2.55	2.52	2.47	2.46	2.44	2.41	2.40	.975	
.99	3.52	3.37	3.29	3.21	3.13	3.08	3.05	2.98	2.96	2.92	2.89	2.87	.99	
.995	4.07	3.88	3.79	3.69	3.59	3.52	3.48	3.39	3.37	3.33	3.29	3.26	.995	
.999	5.54	5.25	5.10	4.95	4.80	4.70	4.64	4.51	4.47	4.41	4.35	4.31	.999	
.9995	6.27	5.93	5.75	5.58	5.40	5.29	5.21	5.06	5.02	4.94	4.87	4.83	.9995	
.0005	.169	.211	.235	.263	.295	.316	.331	.364	.375	.391	.408	.422	.0005	**20**
.001	.191	.233	.258	.286	.318	.339	.354	.386	.395	.413	.429	.441	.001	
.005	.258	.301	.327	.354	.385	.405	.419	.448	.457	.474	.490	.500	.005	
.01	.297	.340	.365	.392	.422	.441	.455	.483	.491	.508	.521	.532	.01	
.025	.363	.406	.430	.456	.484	.503	.514	.541	.548	.562	.575	.585	.025	
.05	.430	.471	.493	.518	.544	.562	.572	.595	.603	.617	.629	.637	.05	
.10	.520	.557	.578	.600	.623	.637	.648	.671	.675	.685	.694	.704	.10	
.25	.708	.736	.751	.767	.784	.794	.801	.816	.820	.827	.835	.840	.25	
.50	.989	1.00	1.01	1.01	1.02	1.02	1.02	1.03	1.03	1.03	1.03	1.03	.50	
.75	1.37	1.36	1.35	1.34	1.33	1.33	1.32	1.31	1.31	1.30	1.30	1.29	.75	
.90	1.84	1.79	1.77	1.74	1.71	1.69	1.68	1.65	1.64	1.63	1.62	1.61	.90	
.95	2.20	2.12	2.08	2.04	1.99	1.97	1.95	1.91	1.90	1.88	1.86	1.84	.95	
.975	2.57	2.46	2.41	2.35	2.29	2.25	2.22	2.17	2.16	2.13	2.10	2.09	.975	
.99	3.09	2.94	2.86	2.78	2.69	2.64	2.61	2.54	2.52	2.48	2.44	2.42	.99	
.995	3.50	3.32	3.22	3.12	3.02	2.96	2.92	2.83	2.81	2.76	2.72	2.69	.995	
.999	4.56	4.29	4.15	4.01	3.86	3.77	3.70	3.58	3.54	3.48	3.42	3.38	.999	
.9995	5.07	4.75	4.58	4.42	4.24	4.15	4.07	3.93	3.90	3.82	3.75	3.70	.9995	
.0005	.174	.218	.244	.274	.309	.331	.349	.384	.395	.416	.434	.449	.0005	**24**
.001	.196	.241	.268	.298	.332	.354	.371	.405	.417	.437	.455	.469	.001	
.005	.264	.310	.337	.367	.400	.422	.437	.469	.479	.498	.515	.527	.005	
.01	.304	.350	.376	.405	.437	.459	.473	.505	.513	.529	.546	.558	.01	
.025	.370	.415	.441	.468	.498	.518	.531	.562	.568	.585	.599	.610	.025	
.05	.437	.480	.504	.530	.558	.575	.588	.613	.622	.637	.649	.659	.05	
.10	.527	.566	.588	.611	.635	.651	.662	.685	.691	.704	.715	.723	.10	
.25	.712	.741	.757	.773	.791	.802	.809	.825	.829	.837	.844	.850	.25	
.50	.983	.994	1.00	1.01	1.01	1.02	1.02	1.02	1.02	1.02	1.03	1.03	.50	
.75	1.35	1.33	1.32	1.31	1.30	1.29	1.29	1.28	1.28	1.27	1.27	1.26	.75	
.90	1.78	1.73	1.70	1.67	1.64	1.62	1.61	1.58	1.57	1.56	1.54	1.53	.90	
.95	2.11	2.03	1.98	1.94	1.89	1.86	1.84	1.80	1.79	1.77	1.75	1.73	.95	
.975	2.44	2.33	2.27	2.21	2.15	2.11	2.08	2.02	2.01	1.98	1.95	1.94	.975	
.99	2.89	2.74	2.66	2.58	2.49	2.44	2.40	2.33	2.31	2.27	2.24	2.21	.99	
.995	3.25	3.06	2.97	2.87	2.77	2.70	2.66	2.57	2.55	2.50	2.46	2.43	.995	
.999	4.14	3.87	3.74	3.59	3.45	3.35	3.29	3.16	3.14	3.07	3.01	2.97	.999	
.9995	4.55	4.25	4.09	3.93	3.76	3.66	3.59	3.44	3.41	3.33	3.27	3.22	.9995	

ν_2, DEGREES OF FREEDOM FOR DENOMINATOR

Table 6. (continued)

ν_1, DEGREES OF FREEDOM FOR NUMERATOR

ν_2, DEGREES OF FREEDOM FOR DENOMINATOR

30

$\mathbb{P}$	1	2	3	4	5	6	7	8	9	10	11	12	$\mathbb{P}$
.0005	$.0^6 40$	$.0^3 50$	$.0^2 50$	.015	.030	.047	.065	.082	.098	.114	.129	.143	.0005
.001	$.0^5 16$	$.0^2 10$	$.0^2 80$	.022	.040	.060	.080	.099	.117	.134	.150	.164	.001
.005	$.0^4 40$	$.0^2 50$	.024	.050	.079	.107	.133	.156	.178	.197	.215	.231	.005
.01	$.0^3 16$	.010	.038	.072	.107	.138	.167	.192	.215	.235	.254	.270	.01
.025	$.0^2 10$	.025	.071	.118	.161	.197	.229	.257	.281	.302	.321	.337	.025
.05	$.0^2 40$	.051	.116	.174	.222	.263	.296	.325	.349	.370	.389	.406	.05
.10	.016	.106	.193	.262	.315	.357	.391	.420	.443	.464	.481	.497	.10
.25	.103	.290	.406	.480	.532	.571	.601	.625	.645	.661	.676	.688	.25
.50	.466	.709	.807	.858	.890	.912	.927	.939	.948	.955	.961	.966	.50
.75	1.38	1.45	1.44	1.42	1.41	1.39	1.38	1.37	1.36	1.35	1.35	1.34	.75
.90	2.88	2.49	2.28	2.14	2.05	1.98	1.93	1.88	1.85	1.82	1.79	1.77	.90
.95	4.17	3.32	2.92	2.69	2.53	2.42	2.33	2.27	2.21	2.16	2.13	2.09	.95
.975	5.57	4.18	3.59	3.25	3.03	2.87	2.75	2.65	2.57	2.51	2.46	2.41	.975
.99	7.56	5.39	4.51	4.02	3.70	3.47	3.30	3.17	3.07	2.98	2.91	2.84	.99
.995	9.18	6.35	5.24	4.62	4.23	3.95	3.74	3.58	3.45	3.34	3.25	3.18	.995
.999	13.3	8.77	7.05	6.12	5.53	5.12	4.82	4.58	4.39	4.24	4.11	4.00	.999
.9995	15.2	9.90	7.90	6.82	6.14	5.66	5.31	5.04	4.82	4.65	4.51	4.38	.9995

40

$\mathbb{P}$	1	2	3	4	5	6	7	8	9	10	11	12	$\mathbb{P}$
.0005	$.0^6 40$	$.0^3 50$	$.0^2 50$	.016	.030	.048	.066	.084	.100	.117	.132	.147	.0005
.001	$.0^5 16$	$.0^2 10$	$.0^2 80$	.022	.042	.061	.081	.101	.119	.137	.153	.169	.001
.005	$.0^4 40$	$.0^2 50$	.024	.051	.080	.108	.135	.159	.181	.201	.220	.237	.005
.01	$.0^3 16$	.010	.038	.073	.108	.140	.169	.195	.219	.240	.259	.276	.01
.025	$.0^3 99$	.025	.071	.119	.162	.199	.232	.260	.285	.307	.327	.344	.025
.05	$.0^2 40$	.051	.116	.175	.224	.265	.299	.329	.354	.376	.395	.412	.05
.10	.016	.106	.194	.263	.317	.360	.394	.424	.448	.469	.488	.504	.10
.25	.103	.290	.405	.480	.533	.572	.603	.627	.647	.664	.680	.691	.25
.50	.463	.705	.802	.854	.885	.907	.922	.934	.943	.950	.956	.961	.50
.75	1.36	1.44	1.42	1.40	1.39	1.37	1.36	1.35	1.34	1.33	1.32	1.31	.75
.90	2.84	2.44	2.23	2.09	2.00	1.93	1.87	1.83	1.79	1.76	1.73	1.71	.90
.95	4.08	3.23	2.84	2.61	2.45	2.34	2.25	2.18	2.12	2.08	2.04	2.00	.95
.975	5.42	4.05	3.46	3.13	2.90	2.74	2.62	2.53	2.45	2.39	2.33	2.29	.975
.99	7.31	5.18	4.31	3.83	3.51	3.29	3.12	2.99	2.89	2.80	2.73	2.66	.99
.995	8.83	6.07	4.98	4.37	3.99	3.71	3.51	3.35	3.22	3.12	3.03	2.95	.995
.999	12.6	8.25	6.60	5.70	5.13	4.73	4.44	4.21	4.02	3.87	3.75	3.64	.999
.9995	14.4	9.25	7.33	6.30	5.64	5.19	4.85	4.59	4.38	4.21	4.07	3.95	.9995

60

$\mathbb{P}$	1	2	3	4	5	6	7	8	9	10	11	12	$\mathbb{P}$
.0005	$.0^6 40$	$.0^3 50$	$.0^2 51$	.016	.031	.048	.067	.085	.103	.120	.136	.152	.0005
.001	$.0^5 16$	$.0^2 10$	$.0^2 80$	.022	.041	.062	.083	.103	.122	.140	.157	.174	.001
.005	$.0^4 40$	$.0^2 50$	.024	.051	.081	.110	.137	.162	.185	.206	.225	.243	.005
.01	$.0^3 16$	.010	.038	.073	.109	.142	.172	.199	.223	.245	.265	.283	.01
.025	$.0^3 99$	.025	.071	.120	.163	.202	.235	.264	.290	.313	.333	.351	.025
.05	$.0^2 40$	.051	.116	.176	.226	.267	.303	.333	.359	.382	.402	.419	.05
.10	.016	.106	.194	.264	.318	.362	.398	.428	.453	.475	.493	.510	.10
.25	.102	.289	.405	.480	.534	.573	.604	.629	.650	.667	.680	.695	.25
.50	.461	.701	.798	.849	.880	.901	.917	.928	.937	.945	.951	.956	.50
.75	1.35	1.42	1.41	1.38	1.37	1.35	1.33	1.32	1.31	1.30	1.29	1.29	.75
.90	2.79	2.39	2.18	2.04	1.95	1.87	1.82	1.77	1.74	1.71	1.68	1.66	.90
.95	4.00	3.15	2.76	2.53	2.37	2.25	2.17	2.10	2.04	1.99	1.95	1.92	.95
.975	5.29	3.93	3.34	3.01	2.79	2.63	2.51	2.41	2.33	2.27	2.22	2.17	.975
.99	7.08	4.98	4.13	3.65	3.34	3.12	2.95	2.82	2.72	2.63	2.56	2.50	.99
.995	8.49	5.80	4.73	4.14	3.76	3.49	3.29	3.13	3.01	2.90	2.82	2.74	.995
.999	12.0	7.76	6.17	5.31	4.76	4.37	4.09	3.87	3.69	3.54	3.43	3.31	.999
.9995	13.6	8.65	6.81	5.82	5.20	4.76	4.44	4.18	3.98	3.82	3.69	3.57	.9995

Table 6. Fisher's *F* distribution **741**

Table 6. (continued)

ν_1, DEGREES OF FREEDOM FOR NUMERATOR

P	15	20	24	30	40	50	60	100	120	200	500	∞	P	
.0005	.179	.226	.254	.287	.325	.350	.369	.410	.420	.444	.467	.483	.0005	30
.001	.202	.250	.278	.311	.348	.373	.391	.431	.442	.465	.488	.503	.001	
.005	.271	.320	.349	.381	.416	.441	.457	.495	.504	.524	.543	.559	.005	
.01	.311	.360	.388	.419	.454	.476	.493	.529	.538	.559	.575	.590	.01	
.025	.378	.426	.453	.482	.515	.535	.551	.585	.592	.610	.625	.639	.025	
.05	.445	.490	.516	.543	.573	.592	.606	.637	.644	.658	.676	.685	.05	
.10	.534	.575	.598	.623	.649	.667	.678	.704	.710	.725	.735	.746	.10	
.25	.716	.746	.763	.780	.798	.810	.818	.835	.839	.848	.856	.862	.25	
.50	.978	.989	.994	1.00	1.01	1.01	1.01	1.02	1.02	1.02	1.02	1.02	.50	
.75	1.32	1.30	1.29	1.28	1.27	1.26	1.26	1.25	1.24	1.24	1.23	1.23	.75	
.90	1.72	1.67	1.64	1.61	1.57	1.55	1.54	1.51	1.50	1.48	1.47	1.46	.90	
.95	2.01	1.93	1.89	1.84	1.79	1.76	1.74	1.70	1.68	1.66	1.64	1.62	.95	
.975	2.31	2.20	2.14	2.07	2.01	1.97	1.94	1.88	1.87	1.84	1.81	1.79	.975	
.99	2.70	2.55	2.47	2.39	2.30	2.25	2.21	2.13	2.11	2.07	2.03	2.01	.99	
.995	3.01	2.82	2.73	2.63	2.52	2.46	2.42	2.32	2.30	2.25	2.21	2.18	.995	
.999	3.75	3.49	3.36	3.22	3.07	2.98	2.92	2.79	2.76	2.69	2.63	2.59	.999	
.9995	4.10	3.80	3.65	3.48	3.32	3.22	3.15	3.00	2.97	2.89	2.82	2.78	.9995	
.0005	.185	.236	.266	.301	.343	.373	.393	.441	.453	.480	.504	.525	.0005	40
.001	.209	.259	.290	.326	.367	.396	.415	.461	.473	.500	.524	.545	.001	
.005	.279	.331	.362	.396	.436	.463	.481	.524	.534	.559	.581	.599	.005	
.01	.319	.371	.401	.435	.473	.498	.516	.556	.567	.592	.613	.628	.01	
.025	.387	.437	.466	.498	.533	.556	.573	.610	.620	.641	.662	.674	.025	
.05	.454	.502	.529	.558	.591	.613	.627	.658	.669	.685	.704	.717	.05	
.10	.542	.585	.609	.636	.664	.683	.696	.724	.731	.747	.762	.772	.10	
.25	.720	.752	.769	.787	.806	.819	.828	.846	.851	.861	.870	.877	.25	
.50	.972	.983	.989	.994	1.00	1.00	1.01	1.01	1.01	1.01	1.02	1.02	.50	
.75	1.30	1.28	1.26	1.25	1.24	1.23	1.22	1.21	1.21	1.20	1.19	1.19	.75	
.90	1.66	1.61	1.57	1.54	1.51	1.48	1.47	1.43	1.42	1.41	1.39	1.38	.90	
.95	1.92	1.84	1.79	1.74	1.69	1.66	1.64	1.59	1.58	1.55	1.53	1.51	.95	
.975	2.18	2.07	2.01	1.94	1.88	1.83	1.80	1.74	1.72	1.69	1.66	1.64	.975	
.99	2.52	2.37	2.29	2.20	2.11	2.06	2.02	1.94	1.92	1.87	1.83	1.80	.99	
.995	2.78	2.60	2.50	2.40	2.30	2.23	2.18	2.09	2.06	2.01	1.96	1.93	.995	
.999	3.40	3.15	3.01	2.87	2.73	2.64	2.57	2.44	2.41	2.34	2.28	2.23	.999	
.9995	3.68	3.39	3.24	3.08	2.92	2.82	2.74	2.60	2.57	2.49	2.41	2.37	.9995	
.0005	.192	.246	.278	.318	.365	.398	.421	.478	.493	.527	.561	.585	.0005	60
.001	.216	.270	.304	.343	.389	.421	.444	.497	.512	.545	.579	.602	.001	
.005	.287	.343	.376	.414	.458	.488	.510	.559	.572	.602	.633	.652	.005	
.01	.328	.383	.416	.453	.495	.524	.545	.592	.604	.633	.658	.679	.01	
.025	.396	.450	.481	.515	.555	.581	.600	.641	.654	.680	.704	.720	.025	
.05	.463	.514	.543	.575	.611	.633	.652	.690	.700	.719	.746	.759	.05	
.10	.550	.596	.622	.650	.682	.703	.717	.750	.758	.776	.793	.806	.10	
.25	.725	.758	.776	.796	.816	.830	.840	.860	.865	.877	.888	.896	.25	
.50	.967	.978	.983	.989	.994	.998	1.00	1.00	1.01	1.01	1.01	1.01	.50	
.75	1.27	1.25	1.24	1.22	1.21	1.20	1.19	1.17	1.17	1.16	1.15	1.15	.75	
.90	1.60	1.54	1.51	1.48	1.44	1.41	1.40	1.36	1.35	1.33	1.31	1.29	.90	
.95	1.84	1.75	1.70	1.65	1.59	1.56	.153	1.48	1.47	1.44	1.41	1.39	.95	
.975	2.06	1.94	1.88	1.82	1.74	1.70	1.67	1.60	1.58	1.54	1.51	1.48	.975	
.99	2.35	2.20	2.12	2.03	1.94	1.88	1.84	1.75	1.73	1.68	1.63	1.60	.99	
.995	2.57	2.39	2.29	2.19	2.08	2.01	1.96	1.86	1.83	1.78	1.73	1.69	.995	
.999	3.08	2.83	2.69	2.56	2.41	2.31	2.25	2.11	2.09	2.01	1.93	1.89	.999	
.9995	3.30	3.02	2.87	2.71	2.55	2.45	2.38	2.23	2.19	2.11	2.03	1.98	.9995	

ν_2, DEGREES OF FREEDOM FOR DENOMINATOR

Table 6. (continued)

ν_1, DEGREES OF FREEDOM FOR NUMERATOR

	$\mathbb{P}$	1	2	3	4	5	6	7	8	9	10	11	12	$\mathbb{P}$
120	.0005	$.0^640$	$.0^350$	$.0^251$	.016	.031	.049	.067	.087	.105	.123	.140	.156	.0005
	.001	$.0^516$	$.0^210$	$.0^281$	.023	.042	.063	.084	.105	.125	.144	.162	.179	.001
	.005	$.0^439$	$.0^250$	.024	.051	.081	.111	.139	.165	.189	.211	.230	.249	.005
	.01	$.0^316$	.010	.038	.074	.110	.143	.174	.202	.227	.250	.271	.290	.01
	.025	$.0^399$	.025	.072	.120	.165	.204	.238	.268	.295	.318	.340	.359	.025
	.05	$.0^239$	.051	.117	.177	.227	.270	.306	.337	.364	.388	.408	.427	.05
	.10	.016	.105	.194	.265	.320	.365	.401	.432	.458	.480	.500	.518	.10
	.25	.102	.288	.405	.481	.534	.574	.606	.631	.652	.670	.685	.699	.25
	.50	.458	.697	.793	.844	.875	.896	.912	.923	.932	.939	.945	.950	.50
	.75	1.34	1.40	1.39	1.37	1.35	1.33	1.31	1.30	1.29	1.28	1.27	1.26	.75
	.90	2.75	2.35	2.13	1.99	1.90	1.82	1.77	1.72	1.68	1.65	1.62	1.60	.90
	.95	3.92	3.07	2.68	2.45	2.29	2.18	2.09	2.02	1.96	1.91	1.87	1.83	.95
	.975	5.15	3.80	3.23	2.89	2.67	2.52	2.39	2.30	2.22	2.16	2.10	2.05	.975
	.99	6.85	4.79	3.95	3.48	3.17	2.96	2.79	2.66	2.56	2.47	2.40	2.34	.99
	.995	8.18	5.54	4.50	3.92	3.55	3.28	3.09	2.93	2.81	2.71	2.62	2.54	.995
	.999	11.4	7.32	5.79	4.95	4.42	4.04	3.77	3.55	3.38	3.24	3.12	3.02	.999
	.9995	12.8	8.10	6.34	5.39	4.79	4.37	4.07	3.82	3.63	3.47	3.34	3.22	.9995
∞	.0005	$.0^639$	$.0^350$	$.0^251$	.016	.032	.050	.069	.088	.108	.127	.144	.161	.0005
	.001	$.0^516$	$.0^210$	$.0^281$	.023	.042	.063	.085	.107	.128	.148	.167	.185	.001
	.005	$.0^439$	$.0^250$	.024	.052	.082	.113	.141	.168	.193	.216	.236	.256	.005
	.01	$.0^316$	.010	.038	.074	.111	.145	.177	.206	.232	.256	.278	.298	.01
	.025	$.0^398$	.025	.072	.121	.166	.206	.241	.272	.300	.325	.347	.367	.025
	.05	$.0^239$	.051	.117	.178	.229	.273	.310	.342	.369	.394	.417	.436	.05
	.10	.016	.105	.195	.266	.322	.367	.405	.436	.463	.487	.508	.525	.10
	.25	.102	.288	.404	.481	.535	.576	.608	.634	.655	.674	.690	.703	.25
	.50	.455	.693	.789	.839	.870	.891	.907	.918	.927	.934	.939	.945	.50
	.75	1.32	1.39	1.37	1.35	1.33	1.31	1.29	1.28	1.27	1.25	1.24	1.24	.75
	.90	2.71	2.30	2.08	1.94	1.85	1.77	1.72	1.67	1.63	1.60	1.57	1.55	.90
	.95	3.84	3.00	2.60	2.37	2.21	2.10	2.01	1.94	1.88	1.83	1.79	1.75	.95
	.975	5.02	3.69	3.12	2.79	2.57	2.41	2.29	2.19	2.11	2.05	1.99	1.94	.975
	.99	6.63	4.61	3.78	3.32	3.02	2.80	2.64	2.51	2.41	2.32	2.25	2.18	.99
	.995	7.88	5.30	4.28	3.72	3.35	3.09	2.90	2.74	2.62	2.52	2.43	2.36	.995
	.999	10.8	6.91	5.42	4.62	4.10	3.74	3.47	3.27	3.10	2.96	2.84	2.74	.999
	.9995	12.1	7.60	5.91	5.00	4.42	4.02	3.72	3.48	3.30	3.14	3.02	2.90	.9995

ν_2, DEGREES OF FREEDOM FOR DENOMINATOR

Table 6. Fisher's *F* distribution **743**

Table 6. (continued)

ν_1, DEGREES OF FREEDOM FOR NUMERATOR

ℙ	15	20	24	30	40	50	60	100	120	200	500	∞	ℙ	
.0005	.199	.256	.293	.338	.390	.429	.458	.524	.543	.578	.614	.676	.0005	**120**
.001	.223	.282	.319	.363	.415	.453	.480	.542	.568	.595	.631	.691	.001	
.005	.297	.356	.393	.434	.484	.520	.545	.605	.623	.661	.702	.733	.005	
.01	.338	.397	.433	.474	.522	.556	.579	.636	.652	.688	.725	.755	.01	
.025	.406	.464	.498	.536	.580	.611	.633	.684	.698	.729	.762	.789	.025	
.05	.473	.527	.559	.594	.634	.661	.682	.727	.740	.767	.785	.819	.05	
.10	.560	.609	.636	.667	.702	.726	.742	.781	.791	.815	.838	.855	.10	
.25	.730	.765	.784	.805	.828	.843	.853	.877	.884	.897	.911	.923	.25	
.50	.961	.972	.978	.983	.989	.992	.994	1.00	1.00	1.00	1.01	1.01	.50	
.75	1.24	1.22	1.21	1.19	1.18	1.17	1.16	1.14	1.13	1.12	1.11	1.10	.75	
.90	1.55	1.48	1.45	1.41	1.37	1.34	1.32	1.27	1.26	1.24	1.21	1.19	.90	
.95	1.75	1.66	1.61	1.55	1.50	1.46	1.43	1.37	1.35	1.32	1.28	1.25	.95	
.975	1.95	1.82	1.76	1.69	1.61	1.56	1.53	1.45	1.43	1.39	1.34	1.31	.975	
.99	2.19	2.03	1.95	1.86	1.76	1.70	1.66	1.56	1.53	1.48	1.42	1.38	.99	
.995	2.37	2.19	2.09	1.98	1.87	1.80	1.75	1.64	1.61	1.54	1.48	1.43	.995	
.999	2.78	2.53	2.40	2.26	2.11	2.02	1.95	1.82	1.76	1.70	1.62	1.54	.999	
.9995	2.96	2.67	2.53	2.38	2.21	2.11	2.01	1.88	1.84	1.75	1.67	1.60	.9995	
.0005	.207	.270	.311	.360	.422	.469	.505	.599	.624	.704	.804	1.00	.0005	**∞**
.001	.232	.296	.338	.386	.448	.493	.527	.617	.649	.719	.819	1.00	.001	
.005	.307	.372	.412	.460	.518	.559	.592	.671	.699	.762	.843	1.00	.005	
.01	.349	.413	.452	.499	.554	.595	.625	.699	.724	.782	.858	1.00	.01	
.025	.418	.480	.517	.560	.611	.645	.675	.741	.763	.813	.878	1.00	.025	
.05	.484	.543	.577	.617	.663	.694	.720	.781	.797	.840	.896	1.00	.05	
.10	.570	.622	.652	.687	.726	.752	.774	.826	.838	.877	.919	1.00	.10	
.25	.736	.773	.793	.816	.842	.860	.872	.901	.910	.932	.957	1.00	.25	
.50	.956	.967	.972	.978	.983	.987	.989	.993	.994	.997	.999	1.00	.50	
.75	1.22	1.19	1.18	1.16	1.14	1.13	1.12	1.09	1.08	1.07	1.04	1.00	.75	
.90	1.49	1.42	1.38	1.34	1.30	1.26	1.24	1.18	1.17	1.13	1.08	1.00	.90	
.95	1.67	1.57	1.52	1.46	1.39	1.35	1.32	1.24	1.22	1.17	1.11	1.00	.95	
.975	1.83	1.71	1.64	1.57	1.48	1.43	1.39	1.30	1.27	1.21	1.13	1.00	.975	
.99	2.04	1.88	1.79	1.70	1.59	1.52	1.47	1.36	1.32	1.25	1.15	1.00	.99	
.995	2.19	2.00	1.90	1.79	1.67	1.59	1.53	1.40	1.36	1.28	1.17	1.00	.995	
.999	2.51	2.27	2.13	1.99	1.84	1.73	1.66	1.49	1.45	1.34	1.21	1.00	.999	
.9995	2.65	2.37	2.22	2.07	1.91	1.79	1.71	1.53	1.48	1.36	1.22	1.00	.9995	

ν_2, DEGREES OF FREEDOM FOR DENOMINATOR

Table 7. The Poisson probability distribution for selected values of μ

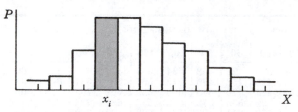

Table entry is height of shaded bar, $P(x = x_i)$.

x	.2	.5	.8	1	1.5	2	2.5
0	.8187	.6035	.4493	.3679	.2231	.1353	.0821
1	.1637	.3033	.3595	.3679	.3347	.2707	.2052
2	.0164	.0758	.1438	.1839	.2510	.2707	.2565
3	.0011	.0126	.0383	.0613	.1255	.1804	.2138
4	.0011	.0016	.0077	.0153	.0471	.0902	.1336
5		.0002	.0012	.0031	.0141	.0361	.0668
6			.0002	.0005	.0035	.0120	.0278
7				.0001	.0008	.0034	.0099
8					.0001	.0009	.0031
9						.0002	.0009
10							.0002

x	3	4	5	6	8	10	15	20
0	.0498	.0183	.0067	.0025	.0003			
1	.1494	.0733	.0337	.0149	.0027	.0005		
2	.2240	.1465	.0842	.0446	.0107	.0023		
3	.2240	.1954	.1404	.0892	.0286	.0076	.0002	
4	.1680	.1954	.1755	.1339	.0573	.1089	.0006	
5	.1008	.1563	.1755	.1606	.0916	.0378	.0019	.0001
6	.0504	.1042	.1462	.1606	.1221	.0631	.0048	.0002
7	.0216	.0595	.1044	.1377	.1396	.0901	.0104	.0005
8	.0081	.0298	.0653	.1033	.1396	.1126	.0194	.0013
9	.0027	.0132	.0363	.0688	.1241	.1251	.0324	.0029
10	.0008	.0053	.0181	.0413	.0993	.1251	.0486	.0058
11	.0002	.0019	.0082	.0225	.0722	.1137	.0663	.0106
12	.0001	.0006	.0034	.0113	.0481	.0948	.0829	.0176
13		.0002	.0013	.0052	.0296	.0729	.0956	.0271
14		.0001	.0005	.0022	.0169	.0521	.1024	.0387
15			.0002	.0009	.0090	.0347	.1024	.0516
16				.0003	.0045	.0217	.0960	.0646
17				.0001	.0021	.0128	.0847	.0760
18					.0009	.0071	.0706	.0844
19					.0004	.0037	.0557	.0888

Table 7. Poisson probabilities   **745**

Table 7. (continued)

| | | | | | | | | μ |
x	3	4	5	6	8	10	15	20
20					.0002	.0019	.0418	.0888
21					.0001	.0009	.0299	.0846
22						.0004	.0204	.0769
23						.0002	.0133	.0669
24						.0001	.0083	.0557
25							.0050	.0446
26							.0029	.0343
27							.0016	.0254
28							.0009	.0181
29							.0004	.0125
30							.0002	0083
31							.0001	.0054
32							.0001	.0034
33								.0020
34								.0012
35								.0007
36								.0004
37								.0002
38								.0001
39								.0001

Probabilities calculated by authors using Microsoft® Excel.

Note on the Poisson distribution. In Chapter 7 we learned how to use the standard normal distribution $Z : N(0, 1)$ to approximate the binomial $X : B(N, p)$ provided $Np \geq 5$ and $N(1 - p) \geq 5$. If one of these inequalities is not satisfied, the Poisson distribution can be used to approximate the binomial. The Poisson (pronounced *pwa · sawn*) distribution is an example of a discrete probability distribution that can take on infinitely many values. Its probability function is given by the formula

$$P(x) = \frac{\mu^x}{x!} e^{-\mu}$$

for $x = 0, 1, 2, \ldots$. Clearly, $P(x)$ will be determined once we know the value of the real number μ, the only parameter in the Poisson. It turns out that the mean and the variance of the Poisson are both equal to its parameter μ.

Table 7 gives probabilities for values of the Poisson distributed random variable X for selected values of μ. To use the Poisson to approximate $X : B(N, p)$ with $Np < 5$, we must also have $p < 0.5$. If $p \geq 0.5$, we can still solve the problem via an approximation of $Y : B(N, q)$. (See Example 2 below.) The procedure is extraordinarily simple. First calculate the binomial mean $\mu = Np$ and then sum the desired values in the column headed μ.

Example 1. What is the probability of obtaining 1, 2, or 3 successes in 100 trials if the probability of success on each trial is only .01? We want $P(1 \leq x \leq 3)$ for $X: B(100, .01)$. The mean is $\mu = Np = (100)(.01) = 1 < 5$ and $p < 0.5$, so the Poisson approximation is appropriate. In Table 7, the required approximations are found in the column headed $\mu = 1$:

$$P(x = 1) \doteq .3679$$

$$P(x = 2) \doteq .1839$$

$$P(x = 3) \doteq .0613$$

Thus, the desired probability is about

$$.3679 + .1839 + .0613 = .6131$$

The exact answer to four decimal places is .6156.

Example 2. What is the probability of obtaining more than 95 successes in 100 trials if the probability of success is .97 on each trial. Now $Nq = 3 < 5$, which makes the normal approximation inappropriate, and $p = .97 > 0.5$, which makes the Poisson approximation to the distribution of X inappropriate. However, if we get more than 95 successes, we must have 0, 1, 2, 3, or 4 failures, where the probability of failure is $1 - .97 = .03$. We therefore let Y be the number of failures and approximate $Y: B(100, .03)$ for $y = 0, 1, 2, 3, 4$. Now,

$$\mu = (100)(.03) = 3$$

so the desired probability is the sum of the appropriate terms in the column headed $\mu = 3$:

$$P(x = 0) \doteq .0498$$

$$P(x = 1) \doteq .1494$$

$$P(x = 2) \doteq .2240$$

$$P(x = 3) \doteq .2240$$

$$P(x = 4) \doteq .1680$$

Thus, the desired probability is about

$$.0498 + .1494 + .2240 + .2240 + .1680 = .8153$$

The exact answer to four decimal places is .8179.

You really don't need a table for the Poisson if you have a calculator that will do exponentials. For example, to compute $P(x = 2)$ for $\mu = 3$,

Table 7. Poisson probabilities **747**

we evaluate

$$P(2) = \frac{3^2}{2!}e^{-3} = \frac{9}{2}e^{-3} \doteq .2240418007$$

to many more decimal places than in Table 7. (Of course, for the purposes of approximation, it makes no sense to use more than four decimal places.)

In [6.8] we presented the binomial probability function,

$$P(x) = {}_NC_x p^x q^{N-x}$$

The Poisson approximation works because if $p < .5$ and $Np < 5$, then it can be shown that

$${}_NC_x p^x q^{N-x} \cong \frac{(Np)^x}{x!}e^{-(Np)}$$

Solutions to odd-numbered exercises

EXERCISES 1.1 (pp. 13–16)

1.

Observations (x)	1	2	3	4	5
(a) Frequency (f)	3	2	3	1	1
(b) Relative Frequency (f/N)	.3	.2	.3	.1	.1
(c) Cumulative f	3	5	8	9	10
(d) Cumulative f/N	.3	.5	.8	.9	1.0

3.

Observations (x)	1	2	3	4	5	6
(a) Frequency (f)	2	3	2	1	1	1
(b) Relative Frequency (f/N)	.2	.3	.2	.1	.1	.1
(c) Cumulative f	2	5	7	8	9	10
(d) Cumulative f/N	.2	.5	.7	.8	.9	1.0

5.

Observations (x)	0.1	0.2	0.3	0.4	0.5
(a) Frequency (f)	3	5	10	5	2
(b) Relative Frequency (f/N)	.12	.20	.40	.20	.08
(c) Cumulative f	3	8	18	23	25
(d) Cumulative f/N	.12	.32	.72	.92	1.0

7. Here is one of many possible correct answers, in which we arbitrarily decided to use six groups.

x	.5 – 5.5	5.5 – 10.5	10.5 – 15.5	15.5 – 20.5	20.5 – 25.5	25.5 – 30.5
(a) f	2	4	7	6	4	2
(b) f/N	.08	.16	.28	.24	.16	.08
(c) Cum. f	2	6	13	19	23	25
(d) Cum. f/N	.08	.24	.52	.76	.92	1.0

9. Here is one of many possible correct answers, in which we arbitrarily decided to use nine groups.

x	31.55 – 34.35	34.35 – 37.15	37.15 – 39.95	39.95 – 42.75	42.75 – 45.55
(a) f	1	2	2	4	7
(b) f/N	.04	.08	.08	.16	.28
(c) Cum. f	1	3	5	9	16
(d) Cum. f/N	.04	.12	.20	.36	.64

45.55 – 48.35	48.35 – 51.15	51.15 – 53.95	53.95 – 56.75
4	1	3	1
.16	.04	.12	.04
20	21	24	25
.80	.84	.96	1.00

11. These data illustrate how an "outlier," a value that differs markedly from the main body of data, can lead to intervals that are so wide that information loss may be unacceptable. If $x_b = 9{,}206$, then even with 20 intervals, the other x-values are concentrated in only 5 groups. For $n = 6$, the first interval includes 45 states. A common procedure in such cases is to calculate class limits exclusive of the outlier and to add an open-ended interval (for the outlier) at the end of the distribution. The distribution given below is one of many possible answers based on this tactic. We arbitrarily used 12 groups, exclusive of Alaska.

x	84.5 – 274.5	274.5 – 464.5	464.5 – 654.5	654.5 – 844.5	844.5 – 1034.5	1034.5 – 1224.5
(a) f	9	10	11	3	5	3
(b) f/N	.18	.20	.22	.06	.10	.06
(c) Cum. f	9	19	30	33	38	41
(d) Cum. f/N	.18	.38	.60	.66	.76	.82

1224.5 – 1414.5	1414.5 – 1604.5	1604.5 – 1794.5	1794.5 – 1984.5	1984.5 – 2174.5	2174.5 – 2364.5	2364.5 or more
2	2	0	0	1	3	1
.04	.04	0	0	.02	.06	.02
43	45	45	45	46	49	50
.86	.90	.90	.90	.92	.98	1.00

13.

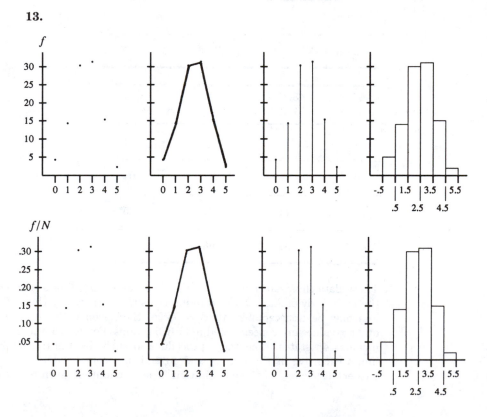

15.

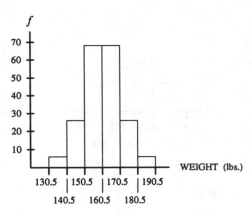
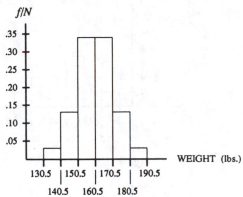

17.

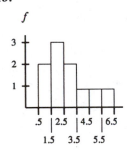
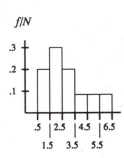
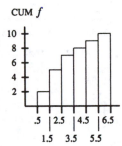
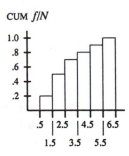

19.

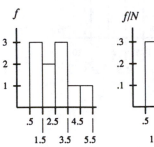
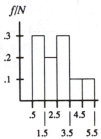
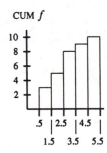
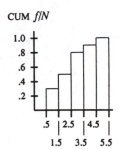

21.

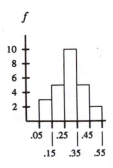
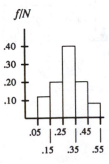
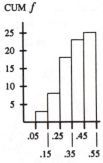
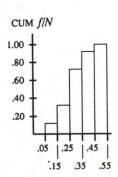

23.

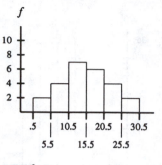 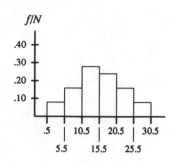

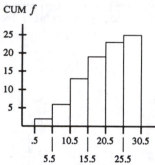 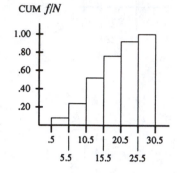

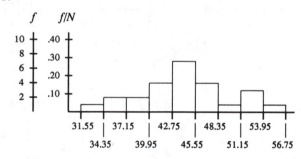

25.

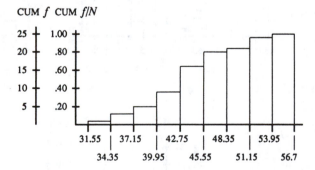

27.

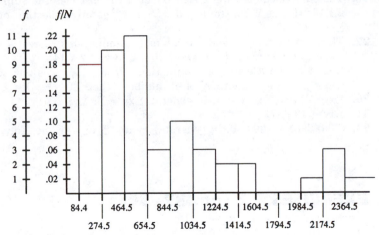

EXERCISES 2.1 (pp. 24–26)

1. $\sum x = 25$; $\dfrac{1}{N}\sum x = 5$

3. $\overset{6}{\sum}x = 12$. Note that this sum is the same as (6) (2) or $N2$.

5. $\overset{N}{\sum}x_i = 21$

7. $\overset{N}{\sum}x_i = 14$; $\overset{n}{\sum}x_i = 9$

9. $\sum x = 31$; $\sum y = 12$; $\sum xy = 74$; $(\sum x)(\sum y) \neq \sum xy$

11. For the first midterm, $N = 15$

13. $\overset{N}{\sum}x = 730$

15. For the first midterm, $n = 10$

17. For the first midterm, $f_3 = 2$

19. $\overset{n}{\sum}f_i x_i = 730$

21. The answers to problems 2.1.13 and 2.1.19 are the same.

23. $\overset{N}{\sum}.80(x_i) = 584$

25. The sum calculated for 2.1.23 is equal to .80 times the sum calculated for 2.1.13. If c is a constant and $x_1, \ldots, x_N$ is any collection of N numbers, $\Sigma cx = c\Sigma x$.

27. The sum of the students' combined examinations scores is 1,126. This is equal to the sum of the answers to 2.1.13 and 2.1.14.
Rule 4 of the Algebra of Summations: If $x_1, \ldots, x_N$ is one set of numbers and $y_1, \ldots, y_N$ is another set of numbers, then $\Sigma(x_i + y_i) = \Sigma x_i + \Sigma y_i$.

29. $\Sigma xy = 2$ and $\Sigma x \Sigma y = 4$. In general, $\Sigma xy \neq \Sigma x \Sigma y$.

31. $\Sigma xy = 19,627$

33. $(730)(396) = 289,080$. In general, Σxy and $\Sigma x \Sigma y$ are not equal.

EXERCISES 2.2 (pp. 37–39)

1. 3   **3.** 10.6   **5.** 1,000; no

7. 4   **9.** 7   **11.** 2.5

13. 4.5   **15.** 157.2 lbs   **17.** 53.4 sec

19. (a) 44.3
(b) For grouping given in answer to Exercise 1.1.9, median is 44.15. *Your* answer depends on how you grouped the data.

21. 3.9   **23.** 104° F

EXERCISES 2.3 (pp. 44–48)

1. $\bar{x} \doteq 4.36$ **3.** $\bar{x} \doteq 6.64$ **5.** $\bar{x} \doteq 3.692$

7. $\bar{x} = 158.5$ lbs   **9.** $\bar{x} = 51.44$ sec

11. Ungrouped, $\bar{x} = 44.336$;
Grouped as in answer to Exercise 1.1.9, $\bar{x} = 44.262$.

13. For $c = 2$, $c\bar{x} = 12$, $\bar{x} + c = 8$, $\bar{x} - c = 4$, $\bar{x} + \bar{y} = 22$.
$c\bar{x} = \overline{cx}$; $\bar{x} + c = \overline{(x + c)}$; $\bar{x} - c = \overline{(x - c)}$; and
$\bar{x} + \bar{y} = \overline{(x + y)}$.

15. Prove: $\bar{c} = c$ **17.** Prove: $\overline{cx} = c\bar{x}$

$$\bar{c} = \frac{\displaystyle\sum_{}^{N} c_i}{N} = \frac{Nc}{N} = c \qquad\qquad \overline{cx} = \frac{\displaystyle\sum_{}^{N} cx_i}{N} = \frac{c\displaystyle\sum_{}^{N} x_i}{N} = c\frac{\displaystyle\sum_{}^{N} x_i}{N} = c\bar{x}$$

19. (a) 203 min $\doteq$ 3.383 hr
(b) 299 min $\doteq$ 4.983 hr
(c) $203 + (40)(.5) = 223$ min; $3.383 + .33333 \doteq 3.716$ hr
(d) $299 + 60 = 359$ min; $4.983 + 1 = 5.983$ hr

21. 75.68   **23.** 67.1

25. Data Set 1: $Mo. = 3$; $\tilde{x} = 3.5$; $\bar{x} = 3.7$
Data Set 2: $Mo. = 3$; $\tilde{x} = 4$; $\bar{x} \doteq 12.5$

27. **29.** $\bar{y} = 0$

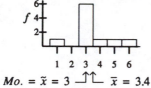

$Mo. = \tilde{x} = 3$ ⟍⟋ $\bar{x} = 3.4$

31. (a) $\sum^{N} \dfrac{(x_i + b)}{N} = \bar{x} + 3 = 43.84$ (b) $\sum^{N} \dfrac{(x_i + b - \bar{x})}{N} = \dfrac{25(3)}{25} = 3$

EXERCISES 2.4 (pp. 59–62)

1. 4   **3.** 5

5.

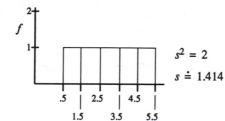

$s^2 = 2$

$s \doteq 1.414$

7.

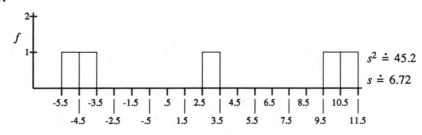

$s^2 \doteq 45.2$

$s \doteq 6.72$

9.

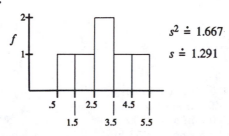

$s^2 \doteq 1.667$

$s \doteq 1.291$

11. $s^2 \doteq 7.14$; $s \doteq 2.67$; $cv \doteq 61.2$
13. $s^2 \doteq 26.05$; $s \doteq 5.10$; $cv \doteq 76.8$
15. $s^2 \doteq 32.39$; $s \doteq 5.69$; $cv \doteq 12.8$
17. $s^2 \doteq 1.29$; $s \doteq 1.14$; $cv \doteq 30.9$

19. $s^2 = 201$; $s \doteq 14.18$ lbs; $cv \doteq 8.9$
21. $s^2 \doteq 48.6$; $s \doteq 6.97$ sec; $cv \doteq 13.0$
23. For $c = 2$:

$$s_{cX}^2 = 20.8; \ s_{cX} \doteq 4.56; \ s_{(X-c)}^2 = s_{(X+c)}^2 = 5.2$$

$$s_{(X-c)} = s_{(X+c)} \doteq 2.28$$

$$s_{cX}^2 = c^2 s_X^2; \ s_{cX} = cs_X$$

$$s_{(X+c)}^2 = s_X^2; \ s_{(X+c)} = s_X$$

$$s_{(X-c)}^2 = s_X^2; \ s_{(X-c)} = s_X$$

$$s_{(X+Y)}^2 = 54.4 > 31.6 = s_X^2 + s_Y^2$$

25. (a) If $x_i = \cdots = x_N = c$, then $\bar{x} = c$ by Exercise 2.3.15. Therefore, the variance of c is

$$\frac{\sum\limits_{}^{N} (c - c)^2}{N} = 0$$

(b) By Exercise 2.3.16, $\overline{x + c} = \bar{x} + c$. Therefore, the variance of $x + c$ is

$$\frac{\sum\limits_{}^{N} [x + c - (\bar{x} + c)]^2}{N} = \frac{\sum\limits_{}^{N} (x - \bar{x})^2}{N} = s_X^2$$

(c) By Exercise 2.3.17, $\overline{cx} = c\bar{x}$. Therefore the variance of cx is

$$\frac{\sum\limits_{}^{N} (cx - c\bar{x})^2}{N} = \frac{\sum\limits_{}^{N} [c(x - \bar{x})]^2}{N}$$

$$= \frac{\sum\limits_{}^{N} c^2 (x - \bar{x})^2}{N} = c^2 \frac{\sum\limits_{}^{N} (x - \bar{x})^2}{N} = c^2 s_X^2$$

27. (a) With Kirk: $s \doteq 186.5$ min or 3.11 hr
   Without Kirk: $s \doteq 49.3$ min or 0.82 hr
(b) With Kirk: $s \doteq 236.0$ min or 3.93 hr
   Without Kirk: $s \doteq 91.8$ min or 1.53 hr
(c) With Kirk: $s^2 \doteq 34{,}792$; 9.66
   Without Kirk: $s^2 \doteq 2{,}431$; 0.68
(d) With Kirk: $s^2 \doteq 55{,}673$; 15.46
   Without Kirk: $s^2 \doteq 8{,}431$; 2.34

EXERCISES 2.5 (pp. 71–73)

1.

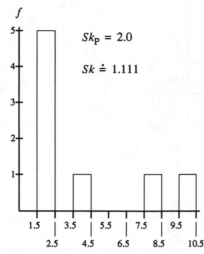

$Sk_P = 2.0$

$Sk \doteq 1.111$

3.

$Sk_P = -.5$ $Sk \doteq 0.0833$

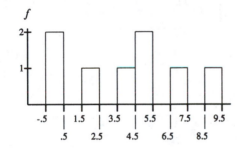

5.

$Sk_P = -1.0$ $Sk \doteq -0.3056$

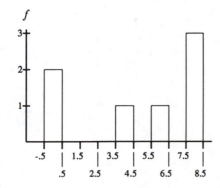

7.

$$Sk_p = 0 \qquad Sk \doteq -0.0833$$

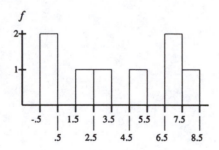

9.

$$Sk_p = -.5 \qquad Sk \doteq 0.4167$$

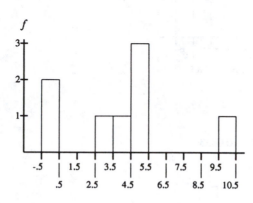

11.

$$Sk_p = .5 \qquad Sk = 0.25$$

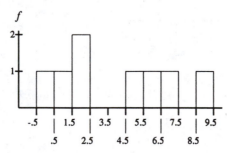

13. $\bar{x} = 5$, $\tilde{x} = 6$ or $5\frac{5}{6}$; $Sk = 0$; $\bar{x} < \tilde{x}$, so $Sk_p < 0$

15. $\bar{x} = 4.5$, $\tilde{x} = 6$ or $5\frac{5}{6}$; $Sk \doteq -.640$; $\bar{x} < \tilde{x}$, so $Sk_p < 0$

17.

$Kur = 5$

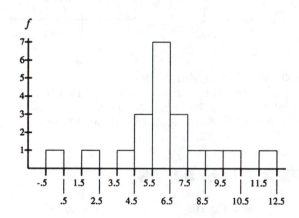

19.

$Kur \doteq 3.75$

21.

$Kur \doteq 3.01$

23.

$Kur \doteq 2.81$

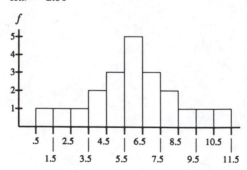

25.

$Kur \doteq 1.91$

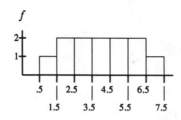

27.

$Kur \doteq 1.40$

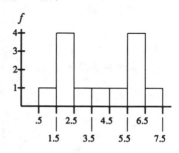

29. $Sk_P \doteq -.839$; $Sk \doteq -.975$; Kur $\doteq 2.95$

31. $Sk_P \doteq .87$; $Sk \doteq 5.08$, Kur $\doteq 32.00$ (Based on 15 intervals of width 152 where $x_a = 86$ and $x_b = 2,366$. Alaska entered as exact value.)

33. $Sk_P \doteq 1.01$, $Sk \doteq 1.91$, Kur $\doteq 6.59$ (Based on 20 intervals of width 130 where $x_a = 32$ and $x_b = 2631$.)

EXERCISES 3.1 (pp. 84–85)

1. (a) 52 (b) 76 (c) 80

3. (a) 36
   (b) The 50th percentile is 48, even though the percentile rank of 48 is about 56. This is only one of the many little oddities that can happen when percentiles for ungrouped data are calculated by interpolation.
   (c) 73 (approximately)

5.

	Percentile	Percent below	Percent above
(a)	38	25	75
(b)	48	47.2	44.4
(c)	75	88.8	5.5

7. (a) 34 (b) 98 (c) 94
9. (a) TN & AL (b) DE & UT (c) CO
11. (a) 55 (b) 96 (c) 98.5
13. (a) 215 min (b) 197 min (c) 153 min

EXERCISES 3.2 (pp. 95–96)

1. For $x = 2$, $z \doteq -.88$; for $x = 7$, $z \doteq .99$
3. For $x = 2$, $z \doteq -.91$; for $x = 11$, $z \doteq .85$
5. For $x = 43.8$, $z \doteq -.09$; for $x = 50.5$, $z \doteq 1.08$
7. (a) $z \doteq -1.59$ (b) $z \doteq -.25$ (c) $z \doteq 1.73$
9. (a) $x \doteq 137.23$ lbs   (b) $x \doteq 158.5$ lbs   (c) $x \doteq 181.19$ lbs
11. For the first applicant, $z = 1.70$.
   For the second applicant, $z = 2.30$.
13. (a) The student received a **B**.
   (b) The lowest C was 1.37 standard deviations from the mean.
15. (a) For $x = 475$, $y \doteq 484$; for $x = 540$, $y \doteq 541$; for $x = 734$, $y \doteq 711$.
   (b) The z-score is the number of standard deviations that the corresponding x-score lies from the mean. The transformation $y = z(100) + 500$ therefore creates a scale with a standard deviation of 100 and a mean of 500.

EXERCISES 4.1 (pp. 104–106)

Blank spaces in all of the following tables are zeros.

1.

(a)   (b)   (c) positive

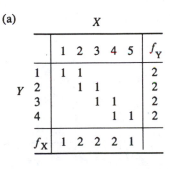

Y \ X	1	2	3	4	5	f_Y
1	1	1				2
2		1	1			2
3			1	1		2
4				1	1	2
f_X	1	2	2	2	1	

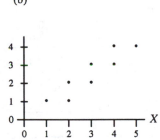

3.

(a)   (b)   (c) negative

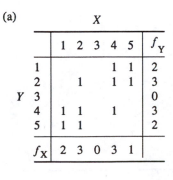

Y \ X	1	2	3	4	5	f_Y
1				1	1	2
2		1		1	1	3
3						0
4	1	1		1		3
5	1	1				2
f_X	2	3	0	3	1	

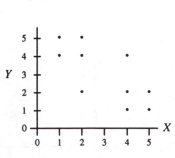

5.

(a)

			X			
	1	2	3	4	5	f_Y
1	1				1	2
2		1		1		2
Y 3			1			1
4			1			1
5	1				1	2
f_X	2	1	2	1	2	

(b)

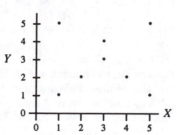

(c) no statistical relation

7.

(a)

		X			
	2	3	4	5	f_Y
3	1				1
4	1	1			2
Y 5	1	1	1		3
6		1	1	1	3
7			1	1	2
8				1	1
f_X	3	3	3	3	

(b)

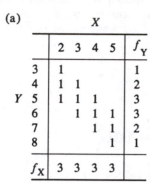

(c) positive

9.

(a)

		X			
	22	23	24	25	f_Y
45	1				1
Y 46	1		1		2
47		1	1	1	3
48			1	1	2
f_X	2	1	3	2	

(b)

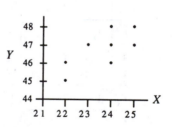

(c) positive

11.

(b)

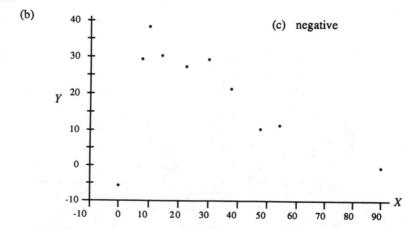

(c) negative

13.

(a)

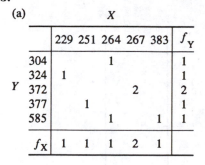

	X					
	229	251	264	267	383	f_Y
304			1			1
324	1					1
Y 372				2		2
377		1				1
585			1		1	1
f_X	1	1	1	2	1	

(b)

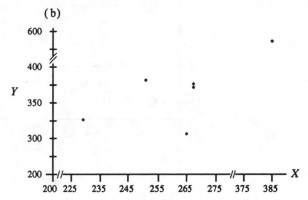

(c) positive

15.

(a)

	X					
	22	23	24	25	26	f_Y
43	1					1
44						0
45	2	1	1	1		5
Y 46		2	2	1		5
47		1	3	2		6
48			1	1		2
49					1	1
f_X	3	4	7	5	1	

(b)

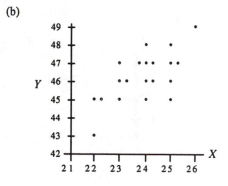

(c) positive

EXERCISES 4.2 (p. 110)

1.

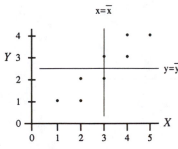

$\bar{x} = 3$
$\bar{y} = 2.5$
$C_{XY} = 1.25$

3.

$\bar{x} = 3$
$\bar{y} = 3$
$C_{XY} = -1.6$

5.

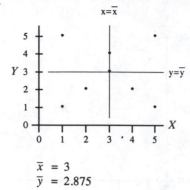

$\bar{x} = 3$
$\bar{y} = 2.875$
$C_{XY} = 0$

7.

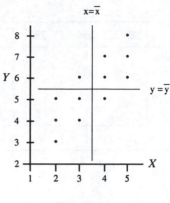

$\bar{x} = 3.5$
$\bar{y} = 5.5$
$C_{XY} = 1.25$

9.

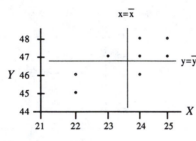

$\bar{x} = 23.625$
$\bar{y} = 46.75$
$C_{XY} = 0.78124$

11.

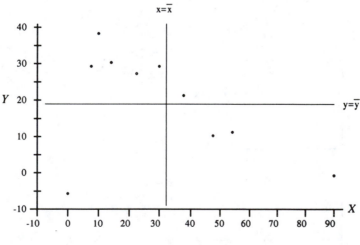

$\bar{x} = 31.1$
$\bar{y} = 18.97$
$C_{XY} = -166.877$

13.

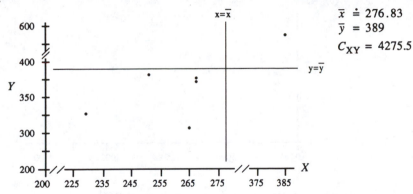

$$\bar{x} \doteq 276.83$$
$$\bar{y} = 389$$
$$C_{XY} = 4275.5$$

15.

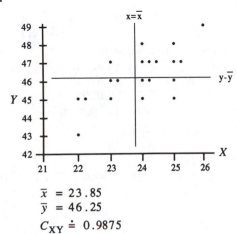

$$\bar{x} = 23.85$$
$$\bar{y} = 46.25$$
$$C_{XY} \doteq 0.9875$$

17. With Alaska: $\bar{x} \doteq 922.44$; $\bar{y} \doteq 21.74$; $C_{XY} \doteq 2,925.31$
Without Alaska: $\bar{x} \doteq 753.39$; $\bar{y} \doteq 21.44$; $C_{XY} \doteq 387.82$
Since $C_{XY} > 0$, the outline of the scatter plot should form a rough ellipse from the lower left to the upper right of the graph. The actual plots are given to compare with your approximations. The vertical axis is salary ($\times \$1,000$) and the horizontal axis is per capita state debt (dollars). Figure (a) is with Alaska; figure (b) is without Alaska.

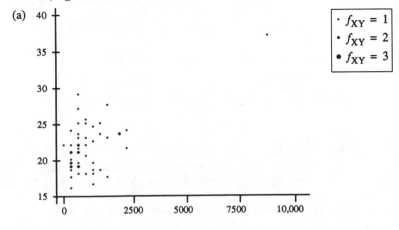

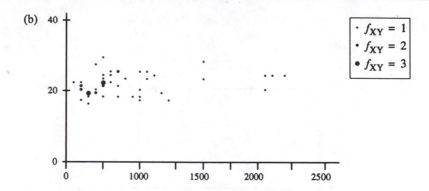

(b)

EXERCISES 4.3 (pp. 121–125)

1. $r \doteq .91$ **3.** $r \doteq -.73$ **5.** $r = 0$
7. $r \doteq .81$ **9.** $r \doteq .73$ **11.** $r \doteq .60$
13. $r = 1$ for persons born on even-numbered days.
   $r = -1$ for persons born on odd-numbered days.
15. $r \doteq .54$ **17.** (a) 200.25
                                                      (b) 16.69
19. $s^2_{Y|x=10} \doteq 243.36$ **21.** $s^2_{Y|x=66} \doteq 662.50$
23. $s^2_{Y|x} \doteq 0.67$; correlation ratio   **25.** $r \doteq .86$
   is about .65
27. $s^2_Y \doteq 9{,}512.24$ for both data sets.

   $s^2_{Y|x} = 500$ for all values of x in both data sets. The conditional means in Data Set 1 are

$$\bar{y}_{x=50} = 45 \qquad \bar{y}_{x=90} = 105 \qquad \bar{y}_{x=130} = 145 \qquad \bar{y}_{x=170} = 205$$

$$\bar{y}_{x=210} = 265 \qquad \bar{y}_{x=240} = 285 \qquad \bar{y}_{x=270} = 325$$

   The conditional means in Data Set 2 are

$$\bar{y}_{x=50} = 45 \qquad \bar{y}_{x=90} = 145 \qquad \bar{y}_{x=130} = 265 \qquad \bar{y}_{x=170} = 325$$

$$\bar{y}_{x=210} = 285 \qquad \bar{y}_{x=240} = 205 \qquad \bar{y}_{x=270} = 105$$

   The correlation ratio for both data sets is

$$\frac{s^2_Y - s^2_{Y|x}}{s^2_Y} \doteq \frac{9{,}512.24 - 500}{9{,}512.24} \doteq .947$$

   For Data Set 1, $r^2_{XY} \doteq .94$. For Data Set 2, $r^2_{XY} \doteq .08$.
   (a) The correlation ratios are equal. The statistical relationship is very strong
      in both data sets. Approximately 95 percent of the uncertainty about Y is
      eliminated if the value of x is specified.
   (b) The coefficients of determination are not equal.

(c) Measures X and Y are related linearly in Data Set 1, but the relationship is nonlinear in Data Set 2.

(d)

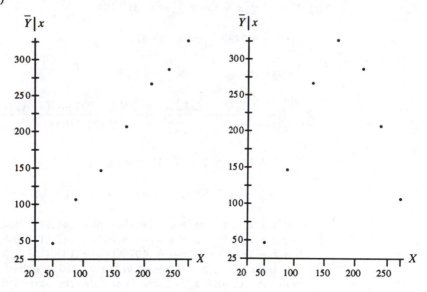

EXERCISES 4.4 (pp. 134–137)

1. (a) $N = 10$ $\sum x = 113$ $\sum x^2 = 1,319$ $\sum y = 105$ $\sum xy = 1,198$
   Therefore,

$$b = \frac{N \sum xy - \sum x \sum y}{N \sum x^2 - (\sum x)^2} = \frac{10(1,198) - (113)(105)}{10(1,319) - (113)^2} \doteq .2732$$

$\bar{x} = 11.3$ $\bar{y} = 10.5$. Therefore,

$$a = \bar{y} - b\bar{x} \doteq 10.5 - (.2732)(11.3) \doteq 7.413$$

For $x = 5$, $\hat{y} = bx + a \doteq .2732(5) + 7.413 \doteq 8.8$.
For $x = 8$, $\hat{y} = bx + a \doteq .2732(8) + 7.413 \doteq 9.6$.
For $x = 9$, $\hat{y} = bx + a \doteq .2732(9) + 7.413 \doteq 9.9$.

(b) $N = 11$ $\sum x = 1,531$ $\sum x^2 = 214,271$

$$\sum y = 1,750 \quad \sum xy = 243,139$$

Therefore,

$$b = \frac{N \sum xy - \sum x \sum y}{N \sum x^2 - (\sum x)^2} = \frac{11(243,139) - (1,531)(1,750)}{11(214,271) - (1,531)^2} \doteq -.3626$$

$\bar{x} = 139.18$ $\bar{y} \doteq 159.09$. Therefore,

$$a = \bar{y} - b\bar{x} \doteq 159.09 - (-.3626)(139.18) \doteq 209.56$$

For $x = 130$, $\hat{y} = bx + a \doteq -.3626(130) + 209.56 \doteq 162.4$.
For $x = 141$, $\hat{y} = bx + a \doteq -.3626(141) + 209.56 \doteq 158.4$.
For $x = 150$, $\hat{y} = bx + a \doteq -.3626(150) + 209.56 \doteq 155.2$.

(c) $N = 12$ $\sum x = 1{,}612$ $\sum x^2 = 217{,}104$

$$\sum y = 2499 \quad \sum xy = 335{,}977$$

Therefore,

$$b = \frac{N\sum xy - \sum x \sum y}{N\sum x^2 - \left(\sum x\right)^2} = \frac{12(335{,}977) - (1{,}612)(2{,}499)}{12(217{,}104) - (1{,}612)^2} \doteq .4976$$

$\bar{x} = 134.33$ $\bar{y} = 208.25$. Therefore,

$$a = \bar{y} - b\bar{x} \doteq 208.25 - (.4976)(134.33) \doteq 141.41$$

For $x = 126$, $\hat{y} = bx + a \doteq .4976(126) + 141.41 \doteq 204.1$.
For $x = 135$, $\hat{y} = bx + a \doteq .4976(135) + 141.41 \doteq 208.6$.
For $x = 145$, $\hat{y} = bx + a \doteq .4976(145) + 141.41 \doteq 213.6$.

(d) For Data Set 1, the observed value $y = 10$ is paired with the observed value $x = 9$, and $s_Y \doteq 1.02$. Therefore, the standardized difference between the observed and predicted values is

$$\frac{10 - 9.9}{1.02} \doteq .10$$

For Data Set 2, the observed $y = 164$ is paired with the observed value $x = 141$, and $s_Y \doteq 6.37$. Therefore, the standardized difference between the observed and predicted values is

$$\frac{164 - 158.4}{6.37} \doteq .88$$

For Data Set 3, the observed value $y = 200$ is paired with the observed value $x = 126$, and $s_Y \doteq 6.52$. Therefore, the standardized difference between the observed and predicted values is

$$\frac{200 - 204.1}{6.52} \doteq -.63$$

The predicted value $\hat{y}$ is nearest the observed value y for $x = 9$ in Data Set 1.

3. (a) $N = 40$ $\sum x = 600$ $\sum x^2 = 11{,}000$

$$\sum y = 4{,}389 \quad \sum xy = 69{,}630$$

Therefore,

$$b = \frac{N\sum xy - \sum x \sum y}{N\sum x^2 - \left(\sum x\right)^2} = \frac{40(69{,}630) - (600)(4{,}389)}{40(11{,}000) - (600)^2} \doteq 1.8975$$

or, $r \doteq .6029$, $s_X \doteq 7.0711$, $s_Y \doteq 22.2564$, so

$$b = r\frac{s_Y}{s_X} \doteq .6029\frac{22.2564}{7.0711} \doteq 1.8976$$

$\bar{x} = 15$, $\bar{y} = 109.725$. Therefore,

$$a = \bar{y} - b\bar{x} = 109.725 - (1.8975)(15) = 81.2625$$

(b) For $x = 35$,

$$\hat{y} = b\bar{x} + a \doteq 1.8975(35) + 81.26 = 147.675$$

5. (a) The values $s_X \doteq 3.6$, $s_Y \doteq 2.9$, and $r \doteq .89$ are calculated in the example. Therefore,

$$b = r\frac{s_Y}{s_X} \doteq .89\frac{2.9}{3.6} \doteq .72$$

The values $\bar{x} = 70$ and $\bar{y} = 66.33$ are also calculated in the example, so

$$a = \bar{y} - b\bar{x} \doteq 66.33 - (.72)70 \doteq 15.93$$

(b) If $x = 73$ in., then

$$\hat{y} = b\bar{x} + a \doteq .72(73) + 15.93 \doteq 68.49$$

From the data, the wife of a man who is 6 ft 1 in. tall should be about 5 feet $8\frac{1}{2}$ in. tall.

7. (a) $N = 11$ $\sum x = 84.7$ $\sum x^2 = 701.21$

$$\sum y = 47.8 \quad \sum xy = 386.12$$

Therefore,

$$b = \frac{N\sum xy - \sum x \sum y}{N\sum x^2 - \left(\sum x\right)^2} = \frac{11(386.12) - (84.7)(47.8)}{11(701.21) - (84.7)^2} \doteq .3684$$

or, $r \doteq .7520$, $s_X \doteq 2.1110$, $s_Y \doteq 1.0343$, so

$$b = r\frac{s_Y}{s_X} \doteq .7520\frac{1.0343}{2.1110} \doteq .3684$$

$\bar{x} \doteq 7.7$, $\bar{y} \doteq 4.3455$. Therefore,

$$a = \bar{y} - b\bar{x} \doteq 4.3455 - (.3684)(7.7) \doteq 1.509$$

(b) For $x = 17$,

$$\hat{y} = b\bar{x} + a \doteq .3684(17) + 1.509 \doteq 7.77$$

In a year of 17 percent unemployment, the data predict a divorce rate of about 7.8 percent.

EXERCISES 5.1 (pp. 163–165)

1. (a) $2/8 = .25$ (b) 0 (c) 1

3. $(1/52)(1/51) = 1/2,652 \doteq .00038$

5. $4/46 \doteq .087$

7. (a) $\{0, 1, 2, 3, 4\}$ (b) $.5$ (c) $(0.5)^4$
(d) $.0625, .25, .375, .25, .0625$

9. It really depends on the food and sleeping accommodations. The traveler's chances of keeping his money are the same in both places. Let E_1 be the probability of *not* being robbed and E_2 be the probability of recovering the money if he *is* robbed. The probability of keeping his money is therefore $P(E_1) + P(E_2)$. At the inn this is $.40 + (.60)(.70) = .82$, and in the village the probability is $.80 + (.20)(.10) = .82$.

11. (a) $P(\text{"hits his opponent"}) = 4/20 = .20$.
(b) $P(\text{"inflicts 5 or more points of damage"}) = 4/8 = .50$
(c) $P(\text{"kills opponent"}) = P(E_1)P(E_2) = (.20)(.50) = .10$
(d) 0; impossible (e) 1; certain

13. (a) $P(\text{"includes two males"}) = 6/15 = .40$
(b) $P(\text{"includes Drusilla"}) = 1/3 \doteq .333$
(c) $P(E_2|E_1) = 0$ (d) mutually exclusive
(e) $P(E_1)P(E_2) = (6/15)(1/3) \doteq .133$
(f) $P(E_1)P(E_2|E_1) = 0$ (g) not independent
(h) Events cannot be both mutually exclusive and independent (unless at least one of them is impossible).

EXERCISES 5.2 (pp. 177–179)

1. (a) $(3)(4)(3)(2)(1) = 72$ (b) $\dfrac{f}{m} = \dfrac{1}{72}$

(c) $m = (3)(2)(3)(2)(1) = 36$ and $f = 1$, so $\dfrac{f}{m} = \dfrac{1}{36}$

3. (a) 720 (b) $479,001,600$ (c) 1 (d) 720 (e) $5,040$
(f) $5,040$ (g) 30 (h) $94,109,400$ (i) $\cong 4 \times 10^{12}$

5. (a) 6 (b) 24 (c) 4 (d) $2,520$ (e) $2,520$ (f) $2,520$

7. (a) $_{52}P_7 = 674,274,182,400$ (or 6.7427×10^{11} on most hand calculators)
(b) There are 4 suits and for any suit there are three ways that one can get the ace, king, queen, jack, and 10 consecutively in a deal of seven cards: cards 1 to 5, cards 2 to 6, or cards 3 to 7. Therefore, there are 12 ways to be dealt the five specified cards. For each of these 12, there are $_{47}P_2 = 2,162$ ways to be dealt the other two cards, so

$$\frac{f}{m} = \frac{12(2,162)}{674,274,182,400} \doteq .0000000385$$

9. (a) $(_{15}P_{15})(_{22}P_{22})(_{11}P_{11})(_{7}P_{7})(_{6}P_{6}) \cong 2.1290455 \times 10^{47}$

(b) The number of arrangements of topics is $(_{5}P_{5}) = 120$. From (a), the number of arrangements of books within topics is approximately 2.1290455×10^{47}. Therefore, the number of ways to arrange all of the books is approximately

$$(120)(2.1290455 \times 10^{47}) \cong 2.5548546 \times 10^{49}$$

(c) Only one of the possible orderings of the 5 topics is alphabetical, so

$$\frac{f}{m} = \frac{1}{_{5}P_{5}} = \frac{1}{120} \doteq .0083$$

(d) From (b), there are approximately 2.5548546×10^{49} possible arrangements. Only one of these has topics and authors within topics in alphabetical order, so

$$\frac{f}{m} \cong \frac{1}{2.5548546 \times 10^{49}} \doteq 3.91 \times 10^{-51}$$

11. (a) 28   (b) 28   (c) 21   (d) 21

13. $1/(_{7}C_{3}) = 1/35 \doteq .03$

15. (a) $_{32}C_{8} = 10{,}518{,}300$

(b) Only one of the 10,518,300 combinations of eight tables includes all eight two-person tables, so

$$\frac{f}{m} = \frac{1}{10{,}518{,}300}$$

(c) There are $_{8}C_{4} = 70$ combinations of four of the eight two-person tables. Each such combination can be combined with any of $_{8}C_{4}$ combinations of four of the eight six-person tables. Therefore, $f = (70)(70) = 4{,}900$ and

$$\frac{f}{m} = \frac{4{,}900}{10{,}518{,}300} \doteq .000466$$

17. (a) $_{11}P_{4} = 7{,}920$ (b) $_{11}C_{4} = 330$

(c) $_{4}P_{4} = 24$ (d) (a) = (b)(c)

EXERCISES 6.1 (pp. 200–203)

1. (a)

x	1	2	3	6	10	$E(X) = 4.4$
P(x)	$\frac{1}{5}$	$\frac{1}{5}$	$\frac{1}{5}$	$\frac{1}{5}$	$\frac{1}{5}$	$V(X) = 10.64$

(b)

x	1	3	5	6	7	8	9	$E(X) \doteq 5.56$
P(x)	$\frac{2}{16}$	$\frac{4}{16}$	$\frac{2}{16}$	$\frac{1}{16}$	$\frac{1}{16}$	$\frac{2}{16}$	$\frac{4}{16}$	$V(X) \doteq 8.12$

(c)

x	−1	5
P(x)	$\frac{9}{10}$	$\frac{1}{10}$

$E(X) = -0.4$

$V(X) = 3.24$

(d)

x	5	100
P(x)	$\frac{4}{5}$	$\frac{1}{5}$

$E(X) = 24$

$V(X) = 1{,}444$

3. (a) $E(X) \doteq 3.67$; $V(X) \doteq 6.22$
  (b) $E(X^2) \doteq 19.67$; $V(X^2) \doteq 440.89$
  (c) They are not equal.   (d) They are not equal.
5. (a) $E(X) = 4.25$ (b) $V(X) = 2.6875$
  (c) $E(X^2) = \mu^2 + \sigma^2 = (4.25)^2 + 2.6875 = 20.75$
7. (a) $k(1) + k(2) + k(3) + k(4) = k(1 + 2 + 3 + 4) = k(10) = 1$
    Therefore, $k = .10$, so

x	1	2	3	4
P(x)	.1	.2	.3	.4

  (b) $E(X) = 1(.1) + 2(.2) + 3(.3) + 4(.4) = 3.0$
  (c) $V(X) = (1 - 3)^2(.1) + (2 - 3)^2(.2) + (3 - 3)^2(.3) + (4 - 3)^2(.4) = 1.0$
9. (a) The student may lay out probability table or give specification statement,
    $X : U(20)$.
  (b) $E(X) = 10.5$ (c) $V(X) = 33.25$
11. Each toss costs $33\frac{1}{3}¢$, so each x-value is the monetary value of the prize—if
    any—minus (approximately) $\$.333$. Therefore,

$$E(X) = (-.333)(.50) + (.417)(.14) + (9.667)(.01) + (.667)(.05)$$

$$+ (-.133)(.30) \doteq -\$.018.$$

13. (a)

x	1	2	3	4	5	6
P(x)	$\frac{1}{6}$	$\frac{1}{6}$	$\frac{1}{6}$	$\frac{1}{6}$	$\frac{1}{6}$	$\frac{1}{6}$

(b) $E(X) = 3.5$

(c) $V(X) \doteq 2.92$

15. (a)

x	2	3	4	5	6	7	8
P(x)	$\frac{1}{16}$	$\frac{2}{16}$	$\frac{3}{16}$	$\frac{4}{16}$	$\frac{3}{16}$	$\frac{2}{16}$	$\frac{1}{16}$

(b) $E(X) = 5$

(c) $V(X) = 2.5$

  This experiment yields 16 possible and equally likely results: 1 on the first
  die, 1 on the second die; 1 on the first die, 2 on the second die; etc. Of
  these 16 results, only one yields $x = 2$, two yield $x = 3$, etc.
17. (a) $E(X + 1) = 6$ (b) If $E(X) = \mu$, then $E(X + c) = \mu + c$ (Rule 2)
  (c) $V(X + 1) = 2.5$ (d) If $V(X) = \sigma^2$, then $V(X + c) = \sigma^2$ (Rule 6)
  (e) 0; 1; the expected value of a standardized random variable is 0 and the
    variance of a standardized random variable is 1.

19. (a) $E(X + Y) = 7$

(b) If $E(X) = \mu_X$ and $E(Y) = \mu_Y$, then $E(X + Y) = \mu_X + \mu_Y$. (Rule 4)

(c) $V(X + Y) \doteq 5.84$

(d) If X and Y are independent random variables and $V(X) = \sigma_X^2$ and $V(Y) = \sigma_Y^2$, then $V(X + Y) = \sigma_X^2 + \sigma_Y^2$. (Rule 9)

21. $E(X) = 0.9$. The probability of inflicting x points of damage is P("hit" on attack die and x on damage die). The probability of a "hit" is $P(17 \cup 18 \cup 19 \cup 20) = 4/20$ and $P(x$ on damage die$) = 1/8$. Since the score on the damage die is independent of the score on the attack die, P("hit" and x) $= (4/20)(1/8) = .025$. Therefore,

$$E(X) = 1(.025) + 2(.025) + \cdots + 8(.025) = .9$$

Another way to look at it is that the expected value of the attack is the probability of a successful attack multiplied by the expected damage,

$$P(\text{"hit"})E(X) = (4/20)(4.5) = .9$$

23. (a) $V(X) = 10.5$ for both characters

(b) If $V(X) = \sigma^2$, then $V(X + c) = \sigma^2$. (Rule 7) Therefore, there is no difference.

(c) $V(2X + 3) = 42$ (d) If $V(X) = \sigma^2$, then $V(cX) = c^2\sigma^2$. (Rule 8)

EXERCISES 6.2 (pp. 214–217)

1. Experiments (a) and (c) are not binomial. In (a) the number of trials is not determined before the experiment begins. (The distribution of this random variable is called the *inverse* binomial and is not discussed in this text.) In (c) the probabilities of success and failure change after each trial, because m changes.

3. $5 \le x \le 7$; $B(15, .82)$; $E(X) = 12.3$; $V(X) = 2.214$

5. $_5C_3(.12)^3(.88)^2 = 10(.001728)(.7744) \doteq .0134$

7. (a) $P(x = 0) = (1/6)^0(5/6)^3 \doteq .579$

(b) $P(x \ge 2) = P(x = 2) + P(x = 3) = 3(1/6)^2(5/6)^1 + (1/6)^3 \doteq .0694 + .0046 = .074$

(c) $E(X) = 3\left(\frac{1}{6}\right) = .5$ (d) $V(X) = 3\left(\frac{1}{6}\right)\left(\frac{5}{6}\right) \doteq .4167$

(e) $E\left(\dfrac{X}{N}\right) = \frac{1}{6}$ (f) $V\left(\dfrac{X}{N}\right) \doteq .046$

9. (a) $E(X) = 2.5$ (b) $V(X) \doteq .4167$ (c) $z \doteq -.7746$

(d) $E\left(\dfrac{X}{N}\right) = \frac{5}{6} \doteq .833$ (e) $V\left(\dfrac{X}{N}\right) \doteq .046$ (f) $z \doteq -.7746$

11. $_{18}C_{13}\left(\frac{17}{22}\right)^{13}\left(\frac{5}{22}\right)^5 + _{18}C_{14}\left(\frac{17}{22}\right)^{14}\left(\frac{5}{22}\right)^4 + _{18}C_{15}\left(\frac{17}{22}\right)^{15}\left(\frac{5}{22}\right)^3 + _{18}C_{16}\left(\frac{17}{22}\right)^{16}\left(\frac{5}{22}\right)^2$

$\doteq .1820 + .2209 + .2003 + .1277 = .7309 \doteq .73$

13. Let X be the number of rejections, so $p = .45$. If $N = 1$, $P(x = 0) = .55$. If $N = 2$, $P(x = 0) = (.55)^2 \doteq .30$. If $N = 3$, $P(x = 0) = (.55)^3 \doteq .17$. Therefore, the answer is 3.

15. Let X be the number of successfully immunized dogs.

$$P(x \le 20) = P(0) + P(1) + \cdots + P(20) \text{ or}$$

$$P(x \le 20) = 1 - P(x \ge 21) = 1 - [P(21) + \cdots + P(25)]$$

$$= 1 - \left[\frac{25!}{21!(25-21)!}(.05)^4(.95)^{21} + \cdots + (.95)^{25} \right] \doteq .007$$

The company's claim would seem to be suspect.

17. $P(4 \text{ or more cases of cancer})$

$$= 1 - P(3 \text{ or fewer cases of cancer})$$

$$= 1 - \left[{}_{100}C_0(.001)^0(.999)^{100} + {}_{100}C_1(.001)^1(.999)^{99} \right.$$

$$\left. + {}_{100}C_2(.001)^2(.999)^{98} + {}_{100}C_3(.001)^3(.999)^{97} \right] \doteq 1 - .999996$$

$$= .000004$$

19. First, find $p = P(\text{success}) = P(3 \text{ sixes in 4 rolls}) = {}_4C_3(1/6)^3(5/6) \doteq .0154$
(a) $P(x = 0) \doteq (1 - .0154)^6 \doteq .911$
(b) $P(x = 1) \doteq 6(.0154)^1(.9846)^5 \doteq .086$
(c) $P(x = 2) \doteq {}_6C_2(.0154)^2(.9846)^4 \doteq .0034$
(d) $P(x \ge 2) \doteq 1 - (.911 + .086) = .003$

EXERCISES 6.3 (pp. 220–221)

1. specification, uniformly distributed, parameter, 1,000, .001

3. (a)

x	1	2	3	4
P(x)	.1	.2	.3	.4

(b) *P*

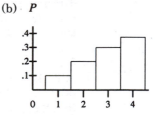

(c) $P(x) = kx$, where $k = P(x = 1)$ or $P(x) = .1x$ for $x = 1, 2, 3, 4$.

5.

(a)

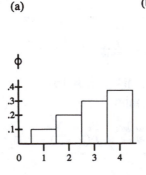

(b) *P*

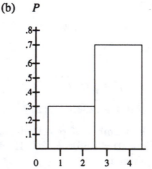

(c) *φ*

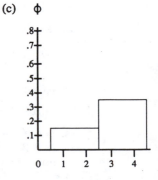

(d) The heights of the bars in Exercise 6.3.5(a) and Exercise 6.3.3(b) are the same.

(e) The heights of the bars in Exercise 6.3.5(b) are twice the heights of the bars in Exercise 6.3.5(c).

EXERCISES 7.1 (pp. 231–232)

1. (a)

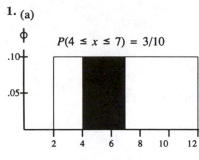

(b)

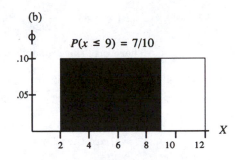

(c)

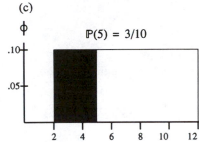

(d)

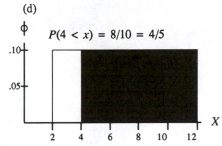

EXERCISES 7.2 (pp. 250–251)

1. (a) .8413   (b) .02275   (c) .93319
     (d) .4090   (e) .9998   (f) .00003
3. (a) .6826   (b) .8990   (c) .083124
     (d) .06612   (e) .72583   (f) .46080
5. (a) .2033   (b) .05050   (c) .02500   (d) .975
     (e) .1587   (f) .009903   (g) .009903   (h) .0455
7. (a) -1.45 (b) 2.65   (c) -2.60 (d) $-.06$
     (e) .01   (f) 3.4   (g) -3.90 (h) -3.05
9. (a) -3.10 (b) -2.58 (c) -1.96
     (d) 3.25   (e) 1.28   (f) 1.96
11. (a) 0   (b) 1.64   (c) 2.05   (d) 2.58   (e) 3.30
13. Approximately 19.15 percent   **15.** Approximately 68.26 percent
17. $[-.67, .67]$, so $P(-.67 \le z \le .67) = .5$
19. For example $[-1.71, 1.01]$ or $[-1.1, 1.53]$

EXERCISES 7.3 (pp. 264–267)

1. (a) .66   (b) .39   (c) .0066   (d) .51   (e) .0149
3. $\underline{50}$ $\underline{49.5}$ $\underline{50.5}$ (a) .0796   (b) 0

5. (a) .397 (b) .383
7. .1114   **9.** .1114   **11.** .4464
13. About 26   **15.** About .00004
17. (a) $X: B(10, .5)$. Using the binomial probability function,

$$P(x = 4) = \frac{210}{1,024}, \ P(x = 5) = \frac{252}{1,024}$$

$$P(x = 6) = \frac{210}{1,024}, \ P(x = 7) = \frac{120}{1,024}$$

Therefore, the probability of obtaining 4 to 7 white balls is

$$\frac{210}{1,024} + \frac{252}{1,024} + \frac{210}{1,024} + \frac{120}{1,024} = \frac{99}{128} = .7734375$$

(b) $V(X) = Npq = 2.5; E(X) = Np = 5$. The standardized limits of the interval are therefore

$$\frac{3.5 - 5}{\sqrt{2.5}} \doteq -.95 \quad \text{and} \quad \frac{7.5 - 5}{\sqrt{2.5}} \doteq 1.58$$

From Table 3 of Appendix VIII, $P(-.95 \leq z \leq 1.58) \doteq .7719$
(c) $X: B(10, 1/3)$. Using the binomial probability function,

$$P(x = 4) = \frac{13,440}{59,049}, \ P(x = 5) = \frac{8,064}{59,049}$$

$$P(x = 6) = \frac{3,360}{59,049}, \ P(x = 7) = \frac{960}{59,049}$$

Therefore, the probability of obtaining 4 to 7 white balls is

$$\frac{13,440}{59,049} + \frac{8,064}{59,049} + \frac{3,360}{59,049} + \frac{960}{59,049} = \frac{25,824}{59,049} = \frac{8,608}{19,683} \doteq .43733170$$

(d) $V(X) = Npq \doteq 2.22$. $E(X) = Np \doteq 3.33$. The standardized limits of the interval are therefore

$$\frac{3.5 - 3.33}{\sqrt{2.22}} \doteq .11 \quad \text{and} \quad \frac{7.5 - 3.33}{\sqrt{2.22}} \doteq 2.80$$

From Table 3 of Appendix VIII, $P(.11 \leq z \leq 2.80) \doteq .4536$.
(e) (b) approximates (a) better than (d) approximates (c) because $X: B(10, 1/3)$ is skewed and because $N = 10$ is not large enough to offset the skew (i.e., $Np < 5$). Nonetheless, the approximation is off only by about .01623.
19. .4464   **21.** Approximately .01616   **23.** .73483
25. (a) .2586   (b) .2126   (c) .1587   (d) .07636
27. $x_i = 4.92$
29. For standardized normally distributed random variables, a cumulative probability of .10 corresponds (approximately) to a score of -1.28. The cutoff score is therefore $-1.28(15) + 100 = 81$ (approximately).

31. (a) For Section 1 the z-score for $x = 25$ is -1.
   $P(z \leq -1) \doteq .16$
   (b) For Section 2 the z-score for $x = 25$ is .83.
   $P(z \geq 25) \doteq .20$.
   (c) The student's score is more probable under the assumption that he or she belongs to Section 2.

EXERCISES 8.1 (p. 281)

1. The population is the incoming freshman class. The sample is the 30 freshmen chosen by the admissions officer. The random variable is final high school grade-point average.

3. The population is all present *and future* applicants *who are hired*. The sample is the current group of applicants—or that subset of applicants who are hired if other selection criteria are used. There are two random variables, test scores and performance evaluation scores. (The civil service is interested in the *population correlation* of these two variables.)

5. There are many possibilities. One way would be to assign each freshman a number from a random number table and pick the lowest 30.

7. (a) If N is the number of family members, $m = {}_NC_2$.
   (b) every pair of family members per student's data
   (c) median and range per student's data
   (d) probably not   (e) probably not   (f) random variables

EXERCISES 8.2 (pp. 292–295)

1.

	$E(\bar{X})$	$V(\bar{X})$
(a)	67.3	2
(b)	67.3	1
(c)	67.3	.16
(d)	67.3	.08

3.

	$E(\bar{X})$	$V(\bar{X})$
(a)	14.29	.39605
(b)	14.29	.07921
(c)	14.29	.015842
(d)	14.29	.007921

5.

	$P \doteq$	Interval Width
(a)	.68	1.78
(b)	.95	1.78
(c)	.997	1.78
(d)	.69	.90
(e)	.69	.60

7. (a) $P \doteq .84$
   (b) $P \doteq .92$
   (c) $P \doteq .98$

9. (a) 3.2, 3.2, 2.6, 2.9, 4.2, 3.9, 4.3, 3.7, 3.0, 3.6

2.45 – 2.85	2.85 – 3.25	3.25 – 3.65	3.65 – 4.05	4.05 – 4.45
1	4	1	2	2

The distribution of sample means exhibits more kurtosis than the distribution of the original population.
   (b) 3.46; $3.5 - 3.46 = .04$
   (c) 0.2924; the variance of the ten sample means is approximately the population variance divided by 10 (sample size).

11. (a) The mean is 0.5 and the standard deviation is about 0.224; the proportion corresponding to z_a is about .30, and the proportion corresponding to z_b is about .70.
   (b) 0.40
   (c) You expect about 63 percent; we actually have 70 percent.
13. (a) uniformly
   (b) The claim is correct by the Central Limit Theorem.
15. (a) 3.5   (b) 3.5

EXERCISES 8.3 (pp. 318–321)

1. (a) $\bar{x} \doteq 472.67$; $s^2 \doteq 1,446.22$ (b) $\bar{x} \doteq 472.67$; $\hat{\sigma}^2 \doteq 1,496.10$
   (c) $\bar{x}$ is unbiased, consistent (and the most relatively efficient of any linear estimator of μ); $\hat{\sigma}^2$ is unbiased and consistent.
   (d) The third observation is an unbiased estimator of the population mean, but it is not consistent since it does not improve with increasing sample size. Nor is it very efficient, since its variance is the same as the population variance.
3. $\hat{p} = .4$
5. (a) [10.4, 49.6]   (b) [20.2, 39.8]   (c) [25.1, 34.9]
7. (a) [27.68, 32.32]   (b) [26.83, 33.17]   (c) [26.03, 33.97]
9. (a) [15.15, 19.45]   (b) [14.85, 19.75]   (c) [14.47, 20.12]
11. (a) 456.9, 488.5

   (b) For $\mathbb{C} = .975$ the confidence coefficients are ± 2.24. That is, $\mathbb{P}(2.24) \doteq .9875$. The upper confidence limit is therefore $2.24\hat{\sigma}_{\bar{X}}$ and is 5 points above the mean. So

$$2.24\hat{\sigma}_{\bar{X}} = 2.24\sqrt{(1,496.1)/N} = 5$$

   Solving for N, a sample of approximately 300 would be needed.
13. (a) .90   (b) .99   (c) .82
15. [2.35, 3.59]   17. [.55, .68]

19. $\dfrac{.025}{\sqrt{(.57)(.43)/1,500}} \doteq 1.96$

   $\mathbb{P}(1.96) \doteq .975$, so $\mathbb{C} = .95$

EXERCISES 9.1 (pp. 341–346)

1. (a) model   (b) test hypothesis   (c) alternative hypothesis

   (d) $B(30, .65)$ (e) 19.5; $\neq$ (f) test statistic

3. (a) $\alpha \doteq .011$ (b) $\alpha \doteq .001$ (c) $\alpha \doteq .059$ (d) $\alpha \doteq .007$
5. (a) $\alpha \doteq .005$ (b) $\alpha \doteq .061$ (c) $\alpha \doteq .121$ (d) $\alpha \doteq .009$
7. (a) $\alpha \doteq .002$ (b) $\alpha \doteq .032$
9. (a) Reject H_0 if $x \geq 8$.   (b) Reject H_0 if $x \geq 16$.
   (c) Reject H_0 if $x \geq 12$.   (d) Reject H_0 if $x \geq 17$.

11. (a) Reject H_0 if $y \geq 12$ $(\alpha \doteq .001)$.
   (b) Reject H_0 if $y \geq 7$ $(\alpha \doteq .018)$.
   (c) Reject H_0 if $y \geq 7$ $(\alpha \doteq .051)$.
   (d) Reject H_0 if $y \geq 8$ $(\alpha \doteq .031)$.
13. (a) $X : B(N, .60)$
   (b) $H_0: p = .60$
      $H_1: p > .60$
   (c) Reject H_0 if $x/N \geq .92$.
   (d) Reject H_0 if $x/N \geq .86$. Using the normal approximation to the binomial distribution gives

$$z_\alpha = 2.33 = \frac{\dfrac{x}{18} - .60}{\sqrt{\dfrac{(.60)(.40)}{18}}}$$

so $x \doteq 15.64$. The nearest integer is $x = 16$, which corresponds to the interval 15.5 to 16.5. Therefore, the critical value is the left boundary of the interval, 15.5, and the critical proportion is

$$\frac{15.5}{18} \doteq .86$$

   (e) Reject H_0 if $x/N \geq .875$.
   (f) Reject H_0 if $x/N \geq .74$. Using the normal approximation to the binomial distribution gives

$$z_\alpha = 1.75 = \frac{\dfrac{x}{36} - .60}{\sqrt{\dfrac{(.60)(.40)}{36}}}$$

so $x \doteq 26.74$. The nearest integer is $x = 27$, which corresponds to the interval 26.5 to 27.5. Therefore, the critical value is the left boundary of the interval, 26.5, and the critical proportion is

$$\frac{26.5}{36} \doteq .74$$

15. (a) $x/N \geq .75$ or $x \geq 38$ (b) $x/N \geq .79$ or $x \geq 40$
   (c) $x/N \geq .725$ or $x \geq 73$ (d) $x/N \geq .755$ or $X \geq 76$
17. (a) The model that is implied by the hunch is $X : B(N, p)$, where $p > \frac{1}{6}$. However, the model that generates the test hypothesis is $X : B(N, \frac{1}{6})$.
   (b) $H_0: p = 1/6$
      $H_1: p > 1/6$
   (c) For $\alpha = .05$, reject H_0 if the number of 1s exceeds 14.
   (d) For $\alpha = .01$, reject H_0 if the number of 1s exceeds 16.
19. (a) Let X be the number of errors.   (b) $X : B(N, .25)$
   (c) If the probability of an error is denoted p, then the test hypothesis is $p = .25$.
   (d) $H_1: p > .25$. This is *not* a misprint. The "obvious" alternative hypothesis is that $p < .25$. However, suppose that one formulates and rejects the

"obvious" alternative in favor of the test hypothesis that $p = .25$. Does the result support the theory or does it weigh against the theory? The result is ambiguous, because the theory implies that $p \leq .25$. This means that $p < .25$ and $p = .25$ are *both* consistent with the theory. The only hypothesis that is unambiguously *inconsistent* with the theory is $p > .25$.

(e) $(X/40)$ distributed approximately $N(.25, .0046875)$.

(f) There is no tabled distribution for $X: N(.25, .0046875)$.

(g) $$\frac{\dfrac{13.5}{40} - .25}{\sqrt{\dfrac{(.25)(.75)}{40}}} \doteq 1.28$$

EXERCISES 9.2 (pp. 360–363)

1. (a) .851   (b) .892   (c) .485   (d) .659

3. (a) .252   (b) .412   (c) .748

    (d) .211   (e) .344   (f) .654

5. (a) For $N = 12$, β is approximately .980

For $N = 30$,

$$z_\alpha = 2.75 = \frac{\dfrac{x}{30} - .50}{\sqrt{\dfrac{(.50)(.50)}{30}}}$$

so $x \doteq 22.53$. The nearest integer is $x = 23$, so

$$\beta = P\left(\frac{x}{N} < \frac{23}{30} \,\middle|\, X: B(30, .60)\right)$$

Using the normal approximation, z_β corresponds to the left boundary of the interval 22.5 to 23.5:

$$z_\beta = \frac{\dfrac{22.5}{30} - .60}{\sqrt{\dfrac{(.60)(.40)}{30}}} \doteq 1.68$$

Under the standard normal curve, $P(z < 1.68) \doteq .95352 = \beta$. (Note that β decreases as N increases.)

(b) For $N = 10$, β is approximately .954

For $N = 25$,

$$z_\alpha = 2.33 = \frac{\dfrac{x}{25} - .50}{\sqrt{\dfrac{(.50)(.50)}{25}}}$$

so $x \doteq 18.33$. The nearest integer is $x = 18$, so

$$\beta = P\left(\frac{x}{N} < \frac{18}{25} \,\middle|\, X : B(25, .60)\right)$$

Using the normal approximation, z_β corresponds to the left boundary of the interval 17.5 to 18.5:

$$z_\beta = \frac{\dfrac{17.5}{25} - .60}{\sqrt{\dfrac{(.60)(.40)}{25}}} \doteq 1.02$$

Under the standard normal curve, $P(z < 1.02) \doteq .8461 = \beta$. (Note that β decreases as N increases.)

7. (a) $\beta \doteq .44$ (b) $\beta \doteq .56$ (c) $\beta \doteq .78$
   $\beta \doteq .23$ $\beta \doteq .37$ $\beta \doteq .63$
   $\beta \doteq .10$ $\beta \doteq .16$ $\beta \doteq .35$

Reading across each row we see that β increases as α decreases, and reading down each column we see that β decreases as N increases.

9. (a) $p \doteq .011$ (b) Type I   (c) $p \doteq .30$ (d) Type II

11. (a) Reject H_0 by binomial $(p \doteq .02)$ or by the normal approximation $(p \doteq .01923)$.
   (b) Reject H_1 by binomial $(p \doteq .24)$ or by normal approximation $(p \doteq .2483)$.
   (c) Reject H_0; $p \doteq .003$.   (d) Reject H_1; $p \doteq .18$.

13. The test hypothesis is H_0: $p = .25$. If H_0 is correct, the expected value of the wager is therefore

$$\$1(.75) - \$2(.25) = \$.25$$

That is, if $P(3 \text{ or } 4) = .25$, Steve can expect to win about 25¢ per roll, or a total of \$250.00. If he incorrectly rejects the hypothesis that $p = .25$ and decides not to proceed with the wager, he will therefore lose about \$250.00. The alternative hypothesis is H_1: $p = .40$. If H_1 is correct, the expected value of the wager is therefore

$$\$1(.60) - \$2(.40) = -\$.20$$

That is, if $P(3 \text{ or } 4) = .4$, Steve can expect to lose about 20¢ per roll. If he incorrectly rejects the hypothesis that $p = .4$ and decides to go ahead with the bet, he will therefore lose \$200.00.

For $N = 24$ and $\alpha = .120$, H_0 must be rejected if the number of 3s and 4s is 9 or more. For this rejection rule,

$$\beta = P[(x < 9) \,|\, X : B(24, .40)] \doteq .328$$

For $\alpha = .053$, H_1 is rejected if the number of 3s and 4s is 10 or more. For this rejection rule,

$$\beta = P[(x < 10) \,|\, X : B(24, .40)] \doteq .489$$

Rejection rule	Expected loss given that	
	H_0 correct	H_1 correct
$\alpha = .120$	$.120(\$250) = \$30.00$	$.328(\$200) = \$65.60$
$\alpha = .053$	$.053(\$250) = \$13.25$	$.489(\$200) = \$97.80$

The maximum expected loss under the first rule is $65.60. The maximum expected loss under the second rule is $97.80. By the minimax criterion, he should use the first rule.

EXERCISES 9.3 (pp. 372–375)

1. (a) $1 - .851 = .149$ (b) $1 - .892 = .108$
   (c) $1 - .485 = .515$ (d) $1 - .659 = .341$

3. There is no x-value that gives us a significance level of exactly .02, but $\mathbb{P}(3) = .017$, so the critical region for the test is $\{0, 1, 2, 3\}$. The power of the test is the probability that x falls in the critical region *if* H_1 is correct, so power is $P(x \leq 3)$ under each of the four alternative hypotheses.
   (a) $P(x \leq 3 | p = .10) \doteq .945$ (b) $P(x \leq 3 | p = .20) \doteq .648$
   (c) $P(x \leq 3 | p = .30) \doteq .298$ (d) $P(x \leq 3 | p = .40) \doteq .090$
   Note that power decreases as the value of p in the alternative hypothesis moves toward $p_0 = .5$.

5. Using the normal approximation to the binomial, the critical value of z is 1.64. The corresponding value of x rounded off to the nearest integer is 68. Correcting for discontinuity, we obtain 67.5, so the critical value of X/N is .675 and H_0 is rejected if $(x/N) \geq .675$. Since these are upper-tailed tests, the most natural way to calculate power is first to calculate β (i.e., the cumulative probability of .675) under each of the alternative hypotheses and then subtract β from 1.
   (a) $1 - P(x/N \leq .675 | p = .70) \doteq .7088$
   (b) $1 - P(x/N \leq .675 | p = .65) \doteq .3015$
   (c) $1 - P(x/N \leq .675 | p = .62) \doteq .1292$
   (d) $1 - P(x/N \leq .675 | p = .61) \doteq .09176$
   Note that power decreases as the value of p in the alternative hypothesis moves toward $p_0 = .6$.

7. (a) Reject H_0 is $x \leq 14$ or if $x = 25$ ($\alpha \doteq .009$).
   (b) Reject H_0 is $x \leq 10$ or if $x = 18$ ($\alpha \doteq .034$).
   (c) $z_{\alpha/2} = 2.58$ and $-z_{\alpha/2} = -2.58$, so

$$\pm 2.58 = \frac{\dfrac{x}{25} - .80}{\sqrt{\dfrac{(.8)(.2)}{25}}}$$

Solving for x, the critical values are about 14.84 and 25.16. Since X is a discrete variable the rejection rule is: Reject H_0 if $x \leq 15$ or if $x = 25$ ($\alpha \doteq .021$).

(d) $z_{\alpha/2} = 2.11$ and $-z_{\alpha/2} = -2.11$, so

$$\pm 2.11 = \frac{\dfrac{x}{18} - .80}{\sqrt{\dfrac{(.8)(.2)}{18}}}$$

Solving for x, the critical values are about 10.82 and 17.98. Since X is a discrete variable the rejection rule is: Reject H_0 if $x \leq 11$ or if $x = 18$ ($\alpha \doteq .069$).

Note: The rule of thumb that the normal approximation to the binomial may be used when $Np \geq 5$ and $Nq \geq 5$ is satisfied by exercise (c), but not by exercise (d). Nevertheless, the value of α obtained by using the normal approximation is more than twice the value of α obtained by using the binomial distribution in *both* exercises. A rule of thumb is a useful guideline, but as p departs from .50, the value of α under the normal approximation is subject to increasing distortion, even if the guideline is satisfied.

9. (a) H_0: $p = .10$ (b) H_1: $p > .10$

(c) $Z_{X/N} = \dfrac{\dfrac{X}{50} - .10}{\sqrt{\dfrac{(.10)(.90)}{50}}}$

(d) $N(0, 1)$

(e) 9 or more. For $x = 9$, the corresponding z-value is

$$\frac{\dfrac{8.5}{50} - .10}{\sqrt{\dfrac{(.10)(.90)}{50}}} \doteq 1.65,$$

and $\mathbb{P}(1.65) \doteq .95053$, so α is just under .05.

11. (a) $X : B(N, \frac{2}{3})$ (b) H_0: $p = \frac{2}{3}$ (c) H_1: $p > \frac{2}{3}$

(d) $Z_{X/N} = \dfrac{\dfrac{X}{50} - \dfrac{2}{3}}{\sqrt{\dfrac{(2/3)(1/3)}{50}}}$

(e) $N(0, 1)$ (f) 42 or more ($\alpha \doteq .007$)

13. (a) The questionnaire used by the independent laboratory forces a choice between the two brands, so the test hypothesis is based on the 15 men in the manufacturer's sample who expressed a preference. Nine of these, or 60 percent, preferred Gillette. Therefore,

$$H_0: p = .60$$

(b) H_1: $p = .50$

(c) Power $\doteq .44$

Solution:

$$z_\alpha = -1.64 = \frac{\dfrac{x_\alpha}{50} - .60}{\sqrt{\dfrac{(.60)(.40)}{50}}}, \text{ so}$$

$$x_\alpha = 50\left[.60 - 1.64\left(\sqrt{\frac{(.60)(.40)}{50}}\right)\right]$$

$$\doteq 24$$

Thus,

$$z_\beta = \frac{\dfrac{24.5}{50} - .50}{\sqrt{\dfrac{(.50)(.50)}{50}}} \doteq -.14 \qquad \text{Power} = \mathbb{P}(z_\beta) \doteq .44$$

EXERCISES 10.1 (pp. 398–401)

1. (a) If $z_{\bar{X}} \le -1.64$, reject H_0. $z_{\bar{X}} \doteq -.53$. Reject H_1.
   (b) If $z_{\bar{X}} \le -2.33$, reject H_0. $z_{\bar{X}} \doteq -.32$. Reject H_1.
   (c) If $z_{\bar{X}} \ge 2.33$, reject H_0. $z_{\bar{X}} \doteq 1.03$. Reject H_1.
   (d) If $z_{\bar{X}} \ge 1.64$, reject H_0. $z_{\bar{X}} \doteq 1.70$. Reject H_0.
   (e) If $z_{\bar{X}} \le -1.96$, reject H_0. $z_{\bar{X}} \doteq -2.23$. Reject H_0.
3. (a) If $z_{\bar{X}} \ge 1.64$, reject H_0. $z_{\bar{X}} = 1.7$. Reject H_0.
   (b) If $z_{\bar{X}} \le -1.96$, reject H_0. $z_{\bar{X}} \doteq -1.27$. Reject H_1.
   (c) If $z_{\bar{X}} \le -1.96$ or ≥ 1.96, reject H_0. $z_{\bar{X}} = 1.7$. Reject H_1.
   (d) If $z_{\bar{X}} \ge 2.33$, reject H_0. $z_{\bar{X}} = 2.8$. Reject H_0.
   (e) If $z_{\bar{X}} \ge 3.10$, reject H_0. $z_{\bar{X}} = 3.09$. Reject H_1.
5. (a) $H_0: \mu = 150$
      $H_1: \mu \ne 150$
   (b) The test statistic is

$$Z_{\bar{X}} = \frac{\bar{X} - \mu_0}{\hat{\sigma}_{\bar{X}}} = \frac{\bar{X} - \mu_0}{\sqrt{\dfrac{s^2\left(\dfrac{N}{N-1}\right)}{N}}}$$

   (c) The rejection region must include all values (equal to or) less than $-z_{\alpha/2}$ or (equal to or) greater than $z_{\alpha/2}$.

   (d) $\dfrac{146.5 - 150}{\sqrt{\dfrac{132.34\left(\dfrac{40}{39}\right)}{40}}} \doteq -1.90$

   (e) $p > .057$ (two-tailed), so reject H_1 if $\alpha \le .057$. Otherwise, reject H_0.

7. (a) $H_0: \mu = .765$
$H_1: \mu \neq .765$

(b) $Z_{\bar{X}} = \dfrac{\bar{X} - \mu_0}{\hat{\sigma}_{\bar{X}}} = \dfrac{\bar{X} - \mu_0}{\sqrt{\dfrac{s^2\left(\dfrac{N}{N-1}\right)}{N}}}$

(c) Reject H_0 if $z_{\bar{X}} \geq z_{\alpha/2}$ or if $z_{\bar{X}} \leq -z_{\alpha/2}$

(d) $z_{\bar{X}} \doteq \dfrac{.769 - .765}{\sqrt{\dfrac{.0000189\left(\dfrac{36}{35}\right)}{36}}} \doteq 5.44$

(e) $p < .0000006$, so reject H_0 if $\alpha \geq .0000006$. Otherwise, reject H_1.

9. (a) $p > .29$; $\beta \doteq .74$; power $\doteq .26$
(b) $p > .37$; $\beta \doteq .67$; power $\doteq .33$
(c) $p > .15$; $\beta \doteq .87$; power $\doteq .13$
(d) $p < .045$; $\beta \doteq .027$; power $\doteq .973$
(e) $p < .013$; $\beta \doteq .76$; power $\doteq .24$

11.

	Test Statistic	Critical Value	Power
(a)	$\dfrac{\bar{X} - 50}{4.743}$	61.05	.42
(b)	$\dfrac{\bar{X} - 50}{3}$	56.99	.83
(c)	$\dfrac{\bar{X} - 50}{2.12}$	54.94	.99
(d)	$\dfrac{\bar{X} - 50}{2.12}$	54.16	.995
(e)	$\dfrac{\bar{X} - 50}{2.12}$	53.48	.998

13.

	Reject H_0 if	Test Statistic	Reject	p	Power
(a)	$z_{\bar{X}} \leq -1.64$	-1.69	H_0	$< .046$	.936
(b)	$z_{\bar{X}} \leq -1.96$	-1.90	H_1	$> .028$	.252
(c)	$z_{\bar{X}} \leq -1.96$ or	-1.964	H_0	$< .05^*$	N/A
	$z_{\bar{X}} \geq 1.96$				
(d)	$z_{\bar{X}} \geq 2.33$	2.33	H_0	$< .01$	N/A

$*p < .025$ if p is reported with respect to $\alpha/2$.

15. (a) Since $\sigma^2 = 10,000$, one standard unit is equal to 100. This smallest scientifically important difference is the synthetic value, δ. Therefore,

$$H_1: \mu = \mu_0 + \delta = 500 + 100 = 600$$

(b) With an upper-tailed test and $\alpha = .01$, $z_\alpha = 2.33$. Under the *alternative* hypothesis, the critical value falls in the lower tail of the distribution of

$Z_{\bar{X}}$, so $Z_\beta = -2.33$. Therefore,

$$N = \left(\frac{\sigma(z_\beta - z_\alpha)}{\mu_0 - \mu_1}\right)^2 = \left(\frac{100(-2.33 - 2.33)}{100}\right)^2 \doteq 21.7 \cong 22$$

EXERCISES 10.2 (pp. 408–410)

1. (a) .75 (b) .995 (c) .60
  (d) .005 (e) .0005 (f) .10
3. (a) $.60 < \mathbb{P} < .75$ (b) $.99 < \mathbb{P} < .995$
  (c) $.10 < \mathbb{P} < .20$ (d) $.995 < \mathbb{P} < .9995$
  (e) $.95 < \mathbb{P} < .975$ (f) $.01 < \mathbb{P} < .025$
5. (a) 1.796 (b) 2.5 (c) .689
  (d) -1.753 (e) -4.604 (f) -2.093
7. (a) .025 (b) .01 (c) .10 (d) .05
  (e) $.005 < p < .01$ (f) $.025 < p < .05$ (g) .99 (h) .95
9. (a) .01 (b) .05 (c) .99 (d) .10
11. (a) .005 (b) .025
13. (a) $.05 < p < .10$ (b) $.005 < p < .01$

EXERCISES 10.3 (pp. 413–417)

1.

	Distribution of Test Statistic	Rejection Region	Value of Test Statistic	Reject	p
(a)	$t(9)$	$t \geq 1.833$	1.85	H_0	$< .05$
(b)	$t(4)$	$t \leq -2.776$	-3.5	H_0	$< .025$
(c)	$t(16)$	$t \leq -2.921$ or $t \geq 2.921$	2.0	H_1	$> .05^*$
(d)	$t(3)$	$t \leq -2.353$	-1.75	H_1	$> .05$
(e)	$t(3)$	$t \leq -3.182$ or $t \geq 3.182$	-3.19	H_0	$< .05^\dagger$

$*p > .025$ if p reported with respect to $\alpha/2$.
$\dagger p < .025$ if p reported with respect to $\alpha/2$.

3. (a) [22.58, 27.42] (b) [21.78, 28.22] (c) [19.98, 30.02]
5. (a) approximately 98 percent
  (b) between 99 and 99.9 percent
  (c) less than 50 percent
7.

	Test Statistic	Distribution of Test Statistic	Rejection Region	Value of Test Statistic	Reject	p
(a)	$Z_{\bar{X}}$	$N(0,1)$	$z \geq 1.64$	1.18	H_1	$> .11$
(b)	t	$t(26)$	$t \geq 1.706$	2.28	H_0	$< .025$
(c)	t	$t(21)$	$t \geq 2.518$	2.05	H_1	$> .025$
(d)	$Z_{\bar{X}}$	$N(0,1)$	$z \geq 2.33$	2.68	H_0	$< .004$
(e)	Nonparametric					

9. (a) H_0: $\mu = 150$
   H_1: $\mu < 150$

(b) The appropriate test statistic is

$$t = \frac{\bar{X} - 150}{s/\sqrt{N - 1}}$$

(c) The rule must identify all values (equal to or) less than t_α as rejection region, where $\mathbb{P}(t_\alpha) = \alpha$ in $t(24)$

(d) $t \doteq -3.58$

(e) $p < .005$, so reject H_0 is $\alpha \geq .005$

11. (a) H_0: $\mu = 58$
   H_1: $\mu < 58$

(b) The test statistic is

$$Z_{\bar{X}} = \frac{\bar{X} - 58}{\hat{\sigma}_{\bar{X}}}$$

(c) The rejection region must include all values (equal to or) less than z_α, where $\mathbb{P}(z_\alpha) = \alpha$.

(d) $\dfrac{55 - 58}{\sqrt{\dfrac{900\left(\dfrac{100}{99}\right)}{100}}} \doteq -.99$

(e) $p \doteq 16$, so if $\alpha \leq .16$, reject H_1.

13. If μ is the mean delay for all flights, the test hypothesis is

$$H_0: \mu = 58$$

and the alternative hypothesis is

$$H_1: \mu = 58 - 10 = 48$$

Variance is estimated on the basis of the data collected by the group in its first study, where $s = 30$ and $N = 100$, so

$$\hat{\sigma} = \sqrt{900\left(\frac{100}{99}\right)} \doteq 30.2$$

Since $\alpha = .05$, $z_\alpha = -1.64$, and for $\beta = .20$, $z_\beta = .84$. Therefore,

$$N = \left(\frac{\hat{\sigma}(z_\beta - z_\alpha)}{\mu_0 - \mu_1}\right)^2 \doteq \left(\frac{30.2(.84 + 1.64)}{10}\right)^2 \cong 56$$

15. (a) With X normally distributed and σ^2 unknown we first try t with the *largest* sample size that requires t instead of $Z_{\bar{X}}$, that is, $N = 29$. We find that t is significant, and we therefore experiment with smaller sample sizes. The smallest sample size for which t is significant is $N = 20$.

(b) With the distribution of X unknown, we must use $Z_{\bar{X}}$. For $\alpha = .01$ the critical value of Z is -2.33. If we set

$$-2.33 = \frac{21 - 33.33}{\sqrt{\dfrac{670}{N-1}}}$$

we find $N \cong 25$. However, $\overline{X}$ cannot be assumed normally distributed unless $N \geq 30$, so the smallest sample size for which the result is significant at $\alpha = .01$ is $N = 30$.

(c) Since X is normally distributed and σ^2 is known, we may use $Z_{\overline{X}}$ as our test statistic. With $\alpha = .005$ the critical value is $z = 2.58$. We set $z_{\overline{X}} = 2.58$, solve for N as in (b), and obtain $N \cong 6$.

(d) Using the same strategy as (a), we first try t with the *largest* sample size that requires t instead of $Z_{\overline{X}}$, that is, $N = 29$. With 28 degrees of freedom and a two-tailed test for $\alpha = .05$, the critical values are $t = \pm 2.048$. Calculation of Student's t with $N = 29$ yields $t \doteq -1.93$. We therefore need a sample larger than 29. At $N \geq 30$, however, we can use $Z_{\overline{X}}$ as our test statistic. With $\alpha = .05$ and a two-tailed test, the critical values are $z = \pm 1.96$. We set $z_{\overline{X}} = 1.96$, solve for N as in (b), and obtain $N \cong 30$.

17. Let μ be the population mean weight of vole nests in rye fields. For $\nu = 10$, the value required to reject $H_0\colon \mu = 405$ in favor of $H_1\colon \mu < 405$ is $t = -2.764$. Therefore,

$$-2.764 = \frac{27 - 405}{s/\sqrt{10}}$$

$s \doteq 432.468$, so $s^2 \doteq 187{,}028.57$.

EXERCISES 10.4 (pp. 424–428)

1. (a) $E(X_1) = 22$; $E(X_2) = 6$
(b) $V(X_1) = 49$; $V(X_2) = 4$

X_1:	15	29
X_2: 4	11	25
8	7	21

(c) $\overline{y} = 16 = E(X_1) - E(X_2)$
(d) $s^2 = 53 = V(X_1) + V(X_2)$

3. (a) .16   (b) .05   (c) .025   (d) .05   (e) .50
5. (a) .33   (b) .00107   (c) .01255   (d) .75   (e) .05

7.

	Reject H_0 if	Value of Test Statistic	p	Reject
(a)	$z_{\overline{X}_1 - \overline{X}_2} \geq 1.64$	1.70	$\doteq .045$	H_0
(b)	$z_{\overline{X}_1 - \overline{X}_2} \leq -2.33$	-2.337	$\doteq .01$	H_0
(c)	$z_{\overline{X}_1 - \overline{X}_2} \begin{cases} \leq -1.96 \text{ or} \\ \geq 1.96 \end{cases}$	-1.99	$\doteq .047$	H_0
(d)	$z_{\overline{X}_1 - \overline{X}_2} \leq -3.10$	-3.00	$> .001$	H_1
(e)	$z_{\overline{X}_1 - \overline{X}_2} \begin{cases} \leq -2.58 \text{ or} \\ \geq 2.58 \end{cases}$	-2.45	$> .014$	H_1

9. (a) If μ_1 is the mean for transfer students from the agricultural college and μ_2 is the mean for junior college transfer students,

$$H_0: \mu_1 - \mu_2 = 0$$

$$H_1: \mu_1 - \mu_2 > 0$$

(b) The appropriate test statistic is

$$Z_{\bar{X}_1 - \bar{X}_2} = \frac{\bar{X}_1 - \bar{X}_2}{\sqrt{\dfrac{\sigma_1^2}{N_1} + \dfrac{\sigma_2^2}{N_2}}}$$

(c) Reject H_0 if $z_{\bar{X}_1 - \bar{X}_2} \geq 1.64$.

(d) $Z_{\bar{X}_1 - \bar{X}_2} \doteq 1.46$ **(e)** Reject H_1.   **(f)** $p > .07$

(g) $\alpha = .07215$

11. (a) $H_0: \mu_{1964} - \mu_{1978} = 0$

$H_1: \mu_{1964} - \mu_{1978} > 0$

(b) The appropriate test statistic is

$$Z_{\bar{X}_1 - \bar{X}_2} = \frac{\bar{X}_{1964} - \bar{X}_{1978}}{\sqrt{\dfrac{\hat{\sigma}_{1964}^2}{N_{1964}} + \dfrac{\hat{\sigma}_{1978}^2}{N_{1978}}}}$$

and will be distributed $N(0, 1)$ if H_0 is correct.

(c) Since the test is one-tailed and the rejection region is in the upper tail of $Z: N(0, 1)$, the rejection region is $z \geq z_\alpha$, where $\mathbb{P}(z_\alpha) = 1 - \alpha$.

(d) $z_{\bar{X}_1 - \bar{X}_2} = \dfrac{822 - 800}{\sqrt{\dfrac{12{,}544\left(\dfrac{400}{399}\right)}{400} + \dfrac{10{,}609\left(\dfrac{275}{274}\right)}{275}}} \doteq 2.63$

(e) $p < .0043$, so H_0 is rejected if $\alpha \geq .0043$.

13. (a) Let μ_1 be the population mean for Catholic college students and let μ_2 be the population mean for Jewish college students.

$$H_0: \mu_1 - \mu_2 = 0$$

$$H_1: \mu_1 - \mu_2 \neq 0$$

(b) The appropriate test statistic is

$$Z_{\bar{X}_1 - \bar{X}_2} = \frac{\bar{X}_1 - \bar{X}_2}{\sqrt{\dfrac{\hat{\sigma}_1^2}{N_1} + \dfrac{\hat{\sigma}_2^2}{N_2}}}$$

and will be distributed $N(0, 1)$ if H_0 is correct.

(c) Since the test is two-tailed the rejection region is $z \geq z_{\alpha/2}$ and $z \leq -z_{\alpha/2}$.

(d) $z_{\overline{X}_1 - \overline{X}_2} = \dfrac{147.4 - 139.5}{\sqrt{\dfrac{900\left(\dfrac{46}{45}\right)}{46} + \dfrac{600.25\left(\dfrac{131}{130}\right)}{131}}} \doteq 1.59$

(e) $p > .11$ (two-tailed), so unless $\alpha > .11$, reject H_1.

EXERCISES 10.5 (pp. 441–447)

1.

	Test Statistic	Distribution of Test Statistic	Value of Test Statistic	p
(a)	$\dfrac{\overline{X}_1 - \overline{X}_2}{\sqrt{\hat{\sigma}^2_{\text{POOLED}}\left(\dfrac{N_1 + N_2}{N_1 N_2}\right)}}$	$t(30)$	2.040	$< .05$ ($\doteq .025$)
(b)	$\dfrac{\overline{X}_1 - \overline{X}_2}{\sqrt{\hat{\sigma}^2_{\text{POOLED}}\left(\dfrac{N_1 + N_2}{N_1 N_2}\right)}}$	$t(12)$	-2.685	$< .01$
(c)	$\dfrac{\overline{X}_1 - \overline{X}_2}{\sqrt{\dfrac{\hat{\sigma}^2_1}{N_1} + \dfrac{\hat{\sigma}^2_2}{N_2}}}$	$t(22)$	-2.079	$< .05$ (2-tailed)
(d)	$\dfrac{\overline{X}_1 - \overline{X}_2}{\sqrt{\dfrac{\hat{\sigma}^2_1}{N_1} + \dfrac{\hat{\sigma}^2_2}{N_2}}}$	$t(10)$	.256	$> .40$
(e)	$\dfrac{\overline{X}_1 - \overline{X}_2}{\sqrt{\dfrac{\hat{\sigma}^2_1}{N_1} + \dfrac{\hat{\sigma}^2_2}{N_2}}}$	$t(38)$ $\nu_W \cong 39$ $N_1 + N_2 - 2 = 38$	-2.423	$< .01$
(f)	$\dfrac{\overline{X}_1 - \overline{X}_2}{\sqrt{\dfrac{\hat{\sigma}^2_1}{N_1} + \dfrac{\hat{\sigma}^2_2}{N_2}}}$	$t(11)$ $\nu_W \cong 11.67$	1.350	$> .10$ (2-tailed)

3. (a) The farmer will have to make at least $75.00 more per cow. At $6.00 for every hundred pounds of milk, Line 2 will have to produce

$$\frac{75}{6} \times 100 = 1{,}250 \text{ lbs}$$

more than Line 1. So, $\delta = 1{,}250$.

(b) Since $\sigma_1 = \sigma_2$, $N_1 = N_2 =$

$$\sigma(\sigma + \sigma)\left(\frac{z_\beta - z_\alpha}{\delta}\right)^2 = 2{,}500(2{,}500 + 2{,}500)\left(\frac{-1.64 - (1.64)}{1{,}250}\right)^2$$

$$\cong 86$$

5. (a) If μ is the population mean for all psychology graduates from this institution, then

$$H_0: \mu = 490$$

$$H_1: \mu = 505$$

(Since we have two simple hypotheses, it doesn't matter from a statistical point of view which is the test and which is the alternative. From a philosophical point of view, we test the hypothesis that the experimenter hopes to reject.)

(b) The test statistic is

$$Z_{\bar{X}} = \frac{\bar{X} - \mu_0}{\sqrt{\dfrac{\hat{\sigma}^2}{N}}} = \frac{\bar{X} - \mu_0}{\sqrt{\dfrac{s^2\left(\dfrac{N}{N-1}\right)}{N}}}$$

(c) The critical region is $z \geq z_\alpha$, where $\mathbb{P}(z_\alpha) = 1 - \alpha$.

(d) The value of the test statistic is

$$\frac{491.4 - 490}{\sqrt{\dfrac{(87.66)^2\left(\dfrac{50}{49}\right)}{50}}} \doteq \frac{1.4}{12.52} \doteq .11$$

(e) The alternative hypothesis is rejected.

7. (a) If μ_H is the population mean for high Mach IV subjects and μ_L is the population mean for low Mach IV subjects, then for both games

$$H_0: \mu_H - \mu_L = 0$$

$$H_1: \mu_H - \mu_L > 0$$

(b) Since variances are assumed equal, the test statistic in both tests is

$$t = \frac{\bar{X}_H - \bar{X}_L}{\sqrt{\hat{\sigma}^2_{\text{POOLED}}\left(\dfrac{N_H + N_L}{N_H N_L}\right)}}$$

(c) For Game 1

$$\hat{\sigma}^2_{\text{POOLED}} = \frac{24(19.9)\left(\dfrac{25}{24}\right) + 28(16.9)\left(\dfrac{29}{28}\right)}{52} \doteq 18.99$$

so $t \doteq 2.00$.

For Game 2

$$\hat{\sigma}^2_{\text{POOLED}} = \frac{24(15.2)\left(\dfrac{25}{24}\right) + 28(11.6)\left(\dfrac{29}{28}\right)}{52} \doteq 13.78$$

so $t \doteq -.88$.

(d) The critical region for both tests is in the upper tail of $t(52)$. There is no row in Table 4 of Appendix VIII corresponding to 52 degrees of freedom, but from the entries for $\nu = 40$ and for $\nu = 60$, it is safe to conclude that $p \cong .025$ for Game 1. For Game 2, the observed value of t is negative, and since our critical region is in the upper tail, $p > .50$.

(e) For Game 1, reject H_0. For Game 2, reject H_1. The experimenter's predictions are supported by both tests.

9. (a) This is a test of differences between correlated pairs. Let x_i be the demerits for wolf pup i on Day 1, let y_i be the demerits for wolf pup i on Day 10, and let $d_i = x_i - y_i$. Then, the hypotheses are

$$H_0: \mu_D = 0$$

$$H_1: \mu_D > 0$$

(b) The appropriate test statistic is

$$t = \frac{\bar{d} - \mu_0}{s/\sqrt{N-1}}$$

(c)

$$t = \frac{-.143 - 0}{1.245/\sqrt{6}} \doteq -.281$$

(d) Since the value of the test statistic is negative and the critical region must lie in the upper tail, it is immediately obvious that $p > .5$, and no further work is necessary. But the mechanics of determining the p-value are nevertheless instructive. The distribution of the test statistic under H_0 is $t(6)$. The probability that $t \geq -.281$ is the same as the probability that $t \leq .281$. In $t(6)$ we find that $\mathbb{P}(.265) = .60$, so the probability of observing a value less than or equal to .281 must be greater than .60. Therefore $p > .60$.

(e) Reject the alternative hypothesis.

11. (a) If μ_H is the population mean of hand-reared wolves and μ_M is the population mean of mother-reared wolves, then

$$H_0: \mu_H - \mu_M = 0$$

$$H_1: \mu_H - \mu_M < 0$$

(b) There is no reason to assume in advance of collecting the data that the two population variances are equal, and the sample sizes are unequal, so the test statistic is

$$t = \frac{\bar{X}_H - \bar{X}_M}{\sqrt{\dfrac{\hat{\sigma}_H^2}{N_H} + \dfrac{\hat{\sigma}_M^2}{N_M}}}$$

(c) $\bar{x}_H \doteq 117.86$ $\quad$ $s_H^2 \doteq 927.55$
$\bar{x}_M \doteq 191.25$ $\quad$ $s_M^2 \doteq 604.69$

so

$$t = \frac{117.86 - 191.25}{\sqrt{\dfrac{927.55\left(\dfrac{7}{6}\right)}{7} + \dfrac{604.69\left(\dfrac{4}{3}\right)}{4}}} \doteq -3.9$$

(d) To determine the p-value, we first determine degrees of freedom for the test statistic.

$$\nu_W = \frac{\left[\left(\hat{\sigma}_H^2/N_H\right) + \left(\hat{\sigma}_M^2/N_M\right)\right]^2}{\dfrac{\left(\hat{\sigma}_H^2/N_H\right)^2}{N_H + 1} + \dfrac{\left(\hat{\sigma}_M^2/N_M\right)^2}{N_M + 1}} - 2$$

$$= \frac{\left[(1,082.14/7) + (806.25/4)\right]^2}{\dfrac{(1,082.14/7)^2}{8} + \dfrac{(806.25/4)^2}{5}} - 2 \doteq 11.41 - 2 \approx 9$$

In the distribution of $t(9)$ we find $\mathbb{P}(3.25) = .995$, so $P(t \geq 3.25) = P(t \leq -3.25) = 1 - .995 = .005$. Therefore, $P(t \leq -3.9) < .005$, and the p-value would be reported as $p < .005$.

(e) Reject the test hypothesis.

13. (a) The experimenters hypothesize that wolves restrained by harness show more improvement than wolves restrained by choke chain. Let D_1 be demerits on Day 1 minus demerits on Day 10 for the harness group and let D_2 be demerits on Day 1 minus demerits on Day 10 for the choke-chain group. If the population means for D_1 and D_2 are denoted μ_{D1} and μ_{D2}, respectively, then the hypotheses are

$$H_0: \mu_{D1} - \mu_{D2} = 0$$

$$H_1: \mu_{D1} - \mu_{D2} > 0$$

(b) It is not assumed that variances are equal and $N_1 \neq N_2$, so the test statistic is

$$t = \frac{\bar{D}_1 - \bar{D}_2}{\sqrt{\dfrac{\hat{\sigma}_{D1}^2}{N_1} + \dfrac{\hat{\sigma}_{D2}^2}{N_2}}}$$

where $\hat{\sigma}_{D1}^2$ is the variance estimate for D_1 and $\hat{\sigma}_{D2}^2$ is the variance estimate for D_2.

(c) $\bar{D}_1 = -.66$ $\quad$ $\hat{\sigma}_{D1}^2 \doteq .333$
$\bar{D}_2 = .25$ $\quad$ $\hat{\sigma}_{D2}^2 \doteq 2.92$

so

$$t = \frac{-.66 - .25}{\sqrt{\dfrac{.333}{3} + \dfrac{2.92}{4}}} \doteq -.99$$

(d) Ordinarily, degrees of freedom ν would be estimated by either Welch's method or the Behren's-Fisher method, but the rejection region is in the *upper* tail of $t(\nu)$ and the calculated value of the test statistic is *negative*, so the exercise is academic; the harness group showed *less* improvement than the choke-chain group, so

(e) the alternative hypothesis is rejected.

15. (a) Let p_1 be the proportion families in Texas with high risk of stroke and let p_2 be the proportion of families in Utah with high risk of stroke.

$$H_0: p_1 - p_2 = 0$$

$$H_1: p_1 - p_2 > 0$$

(b) The appropriate test statistic is

$$Z_{\hat{p}_1 - \hat{p}_2} = \frac{\dfrac{X_1}{N_1} - \dfrac{X_2}{N_2}}{\sqrt{\hat{p}\hat{q}\left(\dfrac{N_1 + N_2}{N_1 N_2}\right)}}$$

which for samples this large will be approximately distributed $N(0, 1)$

(c) $\mathbb{P}(3.70) \doteq .9999$, so the critical region is $z \geq 3.70$.

(d) The value of the test statistic is approximately 4.76.

(e) Reject H_0. (f) .000003

17. (a) If p_1 is the proportion of hyenas' prey consumption that hyenas kill and p_2 is the proportion of lions' prey consumption that lions kill,

$$H_0: p_1 = p_2$$

$$H_1: p_1 < p_2$$

(b) The appropriate test statistic is

$$Z_{\hat{p}_1 - \hat{p}_2} = \frac{\dfrac{X_1}{N_1} - \dfrac{X_2}{N_2}}{\sqrt{\hat{p}\hat{q}\left(\dfrac{N_1 + N_2}{N_1 N_2}\right)}}$$

which for samples this large will be approximately distributed $N(0, 1)$. If the reader is alert, he or she will note that $\hat{p}_1$ is *greater* than $\hat{p}_2$, i.e., that hyenas in this sample killed a *higher* proportion of their prey than did the plains lions, so the p-value of the test statistic will obviously be greater than .50, and there is no need to pursue the problem further.

(c) The critical region is $z \leq z_\alpha$, where $\mathbb{P}(z_\alpha) = \alpha$.

(d) The value of the test statistic is approximately .08.

(e) $p > .53$, so H_1 is rejected.

EXERCISES 11.1 (pp. 467–470)

1. (a) $\mathbb{P} \doteq .025$
   (b) $\mathbb{P} \doteq .75$
   (c) $\mathbb{P} \doteq .01$
   (d) $.01 < \mathbb{P} < .025$
   (e) $\mathbb{P} \doteq .99$

3. (a) $1 - \mathbb{P} \doteq .05$
   (b) $1 - \mathbb{P} \doteq .90$
   (c) $1 - \mathbb{P} \doteq .975$
   (d) $.01 < 1 - \mathbb{P} < .025$
   (e) $.25 < 1 - \mathbb{P} < .50$

5. (a) Critical region is all values greater than or equal to 23.7 in $\chi^2(14)$. The test statistic equals $(25.71)(14)/15 \doteq 24$, so reject H_0.
   (b) Critical region is all values less than or equal to 3.33 in $\chi^2(9)$. The test statistic equals $(51.11)(9)/144 \doteq 3.19$, so reject H_0.
   (c) Critical region is all values less than or equal to 14.3 in $\chi^2(29)$. The test statistic equals $(116.90)(29)/225 \doteq 15.07$, so reject H_1.
   (d) Critical region is all values less than or equal to 4.60 or greater than 32.8 in $\chi^2(15)$. The test statistic equals $(117.33)(15)/350 \doteq 5.03$, so reject H_1.

7. $\dfrac{27,000\left(\dfrac{20}{19}\right)(19)}{45,000} = 12 \qquad (.10 < p < .25)$

9. (a) $\dfrac{\hat{\sigma}^2(N-1)}{\sigma_0^2}$ (b) $\chi^2(6)$ (c) $\dfrac{67.333(6)}{30} \doteq 13.47$
   (d) $.025 < p < .05$ (e) Reject H_0 for $\alpha \geq .05$.

11. (a) $H_0: \sigma_{\text{MAL}}^2 = 6.55$
    $H_1: \sigma_{\text{MAL}}^2 \neq 6.55$

    (b) The appropriate test statistic is $\hat{\sigma}^2(N-1)/\sigma_0^2$.
    (c) Under the test hypothesis the statistic is distributed as $\chi^2(3)$, so reject H_0 if the statistic is less than (or equal to) .216 or if it is greater than (or equal to) 9.35.
    (d) The value of the test statistic is

    $$\dfrac{1.6667(3)}{6.55} \doteq .763$$

    (e) Rejects H_1.

13. For $\nu = 6$, $\mathbb{P}(1.24) \doteq .025$ and $1 - \mathbb{P}(14.4) \doteq .025$. Therefore, $\mathbb{C} = .95$ that

    $$\dfrac{\hat{\sigma}^2(6)}{14.4} \leq \sigma^2 \leq \dfrac{\hat{\sigma}^2(6)}{1.24}$$

    $\hat{\sigma}^2 \doteq 67.33$, so the confidence limits are approximately 28.1 and 325.8.

EXERCISES 11.2 (pp. 474–478)

1. (a) .50   (b) .95   (c) .005   (d) .05   (e) .01
   (f) .90   (g) .05   (h) .01   (i) .95   (j) .02

3.

(ν_1, ν_2)	$(1, 5)$	$(10, 5)$	$(20, 5)$	$(50, 5)$	$(100, 5)$
$E(X)$	1.67	1.67	1.67	1.67	1.67
$Mdn.$	.53	1.07	1.11	1.13	1.14
Difference	1.14	.60	.56	.54	.53

(a) It is positively skewed.
(b) Skew decreases as ν_1 increases.

5. (a) 1.833   .10
(b) 24   1
(c) .05   15
(d) 5.841   1
(e) 3.551   12.6
$t_i^2, t_i, \underline{1}, \nu$

7. (a) .05
(b) .05 (approx.)
(c) .025
(d) .025
(e) ν_2, ν_1

9. (a) $p < .05$, reject H_0
(b) $p < .01$, reject H_0
(c) $p > .05$ (two-tailed), reject H_1
(d) $p < .01$ (two-tailed), reject H_0
(e) $p > .025$, reject H_1

11. (a) Let σ_M^2 be the variance for mother-reared wolves and let σ_H^2 be the variance for hand-reared wolves. Then,

$$H_0: \sigma_M^2 = \sigma_H^2$$

$$H_1: \sigma_M^2 > \sigma_H^2 \quad (\text{or } \sigma_H^2 < \sigma_M^2)$$

(b) $\dfrac{\hat{\sigma}_M^2}{\hat{\sigma}_H^2} \quad \left(\text{or } \dfrac{\hat{\sigma}_H^2}{\hat{\sigma}_M^2}\right)$

(c) The critical region is all values greater than or equal to F_b where $\mathbb{P}(F_b) = 1 - \alpha$ in $F(3, 6)$.

(d) $\dfrac{\hat{\sigma}_M^2}{\hat{\sigma}_H^2} \doteq \dfrac{2775}{278.57} \doteq 9.96$

(e) In $F(3, 6)$, $\mathbb{P}(9.96) > .99$, so $p < .01$. If $\alpha \geq .01$, reject H_0.

13. (a) $H_0: \sigma_{BA}^2 = \sigma_{BS}^2$

$$H_1: \sigma_{BA}^2 > \sigma_{BS}^2 \quad (\text{or } \sigma_{BS}^2 < \sigma_{BA}^2)$$

(b) $\dfrac{\hat{\sigma}_{BA}^2}{\hat{\sigma}_{BS}^2} \quad \left(\text{or } \dfrac{\hat{\sigma}_{BS}^2}{\hat{\sigma}_{BA}^2}\right)$

(c) The critical region is all values greater than or equal to F_b where $\mathbb{P}(F_b) = 1 - \alpha$ in $F(19, 29)$.

(d) $\dfrac{\hat{\sigma}_{BA}^2}{\hat{\sigma}_{BS}^2} \doteq \dfrac{8,394.75}{6,906.94} \doteq 1.22$

(e) There is no table for $F(19, 29)$ in Table 6 of Appendix VIII, but in $F(20, 30)$ we find $p > .25$. For any significance level α, critical values become *more* extreme as degrees of freedom decrease, so the p-value under $F(19, 29)$ would be even greater than the p-value under $F(20, 30)$. The alternative hypothesis should be rejected.

EXERCISES 12.1 (pp. 487–488)

1. $N = 22$ **3.** $N_4 = 6$ **5.** x_{14}

7. $\sum_i^3 x_{i1} + \sum_i^4 x_{i2} + \sum_i^4 x_{i3} = 18 + 101 + 79 = 198$

9. $\bar{x}_3 = 19.75$

11. $\sum_j^J N_j \bar{x}_j = 3(6) + 4(25.25) + 4(19.75) + 6(15) + 5(18.4) = 380$

13. $\dfrac{\sum_i^6 x_{i4}}{6}$ or $\dfrac{\sum_{i=1}^6 x_{i4}}{6}$ **15.** $J = 4$ **17.** $x_{52} = 29$

19. $\sum_i^{N_2} x_{i2} = 29 + 22 + 21 + 27 + 29 + 32 = 160$ **21.** $\bar{x} \doteq 20.20833$

23. $\sum_j^J \sum_i^{N_j} x_{ij} = 94 + 160 + 97 + 134 = 485$ **25.** $\sum_i^N x_i = 485$

27. $\sum_j^J \sum_i^{N_j} 14.3 x_{ij} = 14.3 \sum_j^J \sum_i^{N_j} x_{ij} = 14.3(485) = 6{,}935.5$

EXERCISES 12.2 (pp. 495–498)

1. For Data Set 1:
(a) $s^2 \doteq 10.89$
(b) $\bar{x}_1 = 1$, $\bar{x}_2 = 6$, $\bar{x}_3 = 9$; $s_{\bar{X}}^2 \doteq 10.89$
(c) $s_1^2 = 0$, $s_2^2 = 0$, $s_3^2 = 0$
(d) There is no variability within any of the groups, but there is variability among the group means, so all of the variability tapped by the grand variance s^2 is due to treatment effects.

For Data Set 2:
(a) $s^2 \doteq 10.89$
(b) $\bar{x}_1 \doteq 5.33$, $\bar{x}_2 \doteq 5.33$, $\bar{x}_3 \doteq 5.33$; $s_{\bar{X}}^2 = 0$
(c) $s_1^2 \doteq 10.89$, $s_2^2 \doteq 10.89$, $s_3^2 \doteq 10.89$

(d) There is no variability among group means, but there is variability within each of the groups. Therefore all of the variability tapped by the grand variance s^2 is due to error.

For Data Set 3:

(a) $s^2 \doteq 10.89$

(b) $\bar{x}_1 = 5$, $\bar{x}_2 = 3.5$, $\bar{x}_3 = 7.5$; $s_{\bar{X}}^2 \doteq 2.722$

(c) $s_1^2 = 16$, $s_2^2 = 6.25$, $s_3^2 = 2.25$

(d) There is variability within each group and there is variability among the group means, so the variability tapped by the grand variance s^2 is due to both error and, possibly, treatment effects.

3. (a) 6, 9, 12   (b) N_g (est. $\sigma_{\bar{X}}^2$) = 5(9) = 45

(c) 10, 22.5, 40   (d) 24.17 (approx.)

(e) No. The estimate based on means (b) taps both error and treatment effects. The pooled estimate (d) measures only variability within groups and therefore taps only error. The difference between (b) and (d) suggests the presence of treatment effects.

5. (a) 8.5, 9.5, 11.5   (b) N_g (est. $\sigma_{\bar{X}}^2$) = 6($2\frac{1}{3}$) = 14

(c) 14, 14, 14   (d) 14

(e) Yes. The estimate based on means (b) taps both error and treatment effects. The pooled estimate (d) measures only variability within groups and therefore taps only error. Since (b) and (d) are equal, there can be no treatment effects.

EXERCISES 12.3 (pp. 516–523)

1. (a) $MS_A = 45$. They are equal.   (b) $MS_W \doteq 24.17$. They are equal.

(c) $F(2, 12)$ (d) $MS_A / MS_W \doteq 1.86$

(e) $.75 < \mathbb{P}(1.86) < .90$

(f) The rejection region for analysis of variance always lies in the upper tail of $F(\nu_1, \nu_2)$, so the probability p of making a Type I error is $1 - \mathbb{P}(MS_A/MS_W)$. Therefore, $.10 < p < .25$. This would ordinarily be reported $p > .10$.

3. (a) H_0: $\mu_1 = \mu_2 = \mu_3$

H_1: H_0 incorrect

(b) $$\frac{MS_A}{MS_W} = \frac{SS_A/(J-1)}{SS_W/(N-J)} = \frac{\sum\limits_{j}^{J} N_j (\bar{x}_j - \bar{x})^2/(J-1)}{\sum\limits_{j}^{J} \sum\limits_{i}^{N_j} (x_{ij} - \bar{x}_j)^2/(N-J)}$$

(c) 1.0   (d) larger

(e) The denominator (MS_W) taps *only* error. The numerator (MS_A) taps *both* error *and* systematic differences produced by the different experimental treatments. If the test hypothesis is *incorrect* and there really *are* differences among the three treatment populations, this should inflate MS_A, but it would have no effect on MS_W.

(f) If $MS_A/MS_W \geq 3.89$, reject H_0. Otherwise reject H_1.

(g) By the definitional formulae,

$$SS_A = \sum_j^J N_j(\bar{x}_j - \bar{x})^2 = 5(4-4)^2 + 5(2-4)^2 + 5(6-4)^2$$

$$= 5(0) + 5(4) + 5(4) = 40$$

$$SS_W = \sum_j^J \sum_i^{N_j} (x_{ij} - \bar{x}_j)^2$$

$$= \sum_i^{N_1} (x_{i1} - \bar{x}_1)^2 + \sum_i^{N_2} (x_{i2} - \bar{x}_2)^2 + \sum_i^{N_3} (x_{i3} - \bar{x}_3)^2$$

$$= (10-4)^2 + (0-4)^2 + (8-4)^2 + (1-4)^2 + (1-4)^2$$

$$+ (4-2)^2 + (0-2)^2 + (5-2)^2 + (0-2)^2 + (1-2)^2$$

$$+ (10-6)^2 + (0-6)^2 + (9-6)^2 + (1-6)^2 + (10-6)^2$$

$$= 86 + 22 + 102 = 210$$

By the computational formulae,

$$SS_A = \sum N_j \bar{x}_j^2 - N\bar{x}^2 = 280 - 240 = 40$$

$$SS_W = \sum\sum x^2 - \sum N_j \bar{x}_j^2 = 490 - 280 = 210$$

$$MS_A = SS_A/2 = 20 \qquad \frac{MS_A}{MS_W} \doteq 1.14$$
$$MS_W = SS_W/12 = 17.5$$

(h) Reject H_1.

5. (a) $H_0: \mu_1 = \mu_2 = \mu_3$ (b) $\dfrac{MS_A}{MS_W}$ (c) $F(2, 15)$
     $H_1: H_0$ incorrect

(d) $SS_A = \sum N_j \bar{x}_j^2 - N\bar{x}^2 = 70{,}350 - 68{,}450 = 1900$
    $SS_W = \sum\sum x^2 - \sum N_j \bar{x}_j^2 = 72{,}372 - 70{,}350 = 2{,}022$
    $MS_A = SS_A/2 = 950$
    $MS_W = SS_W/15 = 134.8$ $\dfrac{MS_A}{MS_W} \doteq 7.05$

(e) $.005 < p < .01$

(f) Ordinarily, H_0 would be rejected (unless there is some reason to set $\alpha < .005$).

7. (a) $H_0: \mu_1 = \mu_2 = \mu_3 = \mu_4$ (b) $\dfrac{MS_A}{MS_W}$ (c) $F(3, 11)$

(d) $SS_A = \sum N_j \bar{x}_j^2 - N\bar{x}^2 \doteq 3{,}863{,}526.5 - 3{,}316{,}320.6$

$$= 547{,}206.0$$

$$SS_W = \sum\sum x^2 - \sum N_j \bar{x}_j^2 = 4{,}193{,}651 - 3{,}863{,}526.5$$

$$= 330{,}124.5$$

$$MS_A = SS_A/3 \doteq 182{,}402.0$$
$$MS_W = SS_W/11 \doteq 30{,}011.3 \qquad \frac{MS_A}{MS_W} \doteq 6.08$$

(e) $.01 < p < .025$

(f) The test hypothesis will be rejected for $\alpha \geq .025$.

9. (a) H_0: $\mu_1 = \mu_2 = \mu_3$ (b) $\dfrac{MS_A}{MS_W}$ (c) $F(2, 9)$

(d) $SS_A = \sum N_j \bar{x}_j^2 - N\bar{x}^2 = 52{,}548 - 52{,}272 = 276$

$SS_W = \sum\sum x^2 - \sum N_j \bar{x}_j^2 = 52{,}850 - 52{,}548 = 302$

$MS_A = SS_A/2 = 138$ $\dfrac{MS_A}{MS_W} \doteq 4.11$

$MS_W = SS_W/9 \doteq 33.56$

(e) $.05 < p < .10$

(f) Ordinarily, H_1 would be rejected (unless there is some reason to set $\alpha \geq .10$).

11. (a) H_0: $\mu_1 = \mu_2 = \mu_3 = \mu_4 = \mu_5$ (b) $\dfrac{MS_A}{MS_W}$ (c) $F(4, 100)$

H_1: H_0 incorrect

(d) $SS_A = \sum N_j \bar{x}_j^2 - N\bar{x}^2 = 1{,}478.54 - 1{,}254.94 \doteq 223.60$

$SS_W = \sum\sum x^2 - \sum N_j \bar{x}_j^2 = 2{,}119 - 1{,}478.54 \doteq 640.46$

$MS_A = SS_A/4 \doteq 55.90$ $\dfrac{MS_A}{MS_W} \doteq 8.73$

$MS_W = SS_W/100 \doteq 6.40$

(e) There is no table for $\nu_2 = 100$, but $p < .0005$ in $F(4, 60)$, so $p < .0005$ in $F(4, 100)$.

(f) Reject H_0.

13. (a) H_0: $\mu_1 = \mu_2 = \mu_3$ (b) $\dfrac{MS_A}{MS_W}$ (c) $F(2, 97)$

H_1: H_0 incorrect

(d) SS_A must be calculated using the definitional formula,

$$SS_A = \sum_{}^{J} N_j \left(\bar{x}_j - \bar{x} \right)^2$$

for which we must first calculate the grand mean,

$$\bar{x} = \frac{\overset{N_1}{\sum} x_{i1} + \overset{N_2}{\sum} x_{i2} + \overset{N_3}{\sum} x_{i3}}{N} = \frac{N_1 \bar{x}_1 + N_2 \bar{x}_2 + N_3 \bar{x}_3}{N}$$

$$= \frac{20(56) + 8(52) + 72(49)}{100} = 50.64$$

Therefore,

$$SS_A = 20(56 - 50.64)^2 + 8(52 - 50.64)^2 + 72(49 - 50.64)^2$$

$$\doteq 783.04$$

and

$$MS_A = \frac{SS_A}{2} \doteq 391.52$$

$$MS_W = \hat{\sigma}^2_{\text{POOLED}} = \frac{(146.41)(19) + (118.81)(7) + (110.25)(71)}{97}$$

$$\doteq 117.95$$

$$\frac{MS_A}{MS_W} \doteq 3.32$$

(e) There is no table for $F(2, 97)$, but $p < .05$ in $F(2, 60)$, so $p < .05$ in $F(2, 97)$.

(f) Reject H_0.

EXERCISES 12.4 (pp. 528–532)

1. (a) $\psi_1 = (1/2)\mu_1 + (1/2)\mu_2 + (-1)\mu_3 + (0)\mu_4$
$\psi_2 = (1/2)\mu_1 + (-1)\mu_2 + (1/2)\mu_3 + (0)\mu_4$
$\psi_3 = (1/3)\mu_1 + (1/3)\mu_2 + (1/3)\mu_3 + (-1)\mu_4$

(b) $H_0: \psi_1 = 0 \qquad H_0: \psi_2 = 0 \qquad H_0: \psi_3 = 0$
$H_1: \psi_1 \neq 0 \qquad H_1: \psi_2 \neq 0 \qquad H_1: \psi_3 \neq 0$

(c) $$\frac{\left(\displaystyle\sum_j^J c_j \bar{x}_j \right)^2}{(J-1) MS_W \displaystyle\sum_j^J \frac{c_j^2}{N_j}}$$

(d) $F(3, 11)$

(e) For $\alpha = .05$, the critical value is F_b where $\mathbb{P}(F_b) \doteq .95$ in $F(3, 11)$, so $F_b = 3.59$.

(f) From the answer to 12.3.7,

$$MS_W \doteq 30{,}011.3$$

so for ψ_1. Scheffé's statistic is

$$\frac{\left(\dfrac{1}{2}(247.75) + \dfrac{1}{2}(367.25) + (-1)(545) \right)^2}{(3)30{,}011.3\left(\dfrac{.25}{4} + \dfrac{.25}{4} + \dfrac{1}{3} \right)} \doteq \frac{56{,}406.25}{41{,}265.54} \doteq 1.37$$

For ψ_2, Scheffé's statistic is

$$\frac{\left(\frac{1}{2}(247.75) + (-1)(367.25) + \frac{1}{2}(545)\right)^2}{(3)30{,}011.3\left(\dfrac{.25}{4} + \dfrac{1}{4} + \dfrac{.25}{3}\right)} \doteq \frac{848.27}{35{,}638.42} \doteq 0.024$$

For ψ_3, Scheffé's statistic is

$$\frac{\left(\frac{1}{3}(247.75) + \frac{1}{3}(367.25) + \frac{1}{3}(545) + (-1)(739.5)\right)^2}{(3)30{,}011.3\left(\dfrac{1/9}{4} + \dfrac{1/9}{4} + \dfrac{1/9}{3} + \dfrac{1}{4}\right)}$$

$$\doteq \frac{124{,}491.36}{30{,}844.95} \doteq 4.04$$

(g) $1.37 < 3.59$, so H_1 is rejected for the first contrast. The mean of the pellets group is not significantly different from the average of the means for the two groups of wolves administered choke-chain punishment. $.024 < 3.59$, so H_1 is rejected for the second contrast. The mean of the mother-reared wolves is not significantly different from the average of the means for the two hand-reared groups of wolves. $4.04 > 3.59$, so H_0 is rejected for the third contrast. The mean for the Malamutes is significantly different from the average of the means for the wolves.

3. (a) $\psi = (1/3)\mu_1 + (1/3)\mu_2 + (1/3)\mu_3 + (-1)\mu_4$

(b) $H_0: \psi = 0$ (c) $\dfrac{\left(\sum\limits_{j}^{J} c_j \bar{x}_j\right)^2}{(J-1)MS_W \sum\limits_{j}^{J} \dfrac{c_j^2}{N_j}}$ (d) $F(3, 16)$
   $H_1: \psi \neq 0$

(e) For $\alpha = .05$, the critical value is F_b where $P(F_b) = .95$ in $F(3, 16)$, so $F_b = 3.29$.

(f) $MS_W = \hat{\sigma}^2_{\text{POOLED}} \doteq 41.36$ so Scheffé's statistic is

$$\frac{\left(\frac{1}{3}(35) + \frac{1}{3}(28.5) + \frac{1}{3}(29.33) + (-1)(18.67)\right)^2}{(3)41.36\left(\dfrac{1/9}{5} + \dfrac{1/9}{5} + \dfrac{1/9}{5} + \dfrac{1}{5}\right)} \doteq \frac{150.63}{33.09} \doteq 4.55$$

(g) $4.55 > 3.29$, so reject H_0. The conclusion implies that the backward conditioning group is significantly different from the average of the other three groups.

5. (a) $\psi_1 = (-1/4)\mu_1 + (-1/4)\mu_2 + (-1/4)\mu_3 + (1)\mu_4 + (-1/4)\mu_5$
   $\psi_2 = (-1/4)\mu_1 + (-1/4)\mu_2 + (-1/4)\mu_3 + (-1/4)\mu_4 + (1)\mu_5$
   $\psi_3 = (-1/3)\mu_1 + (-1/3)\mu_2 + (-1/3)\mu_3 + (1/2)\mu_4 + (1/2)\mu_5$

(b) $H_0: \psi_1 = 0$ $H_0: \psi_2 = 0$ $H_0: \psi_3 = 0$
     $H_1: \psi_1 > 0$ $H_1: \psi_2 < 0$ $H_1: \psi_3 \neq 0$

(c) $$\dfrac{\left(\displaystyle\sum_j^J c_j \bar{x}_j\right)^2}{(J-1)MS_W \displaystyle\sum_j^J \dfrac{c_j^2}{N_j}}$$ (d) $F(4, 100)$

(e) For $\alpha = .05$, the critical value is F_b where $P(F_b) = .95$ in $F(4, 100)$. There are no entries for $\nu_2 = 100$, so we have three choices: Use the entry for $\nu_2 = 60$, the next smaller tabled value of ν_2 (which gives $F_b = 2.53$); interpolate the entries for $\nu_2 = 60$ and $\nu_2 = 120$ (which gives $F_b \doteq 2.48$) or use the approximation function that accompanies Table 6 of Appendix VIII (which gives $F_b \cong 2.44$). Since sample sizes are less than 30, the approximation is probably the least desirable choice, so we'll use linear interpolation. Therefore, the critical value of Scheffé's statistic is approximately 2.48.

(f) From the answer to 12.3.11, $MS_W \doteq 6.40$. For ψ_1, Scheffé's statistic is therefore

$$\frac{\left(-\dfrac{1}{4}(3.52) - \dfrac{1}{4}(2.62) - \dfrac{1}{4}(3.19) + (1)(6.14) - \dfrac{1}{4}(1.81)\right)^2}{(4)6.40\left(\dfrac{.0625}{21} + \dfrac{.0625}{21} + \dfrac{.0625}{21} + \dfrac{1}{21} + \dfrac{.0625}{21}\right)} \doteq 7.39$$

For ψ_2, Scheffé's statistic is

$$\frac{\left(-\dfrac{1}{4}(3.52) - \dfrac{1}{4}(2.62) - \dfrac{1}{4}(3.19) - \dfrac{1}{4}(6.14) + (1)(1.81)\right)^2}{(4)6.40\left(\dfrac{.0625}{21} + \dfrac{.0625}{21} + \dfrac{.0625}{21} + \dfrac{.0625}{21} + \dfrac{1}{21}\right)} \doteq 2.78$$

For ψ_3, Scheffé's statistic is

$$\frac{\left(-\dfrac{1}{3}(3.52) - \dfrac{1}{3}(2.62) - \dfrac{1}{3}(3.19) + \dfrac{1}{2}(6.14) + \dfrac{1}{2}(1.81)\right)^2}{(4)6.40\left(\dfrac{1/9}{21} + \dfrac{1/9}{21} + \dfrac{1/9}{21} + \dfrac{1/4}{21} + \dfrac{1/4}{21}\right)} \doteq .74$$

(g) The first contrast supports the hypothesis that the basenjis performed better than the other breeds (reject H_0). The second contrast fails to support the hypothesis that the cockers performed more poorly than the other breeds (reject H_1). The third contrast supports the conclusion that wirehaired terriers, beagles, and shelties fall about midway between cocker spaniels and basenjis (reject H_1).

7. (a) $\psi_1 = (-1)\mu_1 + (1/2)\mu_2 + (1/2)\mu_3$
$\psi_2 = (1/2)\mu_1 + (1/2)\mu_2 + (-1)\mu_3$

(b) $H_0: \psi_1 = 0 \qquad H_0: \psi_2 = 0$
$H_1: \psi_1 \neq 0 \qquad H_1: \psi_2 \neq 0$

(c) $\dfrac{\left(\sum\limits_j^J c_j \bar{x}_j\right)^2}{(J-1)MS_W \sum\limits_j^J \dfrac{c_j^2}{N_j}}$

(d) $F(2, 97)$

(e) For $\alpha = .05$, the critical value is F_b where $P(F_b) = .95$ in $F(2, 97)$. There are no entries for $\nu_2 = 97$, so we have three choices: Use $\nu_2 = 60$, the entry for the next smaller tabled value of ν_2 (which gives $F_b = 3.15$); interpolate the entries for $\nu_2 = 60$ and $\nu_2 = 120$ (which gives $F_b \doteq 3.10$) or use the approximation function that accompanies Table 6 of Appendix VIII (which gives $F_b \cong 3.13$). Since two of the samples are small, the approximation is probably the least desirable choice, so we'll use the value 3.10 obtained by interpolation.

(f) From Exercise 12.3.13, $MS_W = \hat{\sigma}_{\text{POOLED}}^2 \doteq 117.95$. For ψ_1, Scheffé's statistic is therefore

$$\frac{\left((-1)(56) + \dfrac{1}{2}(52) + \dfrac{1}{2}(49)\right)^2}{(2)117.95\left(\dfrac{1}{20} + \dfrac{.25}{8} + \dfrac{.25}{72}\right)} \doteq \frac{30.25}{20.00} \doteq 1.51$$

For ψ_2, Scheffé's statistic is

$$\frac{\left(\dfrac{1}{2}(56) + \dfrac{1}{2}(52) + (-1)(49)\right)^2}{(2)117.95\left(\dfrac{.25}{20} + \dfrac{.25}{8} + \dfrac{1}{72}\right)} \doteq \frac{25}{13.60} \doteq 1.84$$

(g) Neither of the extreme group means differs from the average of the other two group means, which would appear to contradict the significant F test in Exercise 12.3.13. However, Scheffé's statistic for the (pairwise) contrast between extreme groups

$$\psi = (1/2)\mu_1 + (0)\mu_2 - (1/2)\mu_3$$

is approximately 3.26, which is significant at the .05 level. If the omnibus F test is rejected at significance level α, there will be at least one contrast that is significant at that level.

EXERCISES 13.1 (pp. 559–565)

1. (a) $H_0: \tau_j = 0$ for $j = 1, 2, 3, 4$ $H_0: \gamma_k = 0$ for $k = 1, 2, 3$
$H_1: H_0$ incorrect $H_1: H_0$ incorrect

$$H_0: \tau\gamma_{jk} = 0 \text{ for all } j, k$$
$$H_1: H_0 \text{ incorrect}$$

(b)
$$\frac{MS_{ROW}}{MS_{ERROR}} = \frac{\sum_j^R N_j(\bar{x}_j - \bar{x})^2/(R-1)}{\sum\sum\sum(x_{ijk} - \bar{x})^2/(N - RC)} : F(3, 24)$$

$$\frac{MS_{COL}}{MS_{ERROR}} = \frac{\sum_k^C N_k(\bar{x}_k - \bar{x})^2/(C-1)}{\sum\sum\sum(x_{ijk} - \bar{x})^2/(N - RC)} : F(2, 24)$$

$$\frac{MS_{INT}}{MS_{ERROR}} = \frac{\sum_j^R \sum_k^C \left(\bar{x}_{jk} - \bar{x}_j - \bar{x}_k + \bar{x}\right)^2/[(R-1)(C-1)]}{\sum\sum\sum(x_{ijk} - \bar{x})^2/(N - RC)} : F(6, 24)$$

(c) Since all cells have the same number of observations, mean squares can be calculated by the computational formulae.

Basic quantities: $A = 32{,}797$ $D \doteq 32{,}460.03$

$$B_C = 32{,}538.75 \qquad B_R \doteq 32{,}466.33 \qquad G = 32{,}651$$

$$\frac{MS_{ROW}}{MS_{ERROR}} = \frac{(B_R - D)/(R - 1)}{(A - G)/(N - RC)} \doteq \frac{2.10}{6.08} \doteq 0.35 \quad (p > .75)$$

$$\frac{MS_{COL}}{MS_{ERROR}} = \frac{(B_C - D)/(C - 1)}{(A - G)/(N - RC)} \doteq \frac{39.36}{6.08} \doteq 6.47 \quad (p < .01)$$

$$\frac{MS_{INT}}{MS_{ERROR}} = \frac{(G - B_C - B_R + D)/[(R - 1)(C - 1)]}{(A - G)/(N - RC)} \doteq \frac{17.66}{6.08}$$

$$\doteq 2.90 \quad (p < .05)$$

Column effects and interaction are significant.

3. (a) $H_0: \sigma_{\tau\gamma}^2 = 0$ $H_0: \sigma_\tau^2 = 0$ $H_0: \sigma_\gamma^2 = 0$
$H_1: \sigma_{\tau\gamma}^2 > 0$ $H_1: \sigma_\tau^2 > 0$ $H_1: \sigma_\gamma^2 > 0$

The interaction hypothesis should be tested first, because its result will determine whether the appropriate denominator to be used in the F tests for main effects is MS_{INT} or $MS_{\text{POOLED ERROR}}$.

(b) $\dfrac{MS_{\text{INT}}}{MS_{\text{ERROR}}} = \dfrac{\sum\limits_{j}^{R}\sum\limits_{k}^{C}\left(\bar{x}_{jk} - \bar{x}_j - \bar{x}_k + \bar{x}\right)^2/[(R-1)(C-1)]}{\sum\sum\sum\left(x_{ijk} - \bar{x}\right)^2/(N-RC)}$: $F(6, 24)$

(c) $\dfrac{MS_{\text{INT}}}{MS_{\text{ERROR}}} \doteq \dfrac{17.66}{6.08} \doteq 2.90 \quad (p < .05)$

(d) $\dfrac{MS_{\text{ROW}}}{MS_{\text{INT}}} = \dfrac{\sum\limits_{j}^{R} N_j\left(\bar{x}_j - \bar{x}\right)^2/(R-1)}{\sum\limits_{j}^{R}\sum\limits_{k}^{C}\left(\bar{x}_{jk} - \bar{x}_j - \bar{x}_k + \bar{x}\right)^2/[(R-1)(C-1)]}$: $F(3, 6)$

$\dfrac{MS_{\text{COL}}}{MS_{\text{INT}}} = \dfrac{\sum\limits_{k}^{C} N_k\left(\bar{x}_k - \bar{x}\right)^2/(C-1)}{\sum\limits_{j}^{R}\sum\limits_{k}^{C}\left(\bar{x}_{jk} - \bar{x}_j - \bar{x}_k + \bar{x}\right)^2/[(R-1)(C-1)]}$: $F(2, 6)$

(e) $\dfrac{MS_{\text{ROW}}}{MS_{\text{INT}}} \doteq \dfrac{2.10}{17.66} \doteq 0.12 \quad (p > .90)$

$\dfrac{MS_{\text{COL}}}{MS_{\text{INT}}} \doteq \dfrac{39.36}{17.66} \doteq 2.23 \quad (p > .10)$

Column treatments were significant in Exercise 12.5.1 but are not significant in this exercise. The denominator for the tests of main effects in this exercise is MS_{INT}, but in Exercise 12.5.1 the denominator for the tests of main effects was MS_{ERROR}. Mean squares for interaction is much larger ($p < .05$) than MS_{ERROR}, so the F ratios for both main effects are smaller, and $df_{\text{INT}} < df_{\text{ERROR}}$, so larger F ratios are required to achieve significance.

5. (a) H_0: $\tau_j = 0$ for $j = 1, 2, 3, 4$ H_0: $\gamma_k = 0$ for $k = 1, 2, 3, 4$
    H_1: H_0 incorrect H_1: H_0 incorrect

$$H_0: \tau\gamma_{jk} = 0 \text{ for all } j, k$$
$$H_1: H_0 \text{ incorrect}$$

(b) $\dfrac{MS_{\text{ROW}}}{MS_{\text{ERROR}}} = \dfrac{\sum\limits_{j}^{R} N_j\left(\bar{x}_j - \bar{x}\right)^2/(R-1)}{\sum\sum\sum\left(x_{ijk} - \bar{x}\right)^2/(N-RC)}$: $F(3, 32)$

$\dfrac{MS_{\text{COL}}}{MS_{\text{ERROR}}} = \dfrac{\sum\limits_{k}^{C} N_k\left(\bar{x}_k - \bar{x}\right)^2/(C-1)}{\sum\sum\sum\left(x_{ijk} - \bar{x}\right)^2/(N-RC)}$: $F(3, 32)$

$\dfrac{MS_{\text{INT}}}{MS_{\text{ERROR}}} = \dfrac{\sum\limits_{j}^{R}\sum\limits_{k}^{C}\left(\bar{x}_{jk} - \bar{x}_j - \bar{x}_k + \bar{x}\right)^2/[(R-1)(C-1)]}{\sum\sum\sum\left(x_{ijk} - \bar{x}\right)^2/(N-RC)}$: $F(9, 32)$

(c) Since all cells have the same number of observations, mean squares can be calculated by the computational formulae.

Basic quantities: $A \doteq 57.2801$ $D \doteq 56.9917$

$$B_C \doteq 57.0148 \qquad B_R \doteq 57.0798 \qquad G \doteq 57.1793$$

$$\frac{MS_{ROW}}{MS_{ERROR}} = \frac{(B_R - D)/(R - 1)}{(A - G)/(N - RC)} \doteq \frac{.0294}{.00315} \doteq 9.33$$

Since $p < .0005$ in $F(3, 30)$, it must be true that $p < .0005$ in $F(3, 32)$

$$\frac{MS_{COL}}{MS_{ERROR}} = \frac{(B_C - D)/(C - 1)}{(A - G)/(N - RC)} \doteq \frac{.0077}{.00315} \doteq 2.44$$

Since $p > .05$ in $F(3, 40)$, it must also be true that $p > .05$ in $F(3, 32)$

$$\frac{MS_{INT}}{MS_{ERROR}} = \frac{(G - B_C - B_R + D)/[(R - 1)(C - 1)]}{(A - G)/(N - RC)} \doteq \frac{.00850}{.00315} \doteq 2.70$$

Since $.001 < p < .025$ in $F(9, 30)$ and in $F(9, 40)$, it must be true that $.001 < p < .025$ in $F(3, 32)$. Therefore, $p < .025$.

(d) The method of original instruction has no effect on improvement, but remedial method influences improvement, as does the interaction of remedial method and original teaching.

7. (a) This is a random-effects experiment. The experimenters are interested in cumulative dosage *in general* and might have selected many other total dosages for study. Likewise, they are interested in intensity *in general* and might have administered the drug in any of several numbers of treatments.

(b) $H_0: \sigma^2_{\tau\gamma} = 0 \qquad H_0: \sigma^2_{\tau} = 0 \qquad H_0: \sigma^2_{\gamma} = 0$

The interaction hypothesis should be tested first, because its result will determine whether the appropriate denominator to be used in the F tests for main effects is MS_{INT} or $MS_{POOLED\ ERROR}$.

(c) $$\frac{MS_{INT}}{MS_{ERROR}} = \frac{\displaystyle\sum_j^R \sum_k^C \left(\bar{x}_{jk} - \bar{x}_j - \bar{x}_k + \bar{x}\right)^2/[(R - 1)(C - 1)]}{\sum\sum\sum \left(x_{ijk} - \bar{x}\right)^2/(N - RC)} : F(4, 36)$$

(d) Since all cells have the same number of observations, mean squares for all of the hypotheses can be calculated from the computational formulae.

Basic quantities: $A = 197.80$ $D \doteq 183.214$

$$B_C \doteq 184.123 \qquad B_R = 189.396 \qquad G = 190.348$$

$$MS_{INT} = (G - B_C - B_R + D)/[(R - 1)(C - 1)] \doteq .011$$

$$MS_{ERROR} = (A - G)/(N - RC) \doteq .207$$

The F ratio for interaction is less than 1.0 and cannot be significant.

(e) Since $MS_{INT} < MS_{ERROR}$, the sum of squares for error and sum of squares for interaction should be pooled to obtain the estimate of error variance for testing main effects. The test statistics are

$$\frac{MS_{ROW}}{MS_{POOLED\ ERROR}} : F(2, 40) \quad \text{and} \quad \frac{MS_{COL}}{MS_{POOLED\ ERROR}} : F(2, 40)$$

(f) $MS_{POOLED\ ERROR} = \dfrac{A - B_C - B_R + D}{(N - RC) + (R - 1)(C - 1)} \doteq \dfrac{7.495}{40} \doteq .187$

$$\frac{MS_{ROW}}{MS_{POOLED\ ERROR}} = \frac{(B_R - D)/(R - 1)}{MS_{POOLED\ ERROR}} \doteq \frac{3.091}{.187} \doteq 16.53 \quad (p < .0005)$$

$$\frac{MS_{COL}}{MS_{POOLED\ ERROR}} = \frac{(B_C - D)/(C - 1)}{MS_{POOLED\ ERROR}} \doteq \frac{.4545}{.187} \doteq 2.43 \quad (p > .10)$$

(g) The dosage intensity of cyclophosphamide has a significant effect on WBC but cumulative dosage does not.

9. (a) $H_0: \mu_1 = \mu_2 = \mu_3$

(b) $\dfrac{MS_{AT}}{MS_{RESIDUAL}} = \dfrac{\sum_j N(\bar{x}_j - \bar{x})^2 \big/ (J - 1)}{\left[\sum\sum (x_{ij} - \bar{x}_i)^2 - \sum_j N(\bar{x}_j - \bar{x})^2 \right] \big/ [(J - 1)(N - 1)]}$

If there are no differences among treatment groups, the test statistic is distributed $F(2, 18)$.

(c) Using the computational formulae,

$$A = 53,850 \qquad B_S \doteq 53,724.67 \qquad B_T = 53,498 \qquad D = 53,425.2$$

so

$$MS_{AT} = \frac{B_T - D}{J - 1} \doteq \frac{53,498 - 53,425.2}{2} \doteq 36.40$$

$$MS_{RESIDUAL} = \frac{A - B_S - B_T + D}{(N - 1)(J - 1)}$$

$$\doteq \frac{53,850 - 53,724.67 - 53,498 + 53,425.2}{(9)(2)}$$

$$\doteq 2.92$$

so

$$\frac{MS_{AT}}{MS_{RESIDUAL}} \doteq \frac{36.40}{2.92} \doteq 12.47 \quad (p < .001)$$

(d) There are significant differences across preexperimental attitudes, attitudes measured one hour after the experimental treatment, and attitudes measured six weeks after the experimental treatment.

EXERCISES 14.1 (pp. 580–582)

1. (a)

		X		
Y	0	1	2	P(y)
0	0	$\frac{1}{8}$	$\frac{1}{8}$	$\frac{1}{4}$
1	$\frac{1}{8}$	$\frac{1}{4}$	$\frac{1}{8}$	$\frac{1}{2}$
2	$\frac{1}{8}$	$\frac{1}{8}$	0	$\frac{1}{4}$
P(x)	$\frac{1}{4}$	$\frac{1}{2}$	$\frac{1}{4}$	

(b) X and Y are not independent since, for example, $P(x = 0)P(y = 2) = (1/16) \neq P(x = 0, y = 2) = 1/8$.

3. (a)

		X		
Y	2	4	6	P(y)
0	$\frac{1}{9}$	$\frac{1}{9}$	$\frac{1}{9}$	$\frac{1}{3}$
1	$\frac{2}{9}$	$\frac{2}{9}$	$\frac{2}{9}$	$\frac{2}{3}$
P(x)	$\frac{1}{3}$	$\frac{1}{3}$	$\frac{1}{3}$	

(b) X and Y are independent since each of the joint probabilities in the table is the product of the corresponding marginal probabilities.

5. (a)

		X		
Y	2	5	6	P(y)
5	$\frac{4}{48}$	$\frac{1}{48}$	$\frac{1}{48}$	$\frac{6}{48}$
9	$\frac{8}{48}$	$\frac{2}{48}$	$\frac{2}{48}$	$\frac{12}{48}$
11	$\frac{13}{48}$	$\frac{3}{48}$	$\frac{2}{48}$	$\frac{18}{48}$
17	$\frac{7}{48}$	$\frac{2}{48}$	$\frac{3}{48}$	$\frac{12}{48}$
P(x)	$\frac{32}{48}$	$\frac{8}{48}$	$\frac{8}{48}$	

(b) No, since, for example, $P(x = 2)P(y = 11) \neq P(x = 2, y = 11)$

(c)

$X\|(y = 9)$	$P(x\|y = 9)$	$Y\|(x = 5)$	$P(y\|x = 5)$
2	$\frac{2}{3}$	5	$\frac{1}{8}$
5	$\frac{1}{6}$	9	$\frac{1}{4}$
6	$\frac{1}{6}$	11	$\frac{3}{8}$
		17	$\frac{1}{4}$

7.

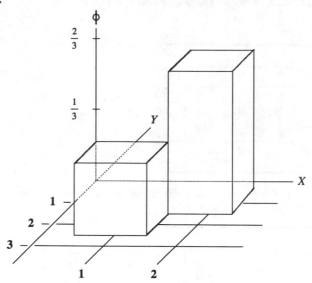

9. (a)

	X		
Y	2	4	6
1	$\dfrac{k}{2}$	$\dfrac{k}{4}$	$\dfrac{k}{6}$
2	$\dfrac{k}{4}$	$\dfrac{k}{8}$	$\dfrac{k}{12}$
3	$\dfrac{k}{6}$	$\dfrac{k}{12}$	$\dfrac{k}{18}$
5	$\dfrac{k}{10}$	$\dfrac{k}{20}$	$\dfrac{k}{30}$

(b) The joint probabilities in (a) must sum to 1, so $k = \dfrac{360}{671}$.

(c)

	X			
Y	2	4	6	$P(y)$
1	$\dfrac{180}{671}$	$\dfrac{90}{671}$	$\dfrac{60}{671}$	$\dfrac{330}{671}$
2	$\dfrac{90}{671}$	$\dfrac{45}{671}$	$\dfrac{30}{671}$	$\dfrac{165}{671}$
3	$\dfrac{60}{671}$	$\dfrac{30}{671}$	$\dfrac{20}{671}$	$\dfrac{110}{671}$
5	$\dfrac{36}{671}$	$\dfrac{18}{671}$	$\dfrac{12}{671}$	$\dfrac{66}{671}$
$P(x)$	$\dfrac{366}{671}$	$\dfrac{183}{671}$	$\dfrac{122}{671}$	

(d)

$X\|(y=3)$	$P(x\|y=3)$	$Y\|(x=6)$	$P(y\|x=6)$
2	$\dfrac{6}{11}$	1	$\dfrac{30}{61}$
4	$\dfrac{3}{11}$	2	$\dfrac{15}{61}$
6	$\dfrac{2}{11}$	3	$\dfrac{10}{61}$
		5	$\dfrac{6}{61}$

11. Since A and B choose their numbers independently, the probability of each pair of numbers $P(x, y)$ is the product of the marginal probabilities $P(x)P(y)$. Each pair of values tabled below is the probability of the pair of choices (x, y) and the monetary outcome of that pair for player A. Losses are tabled as negative values.

			X		
Y	1	2	3	4
1	0, \$0	$\dfrac{210}{2,401}, \$2$	$\dfrac{280}{2,401}, -\$3$	$\dfrac{196}{2,401}, \$2$
2	0, $-\$2$	$\dfrac{300}{2,401}, \$0$	$\dfrac{400}{2,401}, \$2$	$\dfrac{280}{2,401}, -\$3$
3	0, \$3	$\dfrac{225}{2,401}, -\$2$	$\dfrac{300}{2,401}, \$0$	$\dfrac{210}{2,401}, \$2$

The expected value is the sum of the outcomes times their probabilities:

$$-\$3\left(\frac{560}{2,401}\right) + (-\$2)\left(\frac{225}{2,401}\right) + \$0\left(\frac{600}{2,401}\right) + \$2\left(\frac{1,016}{2,401}\right)$$

$$= -\$\frac{2}{49} \doteq -\$.041$$

Player A expects to lose about 4 cents in every round of play.

13. The experiment is "flip six coins," and the random variable X is the number of heads. Since we are concerned with the distribution of Y when $x = 4$, the results of interest are the subset corresponding to the event "obtain 4 heads."

Result	Y	$Y\|(x=4)$	$P(y\|x=4)$
HHHHTT	4	4	$\frac{1}{5}$
HHHTHT	3		
HHHTTH	3	3	$\frac{2}{5}$
HHTHHT	2		
HHTHTH	2		
HHTTHH	2	2	$\frac{2}{5}$
HTHHHT	3		
HTHHTH	2		
HTHTHH	2		
HTTHHH	3		
THHHHT	4		
THHHTH	3		
THHTHH	2		
THTHHH	3		
TTHHHH	4		

15. In general,

$$\mu_{Y|x} = \mu_Y + \rho(\sigma_Y/\sigma_X)(x_j - \mu_X) \quad \text{and} \quad \sigma_{Y|x}^2 = \sigma_Y^2(1 - \rho^2)$$

Therefore, for $\mu_X = 12$, $\mu_Y = 25$, $\sigma_X^2 = 4$, $\sigma_Y^2 = 9$ and $\rho = 0.5$,
(a) $\mu_{Y|8} \doteq 22$ and $\sigma_{Y|8}^2 = 6.75$, so for $y_a = 18$, $z \doteq -1.54$; for $y_b = 22$, $z \doteq 0$.

$$P(18 \le y \le 22 \,|\, x = 8) \doteq P(-1.54 \le z \le 0) \doteq .4382$$

(b) $\mu_{Y|4} = 19$ and $\sigma_{Y|4}^2 = 6.75$, so for $y_a = 18$, $z \doteq -.38$; for $y_b = 22$, $z \doteq 1.15$

$$P(18 \le y \le 22 \,|\, x = 4) \doteq P(-.38 \le z \le 1.15) \doteq .5229$$

(c) $\mu_{Y|12} = 25$ and $\sigma_{Y|12}^2 = 6.75$, so for $y_a = 23$, $z \doteq -.77$; for $y_b = 27$, $z \doteq .77$

$$P(23 \le y \le 27 \,|\, x = 12) \doteq P(-.77 \le z \le .77) \doteq .5588$$

EXERCISES 14.2 (pp. 593–596)

1. (a) $t_r = \dfrac{r\sqrt{N-2}}{\sqrt{1-r^2}} \doteq \dfrac{.42(5.74)}{.908} \doteq 2.66$

$z_r = \dfrac{1}{2} \ln\left(\dfrac{1+.42}{1-.42}\right) \doteq .45$

(b) $t_r = \dfrac{r\sqrt{N-2}}{\sqrt{1-r^2}} \doteq \dfrac{-.78(3.61)}{.626} \doteq -4.50$

$z_r = \dfrac{1}{2} \ln\left(\dfrac{1+(-.78)}{1-(-.78)}\right) \doteq -1.05$

(c) $t_r = \dfrac{r\sqrt{N-2}}{\sqrt{1-r^2}} \doteq \dfrac{.73(2.83)}{.683} \doteq 3.02$

(d) $t_r = \dfrac{r\sqrt{N-2}}{\sqrt{1-r^2}} \doteq \dfrac{.73(3)}{.683} \doteq 3.21$

$z_r = \dfrac{1}{2} \ln\left(\dfrac{1+.73}{1-.73}\right) \doteq .93$

(e) $t_r = \dfrac{r\sqrt{N-2}}{\sqrt{1-r^2}} \doteq \dfrac{.91(2.45)}{.415} \doteq 5.37$

3. (a) $z_r = \dfrac{1}{2} \ln\left(\dfrac{1+.31}{1-.31}\right) \doteq .3205$

$$\dfrac{z_r}{\sqrt{1/(N-3)}} \doteq \dfrac{.3205}{\sqrt{1/29}} \doteq 1.73 \quad (p < .05) \qquad \text{Reject } H_0.$$

(b) $z_r = \dfrac{1}{2} \, ln\left(\dfrac{1 + .43}{1 - .43}\right) \doteq .4599$

$$\dfrac{z_r}{\sqrt{1/(N-3)}} \doteq \dfrac{.4599}{\sqrt{1/12}} \doteq 1.59 \quad (p > .05) \qquad \text{Reject } H_1.$$

(c) $z_r = \dfrac{1}{2} \, ln\left(\dfrac{1 + (-.56)}{1 - (-.56)}\right) \doteq -.6328$

$\zeta_0 = \dfrac{1}{2} \, ln\left(\dfrac{1 + (-.70)}{1 - (-.70)}\right) = -.8673$

$$\dfrac{z_r - \zeta_0}{\sqrt{1/(N-3)}} \doteq \dfrac{-.6328 + .8673}{\sqrt{1/100}} \doteq 2.35 \quad (p < .01) \qquad \text{Reject } H_0.$$

(d) $z_r = \dfrac{1}{2} \, ln\left(\dfrac{1 + .26}{1 - .26}\right) \doteq .2661$

$\zeta_0 = \dfrac{1}{2} \, ln\left(\dfrac{1 + .50}{1 - .50}\right) \doteq .5493$

$$\dfrac{z_r - \zeta_0}{\sqrt{1/(N-3)}} \doteq \dfrac{.2661 - .5493}{\sqrt{1/48}} \doteq -1.96 \quad (p \doteq .05 \text{ two-tailed})$$

Reject H_0.

(e) $z_r = \dfrac{1}{2} \, ln\left(\dfrac{1 + .53}{1 - .53}\right) \doteq .5901$

$$\dfrac{z_r}{\sqrt{1/(N-3)}} \doteq \dfrac{.5901}{\sqrt{1/8}} \doteq 1.67 \quad (p < .05) \qquad \text{Reject } H_0.$$

(f) $t_r = \dfrac{r\sqrt{N-2}}{\sqrt{1-r^2}} \doteq \dfrac{.53(2.83)}{.848} \doteq 1.77 \quad (p > .05) \qquad \text{Reject } H_1.$

(g) $z_r = \dfrac{1}{2} \, ln\left(\dfrac{1 + (-.23)}{1 - (-.23)}\right) \doteq -.2342$

$$\dfrac{z_r}{\sqrt{1/(N-3)}} \doteq \dfrac{-.2342}{\sqrt{1/17}} \doteq -.97 \quad (p > .16) \qquad \text{Reject } H_1.$$

(h) $t_r = \dfrac{r\sqrt{N-2}}{\sqrt{1-r^2}} \doteq \dfrac{.76(2)}{.650} \doteq 2.34 \quad (p > .05 \text{ two-tailed}) \qquad \text{Reject } H_1.$

5. (a) $r \doteq .547$

$t_r \doteq 1.85 \quad (p > .05) \qquad$ Reject H_1.

(b) $r \doteq -.590 \qquad z_r = \frac{1}{2} \ln\left(\frac{1 + (-.590)}{1 - (-.590)}\right) \doteq -.678$

$\rho_0 = -.70 \qquad \zeta_0 = \frac{1}{2} \ln\left(\frac{1 + (-.70)}{1 - (-.70)}\right) \doteq -.867$

$\dfrac{z_r - \zeta_0}{\sqrt{1/(N-3)}} \doteq .535 \quad (p \doteq .30)$

Reject H_1. Note that any *negative* correlation that satisfies the alternative hypothesis implies a weaker statistical relation than does the test hypothesis.

(c) $r \doteq .521 \qquad z_r = \frac{1}{2} \ln\left(\frac{1 + .521}{1 - .521}\right) \doteq .578$

$\rho_0 = .80 \qquad \zeta_0 = \frac{1}{2} \ln\left(\frac{1 + .80}{1 - .80}\right) \doteq 1.099$

$\dfrac{z_r - \zeta_0}{\sqrt{1/(N-3)}} \doteq -1.56 \quad (p > .05) \qquad$ Reject H_1.

7. Since $\eta^2 = \rho^2$, the hypotheses $H_0\colon \eta^2 = .50$ and $H_1\colon \eta^2 < .50$ are equivalent to

$H_0\colon \rho = \sqrt{.50} \doteq .707$

$H_1\colon \rho < \sqrt{.50} \doteq .707$

$r \doteq .569 \qquad z_r \doteq \frac{1}{2} \ln\left(\frac{1 + .569}{1 - .569}\right) \doteq .646$

$\rho_0 \doteq .707 \qquad \zeta_0 = \frac{1}{2} \ln\left(\frac{1 + .707}{1 - .707}\right) \doteq .881$

$\dfrac{z_r - \zeta_0}{\sqrt{1/(N-3)}} \doteq -1.42 \quad (p > .07) \qquad$ Reject H_1.

9. (a) For Data Set 1

$$\frac{s_Y^2 - s_{Y|x}^2}{s_Y^2} \doteq \frac{33.25 - 1.25}{33.25} \doteq .96$$

For Data Set 2

$$\frac{s_Y^2 - s_{Y|x}^2}{s_Y^2} \doteq \frac{6.05 - 1.25}{6.05} \doteq .79$$

There is a strong relationship between X and Y in both populations.

(b) $r_1^2 \doteq .96$ and $r_2^2 \doteq .12$

(c) The correlation ratio taps relatedness, but the correlation coefficient (and the coefficient of determination) tap only linear relatedness. The relationship of X and Y in Population 2 is curvilinear.

EXERCISES 14.3 (pp. 607–609)

1. The basic computational quantities are

$$N = 12 \qquad \sum x = 840 \qquad \sum x^2 = 88{,}200$$

$$\sum y = 204 \qquad \sum y^2 = 4{,}360 \qquad \sum xy = 18{,}720$$

$$\bar{y}_1 = 6 \qquad \bar{y}_2 = 14 \qquad \bar{y}_3 = 22 \qquad \bar{y}_4 = 26$$

$$N_1 = N_2 = N_3 = N_4 = 3$$

(a) $SS_{\text{ERROR}} = \sum y_j^2 - \sum N_j \bar{y}_j^2 = 4{,}360 - 4{,}176 = 184$

$$MS_{\text{ERROR}} = \frac{SS_{\text{ERROR}}}{N - J} = \frac{184}{8} = 23$$

$$b^2 = \left(\frac{N \sum xy - \sum x \sum y}{N \sum x^2 - \left(\sum x \right)^2} \right)^2 = \left(\frac{12(18{,}720) - (840)(204)}{12(88{,}200) - (840)^2} \right)^2$$

$$\doteq .0228$$

$$s_X^2 = \frac{\sum x^2}{N} - \bar{x}^2 = \frac{88{,}200}{12} - \left(\frac{840}{12} \right)^2 = 2{,}450$$

Since $MS_{\text{LINEAR}} = SS_{\text{LINEAR}}$,

$$MS_{\text{LINEAR}} = Nb^2 s_X^2 \doteq 12(.0228)(2{,}450) = 670.32$$

and

$$\frac{MS_{\text{LINEAR}}}{MS_{\text{ERROR}}} \doteq \frac{670.32}{23} \doteq 29.14$$

If $\beta = 0$, test statistic is distributed as $F(1, N - J)$. In Table 6 of Appendix VIII we find that $.999 < \mathbb{P}(29.14) < .9995$ in $F(1, 8)$, so $p < .001$ and H_0 is rejected for $\alpha \geq .001$. The data indicate a significant linear relationship between X and Y.

(b) $MS_{\text{DEV. FROM LINEAR}} = \dfrac{SS_{\text{TOTAL}} - SS_{\text{ERROR}} - SS_{\text{LINEAR}}}{J - 2}$

$$SS_{\text{TOTAL}} = \sum_{}^{N} y^2 - N\bar{y}^2 = 4{,}360 - 12(17)^2 = 892$$

so

$$MS_{\text{DEV. FROM LINEAR}} \doteq \frac{892 - 184 - 670.32}{2} \doteq 18.84$$

and from the answer to (a), $MS_{\text{ERROR}} = 23$. Therefore,

$$\frac{MS_{\text{DEV. FROM LINEAR}}}{MS_{\text{ERROR}}} \doteq \frac{18.84}{23} \doteq .819$$

If $\eta^2 - \rho^2 = 0$, test statistic is distributed as $F(J - 2, N - J)$. In Table 6 of Appendix VIII we find that $.50 < \mathbb{P}(.819) < .75$ in $F(2, 8)$, so $p > .25$, and H_1 is rejected for $\alpha \le .25$. Although the conditional means appear to lie on a curve, the nonlinear component of the relationship of X and Y is not significant at the .25 level.

3. The basic computational quantities are

$$N = 15 \qquad \sum x = 31.2 \qquad \sum x^2 = 68.52$$

$$\sum y = 318 \qquad \sum y^2 = 7{,}562 \qquad \sum xy = 616.5$$

$$\bar{y}_1 = 33 \qquad \bar{y}_2 = 23 \qquad \bar{y}_3 = 21 \qquad \bar{y}_4 = 14 \qquad \bar{y}_5 = 15$$

$$N_1 = N_2 = N_3 = N_4 = N_5 = 3$$

(a) $SS_{\text{ERROR}} = \sum y_j^2 - \sum N_j \bar{y}_j^2 = 7{,}562 - 7{,}440 = 122$

$$MS_{\text{ERROR}} = \frac{SS_{\text{ERROR}}}{N - J} = \frac{122}{10} = 12.2$$

$$b^2 = \left(\frac{N \sum xy - \sum x \sum y}{N \sum x^2 - \left(\sum x \right)^2} \right)^2 = \left(\frac{15(616.5) - (31.2)(318)}{15(68.52) - (31.2)^2} \right)^2$$

$$\doteq 153.78$$

$$s_X^2 = \frac{\sum x^2}{N} - \bar{x}^2 = \frac{68.52}{15} - \left(\frac{31.2}{15} \right)^2 = .2416$$

Since $MS_{\text{LINEAR}} = SS_{\text{LINEAR}}$,

$$MS_{\text{LINEAR}} = Nb^2 s_X^2 \doteq 15(153.78)(.2416) \doteq 557.30$$

and

$$\frac{MS_{\text{LINEAR}}}{MS_{\text{ERROR}}} \doteq \frac{557.30}{12.2} \doteq 45.68$$

If $\beta = 0$, test statistic is distributed as $F(1, N - J)$. In Table 6 of Appendix VIII we find that $\mathbb{P}(45.68) > .9995$ in $F(1, 10)$, so $p < .0005$ and H_0 is rejected for $\alpha \ge .0005$. The data indicate a significant linear relationship between X and Y.

(b) $MS_{\text{DEV. FROM LINEAR}} = \dfrac{SS_{\text{TOTAL}} - SS_{\text{ERROR}} - SS_{\text{LINEAR}}}{J - 2}$

$$SS_{\text{TOTAL}} = \sum^{N} y^2 - N\bar{y}^2 = 7{,}562 - 15(21.2)^2 = 820.4$$

so

$$MS_{\text{DEV. FROM LINEAR}} \doteq \frac{820.4 - 122 - 557.30}{3} \doteq 47.03$$

and from the answer to (a), $MS_{\text{ERROR}} = 12.2$. Therefore,

$$\frac{MS_{\text{DEV. FROM LINEAR}}}{MS_{\text{ERROR}}} \doteq \frac{47.03}{12.2} \doteq 3.85$$

If $\eta^2 - \rho^2 = 0$, test statistic is distributed as $F(J - 2, N - J)$. In Table 6 of Appendix VIII we find that $.95 < \mathbb{P}(3.85) < .975$ in $F(3, 10)$, so $p < .05$, and H_0 is rejected for $\alpha \geq .05$. There is a significant curvilinear relationship between X and Y.

5. (a) $MS_{\text{ERROR}} = \dfrac{\sum y_j^2 - \sum N_j \bar{y}_j^2}{N - J} \doteq \dfrac{501{,}397 - 489{,}178.625}{35} \doteq 349.10$

From Exercise 4.4.3, $b = 1.8975$, so $b^2 \doteq 3.6005$

$s_X^2 = 50$

Therefore,

$$MS_{\text{LINEAR}} = Nb^2 s_X^2 \doteq 40(3.6005)(50) = 7{,}201$$

and

$$\frac{MS_{\text{LINEAR}}}{MS_{\text{ERROR}}} \doteq \frac{7{,}201}{349.10} \doteq 20.63$$

If $\beta = 0$, test statistic is distributed as $F(1, N - J)$. In Table 6 of Appendix VIII there are no entries for $\nu_2 = 35$, but in $F(1, 30)$ we find that $\mathbb{P}(20.63) > .9995$, so $\mathbb{P}(20.63) > .9995$ in $F(1, 35)$. Therefore $p < .0005$, and H_0 is rejected for $\alpha \geq .0005$. Habit strength is linearly related to the number of times a response must be performed to receive a reinforcement.

(b) $MS_{\text{DEV. FROM LINEAR}}$

$$= \frac{\sum^{N} y^2 - N\bar{y}^2 - SS_{\text{ERROR}} - SS_{\text{LINEAR}}}{J - 2}$$

$$= \frac{501{,}397 - 40(109.725)^2 - 12{,}218.375 - 7{,}201}{3} \doteq 131.53$$

and from the answer to (a), $MS_{\text{ERROR}} \doteq 349.10$. Therefore,

$$\frac{MS_{\text{DEV. FROM LINEAR}}}{MS_{\text{ERROR}}} \doteq \frac{131.53}{349.10} \doteq .377$$

If $\eta^2 - \rho^2 = 0$, test statistic is distributed as $F(J - 2, N - J)$. In Table 6 of Appendix VIII there are no entries for $\nu_2 = 35$, but in $F(3, 40)$ we find that $\mathbb{P}(.377) < .25$, so $\mathbb{P}(.377) < .25$ in $F(3, 35)$. Therefore $p > .75$, and H_1 is rejected for $\alpha \le .75$. There is no evidence of a curvilinear component in the relationship of habit strength to the number of times a response must be performed to receive a reinforcement.

7. The basic computational quantities are

$$N = 45 \qquad \sum x = 162{,}000 \qquad \sum x^2 = 626{,}400{,}000$$

$$\sum y = 109 \qquad \sum y^2 = 278.32 \qquad \sum xy = 386{,}400$$

$$\bar{y}_1 = 2.6 \qquad \bar{y}_2 = 2.4 \qquad \bar{y}_3 \doteq 2.26667$$

$$N_1 = N_2 = N_3 = 15$$

(a) $SS_{\text{ERROR}} = \sum y_j^2 - \sum N_j \bar{y}_j^2 \doteq 278.32 - 264.8667 = 13.4533$

$$MS_{\text{ERROR}} = \frac{SS_{\text{ERROR}}}{N - J} \doteq \frac{13.4533}{42} \doteq .3203$$

$$b^2 = \left(\frac{N\sum xy - \sum x \sum y}{N\sum x^2 - \left(\sum x\right)^2} \right)^2 = \left(\frac{45(386{,}400) - (162{,}000)(109)}{45(626{,}400{,}000) - (162{,}000)^2} \right)^2$$

$$\doteq .000000019$$

$$s_X^2 = \frac{\sum x^2}{N} - \bar{x}^2 = \frac{626{,}400{,}000}{45} - \left(\frac{162{,}000}{45} \right)^2 = 960{,}000$$

Since $MS_{\text{LINEAR}} = SS_{\text{LINEAR}}$,

$$MS_{\text{LINEAR}} = Nb^2 s_X^2 \doteq 45(.000000019)(960{,}000) \doteq .8333$$

and

$$\frac{MS_{\text{LINEAR}}}{MS_{\text{ERROR}}} \doteq \frac{.8333}{.3203} \doteq 2.60$$

If $\beta = 0$, test statistic is distributed as $F(1, N - J)$. In Table 6 of Appendix VIII there are no entries for $\nu_2 = 42$, but in $F(1, 60)$ we find that $.75 < \mathbb{P}(2.60) < .90$, so $\mathbb{P}(2.60) < .90$ in $F(1, 42)$. Therefore $p > .10$, and H_1 is rejected for $\alpha \le .10$. The data do not indicate a significant linear relationship between total dosage of cyclophosphamide and average weekly white blood cell count.

(b) $MS_{\text{DEV. FROM LINEAR}} = \dfrac{SS_{\text{TOTAL}} - SS_{\text{ERROR}} - SS_{\text{LINEAR}}}{J - 2}$

$$SS_{\text{TOTAL}} = \sum^{N} y^2 - N\bar{y}^2 \doteq 278.32 - 45(2.42222)^2 \doteq 14.298$$

so

$$MS_{\text{DEV. FROM LINEAR}} \doteq \dfrac{14.298 - 13.4533 - .8333}{1} \doteq .0114$$

and from the answer to (a), $MS_{\text{ERROR}} \doteq .3203$. Therefore,

$$\dfrac{MS_{\text{DEV. FROM LINEAR}}}{MS_{\text{ERROR}}} \doteq \dfrac{.0114}{.3203} \doteq .036$$

If $\eta^2 - \rho^2 = 0$, test statistic is distributed as $F(J - 2, N - J)$. In Table 6 of Appendix VIII there are no entries for $\nu_2 = 42$, but in $F(1, 60)$ we find that $\mathbb{P}(.036) < .25$, so $\mathbb{P}(.036) < .25$ in $F(1, 42)$. Therefore $p > .75$, and H_1 is rejected for $\alpha \leq .75$. There is no curvilinear relationship between total dosage of cyclophosphamide and (average weekly) white blood cell count.

EXERCISES 15.1 (pp. 628–635)

1. (a) $H_0: f_j = \mathbf{F}_j$

   $H_1: H_0$ incorrect

   (b) $\tilde{\chi}^2 = \sum\limits_{j=1}^{3} \dfrac{(f_j - \mathbf{F}_j)^2}{\mathbf{F}_j}$

   (c) $\chi^2(2)$

   (d) $\tilde{\chi}^2 = \dfrac{(5 - 10)^2}{10} + \dfrac{(15 - 20)^2}{20} + \dfrac{(20 - 10)^2}{10} = 13.75$

   (e) In Table 5 of Appendix VIII, we find $.995 < \mathbb{P}(13.75) < .999$ in $\chi^2(2)$, so $p < .005$.

3. (a) The test hypothesis is

   $H_0: f_j = \mathbf{F}_j$

   where $\mathbf{F}_j = NP(x_j)$, and the alternative hypothesis is

   $H_1: H_0$ incorrect

   (b) $\tilde{\chi}^2 = \sum\limits_{j=1}^{5} \dfrac{(f_j - \mathbf{F}_j)^2}{\mathbf{F}_j}$

(c) $\chi^2(4)$

(d) In general, $\mathbf{F}_j = NP(x_j)$. Therefore

$$\mathbf{F}_1 = 600(.178) = 106.8$$

$$\mathbf{F}_2 = 600(.384) = 230.4$$

$$\mathbf{F}_3 = 600(.311) = 186.6$$

$$\mathbf{F}_4 = 600(.112) = 67.2$$

$$\mathbf{F}_5 = 600(.015) = 9$$

(e) $\tilde{\chi}^2 = \dfrac{(94 - 106.8)^2}{106.8} + \dfrac{(209 - 230.4)^2}{230.4} + \dfrac{(199 - 186.6)^2}{186.6}$

$\qquad + \dfrac{(86 - 67.2)^2}{67.2} + \dfrac{(12 - 9)^2}{9} \doteq 10.6$

(f) In Table 5 of Appendix VIII, we find $.95 < \mathbb{P}(10.6) < .975$ in $\chi^2(4)$, so $p < .05$. The test hypothesis is rejected at the .05 level.

5. (a) H_0: $X : U(100,000)$

H_1: H_0 incorrect

(b) $\tilde{\chi}^2 = \displaystyle\sum_{j=1}^{10} \dfrac{(f_j - \mathbf{F}_j)^2}{\mathbf{F}_j}$

(c) $\chi^2(9)$

(d)

0 to 9,999	10,000 to 19,999	20,000 to 29,999	30,000 to 39,999	40,000 to 49,999	50,000 to 59,999	60,000 to 69,999	70,000 to 79,999	80,000 to 89,999	90,0 to 99,9
9	16	10	10	10	8	9	7	6	15

(e)

0 to 9,999	10,000 to 19,999	20,000 to 29,999	30,000 to 39,999	40,000 to 49,999	50,000 to 59,999	60,000 to 69,999	70,000 to 79,999	80,000 to 89,999	90,0 to 99,9
10	10	10	10	10	10	10	10	10	10

(f) $\tilde{\chi}^2 = \dfrac{(9 - 10)^2}{10} + \dfrac{(16 - 10)^2}{10} + \dfrac{(10 - 10)^2}{10}$

$\qquad + \dfrac{(10 - 10)^2}{10} + \dfrac{(10 - 10)^2}{10} + \dfrac{(8 - 10)^2}{10} + \dfrac{(9 - 10)^2}{10}$

$\qquad + \dfrac{(7 - 10)^2}{10} + \dfrac{(6 - 10)^2}{10} + \dfrac{(15 - 10)^2}{10} = 9.2$

(g) In Table 5 of Appendix VIII we find $.50 < \mathbb{P}(9.2) < .75$ in $\chi^2(9)$, so $p > .25$. If $\alpha \leq .25$, reject H_1. The data do not depart significantly from uniformity.

7. (a) The test hypothesis is that frequency of injury is uniformly distributed across categories of plant size. The hypotheses can be stated in several ways. For example, if X is the number of injuries,

H_0: X uniformly distributed

H_1: X not uniformly distributed

Since the total number of injuries was 31,736, the expected frequency of injuries in each group is $31,736/5 = 6,347.2$, so the hypotheses might be stated as

H_0: $f_j = 6,347.2$ for $j = 1$ to 5

H_1: H_0 incorrect

(b) The appropriate test statistic is

$$\tilde{\chi}^2 = \sum_{j=1}^{5} \frac{(f_j - \mathbf{F}_j)^2}{\mathbf{F}_j}$$

(c) $\chi^2(4)$

(d) $\tilde{\chi}^2 = \dfrac{(6,297 - 6,347.2)^2}{6,347.2} + \dfrac{(7,739 - 6,347.2)^2}{6,347.2} + \dfrac{(8,186 - 6,347.2)^2}{6,347.2}$

$+ \dfrac{(7,896 - 6,347.2)^2}{6,347.2} + \dfrac{(1,618 - 6,347.2)^2}{6,347.2} \doteq 4,739.9$

(e) $p < .001$
(f) Reject H_0 for any $\alpha \geq .001$.
(g) No. The frequency of injuries is not uniformly distributed across categories of plant size, but the injury rate does not appear to decrease as plant size increases. The goodness-of-fit test is sensitive to *all* properties of a distribution, so rejection of the test hypothesis does not tell you *how* the observed distribution differs from the model.

9. (a) The test hypothesis can be stated in several ways. For example,

H_0: The ratio of dark to platinum to aleutian to sapphire is $9 : 3 : 3 : 1$

Or, if the subscripts D, P, A, and S represent the various color phases,

H_0: $p_D = \frac{9}{16}$, $p_P = \frac{3}{16}$, $p_A = \frac{3}{16}$, $p_S = \frac{1}{16}$

The alternative hypothesis is

$$H_1: H_0 \text{ incorrect}$$

(b) $\tilde{\chi}^2 = \sum_{j=1}^{4} \frac{(f_j - F_j)^2}{F_j}$

(c) $\chi^2(3)$

(d) For each color phase j, the expected frequency is Np_j, where p_j is the theoretical proportion of the color phase and N is the total number of offspring. Thus, for the dark phase,

$$F_D = Np_D = 102\left(\tfrac{9}{16}\right) \doteq 57.4$$

The observed and expected frequencies are therefore

Color phase	Dark	Platinum	Aleutian	Sapphire
f	53	17	24	8
F	57.4	19.1	19.1	6.4

(e) $\tilde{\chi}^2 = \dfrac{(53 - 57.4)^2}{57.4} + \dfrac{(17 - 19.1)^2}{19.1} + \dfrac{(24 - 19.1)^2}{19.1} + \dfrac{(8 - 6.4)^2}{6.4}$

$\doteq 2.23$

(f) In Table 5 of Appendix VIII we find $.25 < \mathbb{P}(2.23) < .50$ in $\chi^2(3)$, so $p > .50$. If $\alpha \leq .50$ reject H_1. The data are consistent with the predicted ratio of 9 dark to 3 platinum to 3 aleutian to 1 sapphire and support Shackelford's genetic model.

11. (a) The test hypothesis is that frequency of injury is uniformly distributed across years. The hypotheses can be stated in several ways. For example, if X is the number of injuries,

$$H_0: X \text{ uniformly distributed}$$

$$H_1: X \text{ not uniformly distributed}$$

Since the total number of injuries was 30,208, the expected frequency of injuries in each year is $30,208/4 = 7,552$, so the hypotheses might be stated as

$$H_0: f_j = 7,552 \text{ for } j = 1 \text{ to } 4$$

$$H_1: H_0 \text{ incorrect}$$

(b) The appropriate test statistic is

$$\tilde{\chi}^2 = \sum_{j=1}^{4} \frac{(f_j - \mathbf{F}_j)^2}{\mathbf{F}_j}$$

(c) $\chi^2(3)$

(d) $\tilde{\chi}^2 = \dfrac{(7,149 - 7,552)^2}{7,552} + \dfrac{(7,225 - 7,552)^2}{7,552}$

$+ \dfrac{(7,703 - 7,552)^2}{7,552} + \dfrac{(8,131 - 7,552)^2}{7,552} \doteq 83.07$

(e) $p < .001$

(f) Reject H_0 for any $\alpha \geq .001$.

(g) This is a reasonable interpretation. The departure from uniformity is significant, and inspection of the observed frequencies suggests that this departure is in the direction of increase over time.

13. (a) The test hypothesis is that the age distribution of zebra killed by lions fits the age distribution of zebra in the natural population (as estimated by the distribution of zebra shot by humans). If N is the number of zebra killed by lions and p_j is the proportion of the zebra population in age class j, the hypothesis can be stated in several ways.

$$H_0: f_j = N\hat{p}_j$$

where $\hat{p}_j$ is estimated from zebra shot by humans, or

$$H_0: f_j/N = \hat{p}_j$$

where $\hat{p}_j$ is estimated from zebra shot by humans, or

$$H_0: f_j = \mathbf{F}_j$$

where $\mathbf{F}_j = N\hat{p}_j$. In any case, the alternative hypothesis is

$$H_1: H_0 \text{ incorrect}$$

(b) $\tilde{\chi}^2 = \sum_{j=1}^{8} \frac{(f_j - \mathbf{F}_j)^2}{\mathbf{F}_j}$

(c) $\chi^2(7)$

(d) In general,

$$\mathbf{F}_j = N\hat{p}_j$$

where $N = 174$ is the number of zebra killed by lions.

Age category	1	2	3	4	5	6	7	8
F	5.05	13.40	14.44	12.70	18.97	50.63	39.50	19.49

$\Sigma F_j > 174$ because $\Sigma \hat{p}_j > 1.0$ as noted in exercise.

(e) $\tilde{\chi}^2 = \dfrac{(15 - 5.05)^2}{5.05} + \dfrac{(9 - 13.40)^2}{13.40} + \dfrac{(12 - 14.44)^2}{14.44}$

$\qquad + \dfrac{(18 - 12.70)^2}{12.70} + \dfrac{(11 - 18.97)^2}{18.97} + \dfrac{(34 - 50.63)^2}{50.63}$

$\qquad + \dfrac{(20 - 39.50)^2}{39.50} + \dfrac{(55 - 19.49)^2}{19.49}$

$\qquad \doteq 106.81$

(f) In Table 5 of Appendix VIII we find $\mathbb{P}(106.81) > .999$ in $\chi^2(7)$, so $p < .001$. If $\alpha \geq .001$, reject H_0. The age distribution of zebra killed by lions departs significantly from the (estimated) age distribution in the zebra population.

15. (a) Let p_0 be the proportion of injuries in Big Three plants. Then, $p_0 = 3{,}881/308{,}696 \doteq .01257$. If p is the proportion of injuries in supplier plants, the hypotheses are

$\qquad H_0\colon p = .01257$

$\qquad H_1\colon p > .01257$

(b) $Z_{X/N} = \dfrac{\dfrac{X}{N} - p_0}{\sqrt{\dfrac{p_0(1 - p_0)}{N}}}$

(c) $N(0, 1)$

(d) In 1988 supplier plants employed 321,091 workers and recorded 8,084 injuries. Therefore,

$\qquad z_{X/N} = \dfrac{\dfrac{8{,}084}{321{,}091} - .01257}{\sqrt{\dfrac{(.01257)(.98743)}{321{,}091}}} \doteq 64.12$

(e) Reject H_0 ($p < .0000003$).

(f) Let f_1 be the observed frequency of injured workers in supplier plants and let f_2 be the observed frequency of uninjured workers in supplier plants. Let F_1 and F_2 be the expected frequencies of injured and uninjured workers in supplier plants. Then, the hypotheses are

$\qquad H_0\colon f_j = F_j$

$\qquad H_1\colon H_0$ incorrect

(g) The appropriate test statistic is

$$\tilde{\chi}^2 = \sum_{j=1}^{2} \frac{(f_j - \mathbf{F}_j)^2}{\mathbf{F}_j}$$

(h) $\chi^2(1)$

(i) In 1988 supplier plants employed 321,091 workers. Therefore,

$$\mathbf{F}_1 \doteq (.01257)(321,091) \doteq 4,036.1$$

$$\mathbf{F}_2 \doteq (.98743)(321,091) \doteq 317,054.9$$

and

$$\tilde{\chi}^2 = \frac{(8,084 - 4,036.1)^2}{4,036.1} + \frac{(313,007 - 317,054.9)^2}{317,054.9}$$

$$\doteq 4,111.4$$

(j) Reject H_0 ($p < .001$). $\qquad\qquad$ (k) $\tilde{\chi}^2 \doteq (z_{X/N})^2$

EXERCISES 15.2 (pp. 649–655)

1. (a) The test hypothesis is

$$H_0: p_{11} = p_{21} = p_{.1}$$

$$p_{12} = p_{22} = p_{.2}$$

$$p_{13} = p_{23} = p_{.3}$$

$$p_{14} = p_{24} = p_{.4}$$

or, more compactly,

$$H_0: p_{1k} = p_{2k} = p_{.k} \text{ for } k = 1, 2, 3, 4,$$

The alternative hypothesis is

$$H_1: H_0 \text{ incorrect}$$

(b) $\tilde{\chi}^2 = \sum_{k}^{4} \sum_{j}^{2} \frac{(f_{jk} - \mathbf{F}_{jk})^2}{\mathbf{F}_{jk}}$

(c) $\chi^2(3)$

(d) In general,

$$\mathbf{F}_{jk} = \frac{f_j f_k}{N}$$

where f_j is the number of individuals in Population j, and f_k is the number of individuals in category k, and N is the total number of observations. The number of individuals in Population 1 is 91, the number of individuals in Category 1 is 47, and the total number of observations is 172. The expected frequency of individuals in Population 1 classified in Category 1 is therefore

$$\mathbf{F}_{11} = \frac{(91)(47)}{172} \doteq 24.9$$

The number of individuals in Category 2 is 75, so the expected frequency of individuals in Population 1 classified in Category 2 is therefore

$$\mathbf{F}_{12} = \frac{(91)(75)}{172} \doteq 39.7$$

The number of individuals in Category 3 is 39, so the expected frequency of individuals in Population 1 classified in Category 3 is therefore

$$\mathbf{F}_{13} = \frac{(91)(39)}{172} \doteq 20.6$$

The remaining expected frequencies can be obtained by subtraction from either column or row totals:

	Category			
Population	1	2	3	4
1	24.9	39.7	20.6	5.8
2	22.1	35.3	18.4	5.2

(e) $\tilde{\chi}^2 = \dfrac{(31 - 24.9)^2}{24.9} + \dfrac{(40 - 39.7)^2}{39.7} + \dfrac{(17 - 20.6)^2}{20.6} + \dfrac{(3 - 5.8)^2}{5.8}$

$\qquad + \dfrac{(16 - 22.1)^2}{22.1} + \dfrac{(35 - 35.3)^2}{35.3} + \dfrac{(22 - 18.4)^2}{18.4} + \dfrac{(8 - 5.2)^2}{5.2}$

$\qquad \doteq 7.38$

(f) In Table 5 of Appendix VIII we find $.90 < \mathbb{P}(7.38) < .95$ in $\chi^2(3)$, so $p > .05$.

(g) Reject H_1 for any $\alpha \leq .05$.

3. (a) H_0: $P(x_j, y_k) = P(x_j)P(y_k)$

$\qquad H_1$: H_0 incorrect

(b) $\tilde{\chi}^2 = \sum\limits_{k}^{4} \sum\limits_{j}^{3} \dfrac{\left(f_{jk} - \mathbf{F}_{jk}\right)^2}{\mathbf{F}_{jk}}$

(c) $\chi^2(6)$

(d) In general,

$$\mathbf{F}_{jk} = \dfrac{f_j f_k}{N}$$

where f_j is the marginal frequency of x_j, and f_k is the marginal frequency of y_k, and N is the total number of observations. The marginal frequency of x_1 is 80, the marginal frequency of y_1 is 20, and the number of observations is 320. Therefore, expected frequency in cell 1, 1 is

$$\mathbf{F}_{11} = \dfrac{(80)(20)}{320} = 5$$

The marginal frequency of y_2 is 92, so the expected frequency in cell 1, 2 is

$$\mathbf{F}_{12} = \dfrac{(80)(92)}{320} = 23$$

The marginal frequency of y_3 is 140, so the expected frequency in cell 1, 3 is

$$\mathbf{F}_{13} = \dfrac{(80)(140)}{320} = 35$$

The marginal frequency of x_2 is 160, so the expected frequency of cell 2, 1 is

$$\mathbf{F}_{21} = \dfrac{(160)(20)}{320} = 10$$

the expected frequency of cell 2, 2 is

$$\mathbf{F}_{22} = \dfrac{(160)(92)}{320} = 46$$

and the expected frequency of cell 2, 3 is

$$\mathbf{F}_{23} = \dfrac{(160)(140)}{320} = 70$$

The remaining expected frequencies can be obtained by subtraction from either column or row totals:

	y_1	y_2	y_3	y_4
x_1	5	23	35	17
x_2	10	46	70	34
x_3	5	23	35	17

(e) $\tilde{\chi}^2 = \dfrac{(2-5)^2}{5} + \dfrac{(28-23)^2}{23} + \dfrac{(34-35)^2}{35} + \dfrac{(16-17)^2}{17}$

$\quad + \dfrac{(15-10)^2}{10} + \dfrac{(39-46)^2}{46} + \dfrac{(76-70)^2}{70} + \dfrac{(30-34)^2}{34}$

$\quad + \dfrac{(3-5)^2}{5} + \dfrac{(25-23)^2}{23} + \dfrac{(30-35)^2}{35} + \dfrac{(22-17)^2}{17}$

$\quad \doteq 10.7$

(f) In Table 5 of Appendix VIII we find $.90 < \mathbb{P}(10.7) < .95$ in $\chi^2(6)$, so $p > .05$.

(g) Reject H_1 if $\alpha \le .05$.

5. (a) In this data layout, the populations are represented as rows and the categorical variable as columns. The test hypothesis can be written

$$H_0: p_{11} = p_{21} = p_{\cdot 1}$$

$$p_{12} = p_{22} = p_{\cdot 2}$$

$$p_{13} = p_{23} = p_{\cdot 3}$$

or

$$H_0: p_{1k} = p_{2k} = p_{\cdot k}$$

for $k = 1, 2, 3$.

(b) $\tilde{\chi}^2 = \displaystyle\sum_{k}^{3} \sum_{j}^{2} \dfrac{(f_{jk} - \mathbf{F}_{jk})^2}{\mathbf{F}_{jk}}$

(c) $\chi^2(2)$

(d) The observed frequencies in the collapsed intervals are

Performance (no. correct)	Poor (0 – 2)	Moderate (3 – 5)	Good (6 – 8)	Σ
Cocker spaniels	20	6	2	28
Basenjis	3	6	19	28
Σ	23	12	21	56

(e) In general,

$$\mathbf{F}_{jk} = \frac{f_j f_k}{N}$$

where f_j is the total number of observations in row j, and f_k is the total number of observations in column k, and N is the total number of observations. Therefore, the expected frequency of poor performances by cocker spaniels is

$$\mathbf{F}_{11} = \frac{(28)(23)}{56} = 11.5$$

and the expected frequency of moderate performances by cocker spaniels is

$$\mathbf{F}_{12} = \frac{(28)(12)}{56} = 6$$

All of the remaining expected frequencies can be obtained by subtraction from either column or row totals:

Performance (no. correct)	Poor (0 – 2)	Moderate (3 – 5)	Good (6 – 8)
Cocker spaniels	11.5	6	10.5
Basenjis	11.5	6	10.5

(f) $\tilde{\chi}^2 = \dfrac{(20 - 11.5)^2}{11.5} + \dfrac{(3 - 11.5)^2}{11.5} + \dfrac{(6 - 6)^2}{6}$

$\qquad + \dfrac{(6 - 6)^2}{6} + \dfrac{(2 - 10.5)^2}{10.5} + \dfrac{(19 - 10.5)^2}{10.5}$

$\qquad \doteq 26.33$

(g) In Table 5 of Appendix VIII we find that $\mathbb{P}(26.33) > .999$ in $\chi^2(2)$, so $p < .001$. The hypothesis of homogeneity is therefore rejected for $\alpha \geq .001$.

(h) Probably not. It is arguable that the samples are large enough ($N_1 = N_2 = 28$) to assume, under the Central Limit Theorem, that the sampling distribution of means is approximately normal.

7. (a) In the data layout presented with the problem, populations are represented as rows and the categorical variable as columns. The test hypothesis can be written

$$H_0: p_{1k} = p_{2k} = p_{3k} = p_{\cdot k}$$

for $k = 1, 2$. In this particular exercise, the categorical variable happens to be binomial ("Approve" versus "Disapprove"), so it is perfectly acceptable to let p denote the proportion of persons who approve (or the proportion who disapprove) and state the test hypothesis in the form

$$H_0: p_j = p$$

for $j = 1, 2, 3$

(b) $\tilde{\chi}^2 = \sum_k^2 \sum_j^3 \dfrac{(f_{jk} - \mathbf{F}_{jk})^2}{\mathbf{F}_{jk}}$

(c) $\chi^2(2)$

(d) Using the hint to convert percentages to observed frequencies,

| | Response | | |
Political affiliation	**Approve**	**Disapprove**	Σ
Democrats	390	135	525
Republicans	240	360	600
Independents	225	150	375
Σ	855	645	1,500

(e) In general,

$$\mathbf{F}_{jk} = \frac{f_j f_k}{N}$$

where f_j is the total number of respondents in row j, and f_k is the total number of respondents in column k, and N is the total number of respondents. Therefore, the expected frequency of Democrats who approve a tax increase is

$$\mathbf{F}_{11} = \frac{(525)(855)}{1,500} = 299.25$$

and the expected frequency of Republicans who approve a tax increase is

$$\mathbf{F}_{21} = \frac{(600)(855)}{1{,}500} = 342.00$$

and all of the remaining expected frequencies can be obtained by subtraction from either column or row totals:

Political affiliation	Response Approve	Disapprove
Democrats	299.25	225.75
Republicans	342.00	258.00
Independents	213.75	161.25

(f) $$\tilde{\chi}^2 \doteq \frac{(390 - 299.25)^2}{299.25} + \frac{(135 - 225.75)^2}{225.75} + \frac{(240 - 342.00)^2}{342.00} + \frac{(360 - 258.00)^2}{258.00} + \frac{(225 - 213.75)^2}{213.75} + \frac{(150 - 161.25)^2}{161.25} \doteq 136.13$$

(g) In Table 5 of Appendix VIII we find that $\mathbb{P}(136.13) > .999$ in $\chi^2(2)$, so $p < .001$, and the hypothesis of homogeneity is rejected for $\alpha \geq .001$.

9. (a) H_0: $p_{1k} = p_{2k}$ for $k = 1, 2, 3, 4, 5, 6$
H_1: H_0 incorrect

(b) $$\tilde{\chi}^2 = \sum_k^6 \sum_j^2 \frac{(f_{jk} - \mathbf{F}_{jk})^2}{\mathbf{F}_{jk}}$$

(c) $\chi^2(5)$

(d) In general,

$$\mathbf{F}_{jk} = \frac{f_j f_k}{N}$$

where f_j is the number of rats in row j, and f_k is the number of rats in column k, and N is the total number of rats. The number of "inner-city" residents is 50, the number of rats that consumed less than 1.5 percent alcohol is 13, and the total number of rats in the study is 90. The expected frequency of rats in the "inner-city" treatment that consumed less than 1.5 percent alcohol is therefore

$$\mathbf{F}_{11} = \frac{(50)(13)}{90} \doteq 7.22$$

The number of rats that consumed between 1.5 and 4.5 percent alcohol is 23, so the expected frequency of "inner-city" residents in this category of alcohol consumption is

$$F_{12} = \frac{(50)(23)}{90} \doteq 12.78$$

Proceeding in this fashion, we obtain

$$F_{13} = \frac{(50)(24)}{90} \doteq 13.33$$

$$F_{14} = \frac{(50)(18)}{90} = 10$$

and

$$F_{15} = \frac{(50)(8)}{90} \doteq 4.44$$

The remaining expected frequencies can be obtained by subtraction from either column or row totals:

	≤ 1.5	$1.5-4.5$	$4.5-6.5$	$6.5-8.5$	$8.5-10.5$	≥ 10.5
Inner City	7.22	12.78	13.33	10	4.44	2.23
Suburban	5.78	10.22	10.67	8	3.56	1.77

(e) $\tilde{\chi}^2 \doteq \dfrac{(4 - 7.22)^2}{7.22} + \dfrac{(10 - 12.78)^2}{12.78} + \dfrac{(16 - 13.33)^2}{13.33} + \dfrac{(12 - 10)^2}{10}$

$\qquad + \dfrac{(6 - 4.44)^2}{4.44} + \dfrac{(2 - 2.23)^2}{2.23} + \dfrac{(9 - 5.78)^2}{5.78} + \dfrac{(13 - 10.22)^2}{10.22}$

$\qquad + \dfrac{(8 - 10.67)^2}{10.67} + \dfrac{(6 - 8)^2}{8} + \dfrac{(2 - 3.56)^2}{3.56} + \dfrac{(2 - 1.77)^2}{1.77}$

$\qquad \doteq 7.98$

(f) In Table 5 of Appendix VIII we find $.75 < \mathbb{P}(7.98) < .90$ in $\chi^2(5)$, so $p > .10$.

(g) Since $p > \alpha = .05$, the alternative hypothesis is rejected. There is insufficient evidence to conclude that the distribution of alcohol consumption differs across the two groups.

11. (a) The test hypothesis can be formulated in at least three ways. First, we may think of this as a homogeneity of proportions problem in which plant size defines populations and categories are years. If Population 1 is workers in small plants and Population 2 is workers in large plants, then p_{jk} is the proportion of all injuries recorded in year k that occurred in

Population j, and the test hypothesis is

$$H_0: p_{11} = p_{21}$$

$$p_{12} = p_{22}$$

$$p_{13} = p_{23}$$

$$p_{14} = p_{24}$$

or, more compactly,

$$H_0: p_{1k} = p_{2k} \text{ for } k = 1, 2, 3, 4$$

Second, we may think of this as a homogeneity of proportions problem in which year of injury defines populations, and injured workers are categorized according to plant size. That is, rows are categories and columns are populations. If Population 1 is 1986 workers, Population 2 is 1987 workers, etc., then p_{jk} is the proportion of all injuries recorded in plants of size j that occurred in Population k, and the test hypothesis is

$$H_0: p_{11} = p_{12} = p_{13} = p_{14}$$

$$p_{21} = p_{22} = p_{23} = p_{24}$$

or, more compactly,

$$H_0: p_{j1} = p_{j2} = p_{j3} = p_{j4} \text{ for } j = 1, 2$$

Third, we can think of all injured supplier plant workers as a single population, with each worker classified by both plant size (X) and year of injury (Y). We then test the hypothesis that year of injury and plant size are independent variables,

$$H_0: P(x_j, y_k) = P(x_j)(y_k)$$

The test hypothesis one chooses to formulate depends on how one conceives of the underlying issue (e.g., whether one factor is implicitly thought of as the independent variable and the other as the dependent variable). The alternative hypothesis in any case is

$$H_1: H_0 \text{ incorrect}$$

(b) For any of the three scenarios, the appropriate test statistic is

$$\tilde{\chi}^2 = \sum_k^4 \sum_j^2 \frac{(f_{jk} - \mathbf{F}_{jk})^2}{\mathbf{F}_{jk}}$$

(c) $\chi^2(3)$

(d) The first task is to calculate the expected frequencies, $\mathbf{F}_{jk}$. In general,

$$\mathbf{F}_{jk} = \frac{f_j f_k}{N}$$

where f_j is the total number of injuries in Population j, and f_k is the total number of injuries in year k, and N is the total number of injuries, 30,208. Therefore,

$$\mathbf{F}_{11} = \frac{(14{,}036)(7{,}149)}{30{,}208} \doteq 3{,}321.7$$

In the table of expected frequencies given below, the values for small plants in 1986, 1987, and 1988 were calculated in this fashion, and the rest were obtained by subtraction from row and column totals:

	1986	1987	1988	1989
Small	3,321.7	3,357.1	3,579.2	3,778.0
Large	3,827.3	3,867.9	4,123.8	4,353.0

$$\tilde{\chi}^2 = \frac{(3{,}088 - 3{,}321.7)^2}{3{,}321.7} + \frac{(3{,}313 - 3{,}357.1)^2}{3{,}357.1}$$

$$+ \frac{(3{,}528 - 3{,}579.2)^2}{3{,}579.2} + \frac{(4{,}107 - 3{,}778.0)^2}{3{,}778.0}$$

$$+ \frac{(4{,}061 - 3{,}827.3)^2}{3{,}827.3} + \frac{(3{,}912 - 3{,}867.9)^2}{3{,}867.9}$$

$$+ \frac{(4{,}175 - 4{,}123.8)^2}{4{,}123.8} + \frac{(4{,}024 - 4{,}353.0)^2}{4{,}353.0}$$

$$\doteq 86.68$$

(e) $p < .001$
(f) Reject H_0 for any $\alpha \geq .001$.
(g) The substance of the claim is supported. It is clear that the rate (proportion) of injuries exhibits a different profile from year to year in large plants than in small plants, and inspection of the observed frequencies suggests that there is an upward trend in small plants. Note, however, that there is no immediately obvious pattern to the changes across time in large plants.

13. (a) The appropriate test statistic is the fourfold Pearson's *chi*-square.

$$\tilde{\chi}^2 = \frac{N(ad - bc)^2}{(a + b)(c + d)(a + c)(b + d)}$$

(b) $\chi^2(1)$

(c) $\chi^2 = \dfrac{231(1{,}672 - 372)^2}{(158)(73)(214)(17)} \doteq 9.30$

(d) In Table 5 of Appendix VIII we find $.995 < \mathbb{P}(7.72) < .999$ in $\chi^2(1)$, so $p < .005$.

(e) For $\alpha \geq .005$ there is significant association between prey size and predator size.

15. (a) The test hypothesis can be formulated in at least three ways. Given the fundamental issue, it is likely that a researcher would conceptualize this as a homogeneity of proportions problem with two populations, stamping press operators in Big Three plants and stamping press operators in supplier plants. Then, p_{jk} is the proportion of all the injuries recorded in year k that occurred in Population j, and the test hypothesis is

$$H_0: p_{11} = p_{21}$$

$$p_{12} = p_{22}$$

$$p_{13} = p_{23}$$

$$p_{14} = p_{24}$$

or, more compactly,

$$H_0: p_{1k} = p_{2k} \text{ for } k = 1 \text{ to } 4$$

It is just as formally correct, if less intuitively sensible, to conceptualize this as a homogeneity of proportions problem with populations defined by year of injury and injured workers categorized as either Big Three or supplier plant employees. If Population 1 is 1986 workers, Population 2 is 1987 workers, etc., then p_{jk} is the proportion of all injuries recorded in plants of plant category j that occurred in Population k, and the test hypothesis is

$$H_0: p_{11} = p_{12} = p_{13} = p_{14}$$

$$p_{21} = p_{22} = p_{23} = p_{24}$$

or, more compactly,

$$H_0: p_{j1} = p_{j2} = p_{j3} = p_{j4} \text{ for } j = 1, 2$$

Third, we can think of all injured stamping press operators as a single population, with each worker classified as either a Big Three employee or a supplier employee (X) and by year of injury (Y). We then test the hypothesis that X and Y are independent variables.

$$H_0: P(x_k, y_k) = P(x_j)(y_k)$$

The alternative hypothesis in any case is

$$H_1: H_0 \text{ incorrect}$$

(b) The appropriate test statistic is

$$\tilde{\chi}^2 = \sum_{k}^{4} \sum_{j}^{2} \frac{(f_{jk} - \mathbf{F}_{jk})^2}{\mathbf{F}_{jk}}$$

(c) $\chi^2(3)$

(d) The first task is to calculate the expected frequencies, F_{jk}. In general,

$$F_{jk} = \frac{f_j f_k}{N}$$

where f_j is the total number of injuries in Population j, and f_k is the total number of injuries in year k, and N is the total number of injuries, 3,318. Therefore,

$$F_{11} = \frac{(849)(509)}{3,318} \doteq 130.2$$

In the table of expected frequencies given below, the values for Big Three plants in 1986, 1987, and 1988 were calculated in this fashion, and the rest were obtained by subtraction from row and column totals:

Year	1986	1987	1988	1989
Big Three	130.2	125.6	128.4	124.8
Suppliers	718.8	693.4	708.6	688.2

$$\tilde{\chi}^2 = \frac{(165 - 130.2)^2}{130.2} + \frac{(134 - 125.6)^2}{125.6} + \frac{(119 - 128.4)^2}{128.4} + \frac{(91 - 124.8)^2}{124.8}$$

$$+ \frac{(684 - 718.8)^2}{718.8} + \frac{(685 - 693.4)^2}{693.4} + \frac{(718 - 708.6)^2}{708.6} + \frac{(722 - 688.2)^2}{688.2}$$

$$\doteq 23.3$$

(e) $p < .001$

(f) Reject H_0 for any $\alpha \geq .001$.

(g) The most immediately plausible interpretation is that stamper injuries are decreasing in Big Three plants and increasing in supplier plants. This interpretation rests both on rejection of the test hypothesis and inspection of the observed frequencies.

EXERCISES 15.3 (pp. 670–674)

1. (a) If X is puzzle-box score,

$$H_0: X : N(\mu, \sigma^2)$$

$$H_1: H_0 \text{ incorrect}$$

(b) The intervals should be constructed so that they are equally *probable* under the standard normal curve. Since $C = 6$ intervals, one sixth of the total area under the normal curve will fall in each interval. Our first interval includes all values equal to or less than z, where

$$\mathbb{P}(z) = 1/6 \doteq .1667$$

In Table 3 of Appendix VIII we find that $\mathbb{P}(-.97) \doteq .1667$. The first interval is therefore all values less than or equal to $-.97$. The second interval runs from $-.97$ to z, where

$$\mathbb{P}(z) = 2/6 \doteq .3333$$

etc. Using this logic, we obtain the following intervals:

z	$\leq -.97$	$-.97$ to $-.43$	$-.43$ to 0.00	0.00 to $.43$	$.43$ to $.97$	$\geq +.97$

To express the limits of our intervals in natural units, we first calculate the sample mean and the estimate of the population standard deviation:

$$\bar{x} \doteq 3.55 \text{ and } \hat{\sigma} \doteq 2.93$$

The limits of each interval are therefore

$$x_a \doteq z_a(2.93) + 3.55 \text{ and } x_b \doteq z_b(2.93) + 3.55$$

and our six intervals expressed in x-values become

x	$\leq .71$	$.71$ to 2.29	2.29 to 3.55	3.55 to 4.81	4.81 to 6.39	≥ 6.39

(c) If puzzle-box performance is normally distributed, $\frac{1}{6}$ of the 126 dogs should fall in each interval. Therefore,

$$(1/6)126 = 21$$

is the expected frequency in each interval:

x	$\leq .71$	$.71$ to 2.29	2.29 to 3.55	3.55 to 4.81	4.81 to 6.39	≥ 6.39
F	21	21	21	21	21	21

(d) The observed frequencies are

x	$\leq .71$	$.71$ to 2.29	2.29 to 3.55	3.55 to 4.81	4.81 to 6.39	≥ 6.39
f	28	30	9	9	20	30

(e) $\displaystyle \tilde{\chi}^2 = \sum_{j=1}^{6} \frac{(f_j - F_j)^2}{F_j} \doteq \frac{(28-21)^2}{21} + \frac{(30-21)^2}{21} + \frac{(9-21)^2}{21}$

$$+ \frac{(9-21)^2}{21} + \frac{(20-21)^2}{21} + \frac{(30-21)^2}{21}$$

$$\doteq 23.81$$

(f) Since μ and σ^2 were estimated in order to calculate the limits of the intervals, degrees of freedom for this test is $6 - 1 - 2 = 3$. We find $\mathbb{P}(23.81) > .999$ in $\chi^2(3)$, so $p < .001$.

(g) Clearly, the scores depart significantly from normality. Judging from the observed frequencies, it would appear that scores are bimodally distributed.

3. (a) H_0: $p_{\text{PRIMED}} = p_{\text{UNPRIMED}}$

where p is the population proportion at or below the grand median.

(b) Any test of medians calls for an $R \times 2$ Pearson's *chi*-square. Since $R = 2$, the appropriate test statistic is

$$\frac{N(ad - bc)^2}{(a + b)(c + d)(a + c)(b + d)}$$

(c) $\chi^2(1)$

(d) The middle two values for the combined group of 26 are 859 and 872, so the combined median is 865.5. Observed frequencies above and below the median are therefore

	Observed	
	Below $\tilde{x}$	**Above $\tilde{x}$**
Primed	3	10
Unprimed	10	3

$$\frac{N(ad - bc)^2}{(a + b)(c + d)(a + c)(b + d)} = \frac{26(9 - 100)^2}{(13)(13)(13)(13)} \doteq 7.54$$

(e) In Table 5 of Appendix VIII, we find $.99 < \mathbb{P}(7.54) < .995$ in $\chi^2(1)$, so $p < .01$.

(f) Reject H_0 if $\alpha \geq .01$. The median duration of attack is significantly greater for primed subjects.

5. $\varphi = \dfrac{ad - bc}{\sqrt{(a + b)(c + d)(a + c)(b + d)}} \doteq \dfrac{7,109}{44,807.4} \doteq .16$

7. (a) H_0: $p_{11} = p_{21}$

$p_{12} = p_{22}$

$p_{13} = p_{23}$

$p_{14} = p_{24}$

$p_{15} = p_{25}$

or,

H_0: $p_{1k} = p_{2k}$ for $k = 1, 2, 3, 4, 5$

and

H_1: H_0 incorrect

(b) $\mathbf{F}_{jk} = \dfrac{f_j f_k}{N}$

where f_j is the number of kills of species j, and f_k is the number of kills in cover k, and N is the total number of kills.

Prey species	Type of cover				
	1	2	3	4	5
Thomson's gazelle	2.9	13.9	2.2	34.4	25.6
Wildebeest & zebra	1.1	5.1	0.8	12.6	9.4

(c) Since $\nu < 30$ and several of the expected frequencies are smaller than 2, the appropriate statistic is Haldane's statistic:

$$N\left(1 - \dfrac{N\Sigma\left(\dfrac{a_j b_j}{N_j}\right)}{AB}\right)$$

(d) $\chi^2(4)$

(e) $N\left(1 - \dfrac{N\Sigma\left(\dfrac{a_j b_j}{N_j}\right)}{AB}\right) = 108\left(1 - \dfrac{108\left(\dfrac{3}{4} + \dfrac{70}{19} + \dfrac{0}{3} + \dfrac{46}{47} + \dfrac{286}{35}\right)}{2{,}291}\right)$

$\doteq 38.8$

(f) In Table 5 of Appendix VIII we find that $\mathbb{P}(38.8) > .999$ in $\chi^2(4)$, so $p < .001$.

(g) Reject H_0 for $\alpha \geq .001$. The two groups of kills are not homogeneously distributed across types of cover. This might be because the two groups are distributed differently across types of habitat or because the two groups are especially vulnerable to predators in different types of habitat.

Index

GLOSSARY OF SYMBOLS, CONTINUED

CONVENTIONAL MATHEMATICAL SYMBOLS

a	the y-intercept of the line $y = bx + a$
b	the slope of the line $y = bx + a$
$_NC_r$	number of combinations of N things taken r at a time
E	an event or (sub)set of results
e	element of a set; a result in a sample space
ε	(lowercase Greek epsilon) any arbitrarily small positive value
f	function, as in $y = f(x)$
$N!$	N factorial; $N(N-1)(N-2) \cdots 1$
$_NP_r$	number of permutations of N things taken r at a time
S	set; sample space
$\int$	the integral
$\vert$	"given that" in a conditional statement
$\vert\ \vert$	absolute value (e.g., $\vert x \vert$, the absolute value of x)
$[\ ,\ ]$	a closed interval (e.g., $[x_a, x_b]$, the interval x_a to x_b, inclusive)
$\doteq$	equals to the nearest decimal place shown
$\cong$	approximately equal to
$<$	(is) less than (e.g., $5 < 6$)
$\leq$	(is) less than or equal to
$>$	(is) greater than (e.g., $7 > 2$)
$\geq$	(is) greater than or equal to
$\cup$	union of sets (e.g., $E_1 \cup E_2$)
$\cap$	intersection of sets (e.g., $E_1 \cap E_2$)